Enzymes in Nonaqueous Solvents

METHODS IN BIOTECHNOLOGY™

John M. Walker, SERIES EDITOR

METHODS IN BIOTECHNOLOGY™

Enzymes in Nonaqueous Solvents

Methods and Protocols

Edited by

Evgeny N. Vulfson

Institute of Food Research, Norwich, UK

Peter J. Halling

University of Strathclyde, Glasgow, UK

and

Herbert L. Holland

Brock University, St. Catharines, Ontario, Canada

Humana Press ✳ Totowa, New Jersey

© 2001 Humana Press Inc.
999 Riverview Drive, Suite 208
Totowa, New Jersey 07512

This publication is printed on acid-free paper. ∞
ANSI Z39.48-1984 (American Standards Institute) Permanence of Paper for Printed Library Materials.

Cover design by Patricia F. Cleary.

Production Editor: Mark J. Breaugh.

For additional copies, pricing for bulk purchases, and/or information about other Humana titles, contact Humana at the above address or at any of the following numbers: Tel: 973-256-1699; Fax: 973-256-8341; E-mail: humana@humanapr.com, or visit our Website at www.humanapress.com

Printed in the United States of America. 10 9 8 7 6 5 4 3 2 1

Library of Congress Cataloging in Publication Data

Main entry under title: Enzymes in nonaqueous solvents: methods and protocols.

Methods in molecular biology™.

Enzymes in nonaqueous solvents: methods and protocols./edited by Evgeny N. Wulfson, Peter J. Halling, and Herbert L. Holland.
    p. cm.—(Methods in biotechnology; 15)
  Includes bibliographical references and index.
  ISBN 0-89603-929-3 (alk. paper)
  1. Enzymes—Biotechnology. 2. Nonaqueous solvents.   I. Vulfson, Evgeny N.   II. Halling, Peter J.
III. Holland, Herbert L. IV. Series.
  TP248.65.E59 E624 2001
  6608.29423—dc21   00-025881

PREFACE

Enzymatic catalysis has gained considerable attention in recent years as an efficient tool in the preparation of natural products, pharmaceuticals, fine chemicals, and food ingredients. The high selectivity and mild reaction conditions associated with enzymatic transformations have made this approach an attractive alternative in the synthesis of complex bioactive compounds, which are often difficult to obtain by standard chemical routes. However, the majority of organic compounds are not very soluble in water, which was traditionally perceived as the only suitable reaction medium for the application of biocatalysts. The realization that most enzymes can function perfectly well under nearly anhydrous conditions and, in addition, display a number of useful properties, e.g., highly enhanced stability and different selectivity, has dramatically widened the scope of their application to the organic synthesis.

Another great attraction of using organic solvents rather than water as a reaction solvent is the ability to perform synthetic transformations with relatively inexpensive hydrolytic enzymes. It is worth reminding the reader that in vivo, the synthetic and hydrolytic pathways are catalyzed by different enzymes. However, elimination of water from the reaction mixture enables the "reversal" of hydrolytic enzymes and thus avoids the use of the expensive cofactors or activated substrates that are required for their synthetic counterparts. Also, one should bear in mind that water is by no means an ideal solvent for synthesis; it is relatively expensive to remove on a large scale and it often participates in unwanted side reactions. Thus, the use of enzymes in conventional industrial solvents generally makes it easier and cheaper to incorporate a biotransformation step into the overall synthetic sequence.

Indeed, there are numerous examples of the successful application of enzymes in low water media to industrial-scale production of pharmaceuticals, food ingredients, and fine chemicals.

Methods are very important in any area of research, even more so in a field like nonaqueous biocatalysis, where many methods have been developed relatively recently and have not yet been standardized completely in all laboratories. All too often, the format of standard research papers does not allow methods to be fully described. The importance of key details may be known in the originating laboratory, but may not be appreciated in another, because they cannot be stressed enough, nor reasons explained. The prime objective of *Enzymes in Nonaqueous Solvents* is to address this issue because it was compiled to communicate such details. There will also be critical features of methods that are at present not appreciated by anyone, but that may be causing different results in different laboratories. Here again, the fuller presentations in this book should be a basis for the identification of such differences.

For the convenience of the reader, the editors decided to split the submitted material into three parts; broadly, these deal with the biocatalysts, synthetic chemistry, and systems other than just neat organic solvents or solvent mixtures. Those familiar with the subject will no doubt appreciate that such a separation is to a large extent arbitrary and is bound to result in some overlaps. The editors felt, however, that this would provide the book with a certain structure and make it easier for the reader to find specific pieces of relevant information. In addition, each part has a short introduction that surveys the contributions included.

Authors of standard research papers are understandably keen to emphasise their interesting results. Some signs of this can perhaps be detected in contributions to this volume too. As editors, we have tried to encourage authors to include as much detail as possible in describing their methods, and not to dismiss this as rather boring or unnecessary. We hope the result of the authors' efforts will prove valuable to all who are interested in studying or using enzymes in nonaqueous media.

Evgeny N. Vulfson
Peter J. Halling
Herbert L. Holland

CONTENTS

CONTRIBUTORS

PATRICK ADLERCREUTZ • *Department of Biotechnology, Center for Chemistry and Chemical Engineering, Lund University, Lund, Sweden*

MANSOR AHMAD • *Center for Research in Enzyme and Microbial Technology, Fakulti Sains dan Pengajian Alam Sekitar, Universiti Putra Malaysia, Serdang, Malaysia*

MARIA RAQUEL AIRES-BARROS • *Centro de Engenharia Biologica e Quimica, Instituto Superior Tecnico, Lisbon, Portugal*

M. CONCEIÇÃO ALMEIDA • *Instituto de Tecnologia Quimica e Biologica, Universidade Nova de Lisboa, Oeiras, Portugal*

KAMARUZAMAN AMPON • *Universiti Malaysia Sabah, Kota Kinobaler Sabeh, Malaysia*

JOSE A. ARCOS • *Instituto de Catalisis, CSIC, Madrid, Spain*

SUSANA BARREIROS • *Instituto de Tecnologia Quimica e Biologica, Universidade Nova de Lisboa, Oeiras, Portugal*

MAHIRAN BASRI • *Center for Research in Enzyme and Microbial Technology, Fakulti Sains dan Pengajian Alam Sekitar, University Putra Malaysia, Serdang, Malaysia*

GEORGE BELL • *Department of Pure and Applied Chemistry, University of Strathclyde, Glasgow, UK*

PER BERGLUND • *Department of Biotechnology, Royal Institute of Technology, Stockholm, Sweden*

MIKHAIL BORISOVER • *Institute of Soil, Water and Environmental Sciences, The Volcani Center, Bet Dagan, Israel*

EMMANUEL BOURES • *LILICE, Zac "Les Partes de Riom," Riom, France*

MARIE-PIERRE BOUSQUET • *Centre de Bioingenierie Gilbert Durand, Département de Genie Biochimique et Alimentaire, Complexe Scientifique de Rangueil, Toulouse Cedex, France*

JOAQUIM M. S. CABRAL• *Centro de Engenharia Biologica e Quimica, Instituto Superior Tecnico, Lisbon, Portugal*

GIACOMO CARREA • *Istituto di Biocatalisi e Riconoscimento Moleculare, CNR, Milan, Italy*

CRISTINA M. L. CARVALHO • *Centro de Engenharia Biologica Quimica,Instituto Superior Tecnico, Lisbon, Portugal*

SHUI-TEIN CHEN • *Institute of Biological Chemistry, Academia Sinica, Taipei, Taiwan*

DOUGLAS S. CLARK • *Department of Chemical Engineering, University of California, Berkeley, CA*

GIORGIO COLOMBO • *Istituto di Biocatalisi e Riconoscimento Molecolare (CNR), Milan, Italy*

JEAN-STÉPHANE CONDORET • *Institut National des Sciences Appliquees, Departement de Genie Biochimique et Alimentaire, Toulouse Cedex, France*

DAVID R. DODDS • *Biotransformations Group, Schering-Plough Research Institute, Union, NJ*

JONATHAN S. DORDICK • *Department of Chemical Engineering, Rensselaer Polytechnic Institute, Troy, NY*

AMÉLIE DUCRET • *Microbial and Enzymatic Technology Group, Biotechnology Research Institute, Montréal, Québec, Canada*

UWE EICHHORN • *Institute of Biochemistry, Leipzig University, Leipzig, Germany*

JOHAN F. J. ENGBERSEN • *Laboratory of Supramolecular Chemistry and Technology, MESA Research Institute, University of Twente, Enschede, The Netherlands*

MARKUS ERBELDINGER • *Bioscience and Biotechnology Department, University of Strathclyde, Glasgow, UK*

NORHAIZAN M. ESA • *Center for Research in Enzyme and Microbial Technology, Fakulti Sains dan Pengajian Alam Sekitar, University Putra Malaysia, Serdang, Malaysia*

PATRIZIA FERRABOSCHI • *Dipartimento di Chimica e Biochimica Medica, Universitá degli Studi di Milano, Milan, Italy*

NUNO FONTES • *Instituto de Tecnologia Quimica e Biologica, Universidade Nova de Lisboa, Oeiras, Portugal*

HUGO S. GARCIA • *Instituto de Catalisis, CSIC, Madrid, Spain*

LUCIA GARDOSSI • *Department of Pharmaceutical Science, Universitá degli Studi, Piazzale Europa, Italy*

MARION HAENSLER • *Institute of Biochemistry, Leipzig University, Leipzig, Germany*

PETER J. HALLING • *Department of Pure and Applied Chemistry, University of Strathclyde, Glasgow, UK*

JEONG JUN HAN • *Department of Biological Sciences, Korea Advanced Institute of Sciences and Technology, Taejon, South Korea*

NEIL HARPER • *Department of Pure and Applied Chemistry, University of Strathclyde, Glasgow, UK*

ERIK HEDENSTRÖM • *Department of Chemistry and Process Technology, Mid Sweden University, Sundsvall, Sweden*

JOSEPH J. HEIJNEN • *Kluyver Laboratory for Biotechnology, Delft University of Technology, Delft, The Netherlands*
CHARLES G. HILL, JR. • *Instituto de Catalisis, CSIC, Madrid, Spain*
MISAO HIROTO • *Department of Biomedical Engineering, Toin Human Science and Technology Center, Toin University of Yokohama, Yokohama, Japan*
HERBERT L. HOLLAND • *Department of Chemistry, Brock University, St. Catharines, Ontario, Canada*
MICHAEL J. HOMANN • *Biotransformations Group, Schering-Plough Research Institute, Union, NJ*
KARL HULT • *Department of Biotechnology, Royal Institute of Technology, Stockholm, Sweden*
TAKAMITSU IIDA • *Department of Material Science and Technology, Faculty of Engineering, Niigata University, Niigata, Japan*
YUJI INADA • *Department of Biomedical Engineering, Toin Human Science and Technology Center, Toin University of Yokohama, Yokohama, Japan*
HANS-DIETER JAKUBKE • *Institute of Biochemistry, Leipzig University, Leipzig, Germany*
VOLKER KASCHE • *AB Biotechnologie II, Technische Universität Hamburg-Harburg, Hamburg, Germany*
TAKUO KAWAMOTO • *Department of Synthetic Chemistry and Biological Chemistry, Graduate School of Engineering, Kyoto University, Kyoto, Japan*
ROMAS J. KAZLAUSKAS • *Department of Chemistry, McGill University, Montréal, Québec, Canada*
JEFFREY A. KHAN • *Department of Macromolecular Sciences, Institute of Food Research, Reading, UK*
HIDEO KISE • *Institute of Materials Science, University of Tsukuba, Ibaraki, Japan*
NATALIA L. KLYACHKO • *Department of Chemistry, Lomonosov Moscow State University, Moscow, Russia*
YOH KODERA • *Department of Biomedical Engineering, Toin Human Science and Technology Center, Toin University of Yokohama, Yokohama, Japan*
FRAGISKOS N. KOLISIS • *Department of Chemical Engineering, National Technical University, Athens, Greece*
PETER KUHL • *Institute of Biochemistry, University of Technology Dresden, Dresden, Germany*
SEOK JOON KWON • *Department of Biological Sciences, Korea Advanced Institute of Sciences and Technology, Taejon, South Korea*

SYLVAIN LAMARE • *Laboratoire de Genie Proteique, Université de La Rochelle, La Rochelle Cedex, France*

RENÉ LAZARO • *Laboratoire des Aminoacides, Peptides et Proteines LAPP, Université Montpellier 2, Montpellier Cedex, France*

MARIE DOMINIQUE LEGOY • *Laboratoire de Genie Proteique, Université de La Rochelle, La Rochelle Cedex, France*

ANDREY V. LEVASHOV • *Department of Chemistry, Lomonosov Moscow State University, Moscow, Russia*

MIKE J. J. LITJENS • *Kluyver Laboratory for Biotechnology, Delft University of Technology, Delft, The Netherlands*

ROBERT LORTIE • *Microbial and Enzymatic Technology Group, Biotechnology Research Institute, Montréal, Québec, Canada*

DAVID A. MACMANUS • *Institute of Food Research, Norwich, UK*

ALAIN MARTY • *Centre de Bioingenierie Gilbert Durand, Department de Genie Biochimique et Alimentaire, Complexe Scientifique de Rangueil, Toulouse Cedex, France*

AYAKO MATSUSHIMA • *Department of Biomedical Engineering, Toin Human Science and Technology Center, Toin University of Yokohama, Yokohama, Japan*

THIERRY MAUGARD • *Centre de Bioingenierie Gilbert Durand, Departément de Genie Biochimique et Alimentaire, Complexe Scientifique de Rangueil, Toulouse Cedex, France*

LINDSEY MAY • *Department of Pure and Applied Chemistry, University of Strathclyde, Glasgow, UK*

ANNA MILLQVIST-FUREBY • *Institute of Food Research, Norwich, UK*

MAHA M. A. MISBAH • *Unilever Research Laboratory, Vlaardingen, The Netherlands*

PIERRE MONSAN • *Centre de Bioingenierie Gilbert Durand, Departément de Genie Biochimique et Alimentaire, Complexe Scientifique de Rangueil, Toulouse Cedex, France*

BARRY D. MOORE • *Department of Pure and Applied Chemistry, University of Strathclyde, Glasgow, UK*

BRIAN MORGAN • *Biotransformations Group, Schering-Plough Research Institute, Union, NJ*

TOSHIAKI MORI • *Department of Biomolecular Engineering, Tokyo Institute of Technology, Yokohama, Japan*

HIROYUKI NISHIMURA • *Department of Biomedical Engineering, Toin Human Science and Technology Center, Toin University of Yokohama, Yokohama, Japan*

TORBJÖRN NORIN • *Department of Chemistry and Organic Chemistry, Royal Institute of Technology, Stockholm, Sweden*

SHINOBU ODA • *Technical Research Laboratory, Kansai Paint Co., Kanagawa, Japan*

HIROMICHI OHTA • *Technical Research Laboratory, Kansai Paint Co., Kanagawa, Japan*

YOSHIO OKAHATA • *Department of Biomolecular Engineering, Tokyo Institute of Technology, Yokohama, Japan*

CRISTINA OTERO • *Instituto de Catalisis, CSIC, Madrid, Spain*

GIANLUCA OTTLINA • *Instituto di Biocatalisi e Riconoscimento Moleculare, CNR, Milan, Italy*

JENNY OTTOSSON • *Department of Biotechnology, Royal Institute of Technology, Stockholm, Sweden*

JOHANN PARTRIDGE • *Department of Pure and Applied Chemistry, University of Strathclyde, Glasgow, UK*

ALAN D. PEILOW • *Unilever Research Laboratory, Bedford, UK*

SUREE PHUTRAHUL • *Department of Chemistry, Faculty of Science, Chiang Mai University, Chiang Mai, Thailand*

CHE NYONYA A. RAZAK • *Center for Research in Enzyme and Microbial Technology, Fakulti Sains dan Alam Sekitar, University Putra Malaysia, Serdang, Malaysia*

JEFFREY A. REIMER • *Department of Chemical Engineering, University of California, Berkeley, CA*

DAVID N. REINHOUDT • *Laboratory of Supramolecular Chemistry and Technology, MESA Research Institute, University of Twente, Enschede, The Netherlands*

MAGALI REMAUD-SIMEON • *Centre de Bioingenierie Gilbert Durand, Departément de Genie Biochimique et Alimentaire, Complexe Scientifique de Rangueil, Toulouse Cedex, France*

JOON SHICK RHEE • *Department of Biological Sciences, Korea Advanced Institute of Sciences and Technology, Taejon, South Korea*

JOSEPH O. RICH • *EnzyMed Inc., Iowa City, IA*

DONALD A. ROBB • *Department of Pure and Applied Chemistry, University of Strathclyde, Glasgow, UK*

VALÉRIE ROLLAND • *Laboratoire des Aminoacides, Peptides et Proteines LAPP, Université Montpellier 2, Montpellier Cedex, France*

DIDIER ROTTICCI • *Department of Chemistry and Organic Chemistry, Royal Institute of Technology, Stockholm, Sweden*

MICHAEL T. RU • *Department of Chemical Engineering, University of California, Berkeley, CA*

ABU BAKAR SALLEH • *Center for Research in Enzyme and Microbial Technology Fakulti Sains dan Alam Sekitar, University Putra Malaysia, Serdang, Malaysia*

ENZO SANTANIELLO • *Dipartimento di Chimica e Biochimica Medica, Universitá degli Studi di Milano, Milan, Italy*

DOUGLAS B. SARNEY • *Institute of Food Research, Norwich, UK*

FRANCESCO SECUNDO • *Istituto di Biocatalisi e Riconoscimento Moleculare, CNR, Milan, Italy*

VASILIKI SERETI • *Department of Chemical Engineering, National Technical University, Athens, Greece*

VLADIMIR SIROTKIN • *Department of Chemistry, Kazan State University, Kazan, Russia*

BORIS SOLOMONOV • *Department of Chemistry, Kazan State University, Kazan, Russia*

ANDREW J. SMALLRIDGE • *Department of Chemical Sciences, Victoria University of Technology, Melbourne, Australia*

BORIS SOLOMONOV • *Institute of Soil, Water and Environmental Sciences, The Vulcani Center, Bet Dagan, Israel*

BOONYARAS SOOKKHEO • *Institute of Biological Chemistry, Academia Sinica, Tapei, Taiwan*

ANTJE SPIEß • *AB Biotechnologie II, Technische Universität Hamburg-Harburg, Hamburg, Germany*

HARALAMBOS STAMATIS • *Department of Chemical Engineering, National Technical University, Athens, Greece*

ADRIE J. J. STRAATHOF • *Kluyver Laboratory for Biotechnology, Delft University of Technology, Delft, The Netherlands*

TAKESHI SUGAI • *Technical Research Laboratory, Kansai Paint Co., Kanagawa, Japan*

ATSUO TANAKA • *Department of Synthetic Chemistry and Biological Chemistry, Graduate School of Engineering, Kyoto University, Kyoto, Japan*

FRITZ THEIL • *Department of Chemistry, Liverpool University, Liverpool, UK*

MICHAEL TRANI • *Microbial and Enzymatic Technology Group, Biotechnology Research Institute, Montréal, Québec, Canada*

MAURIE A. TREWHELLA • *Department of Chemical Sciences, Victoria University of Technology, Melbourne, Australia*

REIN ULIJN • *Department of Pure and Applied Chemistry, University of Strathclyde, Glasgow, UK*

ROBERT VAIL • *Biotransformations Group, Schering-Plough Research Institute, Union, NJ*

RAO H. VALIVETY • *Department of Pure and Applied Chemistry, University of Strathclyde, Glasgow, UK*

DIRK-JAN VAN UNEN • *Laboratory of Supramolecular Chemistry and Technology, MESA Research Institute, University of Twente, Enschede, The Netherlands*

EVGENY N. VULFSON • *Department of Food Biochemistry and Biotechnology, Institute of Food Research, Norwich, UK*

KUNG-TSUNG WANG • *Institute of Biological Chemistry, Academia Sinica, Tapei, Taiwan*

ERNST WEHTJE • *Department of Biotechnology, Center for Chemistry and Chemical Engineering, Lund University, Lund, Sweden*

ALEXANDRA N. E. WEISSFLOCH • *Chemica Technologies Inc., Bend, OR*

RENÉ-MARC WILLEMOT • *Centre de Bioingenierie Gilbert Durand, Department de Genie Biochimique et Alimentaire, Complexe Scientifique de Rangueil, Toulouse Cedex, France*

ARISTOTELIS XENAKIS • *Industrial Enzymology Unit, Institute of Biological Research and Biotechnology, The National Hellenic Research Foundation, Athens, Greece*

TSUNEO YAMANE • *Laboratory of Molecular Biotechnology, Graduate School of Bio- and Agro-Sciences, Nagoya University, Nagoya, Japan*

WAN MD ZIN W. YUNUS • *Center for Research in Enzyme and Microbial Technology, Fakulti Sains dan Pengajian Alam Sekitar, University Putra Malaysia, Serdang, Malaysia*

DIMITRIY ZAKHARYCHEV • *Department of Chemistry, Kazan State University, Kazan, Russia*

ALEKSEY ZAKS • *Biotransformations Group, Schering-Plough Research Institute, Union, NJ*

PART I

CONTROL OF ENZYME ACTIVITY IN NONAQUEOUS SOLVENTS

Introduction

Peter J. Halling

This section of *Enzymes in Nonaqueous Solvents: Methods and Protocols* covers methods for the control of general conditions affecting the behavior of enzymes in nonaqueous solvents. Just as in aqueous media, the activity, specificity, and stability of enzymes in nonaqueous media are rather strongly dependent on reaction conditions. Some of the important factors, e.g., temperature and substrate concentration, are the same as in conventional aqueous media, although the dependence may be quantitatively very different. Other parameters are somewhat changed: acid–base effects remain important, but medium pH may no longer be a useful concept to describe them. Some completely new factors come into play, like the selection of solvent and the residual water levels.

The way in which an enzyme is prepared for use in nonaqueous media can have very dramatic effects on its behavior, especially the catalytic activity exhibited. So a first set of contributions deal with methods for catalyst preparation. Lyophilization of an enzyme solution remains very commonly used. It can often result in powders of very low catalytic activity, but Ru et al. describe how this can be greatly improved by drying in the presence of salts. Equally important, they note the influence of the precise conditions of lyophilization, which are rarely specified as clearly as they have done. Unfortunately, laboratory freeze dryers do not usually offer the complete control of temperature and pressure–time courses needed to achieve completely

From: *Methods in Biotechnology, Vol. 15: Enzymes in Nonaqueous Solvents: Methods and Protocols*
Edited by: E. N. Vulfson, P. J. Halling, and H. L. Holland © Humana Press Inc., Totowa, NJ

reproducible lyophilization. Rich and Dordick describe another method to improve the activity and specificity of lyophilized powders, by drying in the presence of a suitable imprinting agent. In organic media, activity displayed per enzyme molecule is often better if they are first immobilized, followed by drying of the conjugate (usually by simple air or vacuum drying). Many methods of immobilization can be used, some of them very simple and straight-forward. Three methods are described here: entrapment within a polymer network (Tanaka and Iida), microencapsulation (Khan and Vulfson) and the use of preformed hydrogel supports (Salleh et al). Both lyophilized powders and immobilized conjugates remain as undissolved solids in organic media, but other treatments can give solubilization, or at least very fine dispersion of the enzymes. Four approaches are described in this book: covalent attachment of polyethylene glycol chains (Matsushima et al.) or a variety of hydrophobic modifers (Basri et al.); or noncovalent complexation with free polyethylene glycol (Secundo et al.) or a synthetic lipid (Mori and Okahata). Finally, Moore et al. describe how a different drying method, rinsing with propanol, can give enzyme preparations retaining high activity.

The residual water level has great effects on biocatalyst behavior in nonaqueous media, so methods for measuring and controlling water are important. Bell et al. describe a variety of methods we have used over the years here at Strathclyde. Wehtje and Adlercreutz describe one convenient method for control of water activity by exchange through the walls of suitable tubing. Rhee et al. cover a related membrane-based approach, pervaporation, and also their experience with the use of salt hydrate pairs. Gardossi describes an interesting special form of celite that can absorb and desorb water to buffer water activity, and which can also serve as an immobilization support.

Various measurements on enzymes in nonaqueous media can help us understand what is happening. Most obviously, enzyme activity and selectivity will be measured, but Ducret et al. point out methodological considerations that are important if meaningful values are to be obtained. Calorimetric methods that can tell us about the enzyme and its interaction with water are presented by Borisover et al., while Kise describes fluorescence and CD measurements that probe changes in the protein molecules. van Unen et al. discuss how crown ethers added in various ways to the system can enhance the activity of low–water enzymes. Methods to control acid–base conditions in organic solvents are presented by Partridge et al. Finally, Bousquet et al. describe the acylation of a glycoside to produce a biosurfactant.

1

Salt-Induced Activation of Enzymes in Organic Solvents

Optimizing the Lyophilization Time and Water Content

Michael T. Ru, Jonathan S. Dordick, Jeffrey A. Reimer, and Douglas S. Clark

1. Introduction

Past studies of enzymatic catalysis in nearly anhydrous organic solvents have shown that the amount of water *adsorbed* to an enzyme in organic solvents is a critical determinant of enzyme activity, more important than the water content of the solvent itself *(1,2)*. The exact amount of water associated with the enzyme, however, varies depending on the hydrophobicity of the solvent, with more hydrophilic solvents tending to strip adsorbed water from the enzyme surface *(3)*. In the case of subtilisin Carlsberg, a particularly well-studied enzyme in organic media, a certain population of this essential water is intricately associated with the enzyme and does not exchange with water in the bulk organic solvent *(4)*. Further water bound to the enzyme effects an increase in active-site polarity, which correlates closely with a sharp increase in enzyme activity *(2)*. The correlation between increased activity and increased active-site polarity suggests various strategies for activating enzymes in organic media.

One of the most effective activation methods demonstrated to date involves the inclusion of simple salts during lyophilization, or freeze-drying, of the enzyme prior to use of the enzyme powder as a suspension in an organic solvent. Indeed, Khmelnitsky et al. showed that including excess KCl in an aqueous enzyme solution prior to lyophilization increased the catalytic efficiency of subtilisin-catalyzed transesterification by 3750-fold *(5)*. Activation to a lesser extent was observed for chymotrypsin. More recently, Triantafyllou

From: *Methods in Biotechnology, Vol. 15: Enzymes in Nonaqueous Solvents: Methods and Protocols*
Edited by: E. N. Vulfson, P. J. Halling, and H. L. Holland © Humana Press Inc., Totowa, NJ

et al. demonstrated that adding buffer salts or KCl also increased the catalytic activity of lyophilized *Candida antarctica* lipase fourfold over that with no added salt *(6)*. Moreover, Bedell et al. found that including KCl in the lyophilizate dramatically enhanced thermolysin-catalyzed peptide synthesis in *tert*-amyl alcohol *(7)*. These findings suggest that salt-induced activation of enzyme catalysis in organic solvents may be a general phenomenon, at least for hydrolytic enzymes. Furthermore, Bedell et al. also found that the enhancement in catalytic activity for both thermolysin and subtilisin Carlsberg is *not* a result of reduced substrate diffusional limitations in a suspended enzyme particle, but rather the result of to a mechanism inherent to the biocatalyst itself.

We present here a method to optimize the salt-induced activation of enzymes in organic solvents by optimizing the lyophilization time and water content. Enzymes are first dissolved in an aqueous buffer solution containing excess salt. The solution is subsequently flash-frozen using liquid nitrogen and vacuum-dried using a lyophilizer. The resultant dry powder is then placed in an organic solvent and the catalytic activity assayed using gas chromatography. The water content is measured using Karl Fischer coulometric titration, and the active-site concentration determined from reaction of the biocatalyst with the protease inhibitor, phenylmethylsulfonylfluoride (PMSF). Our findings show that the catalytic efficiencies of salt-induced enzyme preparations depend on both the lyophilization time and the overall water content of the biocatalyst. The catalytic efficiency of lyophilized enzyme powders suspended in organic solvents increased with freeze-drying time and decreasing water content up to an optimum, after which the catalytic efficiency decreased dramatically. The variable catalytic efficiency was the result of the incorporation of salts as well as the lyophilization process. The active-site concentration was also dependent on lyophilization time, with different organic solvents exhibiting opposite trends in this dependence.

2. Materials
2.1. Enzyme Preparation

1. Subtilisin Carlsberg (SC; EC 3.4.21.14; alkaline protease from *Bacillus licheniformis*; specific activity of 10.4 U/mg solid, Sigma Chemical Co., St. Louis, MO; *see* **Note 1**).
2. KCl and K_2HPO_4 (Sigma).
3. Lipase (EC 3.1.1.3; from *Mucor javanicus*; specific activity of 8.7 U/g solid; Fluka, Milwaukee, WI).
4. 50-mL centrifuge tubes (Falcon, Los Angeles, CA), 20 mL of enzyme–salt solution per tube.
5. 100 m*M* HCl and 100 m*M* KOH solution in nanopure water.
6. Lyophilizer and standard lyophilizer jars (>600 mL capacity).

2.2. Kinetic Studies

Transesterification reactions with SC:

1. *N*-acetyl-L-phenylalanine ethyl ester (APEE, Sigma).
2. 1-Propanol (reagent grade, Fisher Scientific Co, Pittsburgh, PA).

Esterification reactions with lipase:

3. 1-Octanol and octanoic acid (Aldrich Chemical Co., Milwaukee, WI).
4. All solvents were purchased from Fisher Scientific and were of the highest grade commercially available.
5. Molecular sieves, 3 Å Linde (Grace, Baltimore, MD).
6. 20-mL glass scintillation vials and Teflon screw caps (VWR Scientific, West Chester, PA).
7. P-5000 and P-1000 pipetmen (Rainin Instruments, Woburn, MA).
8. Vortex mixer (Cole-Palmer, Vernon Hills, IL).
9. Microcentrifuge and 1.5-mL Eppendorf tubes (Eppendorf, Westbury, NY).
10. Shaker–incubator (New Brunswick Scientific, Edison, NJ).

2.3. Active-Site Titration

1. Phenylmethylsulfonyl fluoride (PMSF), succinyl-ala-ala-pro-phe-*p*-nitroanilide (suc-AAPF-*p*Na), and 100 m*M* Tris-HCl buffer solution, pH 7.8 (Sigma).
2. Dimethyl sulfoxide (DMSO, reagent grade, Fisher).
3. UV/Visible spectrophotometer (Beckman Instruments, Fullerton, CA).

3. Methods
3.1. Enzyme Preparation

We have found that the activity of enzyme powders in organic solvents is highly sensitive to the method of preparation prior to use. Directly suspending salt-free subtilisin Carlsberg purchased from the vendor in hexane yielded non-Michaelis–Menten kinetics and extremely slow catalytic rates. The catalytic activity of enzyme powders in organic solvents also depended on the pH of the aqueous buffer solution prior to freeze-drying, hence the pH was adjusted to an optimum for enzyme activity *(8,9)*. The pH optimum for salt-free subtilisin Carlsberg corresponded to the pH optimum for catalytic activity in water. We also determined that the pH optimum prior to lyophilization for salt-activated SC was 7.8, corresponding to the pH optimum for aqueous activity. The enzyme was activated prior to use by lyophilization from an aqueous phosphate buffer containing KCl.

1. Dissolve 50 mg of subtilisin Carlsberg, 50 mg of K_2HPO_4, and 4.90 g of KCl in 200 mL of nanopure H_2O. Upon freeze-drying, the final dry preparation is 98% w/w salt (*see* **Note 2**).

2. At room temperature, adjust the pH of the aqueous solution to 7.8 using a few drops of 100 m*M* KOH solution.
3. The preparation of lipase samples containing 98% (w/w) KCl was identical except that the pH was adjusted to 7.0 using 100 m*M* HCl solution (*see* **Note 3**).
4. Place 20-mL aliquots into ten 50-mL Falcon tubes and flash-freeze for the same duration using liquid nitrogen (typically 5–6 min). Place plastic screw caps drilled with the same number of holes over each tube and place the tubes in glass lyophilization jars.
5. Lyophilization on a Labconco Freeze Dry 5 freeze-dryer was carried out at a condenser temperature of –50°C and a pressure of 200 µm Hg for 98% KCl–lipase, and 180-190 µm Hg and 260–300 µm Hg for 98% KCl–SC samples used in hexane and tetrahydrofuran (THF), respectively (*see* **Note 4**).
6. Successively remove samples at different times for kinetic assays and measurements of residual water content. Typically, assay samples upon removal from the lyophilizer (*see* **Note 5**).

3.2. Kinetic Studies

In studies of enzyme activity as a function of lyophilization time and water content, the catalytic efficiencies of subtilisin Carlsberg and lipase were determined in dry hexane containing less than $0.008\% \pm 0.003\%$ (w/w) H_2O, as determined by the limit of Karl Fischer titration. In all kinetic experiments, eight substrate concentrations spanning suitable concentration ranges were used. The reaction studied for subtilisin Carlsberg was the transesterification of APEE with 1-propanol, and the kinetic parameters were determined with respect to the APEE substrate. The dependence of the catalytic activity of 98% KCl–SC salt-activated samples on lyophilization time and water content is shown in **Fig. 1**.

1. Store all organic solvents over oven-dried (140°C) molecular sieves at least 24 h prior to use. Wash molecular sieves twice with a sequence of ethanol and hexane prior to drying.
2. For a typical experiment in hexane, add 5 mL of hexane containing 1–40 m*M* APEE, 0.85 *M* of 1-propanol, and 1.5 m*M* nonadecane (nonreacting internal standard for gas chromatography assays) to a reaction vial containing 10 mg of lyophilized enzyme powder (with or without salt).
3. Carry out the reaction at 30°C in 20-mL glass scintillation vials with Teflon-lined screw caps by shaking at 250 rpm in a constant-temperature incubator. Upon addition of the solvent to the lyophilized enzyme powder, vortex the solution vigorously for 5 s to achieve a homogeneous suspension and subsequently placed in the shaker–incubator.
4. Perform initial rate measurements over a 40- to 100-min period for lyophilized salt–enzyme samples, 2–6 h for lyophilized pure enzyme, and 5–22 h for unlyophilized enzyme directly out of the bottle, during which time the rates should remain constant (*see* **Note 6**).

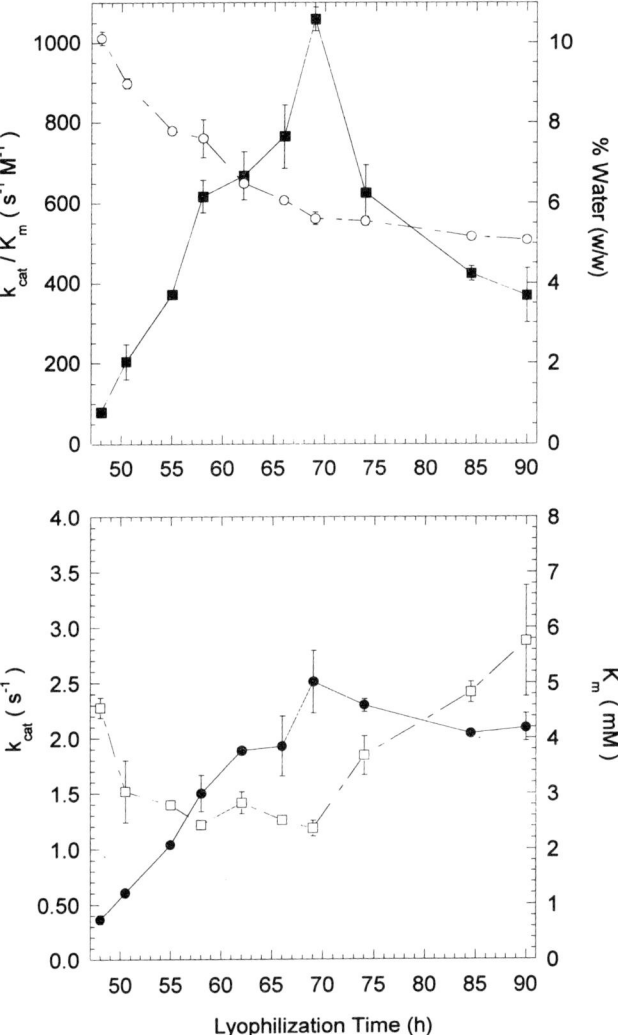

Fig. 1. Kinetic parameters of 98% (w/w) KCl–SC in hexane as a function of lyophilization time. *Top:* k_{cat}/K_m (■) and percent water content of freeze-dried sample (○). *Bottom:* k_{cat} (●) and K_m (□).

5. For the experiments performed in THF, add 10 mg of lyophilized enzyme power to 5 mL of THF containing 20–200 mM APEE and monitor the reaction rate over a 1- to 2-h period, during which time the rates should remain constant.

The reaction studied for lipase is the esterification of octanoic acid with 1-octanol, and the kinetic parameters were determined with respect to the octanoic acid substrate.

1. In a typical experiment in hexane, add 5 mL of hexane containing 1–20 mM (1–40 mM for salt-free lipase) octanoic acid, 45 mM of 1-octanol, and 1.5 mM nonadecane as the internal standard to 10 mg of catalyst (lyophilized 98% [w/w] KCl–lipase or lipase directly from the bottle).
2. Carry out the reaction at 37°C in the same manner as for SC in hexane. Perform initial rate experiments over a 2- to 4-h period for 98% (w/w) KCl–lipase and a 1- to 2-h period for lipase directly out of the bottle, during which time the rates should remain constant.

3.3. Active-Site Titration

It is necessary to determine the concentration of active sites available for catalysis in order to correctly calculate the kinetic constant k_{cat} and the catalytic efficiency k_{cat}/K_m. The fraction of active centers available in water was assumed to be 63 ± 2% for subtilisin Carlsberg, as reported by Wangikar and coworkers *(9)*. Active-site concentrations were determined for the various subtilisin Carlsberg preparations in organic solvents according to the previously published procedure of Wangikar et al. *(9)* with minor modifications. The experimental approach monitors the initial rate of enzyme inhibition using PMSF in the organic solvent and the initial rate of hydrolysis of suc-AAPF-*p*Na. As described by Wangikar and coworkers, a set of six initial rate equations can be reduced to give the ratio of the active-site concentration in organic solvent to that in aqueous buffer. The percentage of active sites in organic solvents can then be calculated by measuring the active-active concentration in aqueous buffer by the procedure of Schonbaum and coworkers *(10)*. The complete active-site titration procedure is as follows:

1. Partition the freshly prepared lyophilized enzyme powder (with or without salt), designated "fresh enzyme," into three equal portions for each organic solvent of interest. Prepare the "inhibited enzyme" by incubating the fresh enzyme (10 mg/mL) in organic solvent containing PMSF, washing the enzyme thoroughly with the neat organic solvent, and vacuum-drying until all solvent is removed. The PMSF concentration is 25 mM in hexane and 100 mM in THF. The lower concentration of PMSF in hexane is required because of the poor solubility of the titrant in the solvent. Prepare the "uninhibited enzyme" by incubating the fresh enzyme (10 mg/mL) in neat organic solvent and then vacuum-dry to remove all solvent. Inhibited and uninhibited enzymes were incubated in solvent with and without PMSF for 24 h at 30°C.
2. Enzyme activity in aqueous buffer was measured for the hydrolysis of suc-AAPF-*p*Na by following the release of *p*-nitroaniline spectrophotometrically at 410 nm. For samples containing salt, hydrolysis was performed in 100 mM Tris-HCl buffer, pH 7.8, using a substrate concentration of 400 μM and catalyst concentrations of 100 μg/mL and 200 μg/mL for enzyme originally inhibited in hexane and THF, respectively. For salt-free enzyme samples, the enzyme concentration was 10 μg/mL.

3. To begin each assay, add appropriate amounts of a 10-mM solution of the suc-AAPF-pNa in DMSO to a 1-mL cuvet containing the enzyme solution. The total reaction volume is 1 mL. Experiments were performed in triplicate.
4. Enzyme activity in organic solvents is determined by monitoring product formation in the transesterification of APEE (10 mM in hexane and 50 mM in THF for samples containing salt, and 20 mM in hexane for the salt-free enzyme) with 1-propanol (0.85 M) using gas chromatography as described in **Subheading 3.4.** All experiments were performed in triplicate. Active-site titration of lipase in hexane was not performed because no titration method was available.

   The active-site concentration of salt-activated SC in organic solvents exhibited a surprising dependence on the solvent type and lyophilization time *(11)*. For the 98% (w/w) KCl–SC catalyst preparation, the active-site concentration *decreased* in hexane from 42% to 20%, whereas it *increased* in THF from 10% to 25% as the lyophilization time increased. The active-site concentration of salt-free SC remained fairly constant at approximately 20%.

3.4. Gas Chromatography Analysis

1. Sample 500-µL aliquots of the homogeneous suspension and centrifuge the suspension in 1.5-mL Eppendorf tubes at 15,300 rpm.
2. Analyze the supernatant by measuring the formation of the transesterification product *N*-acetyl-L-phenylalanine propyl ester using a gas chromatograph (GC). In our analysis, we used a GC (Model 3800, Varian Instruments) equipped with a VA-5MS, 250-µm capillary column (15 m in length and 0.25 µm in inside diameter), a constant He carrier gas pressure of 15 psi (1.3 mL/min), 250°C injection and detection temperatures, and an isothermal column temperature of 215°C.
3. In lipase assays, measure the formation of the esterification product octyl octanoate in the supernatant using a constant He carrier gas pressure of 10 psi and an initial column temperature of 145°C maintained for 1.4 min, ramped to 250°C at 100°C/min, and then held constant for 1.2 min. All GC measurements were performed in triplicate, and initial rates were determined from straight-line fits of the average values.
4. Obtain the kinetic parameters $(V_{max})_{app}$ and $(K_m)_{app}$ by fitting the initial rate data to the Michaelis–Menten equation (Kaleidograph was used in our work). The kinetic parameters were obtained with respect to APEE for subtilisin Carlsberg and octanoic acid for lipase. The intrinsic catalytic efficiencies $(k_{cat}/K_m)_{app}$ were obtained by normalizing $(V_{max}/K_m)_{app}$ by the concentration of active enzyme determined in active-site titration measurements *(12)* (*see* **Note 7**). Reproducibility of kinetic parameters was established from kinetic data obtained from two independent preparations of enzyme powders.

3.5. Measurement of Water Content

The water content of the lyophilized enzyme was determined in each reaction mixture by the Karl Fischer method.

1. Introduce a suspension of catalyst powders (2 mg/mL) in the solvent of choice into capped scintillation vials and equilibrate in a shaker–incubator (250 rpm) for 2 h at 30°C.
2. Prior to introducing the suspension, equilibrate the sealed Karl Fischer titration apparatus with Karl Fischer titrant to react with any water that may be present.
3. Upon equilibration, withdraw the catalyst suspension from the vial using a glass Pasteur pipet (predried in an oven) and immediately transfer the contents into the Karl Fischer titration vessel (*see* **Note 8**).
4. Quickly reseal the vessel with a greased glass stopper and begin the titration. The rapid transfer provides minimal exposure of the sample to environmental water.
5. The water content of the enzyme samples (with or without salt) is calculated as the difference between the total water in the reaction mixture and the water content of the solvent measured prior to adding the enzyme. Values of water content for each lyophilized sample represent averages of eight independent measurements.

4. Notes

1. Other enzymes can also be activated. Enzyme activation by salts has been shown to occur for other proteases such as chymotrypsin and thermolysin, as well as a variety of lipases including those from *Candida antarctica A, B, Mucor miehei,* and *Amano PS.*
2. For SC samples without added salt, freeze the dissolved enzyme solution (pH 7.8, 5 mg/mL) and lyophilize the samples in the same manner as for the salt-activated samples. The phosphate buffer content in the resultant powder should be 1% (w/w) and the enzyme content 99% (w/w).
3. Typically, the pH of the aqueous enzyme solution should be adjusted to the optimal pH for catalytic activity in water. There are exceptions to this rule, however, as noted in **refs.** *13* and *14.*
4. The pressure may vary depending on the types of lyophilizer and vacuum pump used, and pressure variations may potentially result in different optimal conditions specific to the particular drying system.
5. When not in use, store the enzyme powder at $-20°C$ under dry N_2 in sealed (screw caps with no holes) Falcon tubes, and place the tubes over $CaSO_4$ in a desiccator. Before reusing, warm the sealed Falcon tubes to room temperature prior to exposing the contents to air. Avoid exposure of dried samples to ambient moisture by weighing the dried powder into reaction vials under a dry-nitrogen atmosphere.
6. Assay times may vary depending on the activity of the sample. Ideally, rates should remain constant during the assay period.

7. Subtilisin Carlsberg transesterification of APEE with 1-propanol follows the ping-pong, bi-bi mechanism; for further details, see **ref. *13***. Unless otherwise indicated, all subsequent references to the kinetic parameters represent their apparent values $(k_{cat})_{app}$, $(K_m)_{app}$, and $(k_{cat}/K_m)_{app}$.
8. The suspended solids will obstruct syringes and cannulas and therefore cannot be used. Withdraw solids first, then use solvent remaining in the sample to rinse the residual solids adhered to the interior of the pipet.

References

1. Zaks, A. and Klibanov, A. M. (1988) The effect of water on enzyme action in organic media. *J. Biol. Chem.* **263**, 8017–8021.
2. Affleck, R., Xu, Z. F., Suzawa, V., Focht, K., Clark, D. S., and Dordick, J. S. (1992) Enzymatic catalysis and dynamics in low-water environments. *Proc. Natl. Acad. Sci. USA* **89**, 1100–1104.
3. Gorman, L. S. and Dordick, J. S. (1991) Organic solvents strip water off enzymes. *Biotechnol. Bioeng.* **39**, 392–397.
4. Lee, C. S., Ru, M. T., Haake, M., Dordick, J. S., Reimer, J. A., and Clark, D. S. (1998) *Biotechnol. Bioeng.* **57**, 686–693.
5. Khmelnitsky, Y. L., Welsh, S. H., Clark, D. S., and Dordick, J. S. (1994) Salts dramatically enhance activity of enzymes suspended in organic solvents. *J. Am. Chem. Soc.* **116**, 2647,2648.
6. Triantafyllou, A. O., Wehtje, E., Adlercreutz, P., and Mattiasson, B. (1997) How do additives affect enzyme activity and stability in nonaqueous media? *Biotechnol. Bioeng.* **54**, 67–76.
7. Bedell, B. A., Mozhaev, V. V., Clark, D. S., and Dordick, J. S. (1998) Testing for diffusion limitations in salt-activated enzyme catalysts operating in organics solvents. *Biotechnol. Bioeng.* **58**, 654–657.
8. Zaks, A. and Klibanov, A. M. (1985) Enzyme-catalyzed processes in organic solvents. *Proc. Natl. Acad. Sci. USA* **82**, 3192–3196.
9. Schonbaum, G. R., Zerner, B., and Bender, M. L. (1961) The spectrophotometric determination of the operational normality of an a-chymotrypsin solution. *J. Biol. Chem.* **236**, 2930–2935.
10. Wangikar, P. P., Carmichael, D., Clark, D. S., and Dordick, J. S. (1996) Active-site titration of serine proteases in organic solvents. *Biotechnol. Bioeng.* **50**, 329–335.
11. Ru, M. T., Dordick, J. S., Reimer, J. A., and Clark, D. S. (1999) Optimizing the salt-induced activation of enzymes in organic solvents: effects of water content and lyophilization time. *Biotechnol. Bioeng.* **37**, 303–308.
12. Wangikar, P. P., Graycar, T. P., Estell, D. A., Clark, D. S., and Dordick, J. S. (1993) Structure and function of subtilisin BPN' solubilized in organic solvents. *J. Am. Chem. Soc.* **119**, 70–76.
13. Yang, Z., Zacherl, D., and Russell, A. J. (1993) pH dependence of subtilisin dispersed in organic solvents. *J. Am. Chem. Soc.* **115**, 12,251–12,257.
14. Guinn, R. M., Skerker, P. S., Kavanaugh, P., and Clark, D. S. (1990) Activity and flexibility of alcohol dehydrogenase in organic solvents. *Biotechnol. Bioeng.* **37**, 303–308.

2

Imprinting Enzymes for Use in Organic Media

Joseph O. Rich and Jonathan S. Dordick

1. Introduction

Enzymes suspended in nonaqueous media are more rigid than in aqueous media *(1,2)*. This increased rigidity is thought to be the result of increased electrostatic and hydrogen-bonding interactions among the surface residues of the protein in organic solvents *(3)*. Despite this rigidity, enzymes remain active and selective in organic solvents, and this has led to a large number of applications that have impacted the chemical, pharmaceutical, and polymer industries *(4–6)*. Interestingly, because of the rigid structure of biocatalysts when placed in organic media, it has been possible to alter selectivity (e.g., enantioselectivity and regioselectivity) and increase activity through pretreating the enzyme with an inhibitor/substrate analog *(7,8)*. This process, known as "molecular imprinting," locks the enzyme into a conformation that is favorable for catalysis during lyophilization through the addition of the desired substrate or a substrate analog to the enzyme solution prior to freezing (**Fig. 1**). Furthermore, the ligand may prevent the formation of inactive "microconformations" in the active site created during the lyophilization process *(9)*. These hypotheses have been supported by experiments involving the addition of water to the imprinted enzymes. The addition of water to the organic solvent reaction mixture has been shown to strongly depress the activation phenomenon of imprinting because of an increase in enzyme flexibility upon rehydration *(8,10)*.

Enzyme inhibitors *(8)*, substrate analogs *(11)*, and nucleophilic substrates *(10)* have all been successfully used as molecular imprints and have been shown to increase enzyme activity and control enzyme selectivity (both enantioselectivity and substrate selectivity) in organic media *(10,12,13)*. The development of compounds with predetermined molecular recognition

From: *Methods in Biotechnology, Vol. 15: Enzymes in Nonaqueous Solvents: Methods and Protocols*
Edited by: E. N. Vulfson, P. J. Halling, and H. L. Holland © Humana Press Inc., Totowa, NJ

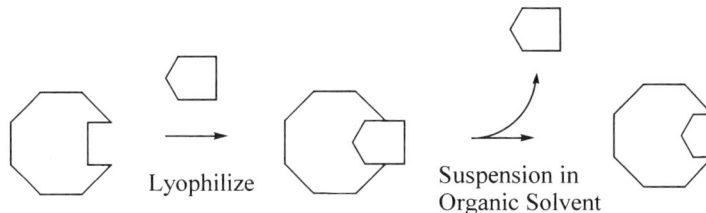

Fig. 1. Representation of the mechanism of enzyme alteration as a result of the addition of an imprinting compound to the lyophilizate. In the lyophilization step, the imprinting compound forces the enzyme active site into a conformation that is more favorable for catalysis. Upon suspension in a solvent in which the imprinting compound is soluble, or rinsing with such a solvent in order to remove the imprinter, the imprinted enzyme is able to accept the substrate of choice. If the imprinting compound is not removed from the active site, the imprinted enzyme is unable to function as a biocatalyst (adapted from **refs.** *8* and *10*).

properties has included the imprinting of synthetic polymers *(14,15)*, hydrogels *(16)*, and proteins *(4)* with a template molecule. Furthermore, the induction of enzymic activity in proteins by lyophilization in the presence of a transition state analog has been recently reported *(17,18)*.

Some important limitations of "molecular imprinting" should be considered when employing this technique. First, the imprinter must be sufficiently soluble in water to obtain the necessary concentration for imprinting, where the effectiveness of a nucleophilic substrate as an imprint is known to be dependent on the concentration of the imprint in the aqueous solution prior to lyophilization *(10)*. Second, the use of imprinted enzymes is limited to nearly anhydrous media where enzymes are sufficiently rigid to maintain the imprint-induced activated conformation. Third, the application of imprinting of enzymes has not been extended beyond hydrolytic enzymes.

The use of additives to the aqueous solution prior to freezing has been shown to prevent the reversible denaturation of proteins during the drying process *(19,20)*. These lyoprotectants, including sucrose and trehalose (and other carbohydrates) and polyethylene glycol, have been shown to increase the activity of many enzymes suspended in anhydrous organic solvents *(11)*. This lyoprotection phenomenon occurs concomitantly with the imprinting, especially when imprinting with nucleophilic substrates such as sugars or nucleosides *(10)*.

Previous studies of imprinting have led to suggestions that the mechanism of the imprinting-induced rate enhancement involves the conservation/alteration of the enzyme active site *(11)*. Similarly, alteration of the substrate specificity via imprinting must also involve the alteration of the enzyme structure, particularly in the vicinity of the transition state. Kinetic observations of altered substrate selectivity show that there must be a structural component to imprinting enzymes *(10)*, and this is consistent with Fourier transform infared (FTIR) studies probing the secondary structure *(21)* and molecular dynamics simulations *(10)* of imprinted proteins.

In conclusion, molecular imprinting of enzymes for use in organic solvents represents a rapid, simple, and often effective methodology to increase enzyme activity and alter enzyme selectivity. Such an approach, in combination with genetic manipulation of enzyme structure (e.g., directed evolution and DNA shuffling) can be used complementarily for the design of biocatalysts with desired properties, particularly for use in extreme environments.

2. Materials
2.1. Imprinting

1. Subtilisin Carlsberg (E.C.3.4.21.62) solution containing 7–15 U/mg solid using casein as the substrate (Sigma): Measure 20 mg of enzyme into a freeze-drying flask.
2. Prepare a solution of 20 mg/mL of water-soluble substrate (e.g., sucrose or thymidine) in 10 mM sodium phosphate buffer, pH 7.8.

2.2. Measurement of Enzyme Activity

1. Imprinted enzyme prepared as detailed in **Subheading 3.1.**
2. Organic solvent dried over 4-Å molecular sieves for 24 h.
3. Prepare a solution of 0.1 M n-butyric acid vinyl ester (TCI America) and 10 mM nucleophile in a suitable organic solvent.
4. High-performance liquid chromatographic (HPLC) grade acetonitrile and water, both containing 0.1% (v/v) trifluoroacetic acid (*see* **Note 1**).

3. Methods
3.1. Imprinting

1. Add 2 mL of substrate solution to 20 mg subtilisin Carlsberg in a glass freeze-drying flask (native enzyme may be prepared by excluding the substrate from the buffer solution).
2. Gently mix the solution for 15 s.
3. Flash-freeze the enzyme–substrate solution in liquid N_2.
4. Lyophilize the mixture for 24 h (*see* **Note 2**).
5. Store the imprinted enzyme in a desiccated environment at –20°C.
6. If necessary, the imprinting compound may be removed from the solid enzyme preparation by washing with a suitable anhydrous organic solvent (*see* **Note 3**).

3.2. Measurement of Enzyme Activity

1. Suspend 1 mg of the imprinted enzyme in 1 mL of anhydrous solvent containing nucleophile and acyl donor in a screw-cap vial.
2. At suitable time intervals (e.g., every 2 h), withdraw samples for analysis by a suitable method (*see* **Note 1**).

4. Notes

1. The analysis of enzyme-catalyzed transesterifications in anhydrous organic solvents typically involves the use of HPLC or GC. For sucrose, GC analysis using a flame ionization detector (FID) and a HP-1 capillary column was employed following precolumn derivitization of the carbohydrates using 1,1,1,3,3,3-hexamethyldisilazane (Sigma Sil-A). The progress of thymidine acylation reactions was followed by reversed-phase high-performance liquid chromatography (RP-HPLC) on a ODS-AQ (C18) column (YMC) using ultraviolet detection at 254 nm and an isocratic mobile phase of 1 mL/min acetonitrile/water (60:40).
2. Lyophilizations were performed using either a Labconco 4.5 or 12 freeze-dryer. The freeze-dryer normally obtained a vacuum of $(10–20) \times 10^{-3}$ torr with a condenser temperature of approx $-50°C$.
3. The removal of a competitive substrate or enzyme inhibitor from the imprinted enzyme preparation may be accomplished by rinsing the enzyme preparation several times (at least three times) with a dry organic solvent in which the imprinting compound is soluble. The nature of the rinsing organic solvent, however, can also affect enzymic activity. If the imprinting molecule is the substrate of interest, it is not necessary to perform this added step.

References

1. Affleck, R., Haynes, C. A., and Clark, D. S. (1992) Solvent dielectric effects on protein dynamics. *Proc. Natl. Acad. Sci. USA* **89,** 5167–5170.
2. Clark, D. S., Creagh, L., Skerker, P., Guinn, M., Prausnitz, J., and Blanch, H. (1989) Enzyme structure and function in water-restricted environments, in *Biocatalysis and Biomimetics* (Burrington, J. D. and Clark, D. S., eds.), American Chemical Society, Washington, DC.
3. Hartsough, D. S. and Merz, K. M., Jr. (1992) Protein flexibility in aqueous and nonaqueous solutions. *J. Am. Chem. Soc.* **114,** 10,113–10,116.
4. Dordick, J. S. (1991) An introduction to industrial biocatalysis, in *Biocatalysts for Industry* (Dordick, J. S., ed.), Plenum, New York.
5. Khmelnitsky, Y. L. and Rich, J. O. (1999) Biocatalysis in nonaqueous solvents. *Curr. Opin. Chem. Biol.* **3,** 47–52.
6. Uhlig, H. (1998) Application of technical enzyme preparations, in *Industrial Enzymes and Their Applications* (translated and updated by Linsmaier-Bednar, E. M.), Wiley, New York.

7. Dabulis, K. and Klibanov, A. M. (1992) Molecular imprinting of proteins and other macromolecules resulting in new adsorbents. *Biotechnol. Bioeng.* **39**, 176–185.
8. Russell, A. J. and Klibanov, A. M. (1988) Inhibitor-induced enzyme activation in organic solvents. *J. Biol. Chem.* **263**, 11,624–11,626.
9. Burke, P. A., Griffin, R. G., and Klibanov, A. M. (1992) Solid-state NMR assessment of enzyme active center structure under nonaqueous conditions. *J. Biol. Chem.* **267**, 20,057–20,064.
10. Rich, J. O. and Dordick, J. S. (1997) Controlling subtilisin activity and selectivity in organic media by imprinting with nucleophilic substrates. *J. Am. Chem. Soc.* **119**, 3245–3252.
11. Dabulis, K. and Klibanov, A. M. (1993) Dramatic enhancement of enzymatic activity in organic solvents by lyoprotectants. *Biotechnol. Bioeng.* **41**, 566–571.
12. Stahl, M., Jeppsson-Wistrand, U., Mansson, M.-O., and Mosbach, K. (1991) Induced stereoselectivity and substrate selectivity of bio-imprinted α-chymotrypsin in anhydrous organic media. *J. Am. Chem. Soc.* **113**, 9366–9368.
13. Stahl, M., Mansson, M.-O., and Mosbach, K. (1990) The synthesis of a D-amino acid ester in an organic media with α-chymotrypsin modified by a bio-imprinting procedure. *Biotechnol. Lett.* **12**, 161–166.
14. Mosbach, K. (1994) Molecular imprinting. *Trends Biochem. Sci.* 9–14.
15. Mosbach, K. and Ramstrom, O. (1996) The emerging technique of molecular imprinting and its future impact on biotechnology. *Bio/technology* **14**, 163–170.
16. Liu, X.-C. and Dordick, J. S. (1999) Sugar acrylate-based polymers as chiral molecularly imprintable hydrogels. *J. Polym. Sci. Part A: Polym. Chem.* **37**, 1665–1671.
17. Ohya, Y., Miyaoka, J., and Ouchi, T. (1996) Recruitment of enzyme activity in albumin by molecular imprinting. *Macromol. Rapid Commun.* **17**, 871–874.
18. Slade, C. J. and Vulfson, E. N. (1998) Induction of catalytic activity in proteins by lyophilization in the presence of a transition state analogue. *Biotechnol. Bioeng.* **57**, 211–215.
19. Hellman, K., Miller, D. S., and Cammack, K. A. (1983) The effect of freeze-drying on the quaternary structure of L-asparaginase from *Erwinia carotovora*. *Biochim. Biophys. Acta* **749**, 133–142.
20. Lee, J. C. and Timasheff, S. N. (1981) The stabilization of proteins by sucrose. *J. Biol. Chem.* **256**, 7193–7201.
21. Griebenow, K. and Klibanov, A. M. (1995) Lyophilization-induced reversible changes in the secondary structure of proteins. *Proc. Natl. Acad. Sci. USA* **92**, 10,969–10,976.

3

Entrapment of Biocatalysts by Prepolymer Methods

Atsuo Tanaka and Takamitsu Iida

1. Introduction

Entrapment of biocatalysts should be carried out under mild conditions, especially when synthetic polymer materials are used as supports. In the case of photo-crosslinkable resin prepolymers having photosensitive functional groups, irradiation of near-ultraviolet (UV) light (maximum intensity at 365 nm) induces radical polymerization of the prepolymers and gel formation completes within a few minutes. Various types of prepolymers with different chain lengths, ionic properties, and/or hydrophilicity–hydrophobicity balance have been developed (1–4). The gels thus formed by photo reaction have a higher mechanical strength than those prepared by thermochemical reaction.

Urethane resin prepolymers with different properties (chain length and/or hydrophilicity–hydrophobicity balance) can form gels by reacting with water. The prepolymers have functional isocyanate groups in the molecules, which can polymerize through urea linkages in the presence of water (2–5).

The features of the prepolymer methods are summarized as follows:

1. The entrapment procedures are simple and mild. Gels can be prepared by only a short-term irradiation of near-UV light to photo-crosslinkable resin prepolymers or by just mixing urethane resin prepolymers with water.
2. The gels entrapping biocatalysts have a great mechanical strength.
3. The properties of the gels formed (network structure, ionic character, and hydrophilicity–hydrophobicity balance) can be regulated by selecting appropriate prepolymers.
4. Prepolymers do not contain monomers, which often have bad effects on biocatalysts to be entrapped.

From: Methods in Biotechnology, Vol. 15: Enzymes in Nonaqueous Solvents: Methods and Protocols
Edited by: E. N. Vulfson, P. J. Halling, and H. L. Holland © Humana Press Inc., Totowa, NJ

This chapter deals with the immobilization of biocatalysts by entrapment with photo-crosslinkable resin prepolymers and urethane resin prepolymers, and applications of these entrapped biocatalysts in organic solvents, especially emphasizing the advantage of hydrophobic gels over hydrophilic gels in these reactions.

2. Materials

2.1. Prepolymers

1. Photo-crosslinkable resin prepolymers (*see* **Note 1**): Model structures and properties of typical photo-crosslinkable resin prepolymers are shown in **Fig. 1** and **Table 1**.

    ENT, ENTP, and ENTB were prepared from hydroxyethyl acrylate (HEA), isophorone diisocyanate (IPDI), and polyethylene glycol (PEG) (in the case of ENT), polypropylene glycol (PPG) (in the case of ENTP), or polybutadiene with hydroxyl groups introduced at both ends (PB) (in the case of ENTB). HEA, PEG, and PPG were purchased from Wako Pure Chemical Industries, Osaka, Japan; IPDI from Kanto Chemicals Co., Tokyo, Japan; PB from Nippon Soda Co., Tokyo, Japan. Dibutyltin dilaurate, a catalyst for urethane linkage formation, and hydroquinone, a radical inhibitor, were obtained from Wako Pure Chemical Industries.

    Equimolar amounts of HEA and IPDI were reacted at 70°C in the presence of 0.05% hydroquinone and 0.03% dibutyltin dilaurate in toluene. After 2 h of reaction, a half-molar ratio of PEG (mol wt, approx 1000, 2000, 4000, or 6000), PPG (mol wt, approx 1000, 2000, 3000, or 4000), or PB (mol wt, approx 1000 having hydroxyl groups at both terminals), 0.01% dibutyltin dilaurate, and hydroquinone (to give 0.2%) were added to the mixture, and the reaction was continued for further 5 h at 60°C. Then, toluene was removed under a reduced pressure until the solvent was not detected by gas chromatography *(4)*. Each prepolymer has a linear skeleton of optional length and both ends are attached to photosensitive groups, acryloyl. The photo-crosslinkable resin prepolymers are stored at 4°C in the dark. ENT containing PEG is water miscible and gives a hydrophilic gel, whereas ENTP having PPG and ENTB containing PB are water immiscible and form hydrophobic gels. These prepolymers are the products of Kansai Paint Co., Osaka, Japan.

2. Urethane resin prepolymers (*see* **Note 2**): A model structure and properties of typical urethane resin prepolymers (PU) are shown in **Fig. 2**.

    Urethane resin prepolymers PU are prepared from toluene diisocyanate and polyether diols composed of polyethylene glycol and polypropylene glycol or polyethylene glycol alone. A typical prepolymer synthesis is as follows *(5)*: 2 mol of toluene diisocyanate (2,4-isomer:2,6-isomer, 80:20) and 1 mol of polyether diol were mixed and reacted at 80°C for 3 h. Prepolymers with a different hydrophilic or hydrophobic character can be obtained by changing the ratio of polyethylene glycol and polypropylene glycol in the polyether diol. For example, PU-3, with a high content of polypropylene glycol, gives a hydrophobic gel, whereas PU-6, with a high content of polyethylene glycol, forms a

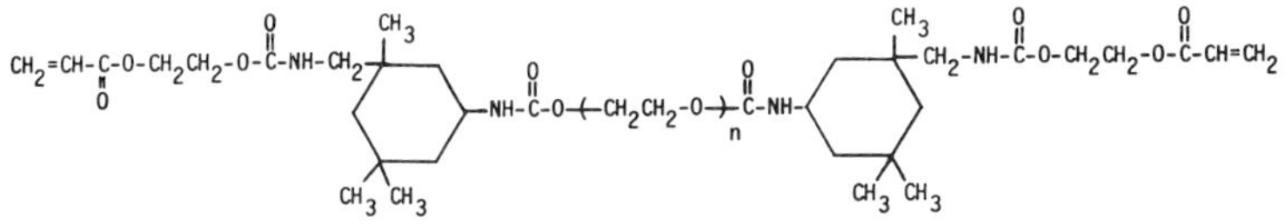

Fig. 1. Model structures of typical photo-crosslinkable resin prepolymers. ENT is water miscible and gives a hydrophilic gel, whereas ENTP and ENTB are water immiscible and give hydrophobic gels. The name "ENT" was derived from the term "entrapment." P and B represent "propylene glycol" and "butadiene", respectively.

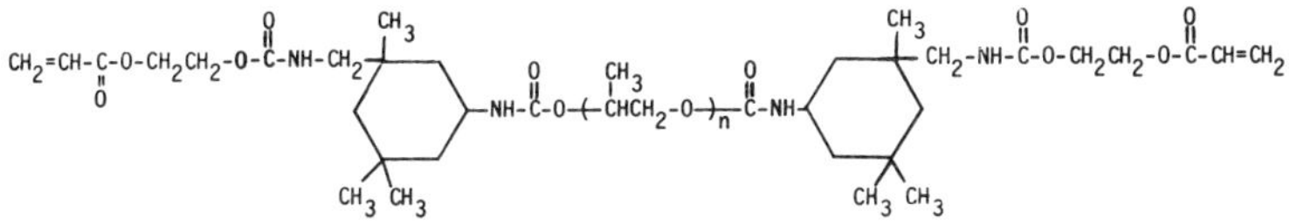

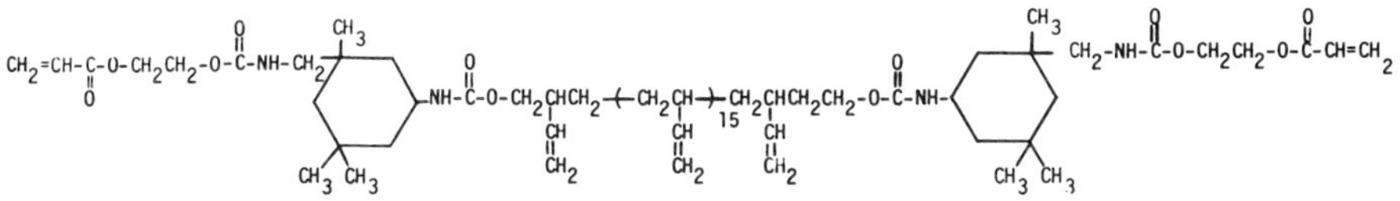

Table 1
Properties of Typical Photo-Crosslinkable Resin Prepolymers

Prepolymer	Main chain of prepolymer	Molecular weight of main chain	Property
ENT-1000	Polyethylene glycol	1000	Hydrophilic
ENT-2000	Polyethylene glycol	2000	Hydrophilic
ENT-4000	Polyethylene glycol	4000	Hydrophilic
ENT-6000	Polyethylene glycol	6000	Hydrophilic
ENTP-1000	Polypropylene glycol	1000	Hydrophobic
ENTP-2000	Polypropylene glycol	2000	Hydrophobic
ENTP-3000	Polypropylene glycol	3000	Hydrophobic
ENTP-4000	Polypropylene glycol	4000	Hydrophobic
ENTB-1000	Polybutadiene	1000	Hydrophobic

Note: Prepolymers are the products of Kansai Paint Co., Osaka, Japan.

hydrophilic gel, although both prepolymers are water miscible. The chain length of the polyether diol can be changed. These prepolymers are stored in a desiccator to avoid moisture. A series of PU are the products of Toyo Tire and Rubber Co., Osaka, Japan, although these are not commercialized.

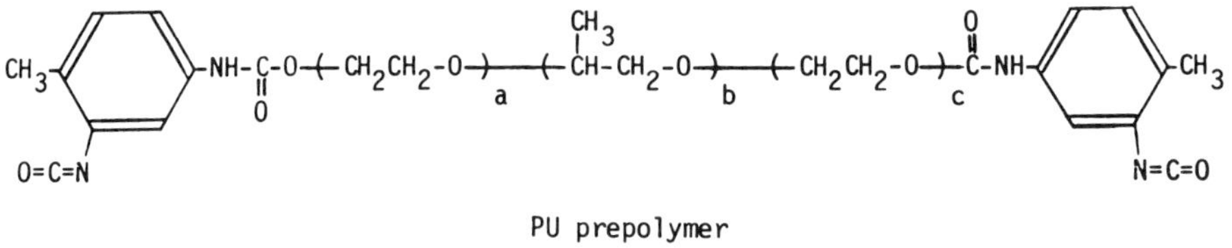

$$CH_3-\underset{\substack{| \\ O=C=N}}{\boxed{}}-NH-\underset{\underset{O}{\parallel}}{C}-O\underset{a}{-(-CH_2CH_2-O-)}\underset{b}{(-\underset{\underset{CH_3}{|}}{CH}-CH_2-O-)}\underset{c}{(-CH_2CH_2-O-)}\underset{\underset{O}{\parallel}}{C}-NH-\underset{\substack{| \\ N=C=O}}{\boxed{}}-CH_3$$

PU prepolymer

Prepolmer	M̄w of diol	NCO content (%)	Ethylene glycol content (%)
PU-3	2529	4.2	57
PU-6	2627	4.0	91

Fig. 2. A model of structure and properties of typical urethane resin prepolymers. Although both PU-3 and PU-6 are water miscible, PU-3 gives a hydrophobic gel, whereas PU-6 gives a hydrophilic gel. Prepolymers are the products of Toyo Tire & Rubber Co.

3. These prepolymers were employed for the entrapment of various biocatalysts (*see* **Note 3**).

2.2. Photoinitiator (see Note 4)

Benzoin ethyl ether (Wako Pure Chemical Industries) was mainly used for photopolymerization as a photoinitiator, at a concentration of 1%, to a photo-crosslinkable resin prepolymer. As this compound is a crystalline solid at room temperature, the initiator should be dissolved completely into the prepolymer by heating at above 60°C prior to the addition of biocatalysts.

2.3. Irradiation Apparatus

The irradiation apparatus consists of UV lamp(s) and a cooling unit, if necessary. The UV lamp used is a low-energy mercury lamp such as a fluorescence chemical lamp or a black-light fluorescence lamp (both are the products of Toshiba Electric Co., Tokyo, Japan). The lamp should be used at a power of 5 W/m^2 or higher.

2.4. Biocatalysts

1. Lipases: Lipase OF 360 from *Candida cylindracea* (360 U/mg powder) was purchased from Meito Sangyo Co., Tokyo, Japan; lipase from *Rhizopus delemar* (6000 U/mg) from Seikagaku Kogyo Co., Osaka, Japan; and lipase P from *Pseudomonas* sp. from Amano Pharmaceutical Co., Nagoya, Japan.

2. *Nocardia rhodocrous* cells: *N. rhodocrous* NCIB 10554 obtained from National Collection of Industrial Bacteria, Torry Research Station, Aberdeen, UK, was maintained on a nutrient agar slant at 4°C. To obtain the bacterial cells, cultivation was carried out at 30°C by shaking (220 rpm) in 500-mL flasks containing 100 mL of a medium of following composition; 10.0 g/L glycerol, 10.0 g/L yeast extract, 2.0 g/L $(NH_4)_2SO_4$, 2.0 g/L K_2HPO_4, 0.1 g/L $MgSO_4 \cdot 7H_2O$, 0.01 g/L $CaCl_2 \cdot 2H_2O$, and 0.01 g/L $FeSO_4 \cdot 7H_2O$ (*6*). The pH was adjusted to 7.0 before sterilization. Seven hours after inoculation (inoculum size, approx 0.2 mg dry cell/mL), a suspension of cholesterol was added to give a concentration of 1.8 mg cholesterol/mL medium, and the cells were harvested by centrifugation after cultivation for an additional 17 h. The time of the inducer addition and the cell harvest corresponded to the beginning of exponential growth phase and of the stationary phase, respectively (*see* **Note 5**). After the harvested cells were washed once with 20 m*M* potassium phosphate buffer (pH 7.0), the cell paste was kneaded thoroughly and stored in sealed sample bottles at –20°C. The frozen cells retained the original activity of 3β-hydroxysteroid dehydrogenase for at least 1 mo. The suspension of cholesterol was prepared under vigorous stirring with 5 g of cholesterol and 833 mg of Tween-80 in 250 mL of water and then sterilized. An aliquot (10 mL) of this sterilized suspension was added to 100 mL of the culture broth.

3. *Rhodotorula minuta* cells: *R. minuta* var. *texensis* IFO 1102 obtained from Institute for Fermentation, Osaka, Japan, was cultivated with shaking (220 rpm) at 30°C in 500-mL flasks containing 100 mL of a medium composed of 5.0 g treacle, 5.0 g corn steep liquor, 0.5 g $(NH_4)_2SO_4$, and 1.0 mL mineral mixture (*7*). The mineral mixture was as follows: 20 g/L $MgSO_4 \cdot 7H_2O$, 5 g/L $FeSO_4 \cdot 7H_2O$, 2 g/L $CaCl_2$, 0.2 g/L $MnCl_2 \cdot 4H_2O$, 0.1 g/L $NaMoO_4 \cdot 2H_2O$, and 0.1 g/L NaCl. The pH of the medium was adjusted to 7.0 with 1*M* NaOH before sterilization. Cells were harvested by centrifugation after 70 h of cultivation, washed twice with 20 m*M* potassium phosphate buffer (pH 7.0), kneaded thoroughly, and stored in sealed sample bottles at –20°C. The frozen cells maintained their original hydrolytic activity for at least 1 mo.

4. Yeast cells: Baker's yeast was obtained from Oriental Yeast Co., Tokyo, Japan. Before use, the yeast was grown by shaking (220 rpm) at 30°C for 20 h in 100 mL medium (pH 5.2) containing 5.0 g glucose, 1.0 g tripotassium citrate·H_2O, 0.2 g citric acid, 0.11 g KH_2PO_4, 0.8 g casamino acids, 8 mg tryptophan, 85 mg KCl, 25 mg $MgSO_4 \cdot 7H_2O$, 19 mg $CaCl_2$, 0.5 mg $FeCl_2$, 0.5 mg $MnSO_4$, 0.75 mg nicotinic acid, 0.5 mg inositol, 0.38 mg calcium pantothenate, 0.15 mg thiamine hydrochloride, 3 μg pyridoxine hydrochloride, and 2.4 μg biotin. Cells were then harvested by centrifugation and washed with 0.9 % NaCl (*8*).

3. Methods

3.1. Entrapment of Enzymes and Microbial Cells

3.1.1. Lipases

Urethane resin prepolymer (0.5 g), melted at 60°C if necessary and cooled to room temperature, was mixed quickly in a beaker with 200 μL of deionized water containing 100 mg of lipase OF 360. After the gelation started, the mixture was kept at 4°C for 1 h to complete the polymerization. The gel thus formed was cut into small pieces (approx 3 × 3 × 3 mm) and used for the enantioselective esterification of menthol *(9)*.

Lipase from *Rhizopus delemar* was entrapped with ENT, ENTP, and PU *(10)*. ENT (0.5 g) was mixed with 5 mg of benzoin ethyl ether and 100 μL of 0.3 *M* *N*-tris(hydroxymethyl)methyl-2-aminoethanesulforic acid (TES)-NaOH buffer (pH 6.5 at 40°C). The mixture was melted by warming at 60°C, cooled to room temperature, and then 5 mg of free lipase in 100 μL of the buffer was added to the molten mixture. This mixture was layered on a sheet of transparent polyester. The layer (thickness, approx 0.5 mm) was covered with another sheet of the same material and then illuminated with near-UV light for 3 min *(10)*. The gel formed was cut into small pieces (approx 3 × 3 mm) and used for the interesterification of triglyceride. In some cases, C-lipase corresponding to 5 mg of free lipase was immobilized by the same procedure. C-lipase was prepared by mixing well 5 mg of lipase in 200 μL of the buffer with 0.25 g of Celite No. 535 (Johns-Manville Co., Denver, CO).

When the hydrophobic prepolymer (ENTP) was used, 0.5 g of the prepolymer and 5 mg of benzoin ethyl ether were dissolved in 400 μL of water-saturated *n*-hexane. Free lipase in 200 μL of the buffer containing 40 mg of Tween-80 or C-lipase (corresponding to 5 mg of free lipase) was added to the mixture and the entrapment was carried out as earlier. Entrapment with PU (0.5 g) was performed as described earlier using 5 mg of lipase dissolved in 200 μL of the buffer.

Lipid-coated lipase P was entrapped by ENTP **(Fig. 3)** *(11)*. An aqueous solution (250 mL) of lipase P (2.0 g) and aqueous dispersion (250 mL) of dioleyl glucosylglutamate (2.0 g) [synthesized according to the method of Goto et al. *(12)*] were mixed under sonication (20 kHz) for 20 min, and the suspension was incubated in a refrigerator for 1 d. The precipitates were collected by centrifugation (18,000*g*, 5 min) and dried *in vacuo* (60 mm Hg) for 12 h. ENTP (1 g) and 10 mg of benzoin ethyl ether were mixed with 1.2 mL of benzene and dissolved by warming at 60°C, cooled to room temperature, and then lipid-coated lipase P (100 mg; enzyme content, 2.8%) was dissolved in the solution. This mixture was spread on a Petri dish, dried for 10 min in air at room temperature (about 15°C), covered with a polyester film, and then irradiated from both sides by UV rays for 10 min. The resulting gel (thickness, approx 0.5 mm) was cut into small pieces (approx 3 × 3 mm), washed thoroughly with benzene, and dried *in vacuo* (60 mm Hg) for 30 min. The entrapped enzyme was used for esterification and transesterification.

3.1.2. Nocardia rhodocrous *Cells* **(6)**

Thawed cells of *N. rhodocrous* (1 g wet cells), suspended in 2.0 mL of a water-saturated mixture of benzene and *n*-heptane (1:1, v/v) under mild soni-

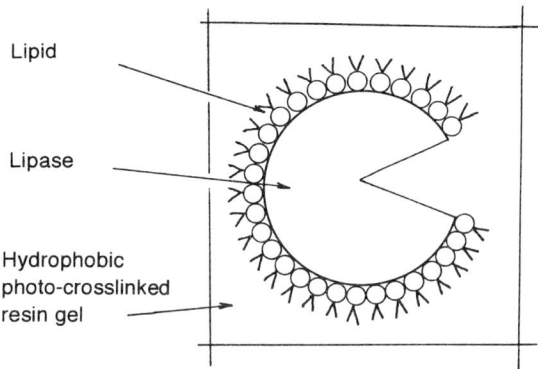

Fig. 3. Schematic model of lipid-coated lipase entrapped in hydrophobic photo-crosslinked resin gel matrix.

cation (10 kHz for several seconds), were mixed with 1.0 mL of the same solvent containing 1.0 g of ENTP or ENTB and 10 mg of benzoin ethyl ether. The mixture was laid on a sheet of transparent polyester and illuminated with near-UV light for 3 min, as described earlier. The resin gel thus formed (thickness, approx 0. 5 mm) was cut into small pieces (approx 1.5 × 1.5 mm) and used for the transformation of 3β-hydroxy-Δ^5-steroids. Mixtures of ENTP and ENT (ENTP content, above 70%) were also employed for entrapment.

The entrapment method with ENT or mixtures of ENT and ENTP (ENT content, above 70%) was the same as that described earlier, except that 20 m*M* potassium phosphate buffer (pH 7.0) was substituted for the organic solvent.

Thawed cells of *N. rhodocrous* (1 g wet cells), suspended in 2.0 mL of the buffer under mild sonication, were mixed quickly in a beaker with 1.0 g of PU-3, PU-6, or their mixtures melted previously at 40°C. After the gelation started, the mixture was kept at 4°C for 30 min, and the formed gel was cut into small pieces (approx 3 × 3 × 3 mm) to use for the reaction.

3.1.3. Rhodotorula minuta *Cells* (7)

Thawed cells (1 g wet cells) suspended in 3 mL of water were mixed with 1 g of ENT dissolving 10 mg of benzoin ethyl ether and entrapped as mentioned in **Subheading 3.1.2.** Thawed cells (1 g wet cells) suspended in 2 mL of water were entrapped with 1 g of PU-3, PU-6, or their mixtures, as described in **Subheading 3.1.2.**

3.1.4. Yeast Cells (8)

Baker's yeast was immobilized by the double-entrapping method. Cultured cells (4.7 g wet cells) were suspended in 3.6 mL of 2.5% (w/v) sodium alginate solution at 40–45°C. The suspension was dropped from a syringe into 0.1 *M* CaCl₂ solution and kept at room temperature for 3 h, and the gel beads formed (diameter, approx 5 mm) were washed with saline (0.9% [w/v] NaCl solution). These calcium-alginate-entrapped cells were further added to a mixture of deionized water (2.0 mL) and PU-6 (2.0 g) and mixed gently for several minutes until the polymerization started. After standing for 1.5 h at 4°C, the gel formed was cut into blocks (approx 2.5 × 2.5 × 1. 0 cm) and used for the reaction. All the materials, except for yeast cells, were sterilized before use and entrapment was carried out aseptically. PU-6 should be autoclaved (121°C, 2 atm), avoiding moisture.

3.2. Enzymatic Reactions in Organic Solvents

3.2.1. Enantioselective Esterification of Menthol by Candida cylindracea Lipase OF 360 (9,13) (see Note 6)

The reaction was carried out at 30°C with shaking (120 strokes/min). The reaction mixture in a 100-mL flask was composed of entrapped lipase OF 360 (corresponding to 100 mg of free lipase) and 10 mL isooctane or cyclohexane containing 130 m*M* d*l*-menthol and 100 m*M* 5-phenylvaleric acid. The solvent was water-saturated before dissolving the substrates. The amounts of the reactants were determined by high-performance liquid chromatography (HPLC) and optical rotation of the product was measured with a polarimeter.

3.2.2. Regiospecific Interesterification of Triglyceride by Rhizopus delemar Lipase (10) (see Note 7)

The reaction was carried out at 40°C with shaking (120 strokes/min). The reaction mixture was composed of entrapped preparations of free lipase or C-lipase (corresponding to 5 mg of free lipase), 0.25 g of olive oil, and 0.25 g of stearic acid in 10 mL of water-saturated *n*-hexane. After development by thin-layer chromatography (TLC), the reformed triglyceride was scraped off from a thin-layer plate and extracted with ethyl ether. After hydrolysis of triglyceride, liberated fatty acids were esterified with BF₃ in methanol, and the methyl esters formed were analyzed by gas chromatography.

3.2.3. Esterification and Transesterification by Pseudomonas Lipase P (11)

The reaction was performed at 30°C with magnetic stirring (220 rpm). The reaction for esterification was composed of 10 mL of benzene, 2.5 mmol lauric acid, 5.9 mmol benzyl alcohol, and 500 mg of entrapped lipid-coated lipase P. The reactants were analyzed using a TSK-GEL ODS-120T column. The reaction mixture for transesterification was composed of 4 mL of organic solvent such as diisopropyl ether, 2 mmol racemic sulcatol, 2 mmol isopropenyl acetate,

1 mmol chlorobenzene as internal standard, and 422 mg of the entrapped lipid-coated lipase P. The composition and enantiomer ratio of the reaction mixture were monitored by gas chromatography using a SUPELCO β-DEX 120 chiral capillary column (60 m).

3.2.4. Transformation of Steroids by N. rhodocrous Cells (6) (see Note 8)

Transformation of 3β-Δ^5-steroids (cholesterol, β-sitosterol, stigmasterol, and dehydroepiandrosterone) to the corresponding 3-oxo-Δ^4-steroids (cholestenone, β-sitostenone, stigmastenone, and 4-androstene-3,17-dione) was carried out by entrapped cells of *N. rhodocrous* at 30°C with shaking (120 strokes/min). The reaction mixture was composed of the entrapped cells corresponding to 1.0 *g* wet cells and 100 mg of a steroid in 10 mL of a water-saturated mixture of benzene and *n*-heptane (1:1, v/v). Each steroid was assayed by gas chromatography.

3.2.5. Enantioselective Hydrolysis of dl-Menthyl Succinate by Rhodotorula minuta Cells (7,13)

dl-Menthyl succinate was synthesized as follows; succinic anhydride (15.7 g) was added dropwise to *dl*-menthol (24.5 g) solution in *o*-xylene with stirring at 80°C. After 3 h, solvent was removed to obtain white crystals, which were recrystallized from benzene. The enzyme reaction was carried out at 30°C with shaking (180 strokes/min). The reaction mixture was composed of 39 m*M* *dl*-menthyl succinate and the entrapped cells corresponding to 1 g wet cells in 10 mL of water-saturated *n*-heptane. The amount of *l*-menthol formed was assayed by gas chromatography and optical rotation of the product was measured with a polarimeter.

3.2.6. Stereoselective Reduction of Ethyl 3-Oxobutanoate by Baker's Yeast (8) (see Note 9)

Reduction of ethyl 3-oxobutanoate to ethyl *(S)*-3-hydroxybutanoate was carried out with the doubly entrapped yeast cells corresponding to 4.7 g wet cells in 20 or 40 mL of isooctane containing 50 m*M* ethyl 3-oxobutanoate at 30°C with shaking (120 strokes/min). Gas chromatography was applied to calculate the conversion ratio and the stereoselectivity of the reaction was determined by measuring the optical purity of the product by HPLC.

Different kinds of yeasts were also applied for the asymmetric reduction of oxo-compounds (*see* Note 10).

4. Notes

1. Various types of photo-crosslinkable resin prepolymers *(2,4)* were synthesized by similar methods and applied to the entrapment of different biocatalysts. When

ENT and ENTP are mixed to control the hydrophilicity–hydrophobicity balance of gels, the mixtures are water miscible up to 30% (w/w) of ENTP.

2. A series of urethane resin prepolymers (PU) having different chain lengths and/or polyethylene glycol contents in polyether diol were prepared by the same method *(2,4,14)*. All the prepolymers are water miscible and can be used in any mixing ratio.

3. Photo-crosslinkable resin prepolymers and urethane resin prepolymers have been used to entrap not only enzymes but also cells of various conditions *(15,16)*.

4. Benzoin isobutyl ether can also be used as a photoinitiator at the same concentration as benzoin ethyl ether.

5. To induce steroid Δ^1-dehydrogenase, 4-androstene-3, 17-dione was applied as an inducer.

6. Although only *l*-menthol was esterified by lipase OF 360 with 5-phenylvaleric acid, enantioselectivity of the reaction was dependent on the kind of acyl donors. Porcine pancreas lipase (19 U/mg) (Wako Pure Chemical Industries) was also applicable to this reaction. PU-3 and PU-6 gave almost the similar results on the enzymatic activity *(9)*.

   Lipase OF 360 entrapped with the prepolymers was used for the ester formation from oleic acid and *n*-heptanol or oleic acid and glycerol in organic solvents *(17)*, esterification of citronellol in water-saturated cyclohexane *(18)*, and hydrolysis of triglyceride in an oil–water emulsion system *(19)*. In these cases, the enzyme entrapped with hydrophobic prepolymers (ENTP and PU-3) showed better activities than that entrapped with hydrophilic prepolymers (ENT and PU-6). As for the esterification of citronellol with 5-phenylvaleric acid, various kinds of lipases and esterases were found to be active *(18)*.

7. In the interesterification of triglyceride by *R. delemar* lipase, entrapment of C-lipase with ENTP gave the highest and the most stable activity *(10)*.

8. *Nocardia rhodocrous* showed the activities of steroid Δ^1-dehydrogenation, 3β-hydroxysteroid dehydrogenase, and 17β-hydroxysteroid dehydrogenation after appropriate induction of the enzyme systems and transformed a variety of steroids *(6,20,21)*.

   In general, the hydrophobic gel-entrapped cells showed higher activities of steroid transformation in organic solvents than the hydrophilic gel-entrapped cells, because of the high partition of substrates into the gels. However, the effect of the hydrophilicity–hydrophobicity balance of gels depends significantly on the hydrophobicity of substrates and the polarity of solvents *(6,20–22)*. The water content of gels is another factor, as shown on esterification by lipase *(17)* and dehydrogenation of cholesterol by *N. rhodocrous* cells *(23)*.

9. When the activity of the doubly entrapped cells becomes insufficient, reactivation is possible by cultivating the entrapped cells at 30°C for 8–12 h with shaking in the culture medium supplemented with 0.3% $CaCl_2$. Thus, the entrapped cells survived and retained the activity for at least 50 d over repeated reactions in isooctane *(8)*. Most of the cells were present in calcium alginate gel but not in polyurethane gel.

10. Prepolymer-entrapped yeasts, such as *Saccharomyces delbrueckii, Saccharomyces fermentati, Candida albicans,* and *Kloeckera saturnus,* were used for the

asymmetric reduction of 2-methyl-3-oxo esters in water-saturated organic solvents (benzene, *n*-hexane, *n*-heptane, and isooctane) *(24)*.

References

1. Fukui, S., Tanaka, A., Iida, T., and Hasegawa, H. (1976) Application of photo-crosslinkable resin to immobilization of an enzyme. *FEBS Lett.* **66,** 179–182.
2. Fukui, S. and Tanaka, A. (1984) Application of biocatalysts immobilized by prepolymer methods. *Adv. Biochem. Eng./Biotechnol.* **29,** 1–33.
3. Fukui, S. and Tanaka, A. (1985) Enzymatic reactions in organic solvents. *Endeavour, New Series* **9,** 10–17.
4. Fukui, S., Sonomoto, K., and Tanaka, A. (1987) Entrapment of biocatalysts with photo-crosslinkable resin prepolymers and urethane resin prepolymers. *Methods Enzymol.* **135,** 230–252.
5. Fukushima, S., Nagai, T., Fujita, K., Tanaka, A., and Fukui, S. (1981) Hydro-philic urethane prepolymers: Convenient materials for enzyme entrapment. *Biotechnol. Bioeng.* **20,** 1465–1469.
6. Omata, T., Iida, T., Tanaka, A., and Fukui, S. (1979) Transformation of steroids by gel-entrapped *Nocardia rhodocrous* cells in organic solvent. *Eur. J. Appl. Microbiol. Biotechnol.* **8,** 143–155.
7. Omata, T., Iwamoto, N., Kimura, T., Tanaka, A., and Fukui, S. (1981) Stereoselective hydrolysis of *dl*-menthyl succinate by gel-entrapped *Rhodotorula minuta* var. *texensis* cells in organic solvent. *Eur. J. Appl. Microbiol. Biotechnol.* **11,** 199–204.
8. Kanda, T., Miyata, N., Fukui, T., Kawamoto, T., and Tanaka, A. (1998) Doubly entrapped baker's yeast survives during the long-term stereoselective reduction of ethyl 3-oxobutanoate in an organic solvent. *Appl. Microbiol. Biotechnol.* **49,** 377–381.
9. Koshiro, S., Sonomoto, K., Tanaka, A., and Fukui, S. (1985) Stereoselective esterification of *dl*-menthol by polyurethane-entrapped lipase in organic solvent. *J. Biotechnol.* **2,** 47–57.
10. Yokozeki, K., Yamanaka, S., Takinami, K., Hirose, Y, Tanaka, A., Sonomoto, K., and Fukui, S. (1982) Application of immobilized lipase to regio-selective interesterification of triglyceride in organic solvent. *Eur. J. Appl. Microbiol. Biotechnol.* **14,** 1–5.
11. Fukunaga, K., Minamijima, N., Sugimura, Y, Zhang, Z., and Nakao, K. (1996) Immobilization of organic solvent-soluble lipase in nonaqueous conditions and properties of the immobilized enzymes. *J. Biotechnol.* **52,** 81–88.
12. Goto, M., Kameyama, H., Goto, M., Miyata, M., and Nakashio, F. (1993) Design of surfactants suitable for surfactant-coated enzymes as catalysts in organic media. *J. Chem. Eng. Jpn.* **26,** 109–111.
13. Fukui, S. and Tanaka, A. (1987) Optical resolution of *dl*-menthol by entrapped biocatalysts. *Methods Enzymol.* **136,** 293–302.
14. Sonomoto, K., Jin, I.-N., Tanaka, A., and Fukui, S. (1980) Application of urethane prepolymers to immobilization of biocatalysts: Δ^1-dehydrogenation of hydrocor-

tisone by *Arthrobacter simplex* cells entrapped with urethane prepolymers. *Agric. Biol. Chem.* **44,** 1119–1126.

15. Fukui, S. and Tanaka, A. (1989) Application of living microbial cells entrapped with synthetic resin prepolymers. *Experientia* **45,** 1055–1061.

16. Tanaka, A. and Nakajima, H. (1990) Application of immobilized growing cells. *Adv. Biochem. Eng./Biotechnol.* **42,** 97–131.

17. Fukui, S., Tanaka, A., and Iida, T. (1986) Immobilization of biocatalysts for bioprocesses in organic solvent media, in *Biocatalysis in Organic Media* (Laane, C., Tramper, J., and Lilly, M. D., eds.), Elsevier, Amsterdam. pp. 21–41.

18. Kawamoto, T., Sonomoto, K., and Tanaka, A. (1987) Esterification in organic solvents: selection of hydrolases and effects of reaction conditions. *Biocatalysis* **1,** 137–145.

19. Kimura, Y., Tanaka, A., Sonomoto, K., Nihira, T., and Fukui, S. (1983) Application of immobilized lipase to hydrolysis of triacylglyceride. *Eur. J. Appl. Microbiol. Biotechnol.* **17,** 107–112.

20. Yamane, T., Nakatani, H., Sada, E., Omata, T., Tanaka, A., and Fukui, S. (1979) Steroid bioconversion in water-insoluble organic solvents: Δ^1 -dehydrogenation by free microbial cells and by cells entrapped in hydrophilic or lipophilic gels. *Biotechnol. Bioeng.* **21,** 2133–2145.

21. Fukui, S., Ahmed, S. A., Omata, T., and Tanaka, A. (1980) Bioconversion of lipophilic compounds in non-aqueous solvent. Effect of gel hydrophobicity on diverse conversion of testosterone by gel-entrapped *Nocardia rhodocrous* cells. *Eur. J. Appl. Microbiol. Biotechnol.* **10,** 289–301.

22. Omata, T., Tanaka, A., and Fukui, S. (1980) Bioconversion under hydrophobic conditions: effect of solvent polarity on steroid transformations by gel-entrapped *Nocardia rhodocrous* cells. *J. Ferment. Technol.* **58,** 339–343.

23. Sonomoto, K., Hosokawa, Y., and Tanaka, A. (1987) Effect of water content in cell-entrapping gels on enzyme activity in organic solvents. *Ann. NY Acad. Sci.* **501,** 343–346.

24. Akita, H., Matsukura, H., Sonomoto, K., Tanaka, A., and Oishi, T. (1987) Asymmetric reduction of α-methyl β-keto esters by microbial cells immobilized with prepolymer. *Chem. Pharm. Bull.* **35,** 4985–4987.

4

Microencapsulation of Enzymes and Cells for Nonaqueous Biotransformations

Jeffrey A. Khan and Evgeny N. Vulfson

1. Introduction

The use of biocatalysts in nonaqueous organic solvents is now a well-established method for the preparation of pharmaceutical products *(1–2)*, food ingredients *(3–5)*, and intermediates used in the fine chemicals industry *(6–10)*. One of the advantages of this approach is that it enables the bioprocess operation at much higher concentrations of poorly water-soluble substrates, thus making both the synthesis and product recovery more attractive from a practical standpoint *(11,12)*. However, many enzymes, and certainly microorganisms, still require a substantial amount of water present to maintain their catalytic activity at synthetically useful levels. In these cases, conventional aqueous–organic two-phase systems are still widely used, although the rates of mass transfer and inactivation of biocatalysts at the interface can be a serious drawback *(13–16)*.

An alternative approach to biotransformations in aqueous-organic two-phase systems involves the entrapment of enzymes and whole cells within permeable polymeric microcapsules containing an aqueous phase. This is achieved using mild methods of interfacial polymerization developed in our laboratory to maintain biocatalyst activity/viability during the immobilization step **(Fig. 1A,B)** followed by the resuspension of these biocatalyst preparations in an organic solvent to perform the required reactions **(Fig. 1D,E)**.

Microencapsulation offers certain advantages over more conventional methods of cell and enzyme immobilization (e.g., calcium alginate gels, κ-carrageenan beads, hydrogels), including (1) improved mass transfer of the substrate and product between the organic and aqueous cell-containing phase resulting from the smaller size of the microcapsules and higher surface-to-volume ratio;

From: *Methods in Biotechnology, Vol. 15: Enzymes in Nonaqueous Solvents: Methods and Protocols*
Edited by: E. N. Vulfson, P. J. Halling, and H. L. Holland © Humana Press Inc., Totowa, NJ

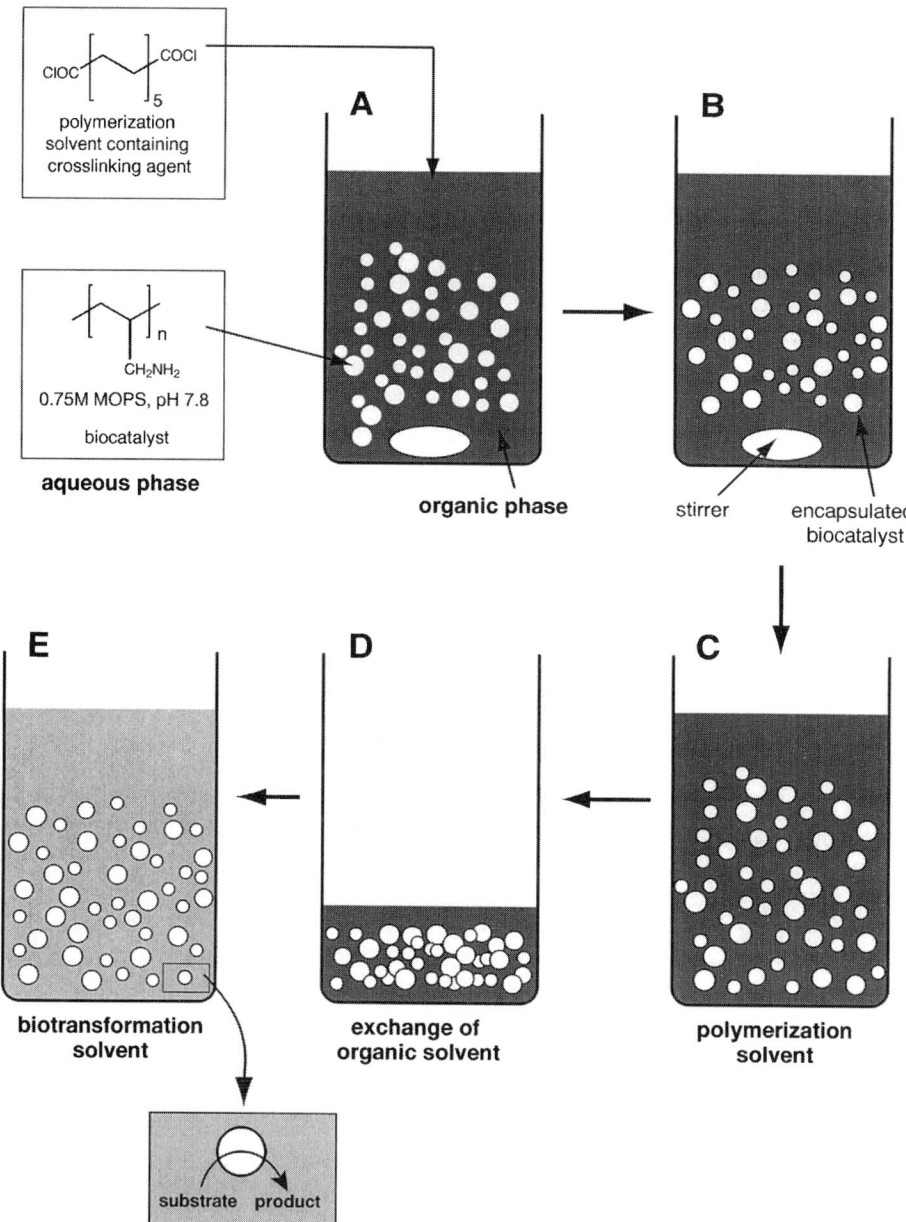

Fig. 1. Preparation and use of microencapsulated biocatalysts in organic solvents. (A) Aqueous buffer containing biocatalyst and a water-soluble amine is dispersed in an organic solvent (e.g., decane, *see* **Notes 1** and **2**) in the presence of an emulsifier; (B) an organic-soluble crosslinker is added to the solvent to initiate interfacial polymerization and stirring is continued to "harden" the microcapsules (*see* **Note 4**); (C) the microcapsules are allowed to sediment; (D) the polymerization solvent is decanted and the microcapsules are rinsed with organic solvent used for the biotransformation (*see* **Note 5**); (E) substrate is solubilized in the organic solvent and the accumulation of product is monitored.

(2) chemical and mechanical stability of polyamide microcapsules in a wide range of organic solvents and under various experimental conditions; (3) the possibility of incorporating water-soluble nutrients, cofactors, and cosubstrates to enhance the activity/viability of the biocatalyst and the rate of biotransformation. The general applicability of this procedure is demonstrated by the utilization of a range of encapsulated biocatalysts, including Baker's yeast for the single-step reduction of a diketone to an alcohol *(17)*, the multistep bioconversion of methyl octanoate to 2-heptanone using fungal spores *(18)* and the preparation of β-[D]-hexylglucopyranoside using immobilized β-glucosidase *(19)*.

2. Materials

α-[D]-Glucose, low-molecular-weight poly(allylamine hydrochloride), sorbitan trioleate (Span 85), 1,6-hexanediamine, methyl octanoate, 2-heptanone, 3-decanone, and dodecanedioyl chloride were obtained from Aldrich Chemical Company Ltd. (Dorset, UK). Baker's yeast (*Saccharomyces cerevisiae*; type II), yeast extract, almond β-glucosidase (G 0395, specific activity 5.2–14 U/mg), sorbitan monolaurate (Tween-20), hexyl β-[D]-glucoside, 2-[*N*-morpholino]-ethanesulfonic (MES) and 3-[*N*-morpholino]propanesulfonic acid (MOPS) were supplied by Sigma Chemical Company Ltd (Dorset, UK). *n*-Decane was purchased from Fluka Chemicals (Dorset, UK) and stored over 4-Å Linde molecular sieves (Aldrich). Potato dextrose agar was obtained from Oxoid (Basingstoke, UK).

2.1. Preparation of Amine Solutions Used During Interfacial Polymerization

1. 0.2 *M* Poly(allylamine) solution in 0.75 *M* MES buffer, pH 6.6: The molecular structure of the polymer is as follows:

Molecular weight = approx 11,500

n = approx 122

The polymer contains one reactive amine group per "residue mole" (molecular weight of repeating unit = 93.5). Hence, in 1 L, a 0.2-*M* solution of poly(allylamine) contains $0.2 \times 93.5 = 18.7$ g of polymer. For the preparation of a 25-mL solution, 468 mg of poly(allylamine) hydrochloride is therefore dissolved in 20 mL of 0.75 *M* MES buffer and the pH is adjusted by the dropwise addition of 5 *M* aqueous sodium hydroxide and 5 *M* aqueous hydrochloric acid to obtain a pH of 6.6. The amine solution is then transferred to a volumetric flask and made up to a final volume of 25 mL by the addition of 0.75 *M* MES buffer, pH 6.6.

2. 0.2 *M* Solution containing poly(allylamine) and 1,6-hexanediamine (10:90 molar ratio) in 0.75 *M* MOPS buffer, pH 7.8: In 1 L, 0.02 mol of poly(allylamine) (0.02 × 93.5 = 1.87 g) and 0.18 moles of 1,6-hexanediamine (0.18 × 116.2 = 20.9 g) are required. Therefore, for the preparation of a 25-mL volume of 0.2 *M* amine solution, 46.8 mg of poly(allylamine) hydrochloride, and 522 mg of 1,6-hexanediamine are dissolved in 20 mL of 0.75 *M* MOPS buffer. The pH is adjusted to 7.8 by the dropwise addition of 5 *M* aqueous sodium hydroxide and 5 *M* aqueous hydrochloric acid. The amine solution is then transferred to a volumetric flask and made up to a final volume of 25 mL by the addition of 0.75 *M* MOPS buffer, pH 7.8.

3. Methods

3.1. Reduction of 1-Phenyl-1,2-Propanedione to 2-Hydroxy-1-Phenyl-1-Propanone by Baker's Yeast (17)

3.1.1. Activation of Baker's Yeast

1. Add 1.25 g of dried Baker's yeast to 25 mL of sterile distilled water containing 10% glucose and 0.1% w/v of yeast extract in a 100-mL conical flask.
2. The culture is incubated at 25°C in a Luckham R300 shaker–incubator at 150 rpm for 30 min.
3. The activated yeast is agitated for 2 min using a laboratory rotamixer (Spinmix, Gallenkamp) to provide a homogenous suspension of cells.
4. A 1.5-mL aliquot of the homogenous suspension of the yeast cells is spun in a lab bench centrifuge at 4000*g* for 5 min.
5. The supernatant is removed using a Pasteur pipet resulting in a sediment of packed yeast cells.

3.1.2. Preparation of Microcapsules Containing Baker's Yeast

1. A 0.5-mL packed volume of yeast cells was added to a 1.5-mL solution of 0.2 *M* poly(allylamine hydrochloride)/1,6-hexanediamine (10:90 molar ratio; *see* **Subheading 2.1.2.** and **Note 6**) in 0.7 5*M* MOPS buffer, pH 7.8.
2. α-[D]-Glucose (5% w/v) was added and the solution stirred for 15 min.
3. The aqueous mixture was added dropwise to 15 mL of ice-cold decane containing 1.4% (v/v) of Span 85 with stirring.
4. The resulting emulsion was stirred continuously for 5 min at 400 rpm (IKA minibar; BDH, Leicestershire, UK) in a 100-mL beaker containing a 2.5-cm stirrer bar (*see* **Note 3**).

5. Microencapsulation was initiated by the dropwise addition of 15 mL dry decane containing 45 µL of dodecanedioyl dichloride over 15 min at 0°C.
6. Stirring was continued for a further 30 min at 0°C.
7. The capsules were allowed to settle under gravity for 15–20 min; the solvent was decanted from the microcapsules and 15 mL of dry decane was added with stirring at room temperature to wash the preparation.
8. The decane was then decanted and the preparation resuspended in 3 mL of the organic solvent used for the biotransformation. Because of swelling of the capsules during rinsing, the resuspension of capsules in the solvent is performed in graduated tubes to obtain accurate final volumes.

3.1.3. Reduction of 1-Phenyl-1,2-Propanedione by Baker's Yeast

1. Microcapsules containing a 0.5-mL packed volume of cells is added to 3 mL of decane containing 27 mM 1-phenyl-1,2-propanedione in a sealed 10-mL vessel. (**Note:** At this stage the solvent used to resuspend the microcapsules can be varied if required.)
2. The reaction is performed at 37°C in a Luckham R300 shaker–incubator at 150 rpm.
3. Aliquots (0.1 mL) of the reaction mixture can be withdrawn at regular time intervals over a period of 4–6 h for product analysis.
4. These samples are centrifuged at 6000g for 5 min and diluted with methanol (50-fold dilution at 27 mM) for high-performance liquid chromatography (HPLC) analysis.

3.1.4. HPLC Analysis

1. This was performed using a Milton Roy CM4000 pump equipped with a Spectra-physics SP8450 UV detector set at 254 nm and a Hichrom RPB (0.46 × 15 cm) reverse-phase column. The substrate (1-phenyl-1,2-propanedione, retention time = 8.4 min) and product (2-hydroxy-1-phenyl-1-propanone, retention time = 4.4 min) are eluted with a binary gradient of 80:20 water / methanol (*A*) against pure methanol (*B*) using a programmed gradient file ($t = 0$ min, 70% A:30% B; $t = 5.5$ min, 70% A:30% B; $t = 9$ min, 20% A:80% B) at a flow rate of 1.0 mL/min.

3.1.5. Microscopic Analysis

1. Light microscopy can be used to determine the distribution of yeast cells within the polyamide microcapsules. A suspension of capsules in solvent are removed from the reaction mixture using a pipet and placed onto a welled glass microscope slide. A cover slip is placed on top of the slide to prevent solvent evaporation and the preparation is examined. The size of the microcapsules can be determined using a standard graticule.

3.2. Bioconversion of Methyl Octanoate to 2-Heptanone by Encapsulated P. roquefortii Spores (18)

3.2.1. Spore Growth and Culture Medium

1. A suspension of *Penicillium roquefortii* spores (ATCC 64383) (200 μL) containing 1×10^9 spores/mL was used to inoculate each potato dextrose agar plate.
2. Mycelial growth and sporulation was initiated by incubation at 25°C over 7 d.
3. The spores were harvested by the addition of 0.05% Tween-80 (10 mL) to each plate followed by gentle agitation to remove spores.
4. The resulting stock solution (containing 1.73×10^{10} spores/mL) was collected and stored at –20°C.
5. The concentration of spores in solution can be determined by serial dilution and cell counting using a hemocytometer (Gelman Hawksley Ltd., Lancing, UK).

3.2.2. Preparation of Microcapsules Containing P. roquefortii *Spores*

1. The stock spore solution (2.8 mL) is centrifuged for 5 min at 2000 rpm (Heraeus, SEPATECH) and the supernatant is removed.
2. Spores are resuspended in 0.3 *M* MOPS buffer, pH 7.2 (2.0 mL) containing 0.1 *M* 1,6-hexanediamine and 0.1 *M* poly(allylamine) hydrochloride (*see* **Subheading 2.1.1.** and **2.1.2.** and **Note 6** for examples on the preparation of amine solutions).
3. The aqueous mixture was added dropwise to decane (20 mL) containing Span 85 (300 μL) and emulsified at room temperature using a conventional magnetic stirrer operating at 400 rpm (IKA RETbasic) and a 2.5-cm stirrer bar (*see* **Note 3**).
4. To the resulting dispersion (30- to 50-μm diameter microspheres), dodecanedioyl chloride (22.5 μL) in dry decane (15 mL) is added dropwise over 10 min.
5. The mixture is stirred for a further 20 min at room temperature.
6. The solvent is decanted from the microcapsules (30–50 μm in diameter), which are then rinsed with water-saturated decane (2 × 20 mL).

3.2.3. Synthesis of 2-Heptanone

1. Encapsulated spores (200 μL) containing 4.84×10^{10} cells are added to 1.8 mL of *n*-decane containing 500 m*M* methyl octanoate in a sealed 10-mL Wheaton vial.
2. The bioconversion is performed at 25°C in a Luckham R300 shaker–incubator at 150 rpm.
3. The reaction is monitored by removal of 50-μL aliquots of the organic layer at regular intervals over a period of 1–14 d. These samples are diluted 10-fold with dichloromethane containing 3-decanone as an internal standard and analyzed by gas chromatography (GC).

3.2.4. GC Analysis

This was performed using an HP 5890 GC (Split) System with a capillary column, methyl silicone gum, HP-1 (25 m × 0.32 mm; film thickness = 1.05 μm) connected to a flame-ionization detector (FID) , and using hydrogen as a carrier gas at a flow rate of 0.7 mL/min. The injection temperature was set to 250°C and the detector was at 280°C. The oven temperature was programmed at 80°C for 2 min, increased to 200°C at 10°C/min, and maintained at 200°C for 2 min. Samples were injected using a split/splitless injector at a split ratio 1/100. The retention time of 2-heptanone was 2.8 min.

3.3. Enzymatic Conversion of Glucose to β-[D]-Hexylglucoside Using Microencapsulated β-Glucosidase (19)

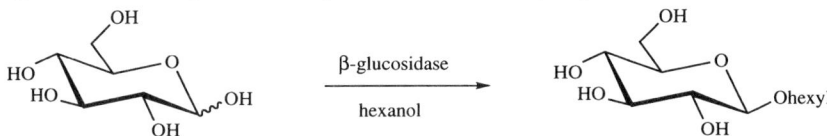

3.3.1. Preparation of Microcapsules Containing β-Glucosidase

1. β-Glucosidase (560 U) and α-[D]-glucose (1600 mg) are dissolved in 1.5 mL of 0.75 *M* MES buffer, pH 6.6 (*see* **Subheading 2.1.1.**) containing 0.2 *M* poly(allylamine) (*see* **Note 6**) with stirring for 15 min at 50°C.
2. This solution is then added dropwise to decane (15 mL) containing Span 85 (220 µL) and emulsified (*see* **Note 3**) at 50°C using a TP 18/10 Ultra-Turrax overhead blender (IKA Labortechnik, Staufen, Germany) fitted with a S-25N-10G dispersion element and accurate control of the mixing speed at 9500 rpm using an in-line variable transformer (Variac, Claude Lyons, Liverpool, UK) set at 120 V.
3. To the resulting dispersion (30- to 50-µm-diameter capsules), dodecanedioyl chloride (45 µL) in dry decane (10 mL) is added dropwise over 10 min and the mixture is stirred for an additional 5 min.
4. Mixing is then continued at room temperature using a stirrer bar (2.5 cm) and conventional magnetic stirrer operating at 400 rpm (IKA RETbasic) for an additional 30 min.
5. The solvent is decanted from the microcapsules (30–50 µm diameter), which are then rinsed with decane (2 × 10 mL) and *n*-hexanol (2 × 10 mL), both presaturated with 50% (w/v) glucose solution.

3.3.2. Enzymatic Conversion of Glucose to β-[D]-Hexylglucoside

1. Allow 1.5 mL of microcapsules (based on the aqueous volume used for their preparation) containing 80 mg of β-glucosidase (560 U) to sediment in a graduated test tube (25 mL).
2. *n*-Hexanol, presaturated with 50% (w/v) glucose solution, is added to a final total volume of 25 mL.
3. The mixture is transferred to a 50-mL Duran bottle and incubated at 50°C in an orbital shaker–incubator at 150 rpm.
4. To monitor the bioconversion, 0.25-mL aliquots of *n*-hexanol are removed from the reaction mixture at regular time intervals over a period of 1–8 h, diluted with methanol (0.75 mL), and analyzed by HPLC.

3.3.3. HPLC Analysis

The analysis was performed using Gilson 305/306 pumps (Anachem, Luton, UK) equipped with an ACS (Macclesfield, UK) or a Sedex 55 (Sedere, Paris, France) evaporative light-scattering detector (evaporation temperatures 70°C and 45°C; nitrogen pressure 1.3 and 1.9 bars, respectively) and a Spherisorb ODS2 250 × 4.6 mm column (Hichrom, Reading, UK) eluted with a linear gradient of acetonitrile/water (from 20 to 70% acetonitrile) that is applied over 7 min at a flow rate of 0.75 mL/min. A calibration curve for hexyl β-[D]-glucopyranoside was constructed for accurate quantitative analysis.

4. Notes

1. It is known that a variety of microorganisms can grow well in the presence of oils and higher hydrocarbons that have a high hydrophobicity index, log *P (20)*. Therefore, the encapsulation of microorganisms is performed in the presence of *n*-decane to minimize cell inactivation.
2. Prior to the initiation of interfacial polymerziation by the addition of acid chloride, it is important to ensure that an adequate dispersion of aqueous droplets is obtained in *n*-decane. This is tested by removing a few drops of the dispersion using a Pasteur pipet during emulsification and analysis using a microscope (*see* **Subheading 3.1.5.**).
3. When the volume ratio of aqueous:organic phase and concentration of emulsifier (Span 85) are kept constant, the size of the capsules produced is dependent on the shear rate applied to produce the emulsion. During enzyme immobilization, a higher shear rate is applied to produce small microcapsules (diameter = 20–50 µm) with a high surface:volume ratio to facilitate mass transfer and rate of the biotransformation. The encapsulation of cells requires less vigorous stirring to produce larger capsules (50–150 µm) for the immobilization of yeast cells (average diameter = 10 µm) and fungal spores (average diameter = 5 µm).
4. Upon complete addition of the acid chloride, stirring is continued to allow the capsules to "harden" using gentle stirring to minimize their breakage. If insufficient time is allowed for this to occur (usually 15–45 min is adequate), the capsules are often unstable and prone to degradation during the rinse and/or biotransformation. Importantly, if excessive stirring times are employed, "clumping" of the capsules may occur because of crosslinking of microcapsules in the presence of excess acid chloride.
5. When the encapsulated biocatalysts have been prepared, excess emulsifier and residual acid chloride is removed by rinsing the microcapsules with the same organic solvent that is to be used for the biotransformation. It is important that this solvent is presaturated with the same aqueous phase present in the microcapsules to prevent breakage of the polymeric biocatalyst during rinsing.
6. In the examples described, defferent molar ratios of poly(allylamine):1,6-hexanediamine amine were used for the encapsulation of biocatalysts to provide cap-

sular membranes of satisfactory composition (i.e., hyrophobicity, charge distribution, and degree of crosslinking in the polymer) to improve mass transfer to substrate/product and hence achieve optimal rates of reaction. These molar ratios were optimized in preliminary experiments.

Acknowledgments

The authors wish to thank the BBSRC and NSERC (Canada) for the financial support of this work. We also thank former workers who have contributed to the development of these methods, especially Professor Q. Yi, Dr. K. D. Green, and Dr. O.-J. Park.

References

1. Margolin, A. L. (1993) Enzymes in the synthesis of chiral drugs. *Enzyme Microb. Technol.* **15,** 266–280.
2. Zaks, A. and Dodds, D. R. (1997) Application of biocatalysis and biotransformations to the synthesis of pharmaceuticals. *Drug Discovery Today* **2,** 513–531.
3. Cheetham, P. S. J. (1995) Biotransformations — new routes to food ingredients. *Chem. Ind.* **7,** 265–268.
4. James, J. and Simpson, B. K. (1996) Application of enzymes in food processing. *CRC Crit. Rev. Food Sci. Nutr.* **36,** 437–463.
5. Vulfson, E. N. (1993) Enzymatic synthesis of food ingredients in low-water media. *Trends Food Sci. Technol.* **4,** 209–215.
6. Dordick, J. S. (1989) Enzymatic catalysis in monophasic organic solvents. *Enzyme Microb. Technol.* **11,** 194–211.
7. Holland, H. L. (1998) Microbial transformations. *Curr. Opin. Chem. Biol.* **2,** 77–84.
8. Klibanov, A. M. (1989) Enzymatic catalysis in anhydrous organic solvents. *TIBS* **14,** 141–144.
9. Klibanov, A. M. (1997) Why are enzymes less active in organic solvents than in water? *TIBTECH* **15,** 97–101
10. Roberts, S. M. (1998) Preparative biotransformations: the employment of enzymes and whole-cells in synthetic organic chemistry. *J. Chem. Soc. Perkin Trans. 1* **1,** 157–169.
11. Cabral, J. M. S., Aires-Barros, M. R., Pinheiro, H., and Prazeres, D. M. F. (1997) Biotransformation in organic media by enzymes and whole cells. *J. Biotechnol.* **59,** 133–143.
12. Leon, R., Fernandes, P., Pinheiro, H. M., and Cabral, J. M. S. (1998) Whole-cell biocatalysis in organic media. *Enzyme Microb. Technol.* **23,** 483–500.
13. Janssens, L., DePooter, H. L., Schamp, N. M., and Vandamme, E. J. (1992) Production of flavours by micro-organisms. *Process Biotechn.* **27,** 195–215.
14. Lilly, M. D. and Woodley, J. M. (1996) A structured approach to design and operating of biotransformation processes. *J. Ind. Microbiol.* **17,** 24–29.
15. Woodley, J. M., Cunnah, P. J., and Lilly, M. D. (1991) Stirred tank 2-liquid phase biocatalytic reactor studies—kinetics, evaluation and modelling of substrate mass transfer. *Biocatalysis* **5,** 1–12.

16. Van Sonsbeek, H. M., Beeftink, H. H., and Tramper, J. (1993) Two-liquid-phase bioreactors. *Enzyme Microb. Technol.* **15,** 722–729.
17. Green, K. D., Gill, I. S., Khan, J. A., and Vulfson, E. N. (1996) Microencapsulation of yeast cells and their use as a biocatalyst in organic solvents. *Biotechnol. Bioeng.* **49,** 535–543.
18. Park, O.-J., Holland, H. L., Khan, J. A., and Vulfson, E. N. (2000) Production of flavour ketones in aqueous-organic two phase systems using free and microencapsulated fungal spores as biocatalysts. *Enzyme Microb. Technol.* **26,** 235–242.
19. Yi, Q., Sarney, D. B., Khan, J. A., and Vulfson, E. N. (1998) A novel approach to biotransformations in aqueous-organic two phase systems: enzymatic synthesis of alkyl β-[D]-glucosides using microencapsulated β-glucosidase. *Biotechnol. Bioeng.* **60,** 385–390.
20. Laane, C., Boeren, S., Vos, K., and Veeger, C. (1987) Rules for optimisation of biocatalysis in organic solvents. *Biotechnol. Bioeng.* **30,** 81–87.

5

Immobilization of Lipases on Hydrogels

**Abu Bakar Salleh, Norhaizan M. Esa, Mahiran Basri,
Che Nyonya A. Razak, Wan Md Zin W. Yunus, and Mansor Ahmad**

1. Introduction
1.1. Background

Immobilization of enzymes can be achieved by methods of varied complexity and efficiency *(1)* on a variety of supports. For example, enzymes can be adsorbed onto insoluble materials, copolymerized with a reactive monomer, encapsulated in gels, crosslinked with a bifunctional reagent, covalently bound to an insoluble carrier *(2)* or entrapped within an insoluble gel matrix of natural or synthetic resin *(1)*.

Numerous methods have been developed for immobilization of enzymes that will be influenced by the activity of the immobilized preparation as well as by the properties of the support matrix *(3)*. Many supports for the immobilization of lipases have been investigated and various natural and synthetic polymers have been used *(1,2,4)*. These supports are often subject to limitations: heterogeneity from batch to batch, and sensitivity to heat and low chemical, physical, and microbiological resistance. As such, the search for new types of immobilization support continues.

The use of biocatalysts in organic media has further enhanced the potential application of immobilized enzymes. The fact that enzymes and cells may be active in an organic solvent has opened a totally new horizon for the enzyme technologists to explore. Enzymes can be used in reaction systems with very low water contents. A low water content favors enzyme stability, mass transfer of hydrophobic substrate to biocatalysts, and the thermodynamic equilibrium in reversed hydrolytic reactions.

From: *Methods in Biotechnology, Vol. 15: Enzymes in Nonaqueous Solvents: Methods and Protocols*
Edited by: E. N. Vulfson, P. J. Halling, and H. L. Holland © Humana Press Inc., Totowa, NJ

The activity of biocatalysts in mainly organic reaction mixtures is usually affected by the amount of water present in the systems. When the biocatalysts are immobilized, the influence exerted by the support on the availability of water to the biocatalysts to maintain its catalytic activity becomes more crucial *(5)*. Apart from water content, water activity (a_w), the intrinsic water content or the water layer around the enzyme molecule, is considered a good measure of the amount of water *(6)*. There was a clear correlation between the aquaphilicity of the support material and the enzymatic activity. Support with low aquaphilicity gave the highest activity. The result can be interpreted as a competition between the enzyme and the support material for the water. Supports with high aquaphilicity adsorb a lot of water, so that the enzyme is insufficiently hydrated. When support materials with low aquaphilicity are used, the enzyme can compete successfully for the water and therefore show high activity.

1.2. Hydrogel as a Matrix for Immobilization

In this chapter, we show how hydrogels can be used as support for immobilization of enzymes. These water-swollen, crosslinked polymeric structures can be produced by the simple reaction of one or more monomers. Hydrogels are polymeric materials made from hydrophilic and/or hydrophobic monomers, which can be a homopolymer or a copolymer. Their major characteristic is that they can imbibe large quantities of water without dissolution of the polymer network. Hydrogels make them interesting supports for immobilization of enzymes. A wide range of hydrophilic polymers have been examined as potential candidates for replacement of soft tissue or for other biomedical applications. The biocompatibility of hydrogel is attributed to their ability to stimulate the natural tissue as a result of their ability to absorb and retain water without dissolving. They also can be made more or less hydrophilic by copolymerization of two or more monomers. This property makes them interesting materials as carriers for immobilization of bioactive compounds.

Hydrogels can be formed with different monomer and crosslinker compositions so as to acquire different capacities for water retention (**Table 1**). Different enzymes require different degrees of hydrophilicity or hydrophobicity of support materials, for optimum protein adsorption and activity expression by the immobilized enzymes.

Aside from providing the water needed for enzyme activity, the hydrogel can also absorb water produced during the esterification reaction, thus increasing the conversions to the products.

2-Hydroxyethylmethacrylate (HEMA) gives marked hydrophilic character to the resulting polymer, very helpful in immobilization experiments. This hydrophilic characteristic gives a somewhat spongy texture to the polymer, thus increasing its surface area. The copolymer can be easily prepared from

Table 1
Percentage of Water Intake by Polymer with Different Monomer Composition (HEMAand MMA) and Percentage of Crosslinkers (DVP and EGDMA) Used

| | Crosslinking agents | | | | | |
| | DVB | | | EGDMA | | |
Poly (HEMA:MMA) ratio	1%	2%	5%	1%	2%	5%
1:2	35.7	32.2	11.7	36.7	33.7	12.5
1:1	40.0	35.6	22.7	45.7	38.1	30.5
2:1	47.0	42.6	39.9	—	—	41.2

cheap monomers, providing good durability against mechanical, chemical, and microbiological agents. Because of its rigid structure, it is not affected by drastic changes in the pH/ionic strength *(7)*.

When choosing the monomer types as well as the polymerization method, one may also control the degree of crosslinking within the hydrogel. Tightly linked gel will immobilize biomolecules well but may inhibit the product and substrate diffusion into and out of the hydrogel matrix. The monomer type is also important because different functional groups may affect the microenvironment of the immobilized species leading to an increase or decrease in its biological activity.

The quantity of enzyme coupled to reactive polymer is dependent on many factors: Particle size, hydrophilic sites, and structure and chemical composition of the polymer. A polymer based on acrylic monomers represent a good choice: mechanically stable and inert to microbial and enzyme degradation *(8)*.

Both the structure and chemical composition of a hydrogel will influence the microenvironment of an immobilized species that eventually would be seen in the activity expressed. Although, the smaller size particles were expected to show high expressed activity owing to the larger surface area, the result showed that particle size in the range of 180–500 μm were preferred *(9)*.

The activities of immobilized lipases so far reported are considerably lower when compared to those reported for other enzymes *(10)*. Lieberman and Ollis *(11)* found that the specific activities of the immobilized lipases were equivalent to only 1–2% that of the free enzymes. The low activity was believed to be the result of the difficulties of the oil droplets in penetrating the pores of the supports. Lipases immobilized to hydrogel would exhibit different hydrolytic and esterification activity with different polymer composition **(Table 2)**.

Table 2
Hydrolytic and Esterification Activity of Lipase Immobilized (Adsorbed) to Hydrogel

Poly(HEMA:MMA) ratio	Crosslinking agents	Hydrolytic activity specific activity (U/mg)	Esterification activity specific activity (U/mg)
1:2	1% DVB	5.87	0.93
	2% DVB	4.92	1.77
	5% DVB	3.82	3.61
1:1	1% DVB	3.71	8.11
	2% DVB	4.20	6.16
	5% DVB	6.33	0.93
	1% DVB	5.24	5.41
2:1	2% DVB	3.89	7.67
	5% DVB	0.81	8.39
1:2	1% EGDMA	8.60	0.72
1:1	1% EGDMA	3.48	3.69
	2% EGDMA	6.10	2.02
	5% EGDMA	7.50	0.08
2:1	1% EGDMA	7.52	0.77
	2% EGDMA	6.26	3.91
	5% EGDMA	4.34	5.20

1.3. Lipases Are Adsorbed or Entrapped

Enzymes can be adsorped onto the hydrogel. Hydrogel is synthesized by copolymerization of a mixture of methyl methacrylate (MMA) and HEMA in various ratio (mole/mole) and enzymes adsorbed onto it. The degree of crosslinking can also affect the swelling behavior of the polymer.

Entrapment of lipases can be achieved by simultaneously mixing enzymes during the copolymerization process of the poly(N-vinyl-2-pyrrolidone-co-2-hydroxyethyl methacrylate) hydrogel. However, it must be noted that owing to the high temperature of the polymerization process, only thermotolerant enzymes can be adopted for this procedure.

2. Materials

Materials for immobilization are as follows: Lipase from *Candida rugosa* (Type VI), monomers; *N*-vinyl-2-pyrrolidone (VP), HEMA, and a crosslinker, ethylene dimethacrylate (EDMA) were obtained from Sigma Chemical Co. (St. Louis, MO). Divinyl benzene (DVB) and the initiator, α,α'-azoisobutyronitrile (AIBN) was from Fluka Chemical, Switzerland. All other reagents were of analytical grade.

3. Methods

3.1. Lipase for Adsorption

Lipase from *Candida rugosa* (Type IV) was obtained from Sigma. The enzyme (500 mg) was dispersed in distilled water (10.0 mL). This mixture was agitated on a vortex mixer and centrifuged at 10,000g for 10 min and the supernatant used for lipase.

3.2. Enzyme Activity Assay

The activities of free and immobilized lipase were determined by monitoring the hydrolysis and esterification reaction. For hydrolysis, the reaction mixture comprises 2.5 mL olive oil emulsion in water (1:1, v/v, initially emulsified by rotax mixing), 20 µL 0.02M CaCl$_2$, and 0.1 g immobilized lipase. The mixture was incubated at 37°C for 30 min in a horizontal water-bath shaker at 200 rpm. For esterification activity, the assay system consisted of immobilized lipase (0.3 g), oleic acid (2.0 mmol), butanol (8.0 mmol), and hexane (2.6 mL). The mixture was incubated at 37°C for 5 h in a horizontal water bath shaker at 150 rpm.

In both cases, the reaction was terminated by dilution with acetone–ethanol (1:1 v/v,). The free fatty acid in the reaction mixture was determined by titration, with aqueous NaOH (0.05M) using an automatic titrator (ABU 90, Radiometer, Copenhagen) to an apparent pH of 9.5 (aqueous calibration). One activity unit (U) is defined as the rate of production or consumption of 1 µmol of free fatty acid per minute under indicated experimental conditions.

3.3. Determination of the Quantities of Protein

Protein concentration was determined both on the initial solution and on the residual supernatant after the coupling procedure using the Bradford method *(12)*.

3.4. Immobilisation of Lipase by Adsorption

3.4.1. Preparation of Poly(HEMA-Co-MMA)

Copolymer beads crosslinked with DVB and ethylene glycol dimethacrylate (EGDMA) are prepared from various ratio of HEMA and MMA. (*See* **Note 1**.) Suspension polymerization is carried out in a 500-mL three-necked flask, equipped with a reflux condenser, a thermometer and an agitator. The suspension medium comprises Na_2SO_4 (30 g), $CaCO_3$ powder (5 g), toluene (60 mL), 4-methylpentan-2-on (20-mL), and gelatin (0.5 g) and 500 mL distilled water. Polymerization is initated as the monomers (HEMA:MMA), the crosslinkers, and 1% benzoyl peroxide as the initiator are added. (*See* **Note 2**.)

The rate of agitation is controlled and with some minor deviation, the stirrer was set at around 250 rpm to ensure the bead size was in the range 180–500 μm. The reaction is carried out for 6 h. The temperature is raised gradually up to 80°C, during the polymerization. The copolymer that is formed as fine particles is collected by filtration and washed thoroughly with a large volume of boiling water and dried overnight.

The polymer is first sieved in order to obtain different classes of particle size. Three samples of hydrogel are obtained; the first from particles sized <180 μm, the second from particles sized between 180 and 350 μm, and the third from particles sized between 350 and 500 μm. The polymer is allowed to swell for 3–4 d in a large volume of distilled water, which is changed every day. The value of equilibrium water content, *Wc* is determined according to the equation

$$Wc = \frac{Ws - Wd}{Ws} \times 100$$

where *Ws* and *Wd* are the weight of the fully swollen support and of the completely dried samples, respectively *(13)*.

3.4.2. Immobilization of Enzyme by Adsorption

The lipase solutions and polymer beads (0.5 g) was mixed at room temperature by shaking at 100 rpm in a sealed vial for 30 min. The polymers were separated from the supernatant by filtration through Whatman No. 1 filter paper and washed three times each with distilled water. The beads were lyophilized, then kept in sealed vials.

3.5. Lipase Immobilization by Entrapment

Purified monomers VP and HEMA of varying wt% composition were mixed together with 1% EGDMA (wt %) in a clean and dry flask (*see* **Note 1**). To these mixtures, a dry initiator, AIBN (10^{-4} mol) was added and the flasks shaken until the AIBN dissolved. The mixtures were then transferred to a polymerization tube and the solutions degassed with nitrogen for 15 min to remove oxygen. The mixtures were incubated to polymerize in a 55–60°C water bath. After the polymer solutions became viscous (1–4 h), the polymers were cooled to 50°C, and the lipase solution (1.0 mL) previously degassed with nitrogen, was added and the polymer solutions shaken until they appeared homogeneous. The solutions in the polymerization flasks were sealed with rubber stoppers and further polymerized at 50°C for about 5 h. The solid polymerized rods were removed from the polymerization tubes. These rods were cut into small pieces (0.2–0.4 cm^3) and stored at –4°C prior to use.

4. Notes

1. The HEMA, VP, and EGDMA are purified by passing through an aluminum oxide (chromatographic grade) column (2.5 cm × 10.0 cm) and used immediately. The organic solvents and substrates are dried over molecular sieves (3Å) before use.
2. To ensure uniformity in bead size, the temperature of the reaction mixture is increased slowly. Protocol for temperature increment is as follows:

Polymerization period	Temperature increment
First hour	Room temperature to 65°C
Second hour	65–75°C
Third hour	75–85°C
Up to the end of process (6 h)	85–90°C

A regular check is essential during polymerization, as temperature fluctuations may occur and adjustment need to be done.

After the polymerization is completed, beads must be washed several time with 1*M* HCl and boiling water to remove the excess $CaCO_3$ and monomers with a final wash with acetone.

Do not forget! Perform the polymerization experiment in a good fume cupboard and take other necessary precautions when handling chemicals and solvents.

References

1. Ramos, M. C., Garcia, M. H., Cabral, F. A. P., and Guthrie, J. T. (1992) Immobilization of lipase from *Mucor miehei* onto poly(ethylene) based graft copolymer. *Biocatalysis* **6(3)**, 223–234.
2. Carta, G., Gainer, J. L., and Benton, A. H. (1991) Enzymatic synthesis of esters using on immobilized lipase. *Biotechnol. Bioeng.* **37**, 1004–1009.
3. Mustranta, A., Forssell, P., and Poutanen, K. (1993) Applications of immobilized lipases to transesterification and esterification reactions in an aqueous system. *Enzyme Microbiol. Technol.* **15(3)**, 133–139.
4. Yunus, W. M. Z., Salleh, A. B., Ismail, A., Ampon, K., Razak C. N. A., and Basri, M. (1992) Poly(Methyl methacrylate) as a matrix for immobilization of lipase. *Appl. Biochem. Biotechnol.* **36**, 97–105.
5. Oladepo, D. K., Halling, P. J., and Larsen, V. F. (1994) Reaction rates in organic media show similar dependence on water activity with lipase catalysts immobilized on different supports. *Biocatalysis* **8**, 283–287.
6. Halling, P. J. and Valivety, R. H. (1992). Physical, chemical nature of low systems for biocatalysis especially phase behaviour, water activity and pH, in *Biocatalysis in Nonconventional Media* (Tramper, J., Vermus, M. H., Beeftink, H. H., and von Stocker, U., eds.), Elsevier, Amsterdam, pp. 13–21.

7. Filippo, P. A., Fadda, D. S., Rescigno, M.B., Rinaldi, A., and Teulada, E. S. D. (1990) A new synthetic polymer as a support for enzyme immobilisation. *Eur. Polym. J.* **26(5),** 545–547.
8. Mosbach, K. (1976) Immobilized enzymes. *FEBS Lett.* **62 (Suppl.),** E80–E95.
9. Esa, N. M. (1996) Studies on the suitability of poly(2-hydroxyethyl methacrylate-co-methyl methacrylate) as a matrix for the immobilisation of lipase, M.S. thesis, Universiti Putra Malaysia.
10. Shaw, J. F., Chang, R., Wang, F. F., and Wang Y. J. (1990) Lipolytic activities of a lipase immobilized on six selected supporting materials. *Biotechnol. Bioeng.* **35,** 132–137.
11. Liebernnan, R. B. and Ollis, D. F. (1975) Hydroloysis of particulate tributyrin in a fluidized lipase reactor. *Biotechnol. Bioeng.* **17,** 1201–1219.
12. Bradford, M. M. (1976) A rapid and sensitive method for the quantitation of microgram quantities of protein utilizing the principle of protein-dye binding. *Anal. Biochem.* **72,** 248–254.
13. Carenza, M., Yoshida, M., Kumakura, M., and Fujimura, T. (1993) Hydrogels obtained by radiation-induced polymerization for yeast cells immobilization. *Eur. Polym. J.* **29(7),** 1013–1018.

6

Polyethylene Glycol-Modified Enzymes in Hydrophobic Media

Ayako Matsushima, Yoh Kodera, Misao Hiroto, Hiroyuki Nishimura, and Yuji Inada

1. Introduction

Chemical modification of proteins with polyethylene glycol (PEG), a nontoxic, nonimmunogenic, and amphipathic polymer, has been extensively studied for the purpose of applying proteins to biomedical and biotechnological processes *(1–9)*. In the biomedical field, the purpose of the chemical modification of enzymes with PEG includes the reduction of their immunoreactivity and immunogenicity as well as the prolongation of the plasma half-life of the modified enzymes *(5,10–13)*. Several PEG enzymes such as PEG–ADA (adenosine deaminase) and L-asparaginase are now approved by the U.S. Food and Drug Administration (FDA). In the biotechnological field, the purpose of the chemical modification of enzymes with PEG includes the preparation of novel catalysts, PEG enzymes, which are soluble and active in organic solvents *(14,15)*.

In 1984, Inada and his co-workers *(16)* happened to find that enzymes modified with PEG became soluble and active in organic solvents. As PEG is an amphipathic macromolecule, its hydrophilic nature makes it possible to modify enzymes in aqueous solution and its hydrophobic nature makes modified enzymes soluble and active in an hydrophobic environment. In fact, PEG enzymes such as PEG catalase *(17)* and PEG peroxidase *(18)* became soluble in organic solvents such as benzene and exhibited markedly high activity in organic solvents. Furthermore, PEG-modified hydrolytic enzymes such as PEG lipase *(16)* and PEG protease *(19)* catalyzed the reverse reaction of hydrolysis effectively in hydrophobic media, namely ester synthesis, ester exchange reaction and acid–amide bond formation (**reactions 1–5**) in transparent organic solvents or in hydrophobic substrates in the absence of organic solvents *(14)*.

From: *Methods in Biotechnology, Vol. 15: Enzymes in Nonaqueous Solvents: Methods and Protocols*
Edited by: E. N. Vulfson, P. J. Halling, and H. L. Holland © Humana Press Inc., Totowa, NJ

$$R_1COOH + R_2OH \longrightarrow R_1COOR_2 + H_2O \qquad (1)$$

$$R_1COOR_2 + R_3OH \longrightarrow R_1COOR_3 + R_2OH \qquad (2)$$

$$R_1COOR_2 + R_3COOH \longrightarrow R_3COOR_2 + R_1COOH \qquad (3)$$

$$R_1COOR_2 + R_3COOR_4 \longrightarrow R_1COOR_4 + R_3COOR_2 \qquad (4)$$

$$R_1COOH + R_2NH_2 \longrightarrow R_1CONHR_2 \qquad (5)$$

Table 1 shows the summarization of PEG hydrolase in relation to their function together with reference *(37–59)*.

Polyethylene glycol is a linear synthetic polymer, of which the general formula is $HO-(CH_2CH_2O)n-CH_2CH_2OH$ and the molecular weight is normally 500–20,000. Although PEG has two reactive hydroxyl groups at the end of the molecule, a monofunctional derivative, monomethoxypolyethylene glycol (mPEG), is often used for protein modification to avoid crosslinking between target protein molecules with PEG modifiers. Various PEG derivatives with high reactivity toward functional groups in a protein molecule have been extensively developed *(9)*.

This chapter deals with the preparation of PEG enzymes and their application in nonaqueous media. Almost all of enzymes are soluble in aqueous solution but are insoluble in hydrophobic media, including organic solvents. We select mainly polyethylene glycol derivatives (PEGs) **(Fig. 1)** and enzymes without a cofactor and/or metal **(Table 1)**. Therefore, coupling reactions between a reactive group of PEG and an amino group (or sulfhydryl group) in the enzyme molecule proceed very smoothly in neutral or weak alkaline aqueous solution at room temperature. As reactive groups of PEG, triazine, acid anhydride or *N*-hydroxysuccinimide is employed. **Figure 1** shows the chemical formula of several PEG derivatives, in which 2,4-bis *(O*-methoxypolyethylene glycol)-6-chloro-*s*-triazine (activated PEG$_2$), methoxypolyethylene glycol succinimidyl succinate (SS-PEG), and a copolymer of polyoxyethylene allyl methyl diether and maleic anhydride (activated PM) are cited in this chapter. Reactions of activated PEG$_2$, SS-PEG and activated PM with amino groups in a protein molecule are shown in **Fig. 2**.

2. Materials

2.1. Oxidoreductase

2.1.1. Modification of Catalase with Activated PEG$_2$ (17)

2,4-Bis *(O*-methoxypolyethylene glycol)-6-chloro-*s*-triazine (activated PEG$_2$) was purchased from Seikagaku Kogyo Co. (Tokyo, Japan) (*see* **Note 1**). Crystalline catalase (E.C. 1.11.1.6) from bovine liver was obtained from Sigma Chemical Co. (St. Louis, MO). It has a molecular weight of 248,000 and con-

Table 1
Application of PEG–Enzymes to Biotechnological Processes.

Enzyme	Origin	Modifier[a]	Purpose	Solvent	Ref.
Lipase	*P. fluorescens*	Activated PEG₂	Ester synthesis, ester exchange reaction	Benzene	(16,27,28, 37–41)
Lipase	*P. fluorescens*	Activated PEG₂	Ester exchange (reformation of oil and fats)	Straight substrate	(27,28)
Lipase	*P. fragi*	Activated PEG₂	Synthesis of terpene alcohol ester and gefarnate, enantioselective esterification	Benzene	(30,42–44)
Lipase	*P. cepacia*	Activated PEG₂	Lactonization, enantioselectivity, regioselectivity	Benzene, trichloroethane, decanol	(29,45,46)
Lipase and others		Activated PEG₂	Effect of water and solvent	Benzene, trichloroethane, dimethylsulfoxide	(47–49)
Lipase	*C. cylindracea*	SS-PEG	Ester synthesis and exchange reactions	Benzene, trichloroethane	(22,50)
Lipase	*C. cylindracea*	SS-PEG	Regioselective deacylation	Trichloroethane	(31)
Chymotrypsin	Bovine pancreas	Activated PEG₂	Formation of oligopeptide	Benzene	(19)
Papain	Papaya latex	Activated PEG₂	Formation of oligopeptide	Benzene trichloroethane	(51–53)
Proteases		Activated PEG₂	Solid-phase peptide synthesis	Methanol	(54)
Lipase	*P. fluorescens*	Activated PM	Stabilization	Benzene	(23)
Trypsin	Bovine pancreas	Activated PM	Stabilization	Aqueous	(24)
Lipase and others		PEG derivatives	Reviews on PEG–enzymes in biotechnological field	Organic solvents	(1–5,14, 15,55–59)

[a]Activated PEG₂, 2,4-*bis*-[*O*-methoxypoly(ethylene glycol)]-6-chloro-*s*-triazine; activated PM,copolymer of poly(oxyethylene) 2-methyl-2-propenyl methyl diether and maleic unhydride; SS-PEG, polyethylene gylcol succinimidyl succinate (*see* **Fig. 1**).

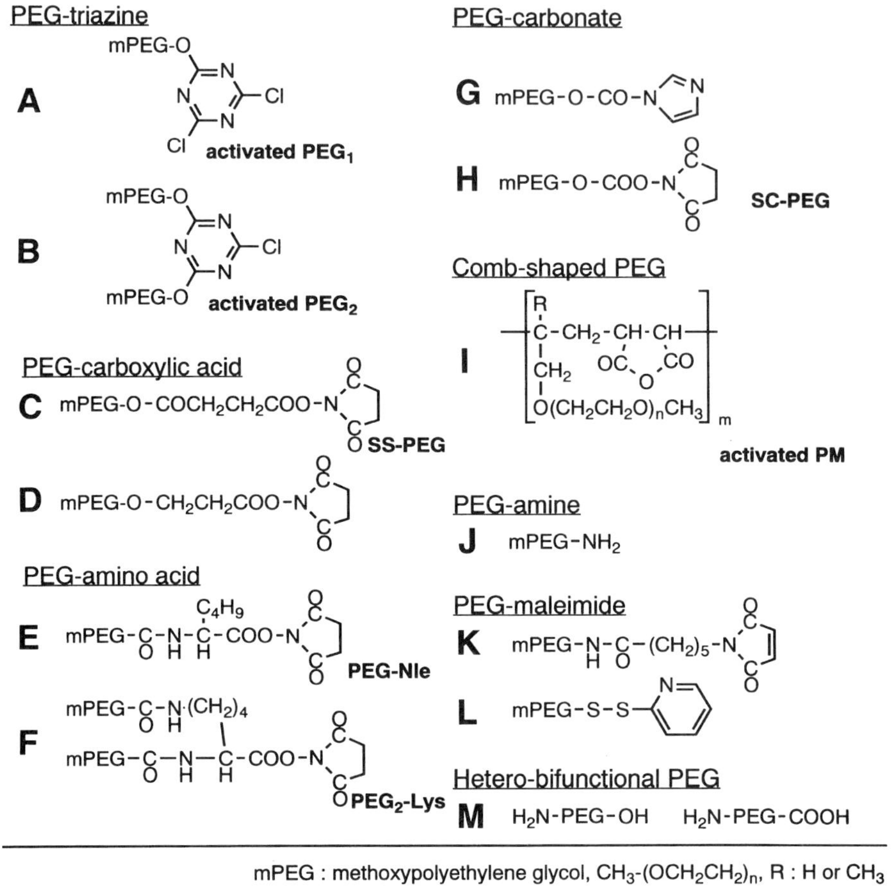

Fig. 1. Polyethylene glycol derivatives, PEGs, frequently used for modification of proteins.

sists of 4 subunits with heme. Enzymic activity of catalase is 4.5×10^4 U/mg in aqueous solution using H_2O_2 as a substrate. Benzene containing hydrogen peroxide was prepared as follows; the mixture of 30% H_2O_2 aqueous solution and benzene are vigorously shaken at room temperature and the benzene layer containing H_2O_2 is prepared. H_2O_2 concentration in benzene is determined by the method of Pobiner *(20)*. The total number of amino groups in the catalase molecule is 112 *(21)*. Phosphate-buffered saline (PBS) consisting of 8 g NaCl, 0.2 g KCl, 2.9 g $Na_2HPO_4 \cdot 12H_2O$, and 0.2 g KH_2PO_4 was dissolved in 1 L of water (pH 7.0); 0.1 *M* Na-borate buffer (pH 9.5) was prepared from boric acid and NaOH.

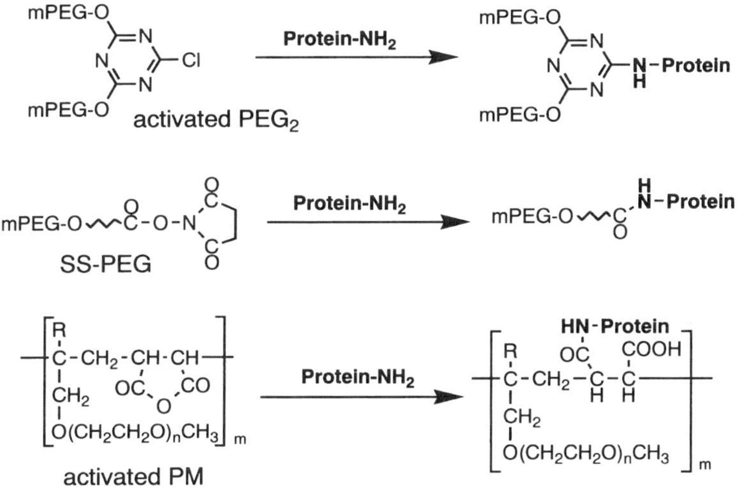

mPEG-O
activated PEG₂ → Protein-NH₂ → mPEG-O ... N–Protein

SS-PEG → Protein-NH₂ → mPEG-O ... N–Protein

activated PM → Protein-NH₂ → [structure with HN-Protein, COOH]

Fig. 2. Reactions of activated PEG₂, SS-PEG, and activated PM with amino groups in a protein molecule.

2.2. Hydrolytic Enzymes

2.2.1. Modification of Lipase from Pseudomonas with Activated PEG₂

Activated PEG$_2$ is available from Seikagaku Kogyo Co. (Tokyo, Japan). A homogeneous lipase PS (E.C. 3.1.1.3) from *Pseudomonas fluorescens* (*P. cepacia*) was purchased from Amano Pharmaceutical Co. Ltd. (Nagoya, Japan) and lipase from *P. fragi* 22.39B was obtained from Sapporo Breweries Ltd. (Yaizu, Japan). Each lipase has a single peptide chain with a molecular weight of 33,000 and catalyzes hydrolysis of triglycerides (2000–3600 U/mg of protein) in an emulsified aqueous system. Na-borate buffers (0.2 M, pH 9.5 and 0.1 M, pH 8.0) are prepared from boric acid and NaOH.

2.2.2. Modification of Lipase from Candida with SS-PEG (3,22)

Methoxypolyethylene glycol succinimidyl succinate (SS-PEG, mol wt 5000) is available from Sigma Chemical Co. (St. Louis, MO) or Shearwater Polymers, Inc. (Hunstville, AL). Lipase MY (E.C. 3.1.1.3, triacylglycerol lipase), a crude preparation, from *Candida cylindracea* was purchased from Meito Sangyo Co., Ltd. (Nagoya, Japan). The *Candida* lipase is unstable in alkaline conditions, so the modification should proceed in a neutral solution such as PBS.

2.2.3. Modification of Lipase with Activated PM (23,24)

A homogeneous lipase from *P. fluorescens* was purchased from Amano Pharmaceutical Co. Ltd. Activated PM$_{13}$ (Sunbright AKM 1510) with molecular weight of 13,000 (m~8, n~33, R = H in **Fig. 1I**) was purchased from NOF (Nippon Oil and Fats) Co. (Tokyo, Japan). Na-borate buffer (0.5 M, pH 8.5) was prepared from boric acid and NaOH. Because of the pH shift caused by hydrolysis of the acid anhydride of activated PM, a strong buffer or a pH-statt should be used for pH control.

3. Methods
3.1. Oxidoreductase
3.1.1. Modification of Catalase with Activated PEG$_2$ (17)

To 2 mL of catalase (6 mg) dissolved in 0.1 M Na-borate buffer (pH 9.5) is added various amounts of activated PEG$_2$ (100~500 mg) and the mixture is kept standing at 37°C for 1 h under gentle stirring. The reaction is stopped by adding 90 mL of cold PBS and then the sample solution is concentrated to one-tenth with an Amicon Diaflo PM-30 ultrafiltration membrane (Beverly, MA). The ultrafiltration procedure is repeated 10 times to remove unreacted modifier sufficiently and the sample solution is lyophilized after dialysis against 5 L of water (three times). The dried sample is stored at –20°C in a box in the presence of silica gel.

3.1.1.1. PROPERTIES OF PEG$_2$ CATALASE **(17)**

3.1.1.1.1. Determination of Protein and Amino Group. The protein concentration in PEG catalase is determined by the biuret method *(25)*. The degree of modification of amino groups in the modified catalase molecule is determined with trinitrobenzenesulfonate *(26)* (*see* **Note 2**) as follows. To 250 µL of catalase solution (0.5 mg protein/mL) are added 250 µL of 0.5 M Na-bicarbonate buffer (pH 8.5) and 250 µL of 0.1% sodium trinitrobenzenesulfonate (TNBS). The reaction mixture is incubated at 40°C for 2 h to complete the reaction. After adding 250 µL of 10% sodium dodecyl sulfate (SDS) and 125 µL of 1 M hydrochloric acid, the absorbance at 335 nm is measured using a quartz cuvet with a light path of 3 mm. A standard curve is obtained with known concentrations of glycine (0–1 mM) instead of a protein solution. The degree of modification is obtained by dividing the number of TNBS-nonreactive amino groups by the total number of amino groups in the catalase molecule.

3.1.1.1.2. Enzymic Activity. The enzymic reaction system (2 mL) consists of 4.5 mM hydrogen peroxide and modified catalase. The amount of enzyme is adjusted so that it will take about 10 s for the decomposition of 1 mM hydrogen peroxide in the reaction system. The concentration of hydrogen peroxide is colorimetrically determined with titanium sulfate by the method of Pobiner *(20)*. One unit (U) of catalase activity is expressed as the decomposition of 1.0 µmol hydrogen peroxide per min.

3.1.1.1.3. Solubility of Enzyme. The modified catalase (1.0 mg) is mixed well with benzene (0.5 mL) at 25°C. After centrifugation (4000g, 10 min), the protein concentration of the supernatant is spectrophotometrically determined.

The modified catalase became soluble in organic solvents by increasing the degree of modification of amino groups in the enzyme molecule. PEG$_2$ catalase, in which 55% of the total amino groups in the molecule were modified, could dissolve in benzene more than 2 mg/mL. The enzymic activity of the modified catalase in benzene, in which 42% of the total amino groups were coupled with the modifier, was unexpectedly high (1.6 times) in comparison with that of unmodified catalase in aqueous system.

3.2. Hydrolytic Enzymes

3.2.1. Modification of Lipase from Pseudomonas with Activated PEG$_2$ (3)

To 100 ml of 0.2 M sodium borate buffer (pH 9.5) containing 1 g of lipase from *P. fragi* is added stepwise 9.1 g of activated PEG$_2$ under stirring (1.5–2.0 g/h). The reaction mixture is stirred at 25°C for 5 h (*see* **Note 3**). Then, the reaction is stopped by adding 10 times the volume of 0.1 M borate buffer (pH 8.0) and then the sample solution is concentrated to one-tenth by an Amicon Diaflo YM100 ultrafiltration membrane (Beverly, MA) to remove unreacted activated PEG$_2$ and its hydrolyzate. The same procedure is repeated 10 times. After dialysis against water and lyophilization, PEG$_2$ lipase is obtained with about 90% recovery by mass. The absence of the unmodified enzyme in the sample preparation is confirmed by electrophoresis on a 5% polyacrylamide gel containing 0.1% SDS and 4.5 M urea.

3.2.1.1. PROPERTIES OF PEG$_2$ LIPASE FROM PSEUDOMONAS (3)

3.2.1.1.1. Ester Synthesis Activity. Water-saturated organic solvents are prepared by mixing the hydrophobic solvent with water at room temperature. The concentration of water in water-saturated benzene is determined to be 30 mM by Karl Fischer titration. Ester synthesis with the modified lipase in benzene is determined as follows. To 300 µL of benzene containing two substrates, a carboxylic acid and an alcohol, are added 100 µL of the modified lipase (about 0.5 U) in benzene, and the reaction mixture is incubated at 25°C. At an appropriate time, the reaction is stopped by the addition of 100 µL of 0.2 N H$_2$SO$_4$ in benzene. To the mixture (500 µL) are added 2 mL of 0.25 N NaOH, 2 mL of petroleum ether, and 0.8 mL of methanol. Then, the mixture is vigorously shaken and centrifuged at 1000g for 10 min. A set of solutions in benzene containing a known concentration of a corresponding ester (0–30 mM) is prepared to make a standard curve. The amount of ester extracted into the upper layer is colorimetrically determined by Hill's method (*36*) as follows. To 1 mL of the upper organic layer containing the extracted ester are added 300 µL of

2.5% NaOH in 95% ethanol saturated with sodium carbonate and 300 µL of 2.5% hydroxylamine hydrochloride in 95% ethanol. The solvent is evaporated off carefully, using a boiling water bath as the solvent contains flammable ether, then the residue is dissolved with 2 mL of the ferrous reagent (*see* **Note 4**) and the absorbance at 520 nm is measured.

For an ester exchange reaction, the products are determined by high-performance liquid chromatography (HPLC) or gas chromatography (GC). The protein concentration in benzene is determined by the biuret method *(25)* after removing the solvent by evaporation.

The PEG$_2$ lipase from *P. fragi* thus prepared had the following properties: the degree of modification of amino groups in the molecule was 49%, and the hydrolytic activity determined with olive oil emulsion (*see* **Note 5**) was 43% of the unmodified lipase. On the other hand, the ester synthesis activity of PEG$_2$ lipase in benzene was 13.6 µmol/min/mg of protein using lauryl alcohol and lauric acid as substrates *(16)*.

3.2.1.1.2. Ester Exchange Activity in Straight Substrates (27,28). Ester synthesis by PEG$_2$-lipase effectively proceeds without benzene when the mixture of two substrates is liquid. PEG$_2$ lipase *(P. fluorescens)* can catalyze the ester exchange reaction (**reaction 3** or **4**) not only in organic solvents but also in straight hydrophobic substrates. A mixture (1.2 mL) of two liquid substrates (e.g., amyl laurate and lauryl alcohol) is added to 1 mg of PEG$_2$ lipase with a trace amount of water (about 1 µL). The transparent sample is incubated at 55–70°C. At a given time, an aliquot is taken out and diluted with acetonitrile to stop the reaction. The products are analyzed by HPLC equipped with a refractive index detector using a Merck LiCrosorb RP-8 column (5 µm, 4 × 250 mm).

In the mixture of two substrates containing 1.35 M amyl laurate and 2.7 M lauryl alcohol, the apparent specific activity of the ester exchange reaction with PEG$_2$ lipase was 2.6 µmol/min/mg of protein at the optimum temperature of 65°C. The optimum temperature for the hydrolysis reaction in an aqueous emulsified state with unmodified lipase was 45°C.

These properties of the modified enzymes make the PEG$_2$ lipase extremely useful for many practical applications such as altering the nature of fats and oil *(28)*.

3.2.1.1.3. Optical Resolution (29,30). Natural lactones which are components of fragrance, are known to be optically active compounds with *(R)*- or *(S)*-form. Although lactones can be synthesized, chiral lactones are rarely obtained by conventional organic synthesis. Optical resolution test of racemic ε-decalactone with PEG$_2$ lipase **(Fig. 3)** is conducted as follows. To 0.1 mL of 1,1,1-trichloroethane containing racemic ε-decalactone (100 mM) and an *n*-alcohol (1 M) are added 0.9 mL of PEG$_2$ lipase (0.02 mM). Hexadecane is used as the internal standard. The reaction mixture is incubated at a temperature ranging from 15°C to 65°C. Amounts of the substrates and the products are analyzed

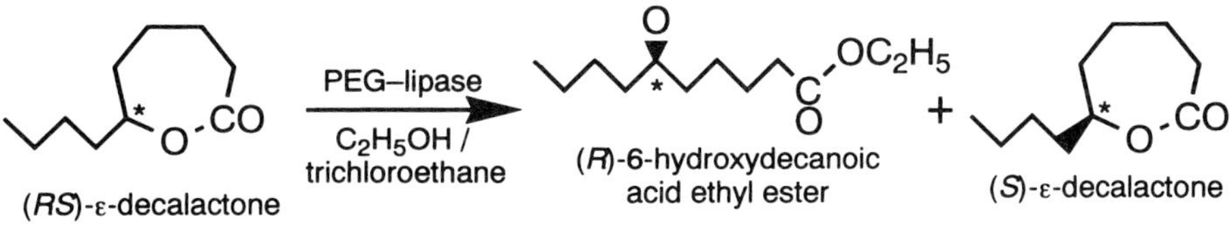

Fig 3. Optical resolution of racemic ε-decalactone in trichloroethane.

with a Shimadzu GC-14A gas chromatograph (Kyoto, Japan) equipped with a flame-ionization detector (FID). A capillary column of Astec Chiraldex B-PH (20 m × 0.25 mm inner diameter, 0.125 μm in film thickness, Madras, OR) is used. The column temperature is increased from 90°C to 150°C in linear gradients of 0.4°C/min for 35 min and then 4°C/min for 11.5 min.

The PEG$_2$ lipase from *P. cepacia* was found to catalyze the alcoholysis of racemic ε-decalactone with ethanol in 1,1,1-trichloroethane to form *(R)*-hydroxydecanoic acid ethyl ester, but no alcoholysis of *(S)*-decalactone took place with the modified lipase *(29)*.

Similarly, PEG$_2$ lipase from *P. fragi* 22.39B was found to catalyze the acylation of *(R)*-alcohols *(30)*.

3.2.2. Modification of Lipase from Candida with SS-PEG (3,22)

A crude lipase from *Candida cylindracea* (560 mg) dissolved in 8 mL of PBS (pH 7.0) is dialyzed against the same buffer followed by centrifugation at 12,000*g* for 10 min to remove insoluble materials. To 10 mL of the enzyme solution (190 mg of proteins) is slowly added 200 mg of SS-PEG. The mixture is incubated for 15 min at 25°C under gentle stirring. The PEG lipase is obtained by dialysis against water followed by lyophilization.

The degree of modification of amino groups in the enzyme molecule was 47% and the hydrolytic activity was retained by 56% of the original activity using emulsified olive oil as the substrate.

3.2.2.1. PROPERTIES OF PEG LIPASE FROM CANDIDA

3.2.2.1.1. Substrate Specificity of Ester Synthesis Activity (22). To a benzene solution containing a fatty acid (0.6 *M*) and an alcohol (0.6 *M*) is added PEG lipase (final concentration 0.5 mg/mL), and the mixture is incubated at 25°C. At a given time, the product is determined by the method described in **Subheading 3.2.1.1.1.**

The substrate specificities of PEG lipase for ester synthesis were tested in benzene containing various combinations of the substrates, alcohols, and acids. The effect of chain length of alcohols and acids on ester synthesis reaction with PEG lipase is shown in **Fig. 4**. In the case of esterification of a series of normal alcohol with pentanoic acid, PEG-modified *Candida* lipase **(Fig. 4A)** preferred

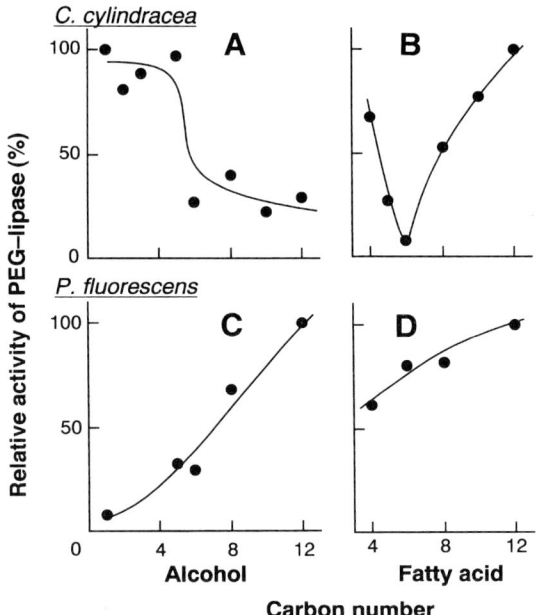

Fig. 4. Ester synthesis activities of PEG lipases from *C. cylindracea* (**A,B**) and from *P. fluorescens* (**C,D**); (**A,C**) alcohols ($C_1 - C_{12}$) + pentanoic acid (C_5). (**B,D**) fatty acids ($C_4 - C_{12}$) + pentanol (C_5).

short-chained alcohols rather than alcohols with more than five carbons. The substrate specificity was quite different with *Pseudomonas* lipase (**Fig. 4C**). Similarly, the substrate specificity of *Candida* lipase for esterification from various fatty acid and *1*-pentanol (**Fig. 4B**) had a different profile from that of the *Pseudomonas* (**Fig. 4D**).

3.2.2.1.1. Regioselective Deacylation of Peracetylated Monosaccharide (31). Peracetylated methyl β-D-xylopyranoside or peracetylated L-serine-β-D-xylopyranoside (10 μmol) is dissolved in 1 ml of 1,1,1-trichloroethane containing 100 μmol of water. To the substrate solution is added PEG lipase (1.15 mg protein), and the reaction mixture is incubated at 50°C for 3 d.

The resultant substances were found to be hydrolyzed only at the C-4 position of xylopyranoside ring in each substrate. Further, L-serine-2,3-di-*O*-acetyl-β-D-xylopyranoside was found to be a source material of the carbohydrate–protein linkage region of proteoglycan.

3.2.3. Modification of Lipase with Activated PM (23,24)

Activated PM, a copolymer of PEG derivative and maleic anhydride, has a comb-shaped form with multivalent reactive groups, as is shown in **Fig. 1I**.

Activated PM$_{13}$ (0–150 mg) is added to 2 ml of lipase (2 mg/mL) in 0.5 *M* borate buffer (pH 8.5) and the mixture is stirred at 4°C for 1 h. Unreacted copolymer is removed by ultrafiltration against 50 m*M* borate buffer with an Amicon Diaflo PM-30 (Beverly, MA) membrane; that is, the reaction mixture is diluted with 100 mL of the borate buffer (pH 8.5) and was concentrated to 5 mL. This procedure is repeated 20 times. PM lipase solution thus obtained is then dialyzed against 3 L of water (three times) and lyophilized.

PM$_{13}$ lipase, in which four out of seven amino groups in the molecule were modified, was soluble and active in hydrophobic media, and catalyzed not only an ester hydrolytic reactions in aqueous solution but also ester synthesis or ester exchange reactions in organic solvents. Furthermore, PM$_{13}$ lipase became heat stable. The stabilization against heat and urea was also obtained for trypsin by the modification with activated PM$_{100}$ *(24)* with a molecular weight of 100,000 (*m*~50, *n*~40, R = CH$_3$ in **Fig. 1I**).

4. Notes

1. In 1977, Abuchowski et al. *(32)* prepared activated PEG$_1$ (**Fig. 1A**), which reacted with amino groups of protein molecule in an aqueous buffered solution at pH 9.2 and 4°C for 1 h. However, activated PEG$_2$ seems to have an advantage in comparison with activated PEG$_1$ as two PEG chains attach to one amino group in the protein molecule.

2. For determination of amino groups in a protein molecule, a fluorometric method with fluorescamine *(33)* is also useful. Although the reaction between fluorescamine and amino group is fast, fluorescent intensity is often influenced with substances dissolved in the reaction system. Instead of the determination of free amino groups with amine-reactive reagents, the number of PEG chains coupled to protein could be directly determined by using "labeled" modifiers (e.g., PEG derivative containing norleucine as a marker [**Fig. 1E, ref. 34**]). Once the modifier coupled to amino groups of target protein, the degree of modification would be clarified by amino acid analysis.

3. When the degree of modification should be checked during the coupling reaction, the following procedure is employed. To 100 mL of 0.2 *M* sodium borate buffer (pH 9.5) containing 1 g of lipase from *P. fragi* is added 1.5–2.0 g of activated PEG$_2$ under stirring. One hour after the reaction at 25°C, an aliquot is taken to determine the degree of modification by fluorescamine *(33)* and the enzymic activity of lipase in aqueous and/or organic solvents *(35,36)*. This modification procedure is repeated until the expected properties of the modified enzyme are obtained (i.e., solubilization into organic solvent and optimization in the catalytic action in the organic solvents). In the case of PEG$_2$ lipase from *Pseudomonas* sp., approx 50% of modification of amino groups in a lipase molecule by activated PEG$_2$ seems to be best because of high solubility and high activity in organic solvents.

4. The ferrous reagent is prepared as follows: 17.5 mL of 60% perchloric acid is added to 1.94 g of ferrous chloride dissolved in 20 mL of HNO$_3$ (10 *M*). The

mixture is heated while being stirred in a water bath under a hood until a vigorous generation of yellow gas is observed. After cooling, 35 mL of water and 10 mL of concentrated HNO_3 are carefully added. The mixture is made up to 100 mL with 60% perchloric acid (about 17.5 mL). The reagent is diluted 19-fold with 95% ethanol prior to use.

5. Hydrolytic activity: The esterase activity of lipase is measured in an emulsified system composed of polyvinyl alcohol and olive oil *(35)*. The olive oil emulsion is prepared by successive sonication of a mixture of 10 mL of olive oil and 30 mL of 2% polyvinyl alcohol. The polyvinyl alcohol used is 9:1 mixture of Poval #117 (degree of polymerization 1750) and Poval #205 (degree of polymerization 550), which are available from Kurare (Kurashiki, Japan). To a mixture of 0.5 mL of emulsified olive oil and 0.4 mL of 0.1 M phosphate buffer (pH 7.0) are added 0.1 mL of lipase solution (about 10 U). In the case of *P. fragi* lipase, 0.2 M Tris-HCl buffer containing 20 mM $CaCl_2$ (pH 9.0), is used instead of phosphate buffer. The reaction mixture is incubated at 37°C for 20 min and then 2 mL of a mixture of acetone and ethanol (1:1, v/v) is added to stop the reaction. The sample solution is titrated with 0.05 M aqueous NaOH in the presence of a few drops of phenolphthalein dissolved in ethanol. One unit of enzyme is defined as the amount which liberates 1 μmol of fatty acid per minute.

References

1. Inada, Y., Yoshimoto, T., Matsushima, A., and Saito, Y. (1986) Engineering physicochemical and biological properties of proteins by chemical modification. *Trends Biotechnol.* **4,** 68–73.
2. Inada, Y., Matsushima, A., Kodera, Y., and Nishimura, H. (1990) Polyethylene glycol (PEG)–protein conjugates; application to biomedical and biotechnological processes. *J. Bioact. Compat. Polym.* **5,** 343–364.
3. Inada, Y., Matsushima, A., Hiroto, M., Nishimura, H., and Kodera, Y. (1994) Modification of proteins with polyethylene glycol derivatives, in *Methods in Enzymology, Volume 242.* (Lee, Y. C. and Lee, R. T., eds.), Academic, San Diego, CA, pp. 65–90.
4. Inada, Y., Furukawa, M., Sasaki, H., Kodera, Y., Hiroto, M., Nishimura, H., and Matsushima, A. (1995) Biomedical and biotechnological applications of PEG- and PM-modified proteins. *Trends Biotechnol.* **13,** 86–91.
5. Inada, Y., Matsushima, A., Hiroto, M., Nishimura, H., and Kodera, Y. (1995) Chemical modification of proteins with polyethylene glycols, *Adv. Biochem. Eng./ Biotechnol.* **52,** 129–150.
6. Topchieva, I. N. (1989) Methods of synthesis and technology of production of drugs. Conjugates of proteins with polyethylene glycol (review). *Pharm. Chem. J.* **23,** 511–517.
7. Delgado, C., Francis, G. E., and Fisher, D. (1992) The uses and properties of PEG-linked proteins. *Crit. Rev. Ther. Drug Carrier Syst.* **9,** 249–304.
8. Harris, J. M. (ed.) (1992) P*oly(ethylene glycol) Chemistry*, Plenum, New York.
9. Kodera, Y., Matsushima, A., Hiroto, M., Nishimura, H., Ishii, A., Ueno, T., and Inada, Y. (1998) Pegylation of proteins and bioactive substances for application to medical and technological processes. *Prog. Polym. Sci.* **23,** 1233–1271.

10. Zalipsky, S. (1995) Chemistry of polyethylene glycol conjugates with biologically active molecules. *Adv. Drug Deliv. Rev.* **16,** 157–182.
11. Smith, R. A. G., Dewdney, J. M., Fears, R., and Poste, G. (1993) Chemical derivatization of therapeutic proteins. *Trends Biotechnol.* **11,** 397–403.
12. Nucci, M. L., Shorr, R., and Abuchowski, A. (1991) The therapeutic value of poly (ethylene glycol)-modified proteins. *Adv. Drug Deliv. Rev.* **6,** 133–151.
13. Fortier, G. (1994) Biomedical applications of enzymes and their polyethylene glycol adducts. *Biotechnol. Genet. Eng. Rev.* **12,** 329–356.
14. Matsushima, A., Kodera, Y., Hiroto, M., Nishimura, H., and Inada, Y. (1996) Bioconjugates of proteins and polyethylene glycol; potent tools in biotechnological processes. *J. Mol. Catal., B: Enzymatic* **2,** 1–17.
15. Inada, Y., Takahashi, K., Yoshimoto, T., Ajima, A., Matsushima, A., and Saito, Y. (1986) Application of polyethylene glycol-modified enzymes in biotechnological processes: organic solvent-soluble enzymes. *Trends Biotechnol.* **4,** 190–194.
16. Takahashi, K., Yoshimoto, T., Ajima, A., Tamaura, Y., and Inada, Y. (1984) Modified lipoprotein lipase catalyzes ester synthesis in benzene; substrate specificity. *Enzyme* **32,** 235–240.
17. Takahashi, K., Ajima, A., Yoshimoto, T., and Inada, Y. (1984) Polyethylene glycol-modified catalase exhibits unexpectedly high activity in benzene. *Biochem. Biophys. Res. Commun.* **125,** 761–766.
18. Takahashi, K., Nishimura, H., Yoshimoto, T., Saito, Y., and Inada, Y. (1984) A chemical modification to make horseradish peroxidase soluble and active in benzene. *Biochem. Biophys. Res. Commun.* **121,** 261–265.
19. Matsushima, A., Okada, M., and Inada, Y. (1984) Chymotrypsin modified with polyethylene glycol catalyzes peptide synthesis reaction in benzene. *FEBS Lett.* **178,** 275–277.
20. Pobiner, H. (1961) Determination of hydroperoxides in hydrocarbon by conversion to hydrogen peroxide and measurement by titanium complexing. *Anal. Chem.* **33,** 1423–1426.
21. Schroeder, W. A., Shelton, J. R., Schelton, J. B., Robberson, B., and Appel, G. (1961) The amino acid sequence of bovine liver catalase: a preliminary report. *Arch. Biochem. Biophys.* **131,** 653–655.
22. Kodera, Y., Takahashi, K., Nishimura, H., Matsushima, A., Saito, Y., and Inada, Y. (1986) Ester synthesis from α-substituted carboxylic acid catalyzed by polyethylene glycol-modified lipase from *Candida cylindracea* in benzene. *Biotechnol. Lett.* **8,** 881–884.
23. Hiroto, M., Matsushima, A., Kodera, Y., Shibata, Y., and Inada, Y. (1992) Chemical modification of lipase with a comb-shaped synthetic copolymer of polyoxyethylene allyl methyl diether and maleic anhydride. *Biotechnol. Lett.* **14,** 559–564.
24. Hiroto, M., Yamada, M., Ueno, T., Yasukohchi, T., Matsushima, A., Kodera, Y., et al. (1995) Stabilization of trypsin by modification with comb-shaped copolymers of poly (ethylene glycol) derivative and maleic anhydride. *Biotechnol. Tech.* **9,** 105–110.
25. Layne, E. (1957) Spectrophotometric and turbidimetric methods for measuring proteins, in *Methods in Enzymology, Volume 3* (Colowick, S. P. and Kaplan, N. O., eds.), Academic, New York, pp. 447–454.

26. Habeeb, A. F. S. A. (1966) Determination of free amino groups in proteins by trinitrobenzenesulfonic acid. *Anal. Biochem.* **14,** 328–336.
27. Takahashi, K., Kodera, Y., Yoshimoto, T., Ajima, A., Matsushima, A., and Inada, Y. (1985) Ester-exchange catalyzed by lipase modified with polyethylene glycol. *Biochem. Biophys. Res. Commun.* **131,** 532–536.
28. Matsushima, A., Kodera, Y., Takahashi, K., Saito, Y., and Inada, Y. (1986) Ester-exchange reaction between triglycerides with polyethylene glycol-modified lipase. *Biotechnol. Lett.* **8,** 73–78.
29. Furukawa, M., Kodera, Y., Uemura, T., Hiroto, M., Matsushima, A., Kuno, H., et al. (1994) Alcoholysis of ε-decalactone with polyethylene glycol-modified lipase in 1,1,1-trichloroethane. *Biochem. Biophys. Res. Commun.* **199,** 41–45.
30. Kikkawa, S., Takahashi, K., Katada, T., and Inada, Y. (1989) Esterification of chiral secondary alcohols with fatty acid in organic solvents by polyethylene glycol-modified lipase. *Biochem. Int.* **19,** 1125–1131.
31. Kodera, Y., Sakurai, K., Satoh, Y., Uemura, T., Kaneda, Y., Nishimura, H., et al. (1998) Regioselective deacetylation of peracetylated monosaccharide derivatives by polyethylene glycol-modified lipase for the oligosaccharide synthesis. *Biotechnol. Lett.* **20,** 177–180.
32. Abuchowski, A., van Es, T., Palczuk, N. C., and Davis, F. F. (1977) Alteration of immunological properties of bovine serum albumin by covalent attachment of polyethylene glycol. *J. Biol. Chem.* **252,** 3578–3581.
33. Udenfriend, S., Stein, S., Bohlen, P., Dairman, W., Leimgruber, W., and Weigle, M. (1972) Fluorescamine. Reagent for assay of amino acids, peptides, and primary amines in the picomole range. *Science* **178,** 871.
34. Yamasaki, N., Matsuo, A., and Isobe, H. (1988) Novel polyethylene glycol derivatives for modification of proteins. *Agric. Biol. Chem.* **52,** 2125–2127.
35. Yamada, K., Ota, Y., and Machida, H. (1962) Studies on the production of lipase by microorganisms. Part II. Quantitative determination of lipase. *J. Agric. Chem. Soc. Japan* **36,** 860–864.
36. Hill, U. T. (1947) Colorimetric determination of fatty acids and esters. *Anal. Chem.* **19,** 932–933.
37. Inada, Y., Nishimura, H., Takahashi, K., Yoshimoto, T., Saha, A. R., and Saito, Y. (1984) Ester synthesis catalyzed by polyethylene glycol-modified lipase in benzene. *Biochem. Biophys. Res. Commun.* **122,** 845–850.
38. Ajima, A., Takahashi, K., Matsushima, A., Saito, Y., and Inada, Y. (1986) Retinyl esters synthesis by polyethylene glycol-modified lipase in benzene. *Biotechnol. Lett.* **8,** 547–552.
39. Yoshimoto, T., Takahashi, K., Nishimura, H., Ajima, A., Tamaura, Y., and Inada, Y. (1984) Modified lipase having high stability and various enzymic activities in benzene, and its re-use by recovering from benzene solution. *Biotechnol. Lett.* **6,** 337–340.
40. Takahashi, K., Yoshimoto, T., Tamaura, Y., Saito, Y., and Inada, Y. (1985) Ester synthesis at extraordinarily low temperature of –3°C by modified lipase in benzene. *Biochem. Int.* **10,** 627–631.

41. Ajima, A., Yoshimoto, T., Takahashi, K., Tamaura, Y., Saito, Y., and Inada, Y. (1985) Polymerization of 10-hydroxydecanoic acid in benzene with polyethylene glycol-modified lipase. *Biotechnol. Lett.* **7,** 303–306.

42. Nishio, T., Takahashi, K., Yoshimoto, T., Kodera, Y., Saito, Y., and Inada, Y. (1987) Terpene alcohol ester synthesis by polyethylene glycol-modified lipase in benzene. *Biotechnol. Lett.* **9,** 187–190.

43. Nishio, T., Takahashi, K., Tsuzuki, T., Yoshimoto, T., Kodera, Y., Matsushima, A., Saito, Y., et al. (1988) Ester synthesis in benzene by polyethylene glycol-modified lipase from *Pseudomonas fragi* 22.39B. *J. Biotechnol.* **8,** 39–44.

44. Mizutani, A., Takahashi, K., Aoki, T., Ohwada, K., Kondo, K., and Inada, Y. (1989) Synthesis of gefarnate with reusable lipase modified with polyethylene glycol. *J. Biotechnol.* **10,** 121–125.

45. Kodera, Y., Furukawa, M., Yokoi, M., Kuno, H., Matsushita, H., and Inada, Y. (1993) Lactone synthesis from 16-hydroxyhexadecanoic acid ethyl ester in organic solvents catalyzed with polyethylene glycol-modified lipase. *J. Biotechnol.* **31,** 219–224.

46. Uemura, T., Furukawa, M., Kodera, Y., Hiroto, M., Matsushima, A., Kuno, H., et al. (1995) Polyethylene glycol-modified lipase catalyzes asymmetric alcoholysis of δ-decalactone in *n*-decanol. *Biotechnol. Lett.* **17,** 61–66.

47. Takahashi, K., Ajima, A., Yoshimoto, T., Okada, M., Matsushima, A., Tamaura, Y., et al. (1985) Chemical reactions by polyethylene glycol-modified enzymes in chlorinated hydrocarbons. *J. Org. Chem.* **50,** 3414,3415.

48. Matsushima, A., Okada, M., Takahashi, K., Yoshimoto, T., and Inada, Y. (1985) Indoxyl acetate hydrolysis with polyethylene glycol-modified lipase in benzene solution. *Biochem. Int.* **11,** 551–555.

49. Takahashi, K., Nishimura, H., Yoshimoto, T., Okada, M., Ajima, A., Matsushima, A., et al. (1984) Polyethylene glycol-modified enzymes trap water on their surface and exert enzymic activity in organic solvents. *Biotechnol. Lett.* **6,** 765–770.

50. Yoshimoto, T., Nakata, M., Yamaguchi, S., Funada, T., Saito, Y., et al. (1986) Synthesis of eicosapentaenoyl phosphatidylcholines by polyethylene glycol-modified lipase in benzene. *Biotechnol. Lett.* **8,** 771–776.

51. Lee, H., Takahashi, K., Kodera, Y., Ohwada, K., Tsuzuki, T., Matsushima, A., et al. (1988) Polyethylene glycol-modified papain catalyzes peptide bond formation in benzene. *Biotechnol. Lett.* **10,** 403–407.

52. Ohwada, K., Aoki, T., Toyota, A., Takahashi, K., and Inada, Y. (1989) Synthesis of N^{α}, N^{δ}-dicarbobenzoxy-L-ornithyl-β-alanine benzyl ester, a derivative of salty peptide, in 1,1,1-trichloroethane with polyethylene glycol-modified papain. *Biotechnol. Lett.* **11,** 499–502.

53. Lee, H.-H., Fukushi, H., Uchino, M., Sato, K., Takahashi, K., Inada, Y., et al. (1989) Substrate specificity of polyethylene glycol-modified papain catalyzed peptide bond synthesis in benzene. *Chem. Express* **4,** 253–256.

54. Sakurai, K., Kashimoto, K., Kodera, Y., and Inada, Y. (1990) Solid phase synthesis of peptides with polyethylene glycol-modified protease in organic solvents. *Biotechnol. Lett.* **12,** 685–688.

55. Inada, Y., Matsushima, A., Takahashi, K., and Saito, Y. (1990) Polyethylene glycol-modified lipase soluble and active in organic solvents. *Biocatalysis* **3,** 317–328.
56. Inada, Y., Takahashi, K., Yoshimoto, T., Kodera, Y., Matsushima, A., and Saito, Y. (1988) Application of PEG-enzyme and magnetite–PEG-enzyme conjugates for biotechnological processes. *Trends Biotechnol.* **6,** 131–134.
57. Takahashi, K., Saito, Y., and Inada, Y. (1988) Lipase made active in hydrophobic media. *J. Am. Oil Chem. Soc.* **65,** 911–916.
58. Inada, Y., Kodera, Y., Matsushima, A., and Nishimura, H. (1992) Enzyme modification with synthetic polymers, in *Synthesis of Biocomposite Materials* (Imanishi, Y., ed.), CRC, London, pp. 85–108.
59. Kodera, Y., Nishimura, H., Matsushima, A., Hiroto, M., and Inada, Y. (1994) Lipase made active in hydrophobic media by coupling with polyethylene glycol. *J. Am. Oil Chem. Soc.* **71,** 335–338.

7

Chemical Modification of Lipase for Use in Ester Synthesis

Mahiran Basri, Kamaruzaman Ampon, Che Nyonya A. Razak, and Abu Bakar Salleh

1. Introduction

Chemical modification studies of proteins originated when interest in quantitative determination of proteins and their various constituent amino acids was started. Later, chemical modification procedures were used to identify the particular amino acid residues required for the biological activity of proteins *(1)*. The increasing interest in the subject during the last decade has been promoted by practical interests related, for example, to possible pharmacological or medical diagnostic application such as to convert a number of protein toxins into toxoids. These toxoids retain some of the original antigenic determinants but are no longer toxic. Other enzymes are modified chemically to alter and improve their native properties and endow them with useful new functions such as to make them more soluble and active and more stable in organic solvents and to change the selectivity of the enzyme *(2)*. Modifications strategies now being developed should soon yield a wide spectrum of novel biomolecules whose activities are optimized for specific industrial processes or therapeutic applications.

Many useful reagents have been developed over the years for chemical modifications to meet the varied and specific needs of protein chemists. These reagents are side-chain selective reagents that react under certain specified conditions, with single or a limited number of side-chain groups. In this work, considerations in selecting the modifiers used for enzyme modification are based on current information available involving the enzyme and the modifiers. These considerations are to select modifiers that do not directly react with

From: *Methods in Biotechnology, Vol. 15: Enzymes in Nonaqueous Solvents: Methods and Protocols*
Edited by: E. N. Vulfson, P. J. Halling, and H. L. Holland © Humana Press Inc., Totowa, NJ

the amino acid groups that are involved in the catalytic activity, and to choose modifiers that will increase the hydrophobicity of the lipase so that it is soluble in organic solvents and suitable to facilitate reactions in organic solvents. The modifiers chosen are imidoesters, aldehydes, and monomethoxypolyethylene glycol (mPEG).

Modification with the chosen modifiers offered several advantages over other modification reactions. The reaction could be carried out readily in aqueous solutions. These modifiers also reacted highly site-specific at the α- and ε-amino groups and the resulting derivatives, involving mainly the numerous side-chain functions of lysine residues, retained the original charge at physiological pH as the native protein. Thus, their physical and biological properties tend to be preserved. The amide bond formed was stable over a wide range of pH, including the strongly acidic region in the absence of a strong nucleophiles *(3)*. Furthermore, homologs of these modifiers are readily available in the case of aldehydes and easily synthesized in the case of imidoesters and activated mPEG.

1.1. Synthesis of Imidoesters

Synthesis of imidoesters are according to the procedure adapted from Hunter and Ludwig *(4)* as described by

$$R\text{-}CN \xrightarrow[\substack{CH_3OH, 0°C, 1.5 \text{ h} \\ \text{nitrile}}]{HCl \text{ (dry gas)}} R\text{-}C\underset{OCH_3}{\overset{NH_2^+ Cl^-}{<}}$$
imidoester hydrochloride

Imidoesters with different hydrophobicity could be synthesized easily from their corresponding nitriles or alcohols in relatively high yield. Methyl acetimidate (imidoester I) methyl benzimidate (imidoester II), methyl 3-phenylpropionimidate (imidoester V), and methyl 4-phenylbutyrimidate (imidoester VI) are very soluble in aqueous 0.1 *M* borate buffer, pH 8.5. The imidoesters are soluble in the buffer (30–40 mg in 5 mL buffer). When these solutions were at room temperature for at least 2 h, the solutions turned cloudy. Methyl 4-phenylbenzimidate (imidoester III) and methyl *n*-dodecaimidate (imidoester IV) are relatively less soluble in aqueous solution (15–20 mg in 5 mL buffer). Imidoesters are soluble in pyridine and dimethylformamide (20–30 mg in 5 mL solvent) but only slightly soluble in benzene, hexane, carbon tetrachloride, and chloroform (10–15 mg in 5 mL solvent).

1.2. Activation of mPEG

Monomethoxypolyethylene glycol (mPEG) of molecular weight 1900 and 5000 were activated by *p*-nitrophenyl chloroformate according to the procedure described by Veronese et al. *(5)*. The reaction is

$$CH_3-O-(C_2H_4O)n-H + Cl-\overset{\overset{\displaystyle O}{\|}}{C}-OR \xrightarrow{28°C,\ 24\ h} CH_3-O-(C_2H_4O)n-\overset{\overset{\displaystyle O}{\|}}{C}-OR$$

CH₃-O-PEG   *p*-nitrophenyl   CH₃-O-PEG-*p*-nitrophenyl
              chloroformate   carbonate

The activated mPEG is extremely soluble in aqueous 0.2 *M* phosphate buffer, pH 8.5. About 50–60 mg of the activated mPEG is soluble in about 5 mL of the buffer. It is also very soluble in all the organic solvents tested (about 40 mg in 5 mL solvent).

1.3. Amidination with Imidoesters

Lipase was amidinated using the procedure adapted from Wofsy and Singer *(6)* as described by

$$\underset{\text{imidoester}}{R-\overset{\overset{\displaystyle NH_3^+}{\|}}{C}-OCH_3 + H_2N\text{-Protein}} \xrightarrow{pH\ 8.5,\ 0°C,\ 2\ h} \underset{\text{amidinated lipase}}{R-\overset{\overset{\displaystyle NH_3^+}{\|}}{C}-HN\text{-Protein}}$$

The reaction mechanism of protein amino groups with imidoesters is already well established *(7)*. Many of the protein amino groups are converted into amidine derivatives directly. Amidine derivatives of proteins are soluble in most buffer systems with low ionic strength and at room temperature. When the ionic strength and temperature increased, they were known to form precipitates, because of aggregation caused by hydrophobic interactions of the covalently attached substituents. However, solutions of the derivatized enzymes cleared upon cooling the solutions.

1.4. Reductive Alkylation with Aldehydes

Reductive alkylation of lipase was carried out as described by Fretheim et al. *(8)*. The reaction is as described by

$$\underset{\text{(aldehyde)}}{R-\overset{\overset{\displaystyle O}{\|}}{C}H + H_2N\text{-Protein}} \xrightarrow{pH\ 9.0,\ 4°C,\ 30\ min} R-CH=N\text{-Protein}$$

$$\downarrow \quad NaCNBH_3$$

$$R-CH_2-HN\text{-Protein}$$
(alkylated lipase)

Reductive alkylation of enzymes with hydrophobic aldehydes and sodium cyanoborohydride to increase the hydrophobicity of several enzymes has been studied by several researchers *(8–10)*. Aldehydes were available with many different chain lengths and can be reacted directly with protein by reductive alkylation. It was found that trypsin modified with dodecyldehyde and sodium borohydride in aqueous solution became very insoluble *(9)*.

1.5. Modification with Activated mPEG

The modification reaction of lipase with activated monomethoxypolyethylene glycol (AmPEG) is

$$CH_3\text{-}(O\text{-}CH_2\text{-}CH_2)n\text{-}\overset{\overset{\displaystyle O}{\|}}{C}\text{-}OR + H_2N\text{-}Protein \xrightarrow[2\,h]{pH\ 8.5,\ 28°C} CH_3\text{-}(O\text{-}CH_2\text{-}CH_2)n\text{-}\overset{\overset{\displaystyle O}{\|}}{C}\text{-}HN\text{-}Protein$$

(Activated mPEG)   (mPEG-Lipase)

mPEGs have been used to modify enzymes by many researchers *(11–14)*. They are linear molecules, uncharged and available in various sizes. They are extremely water soluble by virtue of hydrogen-bonding of three water molecules per ethylene oxide unit. The single terminal hydroxyl group is available for coupling reactions. The amphipathic nature of the mPEG molecule is particularly suitable for lipase reactions. Its hydrophilic nature makes it possible to modify enzymes in aqueous solution and its hydrophobic nature enable modified enzymes to function in hydrophobic environment. mPEG with different molecular weights could be activated easily prior to their reaction with the protein.

The modified lipases in this study were all on soluble in aqueous solution. The degree of solubility however, differed for different enzyme preparations. All lipases modified with mPEG (PL) were very soluble, whereas lipases modified with high-molecular-weight imidoesters and aldehydes were less soluble. The modification reactions with these modifiers were monitored very carefully to minimize insolubilization of the modified enzymes. The total amount of each of them, to be added in the modification reaction mixture to achieve maximum modifications possible without formation of any insoluble derivatives, were determined separately in preliminary experiments

1.6. Catalytic Activity of Amidinated Lipases

The hydrolytic activity of lipase is substantially reduced following its derivatization. The synthetic activities of all the modified lipase preparations, however, are higher compared with the initial native activity. Lipase that was modified with the high-molecular-weight imidoesters such as III, IV, V, and VI show a much higher activity compared with that modified with the low-molecular-weight imidoesters (I and II).

The effects of increasing the degree of lipase modification by imidoester I and imidoester VI, respectively, on the synthetic activity show the maximal activity to be dependent on the degree of modification by the respective imidoesters. To achieve maximum activity, a lower degree of modification (63%) of the enzyme was necessary with the more hydrophobic imidoester VI compared with imidoester I (about 90%). As the degree of modification with imidoester VI was further increased, the modified enzyme showed a decrease in synthetic activity.

1.7. Catalytic Activity of Alkylated Lipases

The effect of reductive alkylation of lipase with different aldehydes on its catalytic activity showed that the modification was accompanied by a reduction in the hydrolytic activity. The hydrolytic activity was highest with lipase modified with acetaldehyde and decreased as the chain length of the aldehydes used in the modification was increased. The synthetic activity of the modified enzyme increased with increasing modification. Generally, increasing the chain lengths of the aldehydes used for modification increases the synthetic activity of the modified lipase. The activity (at about 40% modification) increases slightly with acetaldehyde and by about twofold with dodecyldehyde.

The effect of extensive modification of the enzyme with respect to its esterification reaction is studied. To achieve its maximum activity, a lower degree of modification (40%) of the enzyme was needed with the more hydrophobic dodecyldehyde compared with acetaldehyde (80%).

1.8. Catalytic Activity of mPEG Lipases

Modification of lipase with mPEG was accompanied by a reduction in the hydrolytic activity. However, lipase modified with mPEG1900 (PL1900) at 45% modification showed a threefold increase in synthetic activity, whereas lipase modified with mPEG5000 (PL5000) at 48% modification gave only a slight increase in the synthetic activity compared to the native lipase. The synthetic activity of the PLI900, however, was increased by ninefold as the degree of modification was increased from 0–95%. With PL5000, the activity was increased by sixfold.

2. Materials

Lipases from *Candida rugosa* (Type VI), benzonitrile, 4-biphenylcarbonitrile, undecyl cyanide, hydrocinnamonitrile, 4-phenyl-butyronitrile, sodium cyanoborohydride, acetaldehyde, propionaldehyde, benzaldehyde, octaldehyde, dodecyldehyde, glycine, monomethoxypolyethylene glycol (PEG) of different molecular weights, *p*-nitrophenyl chloroformate (pncf), triethylamine, and trinitrobenzene sulfonate (TNBS) are from Sigma Chemical Co. (St. Louis, MO). All other chemicals used were of the highest purity available.

3. Methods
3.1. Purification of Lipase

Commercial lipase from *Candida rugosa* is purified according to the method of Basri et al. *(15)*. The purification procedures employed are the conventional method and the one-step gel filtration on Superose 6 column using fast performance liquid chromatography (FPLC). The conventional method consists of the following steps: water extraction, ammonium sulfate precipitation, and fractionation through Sephadex G25 and G200. An 11-fold purification with a percent recovery of 42% was obtained with the conventional method. With Superose 6, a 12-fold purification and about 90% recovery was obtained. However, the prepacked Superose 6 column only allows small protein loading (5 mg), whereas the Sephadex G25 and G200 column can separate up to 100 mg protein efficiently. Thus, both methods were used concurrently. The purification procedure was needed because our preliminary experiments showed that unpurified lipase could not be modified efficiently (i.e., only a very low degree of modification was obtained).

3.2. Protein Determination

Samples containing protein (2 mg) and 200 μL of a constant boiling 6 *M* hydrochloric acid (HCl) are introduced into hydrolysis tubes with a side arm and a Rotaflo screw cap (1 × 15 cm). The vials are evacuated to remove the air so as to prevent oxidation of the samples during hydrolysis, which may interfere with amino acid determination. The tubes are then capped tightly and are heated for 1 h at 145°C in a Pierce Reacti-Therm heating module. An aliquot (5 μL) from each hydrolysis tube is transferred to a test tube and dried at 40°C for 5 h in an oven. The amount of protein in the test tube is determined by titration with trinitrobenzene sulfonate (TNBS) *(16)*.

When necessary, protein determination using the method of Lowry et al. *(17)* or Bradford *(18)* was used for comparison. The extent of protein modification is determined by comparing the number of amino acid groups that reacted with TNBS in the modified and unmodified protein *(19)*.

3.3. Synthesis of Imidoesters

Imidoesters methyl acetimidate (I), methyl benzimidate (II), methyl 4-phenyl benzimidate (III), methyl *n*-dodecanimidate (IV), methyl 3-phenylpropionimidate (V), and methyl 4-phenylbutyrimidate (VI) are synthesized from their corresponding nitriles: acetonitrile, benzonitrile, 4-biphenylcarbonitrile, undecyl cyanide, hydrocinnamonitrile, and 4-phenyl-butyronitrile, respectively.

Absolute methanol which has been dried with a 3 Å molecular sieve (0.075 mol) is acidified by bubbling dry hydrochloric acid gas (HCl) for 15 min at –40°C or below using a liquid-nitrogen bath. The corresponding nitrile (0.05 mol) is added to the cold solution and stirred for 30 min. The solution is slowly warmed to 0°C in an ice bath and stirred for an additional hour. Crystallization began after standing overnight at 4°C and continued for 2–3 d. Absolute ether (100 mL) is then added and the white solid is ground with a pestle and mortar with ether. The mixture is filtered and washed with more ether and dried *in vacuo*. In the synthesis of methyl *n*-dodecanimidate hydrochloride, absolute ether (50 mL) is added to the HCl-saturated methanol prior to the addition of *n*-undecyl cyanide in order to prevent immediate solidification of the reaction mixture. The mixture solidified within 5–10 min and was allowed to stand overnight at 4°C, ground, and dried, as described previously.

All products are dried over silica gel and sodium hydroxide pellets *in vacuo*. The melting points of the imidoesters were determined in open capillary tubes using the Electrothermal digital melting point apparatus. Nuclear magnetic resonance (NMR) spectra of the samples were recorded on Bruker WP 80SY nuclear magnetic resonance spectrophotometer using deuterodimethyl sulfoxide-d_6 as solvent. Infrared (IR) spectra were determined on Beckman Acculab 7 infrared spectrophotometer. (*See* **Note 1**.)

3.4. Activation of Monomethoxypolyethylene Glycol mPEG

Monomethoxypolyethylene glycol (mPEG) of molecular weight 1900 and 5000 are activated by *p*-nitrophenyl chloroformate according to the procedure described by Veronese et al. *(4)*. *p*-Nitrophenylchloroformate (0.6 g) is dissolved in 50 mL acetonitrile. mPEG (5 g) and triethylamine (0.29 g) are added and the mixture is stirred for 24 h. The mixture is then filtered to remove the precipitate, which is mostly triethylammonium chloride. Diethyl ether (500 mL) is added to the filtrate and the mixture is left overnight at 4°C to crystallize. The product is filtered, washed with ether, and recrystallized with acetonitrile–ether (1:1) overnight at 4°C. The product is filtered, air-dried in the fume cupboard for about 20 min, put in sealed vials, and kept over silica gel in a vacuum desicator at –20°C prior to use. The products and the reactants were analyzed using a Beckman Acculab 7 infrared spectrophotometer and Bruker WP 80SY nuclear magnetic resonance spectrophotometer using deuterodimethyl sulfoxide-d_6 as solvent. (*See* **Note 2**.)

3.5. Amidination with Imidoesters

The imidoester hydrochloride (0.1–0.3 g total) is added in several small increments (5–10 mg each) at intervals of 10–20 min to a stirred solution of lipase in 0.1 *M* borate buffer ($Na_2B_4O_7 \cdot 10H_2O$) (1%, w/v, protein, 5 mL) at pH 8.5 and 0°C. After each addition, the pH is readjusted to 8.5 using sodium hydroxide solution. Amidination can take place in aqueous solution between pH 7.0 and 10.0 but is more rapid at a high pH *(20)*. The reaction is terminated after 2 h by adjusting the pH to 7.0 and passing the sample through a column (2.5 × 12.5 cm) of Sephadex G25. The eluent used is distilled water. All excess imidoesters and their insoluble hydrolysis products are removed by centrifugation before the sample is passed through the column. The active peak is collected and lyophilized and stored at –20°C prior to use. Lipase is derivatized to different degrees of modification by varying the molar ratio of the imidoester with respect to the enzyme (*see* **Note 3**).

3.6. Reductive Alkylation with Aldehydes

Aldehydes (about 40 μL) are added to lipase solution (0.24 g protein in 20 mL borate buffer, pH 9.0). Sodium cyanoborohydride (NaCNBH$_3$) (4–5 mg) is added as the reducing agent. The mixture is stirred at 4°C for 30 min. The reaction is terminated by adjusting to pH 7.0 and passing the sample through a column (2.5 × 12.5 cm) of Sephadex G25. The active peak is collected and lyophilized and stored at –20°C prior to use. Lipase is derivatized to different degrees of modification by varying the molar ratio of the aldehyde with respect to the enzyme (*see* **Note 4**).

3.7. Modification with Activated mPEG

Activated mPEG (AmPEG) (0.8 g) is added to a solution of purified lipase (58 mg) in phosphate buffer (0.2 *M*, pH 8.5, 4 mL). The reaction mixture is stirred for 2 h at room temperature and terminated by the addition of excess glycine. It is then passed through a Sephadex G25 column (2.5 × 12.5 cm) to remove the unreacted activated mPEG and the active peak is collected and lyophilized. It is kept at –20°C prior to use. Lipase is modified to different degrees by changing the molar ratios of the enzyme and the activated mPEG (*see* **Note 5**).

3.8. Analysis of Modified Lipases Using Hydrophobic Chromatography

Native and modified lipases are passed through a prepacked hydrophobic column, Phenyl Superose HR 5/5 supplied by Pharmacia Fine Chemicals (Uppsala, Sweden) using the Pharmacia FPLC systems. The sample is applied in a solution of high ionic strength, buffer A [50 m*M* phosphate buffer, pH 7.0, and 1.7 *M* ammonium sulfate (NH$_4$)$_2$SO$_4$] to the column, which has been equilibrated with the same buffer. The enzyme is eluted with a linear gradient from 1.7–0 *M* sodium chloride (NaCl) at a flow rate of 0.5 mL/min.

3.9. Synthetic Activity

The reaction mixtures consisting of hexane (0.5 mL), alcohol (2.67 mmol), fatty acid (0.35 mmol), and enzyme (10–20 mg) are put in capped reaction vials. The reactants and solvents were dried with a 3 Å molecular sieve overnight, and the enzyme was kept over silica gel in a vacuum desiccator at –20°C prior to use. The vials are incubated at 28°C for 24 h with continuous shaking at 150 rpm. The reaction is terminated by dilution with 3.5 mL of ethanol:acetone (1:1 v/v). The remaining fatty acid is determined by titrating the free fatty acid present in the reaction mixture with 0.05 *M* aqueous NaOH using an automatic titrator (ABU 90, Radiometer, Copenhagen, Denmark) to an end point of pH 9.5 (calibration with aqueous buffers).

Products of the synthetic reaction are examined periodically on thin-layer chromatography (TLC) using silica gel plates. The mobile phase comprises petroleum ether, diethyl ether, and glacial acetic acid (80:30:1 [v/v]). Appropriate esters are used as standard. The ester (propyl oleate) was detected when both unmodified and modified lipases were used as catalysts.

3.10. Hydrolytic Activity

The reaction system consists of an olive oil emulsion (H_2O:olive oil, 1:1 [v/v], 1% polyvinyl acetate [PVA], 2.5 mL), calcium chloride solution (20 μL, 0.02 M), and enzyme (5–10 mg). The mixture is incubated at 37°C for 30 min with continuous shaking at 150 rpm in a horizontal shaker waterbath. The reaction is terminated by dilution with 3.5 mL of ethanol : acetone (1:1 v/v). The amount of fatty acids produced is determined by titrating the mixture with 0.05 M NaOH using an automatic titrator (ABU 90, Radiometer) to an end point of 9.5. Specific activity of the enzyme is expressed as micromoles free fatty acid produced per minute per milligram of protein.

4. Notes

1. The imidoesters synthesized are crystalline white solids. Synthesis of the imidoesters from their corresponding nitriles are followed by the disappearance of the –C≡N stretch band (2260–2240 cm^{-1}) and the appearance of a new -C=NH absorption band at 1600–1670 cm^{-1} in their IR spectra. The imidoesters show a narrow melting point indicating their relative purity. The –OCH$_3$ protons of the imidoesters show a chemical shift of around 4.10 ppm. It is absent in the NMR spectrum of the corresponding nitrile. The yield is generally about 95–98%. These imidoesters are extremely hygroscopic. They are kept in a vacuum desiccator over sodium hydroxide pellets and silica gel at room temperature. When kept in this manner, the imidoesters are found to be able to amidinate as effectively as freshly prepared samples even after 12 mo of storage. The quality of imidoesters are tested periodically by determining their melting points and IR spectra and followed by their ability to modify amino groups of bovine serum albumin.

2. Activated mPEG1900 and mPEG5000 are white crystals that are very soluble in aqueous solution as well as in organic solvents. The activated compound shows a characteristic peak, –C=0 at 1760–1780 cm^{-1} and the aromatic group at 1550 cm^{-1}, which are absent in the unactivated mPEG. The yield is about 96–98%. These activated mPEGs are extremely hygroscopic. They are kept in a vacuum desiccator over silica gel at 4°C. When stored in this manner, they are found to be able to modify protein amino groups as effectively as freshly prepared samples even after 12 mo of storage. The quality of the activated mPEG is tested periodically by determining their IR spectra, followed by their ability to modify amino groups of bovine serum albumin.

3. Usually, a fourfold molar excess of the imidoesters to the free amino groups of the enzyme produced about 40–50% modification. The imidoesters have a rela-

tively short half-life: for example, methyl acetimidate has a half-life of about 27 min at pH 8.0 and 25°C *(3)*. Hence, the imidoesters are added in small increments (about 10 mg) to the amidination mixture at 10- to 20-min intervals for about 2 h to ensure that they are present at all time to react with the lipase. Amidinated lipases are soluble in aqueous solution (30–20 mg in 5 mL solution) and relatively less soluble in organic solvents (5–10 mg in 5 mL solvent).

4. Aldehydes are used without any further treatment. Usually, about a 10-fold molar excess of the aldehydes to the free amino groups of the enzyme produced about 40–50% modification. All the alkylated lipases are soluble in aqueous solution (about 30–20 mg in 5 mL solution) and only slightly soluble in organic solvents (5–10 mg in 5 mL solvent).

5. Usually, a 25-fold molar excess of the activated mPEG to the free amino groups of the purified enzyme (calculated using 35 free amino groups and a molecular weight of 120,000 per 1000 amino acid residues) produced about 55% modification (i.e., about 19 amino groups were modified per molecule of lipase) under the condition used. mPEG lipases thus formed are also very soluble in aqueous solution (about 40–50 mg in 5 mL solution) and organic solvents (30–40 mg in 5 mL solvent).

References

1. Means, G. E. and Feeney, R. E. (1971) *Chemical Modification of Proteins,* Holden-Day, San Francisco.
2. Basri, M., Ampon, K., Wan Yunus, W. M. Z., Razak, C. N. A., and Salleh, A. B. (1992) Amidination of lipase with hydrophobic imidoesters. *J. Am. Oil Chem. Soc.* **69(5),** 579–583.
3. Dubois, G. C., Robinson, E. A., Inman, J. K., Pernam, R. N., and Appella, E. (1981) Rapid removal of acetimidoyl groups from proteins and peptides. *Biochem. J.* **199,** 335–340.
4. Hunter, M. J. and Ludwig, M. L. (1962) The reaction of imidoesters with proteins and related small molecules. *J. Am. Chem. Soc.* **84,** 3491–3504.
5. Veronese, F. M., Largajolli, R., Boccu, E., Benassi, C. A., and Schiavon, O. (1985) Surface modification of proteins: activation of monomethoxypolyethylene glycol by phenylchloroformates and modification of ribonuclease and superoxide dimutase. *Appl. Biochem. Biotechnol.* **11,** 141–152.
6. Wofsy, L. and Singer, S. J. (1963) Effects of the amidination reaction on antibody-activity and on the physical properties of some proteins. *Biochemistry* **2(1),** 104–115.
7. Inman, J. K., Dubois, G. C., and Appella, E. (1983) Amidination. *Meth. Enzymol.* **91,** 559–569.
8. Fretheim, K., Iwai, S., and Feeney, R. E. (1979) Extensive modification of protein amino groups by reductive addition of different sized substituents. *Int. J. Peptide Protein Res.* **14,** 451–456.
9. Ampon, K., Salleh, A. B., Salam, F., Yunus, W. M., Razak, C. N. A., and Basri, M. (1991) Reductive alkylation of lipase. *Enzyme Microbiol. Technol.* **13,** 597–601.

10. Basri, M., Ampon, K., Yunus, W. M., Razak, C. N. A., and Salleh, A. B. (1997) Enzymatic synthesis of fatty esters by alkylated lipase from Candida rugosa. *J. Mol. Cat. B: Enzymatic* **3,** 171–176.
11. Basri, M., Ampon, K., Yunus, W. M., Razak, C. N. A., and Salleh, A. B. (1995) Synthesis of fatty esters by polyethylene glycol-modified lipase. *J. Chem. Technol. Biotechnol.* **59,** 37–44.
12. Inada, T., Takahashi, K., Toshimoto, T., Kodera, Y., Matsushima, A., and Saito, Y. (1988) Application of PEG-enzyme and magnetite–PEG-enzyme conjugates for biotechnological processes. *Trends Biotechnol.* **6,** 131–134.
13. Basri, M., Salleh, A. B., Ampon, K., Yunus, W. M., and Razak, C. N. A. (1991) Modification of lipase by polyethylene glycol. *Biocatalysis* **4,** 313–317.
14. Baillargeon, M. W. and Sonnet, P. E. (1988) Polyethylene glycol modification of Candida rugosa lipase. *J. Am. Oil Chem. Soc.* **65(11),** 1812–1815.
15. Basri, M., Salleh, A. B., Ampon, K., Yunus, W. M., and Razak, C. N. A. (1990) Studies on the purification of lipase by Fast Performance Liquid Chromatography (FPLC). *Proceedings Malaysian Biochemistry Society Conference* **15,** 159–161.
16. Hazra, A. K., Chock, S. P., and Albers, R. W. (1984) Protein determination with trinitrobenzene sulfonate: a method relatively independent of amino acid composition. *Anal. Biochem.* **137,** 437–443.
17. Lowry, O. H., Rosenberg, N. J., Favo, N. L., and Randall, R. J. (1951) Protein measurement with folin phenol reagent. *J. Biol. Chem.* **193,** 265–275.
18. Bradford, M. M. (1976) A rapid and sensitive method utilising the principly of protein-dye binding. *Anal. Biochem.* **72,** 248.
19. Fields, R. (1972) The measurement of amino acid groups in proteins and peptides. *Biochem. J.* **124,** 581–590.
20. Makoff, A. J. and Malcolm, A. D. B. (1980) Properties of methyl acetimidate and its use as a protein modifying reagent. *Biochem. J.* **193,** 245–249.

8

Preparation and Properties in Organic Solvents of Noncovalent PEG–Enzyme Complexes

Francesco Secundo, Gianluca Ottlina, and Giacomo Carrea

1. Introduction

It is well established that enzymes can be advantageously employed in organic solvents. In fact, there are an enormous number of applications in organic synthesis and numerous examples also in the fields of food-related conversions and analysis. However, enzymes show a lower catalytic efficiency (up to three or four orders of magnitude) when employed in organic solvents rather than in aqueous buffer. It is evident that such behavior represents an obstacle to exploit at industrial level the advantages which come from using enzymes in organic solvents *(1,2)*. One of the reasons that could be responsible for the lower catalytic activity of enzymes in organic solvents can be ascribed to diffusional limitations *(3)*. Generally, when enzymes are employed in organic solvents, they are used as a suspended powder and the dispersion degree of the powder may represent a critical factor for the expression of the catalytic activity. In fact, the activity shown by an enzyme depends on the number of productive encounters that occur between the enzyme molecule and the substrate. Consequently, all methods that may increase the dispersion of the enzyme in organic solvents may be of interest in improving biocatalyst performance in nonaqueous media. Therefore, those methods that allow the dissolution of the enzyme in the organic solvent deserve particular attention because they represent a way to fully disperse the enzyme in the reaction system. Several procedures such as enzyme complexation with ion-pair forming surfactants *(4)* or synthetic amphipatic lipids that coat the enzyme molecule *(5)* or covalent linking of the protein to amphipatic polymers such as poly(ethylene glycol) (PEG) *(6)* have been described. Here, we report on a methodology suitable to prepare subtilisin

From: *Methods in Biotechnology, Vol. 15: Enzymes in Nonaqueous Solvents: Methods and Protocols*
Edited by: E. N. Vulfson, P. J. Halling, and H. L. Holland © Humana Press Inc., Totowa, NJ

Carlsberg and lipase from *Pseudomonas cepacia* (lipase PC) in a form that can be dissolved or highly dispersed in organic solvents, thanks to the formation of stable noncovalent protein–PEG complexes (subtilisin + PEG and lipase PC + PEG). Both enzyme complexes, obtained by the method here described, have been proved *(7,8)* to have a much higher catalytic activity than that of untreated enzymes (**Table 1**). Although the higher catalytic activity shown by the enzymes could, in principle, be attributed also to the lyoprotective properties of PEG, in a recent article *(9)* we have proved that lipase PC + PEG, dissolves in 1,4-dioxane. This suggests that the increase of activity is likely the result of the high dispersion of the enzyme. This holds also for solvents such as carbon tetrachloride, toluene, and benzene, which give clear solutions when they are employed as reaction media for the enzyme–PEG complexes.

If the performance in organic solvents of enzyme–PEG complexes is compared with that of the same enzymes covalently linked to PEG, it can be seen that in the latter case, the activity is slightly higher at least in the case of lipase PC (**Table 1**). However, enzyme–PEG complexes are by far preferable to the covalently modified enzymes because their preparation is much simpler. It should also be emphasized that the complex can be reutilized for several cycles of conversion following a simple procedure that consists of (1) centrifugation of the reaction medium, (2) removal of the solution present in the upper part of the centrifuge tube (no pellet was observed following centrifugation), and (3) addition of a fresh substrate solution.

2. Materials
2.1. Preparation of Subtilisin + PEG

1. Purified subtilisin Carlsberg (Sigma). (*See* **Note 2**.)
2. Buffer: potassium phosphate 0.05 *M*, pH 8.0.
3. Poly(ethylene glycol), *Mr* 5000 (Sigma).

2.2. Preparation of Lipase PC + PEG

1. Purified lipase PC. (*See* **Note 2**). Lipase PS (obtained from Amano) was purified by DEAE chromatography as described by Secundo et al. *(8)*.
2. Buffer: potassium phosphate 0.01 *M*, pH 7.0.
3. Poly(ethylene glycol), *Mr* 5000.

2.3. Activity Determination of Subtilisin

1. 1-Hexanol.
2. Vinyl butyrate.
3. One of the following organic solvents: 1,4-dioxane, carbon tetrachloride, benzene, toluene.

Table 1
Transesterification, Hydrolytic, and Total (Transesterification Plus Hydrolytic) Activity of Subtilisin and Subtilisin + PEG in 1,4-Dioxane, and of Lipase PC and Lipase PC + PEG in 1,4-Dioxane and Carbon Tetrachloride

Enzyme[a]	Transesterification activity	Hydrolytic activity	Total activity
Subtilisin[b]	25	0	25
Subtilisin + PEG[b]	100	0	100
Subtilisin covalently linked to PEG[b]	93	0	93
Crude lipase PC[c]	3	13	16
Lipase PC + PEG[c]	39	61	100
Lipase PC covalently linked to PEG[c]	53	93	146
Crude Lipase PC[d]	4	6	10
Lipase PC + PEG[d]	62	38	100
Lipase PC covalently linked to PEG[d]	88	38	126

Note: See **Note 1.**
[a] The reaction of vinyl butyrate with 1-octanol (or 1-hexanol in the case of subtilisin) was employed as a model and both transesterification (formation of 1-octyl butyrate or 1-hexyl butyrate) and hydrolytic (formation of butyric acid from vinyl butyrate) activities were measured. Total activity is transesterification plus hydrolytic activity. The activities of enzyme + PEG, enzyme covalently linked to PEG, and nontreated enzyme refer to the same amount of enzyme protein.
[b] Activity measured in 1,4-dioxane, $a_w = 0.003$. The activity values were normalized with respect to the total activity value obtained by subtilisin + PEG in dioxane; 100% activity corresponds to 1.5 µmol/min/mg of subtilisin enzyme.
[c] Activity measured in benzene, $a_w = 0.5$. The activity values were normalized with respect to the total activity value obtained by lipase PC + PEG in benzene. 100% activity corresponds to 700 µmol/min/mg of lipase protein.
[d] Activity measured in carbon tetrachloride, $a_w = 0.5$. The activity values were normalized with respect to the total activity value obtained by lipase PC + PEG in carbon tetrachloride; 100% activity corresponds to 1500 µmol/min/mg of lipase protein.
Source: **refs. 7** and **8**.

2.4. Activity Determination of Lipase PC

1. 1-Octanol.
2. Vinyl butyrate.
3. One of the following organic solvents: 1,4-dioxane, carbon tetrachloride, benzene, toluene.

3. Methods
3.1. Preparation of Subtilisin + PEG

1. Dissolve subtilisin in buffer (0.2 mg/mL).
2. Dissolve PEG in water (30 mg/mL). (*See* **Note 3.**)

3. Mix in a vial the subtilisin and the PEG solution in a 1/1 ratio.
4. Freeze and lyophilize. (*See* **Note 4**.)

3.2. Preparation of Lipase PC + PEG

1. Dissolve lipase PC in buffer (0.2 mg/mL).
2. Dissolve PEG in water (30 mg/mL). (*See* **Note 3**.)
3. Mix in a vial the lipase PC and the PEG solution in a 1/1 ratio.
4. Freeze and lyophilize. (*See* **Note 4**.)

3.3. Activity Determination of Subtilisin

1. Add to a proper amount of enzyme a proper volume of the chosen organic solvent.
2. Add vinyl butyrate to obtain a concentration of 0.4 M.
3. Add 1-hexanol to obtain a concentration of 0.8 M.
4. Shake the reaction medium in an orbital shaker at 250 rpm and 25°C. At scheduled times, withdraw an aliquot of the reaction medium and determine the conversion degree. It can be determined by gas-liquid chromatography (GLC) using an HP-1 Cross-Linked Methyl Silicone Gum 25-m, 0.32-mm-inner diameter, 0.52 μm film column (Hewlett-Packard); conditions: oven temperature from 35°C (initial time) to 180°C (final time) with a heating rate of 15°C/min, H_2 as carrier gas.

3.4. Activity Determination of Lipase PC

1. Add to a proper amount of enzyme a proper volume of the chosen organic solvent. (*See* **Note 5**.)
2. Add vinyl butyrate to obtain a concentration of 0.8 M.
3. Add 1-octanol to obtain a concentration of 0.2 M.
4. Shake the reaction medium at 250 rpm and 25°C. At scheduled times, withdraw an aliquot of the reaction and determine the conversion degree. It can be determined by GLC using a HP-1 Cross-Linked Methyl Silicone Gum 25-m, 0.32-mm-inner diameter, 0.52 μm film column; conditions: oven temperature from 35°C (initial time) to 180°C (final time) with a heating rate of 15°C/min, H_2 as carrier gas.

3.5. Reuse of the Enzyme-PEG Complexes

1. Centrifuge the reaction medium containing the enzyme–PEG complex at 10,000g for 5 min.
2. Remove (paying attention not to shake the sample) the upper part of the centrifuged reaction leaving about 10 % of the initial volume.
3. Add fresh substrate solution and shake.

4. Notes

1. The complexes showed good stability in 1,4-dioxane because the enzyme solution remained transparent and active for several days at room temperature.

2. Although this procedure is described for purified enzymes, there are experimental evidences that it can be profitably adopted also in the case of crude enzymes.
3. The recommended PEG/protein ratio to dissolve the enzyme is 150 (w/w). However, beneficial effects on the activity of the enzymes in organic solvents could be obtained also with lower PEG/protein ratios.
4. Freezing of the sample must be as fast as possible.
5. The concentrations of protein and PEG indicated in **Subheading 3**. are the highest tested by us to dissolve lipase PC + PEG in anhydrous 1,4-dioxane.

Acknowledgments

We thank the Biotechnology Programme of the European Commission (BIO5-CT95-0231) and the CNR Target Project on Biotechnology for financial support.

References

1. Carrea, G., Ottlina, G., and Riva, S. (1995) Role of solvents in the control of enzyme selectivity in organic media. *Trends Biotechnol.* **13,** 63–70.
2. Koskinen, A. M. P. and Klibanov, A. M. (eds.) (1996) *Enzymatic Reactions in Organic Media*, Blackie Academic & Professional, London.
3. Klibanov, A. M. (1997) Why are enzyme less active in organic solvent than in water? *Trends Biotechnol.* **15,** 97–101.
4. Paradkar,V. M. and Dordick, J. S. (1994) Mechanism of extraction of chymotrypsin into isooctane at very low concentrations of aerosol OT in the absence of reversed micelles. *Biotechnol. Bioeng.* **43,** 529–540.
5. Okahata, Y., Fujimoto, Y., and Ijiro, K. (1995) A lipid-coated lipase as an enantioselective ester synthesis catalyst in homogeneous organic solvents. *J. Org. Chem.* **60,** 2244–2250.
6. Matsushima, A., Kodera, Y., Hiroto, M., Nishimura, H., and Inada, Y. (1996) Bioconjugates of proteins and polyethylene glycol: potent tools in biotechnological processes. *J. Mol. Catal. B: Enzymatic* **2,** 1–17.
7. Bovara, R., Carrea, G., Gioacchini, A. M., Riva, S., and Secundo, F. (1997) Activity, stability, and conformation of methoxypoly(ethylen glycol)-subtilisin at different concentration of water in dioxane. *Biotechnol. Bioeng.* **54,** 50–57.
8. Secundo, F., Spadaro, S., and Carrea, G. (1999) Optimization of *Pseudomonas cepacia* lipase preparations for catalysis in organic solvent. *Biotechnol. Bioeng.* **62,** 554–561.
9. Secundo, F., Carrea, G., Vecchio, G., and Zambianchi, F. (1999) Spectroscopic investigation of lipase from *Pseudomonas cepacia* solubilized in 1,4-dioxane by non-covalent complexation with methoxypoly(ethylen glycol). *Biotechnol. Bioeng.* **64,** 624–629.

9

Preparation of a Lipid-Coated Enzyme and Activity for Reverse Hydrolysis Reactions in Homogeneous Organic Media

Toshiaki Mori and Yoshio Okahata

1. Introduction

In the last more than two decades, a considerable number of studies have been made on the conduct of enzyme reactions in organic solvents as non-aqueous media *(1–5)*. Merits of employing hydrolytic enzymes in organic solvents are to increase the solubility of lipophilic substrates and to cause reverse reactions such as esterification, transesterification, and transglycosylation by lipases, esterases, and glycosidases, respectively. There have been several approaches to use enzymes as a synthetic catalyst in organic solvents *(6–8)*. In addition to the water-in-oil emulsion and the reversed micellar system containing a small amount of water *(9–11)*, there are two previous reports *(12,13)* of the use of both hydrophobic and hydrophilic organic solvents as a reaction medium for lipase: (1) Klibanov and coworkers reported the direct dispersion of powdered lipase in organic solvents, to produce an ester exchange catalyst for heterogeneous solutions *(2,8,14–16)* and (2) Inada and coworkers prepared a poly(ethylene glycol) (PEG)-grafted lipase that is soluble or swelled in hydrophobic organic solvents and catalyzes simple ester syntheses from aliphatic alcohols and acids *(17–19)*.

The primary significance to the use of enzymes in organic media is the need to avoid enzyme deactivation or denaturation.

In this chapter, we introduce a new method of solubilizing enzymes in hydrophobic organic solvents by coating the enzyme surface with lipid monolayers (**Fig. 1**) *(20–39)*.

From: *Methods in Biotechnology, Vol. 15: Enzymes in Nonaqueous Solvents: Methods and Protocols*
Edited by: E. N. Vulfson, P. J. Halling, and H. L. Holland © Humana Press Inc., Totowa, NJ

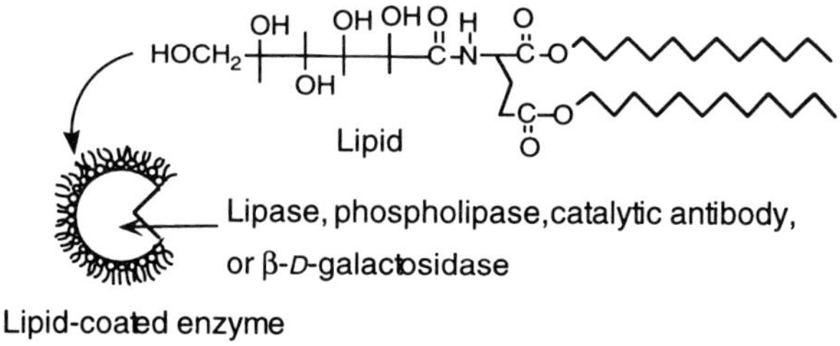

Fig. 1. A schematic illustration of lipid-coated enzymes.

2. Material

The synthetic lipid (**Fig. 1**) was prepared as described *(20–23)*. Lipase B is from *Pseudomonas fragi* 22–39B, Amano, Co., Japan.

3. Methods

3.1. Preparation of Lipid-Coated Enzymes

Most of the lipid-coated enzymes were obtained simply by mixing aqueous solutions of both the enzyme and the lipid *(20–39)*. A typical procedure is as follows. An aqueous buffer solution (50 mL, 0.05 M NaHPO$_4$, pH 5.1) of lipase B (50 mg) was mixed with an aqueous dispersion (50 mL) of the synthetic lipids (50 mg) at 4°C and stirred for 1 d at 4°C. Precipitates were gathered by centrifugation at 4°C (5000 rpm; 15 min) and washed with buffer solution (two times) and distilled water (one time) and then lyophilized. The resulting powders were colorless and insoluble in buffer solution but soluble (optically clear) in most organic solvents such as ethyl acetate, isooctane, benzene, isopropyl ether, dichloromethane, ethyl acetate, ethanol, and dimethyl sulfoxide (DMSO). In the lipid-coated enzyme, the hydrophilic head groups of the lipids interact with the hydrophilic surface of the enzyme and the lipophilic alkyl chains extend away from its surface and solubilize the enzyme in hydrophobic organic solvents. As shown in **Fig. 2**, the molecular weight of a lipid-coated lipase was determined by gel permeation chromatography with dichloromethane elution to be $(13 \pm 2) \times 10^4$, indicating that 150 ± 30 lipid molecules bind per 1 lipase B (mol wt 3.3×10^4) *(25)*. It can be roughly estimated from the lipid molecular area of cross section to the alkyl chains (0.45 nm^2) and the diameter of lipase (approx 3 nm) that 150 ± 50 lipid molecules are required cover the surface of a lipase with a monolayer. The protein content in the complex was estimated to be 8–10 wt%, both from elemental analysis (C, H, and N) and from the ultraviolet (UV) absorption of the aromatic amino acid residues in the protein *(20–25)*. A protein content of 8–10 wt% indicates that 200–300 lipid molecules per enzyme molecule are attached

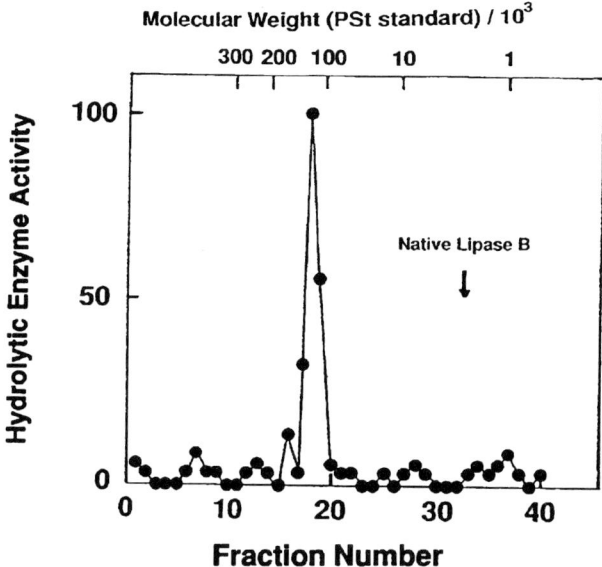

Fig. 2. Gel permeation chromatography of lipid-coated lipase B in CH_2Cl_2.

around the lipase surface. It is likely that a lipase is coated strongly with almost one layer of lipid molecules (200 ± 50). These results confirmed that the coating lipids are stable and are not removed from the enzyme surface in organic media.

In the case of other enzymes including catalytic antibody, similar enzyme–lipid complexes were obtained (protein content: 5–10%, 200 ± 100 lipid molecules per protein).

3.2. Enantioselective Esterification Catalyzed by a Lipid-Coated Lipase in Isooctane (20–25)

Figure 3 shows typical time courses of ester syntheses from *R*-, *S*-, and racemic 1-phenylethanol (50 m*M*) and excess lauric acid (500 m*M*) catalyzed by a lipid-coated lipase B in dry isooctane (60–80 ppm of H_2O in the presence of two pieces of molecular sieves) at 40°C. When the racemic alcohol was used, the esterification reached equilibrium near 50% conversion, within 2 h. The *R*-isomer was completely converted to the ester, but the *S*-isomer scarcely reacted with lauric acid. This indicates clearly that a lipid-coated lipase can recognize the *R*-1-phenylethanol and convert it to ester, but not the *S*-isomer.

Several approaches have been made to use a lipase as a synthetic catalyst in organic solvents. We compared our lipid-coated lipase system with other systems under the same reaction conditions. The time-courses of these reactions are shown in **Fig. 4**. When the lipid-coated lipase B was used (curve a), *R*-1-phenylethanol

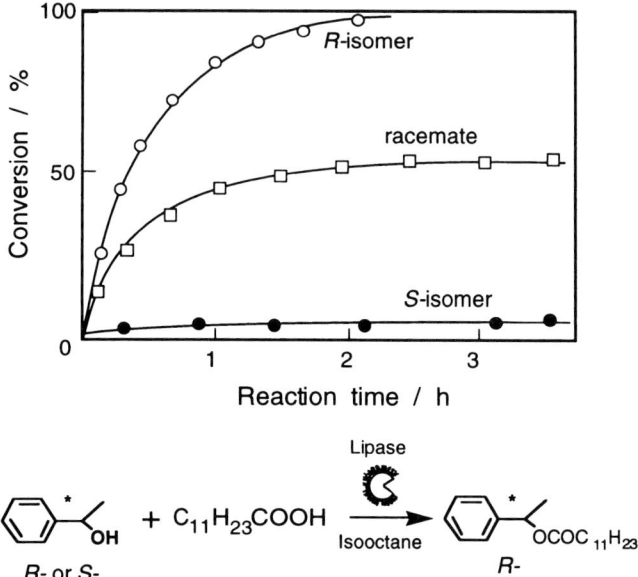

Fig. 3. Typical time-courses of esterification of *R*-, *S*-, or racemic-1-phenylethanol (50 m*M*) with lauric acid (500 m*M*) catalyzed by a lipid-coated lipase B (1 mg of protein) in dry isooctane (5 mL, 60–80 ppm of H_2O) at 40°C.

was completely converted to the ester within 3 h with high enantioselectivity ($v_R = 50$ *M*/s/mg^{-1} protein, $v_R / v_S = 250$) (*see* **Note 1**). When lipase powder was dispersed directly in isooctane, it gave a very slow reaction rate (1/100 times) compared to that of the lipid-coated lipase [$v_R = 0.5$ *M*/s/mg protein, $v_R / v_S = 260$). In the dispersion method (curve d), the enzyme exists as a suspension in the substrate organic solution, therefore, a large amount of enzyme may be required to achieve a fairly high reaction rate. In other words, the homogeneously soluble lipid-coated lipase has a much higher activity than the dispersed lipase when the same amount of enzyme is used. In the water-in-oil system (curve c), in which 1 mg of native lipase B is solubilized in a buffer solution (pH 5.6) and emulsified in isooctane, the rate of ester synthesis was very slow ($v_R < 0.1$ *M*/s/mg^{-1} protein, $v_R / v_S = 80$–120) and decreased with increasing reaction time. It seems that the presence of a small amount of water in the water-in-oil emulsion system caused the reverse hydrolysis reaction. When PEG-grafted lipase B was used (curve b), the esterification proceeded at a fair rate with high enantioselectivity ($v_R = 30$ *M*/s mg^{-1} protein, $v_R / v_S = 280$) in the initial stages. However, the conversion reached a plateau at 70% yield after 40 h in dry isooctane. This suggests that the water produced in the reaction is retained near the amphiphilic PEG chains and causes the reverse hydrolysis.

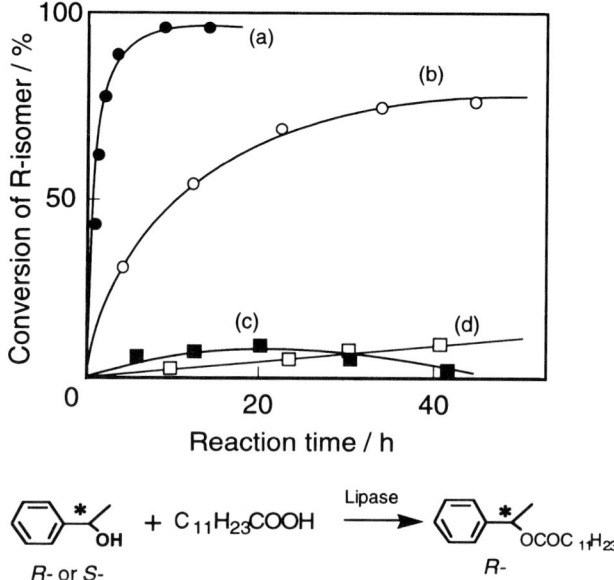

Fig. 4. Comparison of catalytic activities for ester synthesis of *R*-1-phenylethanol and lauric acid in dry isooctane at 40°C catalyzed by (a) lipid-coated lipase B, (b) PEG-grafted lipase B, (c) lipase B in water-in-oil emulsion, and (d) direct dispersion of lipase B powder. Reaction conditions are the same as for **Fig. 3**.

Thus, the lipid-coated lipase can effectively and completely catalyze ester synthesis in dry organic solutions without changing the enzyme enantio-selectivity, because unlike the other enzyme systems, the lipid-coated lipase is homogeneously soluble and stable in dry organic solvents.

3.3. Transphosphatidylation by Lipid-Coated Phospholipase in a Water/Organic Two Phase System (26)

The lipid-coating method can also be applied to phospholipase D (PLD). The PLD from *Streptomyces* sp. was employed and we applied it as the transphosphatidylation catalyst of phosphatidylcholine from *egg yolk* with various compounds containing hydroxyl group to produce various phospho-lipid derivatives. A schematic illustration of the reaction system is shown in **Fig. 5**. Because both the lipid-coated PLD and phosphocholine substrates are soluble only in organic solvents, the reaction was found to proceed smoothly in the homogeneous organic phase, and the water phase was required to remove the produced choline moiety from the organic phase that acts as an inhibitor. Reaction rates or conversions were not affected by agitation speed, because the reaction proceeded only in the organic phase.

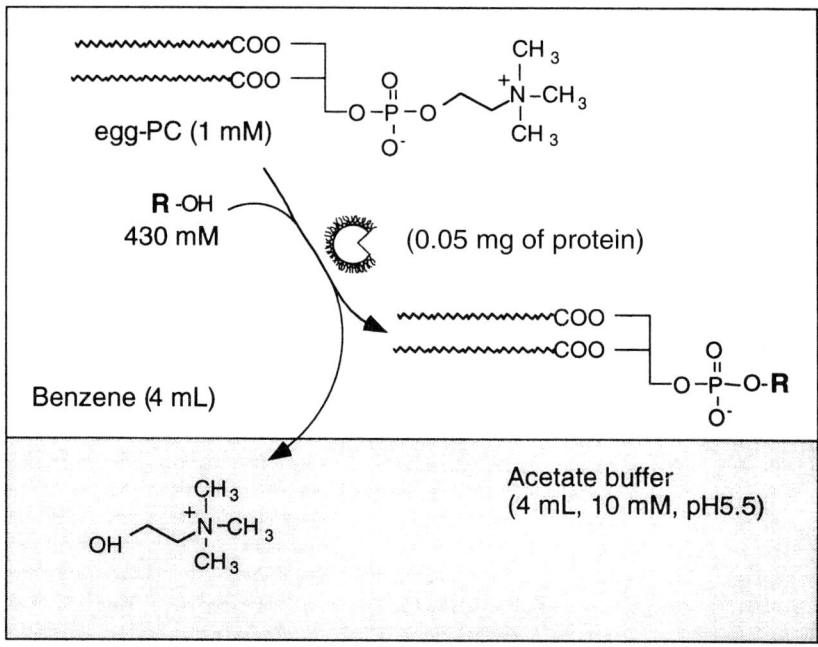

Fig. 5. A schematic illustration of transphosphatidylation of egg-PC with alcohols catalyzed by a lipid-coated PLD in benzene in the presence of aqueous buffer.

Figure 6 shows effects of chemical structures of acceptor alcohols on the transphosphatidylation of egg-PC in the two-phase system. The lipid-coated PLD showed relatively high transphosphatidylation activity with primary aliphatic alcohols having short linear alkyl chains compared with those having long alkyl chains or a phenyl group, or secondary alcohols. When the native PLD was employed under the same condition, a substrate selectivity similiar to that of the lipid-coated PLD was obtained. However, the reaction proceeds more than 300-fold slower than that of the lipid-coated PLD, because the native PLD is solubilized in the aqueous phase and the reaction occurs at the interface of the lipophilic substrates in the organic phase. These results indicate that the coating lipid does not affect the enzyme selectivity but does increase enzyme solubility in the lipophilic phase. A similar tendency was observed in the case of lipid-coated lipase and in the case of β-ᴅ-galactosidase, as noted later.

We have applied this system to produce some practical phospholipid derivatives having amino acids, carbohydrates, and deoxyribonucleotides, as shown in **Fig. 7**. The lipid-coated PLD acts as an efficient catalyst to hydrophobize the water-soluble drug, such as saccharide, amino acid, and nucleic acid, resulting from the introduction as a head group of phospholipids.

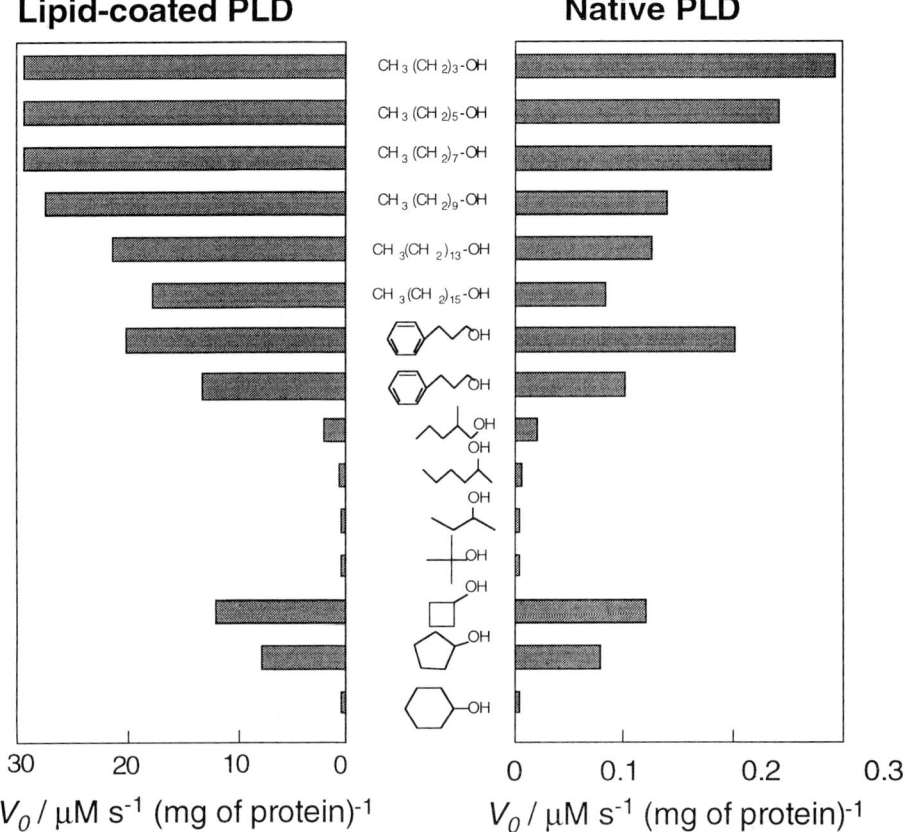

Fig. 6. Substrate selectivity for alcohol moieties of a lipid-coated PLD and a native PLD in the transphosphatidylation of egg-PC in the two-phase mixture of benzene and acetate buffer solution.

3.4. Lipid-Coated Catalytic Antibody in Water-Miscible Organic Solvents (25)

To date numerous publications have appeared on catalytic antibodies. Various reactions, such as hydrolysis of esters and amides, transesterifications, Diels–Alder reactions, decarboxylations, and sigmatropic rearrangements have been reported to be catalyzed by catalytic antibodies (40–46). Catalytic antibodies are induced to compounds (haptens) that mimic the transition states in these chemical reactions. Although substrates in these reactions are generally lipophilic and not very water soluble, the reactions have been always carried out in aqueous buffer solutions because catalytic antibodies are thought to be soluble and stable only in aqueous solution.

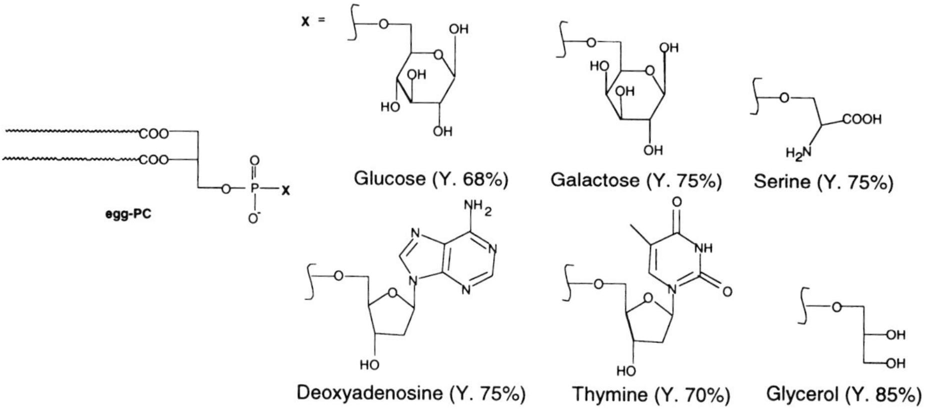

Fig. 7. Chemical structures of the practical compounds obtained with phosphatidylation catalyzed by the lipid-coated PLD in two-phase benzene–acetate buffer solutions.

In an effort to make catalytic antibodies more attractive catalysts from a synthetic standpoint, we applied our lipid-coating method to solubilize catalytic antibodies in water-miscible organic solvents. The catalytic hydrolysis of monoester of chloramphenicol was chosen as an example, because the substrate is lipophilic and the catalytic reaction has been studied in aqueous solution *(45,46)*.

Figure 8 shows the effect of the DMSO content of the buffer solution on hydrolysis rates at 30°C. DMSO must be added to a concentration of at least 20% to solubilize the lipophilic substrate completely. At concentrations of between 20% and 65% DMSO in buffer solution, both substrates and native or lipid-coated antibodies were homogeneously soluble, but the catalytic activity of the lipid-coated antibody was higher than that of the native antibody because the native was denatured by the DMSO. Similar denaturing was observed for other water-miscible organic solvents such as acetone, acetonitrile, and dimethylformamide.

Because catalytic antibodies are designed to catalyze a wide range of organic reactions with lipophilic substrates, it is important to use antibodies in water-miscible or hydrophobic organic solvents. The physical stability of the antibody is also expected to be increased by the coating lipids.

3.5. Transgalactosylation Catalyzed by a Lipid-Coated β-ᴅ-Galactosidase in a Water–Organic Two-Phase System (28–37)

In recent years, glycoside hydrolases have been applied as transglycosylation catalysts to synthesize glycoside compounds or oligosaccharides in aqueous solution containing water-miscible organic solvents, by making use of the

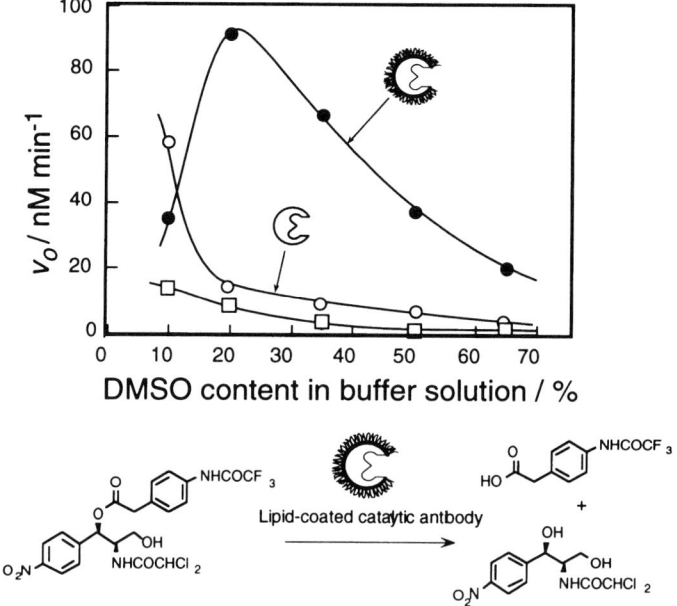

Fig. 8. Effect of DMSO content in Tris buffer solution (0.05 M, pH 8.0) on the hydrolysis of monoester of chloramphenicol (55 mM) at 30°C: (●) catalyzed by a lipid-coated antibody (1 mM, 30 mg of protein), (○) catalyzed by a native antibody (1 mM, 30 mg of protein), and (□) spontaneous hydrolysis.

reverse hydrolysis reaction *(47–50)*. In contrast to chemical synthesis, enzymatic synthesis has the advantage of providing regioselective and stereoselective products in one-step reaction and without using protection groups. However, in enzymatic synthesis using glycoside hydrolases, it has been difficult to obtain high yields of glycosylation products, because in aqueous solution, the hydrolysis reaction predominates over transglycosylation *(47–50)*. If the reaction could be carried out in nonaqueous organic solvents without denaturing the enzymes, the high yields of transglycosylation products would be obtained. We have also used our lipid-coating method to solubilize β-D-galactosidase in organic media. Because galactosyl donors such as lactose are not very soluble in organic media, an organic–water two-phase system was employed, in which both the lipid-coated enzyme and the galactosyl acceptor, alcohol, are in the organic phase and lactose is in the water phase.

Figure 9 shows typical time-courses of transgalactosylation from a 10-fold excess of lactose as a galactosyl donor to 5-phenyl-1-pentanol (PhC$_5$OH) as a galactosyl acceptor in the two-phase system of isopropyl ether and phosphate buffer (10 mM, pH 5.1) at 30°C. When the lipid-coated β-D-galactosidase was

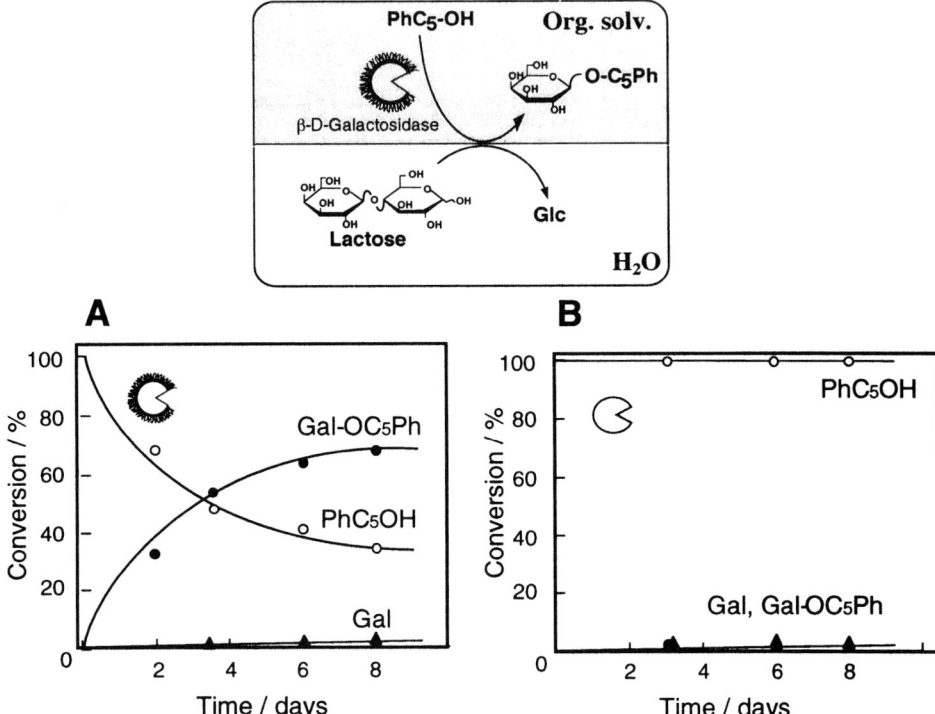

Fig. 9. Typical time courses of transgalactosylation from lactose (10 m*M*) to 5-phenyl-1-pentanol (PhC$_5$OH, 1.0 m*M*) at 30°C catalyzed by **(A)** a lipid-coated β-D-galactosidase and **(B)** a native β-D-galactosidase from *Escherichia coli* in isopropyl ether and 10 m*M* phosphate buffer (pH 5.1). [Enzyme] = 0.1 mg of protein/20 mL. The lower part of the figure shows the transglycosylation catalyzed by a lipid-coated enzyme between galactose acceptor alcohols in the organic solution and lactose in the buffer solution.

solubilized in the organic phase, a yield of 66% of transglycosylated Gal-OC$_5$Ph was obtained, but no lactose was hydrolyzed even after 8 d **(Fig. 9A)**. It was confirmed from ^{1}H- and ^{13}C-NMR (nuclear magnetic resonance) that the chemical structure of the Gal-OC$_5$Ph product remained in the β-configuration of D-galactose. As the amount of PhC$_5$OH consumed corresponds with that of Gal-OC$_5$Ph produced, this means that the carbocation intermediate formed in isopropyl ether was attacked mainly by the alcohol. By contrast, when the native β-D-galactosidase was employed, the starting substrate PhC$_5$OH was completely recovered after 8 d **(Fig. 9B)**. Since neither the galactosylation nor the hydrolysis reaction proceeded in the two-phase system, a native enzyme may be denatured at the interface between the aqueous and organic phases.

3.6. Summary

The lipid-coating system for enzymes is useful for solubilizing enzymes in organic media without denaturation. We have shown that this lipid-coating method can be applied for various enzymes as well as catalytic antibodies. The lipid-coating system has a variety of possibilities for other applications. When the organic solution of the lipid-coated enzyme is cast on an electrode substrate, a thin enzyme film can be obtained as a sensor membrane that is insoluble in aqueous solution *(37,38)*.

4. Note

The activity is expressed as the rate of concentration change divided by the mass of protein in the catalyst used. To allow conversion into other units, the volume of the reaction mixture is stated in each case.

References

1. Whitesides, G. M. and Wong, C.-H. (1985) Enzymes as catalysts in synthetic organic chemistry. *Angew. Chem. Int. Ed. Engl.* **24,** 617–638.
2. Klibanov, A. M. (1983) Immobilized enzymes and cells as practical catalysts. *Science* **219,** 722–729.
3. Klibanov, A. M. (1986) Enzymes that work in organic solvents. *CHEMTECH* **16,** 354–359.
4. Klibanov, A. M. (1990) Asymmetric transformations catalyzed by enzymes in organic solvents. *Acc. Chem. Res.* **23,** 114–120.
5. Jones, J. B. (1986) Enzymes in organic synthesis. *Tetrahedron* **42,** 3351–3403.
6. Chen, C.-S. and Sih, C. J. (1989) General aspects and optimization of enantioselective biocatalysis in organic solvents: the use of lipases. *Angew. Chem. Int. Ed. Engl.* **28,** 695–707.
7. Yokozeki, K., Yamanaka, S., Takinami, K., Hirose, Y., Tanaka, A., Sonomoto, K., and Fukui, S. (1982) Application of immobilized lipase to regio-specific interesterification of triglyceride in organic solvent. *Eur. J. Appl. Microbiol. Biotechnol.* **14,** 1–5.
8. Zaks, A. and Klibanov, A. M. (1984) Enzymatic catalysis in organic media at 100°C. *Science* **224,** 1249–1251.
9. Luisi, P. L. (1985) Enzymes hosted in reverse micelles in hydrocarbon solution. *Angew. Chem. Int. Ed. Engl.* **24,** 439–450.
10. Martinek, K., Levashov, A. V., Klyachko, N., Khmelnitski, Y. L., and Berezin, I. V. (1986) Micellar enzymology. *Eur. J. Biochem.* **155,** 453–468.
11. Hayes, D. G. and Gulari, E. (1990) Esterification reactions of lipase in reverse micelles. *Biotechnol. Bioeng.* **35,** 793–801.
12. Dastoli, F. R., Musto, N. A., and Price, S. (1966) Reactivity of active sites of chymotrypsin suspended in an organic medium. *Arch. Biochem. Biophys.* **115,** 44–47.
13. Dastoli, F. R. and Price, S. (1967) Catalysis by xanthine oxidase suspended in organic media. *Arch. Biochem. Biophys.* **118,** 163–165.

14. Klibanov, A. M. (1989) Enzymatic catalysis in anhydrous organic solvents. *Trends Biochem. Sci.* **14,** 141–144.
15. Russell, A. J. and Klibanov, A. M. (1988) Inhibitor-induced enzyme activation in organic solvents. *J. Biol. Chem.* **263,** 11,624–11,626.
16. Kitaguchi, H., Fitzpatrick, P. A., Huber, J. E., and Klibanov, A. M. (1989) Enzymatic resolution of racemic amines: crucial role of the solvent. *J. Am. Chem. Soc.* **111,** 3094,3095.
17. Takahashi, K., Ajima, A., Yoshimoto, T., Okada, M., Matsushima, A., Tamaura, Y., and Inada, Y. (1985) Chemical reactions by polyethylene glycol modified enzymes in chlorinated hydrocarbons. *J. Org. Chem.* **50,** 3414,3415.
18. Inada, Y., Yoshimoto, T., Matsushima, A., and Saito, Y. (1986) Engineering physicochemical and biological properties of proteins by chemical modification. *Trends Biotechnol.* **4,** 68–73.
19. Matsushima, A., Kodera, Y., Takahashi, K., Saito, Y., and Inada, Y. (1986) Ester-exchange reaction between triglycerides with polyethylene glycol-modified lipase. *Biotechnol. Lett.* **8,** 73–78.
20. Okahata, Y. and Ijiro, K. (1988) A lipid-coated lipase as a new catalyst for triglyceride synthesis in organic solvents. *J. Chem. Soc. Chem. Commun.* 1392–1394.
21. Okahata, Y., Fujimoto, Y., and Ijiro, K. (1988) Lipase–lipid complex as a resolution catalyst of racemic alcohols in organic solvents. *Tetrahedron Lett.* **29,** 5133,5134.
22. Okahata, Y. and Ijiro, K. (1992) Preparation of a lipid-coated lipase and catalysis of glyceride ester synthesis in homogeneous organic solvents. *Bull. Chem. Soc. Jpn.* **65,** 2411–2429.
23. Okahata, Y., Fujimoto, Y., and Ijiro, K. (1995) A lipid-coated lipase as an enantioselective ester synthesis catalyst in homogeneous organic solvents. *J. Org. Chem.* **60,** 2244–2250.
24. Okahata, Y., Hatano, A., and Ijiro, K. (1995) Enhancing enantioselectivity of a lipid-coated lipase via imprinting methods for esterification in organic solvents. *Tetrahedron Asymmetry* **6,** 1311–1322.
25. Tsuzuki, W., Okahata, Y., Katayama, O., and Suzuki, T. (1991) Preparation of organic-solvent-soluble enzyme (lipase B) and characterization by gel permeation chromatography. *J. Chem. Soc., Perkin Trans.* **1,** 1245–1247.
26. Okahata, Y., Niikura, K., and Ijiro, K. (1995) A facile transphosphatidylation of phospholipids catalyzed by a lipid-coated phospholipase D in organic solvents. *J. Chem. Soc. Perkin Trans.* **1,** 919–925.
27. Okahata, Y., Yamaguchi, M., Tanaka, F., and Fujii, I. (1995) A lipid-coated catalytic antibody in water-miscible organic solvents. *Tetrahedron* **51,** 7673–7680.
28. Okahata, Y. and Mori, T. (1996) Effective transgalactosylation catalyzed by a lipid-coated β-D-galactosidase in organic solvents. *J. Chem. Soc. Perkin Trans.* **1,** 2861–2866.
29. Okahata, Y. and Mori, T. (1998) Transglycosylation catalyzed by a lipid-coated β-D-galactosidase in a two-phase aqueous-organic system. *J. Mol. Catal. B: Enzymatic* **5,** 119–123.
30. Mori, T., Fujita, S., and Okahata, Y. (1997) Transglycosylation in a two-phase aqueous-organic system with catalysis by a lipid-coated β-D-galactosidase. *Carbohydr. Res.* **298,** 65–73.

31. Mori, T., Fujita, S., and Okahata, Y. (1997) A facile transglycosylation catalyzed by a lipid-coated β-D-galactosidase in the water–organic two phases. *Chem. Lett.* 73.

32. Mori, T. and Okahata, Y. (1997) A variety of lipid-coated glycoside hydrolases as effective glycosyl transfer catalysts in homogeneous organic solvents, *Tetrahedron Lett.* **38,** 1971–1974.

33. Mori, T. and Okahata, Y. (1998) Effective biocatalytic transgalactosylation in a supercritical fluid using a lipid-coated enzyme. *Chem. Commun.* 2215,2216.

34. Mori, T., Kobayashi, A., and Okahata, Y. (1998) Biocatalytic esterification in supercritical carbon dioxide by using a lipid-coated lipase. *Chem. Lett.* 921,922.

35. Okahata, Y. and Mori, T. (1997) Lipid-coated enzymes as efficient catalysts in organic media. *Trends Biotechnol.* **15,** 50–54.

36. Okahata, Y. and Mori, T. (1997) Catalytic activity of lipid-coated enzymes in organic media. *Proc. Jpn. Acad.* **73,** 210–214.

37. Okahata, Y. and Mori, T. (1998) Biocatalytic reactions in organic media by using lipid-coated enzymes. *J. Synth. Org. Chem. Jpn.* **56,** 931–939.

38. Okahata, Y., Tsuruta, T., Ijiro, K., and Ariga, K. (1988) Langmuir–Blodgett films of an enzyme–lipid complex for sensor membranes. *Langmuir* **4,** 1373–1375.

39. Okahata, Y., Tsuruta, T., Ijiro, K. and Ariga, K. (1989) Preparations of Langmuir–Blodgett films of enzyme–Lipid complex: a glucose sensor membrane. *Thin Solid Films* **180,** 65–72.

40. Lerner, R.A., Benkovic, S. J., and Schultz, P. G. (1991) At the crossroads of chemistry and immunology: catalytic antibodies. *Science* **252,** 659–667.

41. Benkovic, S. (1992) Catalytic antibody. *J. Annu. Rev. Biochem.* **61,** 29–54.

42. Hilvert, D. (1992) Antibody catalysis. *Pure Appl.Chem.* **64,** 1103–1109.

43. Stewart, J. D., Liotta, L. J., and Benkovic, S. J. (1993) Reaction mechanisms displayed by catalytic antibodies. *Acc. Chem. Res.* **26,** 396–404.

44. Hilvert, D. (1993) Antibody catalysis of carbon–carbon bond formation and cleavage. *Acc. Chem. Res.* **26,** 552–558.

45. Miyashita, H., Karaki, Y., Kikuchi, M., and Fujii, I. (1993) Prodrug activation via catalytic antibodies. *Proc. Natl. Acad. Sci. USA* **90,** 5337–5340.

46. Miyashita,H., Hara,T., Tanimura, R., Tanaka, F., Kikuchi, M., and Fujii, I. (1994) A common ancestry for multiple catalytic antibodies generated against a single transition-state analog. *Proc. Natl. Acad. Sci. USA* **91,** 6045–6049.

47. Usui, T., Kubota, S., and Ohi, H. (1993) A convenient synthesis of β-D-galactosyl disaccharide derivatives using the β-D-galactosidase from *Bacillus circulans.* *Carbohydr. Res.* **244,** 315–323.

48. Ooi, Y., Hashimoto, T., Mitsuo, N., and Satoh, T. (1985) Enzymic formation of β-alkyl glycosides by β-galactosidase from *aspergillus oryzae* and its application to the synthesis of chemistry unstable cardiac glycosides. *Chem. Pharm. Bull.* **33,** 1808–1814.

49. Sauerbrei, B. and Thiem, J. (1992) Galactosylation and glucosylation by use of β-galactosidase. *Tetrahedron Lett.* **33,** 201–204.

50. Vulfson, E. N., Patel, R., Beecher, J. E., Andrews, A. T., and Law, B. A. (1990) Glycosidases in organic solvents: I. Alkyl-β-glucoside synthesis in a water–organic two-phase system. Enzyme *Microb. Technol.* **12,** 950–959.

10

Very High Activity Biocatalysts for Low-Water Systems

Propanol-Rinsed Enzyme Preparations

Barry D. Moore, Johann Partridge, and Peter J. Halling

1. Introduction

Depending on the preparation method used, enzymes under low water conditions can exhibit differences in catalytic activity that vary by several orders of magnitude. Because this can make the difference between whether a particular biotransformation is practically useful or not, it is important to control enzyme history carefully. The basic problem is simple: how to transfer the enzyme from an aqueous environment to one where it is dehydrated while ensuring, first, that it remains in a native conformation and, second, that the active site residues are in the correct protonation state. Although the importance of retaining native structure is well recognized, potential detrimental changes to the protonation state are often not considered. Usually, it is assumed that the enzyme will exhibit "pH memory" of the previous aqueous solution. As discussed in detail in Chapter 19, this is not always a valid assumption.

Removal of water from proteins is most often carried out by lyophilization because it is thought to minimize the extent of *irreversible* denaturation. Excipients are added to stabilize the protein during the dehydration process and to provide a convenient powder for storage. Despite these precautions, lyophilization commonly results in a considerable degree of *reversible* denaturation. This is of little consequence where the protein is to be redissolved back into water, but it is of vital importance when native enzyme is required for use as a biocatalyst under low water conditions.

Because of this problem, many alternative forms of enzyme preparation have been explored, some of which are described in other chapters of this volume.

From: *Methods in Biotechnology, Vol. 15: Enzymes in Nonaqueous Solvents: Methods and Protocols*
Edited by: E. N. Vulfson, P. J. Halling, and H. L. Holland © Humana Press Inc., Totowa, NJ

Here, we describe a simple practical procedure for dehydrating enzymes that results in preparations with 1000-fold higher catalytic activities than obtained with conventional lyophilized powders *(1)*. The advantage of this method is that it is rapid and inexpensive and can be implemented in a conventional organic preparative laboratory without any specialized pieces of equipment.

1.1. General Comments on Producing Propanol-Rinsed Enzyme Preparations

Figure 1 illustrates the process involved in the production of propanol-rinsed enzyme preparations (PREPs). The initial step is to immobilize the enzyme on a support material in aqueous buffered solution. This could be via covalent attachment or by an adsorption process. With subtilisin Carlsberg and α-chymotrypsin, we have adopted a very simple method of adsorbing the protein onto chromatography-grade silica gel. For other enzymes, this may not be the best support and materials such as polypropylene, alumina, zeolite, or ion-exchange resin may give better results. The main requirement is for the enzyme to bind efficiently to the support material from aqueous solution without denaturing.

It should be noted that the aqueous immobilization step does not need to be carried out at the pH optimum for enzyme activity, as the protonation state can be altered later, under low-water conditions (*see* Chapter 19). This may be useful if the p*I* of the protein is such that good adsorption to the support is not possible at the pH for maximal activity or if autolysis is a problem.

Following immobilization, most of the aqueous solution should be decanted off from the supported enzyme. However, it is important to ensure that the particles remain fully wetted and do not come into prolonged contact with the air. This is because if supported enzyme is dried by conventional methods, a significant proportion of activity expressed under low-water conditions may be lost. This is certainly the case with silica–enzyme: When dried in air or vacuum it has activity no better than a lyophilized powder.

The next and key step of the process is to carry out a rapid dehydration: This involves rinsing the supported enzyme with a suitable water miscible organic solvent. A particularly good solvent for this process is 1-propanol, but comparable results can be obtained with other solvents such as ethanol *(2)*. It is notable that methanol and ethylene glycol give poor results indicative of irreversible enzyme denaturation.

The dehydration step is best carried out using solvent containing low levels of water (0.5–10% v/v) in order to ensure that essential water molecules remain bound to the protein. For convenience, the level can be chosen to give the same water activity, a_w, required in the subsequent reaction mixture. Control and measurement of a_w is described in detail in Chapter 11. The number of solvent rinses is not critical, but to reach a particular fixed water level, at least three

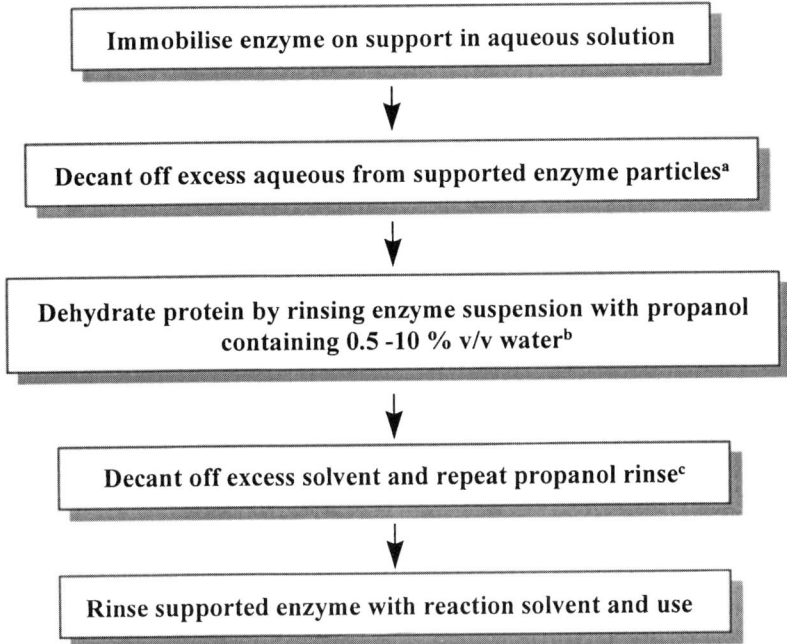

Fig. 1. Production of propanol-rinsed enzyme preparations. [a]The particles should be kept wetted and not dried; [b]The water content can be matched to the water activity required in the subsequent reaction mixture.

rinses of the volume of the original aqueous solution should be used. Karl Fisher titration can be used to ensure that the water content of the rinsing solvent is the same before and after contact with the PREP.

It is convenient to carry out the rinsing step in Eppendorf tubes. Following each addition of solvent, the sample is shaken, allowed to settle (or centrifuged), and the solvent is removed using a pipet. If a larger-scale preparation is required, samples can be prepared in stoppered glass flasks, left to settle under gravity between rinses and the solvent removed by decanting or partial filtration. This means PREPs can easily be prepared in a normal synthetic laboratory because no specialized equipment is required.

If the PREPs are to be stored prior to use they should be kept in suspension in propanol containing >1.5 % water. Their stability will depend on the enzyme and support material used. When required for use in a different reaction solvent, the PREPs may be rinsed once or twice with solvent to remove propanol. In some polar solvents (e.g., acetonitrile), PREPs have much lower stability than in propanol; thus following transfer to this type of solvent it is best for the reaction to be carried out immediately.

2. Materials

Subtilisin Carlsberg protease, Type VIII, from *Bacillus licheniformis*, 13.5 units/ mg solid (P5380) and α-chymotrypsin, Type II, from bovine pancreas, 52 units/ mg solid (C-4129) can be purchased from Sigma Chemical Co. (Poole, UK). *N*-Acetyl-L-tyrosine ethyl ester (A-6751) and silica gel (S-0507) can also be obtained from Sigma. Anhydrous solvents are available from Merck Ltd (Poole, UK).

3. Methods

3.1. Preparation of PREPs

3.1.1. Enzyme Adsorption onto Silica

First the enzyme is dissolved in a suitable buffer. For subtilisin Carlsberg, we have used concentrations of 2 mg/mL and either 20 mM sodium phosphate buffer, pH 7.8, or 20 mM sodium pyrophosphate, pH 5.7. With α-chymotrypsin, good results are obtained with 4 mg/mL enzyme in 25 mM Tris-HCl containing 10 mM CaCl$_2$ (pH 7.8).

Twenty milliliters of the above enzyme solutions held at 4°C are then mixed with 1 g of untreated chromotography-grade silica gel (Sigma S-0507) and the mixture shaken for 1–4 h. A total protein assay (Bio-Rad) of the supernatant can be used to monitor the amount of enzyme adsorbed onto the support as a function of time. Using concentrations of 2–4 mg/mL protein and 50 mg/mL of silica quite high loadings can be achieved with these enzymes. For subtilisin, typically >90% and >80% of protein adsorbs at pH values of 7.8 and 5.7, respectively. For chymotrypsin, adsorption as high as 99% can be obtained. Following rinsing with buffer, the suspension of immobilized enzyme can be stored for weeks at 4°C in aqueous buffer containing azide.

3.1.2. Propanol Rinsing of the Immobilized Enzyme

The stock aqueous mixture of immobilized enzyme is agitated to obtain a homogeneous suspension and a 1-mL aliquot is removed using a pipet, transferred to an Eppendorf tube, and allowed to settle out for a few minutes (in this example, 1 mL contains about 50 mg of silica–enzyme). The excess aqueous buffer over the precipitate can then be removed carefully with the pipet ensuring the silica–enzyme particles are not disturbed. The precipitate is now ready for the key dehydration step using propanol. This is achieved by rapidly adding 1 mL of propanol containing a measured amount of water to the immobilized enzyme and agitating to make a suspension.

Dehydration occurs because most of the water bound to the enzyme is rinsed off and dissolves in the propanol. Having some water in the rinsing solvent ensures that a fraction of essential water molecules remain bound to the enzyme. The most appropriate water level to use depends on what water activity, a_w, the subsequent reaction is going to be performed at (*see* **Table 1**). For example, if the reaction is to be carried out at $a_w = 0.44$, it can be seen from **Table 1** that the water content of the rinsing propanol should be 3.4% v/v.

After the first rinse the enzyme is allowed to settle again and the excess propanol removed. The rinsing process is then repeated at least twice using additional 1-mL aliquots of propanol containing the relevant v/v % water. On the final rinse, the water content of the rinsing solvent should be unchanged after contacting the supported enzyme. By application of this procedure, PREPs can be rapidly prepared for reaction at any particular water activity.

3.1.3. Transferring the PREPs to Another Solvent

Excess propanol is removed from the PREP and 1 mL of the desired reaction solvent containing a certain level of water added. The % v/v water required will depend on the solvent and the desired water activity of the reaction mixture. As an example, we consider transfering to tetrahydrofuran (THF) for a reaction at $a_w = 0.44$. It can be seen from **Table 1** that the water content required for $a_w = 0.44$ is 1.2% in THF compared to 3.4% in propanol. Following agitation and then settling, excess THF is removed and the process repeated. The PREP is now ready for biocatalysis in THF. PREPs of subtilisin have been found to give good results in solvents with a wide range of polarity.

3.2. Comparing Preps with Other Types of Enzyme Preparation

It has been reported that commercially available crosslinked enzyme crystals (CLECs) exhibit very high activities and good stability in organic solvents compared to most other preparations *(4,5)*. It has also been shown by X-ray crystallography that crosslinked crystals of subtilisin Carlsberg retain native protein structure in dry acetonitrile and dioxane *(6)*. CLECs therefore provide a useful benchmark with which to compare other preparations.

It is often difficult to make accurate comparisons between results obtained in different laboratories because of variations in the protocols used for assaying. We have therefore directly compared the activities of PREPs, CLECs, and lyophilized powders of subtilisin Carlsberg under exactly the same reaction conditions. The reaction studied was a simple model transesterification of *N*-acetyl-L-tyrosine ethyl ester with propan-1-ol in acetonitrile.

3.2.1. Typical Protocol for Assaying Activity

N-acetyl-L-tyrosine ethyl ester was dissolved in anhydrous propan-1-ol and added to a suspension of PREP/CLEC/powder in 10 mL of acetonitrile to give 10 m*M* ester and 1 *M* propanol. The acetonitrile also contained fixed amounts of water chosen to give particular water activities, as in **Table 1**. After brief

Table 1
Water Content Required to Attain Selected Water Activities
in Three Solvents

Water activity a_w	Water content % (v/v)		
	1-Propanol	Acetonitrile	Tetrahydrofuran
0.11	0.77	0.50	0.20
0.22	1.60	1.00	0.45
0.44	3.40	2.70	1.20
0.55	4.70	4.50	1.60
0.76	9.80	13.00	3.60

Details on how to calculate values for other solvents can be obtained from **ref. 3**.

mixing, the zero-time sample was removed. The mixture was then incubated at 24°C with constant reciprocal shaking (150 min^{-1}). Samples from the reaction mixture were taken at regular intervals and analyzed by high-performance liquid chromatography (HPLC) on a Gilson 715 equipped with an ODS2 reverse-phase column (Hichrom). The mobile phase consisted of 40% acetonitrile mixed with an aqueous phase adjusted to pH 2 with orthophosphoric acid. The retention times of the acid, ethyl ester, and propyl ester were 2.1, 2.9, and 3.9 min, respectively. The reaction rates are obtained by dividing the number of nanomoles per milliliter of ester converted per minute by the enzyme concentration in milligrams per milliliter, giving convenient units of nanomoles per minute per milligram. Using a discontinuous HPLC sampling procedure the highest rate that can be followed effectively with a 10 mM ester concentration is 50 µM /min. With 1 mg/mL of enzyme, this corresponds to 50 nmol/min/mg and our rates with PREPs and CLECs are generally higher than this. We therefore routinely use enzyme concentrations of less than 0.2 mg/mL in our assays. For the PREPs, the weight of subtilisin in the reaction mixture is calculated from the known loading of enzyme on the silica support.

3.2.2. Comparison of Activities

It can be seen from **Table 2** that PREPs of subtilisin Carlsberg are over 1000 times more active than freeze-dried powders and exhibit comparable activities to crosslinked enzyme crystals. Rate enhancements of the same order have also been observed with α-chymotrypsin PREPs, an enzyme for which CLECs are not commercially available *(1)*. The activity measurements suggest that the propanol-rinsing procedure is much less damaging to protein tertiary structure than conventional lyophilization. In fact, the similarity in rates of PREPs to that of the crosslinked crystals is consistent with high retention of native enzyme conformation during the dehydration.

**Table 2
Variation of Activity in Acetonitrile with Water Content for Preparations
of Subtilisin Carlsberg**

Enzyme preparation	Rate[a] (nmol/mg/min)			
	0% H_2O[b]	1% H_2O[b]	2.7% H_2O[b]	4.5% H_2O[b]
Freeze-dried powder	<0.01	0.13	—	0.28
Crosslinked crystal (CLEC)[c]	278	204	214	156
Propanol rinsed (PREP)	0.82	288	543	530

[a] Initial transesterification rates.
[b] Water content of acetonitrile in % v/v.
[c] As described in **ref 2**, CLECs exhibit an initial 5-min higher burst of activity.

A practical operational advantage of PREPs over CLECs is that at any particular water activity lower levels of hydrolysis are obtained. For example, using acetonitrile containing 2.7% v/v for the above transesterification, the CLEC gives 19% hydrolysis and the PREP 1% hydrolysis. Hydrolysis side reactions reduce product yield and can lead to substantial changes in catalytic rates because acidic or basic side products change the enzyme protonation state in the solvent (*see* Chapter 19). The cause of the increased hydrolysis with CLECs is not clear. It could be that there is a higher density of strong binding sites for water in the crystal lattice, resulting in a higher effective water concentration near the enzyme active site.

The above studies have been carried out with model enzyme systems in which the protein is readily available in high purity. Here, the relative merits of PREPs versus CLECs are not clear-cut. They depend on the availability and relative cost or ease of preparation of the crosslinked crystals. For example, CLECs of subtilisin have now become less expensive and are of comparable cost to other sources of the protein. On the other hand, CLECs of chymotrypsin are not currently available.

By contrast, where the enzyme of interest is only available in small quantities or in a partially purified form, then production of crosslinked crystals is not an option. Here, PREPs offer a major advantage over most other methods of preparing enzymes for low-water systems. They can be made rapidly, on a small scale using simple equipment. Furthermore, if rate enhancements comparable to the above are obtained, 1 mg of PREP will have comparable activity to 1 g lyophilized powder. PREPs are therefore attractive options for applications such as enzyme screening and combinatorial biocatalysis (*7*).

3.2.3. Effect of Water Activity on PREPs

The variation in catalytic activity of PREPs as a function of water availability is very different to that observed for lyophilized powders. With PREPs of subtilisin, the rate rapidly increases from $a_w = 0$ to $a_w = 0.22$ and the catalytic activity reaches a plateau from $a_w = 0.44$ upward *(1)*. By comparison, lyophilized subtilisin generally show very little activity below $a_w = 0.44$ but exhibits a rapid increase from thereon upward *(8,9)*. The practical effect of these differences is that even at quite low water activities, PREPs of subtilisin behave as efficient catalysts. This is of direct interest in applications where hydrolysis is an undesirable side reaction, such as peptide synthesis.

4. Notes

Interestingly, lyophilization of aqueous enzyme solutions containing very high molar ratios of an excipient such as KCl is reported to give preparations with very high activity in organic solvents (*see* Chapter 1). This is probably a consequence of minimizing detrimental protein–protein interactions during the dehydration process so that the native conformation is better retained *(10)*.

References

1. Partridge, J., Halling, P. J., and Moore, B. D. (1998) Practical route to high activity enzyme preparations for synthesis in organic media. *Chem. Commun.* 841,842.
2. Partridge, J., Hutcheon, G. A., Moore, B. D., and Halling, P. J. (1996) Exploiting hydration hysteresis for high activity of cross-linked subtilisin crystals in acetonitrile. *J. Am. Chem. Soc.* **118,** 12,873–12,877.
3. Bell, G., Janssen, A. E. M., and Halling, P. J. (1997) Water activity fails to predict critical hydration level for enzyme activity in polar organic solvents: interconversion of water concentrations and activities. *Enzyme Microb. Technol.* **20,** 471–477.
4. Lalonde, J. (1995) The preparation of homochiral drugs and peptides using cross-linked enzyme crystals. *Chimi. Oggi* **13,** 31–35.
5. Persichetti, R. A., Lalonde, J. J., Govardhan, C. P., Khalaf, N. K., and Margolin, A. L. (1996) Candida–Rugosa lipase: enantioselectivity enhancements in organic-solvents. *Tetrahedron Lett.* **37,** 6507–6510.
6. Schmitke, J. L., Stern, L. J., and Klibanov, A. M. (1997) The crystal structure of subtilisin Carlsberg in anhydrous dioxane and its comparison with those in water and acetonitrile. *Proc. Natl. Acad. Sci. USA* **94,** 4250–4255.
7. Michels, P. C., Khmelnitsky, Y. L., Dordick, J. S., and Clark, D. S. (1998) Combinatorial biocatalysis: a natural approach to drug discovery. *Trends Biotechnol.* **16,** 210–215.
8. Parker, M. C., Moore, B. D., and Blacker, A. J. (1995) Measuring enzyme hydration in non-polar organic solvents using NMR. *Biotechnol. Bioeng.* **46,** 452–458.
9. Partridge, J., Dennison, P. R., Moore, B. D., and Halling, P. J. (1998) Activity and mobility of subtilisin in low water organic media: hydration is more important than solvent dielectric. *Biochim. Biophys. Acta* **1386,** 79–89.
10. Griebenow, K. and Klibanov, A. M. (1997) Can conformational changes be responsible for solvent and excipient effects on the catalytic behavior of subtilisin Carlsberg in organic solvents? *Biotechnol. Bioeng.* **53,** 351–362.

11

Methods for Measurement and Control of Water in Nonaqueous Biocatalysis

George Bell, Peter J. Halling, Lindsey May, Barry D. Moore, Donald A. Robb, Rein Ulijn, and Rao H. Valivety

1. Introduction

It is generally recognized that the small level of remaining water is critical to the behavior of biocatalysts used in mainly non-aqueous (e.g., organic) media. Most biocatalysts are inactive if fully dehydrated, and the reaction rate is stimulated by increasing hydration, at least at first. The rate of the desired reaction can become slower again if water levels increase too much, particularly when a hydrolytic side reaction becomes significant. The water level will also affect the equilibrium position of reactions in which it is a reactant. Most often, the desired synthesis will be the reversal of a hydrolysis or in competition with a hydrolytic side reaction. In such cases, the equilibrium yield will increase as the water level is reduced. Fuller details of these effects of water can be found in recent reviews of the field *(1–7)*. Some further examples are found elsewhere in this volume (Chapters 12 and 13).

It should thus be clear that careful attention to water levels in very important for the best performance of biocatalytic syntheses. This is equally true for fundamental studies of biocatalyst behavior, so that reproducible and interpretable results are obtained. This chapter describes a number of practical methods used at the University of Strathclyde, with special emphasis on the precautions we believe are important.

1.1. Some Basic Definitions and Calculations

The total water content of the system is often known from the recipe used to prepare it. However, it is usually not a particularly useful guide to behavior,

From: *Methods in Biotechnology, Vol. 15: Enzymes in Nonaqueous Solvents: Methods and Protocols*
Edited by: E. N. Vulfson, P. J. Halling, and H. L. Holland © Humana Press Inc., Totowa, NJ

because the water will be partitioned between several different phases (e.g., bound to the biocatalyst or support, dissolved in the organic phase, present in the vapor phase). The water concentration in an individual phase is a more meaningful value. Such water concentrations may be measured by analysis of appropriate samples of one or more of the phases present.

The thermodynamic activity of water is, by definition, equal at equilibrium in all phases present and may be measured by suitable sensors *in situ*. Both water concentrations and water activity have their uses in understanding the behavior of these systems. Fuller details of the theoretical background are presented in **ref. 8**.

Because water activity tends to equality in all phases, it is often most convenient to measure it in the gas phase using a humidity sensor. The amount of water in a gas phase is often presented in terms of its partial pressure, which is defined as the mole fraction of water vapor times the total pressure. Two different partial pressures of water are commonly discussed:

1. The actual value found in a particular gas phase
2. The value found in a gas phase saturated with water at a given temperature (i.e. equilibrated over pure water)

(Confusingly, these two different pressures are both sometimes called the "vapor pressure of water," so this shortened term should be used with care.)

The water activity is given by the ratio of the water partial pressure (definition 1) to the saturated pressure of pure water (definition 2) at the same temperature. Relative humidity given as a percentage is 100 times the water activity. If the humidity sensor reads the partial pressure of water vapor, this can be converted into water activity using standard tables of saturated vapor pressure for water. (These can be found in most scientific data books; e.g., the *CRC Handbook of Physics and Chemistry*. **Figure 1** shows the values graphically.)

Sometimes, such as with saturated salt solutions, tabulated values may directly give relative humidity or water activity. If water partial pressure is required (e.g., for sensor calibration), this is obtained by multiplying the water activity by the saturated vapor pressure of pure water, from tables.

For reaction mixtures at a controlled temperature, it is easiest to think in terms of water activity. However, where a gas phase is in contact with surfaces at different temperatures, it is probably easier to use partial pressures of water. Systems with different temperatures (as in the example presented in the following paragraph) cannot be in complete equilibrium. Using water activity for the whole system involves a difficult choice of the reference temperature. Temperature has a large effect **(Fig. 1)** on the saturated vapor pressure of pure water, involved in the definition of water activity. In some cases, a relevant parameter is the (water) dew point of a gas phase. This is the temperature at

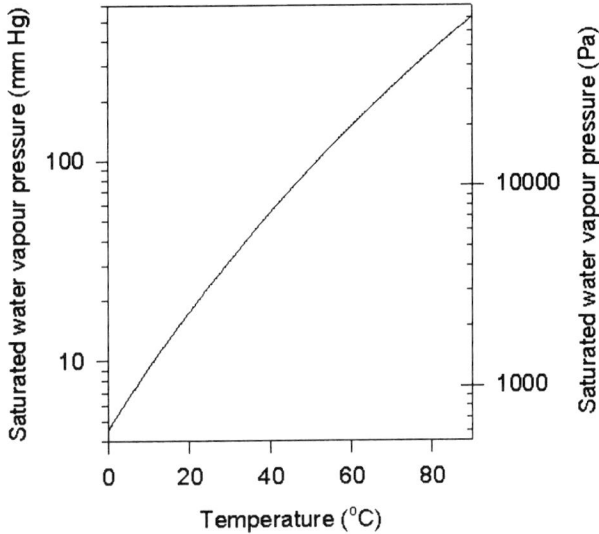

Fig. 1. Saturated vapor pressures of pure water.

which pure liquid water would have a saturated vapor pressure equal to the actual partial pressure of the water vapor present. Cooling the gas phase to this temperature will cause liquid water to start to condense.

To illustrate the various calculations, let us take a hypothetical example. Consider a reaction mixture at 40°C and water activity 0.50. From tables, we read that the saturated vapor pressure of water at 40.00°C is 55.324 mm Hg. Therefore, the partial water vapor pressure in the headspace above our reaction mixture will equilibrate to $0.50 \times 55.324 = 27.7$ mm Hg. The same tables show that this is the saturated vapor pressure of water at 27.7°C. If we were to expose the headspace to a surface at this temperature, the water vapor next to it would be saturated and would just begin to condense to form a dew. Strictly, this system with different temperatures cannot be at equilibrium, but if the temperatures are controlled, it will remain in steady state with a slow transfer of heat to the cooler surface. If we were to have the surface at 20.00°C, the tables show that the saturated vapor pressure of water is now 17.535 mm Hg. Now the water distribution is not at equilibrium either, and there will be continual condensation of water onto the surface. Water will be progressively removed from the liquid reaction mixture, unless the condensed water returns to it immediately. If the water content of the reaction mixture is low, the result may be a significant fall in its water activity. It is possible that an equilibrium for water transfer may be reestablished when the water activity of the reaction mixture (still at 40°C) has fallen to $17.535/55.324 = 0.317$.

2. Experimental Vessels: General Points

2.1. Water Exchange with Atmosphere: Sealing of Vessels

Any reaction mixture will tend to exchange water with the gas headspace above it. When the total water content is low, this can cause significant changes. Therefore, it is usually sensible to seal reaction vessels to prevent unwanted exchange. In some cases, even water exchange when the vessel is opened for sampling may be significant. Leaks resulting from imperfect sealing may also be important, especially over long reaction times. It might be thought that sealing that was sufficient to prevent significant losses of volatile organic solvent would be enough. However, a smaller loss or gain of water may be a large fraction of the total present, and hence lead to a significant change in hydration conditions.

Materials used in sealing reaction vessels must be resistant to the vapors of organic solvents and other components. As well as a possible loss of seal, any attack on the material may leach out compounds that will enter the reaction mixture. We have used the following:

1. Silicone sealants that can be applied in liquid prepolymer form and allowed to set to a solid but highly elastic rubber. They can be reopened by cutting.
2. O-rings made of a solvent-resistant rubber are useful between metal sensors and glass sections.
3. Polytetrafluoro ethylene (PTFE) sleeves give a good seal between ground glass joints.
4. Special glands are required for shafts from external overhead stirrer motors. We have used Quickfit designs. With these, only a very small amount of lubricant should be used, or this will enter the reaction mixture.

2.2. Water Equilibration in Sealed Equipment: Cold Spots

When a reaction mixture is completely sealed inside suitable apparatus, it seems reasonable to expect that water migration will reach equilibrium between the various phases present, resulting in defined hydration conditions. However, unexpected effects can occur if temperature differences exist. For example, as illustrated earlier, if a relatively cold surface in contact with the gas headspace is below the dew point for water, then liquid water will condense. This will tend to reduce the water activity of the reaction mixture, and often it will equilibrate to a value determined by the cold surface. If the temperature difference is large, this can produce a very low water activity (e.g., 0.117 for a cold surface at 20°C above a reaction mixture at 60°C).

Hence, it is generally advisable to ensure that there are no such cold spots on the apparatus used. The simplest approach is to conduct experiments at or near ambient temperature. Where heating is required, complete enclosure by an air bath is possible, but it may present explosion risks from accumulation of organic solvent vapor. Small reaction vials may be completely immersed in water baths if equipped with suitably sealed caps (e.g., "Mininert" valves, from Phase Sep or Supelco). Vessels may be water jacketed or immersed in water baths to above the level of the stopper or other top seal. We have used a specially designed water-jacketed vessel with the top surface electrically heated to above the temperature of the circulated water that controls the reaction temperature *(9)*. It is not normally a problem if surfaces in contact only with the vapor are somewhat hotter than the liquid reaction mixture. A slow transfer of heat via the vapor phase will be easily counteracted by the temperature control of the liquid mixture.

If organic liquid streams pass through regions of different temperature, such as in the feed to a packed-bed reactor, a similar problem is possible. The solubility of water in organic liquids usually falls at lower temperatures, so that an aqueous phase may separate and water can be lost (e.g., trapped in a pump head).

3. Fixing Water Activity by Equilibration
3.1. Saturated Salt Solutions to Provide Controlled Humidity Atmospheres

Equilibration with a saturated solution of an appropriate salt is a convenient method of obtaining vapor phases of known water partial pressure. These are useful for calibrating sensors, pre-equilibrating reaction mixtures, and so forth. The advantage of a saturated solution is that it can gain or lose water without changing vapor pressure, as it should remain at the same (saturated) concentration (provided sufficient excess solid salt is present and equilibrium is maintained).

Care is needed to ensure re-equilibration after temperature changes. These will normally cause changes in the solubility of the salt used, so that the solution becomes temporarily supersaturated or subsaturated. A substantial time may be required before equilibrium is re-established with the solid. During this period, the water-vapor partial pressure in the headspace may be significantly lower or higher than expected.

The water activity of these saturated solutions is usually fairly weakly temperature dependent. Although the water partial pressure above them always rises strongly with temperature, that above pure water does as well; hence, the relative humidity or water activity changes much less. The most reliable and precise values are probably those tabulated by Greenspan *(10)*. A selection of salts we commonly use are shown in **Table 1**, with water-activity values for 25°C.

There may be another volatile component in the system being equilibrated through the vapor phase (e.g., an organic solvent). If so, there is the possibility that this volatile compound will be transferred into the salt solution, affecting its water activity. This is not such a problem as might be expected, because organic molecules will usually be strongly "salted out" by the high concentrations in the saturated solution. The concentration of a low-polarity solvent

Table 1
Saturated Salt Water Activities at 25°C

LiCl	0.113	KI	0.689
KAc	0.225	NaCl	0.753
$MgCl_2$	0.328	KCl	0.843
K_2CO_3	0.432	KNO_3	0.936
$Mg(NO_3)_2$	0.529	K_2SO_4	0.973
NaBr	0.576		

Source: **ref. *10*.**

reached in the salt solution will have a negligible effect on its water activity. A significant effect may be possible in the worse case of very polar solvents, less concentrated salt solutions (high water activity), and "structure-breaking" anions such as iodide and nitrate. (These anions can act to increase solubilities of organic species.)

The saturated solution is normally best used in the form of a solid-rich slush. In other words, the appearance should be of wet solid crystals. This is rather different from most people's immediate picture of a saturated solution, with a few crystals in the bottom of a large liquid volume. Mixtures with a high solid fraction are able to re-equilibrate more quickly after temperature changes, because all the solution remains close to a crystal surface. They also offer a somewhat larger surface area for equilibration with the gas phase and make accidental splashes of solution less likely.

To prepare these mixes, we generally use salts of the best readily available purity. With large excesses of solid, even quite small levels of impurity could reach significant concentrations in the solution phase and depress its vapor pressure. The ratio of salt to water required is often large. This is particularly true for those salts giving the lowest water activities, which have very high solubilities.

The lowest water activity that can be obtained using salt solutions is around 0.05. Values below this require equilibration with appropriate drying agents. Some of these drying agents, notably hydrate-forming salts, give well-defined equilibrium water activities, at least in principle. Some were discussed by Valivety et al. (*11*), although we now have reason to believe that the literature values are in error. Molecular sieves may be used as drying agents to reach very low, although not precisely defined, water activities. Note, however, that to obtain maximal drying efficiency requires very severe reactivation treatment; if heat alone is used, about 350°C is needed.

3.2. Pre-Equilibration of Reaction Mixtures Using Saturated Salt Solutions

Reaction mixture components can be pre-equilibrated by keeping them in a closed vessel with the vapor phase in contact with the appropriate saturated salt solution (*see* **Fig. 2**). We normally use wide-mouth screw-cap jars. The salt-solution slush is placed in the bottom of the jar; it is convenient to keep the experimental material in a double container (e.g., one beaker inside another), so the inner one may be removed clean for weighing. The rate of equilibration is improved by making the surface area of the sample as large as possible. Very careful sealing of the jar is not necessary from considerations of water exchange, but may be needed to prevent loss of volatile solvents during long equilibration.

It is important to check the rates of equilibration for different types of material. For biocatalyst preparations, this is most easily done by weighing. They may require several days; the rate may be increased somewhat by evacuation of the system (in a vacuum desiccator). For organic solutions, equilibration may be monitored by Karl Fischer water analysis on samples. The time required is very dependent on the water capacity of the organic liquid (i.e., the quantity of water that has to be transferred); overnight is sufficient for nonpolar liquids, but not for polar solvents. With polar solvents, it saves a lot of time to add approximately the correct concentration of water before equilibration. The required isotherms of water activity against composition are given by Bell et al. *(12)* for a number of commonly used polar solvents. Stirring the organic liquid phase will usually increase the rate of equilibration. Whatever type of sample is being equilibrated, some form of mixing in the gas phase might increase rates, although we have not studied this.

Saturated salt solutions in deuterium (or oxygen)-labeled water may also be used for subsequent measurement of the label to estimate the water bound *(13,14)*.

3.3. What to Pre-Equilibrate

Thought must be given to the initial hydration state of every component of the reaction mixture. If only some are pre-equilibrated, then water redistribution after they are combined will give a final water activity perhaps quite different from the intention. For example, if a biocatalyst brought to a desired water activity is suspended in a dried solvent, it will probably rapidly lose water. The equilibrium water activity eventually reached may be significantly reduced. Such combination of a biocatalyst and organic phase of different water activity may, in fact, be desirable to obtain high catalytic activity because of hysteresis effects *(15)*. However, if conditions are to be reproducible, all components should still be pre-equilibrated to a defined hydration level. This holds even if different values are chosen for the different phases that will eventually be combined, in order to exploit these hysteresis effects. If the aim is to

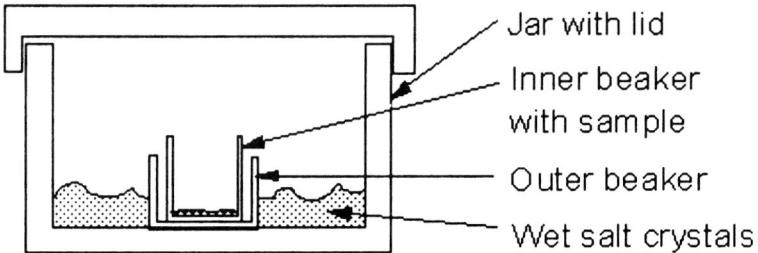

Fig. 2. Equilibration of samples using saturated salt solutions.

pre-equilibrate the entire reaction mixture to the same water activity, a decision must be made as to what to mix at what stage. It is difficult to give clear recommendations here on what combinations of components should be mixed before water pre-equilibration. In principle, mixing afterward may disturb the previously equilibrated hydration conditions. To obtain a system in complete water-distribution equilibrium, the entire reaction mixture should ideally be pre-equilibrated together. However, something must be kept separate until the reaction is to be started, of course. In practice, we have used two options for the choice of the two combinations to be separately pre-equilibrated:

1. (i) The biocatalyst particles and, separately, (ii) an organic solution of all the reactants. Problems may result from the effect of organic solvent on water binding by the biocatalyst. Displacement of water may lead to an increased water activity and aggregation of the biocatalyst *(13)*.
2. (i) The biocatalyst suspended in an organic phase and (ii) one separate reactant. Dissolution or mixing of the missing reactant to start the reaction may alter the pre-equilibrated water activity. To minimize this, the chosen reactant should have a small final concentration. For example, we often include 1 M alcohol nucleophile in the pre-equilibrated organic phase. The reaction will then be started by adding about 20 mM ester.

In all cases, achieving the desired final water activity is easier when the organic solvent used is polar and dissolves a lot of water itself. This will act to buffer the water activity of the whole system.

In general, brief exposure to the laboratory atmosphere during transfer and mixing of pre-equilibrated components does not seem to have major effects. In critical cases, however, especially where the total water content is very small, all such operations can be performed inside a glove box. The interior of the glove box is pre-equilibrated to the desired water activity, by recirculating the air inside it through or over the appropriate salt solution or drying agent; the progress of the glove box interior to the desired water activity can be monitored using one of the sensors described in **Subheading 4.3.**

4. Measurements of Water Concentration or Activity

4.1. Sample Handling for Water Measurements

In some cases, it may be necessary or desirable to make water measurements on samples removed from the reaction mixture. If done correctly, this is a perfectly proper method of analysis. However, when the water content and/or capacity of samples is small, as with nonpolar organic liquids, careful precautions are essential to avoid changes during sample handling. Water vapor is generally present in the environment and may be taken up by dry samples. Equally, water may be lost from wet samples, because (1) it is rather volatile and (2) it adsorbs to many surfaces with which the sample may come into contact. Because we are conditioned to think of water as a dominant component, we tend to neglect these possibilities; but when it is a minor component in nonpolar phases, such exchanges can very quickly lead to major changes in water content.

The possibility of water exchange with a gas phase must always be borne in mind. The capacity of a gas phase for water vapor at room temperature is around 0.02 g/L (or 1 mmol/L), not much less than very nonpolar liquid phases. Allowing a sample of such a liquid phase to equilibrate with many times its volume of gas will substantially change its water content.

This is illustrated by some observations with hexane (saturation solubility of water about 0.1 g/L). If you deliberately breathe out once over a small vial of dry hexane, it will be found to be about half-saturated with water, absorbed from the vapor in expired air. On the other hand, if a few milliliters of water-saturated hexane are poured from one container to another in a dry atmosphere, about half the water will be lost by evaporation into the gas phase.

When samples are taken from systems that are not at ambient temperature, the effect of temperature on water solubility must be remembered. In general, the solubility will fall on cooling, and if saturation is thus exceeded, a separate water phase may form (so that the sample becomes inhomogeneous). To keep the water in solution, it can be convenient to mix the sample with another liquid that increases the water solubility. A more polar solvent is usual, and, obviously, it should be as dry as possible.

4.2. Water Concentrations by Coulometric Karl Fischer

For measurement of low water concentrations, the coulometric Karl Fischer (KF) method is generally the best. This is based on the electrolytic generation of the iodine involved in the Karl Fischer reaction, rather than a volumetric titration. The resulting electrical measurements are more sensitive at low levels. We have used the Metrohm 684KF and Fisher Instruments Accumet KF instruments.

A variety of reagents and recommendations for their use are available from both Riedel de Haen (Hydranal) and Hitachi (Aquamicron). Selection of the best reagents for particular types of sample is important for best performance. These coulometric KF instruments seem to give reasonable reproducibility only over a relatively narrow range of total water contents in the sample. The errors are very large for <20 μg water, probably because of the difficulty of properly correcting for instrument drift. Above 500 μg, the distribution of measurements becomes wider and also skewed to higher water content. For most accurate measurements, between 100 and 300 μg water seems best. From our experience, the reproducibility is better on the Accumet KF machine. With nonpolar solvents like hexane, where the water concentrations are very low, large sample sizes (up to 1 mL) are needed to obtain quantities of water in this range. After a number of such samples have been analyzed, a second phase can be seen to separate in the KF titration vessel, presumably hexane rich. The reagents must then be changed.

Samples are introduced into the instrument by syringe through a rubber septum. With the lowest water concentrations, the syringe must be rinsed with the test solution several times before a correct result can be obtained. Simple transfer from a completely dry organic liquid using a syringe equilibrated with laboratory air will usually reveal a small amount of water that was adsorbed on the surfaces of the syringe. On the other hand, if the syringe has been freshly dried in an oven, it will remove water from a sample with a water concentration near saturation. In some solvents, where even the saturated concentration of water is low, this adsorption may cause a significant reduction. To minimize the risk of changes, we prefer, whenever possible, to transfer samples direct from the reaction mixture to the coulometer, without intermediate holding in any other vessel. A series of samples should be analyzed in rapid succession, until the readings show no further trend, showing that the syringe has equilibrated with the test liquid.

The instrument drift is a major contribution to errors, at least for low water contents. At least part of this is the result of leakage of atmospheric moisture into the titration vessel. This can be minimized by reusing the same hole in the septum for repeat injections, changing the septum regularly, or covering the injection port with parafilm.

4.3. Sensors for Measurement of Thermodynamic Water Activity: General

A major advantage of some humidity sensors is that they can be placed inside a sealed reaction vessel, either in the headspace or even immersed in an organic liquid phase. This avoids all the uncertainties surrounding sample handling discussed earlier. The sensor reading gives a direct indication of the state in the reaction mixture, provided the various phases inside the reaction vessel equilibrate, and the problems of leaks and cold spots noted earlier are avoided.

These sensors will usually be employed to measure the water activity of the system, directly or indirectly. Because water activity is the same in all phases at equilibrium, the water status of the whole system may be characterized by a single value. Furthermore, this can be measured in any convenient phase that is equilibrated with the reaction mixture. For routine use, the water activity can be determined via measurement of the relative humidity or water partial pressure in a gas headspace above the reaction mixture.

A large number of sensors are available for measuring gas humidity. With biocatalysis in organic media, the headspace will usually contain significant levels of volatile organic species (e.g., a solvent) as well as water vapor. This will interfere with many of the possible humidity sensors, notably those that sense the formation of dew on a cooled surface. However, several types are still usable, as discussed in **Subheadings 4.8.–4.10**. Izumoto et al. *(16)* report the use in organic media of a Panametrics sensor, with which we have no experience.

4.4. Sealed Vessels Allowing Sensors to be Placed in the Headspace

A number of approaches are convenient for fitting sensors into the headspace above reaction mixtures. Most start by sealing the sensor inside a Quickfit cone section, using a rubber O-ring or silicone sealant. This may then be fitted into an appropriate socket, either at the mouth of a simple test tube, or in the headpiece of a Quickfit reaction vessel **(Fig. 3)**. It may be convenient to reshape the latter to allow the use of smaller liquid volumes (e.g., **ref. 9**). For sensor calibration, the gas-phase volume may be reduced further by directly sealing the sensor into the mouth of a plain wide test tube, using an O-ring.

4.5. Sensor Calibration

It is important, first, to be clear whether the sensor element responds to partial pressure of water vapor (more common) or to relative humidity/water activity. The relationship between them depends on the temperature of the gas phase being measured. Some sensors incorporate an additional temperature probe, which enables the readout unit to interconvert and perhaps display both parameters.

We have used two methods to obtain gas phases of known relative humidity or water partial pressure for sensor calibration: saturated salt solutions and pure water at a range of temperatures (below the gas-phase temperature). In either case, accurate temperature control of the liquid is critical for precise calibration; temperature has a large effect on the vapor pressure of water. As always, cold spots in the calibration tubes should be avoided. Above ambient temperature, this requires that they be immersed in a constant-temperature bath such that the liquid level is above the level of the O-ring sealing the sensor to the inner wall of the glass tube.

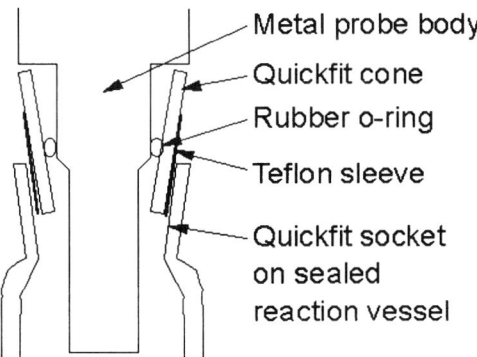

Fig. 3. Method for sealing the metal sensor probe into the headspace of the reaction vessel.

4.6. Testing Gas-Phase Sensors for Organic Solvent Effects

For nonpolar solvents, such tests are simply made by comparing the readings over saturated salt solutions before and after the addition of a drop of the test solvent. More polar solvents may dissolve significantly in the saturated salt solution and, hence, alter its true water activity. In these cases, we prefer to use pairs of solid salt hydrates, again with and without the addition of test solvent. The water activity would only be altered here if the organic solvent reacted chemically with the solid salt (e.g., forming an alternative solvate); this appears to be rare.

4.7. Gas-Phase Sensor Response and Equilibration Times

In discussing the response time of sensors in the gas phase, it is important to distinguish the true response time of the sensor from the equilibration time of the system. The sensor response time may be only a minute or two, as shown by for example removing it from a water-saturated vessel to the laboratory air. However, the reading will take 1 h or more to reach a final value after the sensor is sealed into an unstirred headspace above water or a salt solution; this reflects the time for equilibration of the headspace with the liquid phase.

If a sensor in the gas phase is to be used for control of water activity, this long pre-equilibration time must be remembered. It results in a control loop with a very long response time, suitable only for very slow changes. This is particularly true if water is removed by drying the headspace gases, by pumped circulation through an appropriate absorbent. During a period of drying, the sensor reading will show a low water activity, indicating the adsorbent performance. Afterward, recirculation must clearly be stopped long enough to ensure re-equilibration with the liquid phase, so that the reading from the sensor in the headspace correctly reflects the water activity in the reaction mixture. A laboratory reactor with a stirred headspace to speed up re-equilibration was described *(9)*.

4.8. Weiss (Formerly Philips) LiCl Humidity Sensor

We currently use this sensor for gas-phase humidity measurements. The sensor has an electrically heated LiCl solution that comes into equilibrium for water exchange with the gas phase. The steady-state temperature at which this occurs depends on the partial pressure of water vapor in the gas and is measured by a Pt resistance thermometer. The element temperature and the water partial pressure are related by a calibration curve. A simplified theoretical picture suggests that the calibration curve should be that for the partial pressure of water vapor above saturated aqueous LiCl solutions at various temperatures. The empirically determined calibration curve agrees closely at higher element temperatures (>60°C or so) but deviates significantly below this. The calibration curve also shifts slightly between different sensors and after recharging with LiCl solution. These sensors are now marketed directly by the manufacturer, Weiss Umwelttechnik GmbH, (Reiskirchen, Germany). The sensor we have used is type 1100 H1-F (order no. 7246/001). Besides the sensor, a heater power supply unit is offered, and this appears to be a simple stabilized 24- to 26-V AC supply. The readout unit is definitely a normal 100-Ω PtK9 resistance thermometer module, and we have used one obtained elsewhere. Full information is available from the manufacturers (in German; we have made an unchecked English translation, available on request). The sensor cost about US$ 600 (EUR 500) in the United Kingdom in 1999.

We failed in several attempts to buy from Philips the LiCl solution required to refill the sensor from time to time, or obtain instructions in its use. We have been able to successfully regenerate the sensor as follows:

1. Wash the inner sensor element by soaking overnight, rinse in distilled water, then dry.
2. Apply 150 µL of 50% saturated LiCl evenly distributed over the fiber of the inner sensor element. (This volume agrees with that now indicated in the information from Weiss).
3. Reconnect the electrical supply. The sensor temperature will initially rise to a high value (we usually briefly disconnect the power supply when it exceeds 120°C), but soon stabilizes back to the correct reading.

We have tested this sensor in the presence of the vapors of some 30 solvents, varying in polarity from hexane to methanol. The only one found to significantly affect the reading has been dioxane.

4.9. Sina Instruments

In the late 1970s, one of the authors (P. J. H.) made some measurements on a Sina Equihygroscope instrument and showed that it was not affected by hexane vapor. However, it required samples to be placed in the instrument chamber, risking the problems of sample handling discussed earlier. We believe the company offers a sensor that could be sealed in the headspace of a reactor. Sina instruments have been used much more extensively by Legoy and coworkers *(17,18)*.

4.10. Endress & Hauser Aluminum Oxide Humidity Sensor

We used this sensor extensively in the past, as it is the only one we know of that can be immersed directly in organic liquids (although only the more nonpolar ones). It is also unaffected by most organic vapors when used in the gas phase, but its lower sensitivity and tendency to drift make it no longer our choice for this, compared with the LiCl sensor.

The sensor and corresponding hygrometer are manufactured by the German division of Endress & Hauser (Maulburg). In our studies, we have used the model WMY 270 Hygrometer with type DY20 probes. This Hygrometer is an old model and is no longer advertised by Endress & Hauser. In 1988, it was still possible to obtain replacement DY20 probes from Endress & Hauser UK, on special order (cost about US$ 1200, EUR 1000). Our experience in the use of these sensors in organic reaction mixtures is described in **ref. *19***. Endress & Hauser replaced these models with the Hygrolog WMY 770 Z and DY 43, DY 63, or DY 73 sensors. These operate on the same measurement principle, but the probes now include integral signal processing electronics, so that the meter reads directly in humidity units. We have not tested these sensors ourselves, but believe the primary measuring element would work as well as the old type. There is a problem, however, because we often use the sensors outside their specified range. The manufacturer's literature states a maximum dew point of 20°C, whereas biocatalytic reaction mixtures above ambient temperature will often exceed this. We have found that the older sensor does, nevertheless, give useful readings outside the nominal range. However, we are told that the electronics on the newer types will not pass on any out-of-range signal to the meter.

5. Water Activity Control During Reaction

Even if the entire reaction mixture is successfully pre-equilibrated to the desired water activity, this value can change as the reaction proceeds. Production or consumption of water in the reaction is one clear possibility. A less obvious cause is a change of water solubility because of the conversion of other reactants. For example, acids and alcohols contribute significantly to the solubilization of water in alkanes and other nonpolar solvents, whereas the ester product is much less effective. Hence, as esterification proceeds, there will be a fall in the water solubility and in the dissolved water concentration corresponding to any given water activity. This will contribute a further increase in water activity, in addition to that expected from formation of product water. The water activity can approach 1, leading to the separation of an aqueous phase, which will settle to the bottom of the vial. Its volume may still be very small, and so it may not be noticed. As a result of these effects, it can be desirable to control water activity during the reaction.

Several methods to control water activity during biocatalytic reactions in organic media have been described. Halling *(8)* covers some of the older methods. Other chapters in this volume describe alternative methods (Chapters 12 and 13). Scalable reactors have been described based on either membranes *(20,21)* or headspace evacuation in the absence of volatile organic species *(22)*. Here, we will give details of one method, using salt hydrates.

5.1. Salt Hydrates for Water-Activity Control in Reaction Mixtures

Many salts can form solid hydrates with water molecules present in the crystal lattice. Such salt hydrates can give up water, transforming to a lower hydrate or the anhydrous form. If both forms show ideal crystalline behavior, all the water is given up at a fixed value of water activity. The process is reversible if water is supplied, it being taken up at the same water activity. In practice, many salt pairs come quite close to this ideal behavior. Hence, if a system is prepared so that both forms of the salt remain present at equilibrium, the water activity will come to this fixed value. For a given pair of salt forms, the characteristic water activity is a function of temperature only. Different water activities are obtained by choosing different salt pairs. As an example, in the dissociation,

$$Na_2SO_4 \cdot 10H_2O \rightleftharpoons Na_2SO_4 + 10H_2O$$

equilibrium with both solids present can only be reached at a single value of water activity, 0.80 at 25°C.

In principle, then, salt hydrate pairs can act as perfect "water buffers," giving out or taking up water as required to maintain a fixed water activity. Furthermore, the two solids can be suspended directly in the reaction mixture. This makes the addition of salt hydrates potentially a simple and very attractive method to control water activity in organic reaction mixtures (for laboratory studies). However, it does require a little thought to use the method properly.

5.1.1. Which Salt Pair to Use?

The first task is to choose a suitable salt hydrate pair. In doing so, the following should be borne in mind:

1. Will they give the equilibrium water activity value required? Unfortunately, in many cases, this has been studied much less carefully than for saturated salt solutions. A compilation of literature values has been prepared *(23)*, but it seems that some of these are seriously in error *(24)*. (It may be that equilibration takes

Table 2
Selected Salt Pairs Found Useful for Water
Activity Control in Biocatalysis

Salt pair	Equilibrium water activity at 25°C	Rate of water transfer
NaI.2/0	0.12	Fast
Na_2HPO_4.2/0	0.16	Fast
$LiSO_4$.1/0	0.17	Slow
NaAc.3/0	0.28	Fast
NaBr.2/0	0.35	Slow
$Na_2S_2O_3$.5/2	0.37	Slow
$K_4Fe(CN)_6$.3/0	0.45	Slow
$Na_4P_2O_7$.10/0	0.49	Slow
$CaHPO_4$.2/0	0.50	Slow
Na_2HPO_4.7/2	0.61	Fast
Na_2HPO_4.12/7	0.80	Fast
Na_2SO_4.10/0	0.80	Fast

Note: The pairs used are identified by a shorthand notation: NaI.2/0 means a combination of NaI·$2H_2O$ and anhydrous NaI (i.e., $0H_2O$). "Fast" water transfer indicates equilibration in a few minutes, "slow" indicates that several hours may be needed. There is only limited information on the behavior of hydrate pairs giving lower water activities, although some indication that they generally tend to equilibrate slowly.

an impracticably long time, which will amount to the same thing—the water activity will not be that expected.) **Table 2** gives water-activity values for pairs that we have found can be used successfully.

2. Will water equilibration be fast enough? Some salt pairs transfer water to and from an organic phase very quickly, such that equilibrium will be reached within a few minutes. This is fast enough for most purposes connected with enzyme reactions. In other cases, water exchange will be much slower, taking hours. The relevant hydrate pairs may be suitable for some studies (e.g., of final chemical equilibrium position) but not for others (e.g., of initial rates). In such cases of slow exchange, it may be useful to pre-equilibrate the system with the salt hydrates before starting the reaction. A rough indication of the observed rates of water transfer is given in **Table 2**.

Rates of water transfer are expected to depend on the size, shape and form of the solid salt particles. We have noticed some indications that the rate of water exchange can increase when a salt has been taken through the hydration–dehydration cycle previously. This probably happens because of a reduction in crystal or particle size — often this can be seen by eye, as the new hydrate form develops as a crust of fine powder on the surface of large original crystals.

3. Will the salts interact with the enzyme other than by water exchange? Some transfer of ions from the solid salt to the enzyme particles can be detected *(24)*. We have recently become more aware of the possibility of effects on the protonation state of the enzyme molecules. Many salts have acidic or basic properties, and we have evidence that they can be used deliberately to exchange protons and counterions with the protein (Harper et al., unpublished data; Partridge et al., unpublished data). However, further work is required to understand fully the effects on enzyme protonation, hydration, and their interaction.

4. Will the salts react with substrates, products, or other components of the reaction mixture? Again, a common possibility is acid–base reaction. Thus, Na_2CO_3 and its hydrates, intended for use in controlling water activity, were found to remove butanoic acid from an organic reaction mixture, presumably by forming solid sodium butanoate. Conversely, basic salts used in supercritical CO_2 may be converted to carbonate salts.

5.1.2. How Much to Add?

It is necessary to be sure that excess of both salt forms are added, so that both solids remain at equilibrium, after water exchange with the rest of the system. All the other phases present will need to gain or lose water to reach the equilibrium water activity. The appropriate salt form will have to take up or give out however much water is needed here. It is good practice when using salt hydrates to estimate a "water budget" for the system, to ensure that sufficient salt is being added.

To illustrate, we present an example "water budget." A reaction mixture composed of the phases shown in **Table 3** is to be brought to a target water activity of 0.61 at 25°C, using the hydrate pair of $Na_2HPO_4 \cdot 7H_2O$ plus $Na_2HPO_4 \cdot 2H_2O$. Thus, salt hydrates added initially will have to give up 17.5 mmol of water to the reaction mixture. As the reaction proceeds, 6 mmol will have to be taken up again by the hydrate pair. In principle, adding more than 17.5/5 = 3.5 mmol (0.94 g) of $Na_2HPO_4 \cdot 7H_2O$ should be sufficient to supply the water required, transforming to the $2H_2O$ as it does so. The reaction product water will then be taken up as the transition reverses. In practice, about 1.25 g of $Na_2HPO_4 \cdot 7H_2O$ might be a sensible choice, to ensure excess, together with perhaps 0.1 g of $Na_2HPO_4 \cdot 2H_2O$, to ensure that initial equilibrium is reached as quickly as possible.

Note that this example is an extreme case. With a polar solvent like acetonitrile, the amount of water in the organic phase is much greater than in any other. Thus, the quantity of hydrate required is greatest in such solvents. With less polar solvents, the water in the enzyme particles or the gas headspace may be the largest contribution, but the total will be smaller. Note that the water content in the gas headspace gives a guide to how much may be gained or lost every time a reaction vial is opened for sampling.

In this example, the amount of salt required is close to the maximum that could reasonably be suspended in the liquid phase. However, in polar solvents

Table 3
Sample Water Budget for the Use of Salt Hydrates

Phase	Initial water content (mmol)	Estimated equil. water content (mmol)	Change (mmol)
10 mL acetonitrile	5.5 (1% w/v)	23.5	+18
30 mg immobilized enzyme (on silica)	1.5	1	−0.5
20 mL gas headspace	0.013 (lab air, 50% relative humidity)	0.016	+0.003
Water produced by esterification reaction from 0.6 *M* substrates	—	—	+6

Note: Equilibrium water content in acetonitrile is estimated from the known isotherm *(12)*, showing water mole fraction of 0.115 for water activity 0.61. This is converted to a concentration of 2.35 *M* using the approximation of no volume change on mixing. Equilibrium water content of immobilized enzyme estimated from measured adsorption isotherm (mainly adsorbed by silica support).

like acetonitrile, water-activity control is usually less of a problem. The solvent itself can dissolve or release significant amounts without large change in water activity, so it tends to buffer against changes resulting from reaction, exchange with the surroundings, and so forth.

The safest policy is to add both hydrate forms required, using a water budget estimation to make sure that they remain in excess. In principle, one of the salt forms can be generated *in situ* from the other, during initial water exchange with the rest of the system. This is particularly attractive when one of the salt forms is not available beforehand. However, it requires confidence that at least the direction of net water transfer has been correctly estimated. If this is wrong, then the new salt form produced will not be that intended, but rather that on the "other side" of the added hydrate (perhaps even a saturated solution). Hence, the equilibrium water activity will be totally different from that planned.

5.1.3. Preparing Salt Hydrate Forms Not Available

In some cases, a particular hydrate form required may not be readily commercially available, so it is necessary to prepare it. This can usually be done fairly straightforwardly by incubating the salt in an atmosphere of controlled water activity. (This may be produced by a saturated salt solution; *see* **Subheading 3.1.**). Any given salt form will be the stable end result of equilibration over a characteristic range of water activity. The lower limit will be that found when the target hydrate form is in equilibrium with the next lower hydrate (or anhydrous form). The upper limit will be that generated in the presence of the target salt form and the next higher hydrate (or the saturated solution). At any water activity within the range, the salt should eventually equilibrate to the target hydrate form. (It would be wise not to use a water activity too close to the limits of the range.)

For example, let us take the preparation of $Na_2HPO_4 \cdot 7H_2O$ (although this can be obtained commercially, it is not a usual form). In the case of this salt, all the hydrate transitions are well characterized. Thus we can say, based on the literature values for transition water activities at 25°C *(23)*, that the various forms will be stable over the following ranges: anhydrous salt, 0–0.163; $2H_2O$, 0.163–0.61; $7H_2O$, 0.61–0.80; $12H_2O$, 0.80 up to formation of saturated solution, probably above 0.90. Hence to make $Na_2HPO_4 \cdot 7H_2O$, it will be sufficient to equilibrate the salt at any water activity comfortably within the range 0.61–0.80. Several saturated salt solutions might be used for equilibration through the vapor phase—potassium iodide giving a water activity of 0.689 would be a convenient choice.

Equilibration can be monitored by weighing, with reference to a defined starting state. The anhydrous state is normally most accurate, but a known hydrate form can be used. In either case, be careful that water gain or loss during storage has not altered the starting state. Where large amounts of water have to be transferred, equilibration can be quite slow because of limited rates of diffusion through an unmixed gas phase. It is possible to speed up equilibration by incubating for a period over a drying agent or pure water, monitoring by weight. The addition of liquid water is also possible, but may cause aggregation of a solid powder.

5.1.4. Further Notes

Temperature is a key factor in the use of salt hydrates for water control. First, the equilibrium water activity for a given pair is quite strongly dependent on temperature (usually more so than for saturated salt solutions). Even more importantly, most hydrates have a melting point within the range relevant for biocatalysis. Above this temperature, they will transform into a liquid solution phase, perhaps combined with a solid lower hydrate. Hence, they should never be used in the reaction mixture above the melting temperature, as the liquid will interact with the biocatalyst. As a final warning, do not confuse the use of salt hydrate pairs with the use of saturated salt solutions. Both have their uses for water-activity control, and they have different advantages and limitations. They may often be based on the same salts, but the water-activity values obtained will be quite different. When hydrate pairs are used, the water equilibrium is with two types of solid crystal. With saturated solutions, a single type of crystal is used together with a liquid phase, an aqueous solution saturated with the relevant salt. An example is $SrCl_2$. This forms a saturated solution (in equilibrium with a solid phase of $SrCl_2 \cdot 6H_2O$), which has a water activity of 0.709 at 25°C *(10)*. Alternatively, solid $SrCl_2 \cdot 6H_2O$ may be used together with the lower hydrate $SrCl_2 \cdot 2H_2O$, giving a reported water-activity of 0.35 at 25°C. A second pairing, of $SrCl_2 \cdot 2H_2O$ and $SrCl_2 \cdot H_2O$, gives an even lower water-activity of 0.079. Any of the three combinations may be useful for water-activity control, but clearly if they are confused, the value obtained will be very different from what is intended.

Salt hydrates may be useful for water-activity control in larger-scale reactors for low-water biocatalysis. However, to avoid problems of plugging of packed beds, special approaches are necessary like twin-core design *(25)* or encapsulation *(26)*.

Appendix: How to Calculate Water Activity in Aqueous–Organic Mixtures, Using Wilson Coefficients

This appendix is based on the Appendix to **ref**. *12*. The published version contains several errors (which were somehow introduced after the proofs had been checked!). The following gives the correct equations.

Activities and activity coefficients are obtained from the Wilson coefficients as follows:

1. Calculate the intermediate values:

$$\Lambda_{ws} = \frac{V_s}{V_w} \exp\left[\frac{-(\lambda_{ws} - \lambda_{ww})}{RT}\right], \quad \Lambda_{sw} = \frac{V_w}{V_s} \exp\left[-\frac{(\lambda_{sw} - \lambda_{ss})}{RT}\right]$$

   where Vs and Vw are the molar volumes (molecular weight divided by pure liquid density) of organic solvent and water, respectively. $\lambda ws - \lambda ww$ and $\lambda sw - \lambda ss$ are the Wilson coefficients for that solvent –water system, R is the gas constant (8.314 J/[mol·K]) and T is absolute temperature.

2. Calculate the activity coefficient of water (γw) from

$$\ln \gamma_w = -\ln (x_w + \Lambda_{ws}x_s + x_s) \quad \left[\frac{\Lambda_{ws}}{x_w + \Lambda_{ws}x_s} - \frac{\Lambda_{sw}}{x_s + \Lambda_{sw}x_w}\right]$$

   Table 1 in **ref**. *12* gives the Wilson coefficients for various aqueous–organic mixtures, and also Vs values. Vw is 18.07 mL/mol at 25°C.

6. Acknowledgments

We thank Biotechnology and Biological Sciences Research Council for financial support. We also thank current and former workers at Strathclyde who have contributed to the development of many of these methods, especially Dr. J. M. Cassells, Dr. S. A. Khan, Dr. A. M. Vaidya, Dr. J. Partridge, Dr. M. C. Parker, Ms. L. Curran, and Dr. E. Zacharis.

References

1. De Gomez-Puyou, M. T. and Gomez-Puyou, A. (1998) Enzymes in low water systems. *Crit. Rev. Biochem. Mol. Biol.* **33,** 53–89.

2. Klibanov, A. M. (1997) Why are enzymes less active in organic solvents than in water? *Trends Biotechnol.* **15**, 97–101.
3. Lortie, R. (1997) Enzyme catalyzed esterification. *Biotechnol. Adv.* **15**, 1–15.
4. Vermue, M. H. and Tramper, J. (1995) Interrelations of chemistry and biotechnology 5. Biocatalysis in nonconventional media — medium engineering aspects. *Pure Appl. Chem.* **67**, 345–373.
5. Carrea, G, Ottolina, G., and Riva, S (1995) Role of solvents in the control of enzyme selectivity in organic media. *Trends Biotechnol.* **13**, 63–70.
6. Bell, G., Halling, P. J., Moore, B. D., Partridge, J., and Rees, D. G. (1995) Biocatalyst behaviour in low-water systems. *Trends Biotechnol.* **13**, 468–473.
7. Koskinen, A. and Klibanov, A. M. (1995) *Enzymatic Reactions in Organic Media,* Chapman & Hall, Andover.
8. Halling, P. J. (1994) Thermodynamic predictions for biocatalysis in non-conventional media: theory, tests and recommendations for experimental design and analysis. *Enzyme Microb. Technol.* **16**, 178–206.
9. Cassells, J. M. and Halling, P. J. (1988) Effect of thermodynamic water activity on thermolysin-catalysed peptide synthesis in organic two-phase systems. *Enzyme Microb. Technol.* **10**, 486–491.
10. Greenspan, L. (1977) Humidity fixed points of binary saturated aqueous solutions. *J. Res. Natl. Bur. Stand. A* **81A**, 89–96.
11. Valivety, R. H., Halling, P. J., and Macrae, A. R. (1992) Rhizomucor miehei lipase remains highly active at water activity below 0.0001. *FEBS Lett.* **301**, 258–260.
12. Bell, G., Janssen, A. M. S., and Halling, P. J. (1997) Water activity fails to predict critical hydration level for enzyme activity in polar organic solvents: interconversion of water concentrations and activities. *Enzyme Microb. Technol.* **20**, 471–477. (Because of printing errors in the calculations appendix of this article, a correct version is given as an appendix in this chapter.)
13. Parker, M. C., Moore, B. D., and Blacker, A. J. (1995) Measuring enzyme hydration in nonpolar organic-solvents using NMR. *Biotechnol. Bioeng.* **46**, 452–458.
14. Dolman, M., Halling, P. J., Moore, B. D., and Waldron, S. (1997) How dry are anhydrous enzymes? (Measurement of residual and buried 18-O labelled water molecules using mass spectrometry.) *Biopolymers* **41**, 313–321.
15. Partridge, J., Halling, P. J., and Moore, B. D. (1998) A practical route to high activity enzyme preparations for synthesis in organic media. *J. Chem. Soc. Chem. Commun.* 841,842.
16. Izumoto, E., Fukuda, H., and Nojima, Y. (1992) Feedforward/feedback control of interesterification of fats and oils using a microaqueous bioreactor. *Chem. Eng. Sci.* **47**, 2351–2356.
17. Goldberg, M., Thomas, D., and Legoy, M.-D. (1990) The control of lipase-catalysed transesterification and esterification reaction rates. Effects of substrate polarity, water activity and water molecules on enzyme activity. *Eur. J. Biochem.* **190**, 603–609.
18. Goldberg, M., Thomas, D., and Legoy, M.-D. (1990) Water activity as a key parameter of synthesis reactions: the example of lipase in biphasic (liquid/solid) media. *Enzyme Microb. Technol.* **12**, 976–981.

19. Khan, S. A., Halling, P. J., and Bell, G. (1990) Measurement and control of water activity with an aluminium oxide sensor in organic two-phase reaction systems for enzymic catalysis. *Enzyme Microb. Technol.* **12,** 453–458.
20. Rosell, C. M., Vaidya, A. M., and Halling, P. J. (1996) Continuous in-situ water activity control for organic phase biocatalysis in a packed bed hollow fiber reactor. *Biotechnol. Bioeng.* **49,** 284–289.
21. Ujang, Z., Alsharbati, N., and Vaidya, A. M. (1997) Organic-phase enzymatic esterification in a hollow fiber membrane reactor with in situ gas-phase water activity control. *Biotechnol. Prog.* **13,** 39–42.
22. Napier, P. E., Lacerda, H. M., Rosell, C. M., Valivety, R. H., Vaidya, A. M., and Halling, P. J. (1996) Enhanced organic phase enzymatic esterification with continuous water removal in a controlled air-bleed evacuated-headspace reactor. *Biotechnol. Prog.* **12,** 47–50.
23. Halling, P. J. (1992) Salt hydrates for water activity control with biocatalysts in organic media. *Biotecnol. Technol.* **6,** 271–276.
24. Zacharis, E., Omar, I. C., Partridge, J., Robb, D. A., and Halling, P. J. (1997) Selection of salt hydrate pairs for use in water control in enzyme catalysis in organic solvents. *Biotechnol. Bioeng.* **55,** 367–374.
25. Rosell, C. M. and Vaidya, A. M. (1995) Twin-core packed bed reactors for organic phase enzymatic esterification with water activity control. *Appl. Microbiol. Biotechnol.* **44,** 283–286.
26. Hessbruegge, B. J. and Vaidya, A. M. (1997) Preparation and characterization of salt hydrates encapsulated in polyamide membranes. *J. Membrane Sci.* **128,** 175–182.

12

Water Activity Control in Organic Media by Equilibration Through Membranes

Ernst Wehtje and Patrick Adlercreutz

1. Introduction
1.1. Why Is Water Activity Control Important?

The catalytic activity of an enzyme in an organic medium depends on its hydration state. Of the different possible ways to quantify the amount of water present in the reaction mixture, the water activity has been shown to be preferable because it correlates well with the hydration state of the enzyme (compare Chapter 11). If other ways are used to quantify water in these systems, the results are often more difficult to interpret. In a typical example, the dependence of the enzymatic activity on the solvent was studied. The optimal water concentration varied within a wide range in the solvents studied, but the optimal water activity was virtually the same in all solvents (1). Similarly, the optimal water activity was the same when different support materials were used for immobilization of the enzyme but the optimal water concentration varied depending on the water-absorbing capacity of the support (2).

In order to keep the enzyme working at its maximal rate, it is thus important to keep the water activity at the optimal level. In some cases, it can be sufficient to control the water activity in the reaction mixture before starting the reaction. However, in most cases, continuous water-activity control is beneficial. This is especially so when water is produced or consumed in the reaction catalyzed (e.g., in condensation or hydrolysis reactions). It is important to remember that hydrolytic reactions can occur as a side reaction when hydrolytic enzymes are used for transferase reactions (e.g., transesterification reactions).

From: *Methods in Biotechnology, Vol. 15: Enzymes in Nonaqueous Solvents: Methods and Protocols*
Edited by: E. N. Vulfson, P. J. Halling, and H. L. Holland © Humana Press Inc., Totowa, NJ

Another reason for continuous water-activity control is that the reaction catalyzed changes the composition of the reaction mixture and thus the solubility of water. This can cause a significant change in water activity at fixed water concentration.

1.2. Saturated Salt Solutions for Water Activity Control

Saturated salt solutions have fixed water activity (at fixed temperature). Such solutions have been used extensively for water-activity control of enzymatic reactions in organic media. In most cases, equilibration between the salt solution and the reaction mixture has occurred via the gas phase (compare Chapter 11). Gas-phase equilibration works well in many cases. However, when large amounts of water must be transferred to or from the reaction mixture, this equilibration process is too slow. This mainly happens when water is produced or consumed in relatively fast reactions. The method presented in this chapter describes a possibility to speed up the equilibration between the salt solution and the reaction mixture. The principle of the method is that the gas phase separating the salt solution and the reaction mixture is exchanged for a thin membrane. Thus, the transport distance for water is reduced and the equilibration is accelerated. All membranes that are permeable to water molecules and do not interfere with the rest of the system can be used. The method was developed using silicone tubing as the membrane, but there might be other more suitable materials. No extensive search for the best membrane material has been conducted.

2. Materials

The tubings can be of any material that is permeable to water. In our applications tubings of silicone of various dimensions were used. The tubings in these studies came from Leewood Marketing AB (Stockholm, Sweden). (*See* **Note 1**.)

3. Method

The method was first described in **ref. 3**. In the small-scale laboratory setup (**Fig. 1**), the saturated salt solution (containing excess salt) is kept in a glass vessel. If the reaction to be carried out produces water, there is a need for more excess salt compared to other cases. However, it is important to have enough solution so that it can be pumped through the system. When a water-producing reaction is carried out, the amount of remaining salt crystals in the vessels is a good indication of the remaining water-uptake capacity. The solution returning to the salt vessel (with a slightly changed water activity) is led into the salt-crystal region in order to be equilibrated with the salt (*see* **Note 2**).

The reaction vessel contains the reaction medium, including the substrates and the enzyme preparation. In most cases, the enzyme is immobilized on a

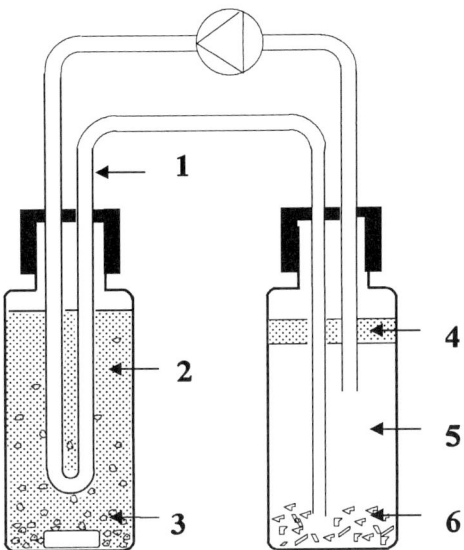

Fig. 1. Small-scale setup for control of water activity by equilibration through silicone tubing. (1) silicone tubing; (2) reaction medium; (3) immobilized enzyme; (4) organic solvent (to keep the salt solution solvent saturated); (5) saturated salt solution; (6) salt crystals.

support material (e.g., Celite or EP-100) prior to use. Mixing in the reaction vessel is normally carried out by magnetic stirring.

Both vessels are equipped with screw caps and rubber septa through which the silicone tubing passes. The salt solution is pumped through the silicone tubing by a peristaltic pump. The flow rate of salt solution is normally not crucial and a moderate rate can be used. Ideally the pump speed should be high enough to ensure that the water activity does not change too much during one pass through the reactor. This is easily established. It is more critical in water-consuming reactions, as there is the risk of salt crystallization in the tubing. (*See* **Notes 3–7.**)

3.1. The Principle of Water Transport

The function of the tubing is to transport water to or from the reaction mixture so that the water activity in the reactor is kept constant or at least within certain limits. The rate of water transport is determined by the reaction catalyzed. Relatively large amounts of water must be transported when water is formed or consumed in the reaction and/or when the reaction causes a large change in solubility of water in the reaction medium. The amount of tubing needed can be determined using straightforward calculations taking into account geometric and chemical considerations. This has been thoroughly described in **refs. 4** and **5**. In the following, the basic principles are presented.

Water transport through the silicone tubing occurs via diffusion. The flux of water through the tubing is thus determined by the geometry of the tubing (length, thickness, etc.), the driving force (water-activity difference between the inside and the outside), and the diffusion coefficient of water in the tubing material:

1. When the tubing is put into the reactor with a reaction medium that is predominantly an organic solvent, it swells. When making calculations on the water transport through the tubing, it is thus important to use the dimensions of the swelled tubing. Swelling influences its inner and outer diameters as well as its length. These parameters are easy to determine by simple size measurements.

2. One can imagine using either the difference in water concentration or water activity as the driving force in the calculations. A comparison of these two possibilities showed that the water activity provided a better correlation with the experimental data *(4)*. (When the water-activity difference was used as driving force, the same diffusion coefficient was obtained irrespective of the magnitude of the driving force, which was not the case when using the concentration difference as driving force.) The rate of water transport is independent of the actual direction water has to be transported. To determine the driving force in a reaction system (in terms of water activity) requires data of how the water content and water activity in the reaction medium are related, because the former is the one that can be readily obtained by standard methods.

3. The diffusion coefficient of water in the tubing material was determined in model experiments in which the driving forces were known (and the dimensions of the tubing). Different values were found depending on which solvent was used in the reactor. A good correlation was found between the diffusion coefficient observed and the swelling observed in the different solvents. A solvent in which the tubing swells substantially thus causes the diffusion coefficient to be high. On the other hand, a high degree of swelling causes the diffusion distance to be long which slows down the water transport.

3.2. A Typical Batch Reaction

In a batch reaction, the water-production rate is highest in the beginning of the reaction. The water-activity difference is close to zero, which means that the water activity in the reactor must increase in the initial phase. The water-activity difference is steadily increasing, which promotes the water transfer, and at a certain stage, the water-production rate (because of the enzymatic reaction) is matched to the water-removal rate (through the membrane) and the water content in the reactor is constant for a period of time. Reaching the end of a batch reaction, the production rate decreases and so does the water-removal rate. At the end, the production rate is zero and the water-activity difference has become zero.

3.3. How to Determine the Amount of Tubing

The required area of tubing (A, m^2) can be calculated using the following equation:

$$A = \frac{F \Delta r}{Da_w \cdot \Delta a_w}$$

where F is the transfer of water given by the water-production/consumption rate in moles per second, Da_w is the diffusion coefficient for water in moles per meter per second, Δr is the thickness of the tubing after swelling, and Δa_w is the accepted maximum difference in water activity between the salt solution and the reaction mixture. Values of Da_w were found to vary between 1.7×10^{-7} and 6.9×10^{-7} mol/m/s when different solvents were used.

The length of tubing (L, m) required is easily determined from A using

$$A = 2\pi L \log (\text{mean radius})$$

The amount of tubing required thus depends on how a large difference in water activity between the inside and outside of the tubing can be accepted. One can, in many cases, accept a rather large water-activity difference across the tubing without losing too much of the reaction rate. It is an advantage here to know the dependence of the catalytic activity on the water activity in detail. In some cases, the enzymatic rate is lowered as the water activity is increased, which can lower the efficiency of the reaction.

3.4. Scaling-Up

The same batch-reactor configuration was scaled up to a working volume of 250 mL reaction medium using 2.5 m silicone tubing. Similar enzymatic and water-transport conditions were obtained and the results were comparable to the smaller scale (a few milliliters). The water-transfer conditions can even be more favorable in larger scale. Longer single tubings were not practical to use; if more surface area is required, a number of shorter and thinner tubings should be used.

3.5. How to Change Water Activity During the Reaction

The water activity influences both the catalytic activity of the enzyme and the equilibrium position. Thus, the two goals to achieve a high reaction rate and a high final yield can come into conflict. Enzymatic esterification reactions are typical examples. The maximal yield is set by the equilibrium position in these thermodynamically controlled reactions. Thus, the maximal yield increases with decreasing water activity. On the other hand, the catalytic activity often decreases with decreasing water activity. At a high water activity, one will thus obtain a high reaction rate but a low yield, and at a low water activity, the reaction rate will be low but the final yield high. A good way around this problem is to start the reaction at a relatively high water activity to obtain a high reaction rate during the first part of the reaction and then shift to a lower water activity to obtain a high final yield. In this way, the final water activity determines the yield and it is obtained much faster compared to when the low water activity is used during the whole time.

The shift in water activity can be obtained by shifting to a new salt vessel equipped with tubing (preferable to avoid mixing of salts) or by changing the contents of the salt vessel used in the first place. A shift in water activity has been shown to be beneficial in a lipase-catalyzed esterification reaction *(6)* as well as in the esterification of a fatty acid and lysophosphatidylcholine catalyzed by phospholipase A_2 *(7)*. In the latter case, the water activity was changed three times during the reaction, but in this case, the reaction was so slow that equlibration via the gas phase was fast enough.

3.6. Other Reactor Configurations

The above-described tubing system is suitable for batch and CSTR (continuous stirred tank reactor) configurations. However, a packed-bed-type reactor can easily be arranged in which the biocatalysts are placed in the tubing and the salt solution is kept on the outside *(5)*. In all these systems, the same type of calculations regarding the water transport (*see* **Subheading 3.3.**) can be used.

A further development of the system in order to increase the available area for transportation is the use of a hollow-fiber unit *(8)*. The salt solution is kept on one side and the biocatalyst on the other, with circulation of the reaction medium.

In addition to saturated salt solutions, there are other means to buffer the water activity. Pairs of salt hydrates (*see* Chapter 11) have been used in connection with a twin-core packed-bed reactor *(9)*. The salt hydrate and reaction medium are physically separated by the shell of the reactor. This system is similar to the one we have described apart from the membrane. The water activity could also be buffered by air of a fixed water activity. This has been demonstrated in **ref. 10**. Our packed bed with the biocatalyst kept inside the tubing can also be used in this manner using air on the outside (unpublished results).

4. Notes

1. Regarding the tubing, the dimensions of the tubing gives the applicability in a given system. Different membrane materials can be used. The most important aspect to consider is the effect of the used reaction medium on the durability of the tubing. In the swelled state, the tubing is considerably weaker than the original; this can easily cause breaks of the tubing in joints and so forth.
2. If part of the silicone tubing is exposed to the surrounding atmosphere, water activity will be equilibrated with the humidity of that atmosphere too. If the

amount of tubing is minimized, this will be a minor problem because it will not affect the transfer in the reactor. However, if a volatile solvent is used, this can be a major cause for solvent leaking from the reactor. The circulating salt solution will be saturated with solvent, and if the transfer is fast enough, solvent can be lost through the tubing into the atmosphere outside. This can be minimized if the salt solution is kept saturated with solvent using a separate solvent layer on top of the salt solution.

3. The membrane is permeable to many types of substances and provided there is a driving force the compound can be transported in any direction across this physical barrier. The compound has to be able to dissolve in the medium on the *other side*. As described above, there can be a transport of solvent and/or substrates from the reactor to the salt. In principle, any of the salt solutions used in normal water-activity equilibration can be used. Some have to be used with care (*see* **Notes 4–7**).

4. Basic salts can decrease the durability of the tubing. If used in reactions containing acids as the substrate, there is a possibility for soap formation.

5. Organic salts (acetates) can penetrate the membrane and be dissolved in the solvent that may affect the pH of the system or be a potential substrate.

6. Lithium salts (e.g., chloride and bromide) have been shown to be impossible to use in the lipase-catalyzed reactions studied. The lithium salts are detrimental for the lipase activity if added directly to the enzyme preparation. It was also found that a lithium salt solution in combination with an acid component in the solvent decreased the enzymatic activity. A possible mechanism is that lithium can ion-pair with the acid and be transferred into the reactor and partitioned to the enzyme, there causing inactivation.

7. Cesium fluoride was also shown to be impossible to use. In this case, it was also a problem connected to the acid in the reaction mixture. Cesium ions in the salt solution made the acid more soluble in the salt solution and a transfer of acid substrate from the solvent to the salt solution was seen, which decreased the substrate concentration.

References

1. Valivety, R. H., Halling, P. J., and Macrae, A. R. (1992) Reaction rate with suspended lipase catalyst shows similar dependence on water activity in different organic solvents. *Biochim. Biophys. Acta* **1118,** 218–222.

2. Oladepo, D. K., Halling, P. J., and Larsen, V. F. (1994) Reaction rates in organic media show similar dependence on water activity with lipase catalyst immobilized on different supports. *Biocatalysis* **8,** 283–287.

3. Wehtje, E., Svensson, I., Adlercreutz, P., and Mattiasson, B. (1993) Continuous control of water activity during biocatalysis in organic media. *Biotechnol. Tech.* **7,** 873–878.

4. Kaur, J., Wehtje, E., Adlercreutz, P., Chand, S., and Mattiasson, B. (1997) Water transfer kinetics in water activity control system designed for biocatalysis in organic media. *Enzyme Microb. Technol.* **21,** 496–501.

5. Wehtje, E., Kaur, J., Adlercreutz, P., Chand, S., and Mattiasson, B. (1997) Water activity control in enzymatic esterification processes. *Enzyme Microb. Technol.* **21,** 502–510.
6. Svensson, I., Wehtje, E., Adlercreutz, P., and Mattiasson, B. (1994) Effects of water activity on reaction rates and equilibrium positions in enzymatic esterifications. *Biotechnol. Bioeng.* **44,** 549–556.
7. Egger, D., Wehtje, E., and Adlercreutz, P. (1997) Characterization and optimization of phospholipase A_2 catalyzed synthesis of phosphatidylcholine. *Biochim. Biophys. Acta* **1343,** 76–84.
8. Rosell, C. M., Vaidya, A. M., and Halling, P. J. (1996) Continuous in situ water activity control for organic phase biocatalysis in a packed bed hollow fiber reactor. *Biotechnol. Bioeng.* **49,** 284–289.
9. Rosell, C. M. and Vaidya, A. (1995) Twin-core packed-bed reactors for organic-phase enzymatic esterification with water activity control. *Appl. Microbiol. Biotechnol.* **44,** 283–286.
10. Ujang, Z. and Vaidya, A. M. (1998) Stepped water activity control for efficient enzymatic interesterification. *Appl. Microbiol. Biotechnol.* **50,** 318–322.

13

Water Activity Control for Lipase-Catalyzed Reactions in Nonaqueous Media

Joon Shick Rhee, Seok Joon Kwon, and Jeong Jun Han

1. Introduction

Lipase can be used for the synthetic reaction if the reaction is carried out in low-water media. The synthetic yield can be enhanced by continuously removing the water produced during the reaction. Many researchers have proposed several methods for water removal, such as headspace evacuation (1,2), pervaporation (3), use of molecular sieve (4), salt hydrate pairs (5,6), saturated salt solution, adsorption (4), and sparging of dry inert gas through the reaction medium (7). However, the continuous water removal might result in the enzyme inactivation when the water content is too low. Therefore, a correct water content must be maintained when biocatalysts are used in synthetic reaction. Especially, synthetic activity of the lipase is greatly affected by the water content or water activity (a_w) (8,9). Generally, it is advantageous to use the water activity (a_w) instead of water content or water concentration to characterize the water effects in the reaction mixture (10). In this section, we described the application of pervaporation and salt hydrate pairs to the a_w control for the lipase-catalyzed esterification.

2. Methods

2.1. Pervaporation

Pervaporation is defined as a separation technique in which a liquid feed mixture is separated by partial vaporization through a nonporous permselective (selectively permeable) membrane (11). Transport phenomena in pervaporation are different when compared to any other membrane processes such as dialysis, reverse osmosis, and ultrafiltration because of multiple interactions

From: *Methods in Biotechnology, Vol. 15: Enzymes in Nonaqueous Solvents: Methods and Protocols*
Edited by: E. N. Vulfson, P. J. Halling, and H. L. Holland © Humana Press Inc., Totowa, NJ

between the feed components and the membrane polymer. The driving force for the mass transfer of permeants from the feed to the permeate is a gradient in chemical potential, which is established by the difference in the partial pressure of the permeants across the membrane. This difference in partial pressure can be created principally by using a condenser and vacuum pump system, or by sweeping an inert gas on the permeate side of the membrane. Pervaporation by using a condenser and a vacuum pump is the more dominant mode of the operation. Recently, pervaporation has been applied as a way to mediate esterification and interesterification reactions that are catalyzed by lipase *(3,12,13)*. The rate, conversion, and the selectivity of lipase are greatly affected by the water concentration in the reaction mixture. Pervaporation can be easily used to control the water concentration of the reaction mixture. As shown in **Fig. 1**, the conversion yield was 65% at 120 min. After equilibrium, the water was removed by the pervaporation process. The conversion yield increased to about 85% after 240 min. Consequently, the thermodynamic equilibrium in the synthesis of *n*-butyl oleate was shifted toward synthesis by water removal.

2.1.1. Removal of Water Produced from Lipase-Catalyzed Esterification

2.1.1.1. MEMBRANE PREPARATION

The membrane was prepared according to the method of Mulder and Smolders *(14)* as follows:

1. Dissolve cellulose acetate in acetone.
2. Cast the cellulose acetate solution upon a glass plate.
3. The acetone was allowed to evaporate. The membrane was completely transparent and the thickness of the membrane was 20 μm.

2.1.1.2. PERVAPORATION

1. Set up a pervaporation reactor *(3)*. The membrane was supported by a filter paper placed on a sintered stainless-steel disk.
2. Add equimolar substrates (40 mmol each of oleic acid and *n*-butanol), Lipozyme® (Novo, 0.4 g), and isooctane (30 mL) into the reactor.
3. Stir at 25°C and 11.5*g*.
4. Apply the vacuum at the permeate side (2 mbar). The water separated from the reaction mixture was condensed in the condensor.

2.1.2. On-Off Dewatering Control by Tubular-Type Pervaporation System

When lipase was used as a biocatalyst in organic synthesis, a_w control was essential. Unfortunately, there is no a_w sensor available that can monitor a_w continuously in the solvent system in a large-scale production system. Therefore, it is impossible to control a_w directly in our pervaporation system.

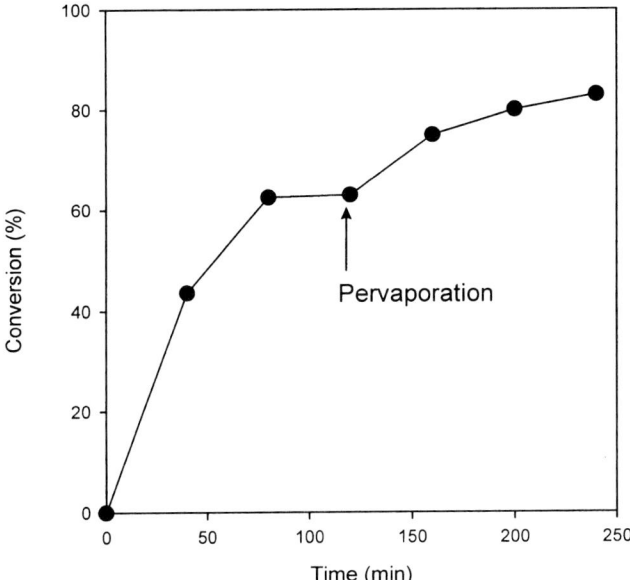

Fig. 1. Equilibrium shift by water removal using pervaporation. The reaction mixture contained 40 mmol of oleic acid, 40 mmol of *n*-butanol, and 0.4 g of Lipozyme® in 30 mL of isooctane at 25°C.

As an alternative, we used a sorption isotherm so that the dissolved water concentration corresponding to the a_w was controlled by an on-off control mode *(13)*; that is, the water formed was removed by pervaporation with the control based on manual analysis of water in the organic phase at the optimal water solubility corresponding to the optimal a_w. A computer-controlled pervaporation system made it possible to control the a_w indirectly in lipase-catalyzed esterification. In this system, a tubular-type membrane was used instead of the sheet-type membrane *(3,15)*. A tubular-type system has several advantages in durability at relatively high temperature and high flow rate, and in the ability to withstand exposure to harsh chemicals *(16)*. Also, this tubular-type pervaporation system was able to increase the membrane area exposed to the reaction mixture by inserting several tubular-type membrane modules when compared to the sheet type system having the membrane area localized on the bottom of the reactor. As shown in **Fig. 2**, the synthesis rate of the controlled reaction was about twice faster than that of the uncontrolled reaction, and the experimental data fitted the calculated ester curve quite well *(13)*. In addition, the water-solubility curve calculated from the dewatering control algorithm fitted the experimental data well (**Fig. 3**). Therefore, the water formed from the enzymatic esterification could be removed while maintaining the optimal a_w range of the reaction mix-

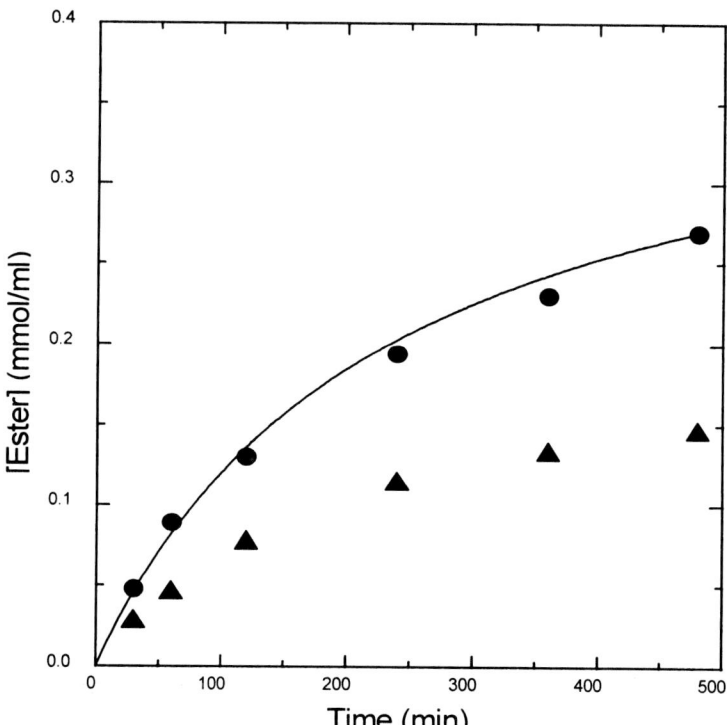

Fig. 2. Time-course for the lipase-catalyzed esterification of *n*-butyl oleate with the on–off dewatering control (●) and without the control (▲). Solid line: Ester concentrations (**Eq. 3**) integrated from the adaptive step-size Runge-Kutta method. The reaction mixtures contained 50 mg *Candida rugusa* lipase, 400 m*M* butanol, and 400 m*M* oleate in 25 mL *n*-hexane.

ture, and the reaction was able to proceed toward synthesis by the irreversible reaction mode. We hope that this method can be used to control the a_w in various enzymatic syntheses of various esters, peptides, and glycosides in the presence of organic solvents where the direct a_w control is impossible.

2.1.2.1. Tubular-Type Membrane Module Preparation

The tubular-type membrane module was prepared according to the method of Song and Hong *(16)* as follows:

1. Prepare the tubular-type porous ceramic support (the average pore diameter was about 0.1 µm.)
2. Coat the porous support with 20 wt% cellulose acetate solution in acetone using dip-coating and rotation-drying techniques. The thickness of the active layer of cellulose acetate was about 30 µm.

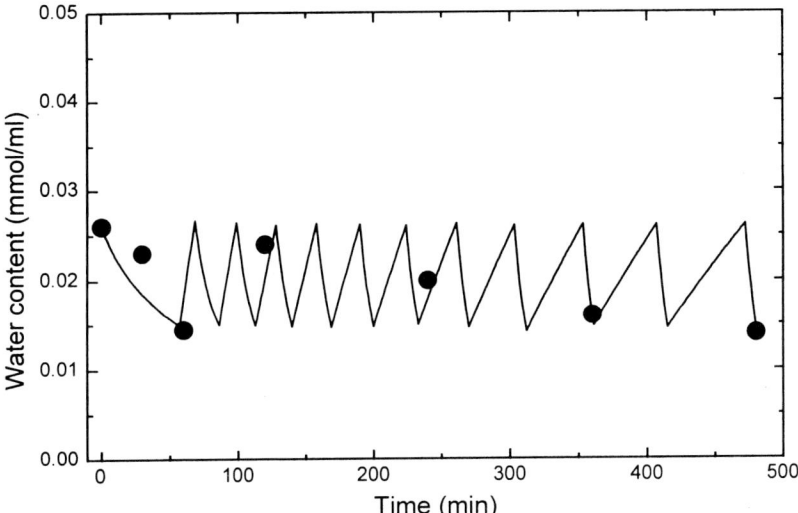

Fig. 3. Time-course for the water solubility in the reaction mixture. Solid line: The change in water calculated from the balance between production in reaction (**Eq. 3**) and removal (**Eq. 1**); ●: experimental value.

2.1.2.2. Determination of Optimal Water Activity

1. Pre-equilibrate the *Candida rugosa* lipase (50 mg, Lipase OF, Meito) with the salt hydrates in a desiccator. [The salt hydtrates used in the following: Na_2HPO_4–2/0 H_2O ($a_w = 0.18$), NaAc–3/0 H_2O ($a_w = 0.3$), $Na_4P_2O_7$–10/0 H_2O ($a_w = 0.52$), Na_2HPO_4–7/2 H_2O ($a_w = 0.85$) (**5**).]
2. Pre-equilibrate 400 mM substrate solution (1:1 molar ratio of oleic acid to butanol in *n*-hexane) with the same salt hydrates with direct contact for 3 d in a cylindrical reactor with screw caps and Teflon-lined septa.
3. Initiate the reaction by adding the 50 mg of lipase into the cylindrical reactor in which the pre-equilibrated substrate solution and 2 g of the salt hydrates were included.
4. Mix vigorously on a shaker (175 strokes/min) at 30°C.
5. Samples taken from the reaction were analyzed by a Hewlett-Packard Model 5890 series II GC with a flame ionization detector (**13**).
6. The optimal water activity was shown to be in the range 0.52–0.65 because the synthetic activity was similar over this range.

2.1.2.3. Determination of Sorption Isotherm Curve

1. Denature the lipase by adding 1N HCl, and dry the denatured lipase.
2. Pre-equilibrate the reaction mixtures including the denatured lipase with 2 g of different salt hydrates at 30°C.

3. Analyze the water contents in the organic phase of the reaction mixtures with a Karl Fischer titrater.
4. The water solubility range corresponding to the optimal a_w range (0.52–0.65) was 0.015–0.026 mmol/mL.

2.1.2.4. DETERMINATION OF DEWATERING RATE OF REACTION MIXTURE BY PERVAPORATION

1. Pre-equilibrate the reaction mixture including the denatured lipase with water ($a_w = 1$).
2. Add the water saturated reaction mixture into the pervaporation system.
3. Set up the tubular-type pervaporation system to remove the water *(13)*.
4. Remove the water in the reaction mixture by pervaporation at 30°C.
5. Take the samples at the predetermined time intervals and analyze the water content of the samples using the Karl Fischer titrater.
6. The dewatering rate was as follows:

$$\log[\text{water (mmol/mL)}] = -1.367 - (0.037)\ \text{Time (min)} \qquad (1)$$

2.1.2.5. STRATEGIES FOR CONTROLLING A_w IN LIPASE-CATALYZED ESTERIFICATION

1. If the water is removed at the same rate that the ester is formed at the optimal water concentration, the original reversible reaction would be shifted predominantly to the following irreversible reaction mode;

$$R_1COOH + R_2OH \xrightarrow{\quad k \quad} R_1COOR_2 \qquad (2)$$

The reaction rate is

$$\frac{d\,[\text{Ester}]}{dt} = k\,[\text{FA}]\,[\text{OH}] \qquad (3)$$

where [FA] is the R_1COOH concentration (mmol/mL), [OH] is the R_2OH concentration (mmol/mL), and [Ester] is the R_1COOR_2 concentration (mmol/mL).
2. The reaction rate constants *(k)* at different water activities were obtained from the experiment by controlling the a_w using salt hydrates (**Table 1**). The k value at the optimal a_w was 0.01075 ± 0.00038 mL/(mmol/min) at 30°C.
3. The ester concentration (**Eq. 3**) was integrated from the fourth-order adaptive step-size Runge-Kutta method. The change in water present was calculated from the balance between production in reaction (**Eq. 3**) and dewatering rate (**Eq. 1**).
4. The water was removed while maintaining the optimal water concentration range corresponding to the a_w range.
5. We put the initial water concentration in solution (*S0*) and the values at the upper (*SU*) and lower (*SL*) limits into the algorithm. The *SU* and *SL* were 0.026 and 0.015 mmol/mL, respectively, which corresponded to the optimal a_w range (0.52–0.65). If the water concentration increased above *SU*, a switch was allowed to be turned on; the switch was turned off if the water was decreased below *SL*. The detailed algorithm was explained by Kwon and Rhee *(13)*.

Table 1
Reaction Rate Constant *(k)* of *C. rugosa* Lipase-Catalyzed Esterification of *n*-Butyl Oleate in *n*-Hexane at Different Water Activities

a_w buffer used	$a_w{}^a$	Rate constant (k) (mL/mmol/min)
CH_3COONa–3/0 H_2O	0.30	0.00375 ± 0.00050
$Na_4P_2O_7$–10/0 H_2O	0.52	0.01070 ± 0.00045
Na_2HPO_4–7/2 H_2O	0.65	0.01075 ± 0.00038
Na_2HPO_4–12/7 H_2O	0.80	0.00742 ± 0.00048

^{a}The a_w value was from **ref. 5**.

2.1.2.6. THE LIPASE-CATALYZED ESTERIFICATION WITH CONTINUOUS A_w CONTROL

1. Pre-equilibrate the lipase (50 mg) with Na_2HPO_4–7/2 H_2O($a_w = 0.65$) in the vapor phase for 3 d.
2. Pre-equilibrate the substrate solution (400 mM) with Na_2HPO_4–7/2 H_2O ($a_w = 0.65$) by direct contact for 3 d.
3. Initiate the reaction by adding 25 mL of the pre-equilibrated substrate solution into the cylindrical reactor in which the pre-equilibrated lipase was included.
4. Mix vigorously on a shaker (175 strokes/min) at 30°C for 480 min. During the reaction, no salt hydrates was included in the reactor.
5. Control the water concentration using the computer-controlled tubular-type pervaporation system.

2.2. Enhancement of Enzyme Activity Using Salt Hydrate Pairs

It is well known that a salt hydrate pair can control the water level in the reaction mixture by taking up or releasing water as required to keep a constant a_w condition during the reaction. Each kind of salt hydrate pair has a typical a_w *(5)*. When a hydrated salt and its corresponding lower hydrate or anhydrous form are present together, ideal behavior implies a fixed equilibrium water vapor pressure and, hence, constant a_w, whatever the relative quantities of the two forms. Although the salts may affect enzyme activity in other aspects of water activity when they are added into the reaction medium directly, many reports have shown that salt hydrate pairs could control the water activity of the reaction system and they increased the reaction rate and yield. We also investigated the effect of various salt hydrate pairs on the lysophospholipid synthesis and sucrose–monoester synthesis in a solvent-free system.

2.2.1. Lysophospholipid Synthesis

For the synthesis of biologically useful lysophospholipids with a desirable fatty acid, an easy and simple synthetic approach was developed by lipase-catalyzed esterification of a glycerophosphoryl group with free fatty acid in a solvent-free system *(17)*. Among the various parameters that govern the lysophospholipid synthesis, a_w was found to be an important factor. Salt hydrate pairs could act to buffer the optimal water level during the reaction. We investigated the effect of various salt hydrate pairs, which were added directly into the reaction medium or suspended in the headspace of the reactor, on the lysophospholipid synthesis in the solvent-free system. Predetermined amounts of the specific fatty acid (10 mmol of decanoic acid) were added to the reactor and the reaction was initiated by adding the lipase and 3 mmol of a glycerophosphoryl group (glycerol-3-phosphate, G-3-P; glycerol-3-phophatidyl choline, G-3-PC; glycerol-3-phophatidyl ethanolamine, G-3-PE) to the melted fatty acid, which had been thermally equilibrated to the predetermined temperature in a water bath and agitated by a magnetic stirrer. Various salt hydrate pairs with an equal amount of each salt form (2.5 g each) were added directly into reaction mixture or they were placed in a tea bag and suspended in the headspace of the closed reactor during the reaction. The salt hydrate pairs used were as follows (the numbers in the parentheses refer to the *aw* values at 50°C, measured by the Novasina a_w-value measuring instrument): Li_2SO_4–1/0 H_2O ($a_w = 0.12$), NaI–2/0 H_2O ($a_w = 0.17$), $BaBr_2$–2/0 H_2O ($a_w = 0.26$), CH_3COONa–3/0 H_2O ($a_w = 0.37$), $NaBr$–2/0 H_2O ($a_w = 0.46$), $Na_4P_2O_7$–10/0 H_2O ($a_w = 0.60$), and $Na_2B_4O_7$–10/0 H_2O ($a_w = 0.80$).

2.2.1.1. DIRECT ADDITION OF SALT HYDRATE PAIR INTO REACTION MEDIUM

In the presence of $Na_4P_2O_7$–10/0 H_2O, the lysophosphatidyl choline (LPC) yield was the best (36.2%) and this is higher than that in an open-reactor system (**Table 2**). In the case of lysophosphatidyl ethanolamine (LPE) synthesis, LPE could not be synthesized readily in either closed- or open- reactor systems. However, when CH_3COONa–3/0 H_2O salt pairs were added, LPE was synthesized readily (yield of 22.9%). In the case of lysophosphatidic acid (LPA) synthesis, however, the yield was not improved by directly adding various salt hydrate pairs to the reaction system (*see* **Note 1**). Increasing the salt content could hardly increase the yield (data not shown). Thus, the direct addition of the salt hydrate pair for higher yield was effective only in LPC and LPE syntheses, but not in LPA synthesis, and the optimal a_w for LPC and LPE syntheses were 0.60 and 0.37, respectively.

2.2.1.2. SUSPENDING SALT HYDRATE PAIR BAG IN THE HEADSPACE OF A CLOSED REACTOR

If the salt hydrate pair is toxic and increases viscosity, its direct addition in the reaction mixture is far from being desirable. Another way of controlling the *aw* of the reaction medium can be possible by suspending the salt hydrate pair bag in the headspace of the closed reactor. By using this method, we can

Table 2
Lysophosphatidyl choline (LPC) and Lysophosphatidyl Ethanolamine (LPE) Yield in Various Reaction System

Reactor system	Controlled $a_w{}^a$	Salt hydrate pair	LPC yield (%)	LPE yield (%)
Closed	0.12	$LiSO_4$–1/0 H_2O	5.2	2.1
	0.17	NaI–2/0 H_2O	7.4	1.9
	0.26	$BaBr_2$–2/0 H_2O	11.9	5.8
	0.37	CH_3COONa–3/0 H_2O	14.2	22.9
	0.46	$NaBr$–2/0 H_2O	13.2	8.7
	0.60	$Na_4P_2O_7$–10/0 H_2O	36.2	9.8
	0.80	$Na_2B_4O_7$–10/5 H_2O	5.0	2.1
	Not controlled	—	15.8	2.2
Open	Not controlled	—	26.1	4.2

^{a} The a_w value was measured by the Novasina a_w-value measuring instrument. The substrate ratio of capric acid to G-3-PC is 10:1.5 (mmol : mmol) and 0.1 g of Lipozyme was added. The reaction was performed at 50°C, 1.0g for 60 h. For a_w control of the reaction system, 0.2 g of a salt hydrate pair with an equal amount of each hydrate form was added.

hopefully reuse the salt hydrates. By definition, a_w is equal in all phases at equilibrium *(5)*. Hence, an organic reaction mixture may be characterized by a single a_w value. The a_w of the reaction medium may be controlled by means of the water vapor pressure in the headspace of the reactor. However, the problem is how fast the a_w of reaction medium can be equilibrated by means of the salt hydrate pair in the headspace. Some lag time was likely to be present. Various salt hydrate pairs (2.5 g of each hydrate form) were suspended in the headspace of the reactor for the control of a_w ranging from 0.08 to 0.80 during the LPA synthesis. As shown in **Table 3**, in the presence of a NaI hydrate pair (2/0), the LPA yield was the highest (45.3%) and higher than that in the open-reaction system (33.8%). The salt hydrate pairs in the headspace could increase the yield of LPA during lipase-catalyzed esterification of G-3-P with fatty acid in a solvent-free system. To find out whether NaI salt hydrate pairs can also increase the LPA synthesis rate in addition to yield, a time-course experiment was performed. Although the lag time was present at the initial stage of the reaction, the LPA synthesis rate was higher than that in the open-reaction system **(Fig. 4)**.

2.2.2. Sucrose-Monoester Synthesis

Sugar monoesters have been widely used in the synthesis of useful intermediates such as food ingredients and chemical and pharmaceutical intermediates *(18)*. There is, therefore, a demand for efficient methods for the synthesis of

Table 3
Lysophosphatidic acid (LPA) Yields in Various Reactor Systems

Reactor system	Controlled $a_w{}^a$	Salt hydrate pair	LPA yield (%)
Closed	0.08	$LiSO_4$–1/0 H_2O	23.8
	0.18	NaI–2/0 H_2O	45.0
	0.37	CH_3COONa–3/0 H_2O	36.8
	0.46	NaBr–2/0 H_2O	34.0
	0.60	$Na_4P_2O_7$–10/0 H_2O	18.6
	0.80	$Na_2B_4O_7$–10/5 H_2O	12.8
	Not controlled	—	21.1
Open	Not controlled	—	33.8

[a]The a_w value was measured by the Novasina a_w-value measuring instrument. The reaction mixture consisted of 6 mmol of G-3-P, 40 mmol of capric acid, and 0.1 g of Lipozyme. The reaction conditions were 50°C, 60 h, and 1.0g. In order to a_w control of reaction system, 5 g of a salt hydrate pair with an equal amount of each hydrate form was put in the headspace of the reactor.

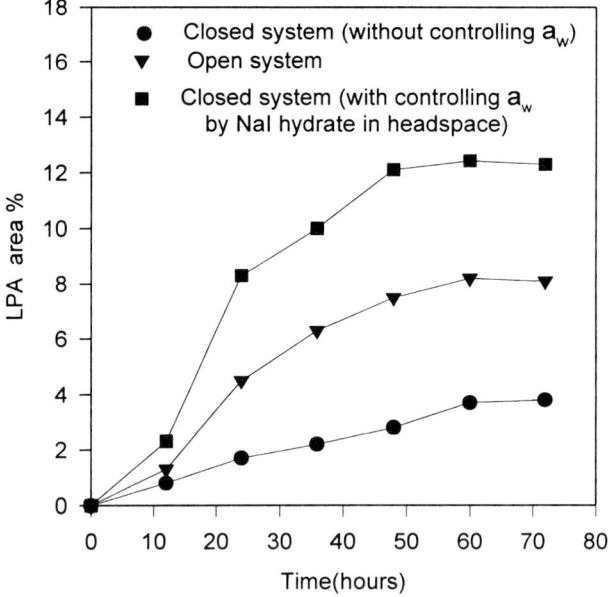

Fig. 4. Time-course data for LPA synthesis. The reaction mixture consisted of 6 mmol of G-3-P, 0.1 g lipase (Lipozyme), and 40 mmol of capric acid. The reaction condition was 50°C, 72 h, and 1.0g.

sugar monoesters. It is difficult to acylate the hydroxyl groups in sucrose regioselectively. Because enzymes have their unique properties such as regioselectivity and stereoselectivity and the ability to catalyze reactions under mild conditions, they can be used for the synthesis of sugar esters. Because of their different solubilities, however, it is not easy to mix sugar and a fatty acid in the reaction medium. In order to solve this problem, a number of different approaches for solubilizing substrates in enzyme reactions have been introduced (e.g., by using solvents *[19,20]* such as pyridine and dimethylformamide [DMF], by using sugar derivatives [alkyl *{21}*, acetal *{22,23}*], or by using solubilizing agents for sugars *[24–26]*, such as organoboronic acids). However, most of the solvents and reagents used for solubilizing sugars are too harmful for the final products to be used as food, cosmetic, and pharmaceutical ingredients. It is also cumbersome to synthesize the sugar derivatives as substrates by chemical methods in order to increase their solubility in solvents. In this study, we examined the effects of various salt hydrate pairs for the synthesis of sugar monoesters by the enzymatic esterification reaction in a solvent-free systems *(27)*. This method shows that the sucrose monoester could be synthesized effectively by adding suitable salt hydrate pairs without using substrate derivatives, solubilizing agents, and solvents (*see* **Note 2**). For the investigation of the effect of initial a_w of both substrate and enzyme on sucrose monoester synthesis, substrates and *Rhizomucor miehei* powder lipase (Novo, hydrolytic activity 121.8 U/mg) were incubated in the container that contains a desired saturated salt solution for 7 d at 20°C in order to equilibrate the materials for the desired a_w. All salts used in this work are as follows: LiBr (a_w = 0.064), LiCl (a_w = 0.11), CH_3COOK (a_w = 0.22), $MgCl_2$ (a_w = 0.33), KCO_3 (a_w = 0.43), NaBr (a_w = 0.57), $CuCl_2$ (a_w = 0.67), NaCl (a_w = 0.75), KCl (a_w = 0.86), and KNO_3 (a_w = 0.94). After equilibration, the sucrose and enzyme were transferred to a stoppered vial, and capric acid and tetracosane (as an internal standard for gas chromatography [GC] analysis) were added. The reaction was started at the predetermined a_w. After 48 h, the reaction was stopped and assayed by GC. To investigate the effect of salt hydrate pairs on sucrose-monoester synthesis, predetermined amounts of substrates and enzyme were added into a stoppered vial. During the reaction, a pair of salt hydrate with an equal amount of each hydrate form was added directly to the reaction mixture. The salts tested were $Na_2B_4O_7$–10/5 H_2O, Na_2HPO_4–12/7 H_2O, Na_2HPO_4–7/2 H_2O, $Na_4P_2O_7$–10/0 H_2O, NaBr–2/0 H_2O, CH_3COONa–3/0 H_2O, $Ba(OH)_2$–8/1 H_2O, $CuSO_4$–5/3 H_2O, Na_2CO_3–2/1 H_2O, $BaCl_2$–1/0 H_2O, and $LiSO_4$–1/0 H_2O. All the reactions were carried out at 350 rpm and for 48 h at various temperatures. After the reaction was stopped, the reaction mixtures were extracted, centrifuged or filtered, and examined by thin-layer chromatography (TLC) and GC. **Table 4** shows that the initial a_w influences the reaction yield significantly and the monoester was formed only with a_w in the range 0.22–0.57. However, the yield was quite low. This phenomenon may result from the increase of water content by the water produced during the esterification. Most

Table 4
Reaction Yield at Different Initial Water Activities

Initial a_w (at 20°C)	Salt-saturated solution	Reaction yield (%)
0.75	NaCl	0
0.67	$CuCl_2$	0
0.57	NaBr	5.1
0.43	KCO_3	6.4
0.33	$MgCl_2$	6.2
0.22	KAc	2.3
0.11	LiCl	0

Note: Enzyme and substrates were pre-equilibrated over saturated salt solution in a closed vessel for 7 d at room temperature before the reaction. The reaction mixture consisted of 2.92 mmol of sucrose, 0.1 g of *M. miehei* lipase, and 14.6 mmol of capric acid in a stoppered vial. The reaction condition was 50°C, 1.4g, and 48 h.

of the product might be produced initially before the significant change of a_w. The data indicate that a_w is also an important factor in this reaction system. **Table 5** shows that monoester was synthesized only in the presence of $Ba(OH)_2$–8/1 H_2O and the yield was 13.8%. We expected that CH_3COONa–3/0 H_2O and $Na_4P_2O_7$–10/0 H_2O would be effective as water buffers for lipase-catalyzed sucrose-monoester synthesis as shown in our initial a_w effect experiment. However, we found that those salt hydrate pairs were actually ineffective (*see* **Note 3**). For the determination of optimal salt content, we investigated the effect of salt hydrate pair concentration at the range of 0–1.0 g per 3.57 g of reaction mixture (**Fig. 5**). The reaction yield significantly depended on the amount of salt hydrate pair added and it was the highest (25.3%) at the salt hydrate pair concentration of 0.6 g of $Ba(OH)_2$–8/1 H_2O per 3.57 g of the reaction mixture. With a salt hydrate pair concentration higher than 0.7 g per 3.57 g of reaction mixture, the yield was sharply reduced. This may have resulted from the mixing problem because of the high viscosity of the reaction mixture at the high content of salt. **Figure 6** shows the time-course change in the a_w controlled closed system and the open system with an initial addition of water (without an initial addition of water, sucrose ester was not synthesized.). The reaction yield in the closed system was significantly higher than that in the open system. Therefore, the product yield was significantly increased by the addition of a suitable pair of solid salt hydrates directly to the reaction mixture in order to control a_w.

Table 5
Yields of Sucrose Ester in the Case of Controlling of Water Activity by Direct Addition of a Salt Hydrate Pair in a Closed Reaction System

Controlled $a_w{}^a$ (at 50°C)	Salt hydrate pair	Reaction yield (%)
0.79	$Na_2B_4O_7$–10/5 H_2O	0
0.60	$Na_4P_2O_7$–10/0 H_2O	n.s.
0.44	$Ba(OH)_2$–8/1 H_2O	13.8 ± 2.4^c
0.37	CH_3COONa–3/0 H_2O	n.s.
0.12	$LiSO_4$–1/0 H_2O	0
Control[b]	$Ba(OH)_2$–8/1 H_2O, no enzyme	0

[a]The a_w value was measured by the Novasina a_w-value measuring instrument. The reaction mixture consisted of 2.92 mmol of sucrose, 0.1 g of *M. miehei* lipase, 14.6 mmol of capric acid, and 0.5 g of various kinds of salt hydrate pair with an equal amount of each hydrate form in a stoppered vial. Reaction condition was 50°C, 1.4*g*, and 48 h.

[b]Control reaction mixture consisted of the same concentration of substrates, $Ba(OH)_2$–8/1H_2O, and no enzyme.

[c]Mean ± SD based on triplicate. n.s.: not significant.

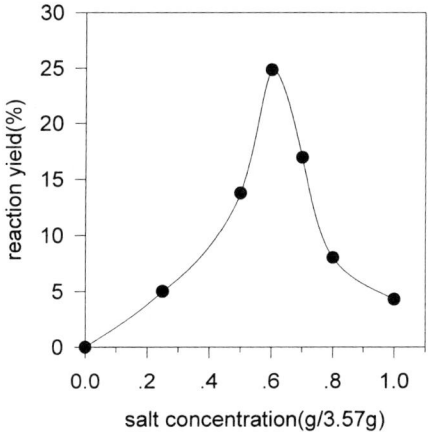

Fig. 5. Yields of sucrose monoester at various salt concentration. The reaction mixture consisted of 2.92 mmol of sucrose, 14.60 mmol of capric acid, and 0.1 g of *M. miehei* lipase. Various amount of $Ba(OH)28/1$ H_2O with an equal amount of each hydrate form was added into the reaction medium. The reaction condition was 50°C, 1.4*g*, and 48h.

3. Notes

1. Unlike sugar acylation, direct addition of salt hydrate pair into the reaction medium for LPA synthesis was not effective for increasing the reaction yield and

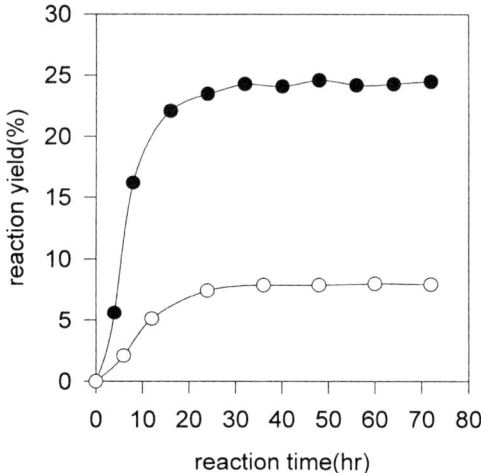

Fig. 6. Time-course of sucrose-monoester synthesis. The reaction mixture consisted of 2.92 mmol of sucrose, 14.60 mmol of capric acid, 0.05 g of tetracosane, and 0.1 g of *M. miehei* lipase. The reaction conditions were 50°C and 1.4*g*. Closed circles indicate reaction yields for the a_w-controlled closed-reaction system by the addition of 0.6 g of the Ba(OH)$_2$–8/1 H$_2$O pair (0.3 g of each hydrate form) to the reaction mixture. Open circles indicate the reaction yields for the open-reaction system. In the open-reaction system, water was added initially in the concentration of 0.5 g of the 3.57 g reaction mixture.

rate in comparison with the open-reaction system. Suspending salt hydrate pairs in the headspace of the reactor could increase both reaction rate and yield. In this case, because salt hydrate pairs were not added directly into the reaction medium, the sole effect of salt hydrate pairs for water-activity control could be monitored. The NaI hydrate pair in the headspace was the most effective for a high reaction rate and yield, and the optimal a_w for LPA synthesis was near 0.18.

2. The salt-addition method was also successfully employed for acylation of primary hydroxyl groups in various unprotected monosaccharides and disaccharides such as glucose, galactose, fructose, trehalose, mannose, maltose, and lactose.

3. When other sugars (glucose, galactose, fructose, mannose, maltose, and lactose) were used as substrates, CH$_3$COONa–3/0 H$_2$O and Na$_4$P$_2$O$_7$–10/0 H$_2$O were effective on lipase-catalyzed monoacylation of these sugars *(27)*. Especially, in fructose monocaprate synthesis, sugar esters could be synthesized with all of the salt hydrate pairs used in this experiment. In the case of glucose, galactose, fructose, and mannose monocaprate syntheses, the CH$_3$COONa–3/0 H$_2$O hydrate pair was most effective, and in the case of maltose monocaprate syntheses, Na$_4$P$_2$O$_7$–10/0 H$_2$O was most effective. In lactose and trehalose monocaprate syntheses, only the Ba(OH)$_2$ hydrate pair was effective, just as in the case of sucrose monocaprate synthesis. Theoretically, CH$_3$COONa–3/0 H$_2$O and Na$_4$P$_2$O$_7$–10/0 H$_2$O must be effective if it acts only as the water buffers for lipase-

catalyzed synthesis of sucrose monocaprate, but, experimentally, sucrose monocaprate was not synthesized in the presence of CH_3COONa and $Na_4P_2O_7$ hydrate pairs as shown in **Table 5**. Therefore, we can conclude that there must be another role of the salt hydrate pair in addition to its a_w control role when it was directly added into the reaction medium. There is a possibility that because some salts are soluble and they could be ionized, it might affect the pH of the reaction medium, resulting in affecting the enzymatic activity. Although much additional work must be required for a complete understanding of the phenomenon, the major role of salt hydrate pairs for enzymatic reaction is water-activity control of reaction system.

References

1. Bloomer, S., Aldercreutz, P., and Mattiasson, B. (1991) Triglyceride interesterification by lipases. *Biocatalysis* **5,** 145–162.
2. Napier, P. E., Lacerda, H. M., Rosell, C. M., Valivety, R. H., Vaidya, A. M., and Halling, P. J. (1996) Enhanced organic-phase enzymatic esterification with continuous water removal in a controlled air-bleed evacuated-headspace reactor. *Biotechnol. Prog.* **12,** 47–50.
3. Kwon, S. J., Song, K. M., Hong, W. H., and Rhee, J. S. (1995) Removal of water produced from lipase-catalyzed esterification in organic solvent by pervaporation. *Biotechnol. Bioeng.* **46,** 393–395.
4. Ergan, F., Trani, M., and Andre, G. (1990) Production of glycerides from glycerol and fatty acid by immobilized lipase in non-aqueous media. *Biotechnol. Bioeng.* **35,** 195–200.
5. Halling, P. J. (1992) Salt hydrates for water activity control with biocatalysts in organic media. *Biotechnol. Tech.* **6,** 271–76.
6. Kvittingen, L., Sjursnes, B., Anthosen, T., and Halling, P. J. (1992) Use of salt hydrates to buffer optimal water level during lipase catalyzed synthesis in organic media: a practical procedure for organic chemists. *Tetrahedron* **48,** 2793–2802.
7. Kosugi, Y. and Azuma, N. (1994) Continuous and consecutive conversion of free fatty acid in rice bran oil to triacylglycerol using immobilized lipase. *Appl. Microbiol. Biotechnol.* **41,** 407–412.
8. Goldberg, M., Thomas, D., and Legoy, M. D. (1990) The control of lipase-catalyzed transesterification and esterification reaction rates. Effects of substrate polarity, water activity and water molecules on enzyme activity. *Eur. J. Biochem.* **190,** 603–609.
9. Goldberg, M., Thomas, D., and Legoy, M. D. (1990) Water activity as a key parameter of synthesis reactions: the example of lipase in biphasic (liquid/solid) media. *Enzyme Microb. Technol.* **12,** 976–981.
10. Halling, P. J. (1994) Thermodynamic predictions for biocatalysis in nonconventional media: theory, tests, and recommendations for experimental design and analysis. *Enzyme. Microb. Technol.* **16,** 178–206.
11. Strathmann, H. and Gudernatsch, W. (1991) Pervaporation in biotechnology, in *Pervaporation Membrane Separation Processes* (Huang, R. Y. M., eds.), Elsevier Science, Amsterdam, pp. 363–389.

12. Van der Padt, A., Sewalt, J. J. W., and Van't Riet, K. (1993) On-line water removal during enzymatic triacylglycerol synthesis by means of pervaporation. *J. Membr. Sci.* **80,** 199–208.

13. Kwon, S. J. and Rhee, J. S. (1998) On-off dewatering control for lipase-catalyzed synthesis of *n*-butyl oleate in *n*-hexane by tubular type pervaporation system. *J. Microbiol. Biotechnol.* **8,** 165–170.

14. Mulder, M. H. V. and Smolders, C. A. (1984) On the mechanism of separation of ethanol/water mixtures by pervaporation I. calculations of concentration profiles. *J. Membr. Sci.* **17,** 289–307.

15. Fleming, H. L. and Slater, C. S. (1992) Design, in *Membrane Handbook* (Ho, W. S. W. and Sirkar, K. K., eds.), Van Nostrand Reinhold, New York, pp. 123–131.

16. Song, K. M. and Hong, W. H. (1997) Dehydration of ethanol and isopropanol using tubular type cellulose acetate membrane with ceramic support in pervaporation process. *J. Memb. Sci.* **123,** 27–33.

17. Han, J. J. and Rhee, J. S. (1998) Effect of salt hydrate pairs for water activity control on lipase-catalyzed synthesis of lysophospholipids in a solvent free system *Enzyme Microb. Technol.* **22,** 158–164.

18. Nishikawa, Y., Okabe, M., Yoshimoto, K., Kurono, G., and Fukuoka, F. (1976) Chemical and biological studies on carbohydrate esters II and III. Antitumor activity saturated fatty acids and their ester derivatives against Ehrich Ascites carcinoma. *Chem. Pharm. Bull.* **24,** 387–393, 756–762.

19. Chopineau, J., McCafferty, F. D., Therisod, M., and Klibanov, A. M. (1988) Production of biosurfactants from sugar alcohols and vegetable oils catalyzed by lipases in nonaqueous medium. *Biotechnol. Bioeng.* **31,** 208–214.

20. Riva, S., Chopineau, J., Kieboom, P. G., and Klibanov, A. M. (1988) Protease-catalyzed regioselective esterification of sugars and related compounds in anhydrous dimethylformamide. *J. Am. Chem. Soc.* **110,** 584.

21. Mutua, C. N. and Akoh, C. C. (1993) Synthesis of alkyl glycoside fatty acid esters in non-aqueous media by *Candida* sp. lipase. *J. Am. Oil Chem. Soc.* **70,** 43–46.

22. Fregapane, G., Sarney, D. B., and Vulfson, E. N. (1991) Enzymic solvent-free synthesis of sugar acetal fatty acid esters. *Enzyme Microb. Technol.* **13,** 796–800.

23. Sarney, D. B., Fregapane, G., and Vulfson, E. N. (1994) Lipase-catalyzed synthesis of lysophospholipids in a continuous bioreactor. *J. Am. Oil Chem. Soc.* **71,** 93–96.

24. Ikeda, I. and Klibanov, A. M. (1993) Lipase-catalyzed acylation of sugars solubilized in hydrophobic solvents by complexation. *Biotechnol. Bioeng.* **42,** 788–791.

25. Oguntimein, G. B., Erdmann, H., and Schmid, R. D. (1993) Lipase-catalyzed synthesis of sugar ester in organic solvents. *Biotechnol. Lett.* **15,** 175–180.

26. Schlotterbeck, A., Lang, S., Wray, V., and Wagner, F. (1993) Lipase-catalyzed monoacylation of fructose. *Biotechnol. Lett.* **15,** 61–64.

27. Kim, J. E., Han, J. J., and Rhee, J. S. (1998) Effect of salt hydrate pair on lipase-catalyzed regioselective monoacylation of sucrose, *Biotechnol. Bioeng.* **57,** 121–125.

14

Immobilization of Enzymes and Control of Water Activity in Low-Water Media

Properties and Applications of Celite R-640 (Celite Rods)

Lucia Gardossi

1. Introduction

Celite R-640 is a chemically inert, silica-based matrix that consists of diatomaceous earth broken up and subsequently recalcined to create porous particles with controlled pore sizes *(1)*. This type of porous Celite differs from Celite powder in its capacity to adsorb water (more than 90% of Celite weight). Recently, it has been demonstrated that Celite R-640 in organic solvent adsorbs and releases water such that water activity (a_w) is maintained constant in a reaction system within defined ranges of water concentrations *(2,3)*. Celite R-640 rods can be used not only as support for enzyme adsorption *(2)* but also as additives in reactions catalyzed by immobilized enzymes. These features make Celite R-640 a practical and simple tool for avoiding some of the problems related to the variation of the water activity/concentration occurring in biotransformations in low-water media. However, Celite R-640 cannot replace hydrated salts *(4)* in a large number of contexts because, at present, there is no established method for fixing the water activity at different values by using Celite R-640. In the following sections the properties of Celite R-640 and their applications to biotransformations in organic solvents are described.

1.1. Celite as a Support for Biocatalysts

Celite is a solid support widely used for the adsorption of biocatalysts. Many examples of improved enzyme performances by means of adsorption onto

From: *Methods in Biotechnology, Vol. 15: Enzymes in Nonaqueous Solvents: Methods and Protocols*
Edited by: E. N. Vulfson, P. J. Halling, and H. L. Holland © Humana Press Inc., Totowa, NJ

Celite powder are reported in the literature *(5–16)*. Porous Celite beads and rods are used primarily as supports for the immobilization of whole cells *(17–19)* and fungi *(1,20,21)*, and, less frequently, of isolated enzymes. For instance, chemically derivatized Celite beads (Celite R-648) have been used for the covalent immobilization of trypsin *(22)*, whereas β-glucosidase and γ-amylase have been adsorbed onto Celite R-640 and used in solvent-free environments *(23)*. Celite R-640 has been employed also for the adsorption of penicillin G amidase *(2,3)* and α-chymotrypsin *(24)*, used in biotransformations in organic solvents. The adsorption of enzymes onto Celite R-640 leads to two major advantages. First, the interfacial area of the catalyst is enlarged and the reaction rate is improved; second, the water activity of biocatalyzed synthesis in organic media is controlled so that the enzyme is provided with the water necessary for preserving its catalytic activity while undesired competing hydrolytic reactions are prevented.

1.2. Adsorption Capacity of Celite R-640

The great capacity of Celite R-640 to adsorb water is determined mainly by its porosity and its surface properties. Studies on the adsorption of water on porous materials indicate that the motion of water molecules in silica-based matrices strongly depends on pores dimensions *(25–28)*. There are two types of water in the pores *(26)*. One type of water behaves like bulk water, probably present in the central region of the pores, and the other type is water perturbed strongly by the silica surface. Consequently, pores of different sizes have different relative amounts of bound and free water *(25–28)*.

Porosimetry studies *(3)* have shown that Celite R-640 has a cumulative pore volume of 0.8 cm^3/g and that small-sized pores prevail (60% of pores have a diameter below 60 nm). Celite powder (30–80 mesh), which has a pore volume more than one order of magnitude smaller (0.06 cm^3/g) *(29)*, is able to adsorb only very small amounts of water (2 mg/g Celite) *(30)*. A further factor which differentiates the Celite R-640 from Celite powder is its surface, which faces the external phase (air or solvent). Celite rods employed in experiments described herein have a cylindrical shape and are 5 mm in height with a diameter of 3 mm. This corresponds to a considerably smaller external surface when compared to the 30- to 80-mesh Celite powder.

Indeed, when Celite rods are ground and reduced to fine powder (>200 mesh), they lose most of their ability to adsorb water *(3)*. In fact, as a result of the grinding the external surface is enlarged and there is partial destruction of the tridimensional organization of the pores, even though most of the pores still remain unbroken.

2. Materials

2.1. Control of Water Activity by Means of Celite R-640: Adsorption Isotherms in Toluene

Celite rods R-640 are from Fluka, who report the following technical information for this product: mean pore diameter 200 nm; pore volume 0.5 cm^3/g; surface area 65 m^2/g.

Other porosimetry studies *(3)*, carried out on various batches of Celite R-640, indicate that the cumulative pore volume is 0.8 cm^3/g and that 55% of pore volume is given by pores having a diameter between 20 and 60 nm. A further 5% of the volume corresponds to pores with smaller diameter, 25% to pores with a diameter between 70 and 1000 nm, and the residual 15% is given by larger pores.

Water activity and relative humidity are measured using a hygrometer (Novasina MS1) equipped with a humidity-temperature sensor (enCR-3). The sensor is calibrated at 25°C at five different a_w values (0.12, 0.33, 0.52, 0.75, 0.90) using standard salt solutions. Measurements are carried out by sealing the sensor into the open end of 5-mL glass vials, thermostatted, until constant reading. The sensor is sealed by means of a stopper that was inserted on the sensor and adapted in order to fit the vial neck and seal it.

Samples were equilibrated in an air-bath-type thermostatted orbital shaker (DARAI, Trieste, Italy). All solvents were dried over molecular sieves (4 Å). Ultrapure water was used in all experiments.

2.2. Hydration Properties of Celite R-640 in Vapor Phases

The saturated solutions are prepared at 25°C using ultrapure water and the following salts (analytical grade): $CaCl_2$ (a_w = 0.29), K_2CO_3 (a_w = 0.45), NH_4NO_3 (a_w= 0.63), NaCl (a_w = 0.75), and KCl (a_w = 0.84).

2.3. Adsorption of Enzymes on Celite R-640 and Their Use in Low-Water Media

Penicillin G amidase (PGA) from *Escherichia coli* (E. C. 3.5.1.11) from Fluka is purchased as phosphate buffer solution (0.1 *M*, pH 7.5) having an activity, assayed in water, of 20 U/mg of protein. The enzyme after lyophilization has a residual activity of 5.88 U/mg of protein and is stored at 4°C. Enzymatic activity of the native and the immobilized PGA is assayed in water by automated titration of the phenylacetic acid formed in the hydrolysis of benzylpenicillin (Aldrich). One enzymatic unit corresponds to the amount of enzyme that hydrolyses 1 µmol of benzylpenicillin in 1 min at pH7.6 at 37°C.

α-Chymotrypsin (Fluka) is a lyophilized powder with an activity of 75 U/mg and is stored at 4°C. Amberlite XAD 2, tyrosine ethyl ester, and Cbz-phenylalanine are from Sigma and methyl phenylacetate is synthesized according to **ref. *31*.**

2.4. Controlling the Hydration of Covalently Immobilized Enzymes by Means of Hydrated Celite Rods

Eupergit-PcA® (PGA immobilized on Eupergit-C®) and PGA from *E. coli* immobilised onto a copolymer of metacrylamide and *N,N'*-methylene-thylene-bis-acrylamide are from Fluka, and batches have variable water content (40–65% w/w) (A. Basso et al., unpublished results). PGA-450 (62.3% w/w of water content) is a generous gift of Boehringer Mannheim GmbH (Germany). It consists of PGA covalently immobilized on a polymer the chemical structure is not disclosed by the supplier.

3. Methods

3.1. Control of Water Activity by Means of Celite R-640: Adsorption Isotherms in Toluene

The following diagrams show how a_w varies as different amounts of water are added to Celite rods in dry toluene. The adsorption isotherms are obtained by mixing 1 mL of dry toluene, three Celite rods (95 mg), and water into 5 mL vials with screw caps and Teflon-lined septa, after which the system is equilibrated for at least 24 h in an air-bath-type thermostatted orbital shaker. After equilibration, water activity is measured by sealing the humidity sensor into the open end of the thermostatted vials, until a constant reading is obtained.

It must be emphasized that the values of water activity reported in the diagrams are always related to measurements carried out under identical conditions, namely by adding three rods (95 mg) with uniform weight (32 ± 2 mg each) and, therefore, uniform surface (1.55 mm^2/mg) (*see* **Note 1**).

Figure 1 illustrates the hydration in toluene of samples coming from three batches of Celite R-640, which, in dry toluene at the equilibrium, have a_w's of 0.54, 0.47, and 0.12 (28°C), respectively. It can be seen clearly that the properties of Celite rods are strongly related to their original hydration, which varies considerably from batch to batch and affects the shape of the adsorption isotherms in toluene, as shown in **Fig. 1.** However, porosimetry studies have demonstrated that batches having different adsorption isotherms are comparable in total pore volume and pore size distribution *(32)*.

The adsorption isotherms indicate that all three batches of Celite R-640 adsorb large amounts of water, keeping the water activity reasonably constant within defined ranges of water concentrations. The first batch adsorbs 100–500 mg water/g of Celite, maintaining the water activity close to a value of 0.78 ± 0.02. Batch 2 shows a similar plateau with values of water activity slightly lower (0.74 $\pm$ 0.04), whereas the driest batch ($a_w = 0.12$) presents a section of constant aw with values close to 0.50 (± 0.02).

It is interesting to note that batch 3 has specific features, as indicated by the adsorption isotherms relative to a broader range of water concentrations (**Fig. 2**).

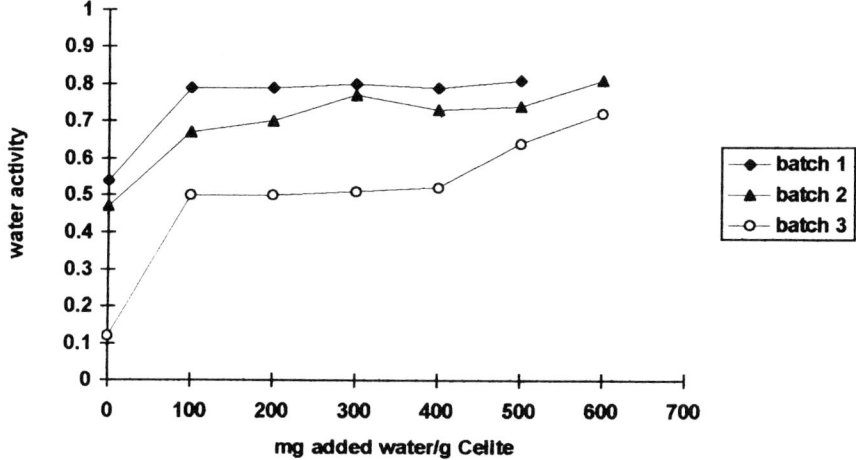

Fig. 1. Adsorption isotherms in toluene (28°C) of three different batches of Celite R-640 (equilibration time = 24 h).

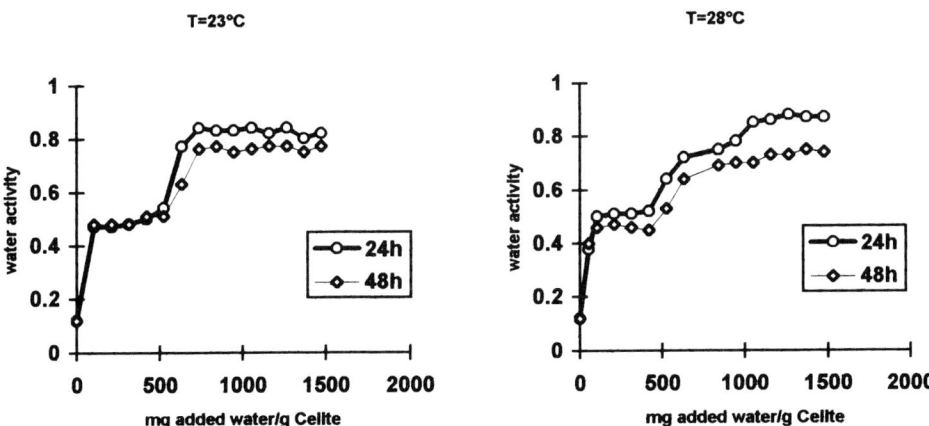

Fig. 2. Adsorption isotherms in toluene of Celite R-640 (batch 3) with an initial value of water activity of 0.12. Samples were equilibrated at 23°C and 28°C for 24 and 48 h.

The profiles indicate the existence of a clearly defined range of water concentrations (100–400 mg/g Celite) with constant a_w (0.50 ± 0.02 at 28°C and 0.48 ± 0.02 at 23°C). At water concentrations between 700 and 1400 mg/g, a second constant section (a_w = 0.82 ± 0.02) can be observed at 23°C. At 28°C, the Celite shows poor capacity of buffering the a_w at water concentrations above 500 mg/g of Celite. However, after 48 h of equilibration, the a_w values undergo a decrement and a second section of the curve, having constant a_w, becomes more defined.

Because Celite rods are hygroscopic and adsorb moisture from the atmosphere their properties may change during the storage. For instance, storing batch 3 over P_2O_5 has proved effective in preserving the dryness of the rods for at least 1 yr by paying attention to avoid any prolonged exposure to the atmosphere humidity, whereas batch 3 after 1 yr of storage but exposed to the atmosphere humidity gives an initial a_w of 0.5 in toluene. By drying the hydrated Celite first at 100°C for 16 h in an oven, and then over P_2O_5 for 24 h, it is possible to remove part of the adsorbed moisture so that the initial a_w in dry toluene decreases to 0.16. More importantly, by drying different batches of Celite with various degrees of hydration, it is possible to obtain Celite rods with homogeneous properties. Adsorption isotherms in **Fig. 3** show that after treatment, the batches have comparable behaviors and they are characterized by a single broad plateau of constant a_w around 0.70 ± 0.02, corresponding to a range of water concentrations of 100–700 mg water/g Celite.

Figure 4 illustrates the adsorption isotherms of batch 3 (after drying) at different equilibration times. In the range of concentrations of 200–700 mg water/g Celite, the system reaches the equilibrium within the first 24 h, and by equilibrating the system for a longer time, the constant section is expanded to a range of 100–900 mg water/g Celite. Adsorption isotherms also demonstrate that the drying treatment is not sufficient to restore the original properties displayed by batch 3 before being exposed to hydration.

Some possible interpretations of the variation of the properties of Celite rods after a first exposure to humidity come from the observation that removing all the water adsorbed on Celite is quite difficult. Thermogravimetry has demonstrated that Celite rods, after having adsorbed 2.5% w/w of water, release at most 1.2% of the adsorbed water at 500°C *(32)*. However, temperatures far above 100°C are not suitable for drying the rods because they tend to become friable and can easily suffer mechanical damage, for instance upon stirring. These observations suggest that some water molecules bind very strongly to the silica-based Celite or deeply penetrate the rods and their removal can be accomplished only with difficulty. Once these molecules are adsorbed on the matrix, it seems that Celite loses the ability to buffer the water activity at a value close to 0.5.

3.2. Hydration Properties of Celite R-640 in Vapor Phase

A general method for performing biotransformations at known water activity consists of exposing each component of the reaction mixture to a vapor phase characterized by a constant known relative humidity *(33,34)*. In principle, the phases, once mixed, reach the equilibrium water activity.

Figures 5 and **6** show the rate of equilibration of dry Celite R-640 (batch 3, **Fig. 1**) exposed to atmospheres generated by saturated salt solutions or pure water

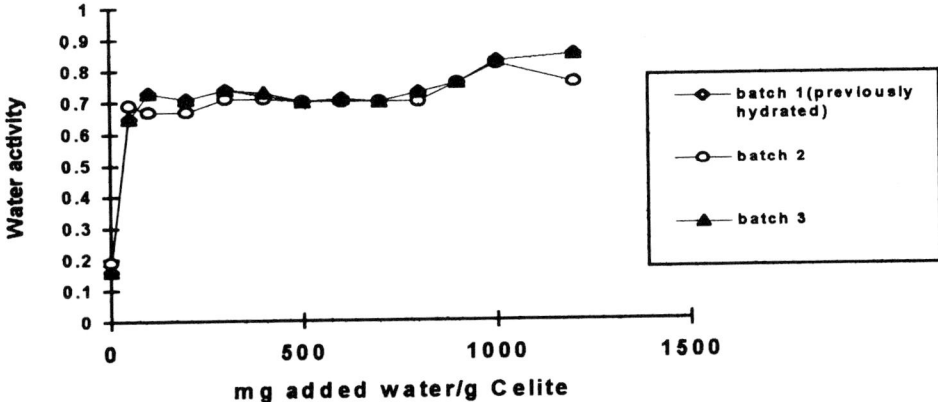

Fig. 3. Adsorption isotherms (equilibration time = 24 h) in toluene (28°C) of Celite batches dried for 16 h in an oven and then over P_2O_5. Batch 1 had been previously exposed at atmospheric humidity.

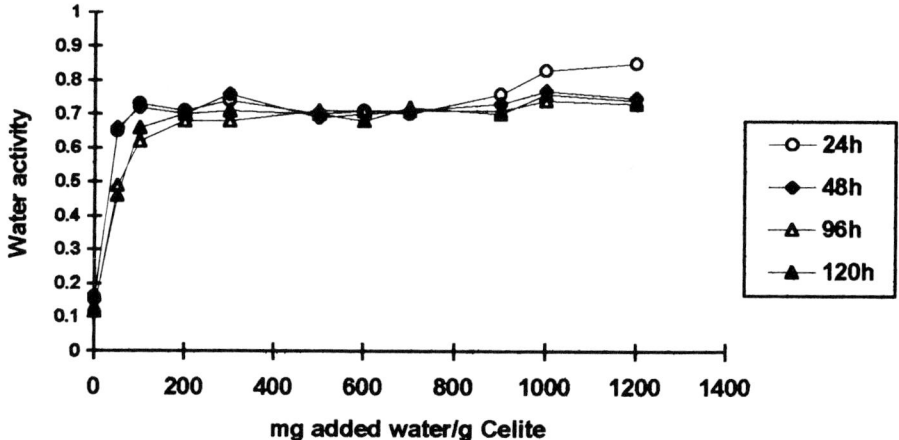

Fig. 4. Adsorption isotherms in toluene of batch 3 after drying, at different equilibration times (28°C).

inside carefully capped vessels. The adsorption isotherms are obtained by keeping three rods (95 mg) in contact with vapor phases characterized by different known relative humidities, and then weighing the Celite every 24 h. **Figure 5** indicates that most of the moisture is adsorbed within the first 24 h of exposure of the rods to the vapor phase; there are negligible increments of weight over subsequent days. In all cases, Celite R-640 has a limited capacity of adsorbing water from vapor phases, ranging from 1% w/w ($a_w = 0.32$) to 4% w/w ($a_w = 0.86$).

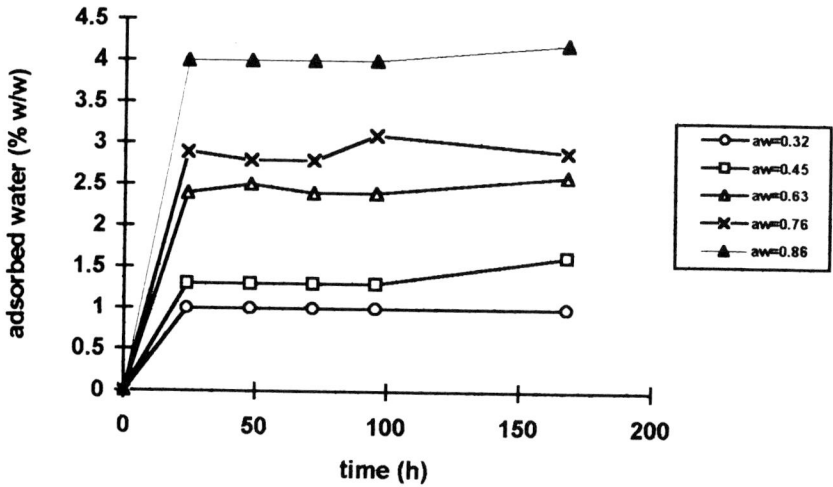

Fig. 5. Water adsorption (25°C) of Celite R-640 versus time at different a_w's generated by saturated salt solutions.

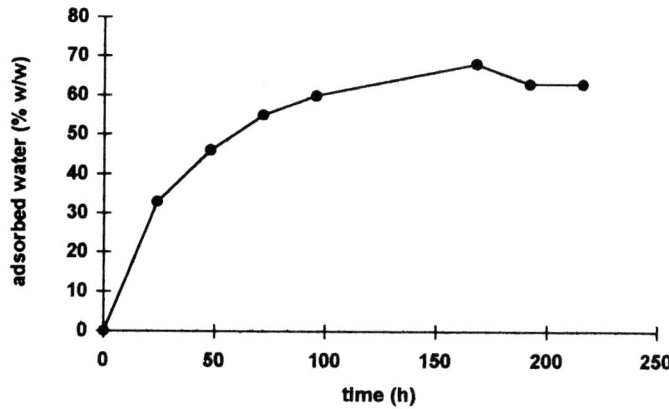

Fig. 6. Water adsorption (25°C) of Celite R-640 versus time at $a_w = 1$ generated by pure water.

The rods equilibrated at $a_w = 1$ behave differently, as they progressively adsorb water up to 60% of their weight **(Fig. 6)**, a result that is quite far, however, from the adsorption capacity exhibited by Celite R-640 in toluene (up to 120% w/w).

The possibility that the plateau reported in **Figs. 5** and **6** represents merely an apparent equilibrium and that the actual equilibrium requires extremely prolonged equilibration times cannot be excluded. Even so, the physical and chemical processes governing the adsorption of vapor water and liquid water in toluene on Celite R-640 appear very different. The difference is even clearer when Celite rods and toluene, previously equilibrated at the same value of water activity, are mixed. **Figure 7** shows how the relative humidity (RH) of the gas phase in the headspace above these mixtures decreases with time, and no equilibrium is reached even after 300 h of incubation (**Fig. 7**). For instance, in the case of the Celite equilibrated at $a_w = 0.86$, the relative humidity after 24 h of incubation is 0.75 and decreases to 0.51 after 300 h of incubation.

Most probably, upon moving from the gas phase to the apolar solvent, the variation of surface forces promotes a redistribution of the water inside the rods, so that the system reaches the equilibrium in the new environment very slowly. Therefore, the equilibration of this porous matrix in vapor phase is not advisable when Celite rods have to be used in apolar medium.

3.3. Adsorption of Enzymes on Celite R-640 and Their Use in Low-Water Media

Enzymes have been adsorbed on Celite R-640 according to different techniques. For example, Celite R-640 has been impregnated with β-glucosidase and γ-amylase together with their substrates for carrying out glycosidations in solvent-free environments *(23)*.

Enzymes adsorbed on Celite-R-640 have been used also in organic solvent. Here examples are reported, describing the advantages deriving from the ability of Celite rods to control the water activity in toluene.

3.3.1. Adsorption of PGA on Hydrated Celite R-640

The use of Celite R-640 as a support for enzyme adsorption has been investigated in a study concerning the activity of penicillin G amidase (PGA) in organic solvent. Penicillin G amidase is an enzyme working in organic media only when sufficiently hydrated, and no enzymatic reaction has been observed by using neither the native *(35)* nor the immobilized PGA in dry apolar solvent *(2,36)*.

As reported earlier, Celite R-640 has high water-binding capacity. Nevertheless, because Celite rods make it possible to carry out biotransformations in low-water media at controlled a_w, they are suitable supports for enzyme adsorption. As previously demonstrated by Halling, when a_w is maintained constant, the enzyme and support are able to take up water independently, in accordance with their individual adsorption isotherm, and water competition effects do not influence the activity/a_w profile *(37)*.

Figure 8 describes the adsorption of PGA on Celite rods (batch 3, **Fig. 2**). The lyophilized enzymatic powder dissolved in water is added to the rods previously rinsed with water (*see* **Note 2**). Excess water is removed under

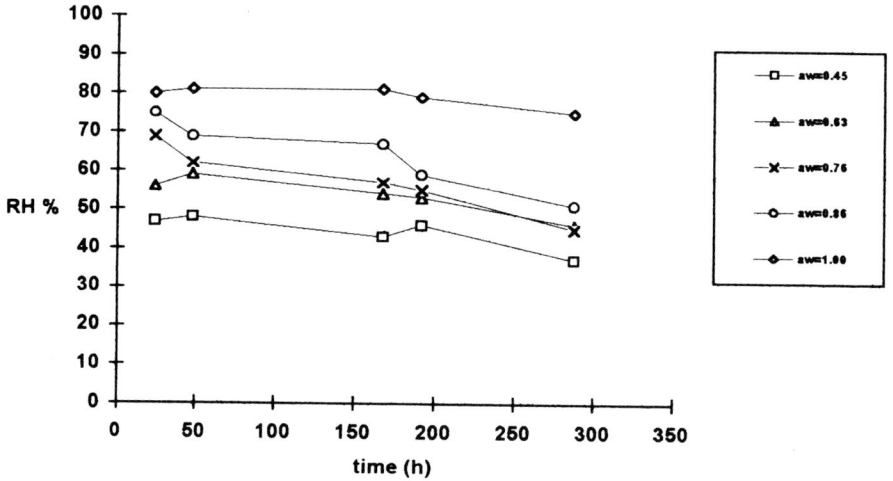

Fig. 7. Relative humidity ($T = 25°C$) of the gas phase in the headspace above toluene (1 mL) and three Celite rods previously equilibrated at the same value of water activity in air using saturated salt solutions and pure water (*see* **Figs. 5** and **6**).

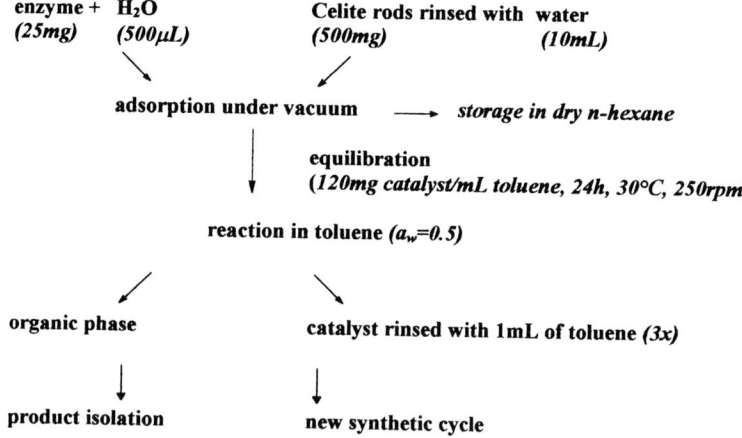

Fig. 8. Adsorption of PGA on hydrated Celite rods and its application to biotransformations in organic medium.

reduced pressure at room temperature for 4 h, obtaining a final water content of 19% (w/w). After drying, all the preparations are stored in dry hexane, a procedure effective in preserving both the hydration and the activity of the catalyst.

The enzymatic preparation (120 mg) after 24 h of equilibration in dry toluene (1 mL in a 5-mL glass vial) has an a_w of 0.50. The equilibration of the enzyme in toluene does not cause any appreciable deactivation of the enzyme *(38)*. It has been demonstrated that the contribution of the enzyme in adsorbing water is negligible in these experimental conditions *(3)* (*see* **Note 3**).

After equilibrating the system, the reaction is started by the addition of the reactants. The a_w remains constant until the complete conversion of the substrates and the low value of a_w helps to prevent competing hydrolytic reactions. In the acylation of the tyrosine derivative **1** (**Fig. 9**), PGA adsorbed on Celite catalyzes in toluene the complete conversion of the substrates, even working with equimolar concentrations of the reactants *(2)*. Therefore, the pure product is isolated simply by removing the organic phase and evaporating the solvent under reduced pressure. Afterward, the catalyst is recovered, washed with dry toluene (120 mg of adsorbed enzymes washed with 1 mL of solvent three times) and recycled. This procedure does not cause any appreciable decrement in the activity of the immobilized catalyst at least after three synthetic cycles, as indicated by initial rates (**Table 1**). Moreover, the a_w of the system is not affected, and no further hydration or equilibration is required.

The method is reproducible, because different preparations of PGA/Celite with similar water content, leading to an a_w close to 0.50 in toluene, give comparable reaction rates.

Because PGA is not active at low values of a_w, the extensive drying of the enzymatic preparation results in the complete loss of the enzymatic activity in toluene (**Table 1**). In this case, the enzymatic activity is restored after increasing the a_w of the system by adding an appropriate volume of water (100 µL/g Celite) or phosphate hydrates ($Na_2HPO_4 \cdot 12H_2O$ and $Na_2HPO_4 \cdot 7H_2O$). It is interesting to note that even in the presence of phosphate hydrates, the water activity remains close to 0.50, indicating that the properties of Celite prevail over the buffering capacity of the pair of hydrated salts, which should, in principle, maintain the water activity at a value of 0.85 at 30°C *(39)*.

It should be mentioned that the same reaction is 25 times slower when catalyzed in benzene by PGA lyophilized in the presence of Na_2HPO_4 and hydrated *in situ* ($a_w = 0.73$, 40°C) *(33)*.

3.3.2. Adsorption of Enzymes on Dry Celite R-640 and Control of Water Concentration in Reaction Producing Water

α-Chymotrypsin is a hydrolase largely employed for peptide synthesis and protection/deprotection of aminoacids *(40,41)*. Its activity in organic media is deeply affected by the degree of hydration. Therefore, controlling the distribution of water among the catalyst, the support, and the reaction medium is of great importance *(5,30)* for preserving the catalytic activity of the enzyme in low-water media.

In the enzymatic esterification of amino acids, the production of water during the reaction limits the reaction yields. Different methods are generally

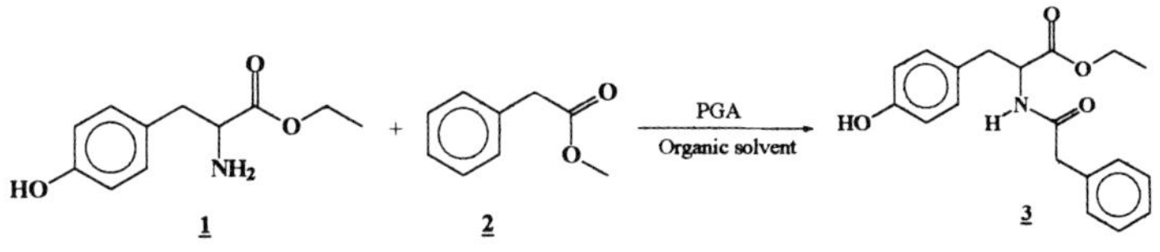

Fig. 9. Acylation of the tyrosine derivative.

Table 1
Activity in Water and in Toluene of PGA Adsorbed on Hydrated Celite R-640

Enzymatic preparation	% PGA (w/w)	% H_2O (w/w)	$a_w{}^a$ (30°C)	Activity in water (U/100 mg)	$v_0{}^b$ (µmol/ min/U)
PGA/Celite	3.4	19	0.49	4.5	0.073
PGA/Celite recycled 3X			0.49		0.070
PGA/Celite	3.7	8.5	0.27	4.9	No reaction
			0.48^c		0.077

[a]Water activity was measured after equilibrating the adsorbed enzyme in 1 mL of dry toluene for 24 h.
[b]Initial rates (reaction reported in **Fig. 9**) are calculated on the basis of the enzymatic units determined in water.
[c]a_w measured after the addition of $Na_2HPO_4 \cdot 12H_2O$ and $Na_2HPO_4 \cdot 7H_2O$ (0.2 M final concentration) and 24 h of equilibration.

employed to control the a_w or remove the formed water, such as hydrated salts *(41)*, saturated salt solutions *(42)*, additives *(43,44)*, ionic exchange resins *(45)*, silica beads *(46)*, pervaporation *(47)*, distillation, or adsorption of water *(48)*.

Because of its high water-binding capacity, Celite R-640 has been used as an alternative method for removing the excess water produced during the esterification of Cbz-phenylalanine *(24)* catalyzed by α-chymotrypsin adsorbed on Celite rods. Here, also, the Celite is employed as a support for the enzyme that is adsorbed, however, on the dry rods in toluene **(Fig. 10)**. By adding the enzymatic aqueous solution to the apolar solvent containing the Celite, a uniform coating of the aqueous phase is formed around the rods, which adsorbs the solution within 24 h of incubation at 30°C in an orbital shaker (250 rpm). After equilibration the adsorbed enzyme in toluene gives an a_w value of 0.73, which is related to the properties of the Celite batch employed, i.e., batch 1 after a drying treatment **(Fig. 3)**.

Enzymatic reactions are started by adding the reactants. The hydrated support provides the enzyme with the hydration necessary for its activity while,

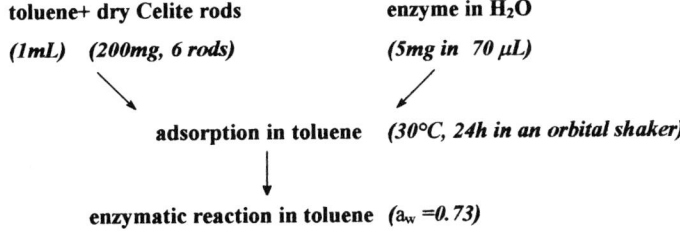

Fig. 10. Adsorption of α-chymotrypsin onto dry rods in toluene.

because of its large capacity of adsorbing water, it removes the excess water produced during the esterification. α-Chymotrypsin adsorbed on Celite R-640 catalyzes the esterification of Cbz-phenylalanine (80 mM) in toluene working in the presence of 0.25 M of methanol. The complete conversion is achieved in 60 h *(24)*. (*see* **Notes 4** and **5**).

3.4. Controlling the Hydration of Covalently Immobilized Enzymes by Means of Hydrated Celite Rods

The properties of Celite R-640 have been applied also in reactions catalyzed by immobilized enzymes in organic medium. In this case, the rods are used as a type of additive that is able to release or adsorb water in a controlled way, such that the water activity of the system is adjusted in accordance to the requirements of the catalyst and the thermodynamics of the reaction.

3.4.1. Rehydration of Dry Immobilized Enzymes by Means of Hydrated Celite Rods

Commercially immobilized enzymes are often available in hydrated form, and the water content may vary significantly among batches. In some cases, they are not suitable for biotransformations in organic media, because an excess of water induces the formation of clusters in apolar solvent, causing a nonhomogeneous dispersion of the catalyst. The water can also affect the thermodynamic equilibrium of the reaction. Halling and Moore recently reported that washing immobilized biocatalysts with anhydrous organic solvents *(49,50)* is a very efficient method for dehydrating different types of immobilized enzyme and this technique does not cause appreciable changes of enzymatic activity. This method has been used also to dehydrate PGA covalently immobilized onto two polymers, Eupergit® and a copolymer of metacrylamide and *N,N'*-methylene-thylene-*bis*-acrylamide *(32)*. Both enzymes are commercially available as hydrated beads that form aggregates in toluene. A large part of the water can be removed by washing 150 mg of the catalyst with 1 mL of 2-propanol three times and by removing the solvent after centrifugation.

Rinsing the catalyst with 2-propanol does not cause any decrement of enzymatic activity (assayed in water) within the first hour of storage, whereas storing the washed enzyme for longer periods of time leads to the progressive decrease of the enzymatic power.

Table 2 reports the activity in toluene of these two enzymes employed straight after the dehydration. Initial rates indicate that the final hydration of the rinsed catalyst is insufficient to maintain the enzymatic activity in toluene. Nevertheless, the catalytic power is completely restored by the presence of Celite rods (95 mg/mL). The rods are hydrated in toluene by adding amounts of water determined on the basis of the adsorption properties of the Celite batch employed (batch 3), so that after 24 h of equilibration, the system has a water activity of 0.5, consistent with the adsorption isotherms reported in **Fig. 2**. The immobilized enzyme is then added to the medium and the hydrated rods provide the catalyst with the water necessary for its activity by controlling the a_w of the system. Kinetic data indicate also that Eupergit-PcA® and PGA/Celite (**Table 1**) have comparable activity in toluene.

3.4.2. Dehydration and Storage of Immobilized PGA by Using Dry Celite R-640

Hydrated enzymatic batches, once exposed to the atmosphere, may undergo microbial contamination, loss of moisture, and the detrimental action of further factors contributing to a progressive reduction of the catalytic activity. Therefore, dehydration and storage of immobilized enzymes are crucial for preserving their catalytic activity. Apolar solvents have proved to be suitable media for the storage of enzymes *(51)*. The adsorption properties of Celite R-640, combined with the use of apolar organic solvents, have been exploited in the pretreatment and storage of immobilized enzymes.

Figure 11 illustrates the dehydration of PGA-450 (covalently immobilized PGA having 62.3% of water content) by means of dry Celite rods in *n*-hexane or petroleum ether. The enzyme and the rods are gently mixed together so that the excess water is adsorbed by the Celite, which is removed after 7 d. The enzyme is stored in the same apolar solvent and samples are taken when required. The batches can be prepared on gram scale and there is no decrement in activity for at least 4 mo. Most of the organic solvent used for the storage can be removed easily from the enzymatic samples at room temperature and atmospheric pressure without causing any detrimental effect to the catalyst. The procedure is particularly fast in the case of petroleum ether because of its high volatility (*see* **Note 6**).

Table 3 compares the activity in water of the immobilized enzyme before and after the dehydration and storage in apolar solvent. The data confirm that there is no decrement of catalytic power resulting from the storage. Moreover,

Table 2
Initial Rates of the Acylation of 1 in Toluene (Fig. 9) Catalyzed
by Immobilized PGA Dehydrated with 2-Propanol

Enzyme	Treatment	$a_w{}^a$ (30°C)	$v_0{}^b$ (μmol/min/U)
Eupergit-PcA®	Washed with 2-propanol	0.35	No reaction
	Washed with 2-propanol + 3 Celite rods + 25 μL H_2O^c	0.50	0.073
PGA-polyacrylamide	Washed with 2-propanol	0.22	No reaction
	Washed with 2-propanol + 3 Celite rods + 25 μL H_2O^c	0.50	0.012

^{a} Water activity of the reaction system after equilibration in the presence of the immobilized enzyme.
^{b} Initial rates are calculated on the basis of the enzymatic units determined in water.
^{c} Reaction volume = 1 mL.

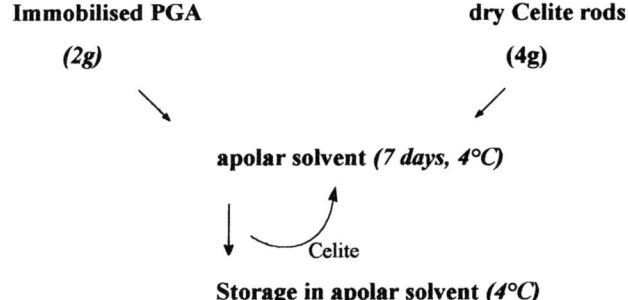

Fig. 11. Dehydration of immobilized PGA by means of dry Celite rods.

the dehydration treatment is effective in removing a large percentage of the water, as indicated by the enhancement of the number of enzymatic units per gram of catalyst. The dehydrated PGA-450 is highly active also in toluene and catalyzes the acylation of tyrosine ethyl ester (**Fig. 9**) with an initial rate of 0.467 μmol/min/U. (*See* **Note 7**.)

It should be noted that Eupergit-PcA®, dehydrated and stored following the same procedure, gives an initial rate of 0.083 μmol/min/U, which is comparable to the activity previously demonstrated by this enzyme (*see* **Table 2**).

3.4.3. Controlling the Hydration of Immobilized PGA in Solvents Having Different Polarity and Water Solubility

The two methods described in **Subheadings 3.4.1.** and **3.4.2.** have been applied, combined, to the acylation of tyrosine ethyl ester (**Fig. 9**) carried out in water-miscible — or partially miscible — solvents (*36*).

Table 3
Activity of PGA-450 Before and After the Dehydration and the Storage in Apolar Solvents

Solvent of storage	Time of storage	a_w in toluene[a] (30°C)	Enzymatic activity in water[b] (U/g)
—	—	>0.95	201
n-Hexane	4 mo	0.85	280
Petroleum ether	1 mo	0.73	290

[a]Water activity measured by equilibrating 62.5 mg of immobilized PGA in 1 mL of toluene for 24 h.
[b]One enzymatic unit corresponds to the amount of enzyme that hydrolyzes 1 µmol of benzylpenicillin in 1 min at pH 8.0 at 37°C.

PGA-450 (62.5 mg, previously dehydrated with Celite rods in apolar solvent) suspended in toluene and equilibrated for 24 h gives an a_w value of 0.73 (**Table 4**) and catalyzes the complete acylation of the substrate **1 (Fig. 9)** in 3 h. The reaction is slower in methylene chloride, although 90% of conversion is achieved in 1 d. The enzyme becomes considerably less efficient, moving to *tert*-amyl alcohol and acetonitrile, whereas it is not active at all in tetrahydrofurane and pyridine.

The very poor activity of the enzyme can be partially explained by the low a_w values measured in these solvents. The a_w can be increased and adjusted at a value close to 0.7 by equilibrating 1 mL of solvent for 24 h with 95 mg of Celite (batch 1, **Fig. 3**) hydrated with 30 µL of water **(Fig. 12)**. In the case of pyridine, the Celite rods are not able to control the hydration of the reaction medium. This could be ascribable to the ability of pyridine to strip and solubilize also the water adsorbed onto the rods.

The increase of the a_w has a considerable positive effect on the reactions performed in *tert*-amyl alcohol and acetonitrile. Furthermore, the presence of the hydrated rods over the course of the reaction prevents the excessive dehydration of the enzyme by controlling the partition of the water between the medium and the catalyst. However, there are effects on the enzyme activity that are intrinsic to the solvents that cannot be avoided. Consequently, tetrahydrofurane and pyridine seem to be incompatible with PGA activity, regardless of the a_w value.

4. Notes

1. The distribution of the same weight of Celite among a different number of rods translates into the variation of the external surface of the matrix. Because the adsorption properties of Celite R-640 are strongly related to its surface, samples

Table 4
Acylation of 1 Catalyzed by PGA-450 Previously Dehydrated in *n*-Hexane with Celite R-640

Reaction medium	Water solubility[a]	Log P^b (24h)	ε^c	a_w (30°C)	Conversion
Toluene	0.046[d]	2.5	2.38	0.73	99.2% (3 h)
Methylene chloride	0.42[d]	1.3	8.93	0.89	90.6%
Tert-amyl alcohol	10.0[d]	1.45	5.8	0.48	4.7%
3 Celite rods (95 mg) + 30 µL water				0.68	58.4%
Acetonitrile	Miscible	−0.3	35.9	0.57	7.8%
3 Celite rods (95 mg) + 30 µL water				0.67	36.4%
Tetrahydrofurane	Miscible	0.49	7.52	0.43	No reaction
3 Celite rods (95 mg) + 30 µL water				0.67	1.14%
Pyridine	Miscible	0.71	12.9	0.28	No reaction
3 Celite rods (95 mg) + 30 µL water		0.34			No reaction
3 Celite rods (95 mg) + 100 µL water		0.53			No reaction

[a] Water solubility taken from **ref. 52**.
[b] Logarithm of partition coefficients taken from **ref. 53**.
[c] Dielectric constant taken from **refs. 53** and **54**.
[d] Percentage (w/w) of water dissolving in a given solvent at 20°C.

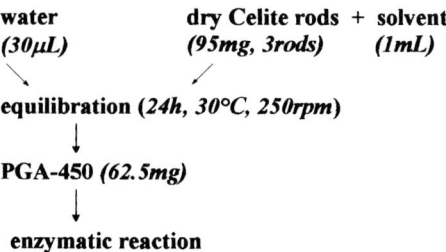

Fig. 12. Equilibration of water-miscible solvents by means of hydrated Celite rods.

of Celite having different surfaces give values of a_w that do not fit in the curves reported here. Further investigations are required to verify whether the size of the surface of the rods can be quantitatively correlated to different "buffering" properties of Celite R-640.

2. The activity of the enzyme is not affected by the use of phosphate buffer (0.1 M, pH 7.0) instead of ultrapure water, for dissolving the enzyme *(3)*.

3. The adsorption isotherm of PGA indicated that 1 g of lyophilized PGA adsorbs up to 300 mg of water and a standard reaction employs about 5 mg of PGA adsorbed on 95 mg of Celite *(3)*.

4. The volume of water that has to be added to the dry rods for the adsorption is related to the solubility of the enzyme in water. Therefore, in order to operate within the range of water concentrations that corresponds to the buffering capacity of Celite R-640, the enzymatic loading can be modified by calculating the proper amount of support on the basis of the adsorption isotherms reported here.

5. Enzymes have been adsorbed also by adding directly an enzymatic aqueous solution to the dry rods in the absence of organic solvent *(3)*. However, this method, in the case of PGA, gives poorer results (activity 40% lower) probably resulting from the nonhomogeneous dispersion of the catalysts on the surface of the support.

6. The final water content of the enzymes cannot be predicted easily, because it is related to many different factors, like the type of solvent, the original hydration of the enzyme, the specific properties of the batch of Celite used, and the amount of enzyme employed. Moreover, because the enzyme is immobilized onto polymers, their own capacity of adsorbing water should be also taken into account to evaluate the partition of water among the polymer, the Celite, and the solvent.

7. When no storage is required, the catalyst can be dehydrated in the reaction medium itself by adding dry Celite rods and equilibrating for 24 h before the addition of the reactants.

Acknowledgments

The author is grateful to those people who have given their precious contribution to the development of these methods, especially Paolo Linda, Cynthia Ebert, Alessandra Basso, and Luigi De Martin.

References

1. Livingston, A. G. (1991) Biodegradation of 3,4-dichloroaniline in a fluidized bed bioreactor and a steady-state biofilm kinetic model. *Biotechnol. Bioeng.* **38**, 260–272.
2. Ebert, C., Gardossi, L., and Linda, P. (1998) Activity of immobilised penicillin amidase in toluene at controlled water activity. *J. Mol. Catal. B: Enzymatic* **5**, 241–244.
3. De Martin, L., Ebert, C., Garau, G., Gardossi, L., and Linda, P. (1999) Penicillin G amidase in low-water media: immobilisation and control of water activity by means of Celite rods. *J. Mol. Catal. B: Enzymatic* **6**, 437–445.
4. Kvittingen, L., Sjursnes, B., Anthonsen, T., and Halling, P. J. (1992) Use of salt hydrates to buffer optimal water level during lipase catalysed synthesis in organic media: a practical procedure for organic chemists. *Tetrahedron* **48**, 2793–2802.
5. Adlercreutz, P. (1991) On the importance of the support material for enzymatic synthesis in organic media. *Eur. J. Biochem.* **199**, 609–614.
6. Valiverty, R., Johnston, G. A., Suckling, C. J., and Halling, P. J. (1991) Solvent effects on biocatysis in organic systems: equilibrium position and rates of lipase catalyzed esterification. *Biotechnol. Bioeng.* **38**, 1137–1143.
7. Kaga, H., Siegmund, B., Neufellner, E., Faber, K., and Paltauf, F. (1994) Stabilization of candida lipase against acetaldehyde by adsorption onto Celite. *Biotechnol. Tech.* **8**, 369–374.
8. Furukawa, S. Y. and Kawakami, K. (1998) Characterization of Candida rugosa lipase entrapped into organically modified silicates in esterification of methanol with butyric acid. *J. Ferment. Bioeng.* **85**, 240–242.
9. Bisht, K. S., Henderson, L. A., Gross, R. A., Kaplan, D. L., and Swift, G. (1997) Enzyme-catalyzed ring-opening polymerization of ω-pentadecalactone. *Macromolecules* **30**, 2705–2711.
10. Johansson, A., Mosbach, K., and Mansson, M. O. (1995) Horse liver alcohol dehydrogenase can accept NADP+ as coenzyme in high concentrations of acetonitrile. *Eur. J. Biochem.* **227**, 551–555.
11. Adlercreutz, P. (1993) Activation of enzymes in organic media at low water activity by polyols and saccharides. *Biochim. Biophys. Acta* **1163**, 144–148.
12. Hanley, A. B., Furniss, C. S. M., Kwiatkowska, K. A., and Mackie, A. R. (1991) The manipulation of DNA with restriction enzymes in low water systems. *Biochim. Biophys. Acta* **1074**, 40–44.
13. Kim, J. and Kim, B. G. (1996) Effect of the hydration state of supports before lyophilization on subtilisin-A activity in organic media. *Biotechnol. Bioeng.* **50**, 687–692.
14. Vyazmensky, M. and Geresh, S. (1998) Substrate specificity and product stereochemistry in the dehalogenation of 2-haloacids with the crude enzyme preparation from Pseudomonas putida. *Enzyme Microb. Technol.* **22**, 323–328.
15. Capellas, M., Benaiges, M. D., Caminal, G., Gonzalez, G., Lopez-Santìn, J., and Clapés, P. (1996) Enzymatic synthesis of a CCK-8 tripeptide fragment in organic media. *Biotechnol. Bioeng.* **50**, 700–708.
16. Lòpez-Fandino, R., Gill, I., and Vulfson, E. N. (1994) Enzymatic catalysis in heterogenous mixtures of substrates: the role of the liquid phase and the effect of adjuvants. *Biotechnol. Bioeng.* **43**, 1016–1023.

17. El-Sayed, A. H., Mahmoud, W. M., and Coughlin, R. W. (1990) Comparative study of production of dextransucrase and dextran by cells of Leuconostoc mesenteroides immobilized on Celite and calcium alginate beads. *Biotechnol. Bioeng.* **36,** 83–91.
18. Wang, S. D. and Wang, D. I. C. (1989) Cell adsorption and local accumulation of extracellular polysaccharide in an immobilized Acinetobacter calcoaceticus. *Biotechnol. Bioeng.* **34,** 1261–1267.
19. Chun, G. T. and Agathos, S. N. (1991) Comparative studies of physiological and environmental effects on the production of cyclosporin A in suspended and immobilized cells of Tolypocladium inflatum. *Biotechnol. Bioeng.* **37,** 256–265.
20. Keshavarz, T., Eglin, R., Walker, E., Bucke, C., Holt, G., Bull, A. T., et al. (1990) The large scale immobilization of penicillium chrysogenum: batch and continuous operation in air-lift reactor. *Biotechnol. Bioeng.* **36,** 763–770.
21. Chun, G.-T. and Agathos, S. N. (1993) Dynamic response of immobilized cells to pulse addition of L-valine in cyclosporin A biosynthesis. *J. Biotechnol.* **27,** 283–294.
22. Huang, X. L., Catignani, G. L., and Swaisgood, H. E. (1997) Comparison of the properties of trypsin immobilized on 2 Celite™ derivatives. *J. Biotechnol.* **53,** 21–27.
23. Gelo-Pujic, M., Guibé-Jampel, E., and Loupy, A. (1997) Enzymatic glycosidations in dry media on mineral supports. *Tetrahedron* **53,** 17,247–17,252.
24. Basso, A., DeMartin, L., Ebert, C., Gardossi, L., and Linda, P. (2000) High isolated in thermodynamically controlled peptide synthesis in toluene catalysed by thermolysin adsorbed on Celite R-640. *Chem. Commun.* 467,468.
25. Clifford, J. (1975) Properties of water in capillaries and thin films, in *Water: A Comprehensive Treatise*, (Franks, F., ed.), Plenum, New York, pp. 75–133.
26. Allen, S. G., Stephenson, P. C. L., and Strange, J. H. (1998) Internal surfaces of porous media studied by nuclear magnetic resonance cryoporometry. *J. Chem. Phys.* **108,** 8195–8198.
27. Overloop, K. and Van Gerven, L. (1993), Exchange and cross-relaxation in adsorbed water. *J. Magn. Reson. A* **101,** 147–156.
28. Takamuku, T., Yamagami, M., Wakita, H., Masuda, Y., and Yamaguchi, T. (1997) Thermal property, structure and dynamics of supercooled water in porous silica by calorimetry, neutron scattering and NMR relaxation. *J. Phys. Chem. B* **101,** 5730–5739.
29. Barros, R. J., Wehtje, E., and Adlercreutz P. (1998) Mass transfer studies on immobilized a-chymotrypsin biocatalysts prepared by deposition for use in organic medium. *Biotechnol. Bioeng.* **59,** 364–373.
30. Reslow, M., Adlercreutz, P., and Mattiasson, B. (1988) On the importance of the support material for bioorganic synthesis. *Eur. J. Biochem.* **172,** 573–578.
31. Baldaro, E., D'Arrigo, P., Pedrocchi-Fantoni, G., Rosell, C. M., Servi, S., and Terreni M. (1993) Pen G acylase catalyzed resolution of phenylacetate esters of secondary alcohols. *Tetrahedron: Asymmetry* **4,** 1031–1034.
32. Basso, A., DeMartin, L., Ebert, C., Gardossi, L., and Linda, P. (2000) Controlling the hydration of covalently immobilised penicillin G amidase in low-water medium: properties and use of Celite R-640. *J. Mol. Catal. B: Enzymatic* **8,** 245–253.

33. Halling P. J. (1994) Thermodynamic predictions for biocatalysis in non conventional media: theory, tests, and recommendations for experimental design and analysis. *Enzyme Microb. Technol.* **16,** 178–206.
34. Zacharis, E., Omar, I. C., Partridge, J., Robb, D. A., and Halling, P. J. (1997) Selection of salt hydrate pairs for use in water control in enzyme catalysis in organic solvents. *Biotechnol. Bioeng.* **55,** 367–374.
35. Ebert, C., Gardossi, L., and Linda, P. (1996) Control of enzyme hydration in penicillin amidase catalysed synthesis of amide bond. *Tetrahedron Lett,* **37,** 9377–9380.
36. Basso, A., DeMartin, L., Ebert, C., Gardossi, L., Linda, P., and Zlatev, V. Activity of covalently immobilised PGA in water miscible solvents at controlled a_w. *J. Mol. Catal. B: Enzymatic,* in press.
37. Valivety, R. H., Halling, P. J., Peilow, A. D., and Macrae, A. R. (1994) Relationship between water activity and catalytic activity of lipases in organic media. *Eur. J. Biochem.* **222,** 461–466.
38. Parker, M. C., Moore, B. D., and Blaker A. J. (1994) In situ hydration of enzymes in non polar organic media can increase the catalytic rate. *Biocatalysis* **10,** 269–277.
39. Halling, P. J. (1992) Salt hydrates for water activity control with biocatalysts in organic media. *Biotechnol. Tech.* **6,** 271–276.
40. Clapés, P. and Adlercreutz, P. (1991) Substrate specificity of α-chymotrypsin-catalysed esterification in organic media. *Biochim. Biophys. Acta* **1118,** 70–76.
41. Kuhl, P. and Halling, P. J. (1991) Salt hydrates buffer water activity during chymotrypsin-catalysed peptide synthesis. *Biochim. Biophys. Acta* **1078,** 326–328.
42. Wehtje, E., Svensson, I., Adlercreutz, P., and Mattiasson, B., (1993) Continuous control of water activity during biocatalysis in organic media. *Biotechnol. Tech.* **7,** 873–878.
43. Otamiri, M., Adlercreutz, P., and Mattiasson, B. (1994) A differential scanning calorimetric study of chymotrypsin in the presence of added polymers. *Biotechnol. Bioeng.* **44,** 73–78.
44. Hyun, C. K., Kim, J. H., and Ryu, D. D. Y. (1993) Enhancement effect of water activity on enzymatic synthesis of cephalexin. *Biotechnol. Bioeng.* **42,** 800–806.
45. Mensah, P., Gainer, J. L., and Carta, G. (1996) Adsorptive control of water in esterification with immobilized enzymes: I. Batch reactor behaviour. *Biotechnol. Bioeng.* **20,** 434–444.
46. Goldberg, M., Parvaresh, F., Thomas, D., and Legoy, M. D. (1988) Enzymatic synthesis with continuous measurement of water activity. *Biochim. Biophys. Acta* **957,** 359–362.
47. Kwon, S. J., Song, K. M., Hong, W. H., and Rhee, J. S. (1994) Removal of water produced from lipase-catalyzed esterification in roganic solvent by pervaporation. *Biotechnol. Bioeng.* **46,** 393–395.
48. Bloomer, S., Adlecreutz, P., and Mattianson, B. (1992) Facile synthesis of fatty acid esters in high yields. *Enzyme. Microb. Technol.* **14,** 546–552.
49. Partridge, J., Hutcheon, G. A., Moore, B. D., and Halling, P. J. (1996) Exploiting hydration hysteresis for high activity of cross-linked subtilisin crystals in acetonitrile. *J. Am. Chem. Soc.* **118,** 12,873–12,877.

50. Partridge, J., Halling, P. J., and Moore, B. D. (1998) Practical route to high activity enzyme preparations for synthesis in organic media. *Chem. Commun.* 841,842.
51. Kaul, R. and Mattiasson, B. (1993) Improving the shelf life of enzymes by storage under anhydrous apolar solvent. *Biotechnol. Tech.* **7,** 585–590.
52. Snyder, L. R. and Kirkland, J. J. (1979) *Introduction to Modern Liquid Chromatography.* Wiley, New York, pp. 248–250.
53. Gorman, L. S. and Dordick, J. S. (1992) Organic solvents strip water off enzymes. *Biotechnol. Bioeng.* **39,** 392–397.
54. Reichardt, C. (1988) *Solvents and Solvent Effects in Organic Chemistry.* VCH, Weinheim.

15

Enzyme Activity and Enantioselectivity Measurements in Organic Media

Amélie Ducret, Michael Trani, and Robert Lortie

1. Introduction

The advent of nonaqueous enzymology has raised many new questions both from theoretical and practical points of view. The influence of the nature of the solvent and the presence of water on enzyme activity and stereoselectivity has been studied extensively *(1–4)*. New tools and methods had to be developed to undertake these studies. In the present chapter, we will describe some methods that were developed in our laboratory to measure the rates of enzymatic reactions in organic media, mostly lipase-catalyzed enantioselective esterification of ibuprofen with fatty alcohols. These methods are based on high-performance liquid chromatography (HPLC) analyses and involve working at constant water activity, the importance of which is emphasized in Chapter 11.

2. Materials

2.1. Reactions at Controlled a_w

The reactions are run at 22°C in 100-mL tissue culture bottles stoppered with a neoprene stopper in which a truncated HPLC autosampler vial is inserted so that samples can be taken with a syringe through a septum (**Fig. 1**). The reaction itself takes place in a 20-mL scintillation vial, surrounded by smaller vials containing the appropriate saturated salt solution or molecular sieves (3 Å). When using molecular sieves, the value of the water activity depends on the rate of the reaction. It is better to measure a_w in a test reaction using a hygrometer such as the Hanna Instruments (Padova, Italy) HI 9065. In such a case, the probe replaces the stopper and is held in place with many layers of parafilm.

From: *Methods in Biotechnology, Vol. 15: Enzymes in Nonaqueous Solvents: Methods and Protocols*
Edited by: E. N. Vulfson, P. J. Halling, and H. L. Holland © Humana Press Inc., Totowa, NJ

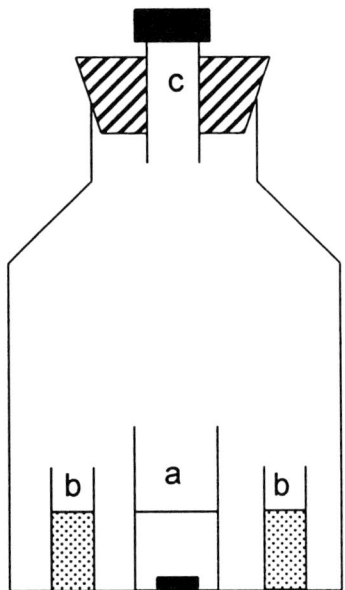

Fig. 1. Experimental setup for reactions in controlled water-activity atmosphere: (a) reaction with stirring bar; (b) vials containing saturated salt solutions or molecular sieve, (c) truncated HPLC vial with screwcap and septum.

2.2. HPLC Analyses

All HPLC analyses were performed using a Waters liquid chromatography system purchased from Waters Scientific (Mississauga, Ontario, Canada). For the measurements of enzyme activity and enantioselectivity in the resolution of ibuprofen presented in this chapter, various types of column were used:

- A reverse-phase column (CSC-S, Spherisorb ODS-2, 5 μm (25 cm × 4.6 mm), CSC, Montréal, Québec, Canada), allowing the separation of ibuprofen, alcohol, and ester.
- Two chiral columns (Chiralcel OD [25 cm × 4.6 mm] or Chiralcel OJ [25 cm × 4.6 mm], Chiral Technologies Inc., Exton, PA, USA), allowing the separation of both enantiomers of ibuprofen and ibuprofen fatty esters, respectively.

During the analysis, the column temperature was kept constant using a Croco-Cil™ column heater (Cluzeau Info-Labo, Sainte-Foy-la-Grande, France) which permits temperature control in the range from –3°C to 99°C as per the manufacturer's claims.

Depending on the method used, the different constituents were detected either with a ultraviolet (UV) spectrophotometer, a refractive index (RI) detector, a photodiode array (PDA) detector (all purchased from Waters Scientific), or an optical rotation (OR) detector (Chiramonitor 2000 Optical Rotation detector purchased from Applied Chromatography Systems Ltd. Cheshire, UK). When different detectors are connected in series, careful attention must be paid to the order of connection; in particular, the Chiramonitor 2000 has to be connected last to avoid excessive back pressure that may damage the flow cell, which is rated to 200 psi working pressure.

3. Methods
3.1. Reactions at Controlled a_w

All the solvents used were previously equilibrated at the proper water activity in a closed chamber containing saturated salt solutions for at least 1 d for apolar solvents such as heptane and up to 1 wk for polar solvents such as acetonitrile. (Equilibration was monitored by coulometric Karl Fischer water analysis on samples.) The reactants are dissolved in 15 mL of the appropriate solvent and allowed to equilibrate for at least 24 h at the proper water activity (*see* **Note 1**).

Various quantities of immobilized enzyme (10–750 mg), depending on the rate of reaction, are equilibrated separately at the same a_w and added to initiate the reaction. The enzyme is maintained in suspension with magnetic stirring. It is important to adjust the rate of reaction so that the water production is not too rapid and the saturated salt solution or molecular sieve can maintain a constant water activity.

Aliquots of the reaction mixture were periodically removed through the septum (without opening the reaction vessel so as not to modify the water activity), filtered through a 0.45-µm pore size filter from Millipore to remove the enzyme and kept at 4°C before analysis (*see* **Notes 1** and **2**).

3.2. HPLC Analyses

Many HPLC analyses based on different principles have been developed in our laboratory for the monitoring of the esterification of ibuprofen and similar products. It is not the aim of this chapter to emphasize typical HPLC details such as the mobile phases used, flow rates, and so forth that can be easily retrieved in the appropriate references if needed. Thus, discussion is essentially based on general observations and problems that we have met and that can be encountered during the HPLC analysis of other chiral compounds.

Each system used has been calibrated on a regular basis (generally every week) using at least five standard solutions of different concentrations of substrates and products. Calibration equations were accepted depending on the correlation coefficients obtained (generally higher than 0.99). It was verified that the obtained equations did not change significantly between each calibration.

Before injection, all samples were diluted in the appropriate mobile phase (*see* **Note 3**).

3.2.1. Use of a Chiral Column Connected with Conventional Nonchiral Detector(s)

The use of a chiral stationary phase permits the physical separation of both enantiomers of a chiral compound (*see* **Note 4**). If all the compounds present in the reaction media can be separated with a chiral column, only one analysis has to be performed. If not, another analysis has to be used so that all species can be quantified. For example, performing the analysis on a Chiralcel OD column allows the separation of the unresolved ester, the alcohol, and the (*R*)- and (*S*)-ibuprofen (*5*). However, at very a high concentration of (*R,S*)-ibuprofen and low concentrations in alcohol, kinetics cannot be monitored accurately. The concentrations of both ester enantiomers have to be measured to obtain reliable results (it is always more precise to measure the appearance of a product rather than the disappearance of a substrate). Thus, a second HPLC method has been developed using a Chiralcel OJ column allowing the separation of both enantiomers of the ester formed (*6*). Under our conditions, ibuprofen remains tightly bound to the column and is not eluted (*see* **Note 5**). The method has not been modified in order to elute ibuprofen to keep analysis times compatible with many tens of analyses required in a complete kinetic study. The analysis performed on the Chiralcel OJ column only gives the molar fraction of both enantiomers of ibuprofen ester, and, consequently, a second short analysis (less than 15 min) has been performed on a C_{18} reverse-phase column in order to quantify the total concentration of ibuprofen, alcohol, and ester. The combination of the obtained results leads to the concentrations of all the components with high accuracy.

3.2.2. Use of a Conventional Nonchiral Reverse-Phase Column Connected with a Chiral Detector

The use of an achiral column does not permit the physical separation of a chiral compound. Thus, it is necessary either to add a chiral additive in the mobile phase in order to form diastereoisomers that can be separated on a nonchiral stationary phase or to use a chiral detector that can detect the presence of enantiomers (like a polarimetric detector, which indicates the specific rotation of a product as a function of the retention time). It is this last technique that has been used in our laboratory. The area under each peak given by the Chiramonitor is a measure of the amplitude of the optical rotation and is, therefore, proportional to the amount and to the rotatory power of the product. When the product is optically pure and has a positive rotatory power, the obtained peak is positive and the area is directly proportional to its quantity. Likewise, with the same amount of the antipodal enantiomer, the peak will have the same area but opposite sign. The positive or negative peak obtained for a mixture of both enantiomers will represent the excess of the major enantiomer present. In the particular case of a racemic mixture, no peak will be detected.

For our experiments, a C_{18} reverse-phase column was connected in series with a refractive index detector (RI) and the Chiramonitor 2000 (OR). Although no physical separation of both enantiomers is obtained, it is possible to quantify the concentration C_1 and C_2 of each enantiomer through simple mathematics using the following equations:

$$C_1 = (C + x)/2 \qquad (1)$$

$$C_2 = (C - x)/2 \qquad (2)$$

with C being the total concentration of the product measured with the RI detector and x the excess concentration of the predominant enantiomer measured with the OR detector. Thus, this method allows the determination of the concentrations of all the components in one short analysis (*see* **Note 6**).

3.2.3. Advantages and Drawbacks of Each Method

Although an initial investment is required for the purchase of a chiral detector, it will rapidly pay for itself with repeated analyses, because of the low cost of achiral columns (generally seven to eight times less expensive than chiral columns) and reduced solvent consumption. In addition, C_{18} reverse-phase columns are very versatile, supporting various mobile phases and can separate many types of compounds compared to chiral columns. Moreover, the analyses performed on a C_{18} reverse-phase column usually lead to the separation of all the compounds in the reaction media and analysis times are often much shorter than on a chiral column. The major drawback with this method is that high specific rotations are required for high sensitivity on the chiral monitor, which limits its application to certain products.

Initially, the first chiral stationary phases sold on the market were expensive and not stable for long periods. This is no longer the case with the introduction of a number of derivatized cellulose-based HPLC chiral stationary phases (many of them commercialized by Diacel [Tokyo, Japan]). However, even if it is always possible to find a column and conditions able to perform the separation of one chiral product of interest, there is not really a "universal" chiral column, and it is often necessary for a laboratory to invest in the purchase of multiple chiral columns, much more expensive than C_{18} reverse-phase ones. In addition, when one wants to combine two HPLC analyses on different columns as presented earlier, two HPLC systems have to be available in the laboratory.

3.3. Rate Calculation

To compensate for possible evaporation of the solvent, all the concentrations obtained directly by HPLC were normalized, by maintaining the total concentration of all ibuprofen species equal to the initial ibuprofen concentration. Calculations of initial rates were done by linear regression using a minimum of five different time-points when conversion was less than 10% (usually less than 5%).

The various kinetic parameters have been determined by fitting the appropriate rate equation to the initial rates obtained from a kinetic study performed at various alcohol and acid concentrations (at least 15 experiments).

Evaluation of the influence of reaction parameters on enzyme activity in nonaqueous media is not always simple. It can be reported as initial rates in given conditions, or in terms of the specificity constant k_{cat}/K_M (7). The latter is, of course, more accurate and general, but to obtain this value, it is necessary to perform a complete kinetic study. Too often, initial rates are measured (sometimes with one single concentration measurement after a given period of time) and divided by the initial substrate concentration to obtain k_{cat}/K_M. This is based on the premise that, in organic solvents, K_M is higher than in an aqueous environment and kinetic measurements are then easily performed in the first-order region (1).

It has been shown that proteases, esterases, and lipases, used in synthetic reactions in organic media, follow a Bi-Bi Ping-Pong mechanism with inhibition by the nucleophile (8–10). It is then obvious that a single activity measurement might not take into account the possible inhibition and, therefore, not give an accurate value of the specificity constant. This situation reflects on the measurement of enantioselectivity. The enantiomeric ratio, as defined by Chen et al. (11) is the ratio of the specificity constants for the two enantiomers:

$$E = \frac{V_{MR}/K_{MR}}{V_S/K_{MS}} \tag{3}$$

The measurements can be made with individual enantiomers or with the racemate. Because resolution is generally applied to isolate one enantiomer from the racemic mixture, the latter is usually preferred. The determination of E can then be made either by measuring the individual rates of reaction for the two enantiomers in the racemate or with one of the relationships between the conversion and the enantiomeric excess, ee, given by Chen et al. (11):

$$E = \frac{\ln[(1-c)(1-eeS)]}{\ln[(1-c)(1-eeS)]} = \frac{\ln[1-c(1+eeP)]}{\ln[1-c(1-eeP)]} \tag{4}$$

where ee$S$ and ee$P$ are the substrate and product enantiomeric excesses, respectively, and c is the conversion defined as:

$$c = 1 - \frac{[R]+[S]}{[R]_0+[S]_0}, \qquad ee_S = \frac{[S]-[R]}{[S]+[R]}, \qquad ee_P = \frac{[P]-[Q]}{[P]+[Q]}$$

It has to be kept in mind that **Eqs. 3** and **4** have been derived for initial rates measurements, without reverse reaction or inhibition by the substrates. In the case of a Bi-Bi Ping-Pong mechanism, the concentration of the second substrate has to be much higher than its Michaelis constant for the equations to be valid. These conditions are usually adequate for hydrolytic reactions. If they are not fulfilled, the value of E obtained with **Eq. 4** may be questionable (*see* **Note 7**).

4. Notes

1. It has been verified that, generally, no chemical reaction occurred during the equilibration of the reactant solutions. When traces of ester were detected, two reactant solutions were prepared and allowed to equilibrate separately, and the proper amounts of each solution were mixed at t_0 just prior to adding the enzyme to initiate the reaction. The rate of the spontaneous chemical reaction has to be negligible compared to that of the enzymatic one. It has also been verified that no reaction occurred during the storage of the samples at 4°C prior to HPLC analysis.

2. Particularly when the reaction volume is low, it is obvious that the total volume removed for analysis has to be minimal when compared to the initial reaction volume so as not to increase significantly the enzyme concentration in the media, which modifies rates of esterification.

3. Depending on the nature of the solvent used during the esterification reaction, traces of this solvent in the sample injected during the HPLC analysis can affect the shape and retention times of the obtained peaks. In these cases, standards containing the same proportion of the reaction solvent have been used for the calibration of the HPLC system.

4. When the solvent peak interfered with the enantiomer peaks, it was removed from the sample under a nitrogen stream or by reduced pressure before redissolving the reaction that remains in the appropriate mobile phase. It was verified that no chemical reaction occurred during the evaporation stage.

5. As ibuprofen was not eluted from the Chiralcel OJ column under our analytical conditions, the column was washed on a regular basis by increasing the polarity of the mobile phase.

6. We have verified that for the esterification of ibuprofen, the results obtained with this method are the same results as those obtained when using a chiral column *(5)*.

7. To illustrate the fact that complete understanding of the stereoselectivity of a protease, esterase, or lipase in a synthetic reaction requires the measurement of the kinetic parameters, we performed some simulations using the reaction in **Fig. 2**.

   In this reaction, both enantiomers, R and S, as well as the nucleophile, B, compete for the active site. The resulting rate equations are given by **Eqs. 5** and **6**. Based on the fact that initial rates are measured, the concentrations of products, including water, were considered to be equal to zero, although it has been shown recently that it can lead to a slightly inacurate model, especially at low substrate concentrations *(12)*:

$$v_S = \frac{V_{mS}[S][B]}{[S][B] + K_{MBS}[S] + K_{MS}[B]\left(1+\dfrac{[B]}{K_{iB}}\right)\left(1+\dfrac{[R]}{K_{MR}}\right) + \dfrac{K_{MBR}K_{MS}}{K_{MR}}[R]} \tag{5}$$

$$v_S = \frac{V_{mR}[R][B]}{[R][B] + K_{MBR}[R] + K_{MR}[B]\left(1+\dfrac{[B]}{K_{iB}}\right)\left(1+\dfrac{[S]}{K_{MS}}\right) + \dfrac{K_{MBS}K_{MR}}{K_{MS}}[S]} \tag{6}$$

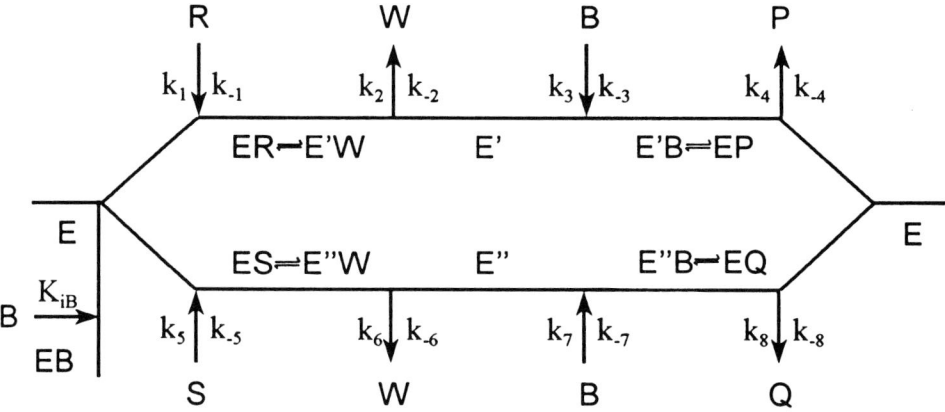

Fig. 2. Reaction scheme for a Ping Pong Bi Bi mechanism in which two enantiomers *(R* and *S)* react with the same nucleophile *(B)*, which is also acting as a competitive inhibitor.

Initial reaction rates were calculated using the parameters indicated in the footnote of **Table 1** in which the results for various substrate concentrations are given. As can be seen, the values obtained for low concentrations correspond to the ratio of the specificity constants given by **Eq. 3**, $E = 36.2$. When the concentrations depart from the first-order region, the ratio of the initial rates no longer corresponds to the ratio of specificity constants.

The rate **Eqs. 5** and **6** were also integrated numerically to calculate the time-course of a resolution reaction. An example is given in **Fig. 3**. As can be seen, the value of E calculated using **Eq. 4** varies with the conversion, even in the early stages of the reaction, and differs from the value calculated from the specificity constants. It is also slightly different from the value calculated from the initial rates for the same conditions ($v_R/v_S = 81$).

Table 1
**Calculation of the Apparent Enantioselectivity for Various
Substrate Concentrations**

$[R]=[S]$ (mM)	$[B]$ (mM)	v_R (mM/d)	v_S (mM/d)	v_R/v_S
0.1	0.5	0.034	9.46×10^{-4}	36.2
0.1	1.0	0.055	1.51×10^{-3}	36.4
0.1	5.0	0.060	1.61×10^{-3}	37.5
0.1	10.0	0.040	1.04×10^{-3}	37.9
0.1	20.0	0.022	5.88×10^{-4}	38.1
0.5	0.5	0.040	1.09×10^{-3}	36.2
0.5	1.0	0.074	2.03×10^{-3}	36.5
0.5	5.0	0.182	4.53×10^{-3}	40.2
0.5	10.0	0.158	3.66×10^{-3}	43.1
0.5	20.0	0.102	2.26×10^{-3}	45.2
5.0	0.5	0.041	1.13×10^{-3}	36.2
5.0	1.0	0.081	2.21×10^{-3}	36.5
5.0	5.0	0.333	7.65×10^{-3}	43.5
5.0	10.0	0.487	8.45×10^{-3}	57.6
5.0	20.0	0.514	6.30×10^{-3}	81.4

*Note:*The values of the parameters used were $V_{mS} = 1.4$ mM/d, $K_{MS} = 20$ mM, $K_{MBS} = 45$ mM, $V_{mR} = 3.8$ mM/d, $K_{MR} = 1.5$ mM, $K_{MBR} = 6.0$ mM, and $K_{iB} = 2.0$ mM.

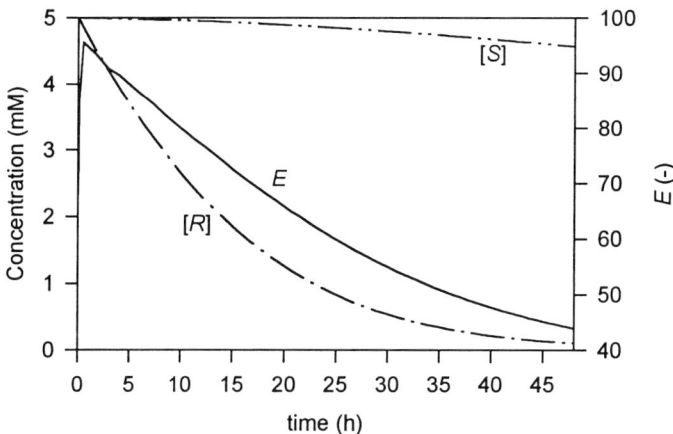

Fig. 3. Concentration of *[R]*- and *[S]*-enantiomers and enantiomeric ratio as a function of time. The concentrations were calculated by integrating the rate equations with an adaptive step-size Runge–Kutta method *(13)*. The values of the parameters were $V_{MS} = 14$ mM/d, $K_{MS} = 20$ mM, $K_{MBS} = 45$ mM, $V_{mR} = 38$ mM/d, $K_{MR} = 1.5$ mM, $K_{MBR} = 6.0$ mM, and $K_{iB} = 2.0$ mM. Initial concentrations of substrates: $[R] = [S] = 5.0$ mM, $[B] = 20$ mM.

In conclusion, it can be said that for the purpose of process development, the measurement of the ratio of the reaction rates for the enantiomers, in conditions as close as possible to those of the process, will give valuable information. If some understanding about the behavior of the enzyme and the influence of the environment on its performance and stereoselectivity is wanted, a complete kinetic study should be undertaken so that all the relevant parameters can be determined.

References

1. Yang, Z. and Russell, A. J. (1995) Fundamentals of non-aqueous enzymology, in *Enzymatic Reactions in Organic Media* (Koskinen, A. M. P. and Klibanov, A. M., eds.), Blackie Academic & Professional, Glasgow, pp. 43–69.
2. Dordick, J. S. (1988) Biocatalysis in nonaqueous media. *Appl. Biochem. Biotechnol.* **19,** 103–112.
3. Halling, P. J. (1994) Thermodynamic predictions for biocatalysis in non-conventional media: theory, tests and recommendations for experimental design and analysis. *Enzyme Microb. Technol.* **16,** 178–206.
4. Bell, G., Halling, P. J., Moore, B. D., Partridge, J., and Rees, D. G. (1995) Biocatalyst behaviour in low-water systems. *Trends Biotechnol.* **13,** 468–473.
5. Ducret, A., Trani, M., Pepin, P., and Lortie, R. (1995) Comparison of two HPLC techniques for monitoring enantioselective reactions for the resolution of (*R,S*)-ibuprofen: chiral HPLC versus achiral HPLC linked to an optical rotation detector. *Biotechnol. Tech.* **9,** 591–596.
6. Ducret, A., Trani, M., Pepin, P., and Lortie, R. (1998) Chiral high performance liquid chromatography resolution of ibuprofen esters. *J. Pharm. Biomed. Anal.* **16,** 1225–1231.
7. Fersht, A. R. (1985) *Enzyme Structure and Mechanism* (1985) W.H. Freeman, New York.
8. Chatterjee, S. and Russell, A. J. (1993) Kinetic analysis of the mechanism for subtilisin in essentially anhydrous organic solvents. *Enzyme Microb. Technol.* **15,** 1022–1029.
9. Chulalaksananukul, W., Condoret, J. S., Delorme, P., and Willemot, R.-M. (1990) Kinetic study of esterification by immobilized lipase in *n*-hexane. *FEBS Lett.* **276,** 181–184.
10. Martinelle, M. and Hult, K. (1995) Kinetics of acyl transfer reations in organic media catalysed by *Candida antarctica* lipase B. *Biochim. Biophys. Acta* **1251,** 191–197.
11. Chen, C. S., Fujimoto, Y., Girdaukas, G., and Sih, C. J. (1984) Quantitative analyses of biochemical kinetic resolutions of enantiomers. *J. Am. Chem. Soc.* **104,** 7294–7299.
12. Janssen, A. E. M., Sjursnes, B. J., Vakurov, A. V., and Halling, P. J. (1999) Kinetics of lipase-catalyzed esterification in organic media: correct model and solvent effects on parameters. *Enzyme Microb. Technol.* **24,** 463–470.
13. Press, W. H., Flannery, B. P., Teukolsky, S. A., and Vetterling, W. T. (1986) *Numerical Recipes: The Art of Scientific Computing.* Cambridge University Press, Cambridge, UK.

16

Calorimetric Methods in Evaluating Hydration and Solvation of Solid Proteins Immersed in Organic Solvents

Mikhail Borisover, Vladimir Sirotkin, Dmitriy Zakharychev, and Boris Solomonov

1. Introduction

Solid proteins immersed in organic solvents represent one class of enzymatic formulations in which structure, properties, and functionality of enzymes may be drastically changed by an organic solvent. Considering the variety of opportunities in use of different nonaqueous media, correct selection of organic environments presents a great challenge to biotechnological scientists and engineers. Such a selection needs the understanding of molecular mechanisms governing the interactions of proteins in nonaqueous media. This understanding should help also in elucidating the mechanisms controlling the conformational stability of the protein macromolecules, protein aggregation in solid state, aging of the protein preparation, and solubility of proteins in nonaqueous media.

Because of the ability to monitor any processes associated with heat evolution, calorimetric methods have a great potential in elucidating interactions in the protein formulations. Thus, the goal of this chapter is to provide calorimetric protocols for examining the interactions occurring in solid proteins immersed in organic solvents. We want to propose here isothermal immersion calorimetry and differential scanning calorimetry (DSC). Fundamentals of calorimetry may be found in **ref. 1**.

1.1. Isothermal Immersion Calorimetry

This involves measuring the heat effects evolved on contacting a certain amount of a solid with a solvent volume. Heat effects determined at a given

From: *Methods in Biotechnology, Vol. 15: Enzymes in Nonaqueous Solvents: Methods and Protocols*
Edited by: E. N. Vulfson, P. J. Halling, and H. L. Holland © Humana Press Inc., Totowa, NJ

temperature and pressure correspond to the change of the enthalpy on formation of the heterogeneous system. Correct analysis of calorimetric data requires defining a system under study. We consider the case when (1) initially both protein sample and organic solvent may contain water, (2) protein immersed in a water–organic mixture forms a two-phase system including the protein phase and liquid solution, (3) the protein phase may contain both water and organic component, and (4) there is no significant dissolution of a protein in the liquid phase. Then, integral change of the enthalpy ΔH corresponding to introducing some amount of protein in a water–organic mixture is given by

$$\Delta H = [\overline{H}_W m_W + \overline{H}_S\, m_S m_S]_{\text{final liquid}} + [\overline{H}_P\, m_P + \overline{H}_W\, m_W + \overline{H}_S + m_S]_{\text{final solid}}$$
$$- \overline{H}_W\, m_W + \overline{H}_S + m_S]_{\text{initial liquid}} + [\overline{H}_P\, m_P + \overline{H}_W\, m_W]_{\text{initial solid}} \tag{1}$$

where $\overline{H}_p$, $\overline{H}_W$, and $\overline{H}_S$ are the partial enthalpies of the protein, water, and organic components, respectively; m_P, m_W, and m_S are mass amounts of the protein, water, and organic components, respectively; phases (liquid or solid) and states (final or initial) are specified by subscripts. The amounts of water (m_W^{tr}) and of the solvent component (m_S^{tr}) transferred from the liquid phase to the protein phase during suspending the protein sample (i.e., sorbed) are defined as

$$m_W^{tr} = [m_{W,\ \text{initial}} - m_{W,\ \text{final}}]_{\text{liquid}} = [m_{W\ \text{final}} - m_{W,\ \text{initial}}]_{\text{solid}} \tag{2}$$
$$m_S^{tr} = [m_{s,\ \text{initial}} - m_{s,\ \text{final}}]_{\text{liquid}} = [m_{s\ \text{final}} - m_{S,\ \text{initial}}]_{\text{solid}} = m_{s,\ \text{final solid}} \tag{3}$$

By introducing the quantities of transferred amounts into **Eq. 1**, the final expression (**Eq. 4**) is obtained as:

$$\Delta H = [m_{W,\ \text{initial liquid}}\, [\overline{H}_{W,\ \text{final liquid}} - \overline{H}_{W,\ \text{initial liquid}}\,]\, m_W^{tr}\, [\overline{H}_{W,\ \text{final solid}} - \overline{H}_{W,\ \text{final liquid}}]$$
$$+ m_{W,\ \text{initial solid}}\, [\overline{H}_{W,\ \text{final solid}} - \overline{H}_{W,\ \text{initial solid}}] + m_P\, [\overline{H}_{P,\ \text{final}} - \overline{H}_{P,\ \text{initial}}] \tag{4}$$
$$+ m_S^{tr}\, [\overline{H}_{S,\text{final solution}} - \overline{H}_{S,\ \text{final liquid}}] + m_{S,\ \text{initial liquid}}\, [\overline{H}_{S,\ \text{final liquid}} - \overline{H}_{S,\ \text{initial liquid}}]$$

Equation 4 accounts for protein interactions in the solid state, the transfer of water and of the solvent from the external liquid phase into the protein solid phase, and the dilution/concentration effects in both phases. When the solid–liquid ratio is small enough, the partial enthalpies of water and organic component in the external liquid phase are not changed during sample immersion. An expression useful for experimental application is given by

$$\Delta H = m_W^{tr}\, [\overline{H}_{W,\ \text{final solid}} - \overline{H}_{W,\ \text{final liquid}}] + m_{W,\ \text{initial solid}}\, [\overline{H}_{W,\ \text{final solid}} - \overline{H}_{W,\ \text{initial solid}}]$$
$$+ [\overline{H}_{P,\ \text{final}} - \overline{H}_{P,\ \text{initial}}] + m_S^{tr}\, [\overline{H}_{S,\ \text{final solid}} - \overline{H}_{S,\ \text{final liquid}}] \tag{5}$$

where ΔH, and m_W^{tr}, $m_{W,\ \text{initial solid}}$, and m_S^{tr} are related now to the unit mass amount of protein.

Detailed thermodynamic analysis based on **Eq. 5** would involve the evaluation of partial enthalpy changes of components in both phases. This analysis needs information concerning the composition of coexisting surface and/or bulk phases. However, useful conclusions may be obtained also when less information is available.

Obtained in the absence of water, the ΔH values demonstrate immediately the energetics of protein–organic solvent interactions. Depending on the specific mechanism, these enthalpy changes may be considered as wetting heats (if solvent molecules solvate the surface of the protein sample) or swelling heats (if the solvent molecules penetrate into the protein bulk).

In the majority of practical cases, both the protein sample and the solvent contain water. In order to distinguish between protein–water interactions (adsorption/desorption, water-stripping effect of organic solvent) and protein–organic solvent interactions, calorimetric data should be compared with water–sorption data obtained independently.

The water amount bound to proteins in organic solvents may be determined using Karl Fischer titration from the change of water concentration in the external phase *(2–5)*. Further, the measured ΔH values should be plotted against the water amount m_W^{tr} removed by the protein sample from the external solution. Nonzero ΔH values obtained at zero water amounts m_W^{tr} will indicate the protein–organic solvent interactions. In fact, any changes of the partial enthalpies for protein, water, and solvent in the protein phase at a fixed water amount have to be related with the solvent uptake (i.e., with protein–solvent interactions). Additional information concerning the energetics of the protein system may be extracted from examining the shape of the obtained dependence. The most significant case is a linear dependence between ΔH and m_W^{tr}. In this case, the differential change ΔH of the enthalpy corresponding to the water sorption $(\partial \Delta H / \partial m_W^{tr}) T_{,p}$ is easily obtained from the constant slope of this dependence. As follows from **Eq. 5** and properties of partial functions [i.e., the generalized Gibbs–Duhem equation *(6)*], this differential change ΔH corresponds to $[\overline{H}_{W,\text{final solid}} - \overline{H}_{w,\text{final liquid}}] + (\partial m_W^{tr}/\partial m_W^{tr}) \, T_{,p} \, [\overline{H}_{S,\text{final solid}} - \overline{H}_{S,\text{ final liquid}}]$, where $(\partial m_S^{tr}/\partial m_W^{tr}) \, T_{,p}$ shows the relation between sorption of water and organic component by the protein sorbent. The linear dependence between ΔH and m^{tr} suggests also that the water partial enthalpy in the solid protein phase is not dependent on the sorbed amount of water. Then, the intercept of the linear ΔH versus m_W^{tr} dependence is $[\overline{H}_{P,\text{ final}} - \overline{H}_{P,\text{ initial}}] + m_S^{tr} [\overline{H}_{S,\text{ final solid}} - \overline{H}_{S,\text{ final liquid}}]$. As estimated for a fixed water amount on the protein, this intercept demonstrates the effect of the organic component penetration into the protein phase. The following analysis of calorimetric data may include (1) examining the solvent nature effect on the differential sorption enthalpy of water, (2) inspecting the solvent nature effect on the enthalpy term related to the protein-solvent interactions, and (3) modeling the water concentration (activity) effect on the ΔH values in different solvents. All together, it may give insight to the energetic quantities characterizing protein–water and protein-solvent interactions *(see* **Note 1***)*.

Discussions of the theoretical background for adsorption thermodynamics and immersion calorimetry may be found in **refs. 7–9**. Experimental illustrations for immersion calorimetry of solid proteins in organic solvents are presented in **refs. 10–14** providing a basis for given protocols.

1.2. Differential Scanning Calorimetry (DSC)

DSC, in which temperature T is changed linearly with time t, shows the thermal behavior of a system in terms of the heat capacity ($\partial Q/\partial T$), or heat flow ($\partial Q/\partial t$), where Q is evolved heat. Measured for solid protein immersed in organic (plus water) solvents, the DSC curve may represent a multiplicity of processes including water desorption from the protein sample into solvent, protein unfolding/denaturation, protein–solvent interactions, and protein destruction and degradation. The DSC curves of solids (which are capable of different transformations [e.g., conformation changes]) are expected to be irreversible.

To minimize the chemical deterioration of the protein, the behavior of protein sample will be considered only at temperatures below 100°C. By changing the water contents of the protein and solvent, it is possible to evaluate the hydration effect on the DSC curve of protein in a suspension. Importantly, temperature-initiated water desorption from protein to the solvent occurred in hermetically closed cell produces a heat effect smeared out in the broad temperature range and does not involve the characteristic peak of heat evolution.

The DSC curve should be recorded for the protein suspension against a some reference material (e.g., solvent). The DSC curve of interest is a difference between the DSC sample curve and the baseline. The latter is evaluated by scanning the volume of solvent as used in the sample experiment against the same reference material. By subtracting the baseline curve from the sample curve, the contribution from the bulk liquid phase is eliminated and the difference is more related with events in the solid phase.

This difference between the DSC sample curve and the baseline is a small difference of great values. Thus, absolute heat capacities are not well-reproducible values. As such, we considered the DSC method mostly as a half-quantitative tool to examine protein interactions in a heterogeneous system. Most attention should be paid to the characteristics of the heat evolution peaks such as peak maximum temperature and peak area. The latter is related to the enthalpy change during the temperature-initiated process. Our main expectation in using the DSC calorimetry was based on the idea that protein–solvent interactions "frozen" partially at ambient temperatures may be seen at higher temperatures. Protocol presented here was applied in **refs. 15** and **16** to the DSC measurements of human serum albumin in organic solvents. An earlier example of use of the DSC was given for solid ribonuclease in organic solvents (**17**). There is a related DSC study of ribonuclease immobilized on Celite in organic solvents *(18)*. Because the immobilization was performed in order to avoid protein aggregation effects, such an immobilized protein should be differentiated from the suspended protein in which aggregation effects may play a significant role in the protein rigidity.

2. Materials

2.1. Protein

The protein used was human serum albumin (HSA), kept hermetically closed at 4°C:

1. Sigma Product No. A1887; essentially fatty acid free (approximately 0.005%).
2. Reanal (Hungary) product N 01092; the electrophoretic purity was >95% and the remainder after burning was <2%. The total concentration of fatty acids (C_{12}–C_{22}) in this protein preparation was found to be 0.2% (*see* **Note 2**).

Initial protein preparations used contained 9–11% water. In each experimental series, the water contents of the initial solid protein samples were measured using Karl Fischer titration. Details of determination are given in **Subheading 3.2.** Dry protein gives no water content according to the Karl Fischer method. (*See* **Note 3**.)

2.2. Solvents

The solvents used were *n*-hexane,1,4-dioxane, acetonitrile, dimethyl sulfoxide (DMSO), pyridine, methanol, ethanol, *n*-propanol, *n*-butanol (reagent grade, purity >99%); the water used for the preparation of water–organic mixtures was bidistillate. The water contents of solvents and water–organic mixtures were determined in each case by Karl Fischer titration.

2.3. Instruments

1. Isoperibolic differential calorimeter (heat flow; two calorimetric vessels; Setaram BT-215; described in **ref. 1** (*see* **Note 4**). The calorimeter should be calibrated (*see* **Note 5**).
2. Differential scanning calorimeter (heat-flow; Setaram DSC-111; described in **ref. 1**).
3. Electrochemical Karl Fischer titrator (*see* **Note 6**).

3. Methods

3.1. Determination of the Enthalpy Changes on Contacting the Protein Sample with Organic Solvent (Water–Organic Mixture)

3.1.1. Blank Experiments

1. Fill *both* calorimetric vessels with solvent (4.0 mL) and place hermetically closed *empty* sample containers in *both* calorimetric vessels. The sample container should be completely immersed in the liquid bulk of the vessel.

2. Close the calorimetric vessels. The vessels should be placed in the corresponding chambers of the calorimeter.
3. Equilibrate the calorimetric system at a chosen temperature until a stabile baseline is seen.
4. With the needle-ended rod, manually punch both Teflon septa of one sample container and then raise this rod up to the initial level. The second calorimetric vessel is the reference vessel.
5. The recorded heat-flow curve represents the heat evolution caused by punching the septa, wetting the septa and the container walls, solvent evaporation, possible friction contributions, and others. Return of the heat flow to the stabile level of the initial baseline indicates the completion of the heat evolution. The usual time for completion of this blank heat evolution is 10–15 min.
6. Determine the full calorimetric effect from the area enclosed by the calorimetric response curve and the straight line connecting the beginning and the end of the heat evolution. The area is found by numerical integration. The coefficient of the proportionality between the heat effect and area of the instrument response should be determined independently (*see* **Note 5**).
7. After the calorimeter comes back to the equilibrium, the Teflon septa of the sample container in the second calorimetric vessel should be punched, and the full calorimetric effect should be determined from the area under heat evolution curve. The first calorimetric vessel is the reference vessel in this experiment.

Thus, blank heat effects related to the punching of the septa in the sample containers are obtained for each calorimetric vessel (*see* **Note 7**). Such a control experiment should be repeated three times. Depending on the organic solvent, reproducibility of the blank effects was 0.005–0.02 J. The averaged values for the blank heat effects are used for the following determination of enthalpy changes at contacting the protein sample with solvents.

3.1.2. Sample Experiments

1. Place 4–10 mg of the protein preparation in *both* sample containers.
2. Fill both calorimetric vessels with solvent (4.0 mL) and install the hermetically closed sample containers in each vessel.
3. Follow **steps 2–4** from **Subheading 3.1.1.**
4. Recorded heat evolution now includes the interactions between the solid sample and the liquid phase. Usually, heat evolution needs 30–40 min. In some cases, the completion of the heat evolution may need 2–3 h. Return of the heat flow to the stabile level of the initial baseline indicates the completion of the heat evolution and the attainment of the equilibrium. The examples of heat evolution curves are given in **Figs. 1** and **2**. (*see* **Note 8**.)
5. Follow **steps 6** and **7** from **Subheading 3.1.1.**
6. Disassemble one calorimetric vessel when the calorimetric experiment is finished. If the sample container was installed well in the calorimetric vessel, then after the completion of the experiment, it will be found below the liquid level and filled by the solvent.

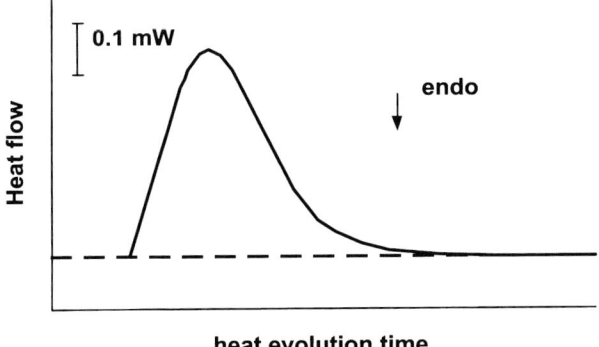

Fig. 1. Typical calorimetric curve recorded for 5.5 mg of HSA contacting with 1,4-dioxane + 3.1 mol/L of water. Full time of the heat evolution is 20 min. Total measured effect is –0.18 J.

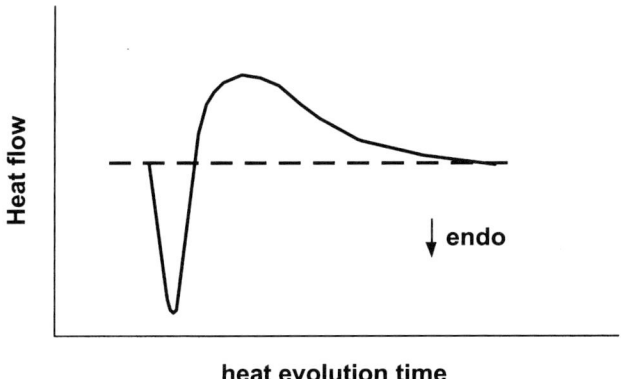

Fig. 2. Shape of calorimetric curve recorded on contacting the solid HSA with DMSO (or ethanol) containing water. The times needed to complete the heat evolution are 1 h in DMSO+7 mol/L of water, 6 h in DMSO + 0.5 mol/L of water, 2 h in ethanol + 0.2 mol/L of water, and 40 min in ethanol + 3.5 mol/L of water.

7. Remove the punched sample container from this calorimetric vessel.
8. Take a certain amount of the solvent from the calorimetric vessel using the syringe. The volume of the solvent aliquot varied from 0.1–0.3 mL at high water contents in solvents to 1.0–1.5 mL at low water contents.
9. Weigh the syringe filled with a solvent aliquot and then transfer the solvent aliquot into the Fischer reagent medium. To obtain an accurate weight of the aliquot, the syringe should be weighed again. Then, the total water amount introduced by the aliquot should be measured electrochemically. This determination as described by **steps 8** and **9** should be at least replicated.

10. Water concentration in the solvent will be found from the measured water amount in the aliquot, the weight of aliquot, and the solvent density.
11. Determine the solvent moisture in the second calorimetric vessel as done in **steps 6–10**.
12. Subtract the blank heat effects from the total heat effects measured in each vessel. The difference is the integral change ΔH of the enthalpy corresponding to the protein contact with a liquid (see **Note 9**)

This effect should be related to the amount of dry protein. The final ΔH value is determined for a given equilibrium water concentration in a solvent (*see* **Note 10**). The dependence of the ΔH values on the equilibrium water concentration is specific for a given organic medium. **Figures 3–5** show the ΔH values plotted against the equilibrium water concentration in three organic media. Smooth dependence is found in acetonitrile **(Fig. 3)**. Such smooth curves were obtained also for HSA in water–methanol and water–ethanol mixtures *(14)*. A more complicated dependence including the sharp changes in a relatively narrow water concentration range is shown in 1,4-dioxane **(Fig. 4)** and pyridine **(Fig. 5)** (also, in *n*-butanol and *n*-propanol *[14]*). At increasing water contents, the ΔH values reach a similar level in different solvents which is close to the solution enthalpy of HSA in water at a concentration of 1 mg/mL *(10,13,14)*. Curves in **Figs. 3–5** correspond to the Langmuir-like trend resulting from water sorption.

3.2. Determination of the Water Sorption by Protein Immersed in Water–Organic Solvent

1. Place 4–10 mg of solid protein preparation and 4.0 mL of organic solvent containing a certain fraction of water in a preweighed thin-bottom glass ampoule and close this ampoule with a preweighed cap (*see* **Note 11**). The solid:liquid ratio is chosen close to the value used in the calorimetric experiment.
2. Maintain the ampoule at a constant temperature (25°C) during some time. *This time period is not less than the time corresponding to the completion of the heat evolution in a given solvent as found by calorimetry* (*see* **Subheading 3.1.2.** and **Note 8**).
3. Remove the aliquot of the solvent from the ampoule with a syringe. The volume of aliquot varied from 0.1–0.3 mL at high water contents in the solvent to 1–1.5 mL at low water contents.
4. Weigh the syringe filled with a solvent aliquot, and then transfer the solvent into the Fischer reagent medium. Then, the total water amount introduced by the aliquot should be measured electrochemically.
5. Weigh the syringe again and obtain the aliquot weight by the difference. The water concentration in the solvent should be calculated from the measured water amount in the aliquot, the weight of aliquot, and the solvent density.
6. Repeat such extraction of the aliquots and the electrochemical measurement two to three times for 40–60 min. Reproducible values of water content in the solvent indicate the attainment of equilibrium (*see* **Note 12**).

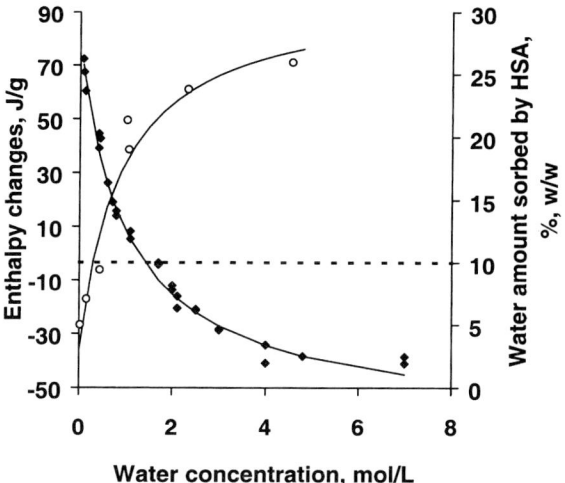

Fig.3. ΔH values (per gram; filled diamonds) and water sorption by HSA (open circles) plotted against water concentration in acetonitrile (298 K). The dashed line shows the initial water amount on the HSA (10%, w/w). (After **ref. *11*.**)

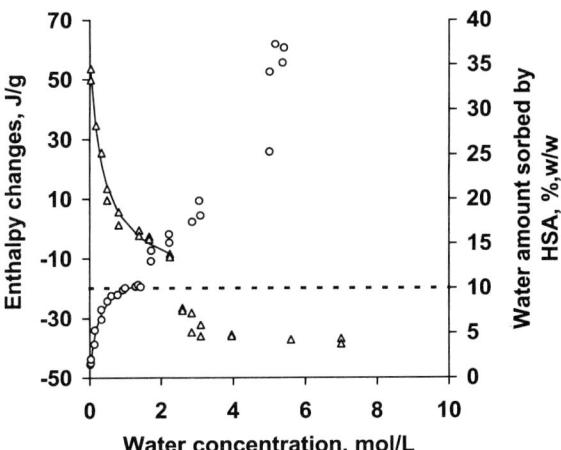

Fig.4. ΔH values (per gram; triangles) and water sorption by HSA (circles) plotted against water concentration in 1,4-dioxane (298 K). The dashed line shows the initial water amount on the HSA (10%, w/w). (After **ref. *13*.**)

7. Withdraw the bulk of the liquid phase from the ampoule with a syringe and weigh the ampoule again. The apparent weight of the remaining liquid phase is calculated as the difference between the final weight of the sealed ampoule and the sum of the weights of the empty ampoule, the cap, and the dry protein.

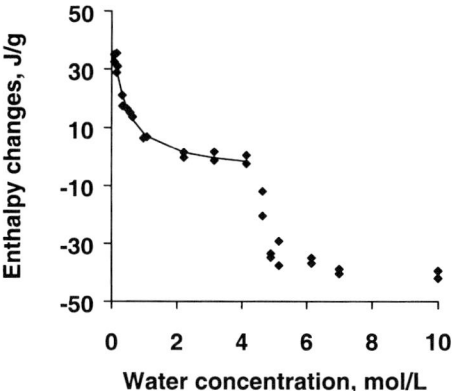

Fig.5. ΔH values (per gram) plotted against water concentration in pyridine (298 K). (After **ref. 13**.)

8. Put this ampoule containing the protein and a small amount of the remaining water–organic mixture into the Fischer apparatus.
9. Break off the bottom of the ampoule and measure the total amount of water on the protein sample (and in the remaining liquid).
10. Weigh the cap after the experiment in order to determine the amount of the solvent that could have been adsorbed by the cap.
11. Subtract the difference in the weight of the cap from the apparent weight of the remaining solvent. The final result is the true weight of the remaining liquid phase.
12. Calculate the amount of water in the remaining liquid using the weight of the remaining liquid phase, its water content measured previously, and the liquid density.
13. The amount of water on the HSA corresponds to the difference between the total measured amount of water in the ampoule and the amount of water in the remaining liquid phase. This bound amount of water is expressed as percentage by weight with respect to the dry protein (%, w/w) (*see* **Note 13**).

Water sorption isotherms are exemplified in **Figs. 3** and **4** (*see* **Note 14**). As is seen, the complicated shape for calorimetric data in 1,4-dioxane is supported by a similar pattern for the water amount bound to HSA. The smooth sorption isotherm in acetonitrile mixtures follows the pattern for calorimetric data. Additional examples combining sorption and calorimetric data for HSA preparation in different solvents are presented in **refs. 12** and **14**.

3.3. Comparison Between Calorimetric and Sorption Data

Figures 6 and **7** show examples in which the ΔH (referred to 1 g of HSA) is plotted against the change in the water amount on the HSA (m_W^{tr}, % w/w) during its suspension. Slopes of the linear dependences in **Figs. 6** and **7** correspond to the differential sorption enthalpy Δh of water (in 100 J/g). The intercepts (i.e.,

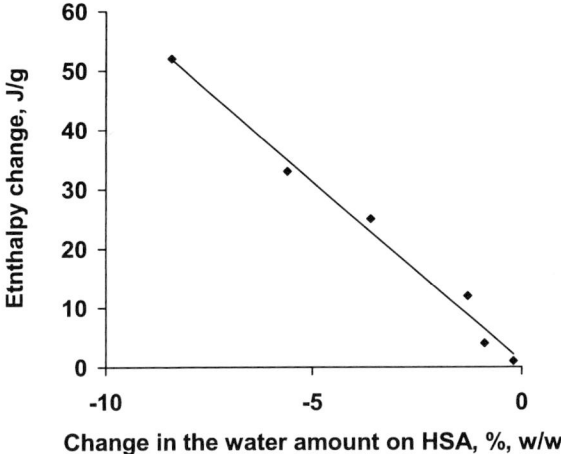

Fig. 6. ΔH values (per gram) plotted against the change in the water amount on HSA (%, w/w) at suspending HSA in water–1,4-dioxane mixtures. The data correspond to the concentration region below 2 mol/L. (After **ref. 5**.)

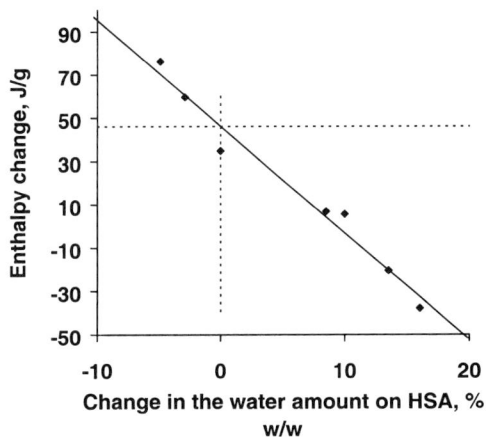

Fig. 7. ΔH values (per gram) plotted against the change in the water amount on HSA (%, w/w) for suspended HSA in water–acetonitrile mixtures (full concentration region available). (Data from **ref. 11**.)

ΔH values at the fixed initial water amount on HSA) represent the term $[\overline{H}_P,$ $_{final} - \overline{H}_{P,\ initial}] + m_S^{tr}\ [\overline{H}_{s,\ final,\ solid} - \overline{H}_{s,\ final\ liquid}\ ]$ of **Eq. 5**, thus indicating the intensity of protein-organic solvent interactions. Intercepts calculated for some solvents (second column) and slopes (expressed in kJ/mol; third column) are exemplified in **Table 1**.

Table 1
Enthalpy Parameters for Various Solvents

Solvent	ΔH at $m_W^{tr} = 0$ (J/g)	Δh (kJ/mol)	ΔH^a (J/g) (dry HSA in dried solvents)
1,4-Dioxane	0.8 ± 1.5	-11.9 ± 1.7	-0.6 ± 1.3
n-Butanol	0.4 ± 0.9	-7.4 ± 1.1	-2.1 ± 1.9
Acetonitrile	44.7 ± 4.9	-9.0 ± 1.5	7.3 ± 3.1
Ethanol	-21.8 ± 0.8	-1.5 ± 0.4	-58.2 ± 0.7
DMSO	-68.0 ± 2.0	~ 0	-89.0 ± 2.0

*a*Equilibrium humidities of solvents are as follows: 0.02 mol/L in 1,4-dioxane , 0.01 mol/L in *n*-butanol, 0.022 mol/L in acetonitrile, 0.21 mol/L in ethanol, and 0.23 mol/L in DMSO. Based on sorption isotherm data, water sorption/desorption contribution to the measured ΔH values is considered insignificant.

The fourth column in **Table 1** shows data on ΔH values measured for dried (0.3 ± 0.5% water) HSA in dried solvents *(19)*. Thus, data in the second and fourth columns of **Table 1** demonstrate different organic solvent–protein interactions for partially hydrated (about 10% w/w) and dried preparations of HSA, respectively (*see* **Note 15**).

3.4. Determination of DSC Curves for HSA Immersed in Organic Solvent

1. Place 5–8 mg of a protein sample and 100 µL of an organic solvent in the calorimetric cell and hermetically close it (*see* **Note 16**).
2. Put the same amount (100 µL) of the solvent in the second cell that is used as a reference.
3. After the 5–10 min equilibration seen as a stabile baseline in the DSC recorder make a scanning in the temperature range 25–120°C with the heating rate 2°C/min (*see* **Note 17**).
4. Cool the DSC cells and repeat the scanning at the same conditions in order to verify the reversibility/irreversibility of observed phenomena.
5. Fill the sample cell by 100 µL of the solvent and determine the baseline by scanning against the same reference material that was used at scanning protein sample.
6. Obtain the final DSC curve by subtracting the baseline curve from the sample curve.
7. In order to determine the peak area, the final subtracted DSC curve is shown on the computer display. The beginning and the end of the observed peak are determined visually and connected by a straight line. The area enclosed by the curve line and this straight line is found by numerical integration. This area is proportional to the enthalpy change, which is obtained from the calibration of the DSC instrument (*see* **Note 18**).

Two examples are given in **Figs. 8** and **9** to illustrate the DSC curves obtained for HSA in different organic solvents. Dried HSA preparation placed in pyridine–*n*-hexane mixtures demonstrates the exothermic heat evolution (curves B–E, **Fig. 8**) in contrast to the DSC curve of this preparation in pure *n*-hexane (curve A). Exothermic heat evolution shows that the state of HSA immersed in pyridine–*n*-hexane mixtures is the nonequilibrium state *(16)*. Increased temperature stimulates protein–solvent interactions "frozen" at near room temperature. Such exothermic peaks were clearly observed on both Sigma and Reanal preparations of HSA (*see* **Note 19**).

HSA in DMSO (**Fig. 9**) also displays its nonequilibrium state resulting in a low-temperature exothermic peak *(15)*.

4. Notes

1. **Equations 1–4** were introduced in terms of the coexistence of two bulk phases (i.e., protein phase and external liquid solution). One may consider these equations applied to the interface solution coexisting with solid protein and external liquid. In this case, the mass quantities referred to the protein phase composition in **Eqs. 1–4** should be reassigned to this interface solution rather than to the protein bulk. Then, spreading pressure π (i.e., change of the surface tension resulted from the change of interface solution composition) has to be considered as one more intensive variable affecting the enthalpy H_S of the interface phase. Then, Δh obtained as $(\partial\Delta H/\partial m_W^{tr})T_{,p}$ from **Eq. 5** and properties of partial functions [generalized Gibbs–Duhem equation *(6)*] will include an additional term $[(\partial HS/\partial\pi)\,(\partial\pi/\partial m_W^{tr})_{T,p}]$, where $(\partial H_S/\partial\pi)$ is considered at fixed temperature, pressure, and amounts of components of the interface solution. When sorption is a nonlinear function of the water concentration in the liquid phase, the surface tension is also related with this water concentration nonlinearly *(7)*. Hence, the $(\partial\pi/\partial m^{tr})_{T,p}$ term should be concentration dependent. As such, in the case of the interface solution, the concentration-independent Δh value (i.e., linear dependence between ΔH and m_W^{tr}) indicates that apparently this π-related contribution to Δh may be neglected.

2. The total concentration of fatty acids in the Reanal HSA preparation was determined by a technique *(20)* that includes the extraction by a chloroform–methanol (2:1) mixture, the washing of the extract by water, the methylation with diazomethane in diethyl ether and gas chromatographic (GC) determination in the concentrated *n*-hexane solution. In fact, isothermal sorption and immersion calorimetry data were measured for the Sigma preparation of HSA. However, some DSC determinations were obtained also using the Reanal HSA preparation.

3. We determined the moisture of the protein preparations also using microthermoanalyzer (Setaram, MGDTD-17S). The water amount on the protein samples was obtained from the weight loss of the protein sample at 298 K and 1×10^{-3} T. These moisture values were in good agreement with the protein moisture measured by Karl Fischer titration.

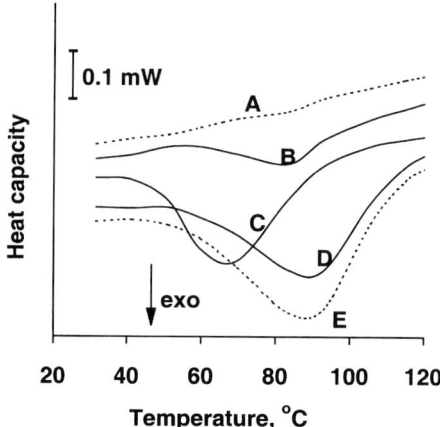

Fig. 8. DSC curves for dried HSA (6 mg; Sigma) immersed in pyridine-*n*-hexane mixtures of different composition. Pure n-hexane (**A**); 0.15 mol fraction of pyridine (**B**); 0.33 mol fraction of pyridine (**C**); 0.62 mol fraction of pyridine (**D**); pure pyridine (**E**). Scanning rate is 2°C/min. (After **ref. *16*.**)

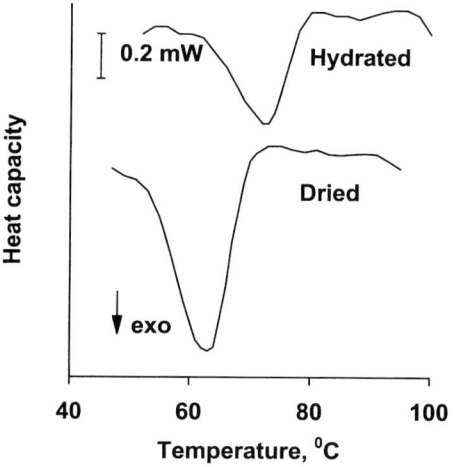

Fig. 9. DSC curves for HSA (6 mg; Reanal) immersed in DMSO. The scanning rate is 2°C/min. The hydrated preparation contained 12% of water. (After **ref. *15*.**)

4. Specifically for given examples, the calorimetric system for introducing the solid sample into solvent was modified (**Fig. 10**). In order to work with corrosive organic solvents, the calorimetric vessel and a sample container were made from titanium. The sample container consisted of the cylinder with the external screw thread and two screw caps with holes. Teflon septa was used in caps as the ends

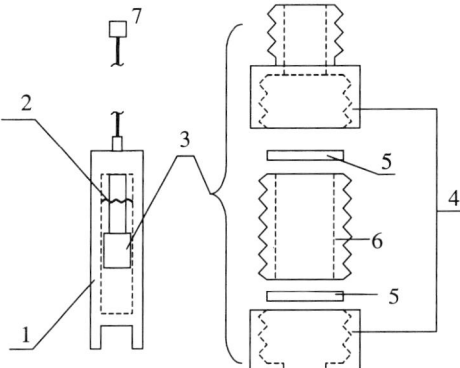

Fig. 10. Scheme of the calorimetric system for introducing the solid sample into the solvent. 1: calorimetric vessel; 2: level of organic solvent; 3: titanium sample container; 4: two screw caps; 5: Teflon septa; 6: external screw; 7: the needle-ended rod coming out of the calorimeter (needle not shown).

of the container. The total volume of this sample container is 0.3 mL. Contact between two compartments (i.e. sample container and vessel) is performed by punching the Teflon septa with a needle 2 mm in diameter. The needle is connected with the long rod coming out of the calorimeter. Punching is carried out manually by pressing on this rod. In many cases, the commercial calorimetric equipment may be quite sufficient for determining the immersion heats of solids (using, for example, glass ampoules as sample containers).

5. The calorimeter is calibrated using the Joule effect and tested by dissolving sodium chloride in water at 0.0347 M according to the recommendations *(21)*. We also tested the calorimeter by measuring the dissolution heats of organic compounds in organic solvents (e.g., naphthalene in benzene). Best testing will be reached by measuring the processes in which the heat evolution kinetics is similar to the kinetics of heat evolution for protein–solvent interactions.

6. The electrochemical Karl Fischer titrator used was home-made. There is no need for additional information concerning this self-designed instrument because the commercial titrators may be effectively used for determination of the water contents in solutions and solids (e.g., Mettler–Toledo titrator).

7. The blank heat effects measured may be significantly positive if the solvent is volatile. The values of blank effects may be also different for each calorimetric vessel.

8. For example, times needed to reach the equilibrium were found to be 30–40 min in methanol, *n*-propanol, *n*-butanol, acetonitrile, and 1–2 h in some ethanol–water mixtures. The type of the calorimetric curve depends on the type of the solid/ solvent systems. The curve may involve endothermic peaks, exothermic peaks, or both together. Endothermic peaks are related to the water desorption from the protein preparation to the solvent. In some cases, evaporation of a solvent will involve a clear endothermic peak (e.g., in methanol and its mixtures, in ethanol).

In cases like acetonitrile, endothermic peak of the solvent evaporation will overlay on the endothermic peak of the protein–solvent interactions. A shift of the final equilibrium baseline relatively to the initial baseline may be observed. If this shift is significant, it may indicate the slowly occurring events. This shift is considered insignificant if it is comparable with that observed in the blank experiment.

9. Depending on the solvent, its water content, and the amount of the HSA sample, experimentally observed heats varied from –0.8 to +0.5 J. Examples of the blank effects (in joules) in different media are as follows: –0.02 in pyridine, –0.01 in *n*-butanol, +0.04 in acetonitrile, +0.01 in 1,4-dioxane, +0.04 in ethanol, and +0.10 in methanol. The presented protocol suggests determining the heat corrections resulting from septa punching, wetting, solvent evaporation, and so forth, in separate control experiments. The often-used alternative protocol for immersion/wetting/dissolution measurements involves (1) the placing of the compound sample only in one sample container and (2) the simultaneous punching of the septa in the container with the substance and the empty container (or simultaneously breaking the glass ampoule with compounds and the empty one). In this way, the measured response is referred directly to the physico-chemical events related to sample interactions. This protocol requires that both calorimetric vessels have to be completely equivalent and give the identical response. However, in many cases, the calorimetric responses obtained on testing two different vessels are not identical. Hence, simultaneous pricking of the containers with substance and without may give erroneous results for the measured heats. Our experience was that the blank corrections determined separately for each vessel in the control experiments (*see* **Subheading 3.1.1.**) were reproducible. Hence, we could correct the total heat effects for each vessel by using the corresponding blank results. The same blank corrections might be used for all experiments with a given solvent, including the mixtures with water additions. Caution in using the same blank corrections for water–organic mixtures differing significantly in composition should be done when the organic component is volatile.

10. As mentioned in **Subheading 1.** (*see* **Eqs. 4** and **5**), the data interpretation is based on the assumption that the change of the liquid composition caused by solid–liquid interactions does not influence strongly the partial enthalpies of water and organic component in the external solution. In other words, the terms $m_{W, \text{ initial liquid}} [\bar{H}_{W, \text{ final liquid}} - \bar{H}_{w, \text{ initial liquid}}]$, and $m_{S, \text{ initial liquid}} [\bar{H}_{S, \text{ final liquid}} - \bar{H}_{S, \text{ initial liquid}}]$ indicating in **Eq. 4** the heat contribution from the dilution of water in the binary mixture may be neglected. For example, for a given solid:liquid ratio (around 1 mg:1 mL) at the lowest equilibrium water concentrations in a solvent, the maximal water desorption from the 10% hydrated protein sample would change the water concentration in 0.005 mol/L. This change is much less than the lowest equilibrium water concentration in solvents under study (i.e., 0.17 in dimethyl sulfoxide, 0.08 in pyridine, 0.07 in acetonitrile). In the worst case of 1,4-dioxane, this water concentration change is about 20% of 0.03 mol/L. Our previous experience on measuring the solution enthalpies of water in different organic solvents showed that partial enthalpies of water do not depend, within

experimental error, on the water concentration up to 0.05 mol/L in these solvents *(22)*. Similar results found in **ref. *23*** show that the partial solution enthalpies for water are constant, within the experimental error, over the water concentrations 0.07–0.24 mol/L in methanol, 0.02–0.16 mol/L in ethanol, 0.06–0.51 mol/L in *n*-propanol, and 0.05–0.32 mol/L in *n*-butanol (concentrations were calculated from the mole fraction compositions). Hence, the contribution of the protein-induced change of the external phase composition to the total measured heat was considered negligible at low water contents. Evidently, at higher water concentrations in solutions, this relative change of the water concentration resulting from the solid–liquid interactions is even less. The following examples will be illustrative. At 6 mol/L of water in 1,4-dioxane, the water sorption by HSA is around 35% w/w. This means that by placing 4 mg of HSA in 4 mL of the mixture, 1.4 mg of water will be extracted from 432 mg of the total water amount in the mixture. The change of the water concentration in the external solution is less than 0.3%. At 2 mol/L of water in acetonitrile, the water sorption by HSA is around 22% w/w. At the same conditions, HSA removes 0.9 mg of water from 144 mg of water (less than 1%). As such, it is concluded that the water dilution in the binary mixture resulting from solid–liquid interactions does not contribute significantly into the integral measured ΔH values in the presented examples *(13,14)*. Thus, the contributions $m_{W, \text{ initial liquid}} [H_{W, \text{ final liquid}} - H_{W, \text{ initial liquid}}]$ and $m_{S, \text{ initial liquid}} [H_{S, \text{ final liquid}} - H_{S, \text{ initial liquid}}]$ may be neglected in **Eq. 4.** In fact, a simple test may be performed in order to verify this assumption. The ratio of experimentally measured heat to the mass of the protein should not depend on the absolute mass of the protein. In such a case, the solid:liquid ratio is small enough in order to neglect the dilution effect. In many of our measurements, the solid:liquid ratio was not constant, and varied by two to three times for a given water–solvent composition. Hence, the amount of the sorbed (desorbed) water was doubled. Heat effects (per gram of protein) measured for different masses of the protein did not show any significant trend exceeding the typical scattering of data. Thus, we considered the solid:liquid ratio 1 mg:1 mL for a given preparation hydration (approx 10%, w/w) and for the mentioned minimal equilibrium humidities of listed solvents as a sufficient condition to use **Eq. 5** in interpretation of data.

11. These thin-bottom ampoules were prepared by blowing out the glass tube.

12. Additional tests were performed in order to examine the effect of the exposure time of the solid HSA preparation in water–organic mixtures on the water sorption. For example, no noticeable variation of the amount of bound water on HSA was observed after maintaining the suspensions for 20–24 h for mixtures finally containing 0.02, and 1.4 m/L of water in 1,4-dioxane, 0.25, 1.6, 2.2, and 4.8 mol/L of water in ethanol, 0.09, 0.65, and 1.2 mol/L of water in *n*-butanol, and 0.16, 0.63, 1.00, 2.2, and 5.1 mol/L of water in dimethyl sulfoxide.

13. In most cases, the ratio of the weight of the remaining liquid phase to the weight of the solid protein was 10–15. The weight change in the cap was not more than 7–10% of the total weight of the remaining liquid phase. The ratio between the water amount sorbed by the protein and water dissolved in the remaining liquid

phase is illustrated by the next examples. At 0.5 mol/L of water in 1,4-dioxane, water sorbed by the protein is around 65% of the water in the remaining liquid phase. At 6 mol/L of water in 1,4-dioxane, this fraction is 30%. At 0.5 mol/L of water in acetonitrile, the ratio is about 50%. At 4 mol/L of water in acetonitrile, it decreases to 30%. Evaluation of this fraction may be easily performed on the basis of the published water sorption isotherms and by using the ratio of the weight of the remaining liquid phase to the weight of the solid protein as 10–15.

14. The described protocol for determination of the water sorption includes estimating the external phase amount remaining in the sample after removal of the solvent bulk (**step 7** in **Subheading 3.2.**). In fact, this amount of the remaining liquid is obtained here *by the difference* between the *total* weight of the remaining sorbet and the *initial* weight of the dry sorbent. In this way, it is neglected that sorbent (i.e., bulk and/or surface) accumulates water and solvent. Hence, use of the dry sorbent weight in this calculation overestimates the weight of the remaining drops of the external liquid phase. The same assumption is used when the total amount of sorbed water is calculated as the change of the water concentration in the external solution multiplied by the *total initial volume* of the liquid phase. In this way, the apparent distribution coefficient K_{app} may be obtained, which shows the ratio between the apparent (calculated by difference) sorbed water amount (weight per weight of dry sorbent) and weight fraction (or any other concentration) of water in the equilibrium liquid phase. Generally, K_{app} is different from K_d, which is the ratio between the true sorbed water amount (weight per weight of dry sorbent) and water concentration in the liquid phase. As it may be easily derived from Eq. 3.6 in **ref. 7**, the weight-fraction-based K_d is the sum of K_{app} and the total weight of water and solvent sorbed by the weight unit of protein. The measured apparent sorption of water approximates the true sorbed amount only if the weight-fraction-based K_{app} exceeds significantly the total weight of water and solvent sorbed by the weight unit of protein. Practically, it may be quite difficult to verify this condition. Still, one may believe that the total amount of sorbed water and solvent (weight per weight of a sorbent) should be less than unity. Then, for practical purposes, we considered that the apparent weight-fraction-based K_{app} for water has to be at least more than unity in order to approximate the true sorption of water by the protein sorbent. Relationships between the water-sorption data and calorimetric enthalpies presented in **Figs. 6** and **7** are considered also as an additional test for correctness of obtained water-sorption data. Still, use of other independent methods enabling the measurement of the amount of water sorbed by the protein may be very helpful.

15. Intercepts obtained from the ΔH versus m_W^{tr} linear analysis for partially hydrated protein cannot be compared directly with ΔH values measured for dried HSA (second and fourth columns in **Table 1**, respectively). In contrast with dry preparation, protein hydration involves occupying the sorption sites and change of the conformational mobility of protein fragments. Correct comparison between these data necessitates building the thermodynamic cycle, including the immersion heats for dried and hydrated protein, heats of water desorption to the solvent and to the gas phase, and solvation enthalpies of water.

16. For multiple uses, we made titanium cells that may be closed hermetically by the screw cap with the Teflon liner. The total volume of this titanium cell was 120 µL.
17. Thermal events in the protein preparation may occur at temperatures quite close to room temperature. Hence, in order to minimize the temperature interval for the initial step of the thermal stabilization of the DSC instrument, a heating rate was chosen to be sufficiently slow.
18. To calibrate the calorimeter, the DSC curve for corundum was measured in the temperature interval of interest.
19. Enthalpy changes corresponding to the exothermic peaks were reproducible within ±4 J/g. The reproducibility of temperature for the maximum of the observed exothermic peaks was ±4°C. Second runs of the HSA suspensions were also performed. No reversible phenomena were observed.

Acknowledgments

Authors gratefully acknowledge financial support from Russian Foundation for Basic Researches (grant N98-03-32102) and from Center for Natural Sciences at St. Petersburg State University (grant N 97-0-9.3-283).

References

1. Hemminger, W. and Hohne, G. (1984) *Calorimetry. Fundamentals and Practice*. Verlag, Chemie, Weinhem.
2. Yamane, T., Kojima, Y., Ichiryu, T., Nagata, M., and Shimizu, S. (1989) Intramolecular esterification by lipase powder in microaqueous benzene: effect of moisture content. *Biotechnol. Bioeng.* **39**, 392–397.
3. Yamane, T., Kojima, Y., Ichiryu, T., and Shimizu, S. (1988) Biocatalysis in a microaqueous organic solvent. *Ann. NY Acad. Sci.* **542**, 282–293.
4. Zaks, A. and Klibanov, A. M. (1988) The effect of water on enzyme action in organic media. *J. Biol. Chem.* **263**, 8017–8021.
5. Borisover, M. D., Sirotkin, V. A., and Solomonov, B. N. (1995) Isotherm of water sorption by human serum albumin in dioxane: comparison with calorimetric data. *J. Phys. Org. Chem.* **8**, 84–88.
6. Munster, A. (1969) *Chemische Thermodynamic*. Akademie-Verlag, Berlin, Chap. 3.
7. Kipling, J. J. (1965) *Adsorption from Solutions of Non-Electrolytes*. Academic, London.
8. Ponec, V., Knor, Z., and Cerny, S. (1974) *Adsorption on Solids*. Butterworth, London.
9. Parfitt, G. D. and Rochester, C. H. (1983) Adsorption of small molecules, in *Adsorption from Solution at the Solid/Liquid Interface* (Parfitt, G. D. and Rochester, C. H., eds.), Academic, London, pp. 3–48.
10. Borisover, M. D., Sirotkin, V. A., and Solomonov, B. N. (1993) Calorimetric study of human serum albumin in the water–dioxane mixtures. *J. Phys. Org. Chem.* **6**, 251–253.
11. Borisover, M. D., Sirotkin, V. A., and Solomonov, B. N. (1995) Thermodynamics of water binding by human serum albumin suspended in acetonitrile. *Thermochim. Acta* **254**, 47–53.

12. Sirotkin, V. A., Borisover, M. D., and Solomonov, B. N. (1995) Heat effects and water sorption by human serum albumin on its suspension in water – dimethyl sulfoxide mixtures. *Thermochim. Acta* **256,** 175–183.

13. Borisover, M. D., Sirotkin, V. A., and Solomonov, B. N. (1996) Interactions of water with human serum albumin suspended in water – organic mixtures. *Thermochim. Acta* **284,** 263–277.

14. Sirotkin V. A., Borisover, M. D., and Solomonov, B. N. (1997) Effect of chain length on interactions of aliphatic alcohols with suspended human serum albumin. *Biophys. Chem.* **69,** 239–248.

15. Zakharychev, D. V., Borisover, M. D., and Solomonov, B. N. (1995) A study of thermal stability of human serum albumin in an anhydrous medium by differential scanning calorimetry. *Russ. J. Phys. Chem. (Engl. Transl.)* **69,** 162–166.

16. Borisover, M. D., Zakharychev, D. V., and Solomonov, B. N. (1999) Effect of solvent composition on DSC exothermic peak of human serum albumin suspended in pyridine–*n*-hexane mixtures. *J. Thermal. Anal. Calorim.* **55,** 85–92.

17. Volkin, D. B., Staubli, A., Langer, R., and Klibanov, A. M. (1991) Enzyme thermoinactivation in anhydrous organic solvents. *Biotechnol. Bioeng.* **37,** 843–853.

18. Battistel, E. and Bianchi, D. (1994) Thermostability of ribonuclease A in organic solvents: a calorimetric and spectroscopic study. *J. Phys. Chem.* **98,** 5368–5375.

19. Sirotkin, V. A., Zinatullin, A. N., Solomonov, B. N., Faizullin, D. A., and Fedotov, V. D. (2000) Study of human serum albumin in anhydrous organic media by isothermal calorimetry and IR-spetroscopy. *Russ. J. Phys. Chem.* (Engl. transl.) **74,** 650–655.

20. Kates, M. (1975) *Techniques of Lipidology*, Mir, Moscow (Russian transl.).

21. Medvedev, V. A. and Efimov, M. E. (1975) Precise calorimeter LKB-8700 for measuring reaction enthalpies in solutions. *Russ. J. Phys. Chem. (Engl. transl.)* **49,** 780–783.

22. Borisover, M. D., Stolov, A. A., Cherkasov, A. R., Izosimova, S. V., and Solomonov, B. N. (1994) Calorimetric and IR-spectroscopic study of intermolecular interactions of water in organic solvents. *Russ. J. Phys. Chem.* **68,** 56–62.

23. Trampe, D. M. and Eckert, C. A. (1991) Calorimetric measurement of partial molar excess enthalpies at infinite dilution. *J. Chem. Eng. Data* **36,** 112–118.

17

Detection of Structural Changes of Enzymes in Nonaqueous Media by Fluorescence and CD Spectroscopy

Hideo Kise

1. Introduction

The ability of enzymes to catalyze synthetic reactions in mainly organic solvents has been well documented. In general, enzymes are insoluble in organic solvents. Therefore, dispersions of hydrated enzymes or immobilized enzymes have been most extensively used for synthetic reactions in organic solvents. In these reaction systems, the nature of organic solvent and water–organic ratio strongly influence the activity and specificity of enzymes. It is reasonably assumed that changes in activity and specificity of enzymes are mainly ascribed to changes in the higher structures of enzymes. Therefore, analysis of enzyme structure is important for the selection of the solvent, but methods to analyze dispersed or immobilized enzymes in organic solvents are rather limited.

Fluorescence spectroscopy can be used for analysis of protein structure either in solutions or in the solid state. The method is based on the fact that, for a tryptophan derivative, fluorescence intensity decreases and the wavelength shifts to a longer wavelength in polar solvents. The intrinsic emission of proteins is mainly the result of tryptophan residues, and hence, a change in higher structure of a protein would result in changes of fluorescence properties of the protein. Therefore, fluorescence technique can be a useful method for detecting the modification of enzyme structure in water-organic solvents.

Circular dichroism (CD) is also a useful method for detecting changes in the higher structures of enzymes. CD spectra are highly sensitive to the folding or unfolding of peptide chains and can be used to detect changes in secondary and

From: *Methods in Biotechnology, Vol. 15: Enzymes in Nonaqueous Solvents: Methods and Protocols*
Edited by: E. N. Vulfson, P. J. Halling, and H. L. Holland © Humana Press Inc., Totowa, NJ

tertiary structures of enzymes. CD spectra are the difference spectra of the absorption of polarized light. Therefore, in contrast to fluorescence measurements, this technique has been limited to analysis of enzyme solutions without precipitates.

In this review, the measurement of fluorescence spectra of α-chymotrypsin and subtilisins dissolved or dispersed in water–organic cosolvent systems will be described. Measurement of CD spectra in water–organic solutions will also be presented. It has been reported that changes in catalytic activity are closely related to changes in fluorescence wavelength or CD spectra for enzymes in solutions or dispersions in cosolvent systems *(1–3)*.

2. Steady-State Fluorescence of α-Chymotrypsin and Subtilisins

Aromatic amino acid residues in proteins emit fluorescence light by excitation with ultraviolet (UV) light at or above 280 nm. The intensity of fluorescence light depends primarily on quantum yield. Fluorescence intensity is also influenced by energy transfer from excited residues to neighboring aromatic amino acid residues or by quenching with other chemical groups. In general, the intensity of fluorescence from phenylalanine and tyrosine residues is negligibly small because of low quantum yield and extensive energy transfer from phenylalanine or tyrosine to tryptophan residues *(4)*. Exceptions are subtilisin Carlsberg and subtilisin BPN' where emission from tyrosine residues is observed by excitation at 280 nm *(5)*. When UV light at or above 295 nm is used, only tryptophan residues are excited.

The emission wavelength of tryptophan or tyrosine residue is a function of solvent polarity. This is illustrated by change of emission wavelength of *N*-acetyl-L-tryptophan ethyl ester in acetonitrile–water **(Fig. 1)**. α-Chymotrypsin has eight tryptophan residues and none of them is accessible to external solvent. However, in acetonitrile–water, the wavelength of fluorescence light of α-chymotrypsin exhibits a characteristic variations with solvent composition. This may be associated with unfolding of the peptide chains resulting in direct contact of trytophan residues with the solvent. Therefore, the emission wavelength would be affected not only by conformational changes of the enzyme but also by the polarity of the solvent that interacts with tryptophan residues. To compensate the effect of polarity change of the solvent on tryptophan emission, a parameter $\Delta\lambda_{em}$ defined by **Eq. 1** was introduced *(2)*:

$$\Delta\lambda_{em} = \lambda^{ATrEE} - \lambda^{E} \tag{1}$$

where λ^{ATrEE} and λ^{E} represent the emission wavelength of *N*-acetyl-L-tryptophan ethyl ester and α-chymotrypsin, respectively. The activity of α-chymotrypsin was well correlated to $\Delta\lambda_{em}$ in acetonitrile–water **(Fig. 2)** *(see* **Note 1)**.

Fluorescence quenching is also useful for probing conformational changes of enzymes. Two kinds of quencher are available for enzymes in solutions:

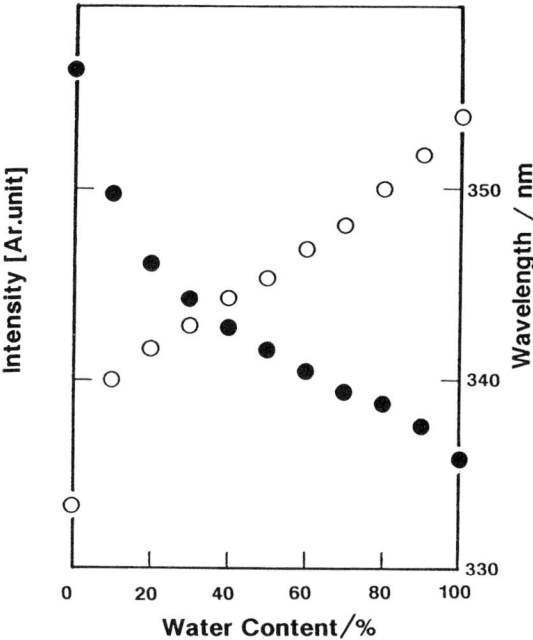

Fig. 1. Wavelength and intensity of maximum fluorescence of *N*-acetyl-L-tryptophan ethyl ester in acetonitrile–water. O: wavelength; ●:intensity.

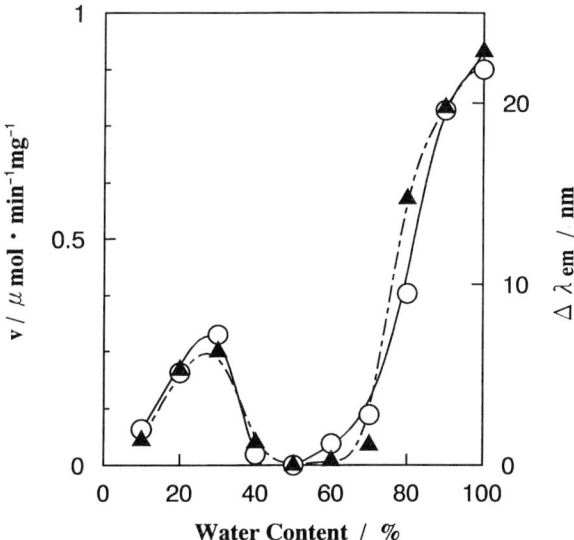

Fig. 2. Hydrolysis rate of *N*-acetyl-L-tyrosine ethyl ester (v) and $\Delta\lambda_{em}$ for α-chymotrypsin in acetonitrile–water ▲: v; O: $\Delta\lambda_{em}$.

ionic and nonionic (organic) quenchers. Examples of the former and the latter are potassium iodide and acrylamide, respectively. The method is based on the fact that a conformational change in enzyme influences the accessibility of fluorophore amino acid residues to a quencher. Iodide ion partitions dominantly in solution, whereas acrylamide partitions between the solvent and enzyme interior. Therefore, acrylamide quenches enzyme fluorescence more effectively than KI, but quenching by KI should be more sensitive to changes in enzyme conformation than acrylamide. However, low solubility of KI in organic solvent limits the water–organic solvent composition for quenching study. Usually, maximum contents of organic solvents are around 60–70%.

Fluorescence quenching for a compound is a function of the concentration of quencher as expressed by the Stern–Volmer equation *(2)*:

$$F_0/F = 1 + K_Q[Q] \qquad (2)$$

where F_0 and F are quantum yields of fluorescence in the absence and presence of a quencher (Q), respectively (**Note 2**). The value of K_Q, which represents quenching efficiency, is obtained as the slope of the plot of F_0/F against [Q]. For most of the proteins that have two or more fluorescent amino acid residues of different accessibility, a modified Stern–Volmer equation is useful for analysis of interactions of quencher with enzyme:

$$\frac{1/f_a}{1/(f_a K_Q[Q])} \qquad (3)$$

where f_a is the fraction of solvent-accessible fluorophore in the enzyme. For the case of $f_a = 1$, **Eq. 3** reduces to **Eq. 2**. The values of f_a and K_Q are obtained from the slope and y-axis intersection of the plot of $F_0/(F_0 - F)$ against [Q]. The value of f_a is a measure of extent of unfolding of peptide chains in enzymes: the higher the f_a, the more unfolded the enzyme.

3. Measurement of Fluorescence Spectra

We have used the Shimadzu RF-5000 instrument with an excitation wavelength of 295 nm and 280 nm for α-chymotrypsin and subtilisin Carlsberg, respectively. The bandwidth was 5 nm. Nonfluorescent quartz cells with a 10-mm path length were used for measurements. Cells were immersed in concentrated sulfuric acid overnight, washed repeatedly by water and acetone, and dried in air. Cells without organic adhesives should be used when organic solvent is used. Organic solvents were dried on molecular sieves. Typically, in a 50-mL vial, 10 mg of enzyme was dissolved in water or the buffer solution. Then the organic solvent was added with gentle stirring at 30°C to the aqueous enzyme solution to obtain a 40-mL solution (1 mg enzyme/4 mL solvent). Solvents had been kept at 30°C before use. Part of the solution was placed in a cell and the solution was stirred magnetically with star-head stirring bars. The temperature of the cell was controlled using a water bath, which circulated water through the cell holder. When the water contents were low (e.g., water contents below 10% in ethanol-water), the mixture was turbid. It was convenient to mix the enzyme solution with organic solvent in the cell, especially when the mixture became turbid (*see* **Note 3**). Fluorescence spectra were measured in side-face geometry. The turbidity was not so high that serious problems occured, as supported by the fact that the plots of maximum wavelength against solvent composition give continuous curves in the solvent concentration range studied.

In solutions or dispersions, enzymes tend to alter their conformation with time. Therefore, it is important to measure the spectrum shortly after the preparation of the sample solutions. Ideally, spectra should be recorded at proper time intervals to follow the changes. The time scale of conformational changes depends on the nature of the enzyme and organic solvent as well as on the organic–water solvent composition. For example, in acetonitrile–borate buffer (90:10 by volume), the maximum emission wavelength (λ_{em}) of α-chymotrypsin shifts from 335 to 343 nm within 1 h. The wavelength and intensity of fluorescence should be determined as an average of more than three measurements.

For measurement of fluorescence quenching, a set of samples with different quencher concentrations in solutions of a specified composition was prepared. Then, an enzyme (1 mg) in the same solvent (1 mL) was added to the quencher solution, and the intensity and the maximum wavelength of emission were measured. The results were analyzed by a modified Stern–Volmer equation. For example, quenching of subtilisin BPN' in mixed solvent of water and ethanol, acetonitrile, dimethyl formamide (DMF), or dimethyl sulfoxide (DMSO) obeyed a modified Stern–Volmer equation for 0.05 M to 1 M KI.

4. Information from CD Spectra

Circular dichroism spectra have been used for probing secondary and tertiary structures of proteins. The measurement is based on the phenomenon that the absorbances, and therefore absorption spectra, of a chiral compound are different for left- and right-circularly polarized lights. In general, the difference in absorbance (ΔA) defined by **Eq. 4** is recorded as a function of wavelength:

$$\Delta A = A_1 - A_r = cl\,(\varepsilon_1 - \varepsilon_r) = cl\Delta\varepsilon \tag{4}$$

where e is the molar extinction coefficient and c and l are molar concentration and path length, respectively. An alternate measure of CD is molar ellipticity ([q]). The ellipticity q (deg) is an angular measure that is related to DA as follows:

$$\theta = 32.98\,\Delta A \tag{5}$$

Most instruments are calibrated in terms of molar ellipticity that is defined as

$$[\theta] = 100 \ \theta/(lc) = 3298 \ \Delta\varepsilon \qquad (6)$$

For proteins, the data are commonly expressed in terms of mean residues rather than molar quantities, and the mean residue ellipticity $[\theta]$ in deg cm^2/decimol) is recorded against wavelength. Ellipticity (in deg), concentration of residue (in decimol/cm^3), and path length (in cm) are used for the calculation.

The CD spectra of polypeptides with high α-helix contents, such as poly(Glu) at pH 4.5, show a typical double minimum at around 222 and 208 nm and a maximum at 190 nm. Polypeptides, which form predominantly β-sheet structure, such as poly(Lys-Leu) with 0.5 M NaF at pH 7, exhibit a minimum band at around 217 nm and a maximum at 195–200 nm. Polypeptides of random structure exhibit a strong minimum band at 195–200 nm.

Several methods have been reported for the determination of a secondary structure with CD. Most of the methods utilize CD spectra of proteins or peptides of known secondary structure determined by X-ray diffraction. A comparison of the results for several methods is summarized in the literature, and a correlation is shown between the results from X-ray and CD methods for a set of proteins *(6,7)*. It should be noted that the CD method is not always accurate for the determination of a secondary structure of specific kinds of proteins. An example is α-chymotrypsin, which is classified as all-β protein (*see* **Note 4**) but does not exhibit typical β-type CD.

Aromatic amino acid residues absorb near-UV light and exhibit CD bands between 250 and 300 nm. Therefore, CD bands in the near-UV region are sensitive to changes in tertiary structure around the aromatic amino acid residues. The intensity of CD bands as a result of aromatic amino acids is generally much smaller than CD bands resulting from peptide chains in the far-UV region. Enzymes with cofactors that absorb visible light exhibit characteristic CD spectra at 380–800 nm. For example, horseradish peroxidase shows a positive band at around 410 nm, which is useful for the detection of conformational changes of the enzyme by organic solvents.

5. Measurement of CD Spectra

The CD spectra of enzymes in water-organic mixed solvents are measured using cells with 1-mm and 10-mm path lengths for far-UV (180–260 nm) and near-UV (240–300 nm) regions, respectively. In our study the concentration of the enzyme is usually 4 μM. Solutions of different enzyme concentrations do not always give the same spectra even if the spectra are normalized for concentration (*see* **Note 5**). Therefore, measurements should be done for solutions of a constant concentration of enzyme for analysis of the structure.

Current commercial instruments can record CD spectra down to about 170 nm. However, for water–organic solutions, the lower limit is higher than 180 nm, depending on the absorption edge of the organic solvent. For example, spectra below 200 nm are not informative for enzymes in mixed solvents of water and acetonitrile, THF, or 1,4-dioxane. In ethanol– or methanol–water, CD can be extended to 190 nm.

The spectra are taken for transparent solutions of enzymes under nitrogen. The temperature of the cell is controlled using a water bath, which circulates water through the cell holder. Warming up of instrument is necessary for 0.5–1 h to measure the spectra under constant conditions. The life of a lamp is about 500–600 h, and the light intensity should be checked before measurement. Calibration of the instrument should be done regularly, especially when a new lamp is used, using a standard solution of 6 mg ammonium *d*-10-camphorsulfonate (ACS) in 10 mL pure water. Spectra are measured on solutions with and without the enzyme, and the difference is recorded as the CD spectrum. It is convenient to measure the absorption spectrum of the sample solution before measuring the CD to check the concentration of the enzyme. Absorbance should not be higher than 2, otherwise CD spectra can be noisy. More than two data accumulations are desirable, but the stability of enzyme in solution during the measurement must be confirmed. As described, time-dependent conformational changes can occur for enzymes in water–organic solvents. Also, rapid autolysis can occur for proteases such as subtilisin Carlsberg. Therefore, the scan speed and the number of accumulations should be decided depending on the time-scale of structural changes of the enzymes.

Circular dicroism is useful for studies on reversibility of conformational changes of enzymes associated with variation of the solvent composition. As an example, **Fig. 3** illustrates a change of CD for α-chymotrypsin by changing the ethanol–water solvent composition. First, the CD spectra was measured for α-chymotrypsin in ethanol–water (50:50) (solution 1). Then, the solvent composition was changed to ethanol–water (10:90) by the addition of water (solution 2). The CD spectrum rapidly changed to a spectrum of nativelike structure, suggesting that folding of peptide chains occurs rapidly and is partly reversible. A problem with this method is that the concentrations of solutions 1 and 2 are different. Therefore, it is advisable to prepare an enzyme solution of the same concentration as solution 2 as a reference. A slower and less complete transition of structure is observed in going form 50:50 to 90:10 ethanol-water.

6. Notes

1. Good correlation was obtained also between activity and $\Delta\lambda_{em}$ for subtilisin Carlsberg and subtilisin BPN' *(1)*.
2. Fluorescence intensity can be used instead of quantum yield.

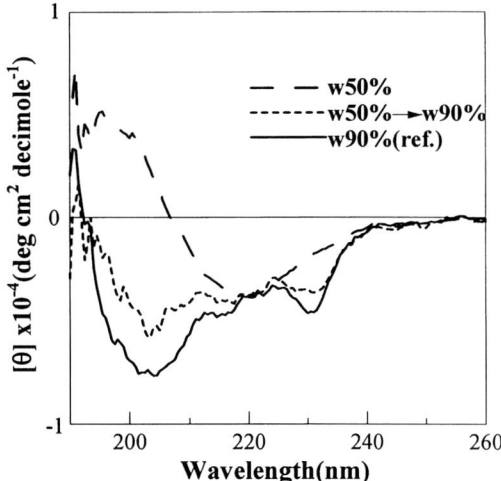

Fig. 3. Changes of CD of α-chymotrypsin by changes of solvent composition (w represents water content).

3. When buffer solutions are used, the salts of buffer components precipitate at high concentrations of organic solvents. In these cases, enzymes form either free suspensions or coprecipitates with salts. Under these conditions, solutions are not buffered, and, therefore, it is advisable to use pure water or buffer solutions of low concentrations (e.g., <0.01 M) to avoid excess turbidity.

4. From X-ray diffraction data, the contents of α-helix and β-sheet structure were determined as 9% and 34%, respectively *(8)*.

5. As stated in the text, CD is basically the difference in absorbance for left- and right-circularly polarized light. Therefore, samples with high absorbance tend to give CD spectra of low reproducibility even if the difference ΔA is small. It is helpful to measure both CD and absorption spectra for a sample solution.

References

1. Kijima, T., Yamamoto, S., and Kise, H. (1994) Fluorescence spectroscopic study of subtilisins as relevant to their catalytic activity in aqueous–organic media. *Bull. Chem. Soc. Jpn.* **67,** 2819–2824.

2. Kijima, T., Yamamoto, S., and Kise, H. (1996) Study on tryptophan fluorescence and catalytic activity of α-chymotrypsin in aqueous-organic media. *Enzyme Microb. Technol.* **18,** 2–6.

3. Sasaki, T., Kobayashi, M., and Kise, H. (1997) Active conformation of α-chymotrypsin in organic solvents as studied by circular dichroism. *Biotechnol. Tech.* **11,** 387–390.

4. Teale, F. W. J. (1960) The ultraviolet fluorescence of proteins in neutral solution. *Biochem. J.* **76,** 381–388.

5. Genov, N., Nicolov, P., Betzel, C., Wilson, K., and Dolashka, P. (1993) Fluorescence properties of subtilisins and related proteinases (subtilases): relation to X-ray models. *J. Photochem. Photobiol. B* **18,** 265–272.

6. Yang, J. T., Wu, C.-S. C., and Martinez, H. M. (1986) Calculation of protein conformation from circular dichroism. *Methods Enzymol.* **130,** 208–290.

7. Woody, R. W. (1995) Circular dichroism. *Methods Enzymol.* **246,** 34–71.

8. Birktoft, J. J. and Blow, D. M. (1972) Structure of crystalline α-chymotrypsin. V. The atomic structure of tosyl-α-chymotrypsin at 2 Å resolution. *J. Mol. Biol.* **68,** 187–240.

18

The Effects of Crown Ethers on the Activity of Enzymes in Organic Solvents

Dirk-Jan van Unen, Johan F. J. Engbersen, and David N. Reinhoudt

1. Introduction

Currently, the applicability of enzymes in synthetic organic chemistry is well recognized. The field of enzyme-catalyzed organic synthesis has been further boosted by the recognition that enzymes can operate in organic solvents. The use of nonaqueous media for enzymatic conversions offers a number of advantages, like enhanced thermal stability of the enzyme, increased substrate solubility, a shift of the equilibrium in favor of synthesis over hydrolysis, and altered selectivity properties of the enzyme *(1)*. However, the most important drawback of the use of enzymes in organic media is the reduced catalytic activity compared to aqueous conditions. Typically, this reduction in activity is two to six orders of magnitude *(2)*. However, studies in our laboratory have revealed that (pre)treatment of the enzymes with crown ethers can enhance enzyme activities in nonaqueous organic solvents up to a level approaching the activity under aqueous conditions.

Crown ethers, discovered in 1967 by Pedersen *(3)*, are cyclic polyethers composed of ethylene-oxy units. They have an extraordinary ability to solvate alkali metal ions by sequestering the metal in the center of the polyether cavity (*see* **Fig. 1**). The selectivity for cation complexation is dependent on the size of the ring (e.g., 18-crown-6 [2] is selective for the K^+ ion, whereas the smaller 15-crown-5 [1] is selective for the Na^+ ion *[4]*). Furthermore, 18-crown-6 is able to complex ammonium ions and water *(5)*. The selectivity of crown ethers cannot only be tuned by variation of the ring size, but also the number and type of donor atoms can be altered, an example being monoaza-18-crown-6 [3].

From: *Methods in Biotechnology, Vol. 15: Enzymes in Nonaqueous Solvents: Methods and Protocols*
Edited by: E. N. Vulfson, P. J. Halling, and H. L. Holland © Humana Press Inc., Totowa, NJ

Crown ether structures

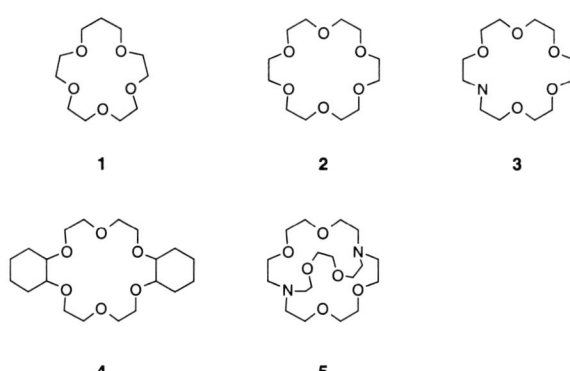

Fig. 1. Various crown ether structures for selective cation binding. 1: 15-crown-5, 2: 18-crown-6, 3: monoasa-18-crown-6, 4:dicyclohexyl-18-crown-6, 5: kryptotix 2.2.2.

Substituents on the ring, like in dicyclohexyl-18-crown-6 [4], may change the flexibility and geometry of the ring, as well as the electronegativity of the donor atoms. Upon introduction of an additional bridge, the name is changed from crown ether into cryptand, like kryptofix 2.2.2 [5].

Interactions between proteins and crown ethers could be expected to occur because proteins possess lysine ammonium groups. Weak association constants between cytochrome C and 18-crown-6 in methanol ($K_s = 1$–$3 \ M^{-1}$) have been reported by Odell and Earlam (6). These interactions did even result in solubilization of some proteins in polar organic solvents.

In this chapter, we present an overview of the studies performed in our laboratories on the effects of crown ethers on the activity of enzymes in nonaqueous organic solvents. Two different types of enzymes, which have shown practical applications in organic solvents (e.g., proteases and lipases), are discussed in this order. Different strategies of (pre)treatment with crown ethers in order to achieve enhancement of enzyme activity are dealt with in **Subheading 3.1.**

2. Materials

Crown ethers: 18-crown-6 (Shell), dibenzo-18-crown-6 (Aldrich), dicyclohexyl-18-crown-6 (Aldrich), monoaza-18-crown-6 (Aldrich), dibenzo-24-crown-8 (Fluka), diaza-18-crown-6 (Merck), 15-crown-5 (Merck), decyl-18-crown-6 (Merck), and kryptofix 2.2.2 didecyl (Merck) are stored at 4°C and used as such. Pentaglyme is synthesized from pentaethylene glycol (Aldrich) by reaction with sodium hydride and methyliodide.

Enzymes. α-Chymotrypsin (E.C. 3.4.21.1), type II, from bovine pancreas (54 U/mg), subtilisin Carlsberg (E.C. 3.4.21.62), type VIII, from *Bacillus licheniformis* (10.4 U/mg), and trypsin (E.C. 3.4.21.4), type III, from bovine pancreas (10,600 U/mg) were from Sigma. Crosslinked crystalline subtilisin Carlsberg is prepared according to the method of Schmitke *(7)*. Lipases (E.C. 3.1.1.3) from *Candida antarctica* (3 U/mg), *Aspergillus niger* (1 U/mg), *Candida cylindracea* (32.3 U/mg), *Mucor miehei* (1.3 U/mg), *Pseudomonas fluorescens* (3414 U/mg), and *Pseudomonas cepacia* (48 U/mg) were obtained from Fluka.

Other materials. The buffers used for the pretreatment of proteases and lipases are subsequently 20 m*M* Na_2HPO_4/NaH_2PO_4, pH 7.8, and 20 m*M* Na_2HPO_4/NaH_2PO_4, pH 8.0. The enzyme preparations are stored at –20°C after pretreatment.

All solvents used are of analytical grade or higher and used as such.

N-Ac-L-Phe-OH, L-Phe-NH$_2$, L-Tyr-NH$_2$, and *N*-Ac-L-Phe-OEt were obtained from Sigma. D-Phe-NH$_2$ and L-Leu-NH$_2$ from Bachem. *N*-Ac-L-Phe-OEtCl is prepared from *N*-Ac-L-Phe-OH and 2-chloroethanol using Amberlite IR-120 as a catalyst. Geraniol was from Aldrich and vinyl acetate from Fluka.

3. Methods

This section describes a number of methods for the treatment of different types of enzyme, viz. proteases and lipases, with crown ethers in order to enhance the enzymatic activity in organic solvents.

3.1. Proteases

Proteases have been intensively studied in water-poor organic solvents using transesterification reactions of *N*-protected amino acid esters *(1)*. Also, peptide bond formation reactions have been catalyzed by proteases in nonaqueous media, which are obviously of greater practical importance. In order to minimize hydrolysis, both types of enzymatic reactions have to be performed under water-poor conditions. This has the drawback that the protease activity is generally low and, therefore, enhancement of the enzymatic activity under these circumstances is of great importance. Crown ether (pre)treatment of the enzymes has proven to be a very effective method for the enhancement of protease activity in organic solvents. Several methods for the pretreatment of normal protease preparations as well as crosslinked crystals of subtilisin Carlsberg will be described.

3.1.1. Crown Ether Addition to Reaction Solvent *(8,9)*

As the structure of the crown ether determines to a large extent the complexation behavior *(vide supra)*, it can be expected that different crown ethers have different effects on the enzyme activity. **Table 1** shows the influence of the crown ether structure on the activity of the serine protease α-chymotrypsin. In

Table 1
Influence of Various Crown Ethers on the
Transesterification Reaction of N-Acetyl-
L-phenylalanine Ethyl Ester with 1-Propanol,
Catalyzed by α-Chymotrypsin in Toluene

Crown ether	Activation[a] $(V_{0(\text{crown ether})}/V_{0(\text{blank})})$
18-Crown-6	19.6
Monoaza-18-crown-6	11.3
Dibenzo-18-crown-6	6.0
Dicyclohexyl-18-crown-6	4.0
2.2 Didecyl kryptofix	2.6
Decyl-18-crown-6	1.9
Dibenzo-24-crown-8	1.8
15-Crown-5	1.8
Pentaglyme	1.4

Conditions: α-Chymotrypsin (5 mg/mL) was lyophilized from 100 mM K$_2$PO$_4$/KH$_2$PO$_4$ buffer, pH 7.8. Model reaction: 2.5 mM ester, 1 M 1-propanol, 2 mM crown ether, 0.5 mg/mL pretreated enzyme, 25°C.
[a]$V_{0(\text{blank})} = 0.50$ μM/min.

this study the transesterification of N-acetyl-L-phenylalanine ethyl ester with 1-propanol in toluene (0.006% H$_2$O) was used as a model reaction.

The crown ethers were simply added to the substrate-containing reaction medium (*see* **Note 1**). **Table 1** clearly indicates that the 18-membered crown ethers are the most effective. The addition of 2 mM 18-crown-6 results in a 20 times higher enzyme activity. The 2-mM crown ether concentration was found to be optimal in case of 18-crown-6 with α-chymotrypsin in toluene. The activation gradually increases until a plateau value for the activation of 20 times is reached at concentrations above 1.5 mM. The linear chain analog of 18-crown-6, pentaglyme (CH$_3$-O-[-CH$_2$-CH$_2$-O-]$_5$-CH$_3$), has hardly any effect on the rate of the reaction. This shows that the acceleration of the enzyme-catalyzed reaction is mainly caused by a *macrocyclic* effect.

As shown in **Table 2** most profound accelerations by crown ether are observed in the more hydrophobic solvents, such as octane, cyclohexane, dibutyl ether, and toluene. The intrinsic enzyme activity (in the absence of crown ether) is already higher in these solvents than in the hydrophilic solvents because of their lower ability to strip of the essential water layer from the enzyme (*1*). Although the crown ether activation effect is lower in hydrophilic solvents also in these solvents, significant enhancements of enzyme activity can be obtained by the simple addition of 18-crown-6 to the reaction solvent.

Table 2
Influence of 18-Crown-6 on the Transesterification Reaction of *N*-Acetyl-L-Phenylalanine Ethyl Ester with 1-Propanol, Catalyzed by α-Chymotrypsin in Various Solvents

Solvent	$V_{0(blank)}$ (μM/min)	$V_{0(2\ mM\ 18\text{-}C\text{-}6)}$ (μM/min)	Activation
Octane	5.4 ± 0.6	154.5 ± 8.0	29
Cyclohexane	4.2 ± 0.8	80.5 ± 9.5	19
Dibutyl ether	0.17 ± 0.01	5.3 ± 0.4	31
Toluene	0.50 ± 0.02	9.8 ± 1.3	20
t-Amylalcohol	0.02 ± 0.00	0.04 ± 0.00	2
THF + 1%H$_2$0	0.18 ± 0.00	1.45 ± 0.05	8

Conditions: α-Chymotrypsin (5 mg/mL) was lyophilized from 100 mM K$_2$PO$_4$/KH$_2$PO$_4$ buffer, pH 7.8. Model reaction: 2.5 mM ester, 1 M 1-propanol, 2 mM 18-crown ether-6, 0.5 mg/mL pretreated enzyme, 25°C.

3.1.2. Pretreatment by Lyophilization in the Presence of Crown Ethers (10–12)

Biocatalysts that are used in organic media are generally lyophilized from an aqueous buffer, adjusted to the optimal pH value for the enzyme activity in water *(13)*. The presence of several additives, such as the inhibitor *N*-Ac-L-Phe-NH$_2$ *(14)*, and the lyoprotectants polyethylene glycol *(14)*, sorbitol *(14)*, and ethyl cellulose *(15)* during the lyophilization process is advantageous for enhancing protease performance in organic solvents. Activation factors up to 150 times have been reported *(14)*.

We found that the addition of 18-crown-6 prior to lyophilization of α-chymotrypsin has a significantly larger effect on the initial activity (**Fig. 2**). In this method the protease (5 mg/mL) is dissolved in a 20 mM Na$_2$HPO$_4$/NaH$_2$PO$_4$ buffer, pH 7.8, containing the indicated amount of crown ether. After incubation for 5 min, the solution is quickly frozen in liquid nitrogen followed by lyophilization for 24 h. After separate equilibration of the enzyme powders, solvents, and substrates in a dessicator above a saturated salt solution at a thermodynamic water activity *(16)* of 0.113, the initial rate of the enzyme reactions is measured. In this case, the transesterification of *N*-Ac-L-Phe-OEt with 1-propanol in cyclohexane was chosen as a model reaction. The presence of 250 equivalents of crown ether with respect to the enzyme results in a 650 times enhanced enzymatic activity (**Fig. 2**). As a result, the second-order rate constant (k_{cat}/K_m) of the suspended enzyme toward *N*-Ac-L-Phe-OEt is 770 M^{-1}/s. This is only 50 times lower than that of the α-chymotrypsin-catalyzed hydrolysis reaction of this substrate in *aqueous solutions*.

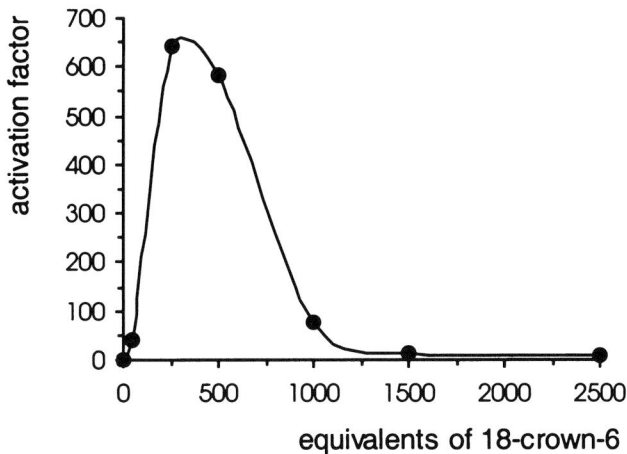

Fig. 2. Crown ether activation ($V_{0(18\text{-crown-}6)}/V_0$) of transesterification catalyzed by α-chymotrypsin as a function of the amount of 18-crown-6 present during lyophilization. α-Chymotrypsin (5 mg/mL) was lyophilized from 20 mM Na$_2$HPO$_4$/NaH$_2$PO$_4$ buffer, pH 7.8, in the presence of the indicated amount of 18-crown-6. Reaction conditions: 2.5 mM N-Ac-L-Phe-OEt, 1 mM 1-PrOH, cyclohexane, a_w = 0.113, 25°C.

The effect of crown ether activation on product yields was investigated for the α-chymotrypsin-catalyzed peptide bond formation between N-Ac-L-Phe-OEtCl and L-Phe-NH$_2$ in acetonitrile. Under initial conditions α-chymotrypsin, pretreated with 50 mol equivalents of 18-crown-6, showed a 450 times higher activity, which corresponds to an enhancement from 1.5×10^{-3} U/mg to 0.7 U/mg of enzyme. After 8 h, almost 70% of the dipeptide was formed with the crown-ether-pretreated enzyme, whereas only 1.5% was formed in case of the nonpretreated enzyme. The crown ether activation effect of protease-catalyzed peptide bond formation was found to be general. Also, other serine proteases, such as subtilisin Carlsberg and trypsin, can be activated by lyophilization in the presence of 18-crown-6 (**11**).

When the crown-ether-pretreated enzyme powders are washed with nonaqueous solvents before application in organic media, the activation is completely lost. Therefore, it can be concluded that the crown-ether-induced activation is not simply a consequence of lyoprotection during the lyophilization process.

3.1.3. Pretreatment of Enzymes by Acetone Precipitation in the Presence of Crown Ethers

Precipitation of proteins from aqueous buffers by trituration with cold acetone is a well-known technique in protein chemistry. After precipitation, followed by centrifugation and decanting of the supernatant, the resulting enzyme powder is washed several times with cold acetone. After solvent evaporation, the dried enzyme powder is obtained. This way of drying proteins is relatively fast compared to lyophilization.

In order to investigate the crown ether activation, 18-crown-6 has been added at three different stages during this precipitation procedure: (1) addition to the aqueous buffer, (2) together with the acetone added for the precipitation, or (3) together with the washing acetone. After equilibration of solvent, substrates, and corresponding enzyme powders at a_w = 0.113, the enzymatic activities were determined for the peptide bond formation reaction with N-Ac-L-Phe-OEtCl and L-Phe-NH$_2$ in acetonitrile. The results are given in **Table 3**. The specific activity of the α-chymotrypsin sample obtained from the acetone precipitation procedure without any addition of crown ether is 4.1 nmol/min/mg of enzyme. This is almost three times higher than the activity of lyophilized α-chymotrypsin powder and shows that acetone precipitation is a fast and useful alternative for pretreatment of enzymes for applications in organic media. A further 10-fold increase in activity is observed when 18-crown-6 is present during the washing procedure, independent of whether 18-crown-6 was present or absent during the actual precipitation. On the other hand, a 10-fold decrease in activity is found in cases where α-chymotrypsin was precipitated in the presence of 18-crown-6 and subsequently washed in the absence of 18-crown-6. This is in accordance with the observation that the crown ether-activated lyophilized α-chymotrypsin powders lose most of their activity upon washing with organic solvent *(vide supra)*. Crown ether activation of α-chymotrypsin by acetone precipitation can therefore only be achieved if the crown ether is present in the final, dry enzyme powder.

3.1.4. Pretreatment of Crosslinked Crystalline Subtilisin Carlsberg (17)

A new and promising technology to improve the operational stability of enzymes in organic solvents is the use of crosslinked enzyme crystals *(18)*. Although such crystals usually have a relatively low catalytic activity in organic media, they are highly resistant against autolysis and thermal inactivation. Therefore, crosslinked enzyme crystals can be reused in many reaction cycles without appreciable loss of activity.

Lyophilization of crosslinked enzyme crystals results in significant loss of the enzymatic activity. Because this method cannot be used, another method of crown ether pretreatment was investigated, that is, soaking of the crosslinked crystalline enzyme in a crown ether solution of acetonitrile, followed by gentle evaporation of the solvent. The enzymatic activities given in **Table 4** for the rates of peptide bond formation of N-Ac-L-Phe-L-Phe-NH$_2$ from N-Ac-L-Phe-OEtCl and L-Phe-NH$_2$ in acetonitrile show that subtilisin Carlsberg crystals can be significantly activated by this procedure. Also in this case, the most effective crown ethers are

Table 3
Influence of 18-Crown-6 Addition at Different Stages During Acetone Precipitation of α-Chymotrypsin

Crown ether addition			V_0	
Buffer	Precipitation acetone	Washing acetone	(nmol/min·mg of enzyme)	$V_0/V_{0(\text{blank})}$
–	–	–	4.1	1
+	–	–	0.3	0.1
–	+	–	0.7	0.2
–	–	+	40.3	10
+	+	–	0.5	0.1
+	–	+	33.2	8
–	+	+	36.6	9
+	+	+	35.7	9

Conditions: 10 mg α-chymotrypsin in 1 mL 20 mM Na$_2$HPO$_4$/NaH$_2$PO$_4$ buffer, pH 7.8, was precipitated with 2 mL cold acetone. The resulting enzyme powder was washed three times with 1 mL acetone, followed by solvent evaporation. If 18-crown-6 was present during the different stages in the process, the concentration was 40 mM. Model reaction: 50 mM N-Ac-L-Phe-OEtCl and L-Phe-NH2, 2.5 mg/mL enzyme powder, acetonitrile, a_w = 0.113, 30°C.

Table 4
Effects of Soaking and Drying of Crosslinked Crystalline Subtilisin Carlsberg from a Solution of Crown Ether in Acetonitrile on the Rate of Peptide Bond Formation

Crown ether	V_0 (nmol/min/mg)	Activation
None	0.43	—
18-Crown-6	4.11	9.6
Monoaza-1-Crown-6	4.82	11.2
15-Crown-5	1.25	2.9
Decyl-18-crown-6	0.54	1.2
Dicyclohexyl-18-crown-6	1.64	3.8
Dibenzo-18-crown-6	0.81	1.9
Dibenzo-24-crown-6	0.38	0.9
Pentaglyme	1.04	2.4

Conditions: Enzyme crystals (1 mg/mL) were dried from acetonitrile containing 15 mM of crown ether. Model reaction: 50 mM N-Ac-L-Phe-OEtCl and L-Phe-NH$_2$, 0.5 mg/mL enzyme crystals, acetonitrile, a_w = 0.113, 30°C.

18-crown-6 and monoaza-18-crown-6, with rate enhancements of about 10. Treatment with more hydrophobic and sterically hindered crown ethers, like dibenzo- and decyl-18-crown-6, is less effective. The macrocyclic effect is also manifest for crosslinked crystalline enzymes, as pretreatment with pentaglyme gives only a twofold rate enhancement.

The crown ether activation of the crosslinked enzyme crystals was found to be very persistent in time. No loss of activity was found for crown-ether-activated crosslinked crystalline subtilisin Carlsberg, even after storage for 4 wk at 4°C.

3.2. Lipases

The use of lipases in nonaqueous media is one of the most promising methods for the enantioselective conversion of a wide variety of hydrophobic acids and alcohols *(19)*. Lipases generally have broad substrate specificity while maintaining a high enantioselectivity. Most of them posses an active site-covering lid and require interfacial activation (i.e., opening of the lid at a water–lipid interface).

Lipases are ubiquitous enzymes and can be of mammalian, fungal, and bacterial origin. The effect of 18-crown-6 on lipase activity has been investigated for different lipases **(Table 5)**. In this case, 18-crown-6 was added prior to lyophilization of the enzymes. The acetylation of geraniol with vinyl acetate in hexane *(20)* was used as a test reaction. The solvent and the pretreated enzyme preparations were equilibrated at $a_w = 0.113$. The results show that crown-ether-induced activation of the enzymatic activity is only significant in cases where the (commercial) lipase preparations are generally of high specific activity (e.g., *C. cylindracea* and *P. fluorescens* lipase). Only minor activation or, sometimes, even a deactivation is observed with less pure lipase samples. The lipase preparations of low specific activity (<10 U/mg) generally contain only a few percent of protein material and the absence of a crown ether effect is probably the result of complexation of 18-crown-6 with the bulk material. After the observation that the purity of the lipase samples is very important for the crown ether activation, additional experiments were perfomed with high-purity *Pseudomonas fluorescens* lipase (3414 U/mg).

Figure 3a–c shows the activity of *P. fluorescens* lipase lyophilized with different amounts of 18-crown-6 in acetonitrile ($a_w = 0.113$), acetonitrile ($a_w = 0.753$), and hexane ($a_w = 0.113$). Although the activity of the enzyme lyophilized in the absence of crown ether is different because of differences in hydrophobicity and thermodynamic water activity of the media, the activation factor is almost identical in the three cases. This suggests that the same activation mechanism is operative under these conditions. As in the case of the proteases, the enzyme activation was also completely lost after washing of the crown-ether-activated enzyme powder with dry ethyl acetate.

3.3. Concluding Remarks

Enzymatic catalysis in nonaqueous solvents is a useful tool for organic synthesis, especially for reactions in which optical resolution steps are involved.

Table 5
Influence of Lyophilization of Various Lipases in the Presence of 18-Crown-6 on the Enzymatic Activity in the Acetylation of Geraniol

Lipase	Conc. 18-crown-6 (mM)	Activity (nmol/min·mg)
Candida antarctica	0	51.4 ± 1.8
(3 U/mg)	5	61.4 ± 2.6
	25	82.8 ± 4.8
Aspergillus niger	0	185 ± 11
(1 U/mg)	5	196 ± 7
	25	34 ± 3
C. cylindracea	0	6.6 ± 1.8
(32.3 U/mg)	5	17.7 ± 2.5
	25	0.05 ± 0.01
Mucor miehei	0	0.15 ± 0.02
(1.3 U/mg)	5	0. 18 ± 0.01
	25	0.05 ± 0.01
Pseudomonas fluorescens	0	1305 ± 94
(3414 U/mg)	5	4713 ± 370
	25	5530 ± 485
P. cepacia	0	309 ± 58
(48 U/mg)	5	550 ± 58
	25	228 ± 76

Conditions: 2 mg/mL lipase in 20 mM Na_2HPO_4/NaH_2PO_4 buffer, pH 8.0, was lyophilized in the presence of the indicated amounts of 18-crown-6. Reaction conditions: 0.4 mg/mL enzyme, 50 μL geraniol, 92 μL vinyl acetate, 5 mL hexane, a_w = 0.113, 30°C.

Because of the intrinsically low catalytic activity of enzymes in these environments, generally two to six orders of magnitude lower than in water, practical methods for enzyme activation are of high importance. We have shown that the (pre)treatment of enzymes with crown ethers is a simple and effective way to increase enzymatic activity in organic media up to a level that is only one to two orders of magnitude lower compared to their activity in water.

The crown ether activation is a macrocyclic effect related with its complexation properties. Highest enzyme activation is generally found for 18-crown-6 and derivatives thereof, which readily form complexes with water, hydronium, ammonium, and alkaline earth metal ions. In previous articles, we have considered a number of mechanisms that may contribute to the enhancement of the enzyme activity by crown ethers *(9,11,17)*. These include complexation of lysine residues to prevent formation of intermolecular and

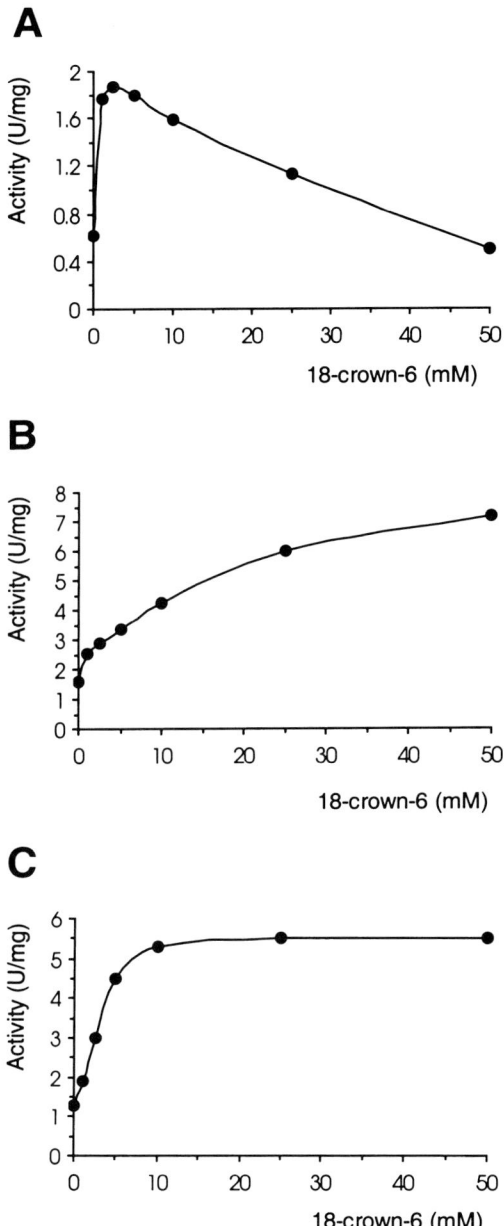

Fig. 3. Effect of 18-crown-6 on the *P. fluorescens* lipase-catalyzed acetylation of geraniol as a function of the presence of 18-crown-6 during lyophilization of the enzyme and reaction conditions. (**A**) Acetonitrile at $a_w = 0.113$. (**B**) Acetonitrile at $a_w = 0.753$. (**C**) Hexane at $a_w = 0.113$. *P. fluorescens* lipase (1 mg/mL) was lyophilized from 20 mM Na$_2$HPO$_4$/NaH$_2$PO$_4$ buffer, pH 8.0, in the presence of the indicated concentration of 18-crown-6. Reaction conditions: 50 mL geraniol, 92 mL vinyl acetate, 5 mL solvent at the indicated thermodynamic water activity, 30°C.

intramolecular salt bridges, which reduce enzyme flexibility, and assistance in transport of tightly bound water molecules from the enzyme active site to improve the binding and orientation of substrate molecules. In addition, crown ether complexes of water or alkali earth metal ions in the active site may contribute to stabilization of the charged transition state *(21)*.

Although the importance of the various factors that can contribute to the crown ether activation are unclear, the crown-ether-induced activation is shown to be generally applicable to various enzymes, different types of enzyme preparation, and a wide variety of solvents, reaction conditions, and reactions. Even the tyrosinase-catalyzed oxidation of *p*-cresol is activated by treatment with crown ethers *(22)*. Therefore, this method gives good prospects for practical applications, as the major problem of enzymatic catalysis in nonaqueous media (i.e., a too low enzymatic activity) can be overcome very effectively by this methodology.

4. Note

1. As most crown ethers are quite hygroscopic and are stored at 4°C, the crown ether-containing vessels should be kept closed during equilibration to ambient temperature.

Acknowledgments

The Council for Chemical Sciences of the Netherlands Organization for Scientific Research (NWO-CW) and the Technology Foundation STW are acknowledged for financial support.

References

1. Koskinen, A. M. P. and Klibanov, A. M. (eds.) (1996) *Enzymatic Reactions in Organic Media*. Blackie Academic & Professional, Glasgow, UK.
2. Dordick, J. S. (1989) Enzymatic catalysts in monophasic organic solvents. *Enzyme Microb. Technol.* **11,** 194–211.
3. Pedersen, C. J. (1967) Cyclic polyethers and their complexes with metal salts. *J. Am. Chem. Soc.* **89,** 2495–2496, 7017–7036.
4. Gokel, G. (1991) *Crown Ethers and Cryptands*. Royal Society of Chemistry, Cambridge, UK.
5. de Jong, F., Reinhoudt, D. N., and Smit, C. J. (1976) On the role of water in the complexation of alkylammonium salts by crown ethers. *Tetrahedron Lett.* 1371–1374.
6. Odell, B. and Earlam, G. (1985) Dissolution of proteins in organic solvents using macrocyclic polyethers: association constants of a cytochrome C–[1,2-^{14}C$_2$]-18-crown-6 complex in methanol. *J. Chem. Soc. Chem. Commun.* 359–361.
7. Schmitke, J. L., Wescott, C. R., and Klibanov, A. M. (1996) The mechanistic dissection of the plunge in enzymatic activity upon transition from water to anhydrous solvents. *J. Am. Chem. Soc.* **118,** 3360–3365.

8. Broos, J., Martin, M. N., Rouwenhorst, I., Verboom, W., and Reinhoudt, D. N. (1991) Acceletation of enzyme-catalyzed reactions in organic solvents by crown ethers. *Recl. Trav. Chim. Pays-Bas* **110,** 222–225.

9. Engbersen, J. F. J., Broos, J., Verboom, W., and Reinhoudt, D. N. (1996) Effects of crown ethers and small amounts of cosolvent on the activity and enantioselectivity of α-chymotrypsin in organic solvents. *Pure Appl. Chem.* **68,** 2171–2178.

10. Broos, J., Sakodinskaya, I. K., Engbersen, J. F. J., Verboom, W., and Reinhoudt, D. N. (1995) Large activation of serine proteases by pretreatment with crown ethers. *J. Chem. Soc. Chem. Commun.* 255,256.

11. van Unen, D. J., Engbersen, J. F. J., and Reinhoudt, D. N. (1998) Large acceleration of α-chymotrypsin-catalyzed dipeptide formation by 18-crown-6 in organic solvents. *Biotechnol. Bioeng.* **59,** 553–556.

12. van Unen, D. J., Engbersen, J. F. J., and Reinhoudt, D. N. (1998) Effects of crown ethers on the activity of enzymes in peptide formation in organic media, in *Progress in Biotechnology, vol. 15: Stability and Stabilization of Biocatalysts* (Ballesteros, A., Plou, F. J., Iborra, J. L., Halling, P. J., eds.), Elsevier, Amsterdam.

13. Zaks, A. and Klibanov, A. M. (1988) Enzymatic catalysts in nonaqueous solvents. *J. Biol. Chem.* **263,** 3194–3201.

14. Dabulis, K. and Klibanov, A. M. (1993) Dramatic enhancement of enzymatic activity in organic solvents by lypoprotectants. *Biotechnol. Bioeng.* **41,** 566–571.

15. Otamiri, M., Aldercreutz, P., and Mattiasson, B. (1992) Complex formation between chymotrypsin and ethyl cellulose as a means to solubilize the enzyme in active form in toluene. *Biocatalysis* **6,** 291–305.

16. Halling, P. J. (1992) Salt hydrates for water activity control with biocatalysts in organic media. *Biotechnol. Tech.* **6,** 271–276.

17. van Unen, D. J., Sakodinskaya, I. K., Engbersen, J. F. J., and Reinhoudt, D. N. (1998) Crown ether activation of cross-linked subtilisin Carlsberg crystals in organic solvents. *J. Chem. Soc. Perkin Trans.* **1,** 3341–3343.

18. St. Clair, N. L. and Navia, M. A. (1992) Cross-linked enzyme crystals as robust biocatalysts. *J. Am. Chem. Soc.* **114,** 7314–7316.

19. Schmid, R. D. and Verger, R. (1998) Lipases: interfacial enzymes with attractive applications. *Angew. Chem. Int. Ed.* **37,** 1608–1633.

20. Gonzalez-Navarro, H. and Braco, L. (1998) Lipase-enhanced activity in flavour ester reactions by trapping enzyme conformers in the presence of interfaces. *Biotechnol. Bioeng.* **69,** 122–127.

21. Xu, Z.-F., Affleck, R., Wangikar, P., Suzawa, V., Dordick, J. S., and Clark, D. S. (1994) Transition state stabilization of subtilisins in organic media. *Biotechnol. Bioeng.* **43,** 515–520.

22. Broos, J., Arend, R., van Dijk, G. B., Verboom, W., Engbersen, J. F. J., and Reinhoudt, D. N. (1996) Enhancement of tyrosinase activity by macrocycles in the oxidation of *p*-cresol in organic solvents. *J. Chem. Soc., Perkins Trans.* **1,** 1415–1417.

19

Control of Acid–Base Conditions in Low-Water Media

Johann Partridge, Neil Harper, Barry D. Moore, and Peter J. Halling

1 Introduction
1.1. Background

The catalytic activity of an enzyme is profoundly affected by its ionization state, whether it is dissolved in aqueous solution or suspended in low-water organic media. In aqueous solution, counterions can freely move around in a solution. Because they are not closely associated with opposite charges, their identity does not effect the protonation state of the enzyme; thus, pH alone governs the protonation state. When a biocatalyst is suspended in a low-water organic solvent, the situation is more complex. In this case, counterions are in closer contact with the opposite charges on the enzyme because of the lower dielectric constant of the medium. Thus, protonation of ionizable groups on the enzyme will be controlled by the type and availability of these ions as well as hydrogen ions. Changes in ionization state of the protein can therefore be described by two equilibria that can, in theory, be controlled independently (*1,2*):

1. Exchange of hydrogen ions and cations with acidic groups of the protein. For example, carboxyl groups require simultaneous exchange of H^+ with a cation such as Na^+.

$$\text{Protein-COOH} + Na^+ \rightleftharpoons \text{Protein-COO}^-Na^+ + H^+$$

Such equilibria can be characterized by the ratio of thermodynamic activities, a_{H^+}/a_{Na^+} (this may also be represented as pH–pNa).

From: *Methods in Biotechnology, Vol. 15: Enzymes in Nonaqueous Solvents: Methods and Protocols*
Edited by: E. N. Vulfson, P. J. Halling, and H. L. Holland © Humana Press Inc., Totowa, NJ

2. Transfer of both a hydrogen ion and an anion onto basic groups of the protein. For example, amino groups bind or release H^+ and an anion such as Cl^- together.

$$\text{Protein-NH}_3^+\text{Cl}^- \rightleftharpoons \text{Protein-NH}_2 + H^+ + Cl^-$$

Such equilibria can be characterized by the product of thermodynamic activities, $a_{H^+}a_{Cl^-}$ (or pH+pCl).

1.2. Need for Buffers That Will Control Acid–Base Conditions

It is well known that the protonation state and subsequent catalytic activity of biocatalysts in low-water organic media are dictated by the pH of the aqueous solution from which they are dried *(3,4)*. The profile of reaction rate in organic solvent for a given enzyme versus pH of aqueous buffer (from which it has been dried) is often found to resemble that for the same enzyme in aqueous solution. Until recently, this phenomenon of "pH memory" has been exploited as the general method for fixing "apparent pH" of enzymes in organic media. Nevertheless, it carries several disadvantages. First, if the effect of the enzyme protonation state on catalytic activity is studied using this approach, it is both tedious and time consuming. Separate samples of enzyme must be prepared by exposing to aqueous buffers of varying pH, followed by subsequent drying. In addition, such a study is further complicated because it has been shown that the identity and concentration of aqueous buffer species also affects the resulting catalytic rate *(5,6)*. More importantly with this procedure, which at best sets the initial enzyme protonation state, there will is no way of buffering against subsequent changes in acid–base conditions that can occur as the reaction proceeds. Such changes have been confirmed by Valivety et al. *(7)* and Brown et al. *(8)*. The authors demonstrated that the pH of an inaccessible aqueous phase was altered significantly during the course of a reaction using hydrophobic indicators. Furthermore, dramatic decreases in enzymatic rate due to formation of acidic by-products have been observed for biocatalysts in organic solvents *(9)*.

The inherent disadvantages of the "pH memory" method have led to the recent development of buffers that can be conveniently added directly to the organic phase and that control acid–base conditions throughout the time-course of the reaction. These buffers function by exchanging protons and counterions with the enzyme, thus setting values for the parameters a_{H^+}/a_{Na^+} and $a_{H^+}a_{Cl^-}$. Depending on the buffer used, either of these parameters will be fixed. They can be added to the reaction mixture in two different forms: organic soluble and solid state (insoluble). The aim of this chapter is to present the practical methods for utilization of these novel buffers with biocatalysts in organic media.

1.3. Organic Soluble Buffers

A useful organic-phase buffer pair must be sufficiently hydrophobic so that each form is soluble in the chosen solvent. In addition, the acid–base strength of the compound must be appropriate to control protein ionization in the range required.

Blackwood et al. *(1)* successfully demonstrated the use of a hydrophobic acid (triphenylacetic acid) and its sodium salt to control the pH–pNa for a transesterification by subtilisin Carlsberg in pentanone. By changing the ratio of buffer acid:base in the reaction mixture, the authors showed that the protonation state of the enzyme and thus its subsequent catalytic activity varied significantly. In addition, they also used a hydrophobic amine (triisooctylamine) with its hydrochloride salt to control the pH+pCl for the transesterification by a lipase in pentanone. A similar more extensive range of soluble buffers for pH–pNa control have been employed *(10)*.

One disadvantage of these buffers is that they are only soluble in fairly polar organic solvents (e.g., pentanone and acetonitrile). However, more recently, a more hydrophobic buffer based on a dendritic polybenzyl ether (composed of acid and sodium salt) has been successfully synthesized and used to control acid–base conditions in more hydrophobic solvents such as toluene *(11)*.

1.4. Solid-State Buffers

These buffer pairs exist as crystalline solids that exchange H^+ with Na^+ (in control of a_{H+}/a_{Na+}) or accept/donate H^+ and Cl^- together (in control of $a_{H+}a_{Cl-}$) in order to control the protonation state of the enzyme. In general, a zwitterion along with the appropriate salt is used because the pair must be insoluble in the organic reaction mixture. Unlike the organic soluble buffers, each buffer pair will set a fixed value of the relevent ionization parameter, regardless of the quantities of each form used. Also, they can be used in a number of solvents because their ionization is independent of the nature of the solvent.

The usefulness of a range of solid-state buffers for control of pH+pCl (e.g., Lys/Lys·HCl) has been shown for immobilized subtilisin in hexane and toluene *(2)*. Similarly, pH–pNa solid-state buffer pairs have recently proved successful for reactions in both polar and nonpolar solvents with different forms of the same enzyme (Harper et al., unpublished results; Partridge et al., unpublished results).

2. Materials

1. *Solvents.* In principle, any solvent should be suitable for use with solid-state buffers. With polar solvents, high water content/water activity should be avoided to prevent the buffer from going into the solution. When organic soluble buffers are used, care should be taken to ensure that the buffer pair is soluble in the

chosen solvent (10 m*M* is an adequate buffer concentration). Most anhydrous solvents can be obtained from Merck and should be dried and stored over molecular sieves prior to use.

2. *Organic soluble buffers.* A number of suitable organic soluble acids and their sodium salts are commercially available from Aldrich (e.g., triphenylacetic acid [T8,120-5], phenylboronic acid [P2,000-9], *p*-toluenesulfonic acid [40,288-5], and *p*-nitrophenol [24,132-6]). Sodium salts can also be prepared simply in the laboratory (*see* **Subheading 3.1.1.**). Similarly, appropriate amines that are soluble in organic solvents can be purchased from Aldrich. Examples of those successfully used in previous work include triisooctylamine (26,149-1) and tridodecylamine (30,613-4). The hydrochlorides of these amines can be prepared (*see* **Subheading 3.1.2.**).

3. *Solid-state buffers.* Generally, biological buffers along with their sodium or potassium salts have been utilized. The following can be obtained from Sigma: AMPSO (3-[(1,1-dimethyl-2-hydroxyethyl)amino]-2-hydroxy-propanesulfonic acid), CAPSO (3-[cyclohexylamino]-2-hydroxy-1-propanesulfonic acid), HEPES (*N*-[2-hydroxyethyl]piperazine-*N*'-[2-ethanesulfonic acid]), HEPPSO (*N*-[2-hydroxyethyl]piperazine-*N*'-[2-hydroxy-propanesulfonic acid]), MOPS (3-[*N*-morpholino]propanesulfonic acid), PIPES (piperazine-*N*,*N*'-*bis*[2-ethane-sulfonic acid]), TAPS (*N*-tris[hydroxymethyl] methyl-3-aminopropanesulfonic acid), and TES (*N*-tris[hydroxymethyl]methyl-2-aminoethanesulfonic acid). The corresponding sodium salts can also be purchased from Sigma.

3. Methods

3.1. Organic Soluble Buffers

3.1.1. Preparation of Sodium Salts of Organic Soluble Acids

1. Mix 0.95 mol equivalent of NaOH with a suspension of the organic soluble acid (e.g., triphenylacetic acid) in water.
2. Remove unreacted acid by filtration.
3. Freeze-dry the filtrate to recover the sodium salt.
4. Dry the resultant powder over molecular sieves and store until required.

3.1.2. Preparation of Hydrochlorides of Organic Soluble Amines

1. Dissolve the organic soluble amine (e.g., triisooctylamine) in a suitable solvent such as dichloromethane to give a final concentration of 100 m*M*.
2. Add a 1.8 mol equivalent of concentrated aqueous HCl to the amine solution.
3. Shake mixture for half an hour.
4. Remove solvent by rotary evaporation (thus ensuring any excess HCl is removed).
5. Dry resultant crystals under vacuum and store over molecular sieves until required.

3.1.3. Preparation of Acid and Sodium Salts of Dendrimers

Synthesis of acid and sodium salts of dendrimers is outlined in **ref. 11**.

3.1.4. Reactions Using Organic Soluble Buffers (see **Note 1**)

The model reaction described here is the transesterification of *N*-acetyl-L-tyrosine ethyl ester (10 mM) with propan-1-ol (1 M) in pentanone using an immobilized form of subtilisin Carlsberg. However, this method is applicable for a wide range of substrates, enzymes, solvents, and so forth.

1. Weigh out a catalytic amount of enzyme (*see* **Note 2**). For convenience do this in the final reaction vial.
2. Dry enzyme over molecular sieves (*see* **Notes 3** and **4**).
3. Equilibrate the enzyme over the relevant saturated salt solution in a sealed jar to give the required a_w (*see* **Note 4**).
4. Weigh out acid and base forms of the chosen organic soluble buffer to give a final total buffer concentration of 10 mM in 12 mL. (Molar ratios from 9:1 to 1:9 can be used depending on buffer ratio that gives good catalytic activity).
5. Dissolve buffer pair in 11.1 mL of dry pentan-3-one.
6. Add 0.9 mL of anhydrous propan-1-ol to the pentanone (containing organic soluble buffer), giving the required final concentration of the alcohol (in this case 1 M).
7. Equilibrate the organic phase (reaction solvent, propanol, and dissolved buffer) to the same fixed a_w over that relevant saturated salt solution in a sealed jar (*see* **Notes 5** and **6**).
8. When equilibration of both phases is complete, add 10 mL of organic phase to the enzyme.
9. Reaction is initiated by adding a solid ester substrate to give the required final concentration in 10 mL (in this case, 10 mM).
10. Shake the reaction mixture at a fixed rate and controlled temperature.
11. Withdraw samples periodically for analysis.

3.2. Solid-State Buffer

3.2.1. Reactions in Polar Solvents

The model reaction described here is the transesterification of *N*-acetyl-L-tyrosine ethyl ester (10 mM) with propan-1-ol (1 M) in acetonitrile using an immobilized form of subtilisin Carlsberg. However, this method should be applicable for a wide range of substrates, enzymes, polar solvents, and so forth.

1. Dry the enzyme using a suitable method.
2. Suspend the enzyme in 9.25 mL of acetonitrile with 1% H_2O by volume, $a_w = 0.22$ (*see* **Notes 7–9**).
3. Add 10 mg of each form of buffer pair per milliliter of organic reaction mixture to the suspension of the enzyme in the reaction solvent (*see* **Note 10**).
4. Dissolve the substrate ester in 1 mL of dry propan-1-ol.

5. To initiate the reaction, immediately add 0.75 mL of the substrate solution to the above reaction mixture, giving the required final substrate concentration (in this case, 10 mM ester, 1 M propanol). (*See* **Note 11.**)
6. Incubate the mixture at a known temperature with constant reciprocal shaking.
7. Withdraw the samples from the reaction mixture at regular intervals for analysis.

3.2.2. Reactions in Hydrophobic Solvents

The model reaction described here is the transesterification of *N*-acetyl-L-tyrosine ethyl ester (10 mM) with propan-1-ol (1 M) in toluene using an immobilized form of subtilisin Carlsberg. However, this method should be applicable to a wide range of substrates, enzymes, hydrophobic solvents, and so forth.

1. Ensure that all water is removed from selected buffer pair prior to use (*see* **Notes 12** and **13**). Store the buffers over molecular sieves until required.
2. Weigh out a catalytic amount of enzyme (*see* **Note 2**). For convenience, do this in the final reaction vial.
3. Dry the enzyme over molecular sieves (*see* **Notes 3** and **4**).
4. Equilibrate enzyme over the relevant saturated salt solution in sealed jar to give the required a_w (*see* **Note 4**).
5. Weigh out 240 mg of each form of solid-state buffer (this is equivalent to 20 mg/mL of each form in organic reaction mixture). (*See* **Note 10.**)
6. To the buffer pair, add 11.1 mL of dry toluene and 0.9 mL of anhydrous propanol (giving the required final concentration of the alcohol substrate, in this case, 1 M).
7. Equilibrate the organic phase (reaction solvent, alcohol substrate, and solid-state buffer) to the same fixed a_w over relevant saturated salt solution in a sealed jar (*see* **Notes 5** and **6**).
8. When equilibration of both phases is complete, add 10 mL of organic phase to the enzyme (*see* **Note 14**).
9. Reaction is initiated by adding the solid ester substrate to give the required final concentration in 10 mL (in this case, 10 mM). (*See* **Note 15.**)
10. Shake the reaction mixture at a fixed rate and controlled temperature.
11. Withdraw 100-µL samples periodically for analysis.

4. Notes

1. There may be solubility problems with these buffers in some solvents. This tends to occur with sodium salts of acids or hydrochlorides of bases. Usually, more polar solvents such as pentanone or acetonitrile must be used. However, some buffer salts (particularly dendrimer salts) can be dissolved in more hydrophobic solvents such as toluene when 1 M of a polar solvent is present. Quite conveniently, this is the substrate alcohol for the model transesterification discussed here.
2. The amount of enzyme used will depend on the form of the enzyme, the reaction solvent, the substrates, and so forth.
3. For reproducible behavior, dry the enzyme first because of possible hysteresis effects (*9*).

4. Monitor the drying/equilibration of the enzyme to a fixed a_w by weighing. When the weight is constant, the drying/equilibration step is complete. Depending on the form of the enzyme this can take between 2 and 7 d.

5. Equilibration can be monitored by measuring changes in water content using Karl Fischer apparatus. Equilibration normally takes between 12 and 24 h.

6. There may be a significant loss of solvent from the vial during equilibration, even though the vessel is sealed. This is particularly noted with volatile solvents such as hexane. To minimize this, a large beaker containing the reaction solvent and propanol (1 *M*) should also be placed in the sealed jar to flood the headspace.

7. For convenient conversions between a_w and water content in polar solvents, see **ref. *14***.

8. Reactions should be carried out at relatively low a_w as a step to minimize solubility of the buffer in the polar solvent. In addition, some of the salts may be capable of forming hydrates at sufficiently high a_w values, particularly in more polar solvents.

9. Some buffer salts have small amounts of water associated with them. However, because this reaction is in polar solvent with 1% (v/v) water, small changes in the system's water content will not significantly alter the a_w. Nevertheless, if these reactions were performed in a hydrophobic solvent such as hexane, it is likely that the low levels of water associated with the buffers would result in significant increases in system's a_w. Therefore, more care would have to be taken to equilibrate the buffers to the required a_w level prior to use in hydrophobic media.

10. Concentrations of >2 mg/mL for each buffer form in organic reaction mixtures were sufficient to control buffering in this system. However, excess concentrations of 10 and 20 mg/mL were used for polar and hydrophobic solvents, respectively.

11. There is no need to allow solid-state buffer to equilibrate in the solvent–enzyme suspension. No lag period in initial rates for reactions in polar solvents has been observed. Thus, it would appear that the transfer of ions between the enzyme and buffers is relatively quick in polar organic media.

12. This can be achieved by drying overnight at 110°C in an oven.

13. In this case, it is essential that the buffers are dry before use, because water will change the system's a_w significantly. In addition, some buffers can readily form hydrates and this should be avoided.

14. Organic phase consists of a solid buffer and a solvent. This suspension should be agitated by magnetic stirring and 10 mL withdrawn to add to the enzyme.

15. Transfer of ions between enzyme and buffer pair may be slower in hydrophobic solvents. As a result, lag periods in the initial rate (of up to 30 min) may be observed. If this is the case, it is advisable to shake the enzyme in the organic phase (solvent, propanol, and buffer) for 30 min prior to initiating the reaction. This allows sufficient time for transfer of ions.

References

1. Blackwood, A. D., Curran, L. J., Moore, B. D., and Halling, P. J. (1994) Organic phase buffers control biocatalyst activity independent of initial aqueous pH. *Biochim. Biophys. Acta* **1206,** 161–165.

2. Zacharis, E., Moore, B. D., and Halling, P. J. (1997) Solid state ionisation control in enzyme catalysis. *J. Am. Chem. Soc.* **119,** 12,396–12,397.
3. Zaks, A. and Klibanov, A. M. (1985) Enzyme-catalysed processes in organic-solvents. *Proc. Natl. Acad. Sci. USA* **82,** 3192–3196.
4. Zaks, A. and Klibanov, A. M. (1988) Enzymatic catalysis in non-aqueous solvents. *J. Biol. Chem.* **263,** 3194–3201.
5. Blackwood, A. D., Moore, B. D., and Halling, P. J. (1994) Are asssociated ions important for biocatalysis in in organic media? *Biocatalysis* **9,** 269–276.
6. Skrika-Alexopoulos, E. and Freedman, R. B. (1993) Factors effecting enzyme characteristics of bilirubin oxidase suspensions in organic solvents. *Biotechnol. Bioeng.* **41,** 887–893.
7. Valivety, R. H., Rakels, J. L. L., Blanco, R. M., Johnston, G. A., Brown, L., Suckling, C. J., et al. (1990) Measurement of pH changes in an accessible aqueous phase during biocatalysis in organic media. *Biotechnol. Lett.* **12,** 475–480.
8. Brown, L. Halling, P. J., Johnston, G. A., Suckling, C. J., and Valivety, R. H. (1990) Water insoluble indicators for the measurement of pH in water-immiscible solvents. *Tetrahedron Lett.* **36,** 5799–5802.
9. Partridge, J., Hutcheon, G. A., Moore, B. D., and Halling, P. J. (1996) Exploiting hydration hysteresis for high activity of cross-linked subtilisin crystals in acetonitrile. *J. Am. Chem. Soc.* **118,** 12,873–12,877.
10. Xu, K. and Klibanov, A. M. (1996) pH control of the catalytic activity of cross-linked enzyme crystals in organic solvents. *J. Am. Chem. Soc.* **118,** 9815–9819.
11. Dolman, M., Halling, P. J., and Moore, B. D. (1997) Functionalized dendritic polybenzylethers as acid/base buffers for biocatalysts in nonpolar solvents. *Biotechnol. Bioeng.* **55,** 278–282.
12. Bell, G., Janssen, A. E. M., and Halling, P. J. (1997) Water activity fails to predict critical hydration level for enzyme activity in polar organic solvents: interconversion of water concentrations and activities. *Enzyme Microb. Technol.* **20,** 471–477.

20

Enzymatic Acylation of α-Butylglucoside in Nonaqueous Media

Marie-Pierre Bousquet, René-Marc Willemot, Pierre Monsan, and Emmanuel Boures

1. Introduction

There is a need to modify many cosmetic or pharmaceutical products so that their properties (bioavailibility, toxicity, liposolubility, and hydrosolubility) are optimized. This can be achieved through transportation by "carrier molecules" which can promote drug intake. The amphiphilic structure of α-butylglucoside makes it particularly efficient for such a purpose. α-Butylglucoside is produced on an industrial scale through enzymatic transglycosylation between maltose and butanol *(1–3)*. This glucose derivative contains several reactive hydroxyl groups and can, therefore, be acylated by means of lipase-catalyzed reactions *(4)*. The advantage of α-butylglucoside over glucose is that the butanol moiety of α-butylglucoside helps the molecule to dissolve in apolar media, which are usually recommended for optimal lipase activity and stability *(5–7)*.

Two types of molecules were grafted onto α-butylglucoside: lactic acid, belonging to the α-hydroxy acid (AHA) family, and linoleic acid (C18: 2*n*-6).

α-Hydroxy acids constitute a class of compounds that exert specific and unique effects on skin structure *(8)*. Their application at low concentration (under 5%) decreases intercorneocyte cohesion and induces skin peeling *(9)*. In particular, lactic acid, which is a constituent of the natural moisturizing factor (NMF), has been described as a very good exfoliating and moisturizing agent *(10)*. Nevertheless, AHAs penetrate too quickly into the deep epiderm and their use at too high a concentration (above 10%) gives rise to irritant effects on the skin *(11,12)*. To prevent these adverse effects, they can be grafted onto an amphiphilic molecule, α-butylglucoside. Such a coupling between an α-hydroxy acid and

From: *Methods in Biotechnology, Vol. 15: Enzymes in Nonaqueous Solvents: Methods and Protocols*
Edited by: E. N. Vulfson, P. J. Halling, and H. L. Holland © Humana Press Inc., Totowa, NJ

α-butylglucoside could provisionally block the acid function, responsible for the harmful effect on skin, and, moreover, reduce the penetration rate of the acid into the deep skin layers.

Preliminary toxicity tests on model skin and on butchered beef cornea showed that α-butylglucoside lactate is significantly less irritant than lactic acid at the same molar concentration *(13)*.

Deficiency of essential fatty acid (EFA) in humans or animals induces morphologic changes characterized by severe scaly dermatosis, extensive percutaneous water loss and hyperproliferation of the epidermis *(14)*. EFAs, in particular linoleic and γ-linolenic (18: 3*n*-6) acids, are essential for the construction and the repair of the skin lipidic barrier *(15)*. Linoleic acid has a crucial role in the synthesis of acylglucosylceramides and, thus, has the essential function of maintaining the integrity of the epidermal water permeability barrier *(16,17)*. For these reasons, linoleic acid has been used for the treatment of *Acne vulgaris (18)*. However, the hydrophobicity of EFAs makes it difficult for them to reach the deep layers of the skin. When grafted onto amphiphilic α-butylglucoside, linoleic acid penetration into the epiderm should be significantly enhanced, thereby improving its dermo-cosmetic effect.

The enzymatic synthesis of α-butylglucoside derivatives through transesterification or direct esterification is described. The reaction is performed under reduced pressure so that the alcohol or water coproduct is removed and the reaction equilibrium is shifted toward synthesis of α-butylglucoside ester *(4,19)*.

2. Materials

1. Enzymes: Chirazyme L2, c.-f., C2 (type B lipase from *Candida antarctica* adsorbed on an acrylic resin) and Chirazyme L9, c.-f. (lipase from *Rhizomucor miehei* grafted onto a macroporous anion-exchange resin) are available from Boehringer Mannheim (Germany).
2. Butyl-lactate is purchased from Fluka (Switzerland). Methanol and hexane are from Prolabo (France). The commercial mixture of linoleic acid (60.5%, mol/mol), oleic acid (32.7%, mol/mol), and linolenic acid (6.8%, mol/mol) is available from Sigma (USA).
3. α-Butylglucoside is produced and purified as described *(2,3)*.
4. The vacuum system is composed of a rotavapor (Büchi, R-114) equipped with an oil bath (Büchi, B-485) and connected to a vacuum controller (Büchi, B-720) and a vacuum pump (KNF, N7263FT18).

3. Methods

3.1. Preparation of α-Butylglucoside Lactate

1. Weigh 369 mg of α-butylglucoside and adjust to 3 mL with butyl-lactate in a glass device for the rotavapor. Subject to rotative agitation, at 60°C and 15 mbars, for 30 min.

2. Add 2 g of Chirazyme L2, c.-f., C2 and return to rotative agitation at 60°C and 15 mbar. Leave the reaction for about 48 h (*see* **Note 1**).
3. When yield (*see* **Note 2**) reaches 95%, remove the biocatalyst by filtration.
4. Remove residual butyl-lactate by liquid–liquid extraction with water and hexane. Mix the reaction mixture vigorously (1 vol) with water (1 vol) and hexane (2 vol) for 30 min, then allow to stand until the organic and aqueous phases separate.
5. Recover the aqueous phase and evaporate the water at 60°C under reduced pressure.
6. Add additional hexane to the syrup at a volumic ratio of hexane:syrup of approx 10:1. Mix thoroughly and take off the upper layer (organic phase) (*see* **Note 3**).
7. Evaporate residual hexane at 40°C under reduced pressure.
8. A syrup containing α-butylglucoside lactate at a purity above 95% (w/w) is obtained (*see* **Note 4**).

3.2. Preparation of α-Butylglucoside Linoleate

1. Weigh 369 mg of α-butylglucoside and 438 mg of the commercial mixture of linoleic acid, oleic acid, and linolenic acid (molar ratio alcohol/acid = 1/1) in a glass device for the rotavapor. Heat to 75°C until the mixture is melted.
2. Add 40.3 mg of Chirazyme L2, c.-f., C2 or Chirazyme L9, c.-f. and place at 65°C, 20 mbars, and under rotative agitation. Leave the reaction for about 48 h (*see* **Note 5**).
3. When the yield (*see* **Note 6**) reaches 95%, add approximately 1 vol of hexane and homogenize.
4. Eliminate the biocatalyst by filtration and evaporate hexane under reduced pressure.
5. A syrup containing a mixture of esters of α-butylglucoside and unsaturated fatty acids containing approximately 60% of α-butylglucoside linoleate is obtained (*see* **Note 7**).

4. Notes

1. Analysis of the mixture composition can be performed by taking samples at intervals. After a 20-fold dilution in water, filter to eliminate the enzyme and analyze by high-performance liquid chromatography (HPLC) on a reverse-phase column (Nucleosil C18, 6 μm, 250 mm × 4 mm inside diameter; I.C.S., France). Eluent: 50/50 methanol/water; flow rate: 0.5 mL/min; temperature: 33°C; injection volume: 10 μL; detection: refractometer (RI-Detector 8110, I.C.S.).
2. The yield is defined as the molar ratio of α-butylglucoside lactate formed to initial α-butylglucoside.
3. Butyl-lactate (boiling point [bp] = 186°C) and hexane (bp = 69°C) can be recovered from the organic phase by distillation.
4. Residual α-butylglucoside, butanol, and butyl-lactate are quantified by HPLC (*see* **Note 1**). The final L-lactic acid concentration is assayed using the L-lactate kit from Boehringer Mannheim (Germany).
5. Analysis of the mixture composition can be performed by taking samples at intervals. After a 10-fold dilution in methanol (for acid analysis) or in 5 mM sulfuric acid (for α-butylglucoside analysis), the sample is filtered to eliminate

the enzyme and analyzed. Linoleic acid, oleic acid, and linolenic acid concentrations are quantified by HPLC chromatography on a reverse-phase column (Nucleosil C18, 6 μm, 250 mm × 4 mm inside diameter, I.C.S., France). Eluent: 87/13/0.3 methanol/water/trifluoroacetic acid, flow rate: 1.0 mL/min; temperature: 50°C; injection volume: 10 μL; detection: refractometer.

α-Butylglucoside is quantified by HPLC chromatography on an ion-exclusion column (PPH 224 [Brownlee Labs], 4.6 mm inside diameter × 220 mm). Elution conditions are 5 mM sulfuric acid, 0.3 mL/min flow rate, 25°C temperature, 10 μL injection volume, and a refractometer for detection.

6. The yield is defined as the molar ratio of total α-butylglucoside esters formed to initial α-butylglucoside (or total acids).
7. Exact concentration of each α-butylglucoside ester (linoleate, oleate, and linolenate) can be quantified by HPLC (*see* **Note 5**). The content of each α-butylglucoside ester is the same as the content of the corresponding acid in the commercial mixture of unsaturated fatty acids.

References

1. Pelenc, V., Paul, F., and Monsan, P. (1993) Enzymatic stereospecific production of α-glucosides from starch etc. by reaction with alcohol in the presence of α-transglucosidase and optionnaly conversion to α-glucoside ester, using a lipase. International Patent WO 93/04185.
2. Monsan, P., Paul, F., Pelenc, V., and Boures, E. (1996) Enzymatic production of α-butylglucoside and its fatty acid esters. *Ann. NY Acad. Sci.* **799,** 633–641.
3. Bousquet, M.-P., Willemot, R.-M., Monsan, P., Paul, F., and Boures, E. (1999) Enzymatic synthesis of α-butylglucoside in a biphasic butanol-water system using the α-transglucosidase from Aspergillus niger, in *Methods in Biotechnology* Vol. 10: *Carbohydrate Biotechnology Protocols* (Bucke, C., ed.), Humana, Totowa. NJ, pp. 291–296.
4. Bousquet M.-P., Willemot, R.-M., Monsan P., and Boures E. (1999) Enzymatic synthesis of α-butylglucoside lactate: a new α-hydroxy acid derivative. *Biotechnol. Bioeng.* **62,** 225–234.
5. Laane, C., Boeren, S., Vos, K., and Veeger, C. (1987) Rules for optimization of biocatalysis in organic solvents. *Biotechnol. Bioeng.* **30,** 81–87.
6. Khmelnitsky, Y. L., Levashov, A. V., Klyachko, N. L., and Martinek, K. (1988) Engineering biocatalytic system in organic media with low water content. *Enzyme Microb. Technol.* **10,** 710–724.
7. Gorman, L. A. S. and Dordick, J. S. (1992) Organic solvents strip water off enzymes. *Biotechnol. Bioeng.* **39,** 392–397.
8. Van Scott, E. J. and Yu, R. J. (1974) Control of keratinisation with alpha-hydroxy acids and related compounds. *Arch. Dermatol* **110,** 586–590.
9. Berardesca, E. and Maibach, H. (1995) AHA mechanisms of action. *Cosmet. Toilet.* **110,** 30,31.
10. Feldmann, J. P. (1994) Les acides alpha-hydroxylés ou alpha-hydroxy acides (AHA). *Cosmétologie* **1,** 31–40.

11. Smith, W. P. (1994) Hydroxy acids and skin aging. *Cosmet. Toilet.* **109,** 41–44.
12. Wolf, B. (1994) A hypoallergenic approach to alpha-hydroxy acids. HBA Global Expo Conference.
13. Bousquet, M.-P., Willemot, R.-M., Monsan, P., and Boures, E. (1998) Enzymatic synthesis of AHA derivatives for cosmetic applications. *J. Mol. Catal. B* **5,** 49–53.
14. Ziboh, V. A. and Chapkin, R. S. (1987) Biologic significance of polyunsaturated fatty acids in the skin. *Arch. Dermatol.* **123,** 1686–1690.
15. Hansen, H. S. and Jensen, B. (1985) Essential function of linoleic acid esterified in acylglucosylceramide and acylceramide in maintaining the epidermal water permeability barrier. Evidence from feeding studies with oleate, linoleate, arachidonate, columbinate and α-linolenate. *Biochim. Biophys. Acta* **834,** 357–362.
16. Hartop, P. J. and Prottey, C. (1976) Changes in transepidermal water loss and the composition of epidermal lecithin after applications of pure fatty acid triglycerides to skin of essential fatty acid-deficient rats. *Br. J. Dermatol.* **95,** 255–264.
17. Prottey, C., Hartop, P. J., Black, J. G., and McCormack, J. I. (1976) The repair of impaired epidermal barrier function in rats by the cutaneous application of linoleic acid. *Br. J. Dermatol.* **94,** 13–21.
18. Gareiß, J., Hoff, E., and Rados-Schaffgans, B. (1996) Liposomes, nanoémulsions de phosphatidylcholine. *Parf. Cosmet. Acta* **131,** 63–66.
19. Bousquet, M.-P., Willemot, R.-M., Monsan, P., and Boures, E. (1999) Enzymatic synthesis of unsaturated fatty acid glucoside esters for dermo-cosmetic applications. *Biotechnol. Bioeng.* **63,** 730–736.

Part II

Synthetic Applications

Introduction

Herbert L. Holland

The ability to form esters and amides by condensation reactions, the absence of hydrolytic side reactions, the manipulation or regio- and stereoselectivity, and the advantages derived from high substrate and product solubilities have all contributed to continued investigation of the synthetic applications of enzymes in nonaqueous solvents. Methodology in this area has focused on the application of isolated enzyme catalysis under nonaqueous or low-water conditions, with free or immobilized enzymes operation in water-miscible solvent mixtures (e.g., methanol/water), water-immiscible solvents (e.g., hexane), and reverse micelle and encapsulated environments. Although water-miscible solvents are generally used only as a means to increase substrate solubility, water-immiscible solvents can also be used to alter and enzyme's reactivity.

Whole-cell biocatalysts can also be used in nonaqueous media, water/solvent two-phase systems, and reverse micelles. Such methods are particularly useful when dealing with the biotransformations of water-immiscible substrates that can act as the second phase (e.g., toluene), or for microbial conversions of those substrates possessing an appropriate partition coefficient between the two solvents. They are also valuable for biotransformations for which extensive cofactor recycling and thus cell viability is not a requirement, and for conversions involving microorganisms (such as Rhodococcus) that maintain high viability in the presence of organic solvents. In some instances, such as cases where competing enzymes may have different activities in a range of solvents, the nature of the solvent may also control the regio- or stereoselectivity of the biotransformation of a single substrate by whole cells.

From: *Methods in Biotechnology, Vol. 15: Enzymes in Nonaqueous Solvents: Methods and Protocols*
Edited by: E. N. Vulfson, P. J. Halling, and H. L. Holland © Humana Press Inc., Totowa, NJ

The contributions presented in this section cover a wide range of synthetic applications. The bulk of the work in this area has focused on the use of hydrolytic enzymes for the formation of esters and amides, and this is reflected in the following section, beginning with a discussion of the empirical rules available for the selection of such enzymes for enantioselective reactions. This is followed by a range of specific applications for the preparation of chiral alcohols by transesterification reactions using various lipase enzymes, together with an example of how such processes can be optimized for enantioselectivity. The selective formation of amides of amino-polyols is the presented, followed by a discussion of the use of lipases for esterification reactions in water–oil emulsions. Several reactions of water-insoluble organosilicon substrates catalyzed by hydrolytic enzymes are then discussed, followed by two contributions covering the use of a wider range of enzymes under a variety of conditions for hydrolytic and other reactions of multiple substrates types.

Whole-cell applications are exemplified by discussions of microbial reactions using an interface bioreactor, and of yeast-mediated reactions in organic solvents, and finally a case study of the development of a method for the preparation of the antifungal agent SCH56592 is presented, in which the use of nonaqueous biocatalysis is played a significant role.

The contributions in this section thus cover the scope of synthetic applications, from the initial concept of an idea and the choice of a suitable catalyst to the application of nonaqueous enzyme technology for chemical production.

21

Choosing Hydrolases for Enantioselective Reactions Involving Alcohols Using Empirical Rules

Alexandra N. E. Weissfloch and Romas J. Kazlauskas

1. Introduction

With over 100 hydrolases available commercially, researchers need to choose the best hydrolase for their problem. One good way to choose a hydrolase is to use empirical rules or models that summarize earlier results.

Some rules—substrate rules—define what the substrate looks like. The simplest rules specify only relative sizes of substituents (e.g., small, medium and large). The advantage and disadvantage of these rules are their simplicity. They are very easy to use and work for a wide range of substrates, but they do not provide detailed information. Other substrate rules are more detailed, but apply only to a specific type of molecule.

Other rules—active-site models—define what the binding site of hydrolases looks like. These are usually two- or three-dimensional box-type models that give more detail about the shape, size, hydrophobicity, and, possibly, electronic character of the binding site. These models are harder to apply to a new substrate and are less general. Each hydrolase needs its own model.

This chapter summarizes the empirical rules and models developed for lipases and proteases primarily over the last 10 years. The aim of these models is to guide the choice of hydrolases and also to identify what features of the substrate are most important for high enantioselectivity. Most of the models presented focus on the size and shape of the substituents; thus, these features appear to be the most important for high enantioselectivity. However, many of the models also mention the polar or nonpolar character of substituents, indicating that size and shape are not the only features important for high enantioselectivity. Naemura recently reviewed empirical rules and models for hydrolases (1).

From: *Methods in Biotechnology, Vol. 15: Enzymes in Nonaqueous Solvents: Methods and Protocols*
Edited by: E. N. Vulfson, P. J. Halling, and H. L. Holland © Humana Press Inc., Totowa, NJ

To limit the size of this chapter, we focus only on chiral alcohols and amines and omit chiral acids. Lipases are usually more enantioselective toward alcohols and amine; thus, the chapter covers most of the models developed for lipases.

Although researchers developed these rules and models empirically, recent X-ray crystal structures provide a molecular-level rationalization for many of these models. The binding site of hydrolases often shows a remarkable similarity to the active-site model.

2. Materials

The proposed substrate binding sites for synthetically useful hydrolytic enzymes presented in **Figs. 6** and **7** were generated using RasMac v2.6, starting from available X-ray crystal-structure coordinates of the enzymes.

3. Method

3.1. Secondary Alcohols and Primary Amines

Based on the observed enantioselectivity of lipases and subtilisins, researchers proposed rules to predict which enantiomer reacts faster (**Fig. 1**). These rules are based on the size of the substituents and apply to both hydrolysis and acylation reactions (*see* **Note 1**). For acylation, the enantiomer shown reacts faster; for hydrolysis, the ester of enantiomer shown reacts faster.

The advantage of this rule is that it is simple to apply. First, draw the substrate so that the alcohol (or ester) group points out of the page toward the reader. Second, imagine a line extending the C—O bond so that it divides the molecule in two parts. For lipases, the enantiomer with the larger group on the right side of the line is the one that will react faster. Keep the flexibility of molecular structures in mind when comparing the sizes of the substituents. For instance, an alkyl chain can fold so that its effective size is smaller than a substituent that cannot fold such as phenyl.

The rule for secondary alcohols applies to all lipases tested so far. These include cholesterol esterase (CE) *(2)*, lipases from several *Pseudomonas* species, including *Ps. cepacia* (PCL) *(2)*, *Ps. fluorescens* (PFL) *(3,4)*, *Ps. aeruginosa* (PAL) *(5)*, lipase from *Rhizomucor miehei* (RML) *(6)*, lipase B from *Candida antarctica* (CAL-B) *(7)*, and porcine pancreatic lipase *(8)*. The rule also works for lipase from *Candida rugosa* (CRL), but only for cyclic secondary alcohols *(2)*.

This rule also applies to primary amines of the type RR'CHNH$_2$ because their shape is similar to that of secondary alcohols *(9)*. The lipases tested include CAL-B, PCL, and PAL. Lipases are very poor catalysts for the hydrolysis of amides, but these lipases catalyze the reverse reaction—acylation of amines.

Subtilisins favor the enantiomer opposite to the one favored by lipases for both secondary alcohols and primary amines of the type RR'CHNH$_2$ *(9)*. Thus, subtilisins and lipases are complementary reagents. The enantioselectivity of subtilisins is usually lower than that of lipases.

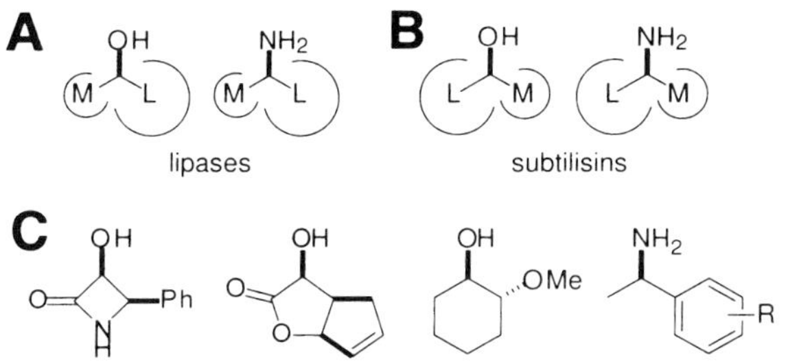

Fig. 1. Empirical rules predict the fast reacting enantiomer for lipases and subtilisins. Both secondary alcohols and amines of the type RR'CHNH$_2$ have similar structures so similar rules apply. (**a**) Rules to predict the fast reacting enantiomer for lipase-catalyzed reactions. M represents a medium-sized substituent such as methyl; L represents a large substituent such as phenyl. In acylation reactions, the enantiomer shown reacts faster; in hydrolysis reactions, the ester of the enantiomer shown reacts faster. (**b**) Rules to predict the fast reacting enantiomer in subtilisin-catalyzed reactions. Note that lipases and subtilisins favor opposite enantiomers; thus, they are complementary reagents. Subtilisins are usually less enantioselective than lipases. (**c**) Examples of pure enantiomers prepared by lipase-catalyzed reactions showing the fast reacting enantiomer.

These empirical rules suggest that lipases and subtilisins distinguish between enantiomeric secondary alcohols primarily by the relative sizes of the two substituents. Changing the relative sizes of the substituents might change the selectivity. Indeed, a number of researchers increased the enantioselectivity of lipase-catalyzed reactions by modifying the substrate to increase the size of the large substituent (examples can be found in the articles cited in **refs. *2,10–14***. Similarly, Shimizu et al. *(15)* reversed the enantioselectivity by converting the medium substituent into the large one.

However, sometimes, small changes in the structure of the secondary alcohol alter the enantioselectivity. For example, CAL-B shows high enantioselectivity toward 3-nonanol ($E > 300$), but low enantioselectivity toward 1-bromo-2-octanol ($E = 7.6$) under the same conditions **(Fig. 2.)** *(16)* Both an ethyl and a—CH$_2$Br group have similar sizes, so the difference suggests that an electronic effect lowers the enantioselectivity. Similarly, changing the position of a methyl substituent in the aryl ring of the large substituent drastically altered the selectivity of PCL *(17)*. The rules in **Fig. 1** cannot account for these changes in selectivity.

To map more precisely how structure affects enantioselectivity, researchers have developed more detailed substrate models. These models apply only to substrates similar to those used to develop the model. For example, Oberhauser et al. developed a qualitative *(18)* and later quantitative *(19)* model for CRL-

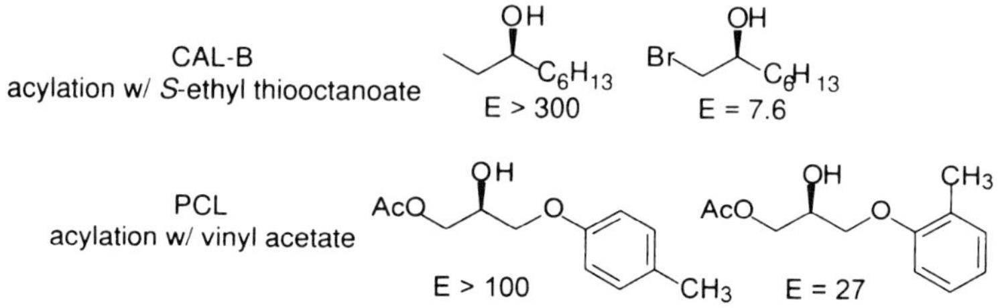

Fig. 2. Small changes in structure can alter the selectivity. Nevertheless, all these examples follow the secondary alcohol rule.

catalyzed resolutions of bicyclic secondary alcohols. This model predicts the enantioselectivity for various bicyclo[2.2.1]heptanols and bicyclo[2.2.2] octanols. This model identifies regions of the substrate that tolerate substituents, as well as positive and negative charges.

Another way of adding detail to the simple rules in **Fig. 1** is to define the maximum size for each substituent. For example, Exl et al. *(20)* found a maximum size of 9.2 Å for the large substituent in CRL, but a maximum size of only 7.1 Å for PCL. A natural extension of these limits is the two-dimensional box-type models **(Fig. 3)**. The advantage of these active-site models is that they define limits for the size of each substituent. The disadvantages are that they are harder to use and each hydrolase requires a separate model.

To use these models, first draw the secondary alcohol using the same scale as the model. A computer drawing program that permits scaling of the drawing is helpful. Next, fit the hydroxyl group into the catalytic site (the OH pointing out from the page). If part of a molecule goes beyond the boundaries of the model, it is not a substrate. The enantiomer with the best fit is predicted to be the major product. Not surprisingly, the overall shape of these models is similar to those in **Fig. 1**.

Burgess and Jennings developed the first model of this type for lipases *(21)*. They tested a range of unsaturated alcohols with lipase from *P. fluorescens* (lipase AK). Bornscheuer et al. also used this model to explain PCL- catalyzed and *Chromobacterium viscosum* (*P. glumae*)-catalyzed reactions *(22)*. Naemura et al. proposed a similar model for another lipase from *P. fluorescens* (lipase YS) *(23)* using acyclic, cyclic, and bicyclic secondary alcohols. Naemura et al. later proposed a similar model for a lipase from *Alcaligenes* sp. (lipase QL) *(24)*.

To better account for the three-dimensional nature of molecules, several groups proposed three-dimensional models of the enzyme active site. Researchers overlaid energy-minimized structures of substrates and nonsubstrates to define the size limits and polar or nonpolar character of each pocket. Jones'

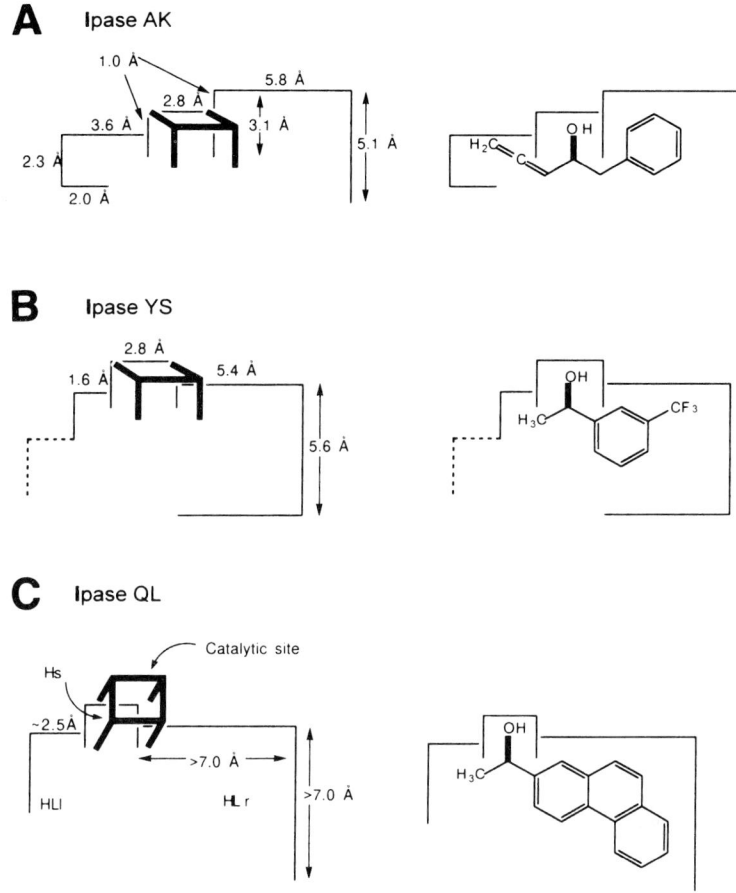

Fig. 3. Two-dimensional active-site models for several lipases predict which secondary alcohol reacts faster. Examples show good substrates for each enzyme. (**A**) Model for lipase from *Pseudomonas fluorescens* (lipase AK). (**B**) Model for another lipase from *Pseudomonas fluorescens* (lipase YS). (**C**) Model for a lipase from *Alcaligenes* sp. (lipase QL).

group proposed a model for pig liver esterase (PLE) to account for the stereoselectivity of this hydrolase toward chiral and prochiral acids (**Fig. 4**) *(25,26)*. This model can account for the reversal in stereoselectivity of PLE toward the meso diacids shown. To use this model, build a three-dimensional model of the molecule using molecular models. Next, build a three-dimensional cardboard or plastic model. (Framework molecular models or children's Lego™ toys are convenient.) according to the dimensions in **Fig. 4**. Use the same scale for both models. Use the model as follows:

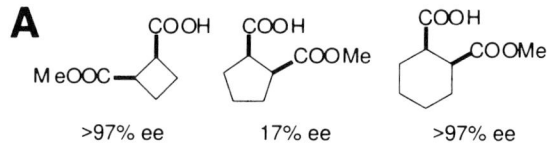

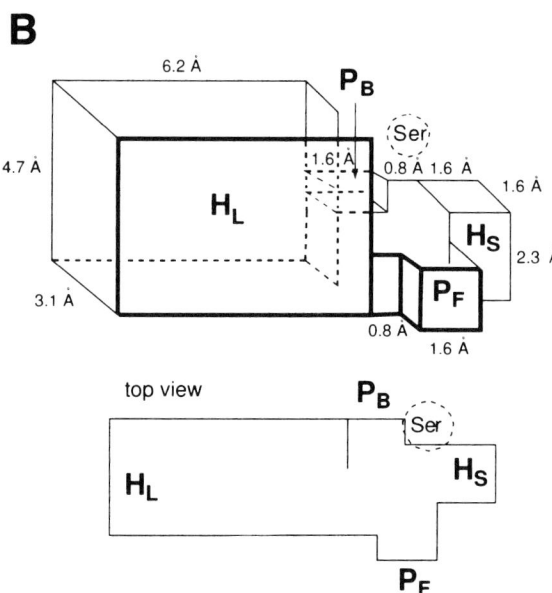

Fig. 4. Three-dimensional active-site model accounts for the enantioselectivity of PLE toward chiral acids. **(A)** PLE shows a reversal in enantioselectivity in the hydrolysis of a series of *meso* diester. The product monoester has high enantiomeric purity, but opposite configuration for the four- and six-membered rings and low enantiomeric purity for the five-membered ring. **(B)** The active-site model contains four regions: large hydrophobic, H_L, small hydrophobic, H_S, front polar, P_F, and back polar, P_B. The reactive ester group must lie near the serine sphere. The model accounts for the reversal in stereoselectivity as follows. The hydrophobic portion of the cyclobutane binds in H_S because it fills the hydrophobic site. The hydrophobic portion of the cyclohexane is too large to bind in H_S, so it binds in H_L, causing reversed enantioselectivity.

1. Fit the carbon atom of the reacting ester carbon within 1.4 Å (a C—O bond length) of the serine sphere.
2. Fit the hydrophobic groups into either hydrophobic pocket to fill the pocket if possible. Thus, smaller groups, usually aliphatic, bind in the small hydrophobic pocket, whereas larger groups, usually aromatic, bind in the larger pocket. This requirement accounts for size-induced reversals of stereoselectivity. Nearby nonreacting ester groups bind in P_F, and hydrogen bonding groups will bind in P_B.

3. After comparing the fit for each enantiomer, if one enantiomer fits much better, then the model predicts high stereoselectivity, but if both fit similarly, then the model predicts low stereoselectivity.

Although Jones' group developed this model for chiral acids, not alcohols, Naemura et al. used this model to account for PLE's enantioselectivity toward several secondary alcohols *(27)*. It may also be useful for other secondary alcohols.

Two groups independently proposed three-dimensional models for PCL **(Fig. 5)** *(28,29)*. Lemke et al. found one spherical hydrophobic pocket and another tubelike pocket. The spherical pocket accommodates bulky groups such as phenyl, phenoxymethyl, and substituted aryl derivatives, whereas the tube accepts only long groups such as acetoxymethyl and (*n*-hexadecanoyloxy)methyl. Therefore, Lemke et al. suggested that not just size but also shape of the substituents determines enantioselectivity. Grabuleda et al. proposed a similar, but slightly smaller, model. This difference is likely because the two groups examined different types of substrates.

The X-ray crystal structures support the aforementioned models. The alcohol binding site of lipases indeed contains a large hydrophobic pocket and a smaller medium-sized pocket (or stereoselectivity pocket) **(Fig. 6)**. The structure of PCL also supports the three-dimensional model because the binding site indeed contains a spherical pocket and a tubelike pocket. The spherical pocket is the large and hydrophobic pocket, whereas the tubelike pocket is a crevice below the serine in **Fig. 6**. The medium-sized pocket is the top part of this crevice.

The X-ray structure of subtilisin provides a rationalization for why it favors the enantiomer of secondary alcohols opposite to the one favored by lipases **(Fig. 7)** *(9)*. The catalytic triad has the opposite chirality in the two classes of enzyme, but the hydrophobic pocket has approximately the same location. Thus, these two classes of hydrolases favor opposite enantiomers of secondary alcohols and primary amines.

Computer modeling of hydrolase–substrate interactions also promises to be a useful method for predicting enantioselectivity. However, given the current state of computer modeling *(35)*, the empirical rules are much faster to use and do not require specialized software or training.

Changing from water to organic solvent often changes lipase enantioselectivity, as does changing from one organic solvent to another. Empirical rules and active-site models do not explain these changes (*see* **Note 3**).

3.2. Primary Alcohols

Lipases usually show lower enantioselectivity toward primary alcohols than toward secondary alcohols. Only porcine pancreatic lipase (PPL) and PCL show high enantioselectivity toward a wide range of primary alcohols. An empirical rule can predict some of the enantiopreference of PCL toward pri-

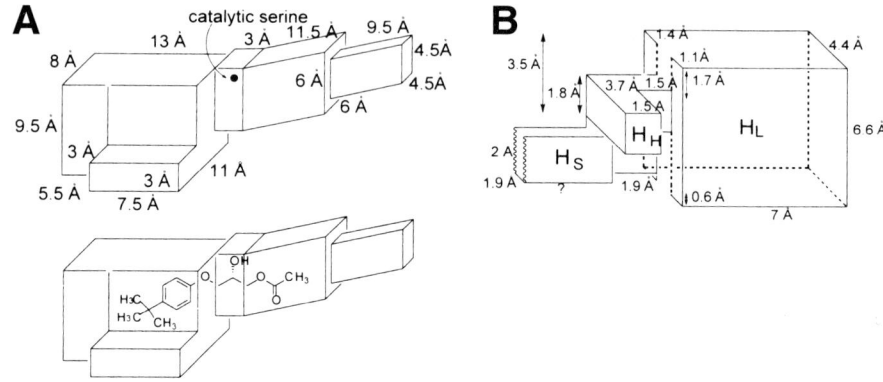

Fig. 5. Three-dimensional active-site models for PCL proposed after overlaying energy-minimized structures of both good and bad substrates. (a) Lemke et al.'s model contains a spherical hydrophobic pocket and a tubelike hydrophobic pocket. (b) Grabuleda et al.'s model is slightly smaller. The reacting hydroxyl group goes in the H_H site (hydrophilic), whereas the substituents go into the two hydrophobic sites, H_S and H_L. Both models predict the same absolute configuration. The drawing shows Grabuleda et al.'s model rotated by 180° about the vertical axis relative to Lemke et al.'s model.

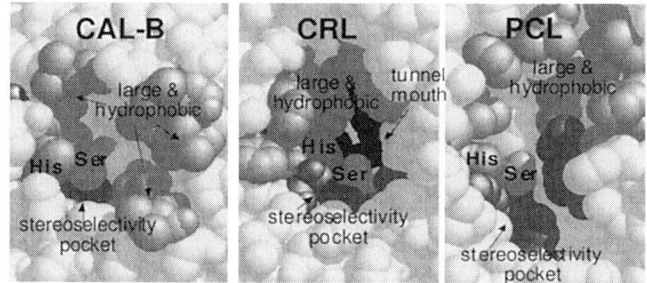

Fig. 6. Proposed substrate binding site in three synthetically useful lipases. The catalytic Ser lies at the bottom of a crevice with the catalytic His on the left. Although the details of this crevice differ for each lipase, each crevice contains a large hydrophobic pocket (light gray) and smaller pocket (medium gray), labeled stereoselectivity pocket. This crevice is the alcohol binding site and the two pockets resemble the empirical rule in **Fig. 1**. The structure for CRL also shows the mouth of a tunnel that binds the acyl chain of an ester. Pictures were drawn with RasMac v2.6 *(30)* starting from the X-ray crystal structures for each lipase: CAL-B *(31)*, CRL *(32)*, and PCL *(33)*.

mary alcohols **(Fig. 8)** *(36)*. Like the foregoing secondary alcohol rule, the primary alcohol rule is based on the size of the substituents, but, surprisingly, the sense of enantiopreference is opposite; that is, the —OH of secondary alco-

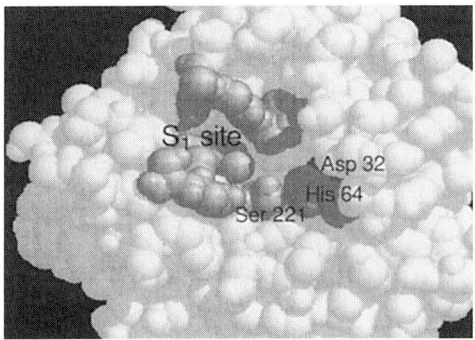

Fig. 7. Substrate binding site in subtilisin Carlsberg. The labels and coloring show the amino acid residues of the catalytic triad and the residues forming the S_1 binding site. This binding site is a shallow groove lined with nonpolar amino acid residues. The large hydrophobic substituent of secondary alcohols probably binds in this pocket. Note that the histidine of the catalytic triad lies to the right of the catalytic serine, whereas in lipases (**Fig. 6**), it lies to the left of the serine. This opposite chirality of the catalytic triad in part accounts for the opposite enantiopreference of lipases and subtilisins. Coordinates are from Brookhaven protein data bank file 1sbc *(34)* and the figure was created using RasMac v 2.6 *(30)*.

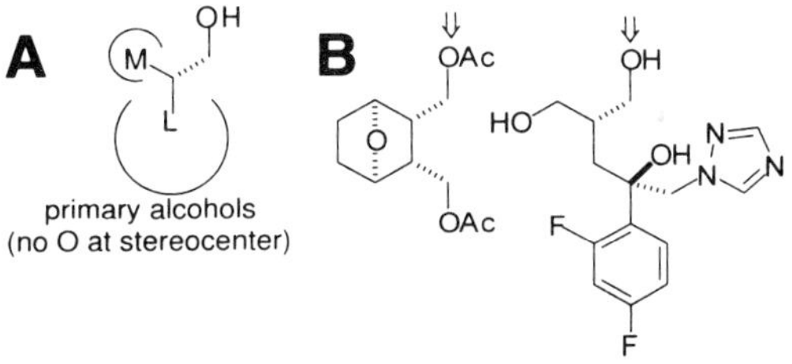

Fig. 8. Empirical rule predicts the enantiopreference of PCL toward primary alcohols. (**A**) The empirical rule is based on the relative sizes of the substituents at the stereocenter. This rule is reliable only when there is no oxygen directly bonded at the stereocenter. Computer modeling (and the drawing) suggests that the large substituent, L, binds to different regions of the active site. Comparing this rule to the one in **Fig. 1** shows that PCL favors opposite enantiomers in the case of primary and secondary alcohols. (**B**) PCL shows high enantioselectivity toward the primary alcohols shown. Reaction (hydrolysis of the diacetate in the first case, acylation of the diol in the second case) occurs at the arrow.

hols and the —CH$_2$OH of primary alcohols point in opposite directions in the two models. Computer modeling suggests that the large substituent of primary alcohols does not bind in the same pocket as secondary alcohols *(37)*.

Not all primary alcohols fit this rule. In particular, primary alcohols that have an oxygen at the stereocenter (e.g., glycerol derivatives) do not fit this rule. An example of how the empirical rule does not apply to primary alcohols with an oxygen at the stereocenter is the γ-butyrolactones in **Fig. 9** *(38)*. For the trans isomer, PCL favors one enantiomer, but for the cis isomer, PCL favors the other. Computer modeling on other substrates containing an oxygen at the stereocenter suggested that Tyr29 might form a hydrogen bond to this oxygen *(37)*. The formation of this hydrogen bond and its influence on enantioselectivity likely depends on differences in structure more subtle than just size of the substituents. For those primary alcohols without an oxygen at the stereocenter, the rule in **Fig. 8** showed 89% reliability (correct for 54 of 61 examples). For primary alcohols that did contain an oxygen at the stereocenter, the rule showed only 37% reliability (10 of 27 examples).

For secondary alcohols, increasing the difference in the size of the substituents often increased the enantioselectivity of PCL and other lipases as discussed in **Subheading 3.1**. However, for primary alcohols, this strategy was not reliable. Upon adding large substituents, the enantioselectivity sometimes increased *(39)*, sometimes decreased, and sometimes remained unchanged *(36)*.

Researchers have had difficulties finding a reliable rule for PPL-catalyzed reactions of primary alcohols *(40–43)*. Two proposed rules, based on different sets of substrates, even predict opposite enantiomers. An example of the difficulty is shown in **Fig. 10**. Enantiopreference of PPL reversed upon changing from a trans to a cis configuration of the double bond in the 2-substituted 1,3-propandiol derivatives *(44)*. PPL favored the (*S*)-enantiomer with high enantioselectivity for the trans isomer, the (*S*)-enantiomer with moderate enantioselectivity for the saturated analog, but the (*R*)-enantiomer with low to moderate enantioselectivity for the cis isomer. This reversal is difficult to explain using only the relative sizes of the substituents. Note that for secondary alcohols, the enantioselectivity of PPL also varied with the configuration of double bonds in the large substituent, but the enantiopreference remained the same *(45)*. A similar division of primary alcohols into two groups, those with and without oxygens at the stereocenter, may resolve dilemma of enantiomeric rules for PPL. The rule in **Fig. 8** was reliable for primary alcohols without an oxygen at the stereocenter (27 out of 31 substrates, 87% reliability), but not for those with an oxygen at the stereocenter (3 out of 9, 33% reliability).

Naemura et al. *(23)* used a box-type model to predict the fast reacting enantiomer of primary alcohols in PFL (lipase YS)-catalyzed reactions (**Fig. 11**). None of the primary alcohols contained oxygen at the stereocenter. They used the same model to predict the fast reacting enantiomer for secondary alcohols. This model, unlike the computer modeling mentioned earlier, predicts that the large substituent of both primary and secondary alcohols binds in the same region of the lipase.

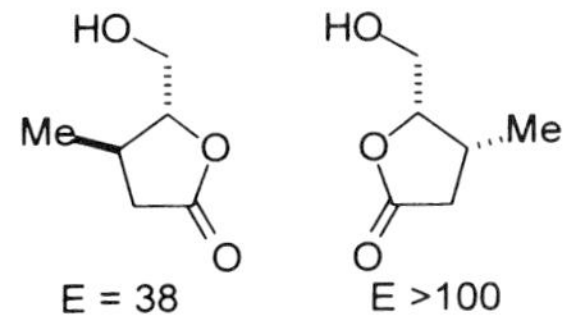

E = 38   E >100

Fig. 9. Examples of primary alcohols with oxygen directly bonded to the stereocenter. The primary alcohol rule does not apply to these types of alcohols. In the hydrolysis of the corresponding acetates, the absolute configuration of the favored enantiomer reverses in the cis and trans diastereomers. Thus, a rule based only on the size of the substituents cannot predict which enantiomer reacts faster.

trans isomer: 97% ee, 75% yield
saturated analog: 72% ee, 47% yield
cis isomer: 21% ee (ent), 25% yield

trans isomer: 95% ee, 63% yield
saturated analog: 70% ee, 56% yield
cis isomer: 55% ee (ent), 44% yield

Fig. 10. The presence and configuration of a double-bond change and even reverse the enantioselectivity in PPL-catalyzed hydrolyses. The ent indicates that the enantiomer of the structure drawn reacts faster.

Tanaka et al. *(46)* proposed a box-type model with a single binding site for lipase from *Rhizopus delemar* (not shown). They used this model to rationalize RDL-catalyzed hydrolyses of *meso*-bis(acetoxymethyl)cyclopentanes (*meso* diacetates of primary alcohols).

3.3. Primary Alcohols of the Type RR'OHCCH₂OH

Pseudomonas fluorescens (lipase AK) efficiently resolves α,α-disubstituted 1,2-diols, a group of primary alcohols having a quaternary stereocenter. Acylation occurs at the primary alcohol and the hindered quaternary stereocenter lies far enough away from it that lipase-catalyzed reactions remain fast. Two groups proposed models to predict the fast reacting enantiomer (**Fig. 12**). Hof and Kellog *(47)* proposed a box-type model with three sites, whereas Chen and Fang *(48)* proposed a simplified version. Although the S (small) site of Hof and Kellog corresponds to the L (large) site of Chen and Fang, this difference may reflect the different types of substrate tested by the two groups or it may reflect problems caused by the oxygen at the stereocenter, as mentioned earlier.

4. Notes

1. Although the alcohol portion of a lactone is a secondary alcohol, the secondary alcohol rule does not apply to lactones because the stereocenter lies in a different

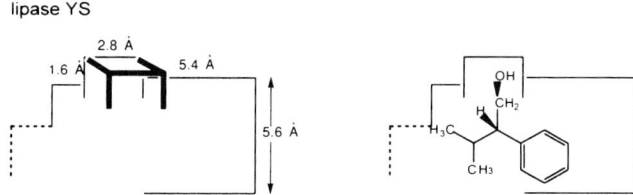

Fig. 11. A box-type model to predict the fast reacting enantiomer in PFL (lipase YS)-catalyzed reactions. This model applies to both primary and secondary alcohols. The example in the figure is a primary alcohol. A secondary alcohol example is given in **Fig. 3B**.

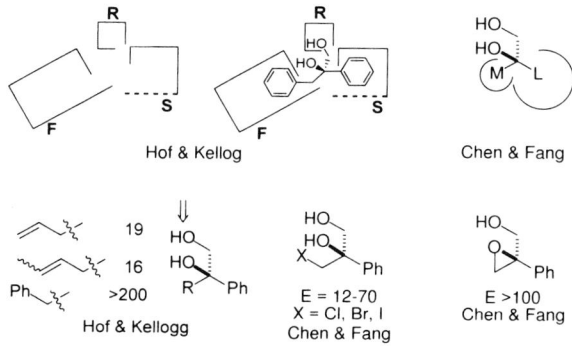

Fig. 12. Model and several examples of α,α-disubstituted 1,2-diols resolved by PFL (lipase AK). Hof and Kellog's model proposes three sites: R (reaction), F (flat), and S (small). The primary hydroxyl lies in the R site pointing into the page. The tertiary hydroxyl group points out of the page. The F site accepts only flat, nearly planar, side chains (e.g., allyl group). Aliphatics do not fit well. The S site accepts phenyl, but not 4-substituted phenyl. Chen and Fang proposed a simplified version of this model. The medium substituent corresponds to the flat site and the large substituent surprisingly corresponds to the S (small) site. Several examples include α,α-disubstituted 1,2-diols as well as the structurally similar epoxides.

position. Esters adopt a syn conformation along the carbonyl C—alcohol O bond. The crystal structure of transition state analogs bound to lipases suggests that this conformation persists in the active site. On the other hand, the lactone ring forces an anticonformation along the carbonyl C—alcohol O bond, which places the stereocenter in a different part of the enzyme. In particular, the lactone stereocenter appears to lie entirely within the L-pocket of the alcohol-binding crevice **(Fig. 13)**. Indeed, many lactones do not follow the secondary alcohol rule.

2. Although the secondary alcohol rule is valid for crude PPL, pure PPL does not catalyze hydrolysis of secondary alcohol esters *(49)*. Thus, an impurity in the

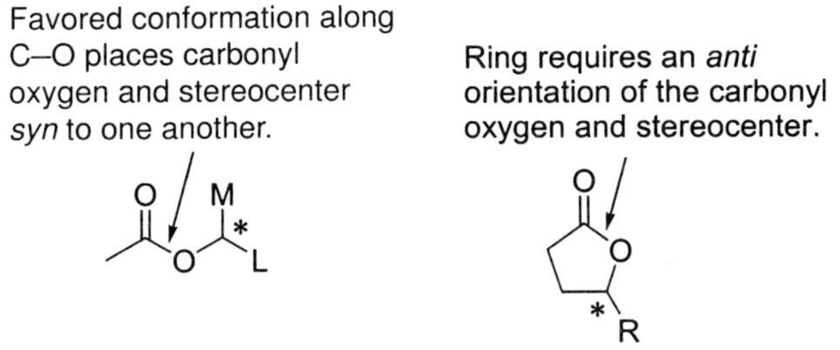

Favored conformation along C–O places carbonyl oxygen and stereocenter *syn* to one another.

Ring requires an *anti* orientation of the carbonyl oxygen and stereocenter.

Fig. 13. The secondary alcohol rule does not apply to lactones because the stereocenter (marked with an asterisk) lies in a different position. M and L represent the medium-sized and large substituents of the secondary alcohol, respectively, R represents a substituent of the lactone.

crude PPL, possibly cholesterol esterase, is the true catalyst. The difficulties in finding a reliable rule for PPL-catalyzed reactions of primary alcohols may also stem from contaminating hydrolases.

3. Changing from water to organic solvent often changes lipase enantioselectivity, as does changing from one organic solvent to another. Empirical rules and active-site models do not explain these changes. Many researchers used this "medium engineering" to optimize reactions in organic solvents. For example, Mori et al. *(50)* reported that PPL showed no enantioselectivity in the hydrolysis of seudenol acetate, but Johnston et al. *(51)* reported moderate enantioselectivity ($E = 17$) in the acetylation of seudenol with trifluoroethyl acetate in ethyl ether. The enantioselectivity of the CAL-B-catalyzed acetylation of seudenol with vinyl acetate varied from 8 to 32, depending on the solvent and the water content. The highest enantioselectivity was in dry benzene *(52)*. Occasionally, the enantioselectivity even reverses upon changing the solvent. PCL acetylated the (*R*)-enantiomer of a secondary alcohol, methyl 3-hydroxyoctanoate, with vinyl acetate in methylene chloride ($E = 5$), but the (*S*)-enantiomer in hexane ($E = 16$) *(53)*. In the most dramatic example, Amano lipase AH hydrolyzed the *pro-R* ester of a dihydropyridine derivative in cyclohexane ($E \sim 20$), but the *pro-S* ester in diisopropyl ether ($E > 100$) *(54)*.

References

1. Naemura, K. (1994) Stereoselectivity of enzymatic hydrolyses and acylations. *J. Synth. Org. Chem. Jpn.* **52,** 49–58. (in Japanese)
2. Kazlauskas, R. J., Weissfloch, A. N. E., Rappaport, A. T., and Cuccia, L. A. (1991) A rule to predict which enantiomer of a secondary alcohol reacts faster in reactions catalyzed by cholesterol esterase, lipase from *Pseudomonas cepacia*, and lipase from *Candida rugosa*. *J. Org. Chem.* **56,** 2656–2665.
3. Burgess, K. and Jennings, L. D. (1991) Enantioselective esterifications of unsaturated alcohols mediated by a lipase prepared from *Pseudomonas* sp. *J. Am. Chem. Soc.* **113,** 6129–6139.

4. Naemura, K., Ida, H., and Fukuda, R. (1993) Lipase YS-catalyzed enantioselective transesterification of alcohols of bicarbocyclic compounds. *Bull. Chem. Soc. Jpn.* **66,** 573–577.

5. Kim, M. J. and Cho, H. (1992) *Pseudomonas* lipases as catalysts in organic synthesis: specificity of lipoprotein lipase. *J. Chem. Soc., Chem. Commun.* 1411–1413.

6. Roberts, S. M. (1989) Use of enzymes as catalysts to promote key transformations in organic synthesis. *Philos. Trans. R. Soc. Lon. B* **324,** 577–587.

7. Orrenius, C., Öhrner, N., Rotticci, D., Mattson, A., Hult, K., and Norin, T. (1995) Candia antarctica lipase B catalyzed kinetic resolutions: substrate structure requirements for the preparation of enantiomerically enriched secondart alcohols. *Tetrahedron: Asymmetry* **6,** 1217–1220.

8. Janssen, A. J. M., Klunder, A. J. H., and Zwanenburg, B. (1991) Resolution of secondary alcohols by enzyme-catalyzed transeserification in alkyl carboxylates as the solvent. *Tetrahedron* **47,** 7645–7662.

9. Kazlauskas, R. J. and Weissfloch, A. N. E. (1997) A structure-based rationalization of the enantiopreference of subtilisin toward secondary alcohols and isosteric primary amines. *J. Mol. Catal. B Enzymes* **3,** 65–72.

10. Scilimati, A., Ngooi, T. K., and Sih, C. J. (1988) Biocatalytic resolution of (–)-hydroxyalkanoic esters. A strategy for enhancing the enantiomeric specificity of lipase-catalyzed ester hydrolysis. *Tetrahedron Lett.* **29,** 4927–4930.

11. Johnson, C. R., Golebiowski, A., McGill, T. K., and Steensma, D. H. (1991) Enantioselective synthesis of 6-cycloheptene-1,3,5-triol derivatives by enzymatic asymmetrization. *Tetrahedron Lett.* **32,** 2597–2600.

12. Kim, M. J. and Choi, Y. K. (1992) Lipase-catalyzed enantioselective transesterification of O-trityl 1,2-diols. Practical synthesis of (R)-tritylglycidol. *J. Org. Chem.* **57,** 1605–1607.

13. Gupta, A. K. and Kazlauskas, R. J. (1993) Substrate modification to increase the enantioselectivity of hydrolases. A route to optically-active cyclic allylic alcohols. *Tetrahedron: Asymmetry* **4,** 879–888.

14. Adam, W., Mock-Knoblauch, C., and Saha-Möller, C. R. (1997) Kinetic resolution of hydroxy vinylsilanes by lipase-catalyzed enantioselective acetylation. *Tetrahedron: Asymmetry* **8,** 1441–1444.

15. Shimizu, M., Kawanami, H., and Fujisawa, T. (1992) A lipase mediated asymmetric hydrolysis of 3-acyloxy-1-octynes and 3-(E)-acyloxy-1-octenes. *Chem. Lett.* 107–110.

16. Rotticci, D., Orrenius, C., Hult, K., and Norin, T. (1997) Enantiomerically enriched bifunctional *sec*-alcohols prepared by *Candida antarctica* lipase B catalysis. Evidence of non-steric interactions. *Tetrahedron: Asymmetry* **8,** 359–362.

17. Theil, F., Lemke, K., Ballschuh, S., Kunath, A., and Schick, H. (1995) Lipase-catalysed resolution of 3-(aryloxy)-1,2-propanediol derivatives—towards an improved active site model of *Pseudomonas cepacia* lipase (Amano PS). *Tetrahedron: Asymmetry* **6,** 1323–1344.

18. Oberhauser, T., Faber, K., and Griengl, H. (1989) A substrate model for the enzymic resolution of esters of bicyclic alcohols by *Candida cylindracea* lipase. *Tetrahedron* **45,** 1679–1682.

19. Faber, K., Griengl, H., Hoenig, H., and Zuegg, J. (1994) On the prediction of the enantioselectivity of *Candida rugosa* lipase by comparative molecular field analysis. *Biocatalysis* **9,** 227–239.

20. Exl, C., Hoenig, H., Renner, G., Rogi-Kohlenprath, R., Seebauer, V., and Seufer-Wasserthal, P. (1992) How large are the active sites of the lipases from *Candida rugosa* and from *Pseudomonas cepacia*? *Tetrahedron: Asymmetry* **3,** 1391–1394.

21. Burgess, K. and Jennings, L. D. (1991) Enantioselective esterifications of unsaturated alcohols mediated by a lipase prepared from *Pseudomonas* sp. *J. Am. Chem. Soc.* **113,** 6129–6139.

22. Bornscheuer, U., Herar, A., Kreye, L., Wendel, V., Capewell, A., Meyer, H. H., et al. (1993) Factors affecting the lipase catalyzed transesterification reactions of 3-hydroxy esters in organic solvents. *Tetrahedron: Asymmetry* **4,** 1007–1016.

23. Naemura, K., Fukuda, R., Murata, M., Konishi, M., Hirose, K., and Tobe, Y. (1995) Lipase-catalyzed enantioselective acylation of alcohols: a predictive active site model for lipase YS to identify which enantiomer of an alcohol reacts faster in this acylation. *Tetrahedron: Asymmetry* **6,** 2385–2394.

24. Naemura, K., Murata, M., Tanaka, R., Yano, M., Hirose, K., and Tobe, Y. (1996) Enantioselective acylation of alcohols catalyzed by lipase QL from *Alcaligenes* sp.: a predictive active site model for lipase QL to identify the faster reacting enantiomer of an alcohol in this acylation. *Tetrahedron: Asymmetry* **7,** 1581–1584.

25. Toone, E. J., Werth, M. J., and Jones, J. B. (1990) Active site model for interpreting and predicting the specificity of pig liver esterase. *J. Am. Chem. Soc.* **112,** 4946–4952.

26. Provencher, L. and Jones, J. B. (1994) A concluding specification of the dimensions of the active site model of pig liver esterase. *J. Org. Chem.* **59,** 2729–2732.

27. Naemura, K., Takahashi, N., Ida, H., and Tanaka, S. (1991) Pig liver esterase-catalyzed hydrolyses of racemic diacetates of bicyclic compounds and interpretation of the enantiomeric specificity of PLE. *Chem. Lett.* 657–660.

28. Lemke, K., Lemke, M., and Theil, F. (1997) A three-dimensional predictive active site model for lipase from *Pseudomonas cepacia*. *J. Org. Chem.* **62,** 6268–6273.

29. Grabuleda, X., Jaime, C., and Guerrero, A. (1997) Estimation of the lipase PS (*Pseudomonas cepacia*) active site dimensions based on molecular mechanics calculations. *Tetrahedron: Asymmetry* **8,** 3675–3683.

30. Sayle, R. A. and Milner-White, E. J. (1995) RASMOL: biomolecular graphics for all. *Trends Biochem. Sci.* **20,** 374–376.

31. Uppenberg, J., Öhrner, N., Norin, M., Hult, K., Patkar, S., Waagen V., et al. (1995) Crystallographic and molecular modeling studies of lipase B from *Candida antarctica* reveal a stereospecificity pocket for secondary alcohols. *Biochemistry* **34,** 16,838–16,851.

32. Grochulski, P., Li, Y., Schrag, J. D., Bouthillier, F., Smith, P., Harrison, D., et al. (1993) Insights into interfacial activation from an open structure of *Candida rugosa* lipase. *J. Biol. Chem.* **268,** 12,843–12,847.

33. Schrag, J. D., Li, Y. G., Cygler, M., Lang, D. M., Burgdorf, T., Hecht, H. J., et al. (1997) The open conformation of a *Pseudomonas* lipase. *Structure* **5,** 187–202.

34. Neidhart, D. J. and Petsko, G. A. (1988) The refined crystal structure of subtilisin Carlsberg at 2.5 resolution. *Protein Eng.* **2,** 271–276.

35. Kazlauskas, R. J. (2000) Molecular modeling in biocatalysis: explanations, predictions, limitations, and opportunities. *Curr. Opin. Chem. Biol.* **4,** 81–88.

36. Weissfloch, A. N. E. and Kazlauskas, R. J. (1995) Enantiopreference of lipase from *Pseudomonas cepacia* toward primary alcohols. *J. Org. Chem.* **60,** 6959–6969.

37. Tuomi, W. V. and Kazlauskas, R. J. (1999) Molecular basis for enantioselectivity of lipase from *Pseudomonas cepacia* toward primary alcohols. Modeling, kinetics and chemical modification of Tyr29 to increase or decrease enantioselectivity. *J. Org. Chem.* **64,** 2638–2647.

38. Ha, H.-J., Yoon, K.-N., Lee, S.-Y., Park, Y.-S., Lim, M.-S., and Yim, Y.-G. (1998) Lipase PS (*Pseudomonas cepacia*) mediated resolution of γ-substituted γ-((acetoxy)methyl)-γ-butyrolactones: complete stereochemical reversion by substituents. *J. Org. Chem.* **63,** 8062–8066.

39. Lampe, T. F. J., Hoffmann, H. M. R., and Bornscheuer, U. T. (1996) Lipase mediated desymmetrization of *meso*-2,6-di(acetoxymethyl)tetrahydropyran-4-one derivatives. An innovative route to enantiopure 2,4,6-trifunctionalized *C*-glycosides. *Tetrahedron: Asymmetry* **7,** 2889–2900.

40. Ehrler, J. and Seebach, D. (1990) Enantioselective saponification of substituted achiral 3-acyloxypropyl esters with lipases. Preparation of chiral derivatives of tris(hydroxymethyl)methane. *Liebigs Ann. Chem.* 379–388.

41. Wimmer, Z. (1992) A suggestion to the PPL active site model dilemma. *Tetrahedron* **48,** 8431–8436.

42. Guanti, G., Banfi, L., and Narisano, E. (1992) Chemoenzymic preparation of asymmetrized tris(hydroxymethyl)methane (THYM) and of asymmetrized bis(hydroxymethyl)acetaldehyde (BHYMA) as new highly versatile chiral building blocks. *J. Org. Chem.* **57,** 1540–1554.

43. Hultin, P. G. and Jones, J. B. (1992) Dilemma regarding the active site model for porcine pancreatic lipase. *Tetrahedron Lett.* **33,** 1399–1402.

44. Guanti, G., Banfi, L., and Narisano, E. (1990) Enzymes in organic synthesis: remarkable influence of a system on the enantioselectivity in PPL catalyzed monohydrolysis of 2-substituted 1,3-diacetoxypropanes. *Tetrahedron: Asymmetry* **1,** 721–724.

45. Morgan, B., Oehlschlager, A. C., and Stokes, T. M. (1992) Enzyme reactions in apolar solvent. 5. The effect of adjacent unsaturation on the PPL-catalyzed kinetic resolution of secondary alcohols. *J. Org. Chem.* **57,** 3231–3236.

46. Tanaka, M., Yoshioka, M., and Sakai, K. (1993) Highly asymmetric enzymic hydrolysis and transesterification of *meso*-bis(acetoxymethyl)- and bis(hydroxymethyl)cyclopentane derivatives: an insight into the active site model of *Rhizopus delemar* lipase. *Tetrahedron: Asymmetry* **4,** 981–996.

47. Hof, R. P. and Kellogg, R. M. (1996) Lipase AKG mediated resolutions of α,α-disubstituted 1,2-diols in organic solvents: remarkably high regio- and enantio-selectivity. *J. Chem. Soc., Perkin Trans. 1* 2051–2060.

48. Chen, S. T. and Fang, J. M. (1997) Preparation of optically active tertiary alcohols by enzymatic methods. Application to the synthesis of drugs and natural products, *J. Org. Chem.* **62,** 4349–4357.

49. Cotterill, I. C., Sutherland, A. G., Roberts, S. M., Grobbauer, R., Spreitz, J., and Faber, K. (1991) Enzymatic resolution of sterically demanding bicyclo[3.2.0]heptanes: evidence for a novel hydrolase in crude porcine pancreatic lipase and the advantages of using organic media for some of the biotransformations. *J. Chem. Soc., Perkin Trans. I* 1365–1368.

50. Mori, K., Hazra, B. G., Pfeiffer, R. J., Gupta, A. K., and Lindgren, B. S. (1987) Synthesis and bioactivity of optically-active forms of 1-methyl-2-cyclohexen-1-ol, an aggregation pheromone of *Dendroctonus pseudotsugae*. *Tetrahedron* **43,** 2249–2254.

51. Johnston, B. R., Morgan, B., Oehlschlager, A. C., and Ramaswamy, S. (1991) A convenient synthesis of both enantiomers of seudenol and their conversion to 1-methyl-2-cyclohexen-1-ol. *Tetrahedron: Asymmetry* **2,** 377–380.

52. Orrenius, C., Norin, T., Hult, K., and Carrea, G. (1995) The *Candida antarctica* lipase B catalysed kinetic resolution of seudenol in non-aqueous media of controlled water activity. *Tetrahedron: Asymmetry* **6,** 3023–3030.

53. Bornscheuer, U., Schapoehler, S., Scheper, T., and Schuegerl, K. (1991) Influences of reaction conditions on the enantioselective transesterification using *Pseudomonas cepacia* lipase. *Tetrahedron: Asymmetry* **2,** 1011–1014.

54. Hirose, Y., Kariya, K., Sasaki, I., Kurono, Y., Ebiike, H., and Achiwa, K. (1992) Drastic solvent effect on lipase-catalyzed enantioselective hydrolysis of prochiral 1,4-dihydropyridines. *Tetrahedron Lett.* **33,** 7157–7160.

22

Candida antarctica Lipase B

A Tool for the Preparation of Optically Active Alcohols

Didier Rotticci, Jenny Ottosson, Torbjörn Norin, and Karl Hult

1. Introduction

Lipases (E.C. 3.1.1.3) have proved to be efficient catalysts for the preparation of enantiomerically enriched compounds. Among them, *Candida antarctica* lipase B (CALB) has been found to be a particularly useful biocatalyst for the asymmetric transformation of *sec*-alcohols and related compounds. Indeed, about 200 compounds have already been successfully resolved using this enzyme. This number includes some chiral acids, but other lipases have proved to be superior to CALB for the resolution of most chiral acids. This chapter will, therefore, focus on the resolution of chiral alcohols. CALB is supplied as a recombinant protein patented by Novo-Nordisk. Despite its origin in the Antarctics, CALB is stable at 60–80°C for extended periods of time, once it is immobilized *(1)*. Furthermore, CALB retains most of its activity and robustness in nonaqueous media. An example is that Glaxo selected CALB out of two lipases suitable for a multikilo resolution, because of its stability over multiple-use cycles in nonaqueous media *(2)*.

A general review about the application of CALB in organic synthesis has recently been published by Anderson et al. *(1)*. Kazlauskas and Bornscheuer wrote a comprehensive chapter on biotransformations with lipases, which thoroughly covered CALB applications *(3)*.

The aim of present chapter is to help the reader successfully prepare optically active alcohols using CALB. A simple model for estimating the enantioselectivity of CALB toward *sec*-alcohols is presented in the Notes. The model is illustrated by a selection of literature data on enantioselectivity of

From: *Methods in Biotechnology, Vol. 15: Enzymes in Nonaqueous Solvents: Methods and Protocols*
Edited by: E. N. Vulfson, P. J. Halling, and H. L. Holland © Humana Press Inc., Totowa, NJ

CALB toward *sec*-alcohols. The reader can perceive the potentials and limitations of CALB in asymmetric transformations from these and other selected examples (*see* **Notes 1–5**). We also present a review of parameters influencing the enantioselectivity of CALB and suggest actions that may improve its selectivity. A basic reaction scheme for the CALB-catalyzed kinetic resolution of an alcohol is presented in **Fig. 1**. In organic solvents, lipases catalyze the acyl-group transfer from a suitable acyl-donor to an acyl acceptor (e.g., a chiral alcohol). Most alcohol enantiomers react with unequal rates in reactions catalyzed by CALB, allowing the preparation of optically active alcohols and esters by kinetic resolution.

2. Materials
2.1. Preparation of Enzyme and Reagents

1. Immobilized enzyme preparation (*see* **Notes 6–9**).
2. Chiral alcohol to be resolved.
3. Vinyl acetate or an other suitable acyl donor (*see* **Notes 10–16**). A variety of vinyl alkanoates are available from TCI Japan. *S*-Methyl thioacetate and thiobutyrate are sold by Aldrich.
4. Reaction vessel with lid or penetrable septum to enable sampling.
5. Solvent pro analysi (p.a.) grade (*see* **Note 17**).
6. Desiccator, saturated salt solutions, or molecular sieves (*see* **Notes 18** and **19**) (optional).
7. Internal standard such as dodecane, p.a. grade (*see* **Notes 20** and **21**) (optional).

2.2. Kinetic Resolution

1. Stirring device: propeller, end-over-end shaker, or magnetic stirrer (*see* **Note 22**).
2. Thermostated bath (*see* **Note 23**) (op).

2.3. Sampling, Monitoring, Analysis

1. Syringe, 10–100 µL, or pipet, 10–100 µL.
2. Sample vials.
3. Solvent, p.a. grade.
4. Chromatographic system, gas chromatography (GC) or high-performance liquid chromatography (HPLC), with chiral column or columns for the analysis of enantiomeric excess of substrate (ee_S), product (ee_P), or conversion (*see* **Note 20**).

3. Methods
3.1. Preparation of Enzyme and Reagents

1. Weigh the lipase preparation (e.g., 20 mg/mmol alcohol [*see* **Note 24**]) in the reaction vessel.
2. If you wish to control the water activity, equilibrate the lipase and solutions (*see* **Note 18**).

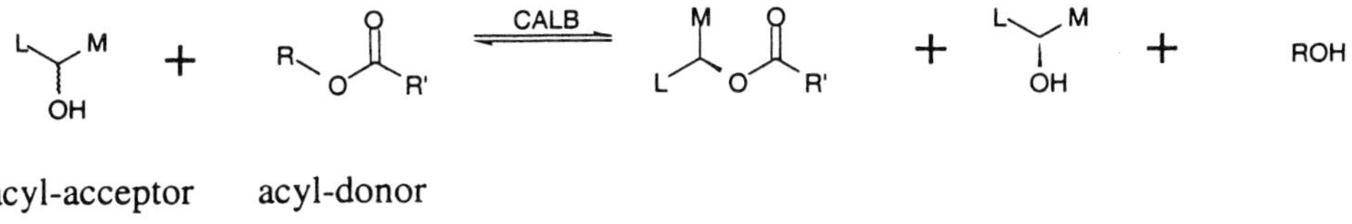

acyl-acceptor   acyl-donor

Fig. 1. Reaction scheme.

3. Add alcohol (e.g., 0.5 M [*see* **Note 25**]) and solvent (e.g., hexane [*see* **Note 17**]) to the reaction vessel. If the ee cannot be determined for both product and substrate, add an internal standard (e.g., dodecane 0.05 M [*see* **Notes 20** and **21**]) in order to enable the determination of the conversion.
4. Close the vessel and let the reaction reach the desired temperature (e.g., 25°C [*see* **Note 10**]) in the presence of an efficient stirring (*see* **Note 22**).

3.2. Kinetic Resolution of Alcohol

Start the reaction by adding the acyl donor (e.g., 0.5 M vinyl acetate [*see* **Notes 10–16** and **26**]).

3.3. Sampling, Monitoring, Analysis

1. To monitor the reaction as it proceeds, take samples regularly with a syringe or a pipet.
2. Dilute each sample to a suitable concentration (2 mg/mL for GC analysis).
3. Derivatize samples if necessary for analysis.
4. Analyze the samples on a chiral GC or HPLC column and determine two of the following three parameters: conversion, ee_S, and ee_P (*see* **Note 20**).
5. Calculate the E ratio and determine at which conversion to stop the reaction (*see* **Note 27**).
6. Stop the reaction by removing the immobilized lipase by filtration. Rinse the carrier beads (*see* **Note 28**).

3.4. Workup

The workup procedure depends on the properties of the starting materials and products. No process can be recommended for general use (*see* **Note 29**).

4. Notes

1. *Substrate structure requirements and enantioselectivity:* Concerning *sec*-alcohols, CALB follows the rule described by Kazlauskas et al. *(4)*. By comparing the size of the substituents at the stereocentre, this empirical rule **(Fig. 2)** predicts which enantiomer will react the fastest.
2. *Acyclic sec-alcohols:* A simple model based on substrate structure requirements can be used to estimate the CALB enantioselectivity toward *sec*-alcohols. A high selectivity (generally $E > 100$) and a relatively high reaction rate can be expected

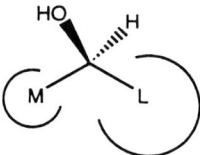

Fig. 2. Empirical rule for predicting the fast reacting enantiomer of *sec*-alcohols.
M = medium-sized substituent; L = large-sized substituent.

for substrates that fulfill the following requirements: The medium-sized substituent should be smaller than a propyl group (**Table 1**, **1–8** and **10–11**). Furthermore, halogen atoms placed on this substituent tend to decrease the enantioselectivity (substrate **14**). As predicted by the model, **13** and **15** are not suitable substrates for CALB because their medium-sized substituents are too large. Therefore, **13** reacts very slowly and has a low *E* value, whereas **15** does not react at all. The large-sized substituent should be at least as large as a propyl group. Otherwise, the selectivity is poor as for **12**. Halogen atoms placed on this substituent tend to increase the enantioselectivity (substrate **4**). One can see that the large substituent can be of almost any size (substrates **1–8**, **10**, and **11**), but the rate goes down dramatically for the bulky compounds **10** and **11**. A steric clash is reached in **9**, showing some limitations with regards to the size for the large substituent.

3. *Cyclic alcohols:* The aforementioned model can generally be applied to cyclic alcohols. Here, the ring replaces the large and medium substituents. Carbons linked to the stereogenic center should be regarded as part of the hypothetical large and medium substituents. The size of these substituents will be assigned according to the sizes of the ligands placed on the α- and β-carbons as exemplified by 2-methylcyclohexan-1-ol in **Fig. 3**.
Some examples of cyclic alcohols are shown in **Table 2**. One can notice that the *trans*-α-substituted alcohols (substrates **21** and **22**) react faster than their *cis*-stereomers (substrates **23** and **24**). The enantioselectivity decreases with the non-α-substituted alcohols as well as with those substituted on both α-carbons (substrates **25–29**).

4. *Prochiral or meso compounds:* **Table 3** shows some examples of asymmetrization of prochiral or meso compounds catalyzed by CALB. One advantage of this type of asymmetrization processes is the theoretical 100% yield, compared to the maximum of 50% in a kinetic resolution of a racemic mixture. The outcome has been reported for only few substrates.

5. *Other types of alcohols and masked alcohols:* A great variety of alcohols, which do not belong to the aforementioned classes, have been resolved by CALB (**Table 4**, substrates **35–43**). The model for *sec*-alcohols can neither be applied to the lactones **44** and **45** nor to the carbonates **46–48**. The chiral center of the vinyl carbonates can be regarded as part of the acyl-chain. CALB shows opposite enantiopreference for vinyl 2-butyl carbonate **46** and 2-butanol **12**. Furthermore,

Table 1
Examples of Acyclic *sec*-Alcohols, Most of Which Are Resolved by CALB

Substrate	Structure	R. t.[a]	C. (%)[b]	E[c]	F. r.[d]
1[f]	R=Pr	0h20	34	>500[e]	R
2[g]	R=*i*-Pr	2h48	35	>500[e]	R
3[f]	R=*n*-hexyl	0h21	28	>900[e]	R
4[h]		3h40	47	370[e]	R
5[i]	R=Me	4h00	49	>500	R
6[i]	R=Et	14h00	50	>500	R
7[j]	R_1=CHOHCH$_3$ R$_{2,3}$=H	3h30	50	>500	
8[j]	R_2=CHOHCH$_3$ R$_{1,3}$=H	24h00	45	>500	
9[j]	R_3=CHOHCH$_3$ R$_{1,2}$=H		n. r.[m]		
10[k]	M=H$_2$	9h00	50	>458	R
11[k]	M=Zn	48h00	20	>126	R
12[g]		2h50	37	8[e]	R
13[l]		123h00	51	10	
14[h]		23h00	46	7[e]	S
15[l]			n. r.[m]		

[a]R. t. = reaction time.
[b]C = conversion.
[c]E = enantiomeric ratio.
[d]F. r. = fast reacting enantiomer.
[e]Based on at least five measurements at different conversions.
[f](Ottoson, J., unpublished data) Novozyme 435 (4–5 mg/mL, a_w = 0.1), alcohol (0.43 *M*), vinyl octanoate (0.43 *M*), hexane, solutions predried over molecular sieves. **1**: 30°C; **2**: 39°C.
[g](**5**) Novozyme 435 (2–5 mg/mL, a_w = 0.1), alcohol (0.43 *M*), vinyl octanoate (0.43 *M*), hexane, solutions predried over molecular sieves, 30°C.

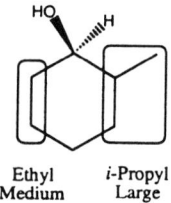

Ethyl   *i*-Propyl
Medium   Large

Fig. 3. Determination of the hypothetical medium and large substituents of cyclic alcohols.

CALB displays high selectivity for **46**, whereas its counterpart **12** is poorly resolved. Thus, the kinetic resolution of carbonates may be regarded as an attractive alternative for the preparation of optically active alcohols.

6. *CALB preparations:* Immobilized preparations are preferable for reactions in nonaqueous media. Some of the advantages of the immobilized lipases are the easy enzyme recovery during workup, the decreased diffusion limitation in organic solvent, and the increase in thermostability.

7. *Availability:* CALB is available from Novo-Nordisk A/S (Denmark) and Boehringer Mannheim (Germany). The former firm sells CALB under the trade name Novozyme 435, which is a recombinant enzyme, immobilized on a macroporous acrylic resin. Polyacrylamide gel electrophoresis of CALB desorbed from the Novozyme 435 support exhibited a single band of 33 kDa *(26)*. Earlier, Novo-Nordisk distributed CALB as special preparations (SP). SP 435 is equivalent to the Novozyme 435. SP 525 is a powder containing the component B only, whereas SP 382 is an immobilized mixture of the components A and B (component A is a lipase quite different from component B). Boehringer Mannheim sells CALB under the trade name Chirazyme® L-2, which is available either as the free enzyme or immobilized. The three immobilized products are all characterized by the abbreviation c.-f. (carrier-fixed enzyme). According to the product information, Chirazyme L-2, c.-f., lyo is covalently bound to the carrier and has been developed for hydrolysis reactions. Chirazyme L-2, c.-f., C-3, lyo, on the other hand, is a noncovalently carrier-fixed version that is especially suited for

---

[h]*(6)* Novozyme 435 (**4:** 7 mg/mL; **14:** 25 mg/mL), alcohol (0.5 M), vinyl butyrate (0.5 M), hexane, 23°C.

[i]*(7)* Novozyme 435 (30 mg), alcohol (100 mg), vinyl acetate (0.2 mL), dry *i*-Pr$_2$O (20 mL), molecular sieves, room temperature.

[j]*(8)* Immobilized CALB (40 mg, Novo), alcohol (10 mg), vinyl acetate (100 mg), benzene (5 mL), molecular sieves, room temperature.

[k]*(9)* Chirazyme® L-2 (900 mg), alcohol (42 μmol), vinyl acetate (1.3 mmol), *i*-Pr$_2$O (30 mL), 30°C.

[l]*(10)* Novozyme 435 (130 mg, a_w = 0.1), alcohol (2 mmol), vinyl acetate (4 mmol), hexadecane (1 mmol).

[m]n.r. = no reaction.

Table 2
Examples of Cyclic Alcohols Resolved by CALB

Substrate	Structure		R. t.[a]	C. (%)[b]	E[c]	F. r.[d]
16[f]		n=1 X=I	1h30	49	>700	R
17[f]		n=2 X=I	18h00	50	>450	R
18[f]		n=2 X=Br	18h30	48	>110	R
19[f]		n=2 X=H	1h30	48	1.6	R
20[f]		n=3 X=I	96h00	49	>240	R
21[g]		X=NMe$_2$	92h00	45	>200	R
22[g]		X=Piperidine	142h00	48	>200	R
23[g]		X=NMe$_2$	116h00	9	>200	R
24[g]		X=Piperidine	144h00	4	200	R
25[h]		R=cis-N-(benzylcarbamoyl)	31h00	39	34	S
26[i]		R=cis-O-TBS	5h30	63	15	S
27[j]			1h00	52	62[e,j]	R
28[k]			120h00	30	1.3	S
29[l]			48h00	49	90	S

[a]R. t. = reaction time.
[b]C = conversion.
[c]E, enantiomeric ratio.
[d]F. r. = fast-reacting enantiomer.
[e]Based on at least five measurements at different conversions.
[f](11) Novo SP-435 (25 wt% alcohol), isopropenyl acetate (4 mL/g alcohol), hexane (16 mL/g alcohol).
[g](12) Novozyme 435 (30 mg/mL), alcohol (0.1 M), vinyl acetate (0.2 M), Et$_2$O, room temperature.
[h](13) CALB (5.2 mg), alcohol (0.12 mmol), vinyl acetate (0.43 mmol), CH$_2$Cl$_2$ (0.5 mL).
[i](14) Sp 435 (0.3 equiv. wt.), isopropenyl acetate (3 equiv.), TBME, room temperature.
[j]See **Note 2**, **item 3**.
[k](15) CALB, vinyl acetate, hexane/ether, 40°C.
[l](16) immobilized CALB (1 g, Novo), alcohol (10 mmol), vinyl acetate (10 mmol), i-Pr$_2$O (50 mL), 40°C.

organic solvents. Chirazyme L-2, c.-f., C-2, lyo corresponds to Novozyme 435. Information about Chirazyme products can be found on Boehringer Mannheim's web page (http://biochem.boehringer-mannheim.com).

Table 3
Asymmetrization of Prochiral or *meso* Alcohols

Substrate	Structure	R. t.[a]	Yield	ee
30[b]		6h00	71	98.2
31[c]	n=1	72h00	48	>99
32[c]	n=2	89h00	25	59
33[c]	n=3	89h00	81	>99
34[d]		23h00	74	>98

[a]R. t. = reaction time.

[b]*(17)* Novo SP435 (0.26 g), diol (21.95 mmol), vinyl acetate (43.4 mmol), MeCN (25 mL), 0°C, reaction was scaled up to 30 kg.

[c]*(18)* SP-435, $n = 1$ isopropenyl acetate (neat), 50°C; $n = 2$ isopropenyl acetate (5 parts), TBME (3 parts), 50°C; $n = 3$ isopropenyl acetate (1 part), TBME (4 parts), 50°C; **Note:** CALB was more efficient in the hydrolysis of the diacetates.

[d]*(19)* CALB (30.3 mg), diol (505 µmol), vinyl acetate (2.53 mmol), benzene (50 mL), room temperature.

8. *Storage:* CALB preparations should be stored at 4°C in a dry place when not used. Using these conditions for years, we have not noted any significant decrease in the activity of the Novozyme 435 preparation.

9. *Comparison of immobilized CALB preparations:* We compared the reaction rates and enantioselectivity for the three commercially available preparations in transesterification reactions (**Table 5**).

   Noticeable differences among the preparations were found for the substrates tested. Novozyme 435 turned out to be the best catalyst for these particular substrates under the conditions used. This demonstrates that differences in rate and enantioselectivity can be obtained using different preparations under identical conditions. However, one should not conclude that the Novozyme 435 is the best under other conditions or for other substrates.

10. *Acyl-donors:* In organic solvents, lipases catalyze acyl-transfer reactions such as esterifications (alcohol and acid) and transesterifications (alcohol and ester). The reactions can be either reversible or irreversible, depending on the leaving group of the acyl donor and the reaction conditions. The equilibrium constant K affects the optical yield of the reaction because of the fact that the reverse reaction lowers the ee. Hence, the best results will very often be obtained under irreversible conditions as described in **Notes 11–15**. The acyl part of the acyl donor influences the enantioselectivity as well as the reaction rate (*see* **Note 16**).

11. *Leaving groups:* Irreversible conditions can be obtained by using various activated esters.

Table 4
Examples of Other Types of Alcohols and Masked Alcohols Resolved by CALB

Substrate	Structure			R. t.[a]	C. (%)[b]	E[c]	F. r.[d]
35[e]	R=Ph	X=O		240h00	44	>200	S
36[e]	R=4-BrPh	X=O		20h00	38	>100	S
37[e]	R=Me	X=O		20h00	60	21	S
38[e]	R=Ph	X=NMe		72h00	54	>50	S
39[f]	R=Me			3h00	60	10	R
40[f]	R=CH₂Ph			48h00	5	49	S
41[g]	R=Me			0h54	32	30	R
42[g]	R=Ph			1h00	42	15	R
43[h]				7h00	59	20	P
44[i]	R=Me			11h00	51	>200	R
45[i]	R=i-Pr			72h00	50	>200	S
46[j]	R=Et			5h00	45	118	S
47[j]	R=n-Hexyl			4h00	40	75	S
48[j]	R=Ph			9h00	50	96	S

[a]R. t. = reaction time.
[b]C = conversion.
[c]E = enantiomeric ratio.
[d]F. r. = fast-reacting enantiomer.
[e](20) CALB (20 mg, from Novo), alcohol (0.3 mmol), vinyl acetate (0.2 mL), i-Pr₂O (1 mL), room temperature.
[f](21) Novozyme 435 (50 mg), alcohol (0.4–0.6 mmol), vinyl acetate (5 equiv, neat), room temperature.
[g](22) Novozyme 435 (40 mg/mL), alcohol (20 mg/mL), vinyl acetate (10 equiv), i-Pr₂O.
[h](23) Immobilized CALB (5 mg, Novo), alcohol (2 mg), vinyl acetate (40 μL), CH₂Cl₂ (2 cm₃), 30°C.
[i](24) Chirazyme® L-2 (44: 15 mg; 45: 110 mg), lactone (44: 0.19 mmol; 45: 3.17 mmol), benzyl alcohol (44: not stated, 45: 12.7 mmol), MTBE (44: not stated, 45: 12 mL), 20°C.
[j](25) SP435A (100 mg), carbonate (1 mmol), benzyl alcohol (0.6 mmol), hexane (15 mL), molecular sieves.

12. *Vinyl esters:* The acyl donors by far most used are enol esters, such as vinyl acetate and isoprenyl acetate *(27)*. The leaving enol tautomerizes to non-nucleo-

Table 5
Comparison of CALB-Immobilized Preparations

CALB preparation	LU/g carrier[a]	E^b + S.D.	Fast-reacting enantiomer[b]
Chirazyme L-2, c.-f., lyo	3200	34 ± 2	R
Novozyme 435	20,100	68 ± 3	R
Chirazyme L-2, c.-f., C-3, lyo	18,900	62 ± 3	R

[a]Lipase unit measurements in organic solvents (LU): The procedure is described in **Subheading 3**. Reaction conditions: immobilized enzyme (typically 10 mg, $a_w = 0.1$), 1-octanol (9 mmol, 1.41 mL), vinyl acetate (18 mmol, 1.66 mL), hexane (5.93 mL, predried over molecular sieves), 20°C. Samples (20 µL) were withdrawn and conversions were determined by gas chromatography (GC) (column: Chrompack CP-WAX 58CB, 25 m × 0.32 mm, He, 120°C isothermal, response factor 1.2). The activity was calculated at low conversion (<10%) to minimize product inhibition. One LU was defined as micromoles of product formed per minute under the above mentioned conditions.
[b]Kinetic resolution of 3-methyl-2-cyclohexen-1-ol **Table 2**, substrate **27** [a sex pheromone of the Douglas-fir beetle): Procedure is described in **Subheading 3**. Reaction conditions: enzyme (30 mg), alcohol (0.128 mL, 1.1 mmol), vinyl acetate (0.205 mL, 2.2 mmol), hexane (4.67 mL), 23°C, $a_w = 0.1$. Analysis: ee_S and ee_P were determined on a Chrompack CP chirasil-dex CB, 25 m × 0.32 mm GC column. The E values are based on six measurements at different conversions. S.D. = standard deviation.

philic volatile aldehydes or ketones, thus driving the equilibrium toward the product side. Aldehydes have been shown to deactivate some lipases by forming Schiff's bases with exposed lysine residues, but CALB has proven to be little affected *(28)*. If the substrate cannot stand the presence of aldehydes or ketones, an alternative acyl donor has to be used (*see* **Notes 13–15**).

13. *Thioesters:* Our group has used *S*-ethyl thiooctanoate for many transesterification reactions *(29)*. This acyl donor drives the reaction toward the product side because the thiol is a good leaving group, a bad nucleophile, and easily removable by evaporation above 35°C. *S*-Methyl thioacetate has also been successfully used in our laboratory *(30)*. Advantages of *S*-methyl thioalkanoates are their commercial availability and the low boiling point of the leaving group (6°C). However, work with thiols require extra care, such as working in well-ventilated fume hoods. In order to drive the equilibrium to the right, the system should be open. If dry reaction conditions are needed, the vessel can be covered by a drying tube, containing a drying agent such as calcium chloride to prevent moisture from entering the system.

14. *Ethyl esters:* Ethyl alkanoates are inexpensive and easily accessible acyl donors, but a low product equilibrium is a major problem. This can be circumvented by continuous evaporation of the coproduct ethanol under reduced pressure *(31)*. The drawbacks of this method are complicated experimental setup, limitations in the choice of solvents, and nonapplicability for volatile substrates.

15. *Acid anhydrides:* Acid anhydrides also allow practically irreversible reactions *(32)*, but the acid formed may have a negative effect on the enantioselectivity. Furthermore, the spontaneous reaction may degrade ee_P in a polar solvent.

16. *Acyl part:* The enantioselectivity is influenced by the length and structure of the acyl part of the acyl donor. In general, we find that E is higher for the longer and larger acyl groups, as exemplified in **Table 6**. We suggest that the inexpensive, easily accessible vinyl acetate be tried at first and if the enantioselectivity is too low, a longer or larger acyl part then tried.

17. *Solvent:* General recommendation as regards the choice of organic solvents is to use a hydrophobic solvent ($\log P > 1.5$) if possible *(34)*. CALB is somewhat of an exception being rather stable and active even in relatively polar solvents, such as acetonitrile *(1,35)*. This can be advantageous when the substrate or the product is soluble only in polar solvents. The enantioselectivity of CALB has been shown to vary with the choice of solvent *(12,36,37)*. Arroyo and Sinisterra have even reported a change in enantiopreference when changing the solvent from R- to S-carvone *(38)*. Generally, the variations in E due to a solvent change are not very dramatic *(36,39,40)*, but a solvent change can sometimes make all the difference for a successful kinetic resolution *(12,41)*. There have been attempts to correlate these variations with physical parameters, such as $\log P$. Some studies disclose no correlation between E and $\log P$ *(12,41)*, whereas others show an increase in E with decreasing $\log P$ *(39,40,42)*. These investigations usually show a decreased reaction rate in more polar solvents. The reaction can also be run without a solvent or in a large excess of acyl-donor.

18. *Water activity/content:* If you choose to control the water activity, you start by equilibrating the reaction vessel containing the lipase and the solutions through the vapor phase in a sealed container with the desired saturated salt solution for 24 h. The water activity has been shown to influence the enantioselectivity of some lipase-catalyzed kinetic resolutions, but no general trend has been identified. Some reports in the literature show an increase in E with increasing water activity *(43,44)*, others a decrease in E *(36,45)*, and yet others no effect at all *(46,47)*. In a study of several solvents, Orrenius et al. found a tendency of increased selectivity of CALB in the drier systems *(36)*. Wehtje et al. found that the enantioselectivity of CALB was unaffected by the water activity in the esterification of 2-octanol with decanoic acid *(47)*. If the water content in the system is to high, water can compete with the alcohol as a nucleophile, producing acid too.

19. *Molecular sieves:* Molecular sieves (without indicator) should be activated before use. Once activated, they should be stored in a closed flask, in a dry desiccator or in an oven at 120°C. To activate the molecular sieves, place them at 300°C under vacuum (approx 1 mmHg) for 2 h.

20. *Calculation of the enantiomeric ratio* E: The ability of an enzyme to distinguish between two competing enantiomers is defined by the enantiomeric ratio E. This is the ratio between the specificity constants (k_{cat}/K_m) of the enzyme for the two competing enantiomers *(48)*.

   The E of the reaction can be determined from two of the three following parameters: the enantiomeric excess of the substrate, ee_s, the enantiomeric excess of the product, ee_P, and the conversion, c, determined by the internal standard method (**Eqs. 1–3**) *(33,48)*. The determination of E from ee_S and ee_P gives a more accurate value of E than a determination from c and ee_S or ee_P and is to be used if possible.

Table 6
Effect of Acyl Length on *E*

Substrate	Acyl donor	E^a
3-Methyl-2-butanol	Vinyl propionate	300
3-Methyl-2-butanol	Vinyl butanoate	250
3-Methyl-2-butanol	Vinyl hexanoate	490
3-Methyl-2-butanol	Vinyl octanoate	600

^{a}*E*-calculated from ee$_s$ and ee$_p$ as the average of 6–11 measurements at conversions between 3% and 47% *(33)*. The reaction performed as described under in **Subheading 3**. Reaction conditions: Novozyme 435 ($a_w = 0.1$), alcohol (0.43 *M*), vinyl alkanoate (0.43 *M*), hexane, solutions predried over molecular sieves, 40°C.

$$E = \frac{\ln[1 - c(1 + ee_P)]}{\ln[1 - c(1 - ee_P)]} \tag{1}$$

$$E = \ln\left[\frac{1 - ee_S}{1 + ee_S/ee_P}\right]\Bigg/\left[\frac{1 + ee_S}{1 + ee_S/ee_P}\right] \tag{2}$$

$$E = \frac{\ln[1 - c(1 + ee_S)]}{\ln[1 - c(1 - ee_S)]} \tag{3}$$

For the systems described herein, *E* should be calculated from the equation for an irreversible reaction. However, a declining *E* at a conversion higher than 50% is sometimes found. This often means that the backward reaction is becoming significant. A trend in *E* can be a sign of contamination, faulty ee measurements, side reactions, and so forth.

21. *Selection of the internal standard:* The internal standard should be inert and of a hydrophobicity similar to that of the solvent. The accuracy of the conversion calculated with the internal standard method is often low. To keep the errors in *E* down, it should be calculated from ee$_p$ and *c*; this should be done at low conversions.

22. *Stirring:* Stirring should ensure a homogeneous dispersion of the carrier beads in the medium. Magnetic stirring bars tend to disintegrate the carrier beads and should be avoided in preparative scale syntheses.

23. *Temperature:* The temperature influences the efficiency of enzymatic kinetic resolutions. The enantioselectivity can generally be improved by a decrease in the reaction temperature. For CALB, the magnitude of the effect varies but is significant in all systems known to us *(5,49,50)*. A decreased reaction temperature will also result in a decrease in reaction rate. The choice of temperature will, therefore, always be a compromise between optical yield and reaction time. The reaction time may be decreased for substrates with high *E* by increasing the temperature. Substrates with an insufficient value of *E* at high temperature may still be successfully resolved, at the expense of reaction rate, by lowering the temperature.

24. *Amount of lipase:* CALB displays large differences in reaction rate between different alcohols, so that the amount of catalyst must be adjusted to compensate for these differences. Usually, the lipase amounts vary from 5 to 300 mg/mmol of alcohol. One can start with the amount used for a similar alcohol presented in one of the **Figs. 2** and **3** and **Tables 1**, **3**, and **4**.
25. *Alcohol concentration:* The alcohol concentration will generally be kept low for screening experiments, but be increased for the preparative work. Usually, the concentration varies between 0.01 and 2 *M*.
26. *Acyl-donor concentration:* The acyl donors have been used in very different amounts going from one equivalent to bulk solvent.
27. The ee of the product or the remaining alcohol will be a compromise between optical purity and yield. If the reaction is monitored as it proceeds, the choice can be made. The program Selectivity for Mac or Windows (Faber et al. http://borgc185.kfunigraz.ac.at) allows calculating E for irreversible reactions as well as to plot ee_s and ee_p versus the conversion for the E obtained. This plot should help you to determine when to stop the reaction. If the E is low, a double resolution can be considered to improve the ee.
28. Boehringer Mannheim recommends that the rinsing of the carrier be finished with an apolar solvent.
29. *Workup:* In syntheses of up to 10 g, we purify the optically active ester and alcohol by medium pressure chromatography or by flash chromatography.

Acknowledgments

Financial support from the Swedish Research Council for Engineering Sciences (TFR) and the European Union is gratefully acknowledged. Generous supplies of Novozyme 435 by Novo-Nordisk and Chirazyme® L-2 by Boehringer Mannheim are also acknowledged. We thank Gunhild Aulin-Erdtman for valuable linguistic comments.

References

1. Anderson, E. M., Larsson, K. M., and Kirk, O. (1998) One biocatalyst — many applications: the use of *Candida antarctica* B-lipase in organic synthesis. *Biocatal. Biotransform.* **16**, 181–204.
2. Stead, P., Marley, H., Mahmoudian, M., Webb, G., Noble, D., To Ip, Y., et al. (1996) Efficient procedures for the large-scale preparation of (*1S,2S*)-*trans*-2-methoxycyclohexanol, a key chiral intermediate in the synthesis of tricyclic β-lactam antibiotics. *Tetrahedron: Asymmetry* **7**, 2247–2250.
3. Kazlauskas, R. J. and Bornscheuer, U. T. (1998) Biotransformations with lipases, in *Biotechnology, 2nd ed., vol. 8a: Biotransformation* I (Kelly, D. R., ed.), Wiley–VHC, Weinheim, pp. 37–191.
4. Kazlauskas, R. J., Weissfloch, A. N. E., Rappaport, A. T., and Cuccia, L. A. (1991) A rule to predict which enantiomer of a secondary alcohol reacts faster in reactions catalyzed by cholesterol esterase, *Pseudomonas cepacia* and *Candida rugosa* lipase. *J. Org. Chem.* **56**, 2656–2665.

5. Overbeeke, T., Ottosson, J., Hult, K., Jongejan, J., and Duine, J. (1999) The temperature dependence of enzymatic kinetic resolutions reveals the relative contribution of enthalpy and entropy to enzymatic enantioselectivity. *Biocatal. Biotransform.* **17**, 61–79.

6. Rotticci, D., Orrenius, C., Hult, K., and Norin, T. (1997) Enantiomerically enriched bifunctional sec-alcohols prepared by *Candida antarctica* lipase B catalysis. Evidence of non-steric interactions. *Tetrahedron: Asymmetry* **8**, 359–362.

7. Uenishi, J. I., Hiraoka, T., Hata, S., Nishiwaki, K., Yonemitsu, O., Nakamura, K., et al. (1998) Chiral pyridines: optical resolution of 1-(2-pyridyl)- and 1-[6-(2,2'-bipyridyl)]ethanols by lipase-catalyzed enantioselective acetylation. *J. Org. Chem.* **63**, 2481–2487.

8. Kano, K., Negi, S., Kawashima, A., and Nakamura, K. (1997) Optical resolution of 1-arylethanols using transesterification catalyzed by lipases. *Enantiomer* **2**, 261–266.

9. Ema, T., Jittani, M., Sazkai, T., and Utaka, M. (1998) Lipase-catalyzed kinetic resolution of large secondary alcohols having tetraphenylporphyrin. *Tetrahedron Lett.* **39**, 6311–6314.

10. Orrenius, C., Hæffner, F., Rotticci, D., Öhrner, N., Norin, T., and Hult, K. (1998) Chiral recognition of alcohol enantiomers in acyl transfer reactions catalysed by *Candida antarctica* lipase B. *Biocatal. Biotransform.* **16**, 1–15.

11. Johnson, C. R. and Sakaguchi, H. (1992) Enantioselective transesterifications using immobilized, recombinant *Candida antarctica* lipase B: resolution of 2-iodo-2-cycloalken-1-ols. *Synlett* **10**, 813–816.

12. Forro, E., Kanerva, L. T., and Fulop, F. (1998) Lipase-catalyzed resolution of 2-dialkylaminomethylcyclohexanols. *Tetrahedron: Asymmetry* **9**, 513–520.

13. Mulvihill, M. J., Gage, J. L., and Miller, M. J. (1998) Enzymic resolution of aminocyclopentenols as precursors to D- and L-carbocyclic nucleosides. *J. Org. Chem.* **63**, 3357–3363.

14. Curran, T. T. and Hay, D. A. (1996) Enzymic resolution of cis-4-O-TBS-2-cyclopenten-1,4-diol. *Tetrahedron: Asymmetry* **7**, 2791,2792.

15. Mitrochkine, A., Gil, G., and Réglier, M. (1995) Synthesis of enantiomerically pure *cis* and *trans*-2-amino-1-indanol. *Tetrahedron: Asymmetry* **6**, 1535–1538.

16. Igarashi, Y., Otsutomo, S., Harada, M., Nakano, S., and Watanabe, S. (1997) Lipase-mediated resolution of indene bromohydrin. *Synthesis* **5**, 549–552.

17. Saksena, A. K., Girijavallabhan, V. M., Lovey, R. G., Pike, R. E., Wang, H., Ganguly, A. K., et al. (1995) Highly stereoselective access to novel 2,2,4-trisubstituted tetrahydrofurans by halocyclization: practical chemoenzymic synthesis of SCH 51048, a broad-spectrum orally active antifungal agent. *Tetrahedron Lett.* **36**, 1787–1790.

18. Johnson, C. R. and Bis, S. J. (1992) Enzymic asymmetrization of meso-2-cycloalken-1,4-diols and their diacetates in organic and aqueous media. *Tetrahedron Lett.* **33**, 7287–7290.

19. Chênevert, R. and Rose, Y. S. (1998) A chemoenzymic synthesis of both enantiomers of a *cis*-lignan lactone. *Tetrahedron: Asymmetry* **9**, 2827–2831.

20. Hof, R. P. and Kellogg, R. M. (1996) Synthesis and lipase-catalyzed resolution of 5-(hydroxymethyl)-1,3-dioxolan-4-ones: masked glycerol analogs as potential building blocks for pharmaceuticals. *J. Org. Chem.* **61,** 3423–3427.

21. Gais, H.-J. and von der Weiden, I. (1996) Preparation of enantiomerically pure α-hydroxymethyl *S*-tert-butyl sulfones by *Candida antarctica* lipase catalyzed resolution. *Tetrahedron: Asymmetry* **7,** 1253–1256.

22. Patti, A., Lambusta, D., Piattelli, M., Nicolosi, G., McArdle, P., Cunningham, D., et al. (1997) Lipase-assisted preparation of enantiopure ferrocenyl sulfides possessing planar chirality and their use in the synthesis of chiral sulfoxides. *Tetrahedron* **53,** 1361–1368.

23. Tanaka, K., Osuga, H., Suzuki, H., Shogase, Y., and Kitahara, Y. (1998) Synthesis, enzymic resolution and enantiomeric enhancement of *bis*(hydroxymethyl)[7]thiaheterohelicenes. *J. Chem. Soc., Perkin Trans. 1* **5,** 935–940.

24. Adam, W., Groer, P., and Saha-Möller, C. (1997) Enzyme preparation of optically active α-methylene β-lactone by lipase-catalyzed kinetic resolution through asymmetric transesterification. *Tetrahedron: Asymmetry* **8,** 833–836.

25. Pozo, M. and Gotor, V. (1993) Kinetic resolution of vinyl carbonates through a lipase-mediated synthesis of their carbonate and carbamate derivatives. *Tetrahedron* **49,** 10,725–10,732.

26. Parker, M.-C., Brown, S. A., Robertson, L., and Turner, N. J. (1998) Enhancement of Candida antarctica lipase B enantioselectivity and activity in organic solvents. *J. Chem. Soc. Chem. Commun.* 2247,2248.

27. Wang, Y.-F., Lalonde, J. J., Momongan, M., Bergbreiter, D., E., and Wong, C.-H. (1988) Lipase-catalyzed irreversible transesterifications using enol esters as acylating reagents: preparative syntheses of alcohols, glycerol derivatives, sugars, and organometallics. *J. Am. Chem. Soc.* **110,** 7200–7205.

28. Weber, H. K., Stecher, H., and Faber, K. (1995) Sensitivity of microbial lipases to acetaldehyde formed by acyl-transfer reactions from vinyl esters. *Biotechnol. Lett.* **17,** 803–808.

29. Frykman, H., Öhrner, N., Norin, T., and Hult, K. (1993) *S*-Ethyl thiooctanoate as acyl donor in lipase catalysed resolution of secondary alcohols. *Tetrahedron Lett.* **34,** 1367–1300.

30. Trollsås, M., Orrenius, C., Sahlen, F., Gedde, U. W., Norin, T., Hult, A., et al. (1996) Preparation of a novel cross-linked polymer for second-order nonlinear optics. *J. Am. Chem. Soc.* **118,** 8542–8548.

31. Öhrner, N., Martinelle, M., Mattson, A., Norin, T., and Hult, K. (1992) Displacement of the equilibrium in lipase catalyzed transesterification in ethyl octanoate by continuous evaporation of ethanol. *Biotechnol. Lett.* **14,** 263–268.

32. Bianchi, D., Cesti, P., and Battistel, E. (1988) Anhydrides as acylating agents in lipase-catalyzed stereoselective esterification of racemic alcohols. *J. Org. Chem.* **53,** 5531–5534.

33. Rakels, J. L. L., Straathof, A. J. J., and Heijnen, J. J. (1993) A simple method to determine the enantiomeric ratio in enantioselective biocatalysis. *Enzyme Microb. Technol.* **15,** 1051–1056.

34. Reslow, M., Adlercreutz, P., and Mattiasson, B. (1987) Organic solvents for bioorganic synthesis 1. Optimization of parameters for a chymotrypsin catalyzed process. *Appl. Microbiol. Biotechnol.* **26,** 1–8.
35. Martinelle, M. and Hult, K. (1995) Kinetics of acyl transfer reactions in organic media catalysed by *Candida antarctica* lipase B. Biochim. *Biophys. Acta* **1251,** 191–197.
36. Orrenius, C., Norin, T., Hult, K., and Carrea, G. (1995) The *Candida antarctica* lipase B catalyzed kinetic resolution of seudenol in non-aqueous media of controlled water activity. *Tetrahedron: Asymmetry* **6,** 3023–3030.
37. Iglesias, L., Sanchez, V., Rebolledo, F., and Gotor, V. (1997) *Candida antarctica* B lipase catalysed resolution of 1-(heteroaryl)ethylamines. *Tetrahedron: Asymmetry* **8,** 2675–2677.
38. Arroyo, M. and Sinisterra, J. V. (1995) Influence of chiral carvones on selectivity of pure lipase-B from *Candida antarctica*. *Biotechnol. Lett.* **17,** 525–530.
39. Morrone, R., Nicolosi, G., Patti, A., and Piattelli, M. (1995) Resolution of racemic flurbiprofen by lipase-mediated esterification in organic solvent. *Tetrahedron: Asymmetry* **6,** 1773–1778.
40. Roure, F., Ducret, A., Trani, M., and Lortie, R. (1997) Enantioselective esterification of racemic ibuprofen in solvent media under reduced pressure. *J. Chem. Technol. Biotechnol.* **69,** 266–270.
41. Morrone, R., Nicolosi, G., and Patti, A. (1997) Resolution of racemic 1-hydroxyalkylferrocenes by lipase B from *Candida antarctica*. *Gazz. Chim. Ital.* **127,** 5–9.
42. Ducret, A., Pepin, P., Trani, M., and Lortie, R. (1996) Lipase-catalyzed selective esterification of ibuprofen. *Ann. NY Acad. Sci.* **799,** 747–751.
43. Kitaguchi, H., Itoh, I., and Ono, M. (1990) Effects of water and water-mimicking solvents on the lipase-catalyzed esterification in an apolar solvent. *Chem. Lett.* 1203–1206.
44. Högberg, H.-E., Edlund, H., Berglund, P., and Hedenström, H. (1993) Water activity influences enantioselectivity in a lipase-catalysed resolution by esterification in an organic solvent. *Tetrahedron: Asymmetry* **4,** 2123–2126.
45. Bodnár, J., Gubicza, L., and Szabó, L.-P. (1990) Enantiomeric separation of 2-chloropropionic acid by enzymatic esterification in organic solvents. *J. Mol. Catal.* **61,** 353–361.
46. Bovara, R., Carrea, G., Ottolina, G., and Riva, S. (1993) Water activity does not influence the enantioselectivity of lipase PS and lipoprotein lipase in organic solvents. *Biotechnol. Lett.* **15,** 169–174.
47. Wehtje, E., Costes, D., and Adlercreutz, P. (1997) Enantioselectivity of lipases: effects of water activity. *J. Mol. Catal. B: Enzym.* **3,** 221–230.
48. Chen, C.-S., Fujimoto, Y., Girdaukas, G., and Sih, C. J. (1982) Quantitative analysis of biochemical kinetic resolutions of enantiomers. *J. Am. Chem. Soc.* **104,** 7294–7299.
49. Heinsman, N. W. J. T., Orrenius, S. C., Marcelis, C. L. M., De Sousa Teixeira, A., Franssen, M. C. R., Van Der Padt, A., et al. (1998) Lipase mediated resolution of γ-branched chain fatty acid methyl esters. *Biocatal. Biotransform.* **16,** 145–162.
50. Schieweck, F. and Altenbach, H.-J. (1998) Preparation of S-(−)-2-acetoxymethyl-2,5-dihydrofuran and S-(−)-*N*-Boc-2-hydroxymethyl-2,5-dihydropyrrole by enzymatic resolution. *Tetrahedron: Asymmetry* **9,** 403–406.

23

Enantioselective Lipase-Catalyzed Transesterifications in Organic Solvents

Fritz Theil

1. Introduction

Among the biocatalysts used in organic synthesis, lipases (triacylglycerol acyl hydrolases, E.C. 3.1.1.3) are the most frequently used. Because of their ability to discriminate between enantiomers and enantiotopic groups, they are utilized in kinetic resolutions of racemates and desymmetrizations of prostereogenic or *meso*-compounds to provide an easy access to enantiomerically pure building blocks *(1–6)*. Protecting group techniques take advantage of the regioselectivity and chemoselectivity of lipase-catalyzed reactions *(7–9)*.

Lipases are inexpensive, available from many sources, and easy to handle. As induced-fit enzymes, lipases accept a broad range of substrates, which are very different from the natural ones. The natural function of lipases is to catalyze the hydrolysis of triacylglycerols. Therefore, they have been used in many cases to perform enantioselective or regioselective hydrolyses. However, lipases are active in practically water-free organic solvents to catalyze the formation of carboxylic esters and amides. This behavior of lipases offers some advantages in their practical use, such as higher substrate concentration compared with hydrolytic reactions, simpler workup procedures, suppression of undesired side reactions, and, in many cases, higher selectivities. Lipases are available as freeze-dried powders, immobilizates on different kinds of solid supports, and, now, immobilized in sol-gel materials *(10)* (Fluka Chemie, AG, Catalogue 1997/98).

According to X-ray analyses of some lipases, they show a common catalytic machinery based on their catalytic triad consisting of the amino acids serine, histidine, and aspartic acid *(11)*. In some cases, the latter amino acid is replaced

From: *Methods in Biotechnology, Vol. 15: Enzymes in Nonaqueous Solvents: Methods and Protocols*
Edited by: E. N. Vulfson, P. J. Halling, and H. L. Holland © Humana Press Inc., Totowa, NJ

by glutamic acid. The crucial mechanistic step of lipases in their action on carboxylic esters is the nucleophilic attack of the primary hydroxy group of the serine residue of the catalytic triad onto the ester carbonyl group to form an acyl-enzyme intermediate. The acyl-enzyme reacts further with nucleophiles such as water, alcohols, ammonia, amines, or even hydrogen peroxide to yield carboxylic acids, carboxylic esters, amides, or peroxycarboxylic acids, respectively.

Because this chapter describes procedures using lipase-catalyzed transesterifications for the synthesis of enantiomerically pure building blocks, a short general discussion of the experimental parameters should be helpful for chemists trying this type of reaction for the first time.

Figure 1 represents the main methods for the formation of an ester bond catalyzed by lipases.

Method 1, the direct esterification between a carboxylic acid and an alcohol, requires the removal of water formed in order to shift the equilibrium to the formation of the ester. Much more convenient is Method 2 (i.e., the transesterification of an ester with an alcohol either by reaction of an alcohol with an excess of an ester, or by reaction of an ester with an excess of an alcohol [alcoholysis]). The former transformation represents the main method used for the kinetic resolution of chiral alcohols and for the desymmetrization of prostereogenic or *meso*-alcohols. Method 2 also encompasses the formation of γ- or δ-lactones by intramolecular esterification of hydroxy esters.

Method 3, also convenient but used less often, represents the ester formation by alcoholysis of aliphatic or cycloaliphatic carboxylic acid anhydrides. The interesterification between an ester and a carboxylic acid or between two esters, not mentioned in **Fig. 1**, are very rarely used in organic synthesis.

The most convenient way for the lipase-catalyzed formation of carboxylic esters according to Method 2 is the transesterification of alcohols using enol esters such as vinyl acetate as acyl donor **(Fig. 2)**. This reaction is irreversible due to the formation of vinyl alcohol, which tautomerizes spontaneously to acetaldehyde: this does not compete as a nucleophile with the acyl-enzyme. Furthermore, vinyl acetate is a cheap product with a low boiling point and therefore easily removed by distillation.

Further useful transesterification agents are activated esters such as 2,2,2-trichloro- or 2,2,2-trifluoroethyl esters, phenylesters, oximesters, and cyanomethyl esters. These activated esters release weakly nucleophilic hydroxy compounds; the reactions can be regarded as quasi-irreversible.

Typical solvents for lipase-catalyzed acylations are aliphatic and cycloaliphatic hydrocarbons, aromatic hydrocarbons, acyclic and cyclic ethers, halogenated alkanes, tertiary alcohols, and the acylating agent itself.

1. R^1-COOH + R^3-OH $\xrightarrow{\text{lipase}}$ $R^1-CO\text{-}OR^3$ + H_2O

2. $R^1-CO\text{-}OR^2$ + R^3-OH $\xrightarrow{\text{lipase}}$ $R^1-CO\text{-}OR^3$ + R^2-OH

3. $R^1-CO)_2O$ + R^3-OH $\xrightarrow{\text{lipase}}$ $R^1-CO\text{-}OR^3$ + R^1-COOH

Fig. 1. Main methods for the formation of an ester bond catalyzed by lipases.

$R-OH$ + $H_2C=CH\text{-}OAc$ $\xrightarrow[\text{solvent}]{\text{lipase}}$ $R-OAc$ + H_3C-CHO

Fig. 2. Transesterification of alcohols using enol esters.

In order to optimize a lipase-catalyzed kinetic resolution or desymmetrization of a given hydroxy compound, certain parameters of the transesterification reaction can be changed. One should choose one of the above-mentioned solvents that dissolves the alcohol and the acylating agent. The next step is a screening of available lipases with respect to, first, high stereoselectivity and, second, high reaction rate. According to the author's experience, the most powerful lipases for transesterifications regarding substrate acceptance and selectivity according to **Fig. 2** are the lipases from *Pseudomonas cepacia*, *Ps. fluorescens*, *Ps.* sp., porcine pancreas, and the lipase B from *Candida antarctica*. Having selected the appropriate lipase, solvent engineering very often improves the selectivity and/or reaction rate.

Changing the acylating agent can be a further possibility to improve the outcome of the reaction. In general, it is difficult to predict which combination of solvent, lipase, and acylating agent for a given alcohol furnishes optimal results although predictive models, particularly for the lipase from *Ps. cepacia* have been developed *(12–14)*.

Finally, if all attempts fail to improve the outcome of the reaction by modifying the reaction conditions, the substrate structure can be changed as well. A simple way is, for example, the exchange of a protecting group by another one with different steric or electronic properties, but this requires a return to the initial starting point of the optimization procedure.

The following methods represent typical optimized procedures for lipase-catalyzed transesterifications using vinyl acetate with the aim of preparing building blocks of high enantiomeric purity. The four substrates used are diols of different structure: the *meso*-diol **1 (Fig. 3)**, the C_2-symmetric diol *rac*-**4 (Fig. 4)**, the diol *rac*-**7 (Fig. 5)** in which the primary hydroxy group is protected, and the unprotected diol *rac*-**9 (Fig. 6)** with a primary and a secondary hydroxy group.

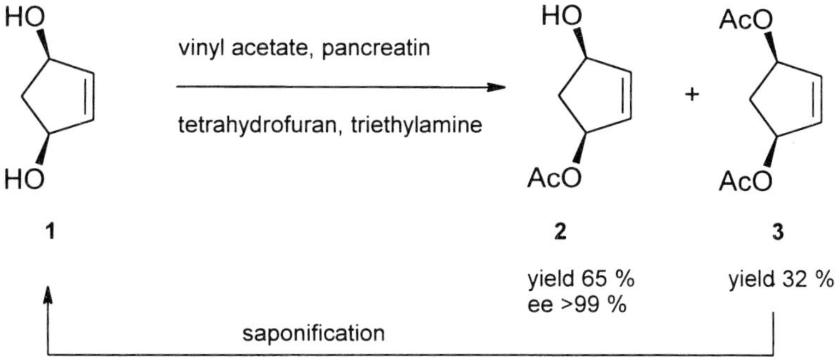

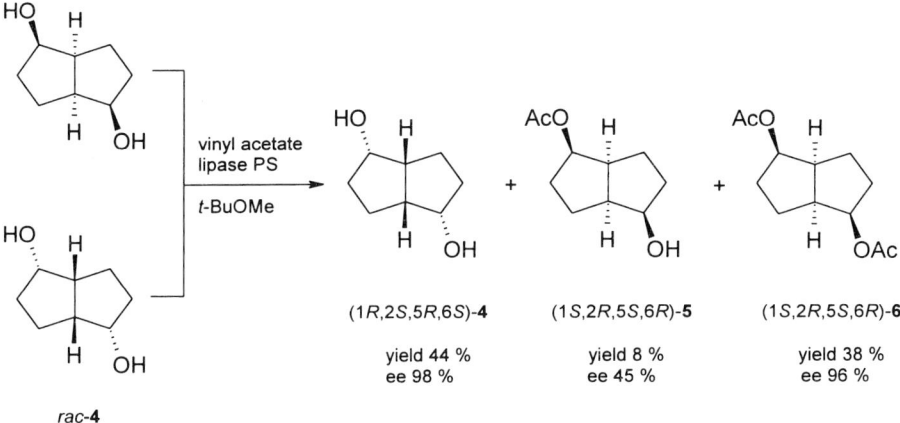

Fig. 3. Desymmetrization of *cis*-2-cylcopeptene-1,4-diol *(1)*.

Fig. 4. Kinetic resolution of *endo-endo-cis*-bicyclo[3.3.0]octane-2,6-diol (*rac*-**4**).

2. Materials

All reactions were carried out in round-bottom flasks under magnetic stirring with Teflon® coated stirring bars. Vinyl acetate was freshly distilled. The solvents in which the reactions were carried out were dried with and stored over sodium wire. All lipases were used as obtained from the supplier.

2.1. Synthesis of (1S,4R)-(–)-4-Hydroxy-2-cyclopentenyl Acetate (2) by Desymmetrization of cis-2-Cyclopentene-1,4-diol (1) (Fig. 3)

1. *cis*-2-Cyclopentene-1,4-diol.
2. Vinyl acetate.
3. Pancreatin 6 × NF (crude porcine pancreatic lipase, activity 820 U/g with triolein as substrate, water content 5.4%).

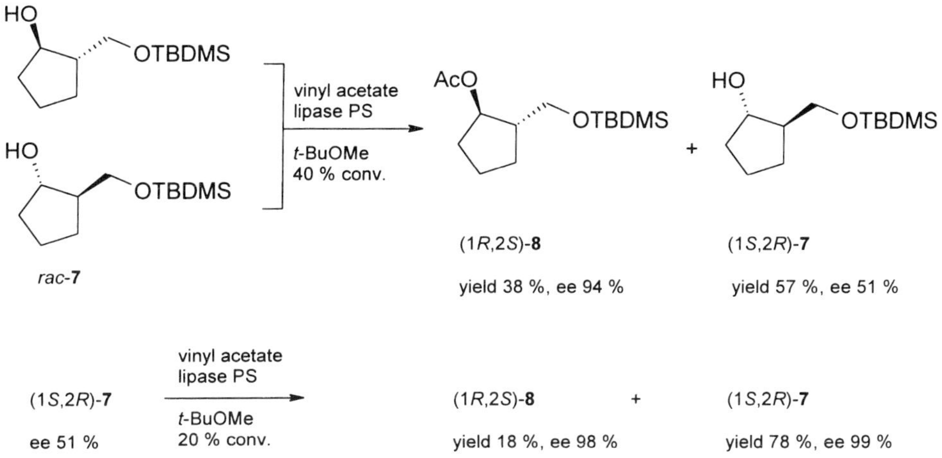

Fig. 5. Kinetic resolution of *trans*-2-(*tert*-butyldimethylsiyloxymethyl)cyclopentanol (*rac*-**7**).

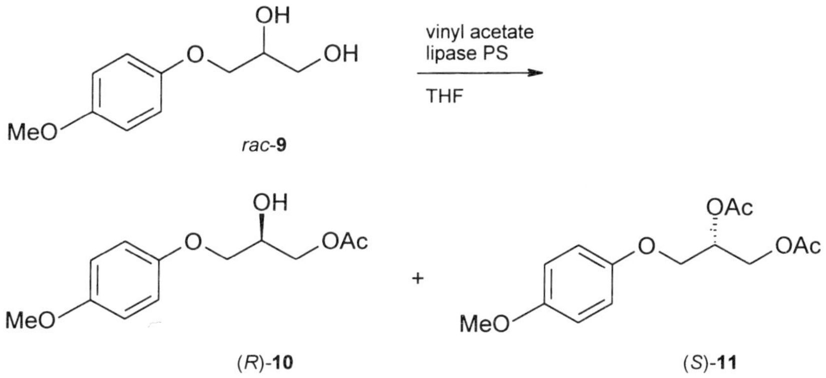

Fig. 6. Kinetic resolution of (RS)-3-(4-methoxyphenoxy)propane-1, 2-diol (rac-9).

4. Triethylamine (distilled from and stored over KOH pellets).
5. Tetrahydrofuran.
6. Celite®, silica gel 60 (0.063 – 0.040 mm), ethyl acetate, *n*-hexane.

2.2. Kinetic Resolution of Endo-Endo-Cis-*Bicyclo[3.3.0]Octane-2,6-Diol* (Rac-4) (Fig. 4)

1. *endo-endo-cis*-Bicyclo[3.3.0]octane-2,6-diol (*rac*-**4**).
2. Vinyl acetate.
3. *tert*-Butyl methyl ether.

4. Lipase from *Ps. cepacia* (lipase PS from Amano Enzyme Europe Ltd., Milton Keynes, UK).
5. Celite, silica gel 60 (0.063 – 0.040 mm), ethyl acetate, *n*-hexane.

2.3. Kinetic Resolution of Trans-2-(Tert-Butyldimethylsilyl-oxymethyl)Cyclopentanol (Rac-7) (Fig. 5)

1. *trans*-2-(*tert*-Butyldimethylsilyloxymethyl)cyclopentanol (*rac*-**7**).
2. Vinyl acetate.
3. *tert*-Butyl methyl ether.
4. Lipase from *Ps. cepacia* (lipase PS from Amano Enzyme Europe Ltd., Milton Keynes, UK).
5. Celite, silica gel 60 (0.063 – 0.04 mm), *n*-hexane.

2.4. Kinetic Resolution of (RS)-3-(4-Methoxyphenoxy)Propane-1,2-Diol (Rac-9) (Fig. 6)

1. (*RS*)-3-(4-Methoxyphenoxy)propane-1,2-diol (*rac*-**9**).
2. Vinyl acetate.
3. Tetrahydrofuran.
4. Lipase from *Ps. cepacia* (lipase PS from Amano Enzyme Europe Ltd., Milton Keynes, UK).
5. Celite, silica gel 60 (0.063 – 0.04 mm), ethyl acetate, *n*-hexane.

3. Methods
3.1. General Method

1. Dissolve the substrate in an appropriate solvent.
2. Add, in sequence, the acylating agent and the lipase.
3. Stir at ambient temperature.
4. Follow the conversion of the substrate by an appropriate chromatographic method.
5. Filter off the lipase when the desired conversion is reached.
6. Separate the products by flash chromatography.

The detailed procedures for the individual compounds follow.

3.2. (1S,4R)-(–)-4-Hydroxy-2-Cyclopentenyl Acetate (2) (15–17)

1. A solution of *cis*-2-cyclopentene-1,4-diol (**1**) (5.0 g, 50 mmol) in tetrahydrofuran (125 mL) (*see* **Note 1**) is treated in sequence with triethylamine (5 mL) (*see* **Note 2**), vinyl acetate (32 mL, 50 mmol), and pancreatin (25 g) (*see* **Notes 3** and **4**).
2. The reaction mixture is stirred at room temperature for about 2.5 h (*see* **Note 5**). The lipase is subsequently filtered off through a sintered glass funnel covered with a pad of Celite.

3. The filter cake is washed with ethyl acetate (3 × 20 mL). The combined filtrates were concentrated under reduced pressure.
4. The products are separated by flash chromatography (*see* **Note 6**) with *n*-hexane/ethyl acetate (2:1 and 1:1) (*see* **Note 7**) as eluent to give in the order of elution the diacetate **3** (2.8 g, 32%) as a colorless liquid and the monoacetate **2** (4.61 g, 65%) as a colorless crystalline material (melting point 39–47°C) with an enantiomeric excess >99% (*see* **Note 8**). Recrystallization of **3** from diethyl ether/*n*-hexane gives the analytically pure sample.

2: melting point: 46–48 °C; $[\alpha]_D^{20}$: –66.5° (*c* 1.0, CHCl$_3$); ^{1}H-NMR (nuclear magnetic resonance) (CDCl$_3$) 1.68 (1 H, ddd, J = 15, 4, and 4 Hz), 1.99 (1 H, d, J = 4 Hz, with D$_2$O exchangeable), 2.04 (3 H, s), 2.80 (1 H, ddd, J = 15, 8, and 8 Hz, 4.71 (1 H, m), 5.49 (1 H, m), 5.98 (1 H, m), 6.12 (1 H, m); ^{13}C-NMR (CDCl$_3$) 21.12, 40.54, 74.84, 77.06, 132.61, 138.59, 170.83.

3.3. Kinetic Resolution of Endo-Endo-Cis-*Bicyclo[3.3.0]Octane-2,6-Diol (*Rac-4*) (18,19)

Monitoring this reaction by a chromatographic method showed the formation of the monoacetate **5**; subsequently, the diacetate (1*S*,2*R*,5*S*,6*R*)-**6** formed and the diol (1*R*,2*S*,5*R*,6*S*)-**4** and only a trace of the monoacetate (1*R*,2*S*,5*R*,6*S*)-**5** remained.

1. A solution of *endo-endo-cis*-bicyclo[3.3.0]octane-2,6-diol (*rac*-**4**) (4.0 g, 28.5 mmol) in *tert*-butyl methyl ether (80 mL) is treated in sequence with vinyl acetate (18.4 mL, 200 mmol) and lipase from *Ps. cepacia* (0.80 g) (*see* **Note 9**).
2. The reaction mixture is stirred at room temperature for about 200 h (*see* **Note 10**). The lipase is subsequently filtered off through a sintered glass funnel covered with a pad of Celite.
3. The filter cake is washed with *tert*-butyl methyl ether (3 × 20 mL). The combined filtrates are concentrated under reduced pressure.
4. The products are separated by flash chromatography (*see* **Note 11**) to afford, in the order of elution, the diacetate (1*S*,2*R*,5*S*,6*R*)-**6** (2.45 g, 38%), the monoacetate (1*S*,2*R*,5*S*,6*R*)-**5** (0.43 g, 8 %), and the diol (1*R*,2*S*,5*R*,6*S*)-**4** (1.78 g, 44%).

(1*R*,2*S*,5*R*,6*S*)-**4**: colorless oil; $[\alpha]_D^{20}$: +42.5° (*c* 1.0, CHCl$_3$); enantiomeric excess 98% (*see* **Note 12**); ^{1}H-NMR (CHCl$_3$) 1.55–1.94 (8 H, m), 2.52 (2 H, m), 3.54 (2 H, br s), 3.9 (2 H, s); ^{13}C-NMR (CDCl$_3$) 20.30, 38.98, 48.92, 72.80.

(1*S*,2*R*,5*S*,6*R*)-**6**: colorless crystals; melting point: 37–39°C; $[\alpha]_D^{20}$: +104.3° (*c* 1.0, CHCl$_3$); enantiomeric excess 96% (*see* **Note 13**); ^{1}H-NMR (CDCl$_3$) 1.46–1.86 (8 H, m), 2.05 (6 H, m), 2.73 (2 H, m), 5.10 (2 H, m); ^{13}C-NMR (CDCl$_3$) 21.07, 22.82, 32.98, 44.74, 77.22, 170.81.

3.4. Kinetic Resolution of Trans-2-(Tert-Butyldimethylsilyl-oxymethyl)Cyclopentanol (*Rac-7*) (20,21)

The value of the enantiomeric ratio (*E* value) (a measure for the selectivity of a kinetic resolution [*see* **Note 14**]) of this reaction under the below-described optimal conditions is 57. In order to obtain both enantiomers with an enantiomeric excess higher than 90%, the reaction has to be stopped when 40% of the product was formed. After separation of the products, a second resolution step of the enantiomerically enriched starting material was applied.

1. A solution of *trans*-2-(*tert*-butyldimethylsilyloxymethyl)cyclopentanol (*rac*-**7**) (3.6 g, 15.5 mmol) in *tert*-butyl methyl ether (110 mL) (*see* **Note 15**) is treated in sequence with vinyl acetate (10.8 mL, 117 mmol) and lipase from *Ps. cepacia* (0.72 g) (*see* **Note 16**).
2. The reaction mixture is stirred at room temperature for about 4 h until the conversion of the starting material *rac*-**7** reached 40% (*see* **Note 17**).
3. The lipase is subsequently filtered off through a sintered glass funnel covered with a pad of Celite. The filter cake is washed with *tert*-butyl methyl ether (3 × 20 mL).
4. The combined filtrates are concentrated under reduced pressure. The products are separated by flash chromatography (*see* **Note 18**) to afford in the order of elution the acetate (1*R*,2*S*)-**8** (1.60 g, 38%) with an enantiomeric excess of 94% (*see* **Note 19**) and the alcohol (1*S*,2*R*)-**7** (2.06 g, 57%) with an enantiomeric excess of 51% (*see* **Note 19**).

The enantiomerically enriched alcohol (1*S*,2*R*)-**7** (2.06 g) was subjected to a second lipase-catalyzed transesterification: A solution of the enantiomerically enriched alcohol (1*S*,2*R*)-**7** (100-mL-round-bottom flask) in *tert*-butyl methyl ether (60 mL) was treated in sequence with vinyl acetate (6.0 mL, 65 mmol) and lipase from *Ps. cepacia* (0.575 g). The reaction mixture was stirred at room temperature for about 24 h until the conversion of the starting material reached 20% (for conditions, *see* **Note 17**). The reaction mixture was filtered through a sintered glass funnel covered with a pad of Celite. The filter cake was washed with *tert*-butyl methyl ether (3 × 10 mL). The combined filtrates were concentrated under reduced pressure. The products were separated by flash chromatography (*see* **Note 20**) to afford, in the order of elution, the acetate (1*R*,2*S*)-**8** (0.44 g, 18%) with an enantiomeric excess of 98 % and the alcohol (1*S*,2*R*)-**7** (1.56 g, 78%) with an enantiomeric excess of >99 %.

(1*R*,2*S*)-**8**: colorless liquid; $[\alpha]D^{20}$: –6.24° (*c* 1.0, CHCl$_3$); ^{1}H-NMR (CDCl$_3$): 0.02 (3 H, s), 0.03 (3 H, s), 0.87 (9 H, s), 1.38 (1 H, m), 1.56–1.74 (3 H, m), 1.76–1.95 (2 H, m), 2.00 (3 H, s), 2.09 (1 H, m), 3.57 (2 H, dd, J = 6.0 and 4.0 Hz), 4.94 (1 H, dd, J = 6.7 and 3.7 Hz); ^{13}C-NMR (CDCl$_3$) –5.46, –5.45, 18.25, 21.35, 23.38, 25.87, 27.20, 32.68, 47.89, 64.29, 78.66, 170.87.

(1*S*,2*R*)-**7**: colorless liquid; $[\alpha]_D^{20}$: +2.0° (*c* 1.0, CHCl$_3$); ^{1}H-NMR (CDCl$_3$): 0.00 (6 H, s), 0.83 (9 H, s), 1.10 (1 H, m), 1.42–1.56 (2 H, m), 1.62–1.74 (2 H, m), 1.83–1.92 (2 H, m), 2.53 (1 H, br s), 3.41 (1 H, dd, J = 2 × 9.5 Hz), 3.73 (1 H, dd, J = 9.5 and 5.2 Hz), 3.91 (1 H, m); ^{13}C-NMR (CDCl$_3$) –5.56, –5.50, 18.17, 21.56, 25.87, 33.81, 49.24, 63.18, 66.86, 78.35.

3.5. Kinetic Resolution of (RS)-3-(4-Methoxyphenoxy)Propane-1,2-Diol (Rac-9) (Fig. 6) (23)

The first step of this one-pot reaction is the fast formation of the racemic primary monoacetate, which, in the subsequent second acylation step undergoes an efficient kinetic resolution. The E value (*see* **Note 14**) of this reaction under the optimal conditions described as follows is >100. Therefore, the reaction can be run to 50% conversion to obtain both enantiomers in an enantiomeric excess of ≥95%.

1. A solution of (*RS*)-3-(4-methoxyphenoxy)propane-1,2-diol (*rac*-**9**) (4.95 g, 25 mmol) in tetrahydrofuran (60 mL) (*see* **Note 21**) is treated in sequence with vinyl acetate (15.4 mL, 175 mmol) and lipase from *Ps. cepacia* (1.5 g) (*see* **Note 22**).
2. The reaction mixture is stirred at room temperature for about 64 h until the conversion of the starting material *rac*-**9** is reached 50% (*see* **Note 23**).
3. The lipase is subsequently filtered off through a sintered glass funnel covered with a pad of Celite. The filter cake is washed with tetrahydrofuran (3 × 20 mL). The combined filtrates are concentrated under reduced pressure.
4. The products are separated by flash chromatography (*see* **Note 24**) to afford in the order of elution the diacetate (*S*)-**11** (3.5 g, 50%) with an enantiomeric excess of 95% (*see* **Note 25**) and the monoacetate (*R*)-**10** (3.0 g, 50%) with an enantiomeric excess of 98% (*see* **Note 25**).

(*R*)-**10**: colorless oil; ^{1}H-NMR (CDCl$_3$) 2.05 (3 H, s), 2,55 (1 H, br s), 3.71 (3 H, s), 3.81–4.23 (5 H, m), 6.78 (4 H, s); ^{13}C-NMR (CDCl$_3$) 20.80, 55.68, 65.41, 68.52, 69.48, 114.69, 115.58, 152.47, 154.23, 171.18.

(*S*)-**11**: colorless oil, boiling point: 250°C (1 Pa, Kugelrohr); 2.00 (3 H, s), 2.03 (3 H, s), 3.69 (3 H, s), 4.00 (2 H, d, $J = 4$ Hz), 4.20 (1 H, dd, $J = 12$ and 6 Hz), 4.36 (1 H, dd, $J = 12$ and 4 Hz), 5.27 (1 H, m), 6.77 (4 H, s); ^{13}C-NMR (CDCl$_3$) 20.71, 20.94, 55.65, 62.57, 66.81, 69.83, 114.64, 115.65, 152.41, 154.26, 170.26, 170.57.

4. Notes

1. Tetrahydrofuran is the solvent of choice because in other suitable solvents the diol **1** is insoluble.
2. The addition of triethylamine is essential to achieve a high rate of conversion *(15,16)*. However this effect is not understood.
3. This is a crude preparation of porcine pancreatic enzymes with lipase, amylase, and protease activity. Pancreatin can be replaced by a more purified porcine pancreatic lipase. However, there are no advantages compared with the crude preparation.
4. Pancreatin can be replaced by lipase SP 382 from Novo-Nordisk (Denmark), which is mixture of the lipases A and B from *Candida antarctica* or as found later by the pure B lipase from *C. antartica* (Novozyme 435 from Novo-Nordisk): however, this is at the expense of the yield of **2**.

5. The reaction was monitored by thin-layer chromatography until the starting material **1** was completely consumed. There was no need for an immediate workup. Even keeping the reaction mixture for 24 h had no influence on the yield and the enantiomeric excess of **2**. Conditions: TLC plates, silica gel, 60 F_{254}, ethyl acetate; visualizing by spraying with a 2.5% solution of molybdatophosphoric acid in ethanol followed by heating.

6. One hundred forty grams of silica gel 60 (0.063–0.040 mm); column dimensions: 25×4 cm; flow rate: 60 mL/min; size of the fractions: 20 mL.

7. When the nonpolar diacetate **3** was completely eluted with *n*-hexane/ethyl acetate 2:1, the eluent was changed to *n*-hexane/ethyl acetate 1:1 in order to decrease the elution time of **2**.

8. The enantiomeric excess of **2** has been determined either by ^{19}F-NMR spectroscopy of the corresponding Mosher ester *(22)* or by high-performance liquid chromatography (HPLC) on Chiralpak® AD (250×4.6 mm): ultraviolet (UV) detection at 220 nm; mobile phase: *n*-hexane/ethanol/methanol (80:10:10); flow rate: 1 mL/min.

9. The reaction was less efficient using 2,2,2-trichloroethyl acetate and pancreatin in tetrahydrofuran/triethylamine *(18)*.

10. The progress of the reaction was monitored by HPLC: refractive index detector; column: Lichrosorb Si60® (120×4.6 mm, 7 μm); mobile phase: *n*-heptane/ethyl acetate (3:1); flow rate: 1 mL/min. Alternatively, the progress of the reaction can be monitored by thin-layer chromatography, but this is less accurate.

11. Five hundred grams of silica gel 60 (0.063–0.040 mm); column dimensions: 20×7 cm; eluent: *n*-hexane/ethyl acetate 1:1; flow rate, 100 mL/min; size of the fractions: 20 mL.

12. The enantiomeric excess was determined by HPLC on Chiralpak® AD (250×4.6 mm): refractive index detector; mobile phase: *n*-heptane/ethanol (90:10); flow rate: 1 mL/min.

13. The enantiomeric excess was determined after converting (1*S*,2*R*,5*S*,6*R*)-**6** into the corresponding diol according to **Note 12**. The diacetate was deacetylated by treatment with methanol in the presence of the strong basic ion-exchange resin Dowex® 1×2-100 (OH⁻-form).

14. For the definition of the *E* value and for the general relationship among *E*, the enantiomeric excess of the product, and the remaining substrate from the degree of conversion, see **ref. 1**.

15. *n*-Hexane gave similar results. In 3-methyl-3-pentanol, the selectivity dropped.

16. Lipase B from *Candida antarctica* (Novozym 435) gave similar results in tetrahydrofuran and *tert*-butyl methyl ether as solvents.

17. The progress of the reaction was monitored by HPLC: refractive index detector; column: Lichrosorb Si60® (120×4.6 mm, 7 μm); mobile phase: *n*-heptane/*tert*-butyl methyl ether (5:1); flow rate: 1 mL/min. Careful control of the conversion was essential; otherwise, the enantiomeric excess of the product (1*R*,2*S*)-**8** dropped. Therefore, the reaction had to be stopped immediately when 40% of conversion was reached, by filtering off the lipase.

18. One hundred grams of silica gel 60 (0.06 –0.040 mm); column dimensions: 20 × 3.6 cm; eluent: *n*-hexane/*tert*-butyl methyl ether (5:1); flow rate: 60 mL/min; size of the fractions: 20 mL.

19. The enantiomeric excess was determined by GC on FS-Lipodex® E (25 m, Macherey-Nagel) after conversion of the products into the corresponding diols as follows: (1*R*,2*S*)-**8** was deacetylated with the strong basic ion-exchange resin Dowex® 1 × 2-100 (OH⁻-form). The alcohols were desilylated by hydrolysis with acetic acid/water in tetrahydrofuran.

20. Fifty grams of silica gel 60 (0.063–0.040 mm); column dimensions: 20 × 2.2 cm; eluent: *n*-hexane/*tert*-butyl methyl ether (5:1); flow rate: 20 mL/min; size of the fractions: 20 mL.

21. Instead of tetrahydrofuran, 1,4-dioxane can be used. Other solvents such as diethyl ether, *tert*-butyl methyl ether, toluene, 3-methyl-3-pentanol, and *tert*-amyl alcohol resulted in a remarkably lower selectivity.

22. Other lipases such as pancreatin, lipases A and B from *Candida antarctica* (SP 382 from Novo Nordisk), or lipase from *Mucor miehei* (Lipozyme® from Novo Nordisk) are not useful.

23. The progress of the reaction was monitored by HPLC: UV detection at 254 nm; column: Lichrosorb Si60® (120 × 4.6 mm, 7 mm); mobile phase: *n*-heptane/2-propanol (80:20); flow rate: 1 mL/min. However, because of its high selectivity, this reaction can also be followed by thin-layer chromatography. The reaction practically stops after reaching 50% conversion.

24. Two hundred grams of silica gel 60 (0.063–0.040 mm); column dimensions: 20 × 4.5 cm; eluent: *n*-hexane/ethyl acetate (2:1); flow rate: 100 mL/min; size of the fractions: 20 mL.

25. For the determination of the enantiomeric excess mono- and diacetate were converted into the corresponding diols by deacetylation with methanol in the presence of the strong basic ion-exchange resin Dowex® 1 × 2-100 (OH⁻-form). The enantiomeric excess of the resulting diols was determined by HPLC on Chiralpak® AD (250 × 4.6 mm): UV detector; mobile phase: *n*-heptane/2-propanol (80:20); flow rate: 1 mL/min.

References

1. Chen, C.-S. and Sih, C. J. (1989) General aspects and optimization of enantioselective biocatalysis in organic solvents: the use of lipases. *Angew. Chem. Int. Ed. Engl.* **28,** 695–707.

2. Boland, W., Frößl, C., and Lorenz, M. (1991) Esterolytic and lipolytic enzymes in organic synthesis. *Synthesis* 1049–1072.

3. Faber, K. and Riva, S. (1992) Enzyme-catalyzed irreversible acyl transfer. *Synthesis* 895–910.

4. Santaniello, E., Ferraboschi, P., and Grisenti, P. (1993) Lipase-catalyzed transesterification in organic solvents: application to the preparation of enantiomerically pure compounds. *Enzyme Microb. Technol.* **15,** 367–382.

5. Theil, F. (1994) Diols as substrates in lipase-catalyzed enantioselective acylations—a brief review. *Catal. Today* **22,** 517–536.

6. Theil, F. (1995) Lipase-supported synthesis of biologically active compounds. *Chem. Rev.* **95,** 2203–2227.

7. Reidel, A. and Waldmann, H. (1993) Enzymatic protecting group techniques in bioorganic synthesis. *J. Prakt. Chem.* **335,** 109–127.

8. Waldmann, H. and Sebastian, D. (1994) Enzymatic protecting group techniques. *Chem. Rev.* **94,** 911–937.

9. Bashir, N. B., Phythian, S. J., Reason, A. J., and Roberts, S. M. (1995) Enzymatic esterification and de-esterification of carbohydrates: synthesis of naturally occurring rhamnopyranoside of *p*-hydroxybenzaldehyde and a systematic investigation of lipase-catalyzed acylation of selected aryl pyranosides. *J. Chem. Soc. Perkin Trans.* 1, 2203–2222.

10. Reetz, M. T., Zonta, A., and Simpelkamp, J. (1995) Efficient heterogeneous biocatalysis by entrapment of lipases in hydrophobic sol-gel materials. *Angew. Chem. Int. Ed. Engl.* **34,** 301–304.

11. Cygler, M., Grochulski, P., Kazlauskas, R. J., Schrag, J. D., Bouthillier, F., Rubin, B., et al. (1994) A structural basis for the chiral preference of lipases. *J. Am. Chem. Soc.* **116,** 3180–3186, and references cited therein.

12. Kazlauskas, R. J., Weissfloch, A. N. E., Rappaport, A. T., and Cuccia, L. A. (1991) Rule to predict which enantiomer of a secondary alcohol reacts faster in reactions catalyzed by cholesterol esterase, lipase from *Pseudomonas cepacia*, and lipase from *Candida rugosa*. *J. Org. Chem.* **56,** 2656–2665.

13. Weissfloch, A. N. E. and Kazlauskas, R. J. (1995) Enantiopreference of lipase from *Pseudomonas cepacia* toward primary alcohols. *J. Org. Chem.* **56,** 6959–6969.

14. Lemke, K., Lemke, M., and Theil F. (1997) A three-dimensional predictive active site model for lipase from *Pseudomonas cepacia*. *J. Org. Chem.* **62,** 6268–6273.

15. Theil, F., Ballschuh, S., Schick, H., Haupt, M., Häfner, B., and Schwarz, S. (1988) Synthesis of (1*R*,4*S*)-(–)-4-hydroxy-2-cyclopentenyl acetate by a highly enantioselective enzyme-catalyzed transesterification in organic solvents. *Synthesis* 540–541.

16. Theil, F., Schick, H., Lapitskaya, M. A., and Pivnitsky, K. K. (1991) Investigation of the pancreatin-catalyzed acylation of *cis*-cyclopent-2-ene-1,4-diol with various trichloroethyl and vinyl alkanoates. *Liebigs Ann. Chem.* 195–200.

17. Theil, F., Schick, H., Winter, G., and Reck, G. (1991) Lipase-catalyzed transesterification of *meso*-cyclopentane diols. *Tetrahedron* **47,** 7569–7582.

18. Djadchenko, M. A., Pivnitsky, K. K., Theil, F., and Schick, H. (1989) Enzymes in organic synthesis. Part 3. Synthesis of enantiomerically pure prostaglandin intermediates by enzyme-catalyzed transesterification of (1*SR*,2*RS*,5*SR*,6*RS*)-bicyclo[3.3.0]octane-2,6-diol with trichloroethyl acetate in an organic solvent. *J. Chem. Soc. Perkin Trans 1* 2001,2002.

19. Lemke, K., Ballschuh, S., Kunath, A., and Theil, F. (1997) An improved procedure for the lipase-catalyzed kinetic resolution of *endo-endo-cis*-bicyclo(3.3.0)octane-2,6-diol – synthesis of potential C_2-symmetric enantiomerically pure bidentate auxiliaries. *Tetrahedron: Asymmetry* **8,** 2051–2055.

20. Weidner, J., Theil, F., and Schick, H. (1994) Kinetic resolution of (1*RS*,2*SR*)-2-(hydroxymethyl)cyclopentanol by a biocatalytic transesterification using lipase PS. *Tetrahedron: Asymmetry* **5,** 751–754.

21. Theil, F. and Ballschuh, S. (1996). Chemoenzymatic synthesis of both enantiomers of cispentacin. *Tetrahedron: Asymmetry* **7,** 3565–3572.

22. Theil, F., Costisella, B., and Schick, H. (1992) A correlation of configuration and ^{19}F-NMR chemical shifts of (*R*)-(+)-Mosher esters of chiral cyclopentanediol derivatives. *J. Prakt. Chem.—Chemiker-Zeitung* **334,** 85,86.

23. Theil, F., Weidner, J., Ballschuh, S., Kunath, A., and Schick, H. (1994) Kinetic resolution of acyclic 1,2-diols using a sequential lipase-catalyzed transesterification in organic solvents. *J. Org. Chem.* **59,** 388–393.

24

Pseudomonas cepacia Lipase-Catalyzed Enantioselective Acylation of 2-Substituted-1-alkanols in Organic Solvents

Patrizia Ferraboschi and Enzo Santaniello

1. Introduction

Lipase-catalyzed acylation is an enzymatic process that has recently gained wide popularity in organic synthesis, because the reaction can be carried out efficiently in organic solvents and often proceeds with a high degree of regiocontrol and stereocontrol. (For extensive literature, see **ref. *1***; among the many specialized texts, see **refs. *2–5*.**) Several methods of acylation are compatible with the activity of the enzyme so that the biocatalyst can be used in solvents of different polarity, revealing a great flexibility toward a great variety of hydroxy compounds as substrates (*6*). Several acylation procedures such as esterification or transesterification have been reported in the literature, but, from a preparative point of view, the transesterification of alcohols is by far the most useful and applied method (*7*).

$$R\text{-COOH} + R^1OH = R\text{-COOR}^1 + H_2O$$
$$R\text{-COOR}^1 + R^2OH = R\text{-COOR}^2 + R^1OH$$

An additional advantage of the method is the fact that the transesterification reaction can also be carried out under irreversible conditions, so that the equilibrium is shifted almost completely to the right. The use of vinyl esters as acylating reagents are specially indicated for preparative purposes, because the vinyl alcohol that is formed completely tautomerizes to acetaldehyde in the reaction conditions and the back reaction is almost totally prevented (*8,9*).

$$R\text{-COOCH} = CH_2 + R^1OH = R\text{-COOR}^1 + HOCH = CH_2$$
$$HOCH = CH_2 \rightarrow CH_3\text{-CHO}$$

From: *Methods in Biotechnology, Vol. 15: Enzymes in Nonaqueous Solvents: Methods and Protocols*
Edited by: E. N. Vulfson, P. J. Halling, and H. L. Holland © Humana Press Inc., Totowa, NJ

The stereochemical outcome of the lipase-catalyzed acylation of alcohols in organic solvents can be generalized and has been interpreted through a model of the active site and in terms of the reaction mechanism *(10–13)*. From these models, it is possible to propose the configuration of the reacting substrate and, in the case of secondary alcohols, the model can be adopted as a predictive rule *(10)* (*see* **Fig. 1**). The previous observations have been extended also to primary alcohols, although, in this case, it is more difficult to predict the stereopreference and it is not easy to generalize the obtained stereochemical results *(11)*.

Primary alcohols containing a stereogenic center at the 2 position (i.e., racemic 2-methyl-alkanols [**Fig. 2**]) have been studied by our group as substrates of the lipase-catalyzed irreversible transesterification with vinyl acetate in organic solvents. From a synthetic point of view, the preparation of optically pure compounds was the main achievement and, generally, we were able to prepare the acetate at the highest optical purity, stopping the reaction at approx 40% conversion. The unreacted alcohol could be recovered at the highest enantiomeric excess (ee) when the reaction was carried out at 60% conversion to the acetate. The results collected in **Table 1** show that the unreacted (*R*)-alcohols **1a–f** and the (*S*)-esters **2a–f** can be obtained nearly enantiomerically pure *(14–17)*. The only exception is constituted by 2-phenyl-1-propanol **1g**, which is converted to the acetate **2g** with poor enantioselectivity (20% ee).

The foregoing results were obtained using as a biocatalyst the lipase from *Pseudomonas fluorescens* (PFL) (*see* **Note 1**) in chloroform or dichloromethane. 3-Phenyl-2-methyl-1-propanol **1f** has been considered a model substrate for 2-substituted alkanols, because a high value of enantiomeric ratio ($E = 172$) has been calculated for this compound in the enzymatic process when the reaction is carried out in dichloromethane *(18)*. We have used the alcohol **1f** as the substrate for further studies, such as the influence of the nature of the solvent (**Fig. 3**) on the activity of *Pseudomonas cepacia* lipase (PCL) (*see* **Note 1**) in an irreversible transesterification reaction with vinyl acetate *(19)*.

In another study, it has been shown that three to five cycles of reactions with the same enzyme does not significantly affect the enantioselectivity of the reaction *(20)*. The results of the recycling in benzene and chloroform are collected in **Table 2**, whereas **Table 3** shows the values of the enantiomeric ratio E in different solvents *(19)*. It is worthy of note that the highest E has been observed in chloroform and that this value matches the result obtained in dichloromethane. Therefore, in the following enzymatic reactions, the choice between the two chlorinated solvents relied only on their different boiling points, and dichloromethane was preferred when products were volatile.

2-Methyl-1-alkanols bearing a naphthyl moiety in the structure such as the racemic compounds **3–6** have also been enzymatically acylated in the presence of PCL (**Fig. 4**). The enantioselectivity of the process was influenced by

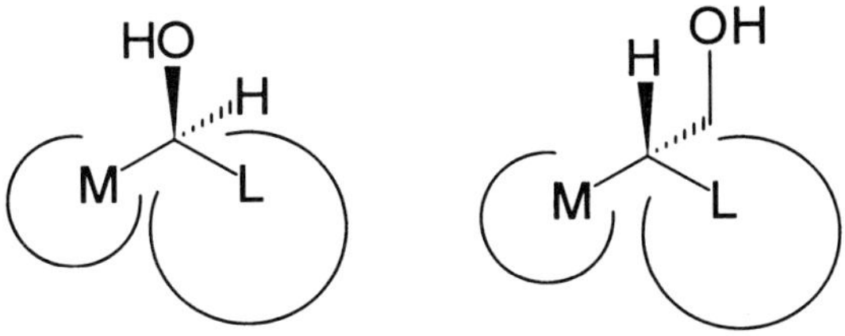

Fig. 1. Proposed rules for the enantiopreferences of a lipase-catalyzed reaction (size of groups: M = medium, L = large).

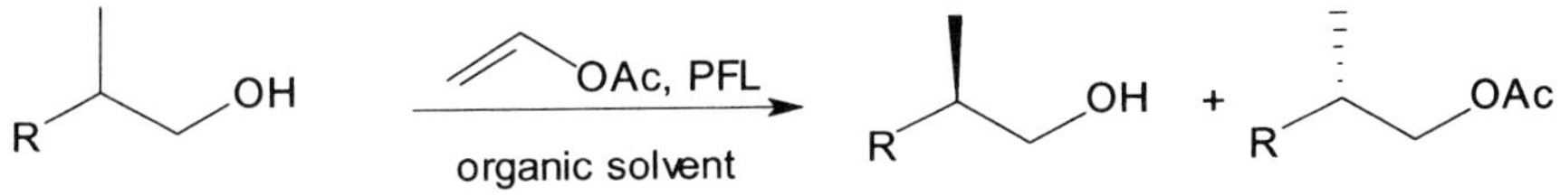

a. R =PhSCH$_2$CH$_2$

b. R =PhSO$_2$CH$_2$CH$_2$

c. R = PhSeCH$_2$CH$_2$

d. R = CH$_3$CH=CHCH$_2$

e. R = CH$_3$(CH$_2$)$_7$

f. R = PhCH$_2$

g. R = Ph

Fig. 2. Racemic 2-methyl-alkanols containing a stereogenic center at the 2 position.

the position of the alkyl chain with respect to the 1' versus the 2' position of the naphthyl ring and by the distance of the stereogenic center from the aromatic system (*21*), as shown in **Table 4**. The consistent differences in reaction times and ee observed within compounds such **3** and **5** or **4** and **6** are reminiscent of the marked difference recorded for the phenyl analogs **1f** and **1g**.

Starting from the consideration that the hydrophobic structure of a steroid should be mostly suitable to the transesterification procedure in organic solvents, the enzymatic reaction has been applied to steroid side chains containing a primary alcohol bearing a methyl group at the stereogenic center corresponding to the α-position (*22,23*). **Figure 5** describes the results obtained from the PCL-catalyzed transesterification procedure of the 22- and 26-

Table 1
**PFL - Catalyzed Transesterification of 2-Methylalkanols
in Chloroform (Entries a–c) and Dichloromethane (Entries d–g)**

Substrate 1	Products Alcohol-1 (reaction time, %ee)	Acetate-2 (reaction time, %ee)	Ref.
a	18 h, 98	7 h, 98	*14*
b	16 h, 98	7 h, 98	*14*
c	15 h, >98	7 h, 98	*15*
d	5.5 h, >98	3.5 h, >98	*16*
e	4 h, >98	1 h, 98	*17*
f	2 h, >98	1 h, >98	*18*
g	8 h, 20	3 h, 20	*18*

hydroxysteroids, compounds **7** and **9**, with a clear indication of the stereopreference of the enzymatic process. It is worth mentioning the slow reaction of the more hindered 22-hydroxy compound **7** with respect to the 26-hydroxycholesterol **9**.

Epoxy alcohols such as 2-substituted-oxiranemethanols **11** were selected as a special class of 2-substituted alkanols to extend the studies on the enantioselectivity of the enzymatic transacylation procedure. Racemic compounds **11a–e** (**Fig. 6**) were synthesized and efficiently and enantio-selectively resolved by the PFL-catalyzed transesterification procedure *(24–26)*. The results are collected in **Table 5** and it should be noted that the unreacted epoxyalcohols **11a–e** and the obtained esters **12a–e** are enantiomers but have the same (*S*) configuration in accord with the Cahn–Ingold–Prelog rules.

In addition, several 2-substituted-1,2-propanediols **13a–c** have been prepared and subjected to the PFL-catalyzed reaction in dichloromethane (**Fig. 7**) and the results *(27)* show high enantioselectivity for diols **13a,b**, whereas from the 2-phenyl analog **13c**, nearly racemic unreacted alcohol **13c** and acetate **14c** have been obtained (**Table 6**). This result is in agreement with the resolution of the alcohol **1g** and in contrast with the enantioselective acyla-tion of the epoxyalcohol **11e**.

2. Materials

2.1. Synthesis of Substrates

1. Reagents were from Fluka (Buchs, Switzerland) and Sigma-Aldrich (Italy) (*see* **Note 2**).
2. Solvents were of analytical grade and were purchased from Fluka and Mallinckrodt Baker Italia (Italy).

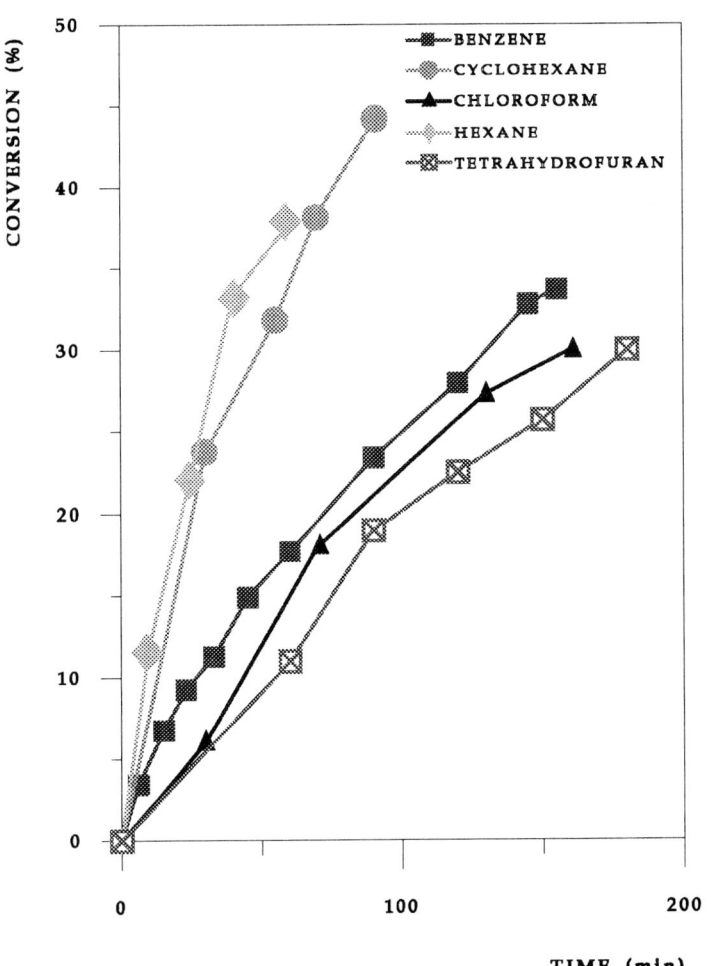

Fig. 3. Activity of PCL-catalyzed transesterification of alcohol 1f in solvents of different polarity.

Table 2
Effect of PCL Recycling on the Enantioselectivity
of the Transesterification of Alcohol 1f

| Use | Benzene | | Choloroform | |
	Time (h)^a	ee (%)^b	Time (h)^a	ee (%)^b
I	2	>98	2	>98
II	2	96	2.3	96
III	2	92	7.5	95
IV	2	91	7.5	92

^{a}Coversion to 60% acetate **2f**.
^{b}From ^{1}H-NMR (500 MHz) analysis of (R)-MTPA Ester Alcohol **1f**.

Table 3
Enantiomeric ratio (E) of the PCL-Catalyzed acylation of
Alcohol 1f **in Different Solvents**

Solvent	Acetate **2f** (%)	Alcohol **1f** (%)	Time (h)	E
Choloroform	36	64	2	172
Benzene	34	66	2.5	46
Cyclohexane	44	56	1.5	32
Hexane	38	62	1	22
Tetrahydrofuran	22	78	2	18

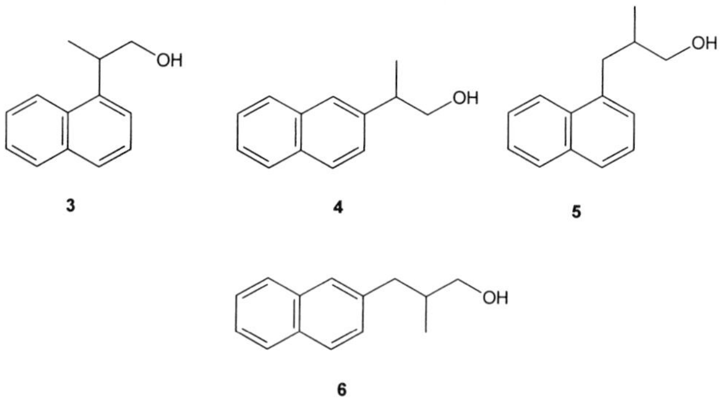

Fig. 4. 2-Methyl-1-alkanols bearing a naphthyl moiety in racemic compounds **3–6**.

2.2. Enzymatic Reaction

1. Lipase from *P. fluorescens* (SAM-2, E.C. 3.1.1.3) from Fluka.
2. Lipase from *P. cepacia* (Amano lipase PS) from Amano Pharmaceutical Co., Japan (*see* **Note 3**).

Table 4
PCL-Catalyzed Transesterification of 1-Naphthyl and 2-Naphthyl Alcohols in Chloroform

Substrates 3–6	Products		E
	Alcohol (reaction time, %ee, config)	Acetate (reaction time, %ee, config)	
3	97 h, 70, R	86 h, 72, S	9
4	22 h, 65, R	16 h, 62, S	6
5	23 h, >98, R	4 h, >98, S	150
6	7 h, 85, R	3 h, 92, S	45

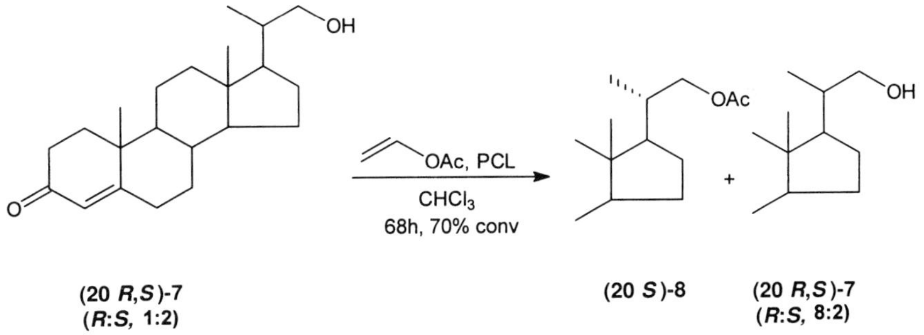

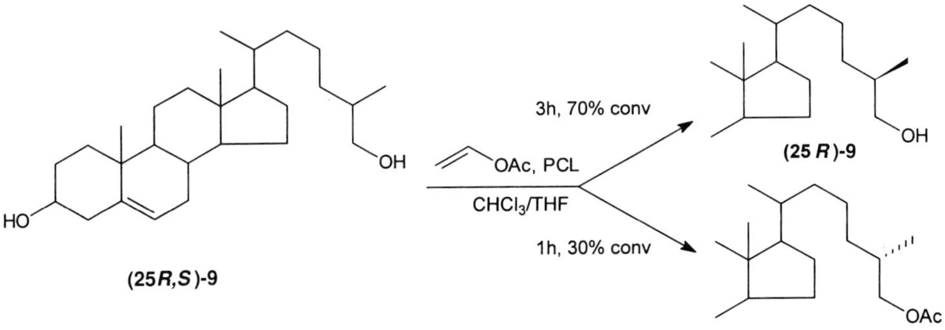

Fig. 5. PCL-catalyzed acylation of steroid side chain.

3. Chloroform (Fluka, Switzerland), analytical grade (for ultraviolet [UV] spectroscopy, stabilized with 1% ethanol, water content <0.01%).
4. Dichloromethane (Fluka), analytical grade (for UV spectroscopy, stabilized with cyclohexene).
5. (S)-2-Methoxy-2-trifluoromethylphenylacetic acid chloride (S-MTPA-Cl), ee >99.5% from JPS (Bevaix).

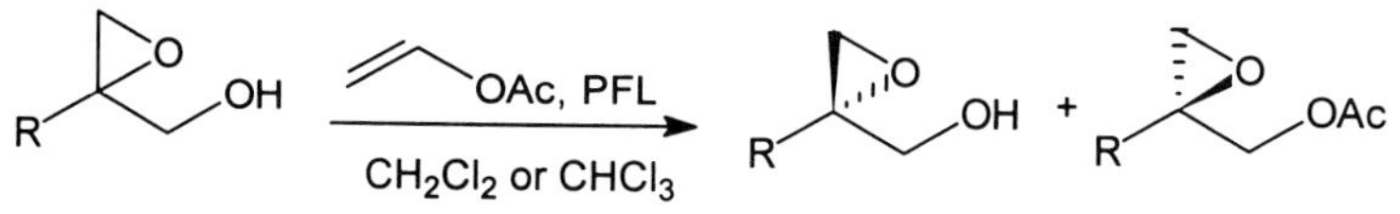

a. $R = PhCH_2$ c. $R = (CH_3)_2C=CH-CH_2$

b. $R = C_9H_{19}$ d. $R = CH_2=CH-(CH_2)_3$

e. $R = Ph$

Fig. 6. Racemic compounds 11a–e synthesized and efficiently and enantioselectivity resolved by the PFL-catalyzed transesterification procedure.

2.3. Analytical Procedures

1. Thin-layer chromatography (TLC) performed with silica gel plates Merck 60 F254 purchased from Merck, Darmstadt (Italian dealer: Bracco, Milano).
2. Gas chromatography analyses performed on a capillary column (HP-5) with a Hewlett Packard gas chromatograph (model 5890/II). The samples were injected as solutions in ethyl acetate.
3. Routine control of the reactions was realized by ^{1}H-nuclear magnetic resonance (NMR) spectroscopy, recording the spectra at 360 MHz on a Varian EM 360 L spectrometer. Samples were analyzed as solutions in $CDCl_3$ and resonances (δH) are expressed in parts per million from tetramethylsilane (TMS) used as the internal standard.
4. Structural definition of the substrates were performed on solutions in $CDCl_3$ by ^{1}H-NMR spectroscopy recording the spectra at 500 MHz on a Brucker AM 500.
5. Optical rotations of enantiomerically enriched or pure compounds were measured at 25°C on a Perkin-Elmer polarimeter (model 241).
6. Infrared (IR) spectra were recorded on a 1420 Perkin-Elmer spectrometer for solution in chloroform.
7. Mass spectra were recorded on a Hewlett-Packard instrument (model 5988) by direct inlet probe and electronic impact technique (electron energy at 70 eV and ion source at 270°C).

2.4. Enzyme Activity

1. Lipase from Fluka (SAM-2) had an activity of 32–42 U/mg of enzyme preparation, established titrimetrically by hydrolysis of triolein the (Fluka) as the substrate. One unit corresponds to the amount of lipase that liberates 1 μmol of oleic acid per minute from triolein at pH 7.4 at 37°C (*see* **Note 4**).
2. Lipase from Amano (Lipase PS) had an optimum activity (*see* **Note 5**) of 30 U/mg of enzyme preparation, determined titrimetrically by hydrolysis of triolein (Fluka), as established for the *Ps. fluorescens* lipase (SAM-2).

Table 5
PFL-Catalyzed Acylation of 2-Oxiranemethanols in Chloroform (Entries a–c) and in Dichloromethane (Entries d and e)

Substrate 11	Products		Ref.
	Alcohol-11 (reaction time, %ee)	Acetate-12 (reaction time, %ee)	
a	6 h, >98	3 h, >98	*24*
b	2 h, 96	1 h, 96	*24*
c	5.5 h, 98	3 h, 98	*25*
d	2 h, 98	1 h, >98	*26*
e	3 h, 90	0.5 h, >98	*18*

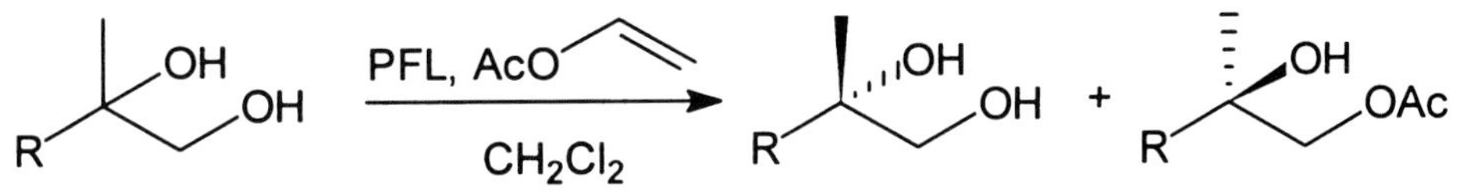

(R,S)-13   **S-13**   **R-14**

a. R = CH₂Ph   b. R = (CH₃)₂C=CHCH₂

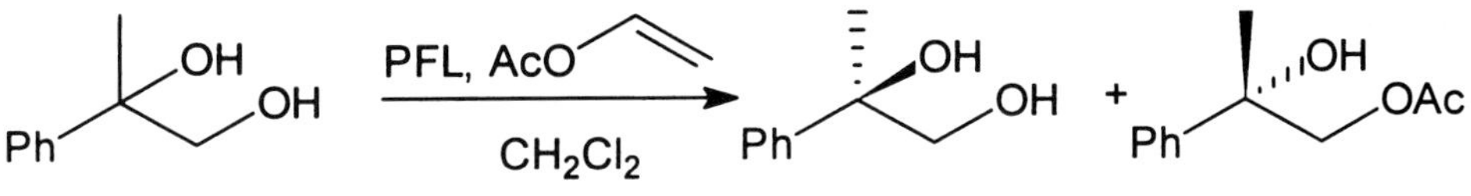

(R,S)-13c   **R-13c**   **S-14c**

Fig. 7. 2-Substituted-1,2-propanediols 13a–c prepared and subjected to the PFL-catalyzed reaction in dichloromethane.

Table 6
PFL-Catalyzed Acylation of 2-Substituted-1, 2-Propanediols

Substrate 13	Products	
	Alcohol-13 (reaction time, %ee)	Acetate-14 (reaction time, %ee)
a	66 h, >98	8 h, 92
b	15 h, 92	4 h, 92
c	69 h, 10	31 h, 10

3. Methods

3.1. Synthesis of Racemic Substrates

1. The methods for the chemical synthesis of the substrates have been described in the experimental sections of the original papers.
2. The required substrate was purified by conventional column chromatography with silica gel Merck 60 (230–400 mesh) (*see* **Note 6**) to a purity, that was considered optimal by a series of analytical controls.
3. Distillations for purification and analytical purposes were carried out under vacuum in a glass tube oven (Buchi GKR-50).

3.2. Enzymatic Reactions–General Procedure

To a solution of the alcohol (10 mmol) in the suitable solvent (20 mL), vinyl acetate (40 mmol, 3.7 mL) and lipase (400 U/mmol) were sequentially added and the mixture was kept under stirring (*see* **Note 7**) for the time required for the conversion of the substrate. The progress of the reaction can be monitored by thin-layer chromatography (TLC) (toluene/ethyl acetate with variable ratios, depending on the substrates), gas–liquid chromatography (GLC), and by integration of suitable signals of 60-MHz NMR spectra. The products mixture (unreacted alcohol and produced ester) is recovered by filtration of the enzyme (*see* **Note 8**); the solvent and vinyl acetate are evaporated at reduced pressure and the mixture purified by column chromatography.

3.3. Isolation of Products

The mixture of the unreacted alcohol and the formed ester at the required ratio (*see* **Note 9**) are purified by column chromatography (silica gel:products ratio, 40:1), and the less polar ester is generally eluted with hexane/ethyl acetate between 9:1 and 8:2, whereas the alcohol requires an eluants ratio from 7:3 to 1:1. Evaporation of the collected fractions of the required compound affords the product that can be eventually distilled under vacuum.

3.4. Determination of Absolute Configurations

1. The absolute configuration of the unreacted alcohol was established by the measure of the optical rotation of the purified sample that was compared with the reported value, if the absolute configuration of the alcohol was known. (Literature about optical rotation values are available in the original papers cited for each enzymatically prepared compound.)
2. If the absolute configuration of the alcohol was unknown, it was transformed into a compound of known configuration and the configuration established by the optical rotation of the compound prepared for the chemical correlation.
3. Because the configuration of acetates is less frequently reported in literature, generally these esters were converted into the corresponding alcohols by reaction with lithium aluminum hydride (*see* **Note 10**). The configuration of the alcohols

so obtained was determined by the measure of their optical rotation, which also established the configuration of the corresponding acetate.

4. To a solution of the acetate (1 mmol) in dry ethyl ether (2 mL; *see* **Note 11**), excess solid lithium aluminum hydride (0.38 g, 10 mmol) was added and the mixture was magnetically stirred for the time required for the conversion, as monitored by TLC or GLC. For the workup, water (0.4 mL), 15% sodium hydroxide (0.4 mL), and water (1.2 mL) were sequentially added; the solid was removed by filtration on a Celite® pad and the final solution was evaporated under reduced pressure. A practically pure alcohol (90% yield) was obtained that could be used as such for optical rotation measurement or for the evaluation of the enantiomeric excess.

3.5. Evaluation of Enantiomeric Excess

The enantiomeric excesses of the unreacted alcohol and the formed acetate as products of the enzymatic reaction were evaluated by converting the purified unreacted alcohol into the corresponding ester with optically pure (*S*)-MTPA-Cl. The MTPA esters were prepared according the method presented in **ref. 28** and used as such without purifications for the NMR analysis. The enzymatically formed acetates were converted into the corresponding alcohols that were transformed into the MTPA esters for the NMR determination of the enantiomeric excess of the original acetate. For this purpose, it was also necessary to prepare by the same procedure the MTPA ester of the racemic alcohol. In a typical ^{1}H-NMR (500 MHz) analysis of a the MTPA ester, the resonances of the CH_2-OCO or *CH*-OCO groups were considered. The signals corresponding to MTPA ester of the (*R*)- and (*S*)-alcohol are well resolved in the spectra of the racemic compound. The integration of the signals enabled the quantitative determination of the enantiomeric excess of enantiomerically enriched alcohol.

1. To a solution of the alcohol (0.1 mmol) in carbon tetrachloride/pyridine mixture (1:1, 1 mL) (*S*)-MTPA-Cl (30.3 mg, 0.12 mol) was added under nitrogen. The solution was kept at room temperature overnight and then 3-dimethylamino-1-propylamine (0.02 mL) was added. The solution was poured into water (2 mL), the product extracted with dichloromethane (3 × 1 mL), and the organic solution washed with saturated ammonium chloride and sodium hydrogencarbonate solution and then water, and finally dried with sodium sulfate. Evaporation of the solvent at reduced pressure afforded the required derivative for NMR analysis.

2. Typically, for the MTPA ester of racemic 3-phenyl-2-methyl-1-propanol, the signals for the CH_2-OCO grouping consisted of three multiplets centered at 4.00, 4.15, and 4.25 ppm corresponding to 0.5, 1, and 0.5 H, respectively. In the sample prepared from enzymatically prepared (*R*)-(+)-3-phenyl-2-methyl-1-propanol, only the multiplet centered at 4.15 ppm (1 H) was detectable and a >98% ee was assigned to the compound. In the case of the (*S*)-(–)-3-phenyl-2-methyl-1-propanol derived from the hydrolysis of the enzymatically prepared acetate, only the

two multiplets at 4.00 and 4.25 ppm (0.5 and 0.5 H) were present and a >98% ee was assigned to the *S*-alcohol.

3.6. Evaluation of the Enantiomeric Ratio

The enantiomeric ratio *E* was calculated using the relation introduced by Sih and coworkers *(29,30)* that correlates *E* with the conversion *c* expressed as a function of the ee of the substrate (ee$_s$) and ee of product (ee$_P$):

$$E = \frac{\ln[1 - c(1 + ee_P)]}{\ln[1 - c(1 - ee_P)]}$$

$$c = \frac{ee_S}{ee_S + ee_P}$$

The enantiomeric excess of the unreacted alcohol and obtained acetate at a <50% conversion was evaluated and these values were used to resolve the above equation. Typically, for the enzymatic resolution of racemic 3-phenyl-2-methyl-1-propanol **1f** in chloroform or dichloromethane, the value of *E* was calculated from the ee of the unreacted alcohol **1f** (56% ee) and of the formed acetate **2f** (98% ee) at a 40% conversion:

$$E = \frac{\ln[1 - 0.36(1 + 0.98)]}{\ln[1 - 0.36(1 - 0.98)]} = 172$$

$$c = \frac{0.56}{0.56 + 0.98} = 0.36$$

4. Notes

1. The original name of *Pseudomonas fluorescens* lipase had been later changed to *P. cepacia* lipase because the microorganism was reclassified. We have used the names of the enzyme as given by the commercial dealer.
2. Reagents were purchased at the highest quality offered by the more qualified dealer and were generally used without any further purification.
3. The Amano PS lipase was a generous gift of the Amano Pharmaceuticals Co. through the Italian subsidiary (Mitsubishi Italia, Milano, Italy). The Amano lipase PS is one of the many types (PS, P, LPL-80, SAM-II) sold by the company; these types all come from the same micro-organism (*Ps. cepacia*, formerly *Ps. fluorescens*) and differ only in the purification method and types of stabilizers (*see* **ref. 11**).
4. The activity of the SAM-2 samples (100 or 500 mg powder) consistently was in the range of the reported value and mantained stable for a few weeks in the refrigerator.
5. Different values of the enzyme activity were found in different lipase PS samples, supplied in 10- to 20-g amounts as a powder. The samples of the enzymes were kept at 0–4°C and the activity diminished with time.
6. All substrates present chemico-physical properties in agreement with their structure.

7. Magnetic or gentle mechanical stirring ensures an efficient dispersion of the heterogeneous mixture, as required for reproducible conversions of the substrate. Shaking of the reaction does not furnish comparable results with the above stirring systems.
8. In some case, it is possible to recycle the enzyme and the activity is not significantly diminished after three to five cycles.
9. In order to obtain the highest ee for the products, it is advisable to carry out two separate reactions. If the optically pure ester is required, a <50% conversion has to be reached, whereas at >50% transformation, the unreacted alcohol can be recovered enantiomerically pure.
10. The conversion of the acetate into the corresponding alcohol was preferentially realized by reaction with LiAlH$_4$ because the final workup did not require extraction of the product from the final mixture. By this method, the loss of material that invariably might accompany the workup in the traditional alkaline hydrolysis of an ester to alcohol (neutralization of the final aqueous mixture, extraction with solvents, drying with sodium sulfate, filtration, evaporation) could be avoided.
11. For small-scale preparation, dry ethyl ether could be prepared by pouring the solvent through an alumina (neutral, grade I, Fluka) column (3 g alumina/25 mL of ether). The required volume of the solvent can be directly collected into the reaction vessel.

References

1. Schoffers, E., Golebiowski, A., Johnson C. R. (1996) Enantioselective synthesis through enzymatic asymmetrization. *Tetrahedron* **52**, 3769–3826.
2. Wong, C.-H. and Whitesides, G. M. (1994) *Enzymes in Synthetic Organic Chemistry*, Tetrahedron Organic Chemistry Series vol. 12, Elsevier Science, Oxford.
3. Faber, K. (1995) *Biotransformations in Preparative Organic Chemistry,* Springer-Verlag, Berlin.
4. Drauz, K. and Waldmann, H. (eds.) (1995) *Enzyme Catalysis in Organic Synthesis*, Vols. 1 and 2, VCH, Weinheim.
5. Koskinen, A. M. P. and Klibanov, A. M. (eds.) (1996) *Enzymatic Reactions in Organic Media*, Blackie Academic & Professional, London.
6. Santaniello, E., Ferraboschi, P., and Rezahelahi, S. (1999) Chiral synthons by enzymatic acylation and esterification reactions, in *Stereoselective Biocatalysis* (Patel, R., ed.), Dekker, New York.
7. Santaniello, E., Ferraboschi, P., and Grisenti, P. (1993) Lipase-catalyzed transesterification in organic solvents: applications to the preparation of enantiomerically pure compounds. *Enzyme Microb. Technol.* **15**, 367–382.
8. Degueil-Castaing, M., De Jeso, B., Drouillard, S., and Maillard, B. (1987) Enzymatic reactions in organic synthesis. II. Ester interchange of vinyl esters. *Tetrahedron Lett.* **28**, 953–954.
9. Wang, Y. F., Lalonde, J. J., Momongan, M., Bergbreiter, D. E., and Wong, C.-H. (1988) Lipase-catalyzed irreversible transesterifications using enol esters as acylating reagents: preparative enantio- and regioselective syntheses of alcohols, glycerol derivatives, sugars, and organometallics. *J. Am. Chem. Soc.* **110**, 7200–7205.
10. Kazlauskas, R. J., Weissfloch, A. N. E., Rappaport, A. T., and Cuccia, L. A. (1991) A rule to predict which enantiomer of a secondary alcohol reacts faster in reac-

tions catalyzed by cholesterol esterase, lipase from *Pseudomonas cepacia*, and lipase from *Candida rugosa. J. Org. Chem.* **56**, 2656–2665.

11. Weissfloch, A. N. E. and Kazlauskas, R. J. (1995) Enantiopreference of lipase from *Pseudomonas cepacia* toward primary alcohols. *J. Org. Chem.* **60**, 6959–6969.

12. Lemke, K., Lemke, M., and Theil, F. (1997) A three-dimensional predictive active site model for lipase from *Pseudomonas cepacia. J. Org. Chem.* **62**, 6268–6273.

13. Lang, D. A., Mannesse, M. L. M., De Haas, G. H., Verheu, H. M., and Dijkstra, B. W. (1998) Structural basis of the chiral selectivity of *Pseudomonas cepacia* lipase. *Eur. J. Biochem.* **254**, 333–340.

14. Ferraboschi, P., Grisenti, P., Manzocchi, A., and Santaniello, E. (1990) New chemoenzymatic synthesis of (R)- and (S)-4-(phenylsulfonyl)-2-methyl-1-butanol: a chiral C5 isoprenoid synthon. *J. Org. Chem.* **55**, 6214–6216.

15. Ferraboschi, P., Grisenti, P., and Santaniello, E. (1990) Lipase-catalyzed resolution of (R,S)-2-methyl-4-phenylseleno-1-butanol: synthesis of enantiomerically pure 2-methyl-1,3-propanediol derivatives. *Synthesis* 545–546.

16. Ferraboschi, P., Brembilla, D., Grisenti, P., and Santaniello, E. (1991) Enzymatic synthesis of enantiomerically pure chiral synthons: lipase-catalyzed resolution of (R,S, 4E)-2-methyl-4-hexen-1-ol. *Synthesis* 310–312.

17. Ferraboschi, P., Grisenti, P., Manzocchi, A., and Santaniello, E. (1992) A chemoenzymatic synthesis of enantiomerically pure (R)- and (S)-2-methyldecan-1-ol. *J. Chem. Soc., Perkin Trans. I* 1159–1161.

18. Ferraboschi, P., Casati, S., De Grandi, S., Grisenti, P., and Santaniello, E. (1994) An insight into the active site of *Pseudomonas fluorescens (P. cepacia)* lipase to define the stereochemical demand for the transesterification in organic solvents. *Biocatalysis* **10**, 279–288.

19. Casati, S. (1996) Native and modified lipases: study on the activity in water and organic solvents. Thesis of Doctorate in Biochemistry, University of Milan, Italy.

20. Casati, S., Caporali, M., Ferraboschi, P., Manzocchi, A., and Santaniello, E. (1997) Enzyme recycling does not influence the enantioselectivity of the *Pseudomonas cepacia* lipase-catalyzed acylation of racemic alcohol in organic solvents. *Biotechnol. Tech.* **11**, 81–83.

21. Ferraboschi, P., Casati, S., Manzocchi, A., Grisenti, P., and Santaniello, E. (1995) Studies on the regio- and enantioselectivity of the lipase-catalyzed transesterification of 1'- and 2'-naphthyl alcohols in organic solvent. *Tetrahedron: Asymmetry* **6**, 1521–1524.

22. Ferraboschi, P., Molatore, A., Verza, E., and Santaniello, E. (1996) The first example of lipase-catalyzed resolution of a stereogenic center in steroid side chains by transesterification in organic solvent. *Tetrahedron: Asymmetry* **7**, 1551–1554.

23. Ferraboschi, P., Rezaelahi S., Verza, E., and Santaniello, E. (1998) Lipase-catalyzed resolution of stereogenic centers in steroid side chains by transesterification in organic centers: the case of a 26-hydroxycholesterol. *Tetrahedron: Asymmetry* **9**, 2193–2196.

24. Ferraboschi, P., Brembilla, D., Grisenti, P., and Santaniello, E. (1991) Enzymatic resolution of 2-substituted oxiranemethanols, a class of syntethically useful building blocks bearing a chiral quaternary center. *J. Org. Chem.* **56**, 5478–5480.

25. Ferraboschi, P., Grisenti, P., Casati, S., and Santaniello, E. (1994) A new synthesis of (R)- and (S)-mevalonolactone from the enzymatic resolution of (R,S)-2-(3-methyl-2-butenyl)-oxiranemethanol. *Synthesis* 754–756.
26. Ferraboschi, P., Casati, S., Grisenti, P., and Santaniello, E. (1993) A chemoenzymatic approach to enantiomerically pure (R)- and (S)-2,3-epoxy-2-(4-pentenyl)-propanol, a chiral building block for the synthesis of (R)- and (S)-frontalin. *Tetrahedron: Asymmetry* **4,** 9–12.
27. Ferraboschi, P., Casati, S., Grisenti, P., and Santaniello, E. (1994) Enantioselective *Pseudomonas fluorescens (P. cepacia)*-lipase-catalyzed irreversible transesterification of 2-methyl-1,2-diols in an organic solvent. *Tetrahedron: Asymmetry* **5,** 1921–1924.
28. Dale, J. A., Dull, D. L., and Mosher, H. S. (1969) α-Methoxy-α-trifluoromethyl-phenylacetic acid, a versatile reagent for the determination of enantiomeric composition of alcohols and amines. *J. Org. Chem.* **34,** 2543–2549.
29. Chen, C.-S., Fujimoto, Y., Girdaukas, G., and Sih, C. J. (1982) Quantitative analyses of biochemical kinetic resolutions of enantiomers. *J. Am. Chem. Soc.* **104,** 7294–7299.
30. Sih, C. J. and Wu, S.-H. (1989) Resolution of enantiomers via biocatalysis. *Topics Stereochem.* **19,** 63–125.

25

Preparation of 2-, 3-, and 4-Methylcarboxylic Acids and the Corresponding Alcohols of High Enantiopurity by Lipase-Catalyzed Esterification

Per Berglund and Erik Hedenström

1. Introduction

Chiral methyl-branched carboxylic acids or the corresponding alcohols of high enantiomeric purity are valuable intermediates for the synthesis of many pharmaceuticals, pesticides, and natural substances such as pheromones. Often, compounds of a very high enantiomeric excess (>99.5% ee) are needed. Various methyl-branched carboxylic acids have been synthesized by chemical methods such as diastereoselective alkylation of alkylamide enolates bearing a chiral auxiliary *(1)*. However, a loss in enantiomeric excess of about 2% ee is often associated with the final hydrolytic removal of the chiral auxiliary to release the chiral alkylated acid. This has been noted, for instance, with the currently popular auxiliary pseudo-ephedrine *(2)*, and much work has been devoted to overcome this obstruction, albeit with limited success so far *(3)*.

Biocatalyzed procedures are remarkably powerful in producing methyl-branched carboxylic acids of very high enantiomeric excess. For example, 2-methyloctanoic acid has been prepared in >99.5% ee *(4)*. The use of a kinetic resolution procedure catalyzed by a lipase in organic media gives both enantiomers in high ee in a simple process without any risk of racemization. The racemic starting material for the resolution can easily be obtained either by alkylation of a malonic acid ester *(5)* or by methyl iodide alkylation of a carboxylic acid *(6)*. The advantage of a kinetic resolution process is that it provides both enantiomers from a racemate. This chapter describes the protocol for kinetic resolution of 2-, 3-, and 4-methyl carboxylic acids by esterification

From: *Methods in Biotechnology, Vol. 15: Enzymes in Nonaqueous Solvents: Methods and Protocols*
Edited by: E. N. Vulfson, P. J. Halling, and H. L. Holland © Humana Press Inc., Totowa, NJ

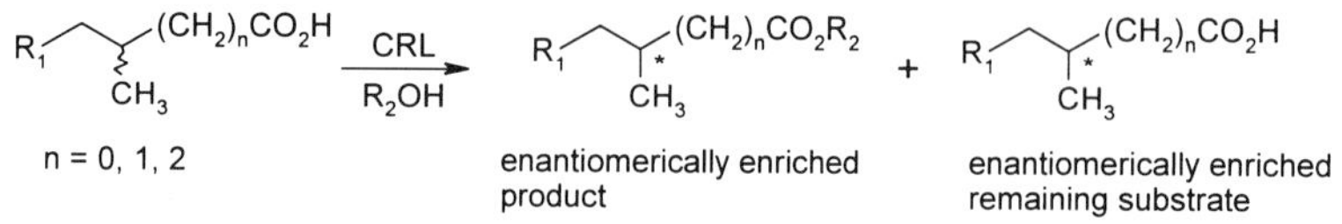

Fig. 1. Resolution of 2-, 3-, or 4-methylcarboxylic acids by esterification catalyzed by *Candida rugosa* lipase (CRL). The asterisk indicates enantiomerically enriched compounds of either *R*- or *S*-configuration.

catalyzed by a lipase in order to produce both enantiomers in high enantiomeric excess (**Fig. 1**). The protocol also involves a reduction step, which provides the corresponding alcohols as an alternative.

The theoretical yield of a product from a kinetic resolution process is never more than 50%. In the cases where only one of the enantiomers are needed, a dynamic resolution process, where the substrate is racemized, can be used. Racemization of esters of 2-methylalkanoic acids is possible in alkaline solution *(4)* or with non-nucleophilic bases *(7)*, whereas sodium has been used for racemization of methyl-substituted alcohols *(8)*. Dynamic resolution is beyond the scope of this chapter and can be found in recent review articles elsewhere *(9,10)*.

2. Materials

2.1. Enantioselective Esterification of Methyl-Branched Carboxylic Acids

1. Commercial extracellular lipase (E.C. 3.1.1.3) from *Candida rugosa* (CRL) with a specific activity stated as 900 U/mg solid and 4865 units/mg protein (obtained from Sigma, St. Louis, MO). One unit is the amount hydrolyzing 1.0 µmol/h of olive oil at pH 7.2 and 37°C. Store the lipase at 4°C and dry conditions.
2. Macroporous polypropylene, Accurel EP100, 350–1000 µm (obtained from Akzo Faser AG, Obernburg, Germany).
3. Cyclohexane proanalysi (p.a.), (analytical grade, >99.5% purity).
4. A nucleophile; for example, a long-chained alcohol of p.a. grade.
5. A racemic or enantiomerically enriched methyl-branched carboxylic acid.
6. An internal standard of p.a. grade.
7. Na_2SO_4 and $Na_2SO_4 \times 10\ H_2O$, both of p.a. grade, to obtain a water activity of 0.8.
8. A saturated aqueous solution of NaCl of puriss grade (99% purity) to obtain a water activity of 0.76.

2.2. Isolation of the Remaining Enantiomerically Enriched Substrate from the Esterification

1. 10% Na_2CO_3 (aq), puriss grade.
2. Saturated NaCl (aq), puriss grade.
3. 6 *M* HCl (aq), puriss grade.
4. Et_2O, puriss grade.

2.3. Isolation of the Enantiomerically Enriched Product from the Esterification

2.3.1. Chemical Hydrolysis Procedure of the Produced Enantiomerically Enriched Ester

1. 10% KOH (p.a. grade) in EtOH (95%).
2. 6 M HCl (aq), puriss grade.
3. Et$_2$O, puriss grade.
4. 10% Na$_2$CO$_3$ (aq), puriss grade.

2.3.2. Chemical Reduction Procedure of the Produced Enantiomerically Enriched Ester

1. LiAlH$_4$, synthesis grade.
2. Et$_2$O, anhydrous grade.
3. 15% NaOH (aq), p.a. grade.

2.3.3. Oxidation Procedure of the Alcohol Obtained from the Chemical Reduction of the Produced Enantiomerically Enriched Ester

1. Acetone, p.a. grade.
2. CrO$_3$, p.a. grade.
3. Concentrated H$_2$SO$_4$, puriss grade.
4. 6 M HCl (aq), puriss grade.
5. Pentane, purum grade.
6. Saturated Na$_2$S$_2$O$_5$ (aq), p.a. grade.
7. Et$_2$O, puriss grade.
8. 10% Na$_2$CO$_3$ (aq), puriss grade.

3. Methods

3.1. Enantioselective Esterification of Methyl-Branched Carboxylic Acids

1. Procedure for the immobilization of the commercial lipase (*see* **Note 1**): Dissolve the lipase from *C. rugosa* (200 mg) in sodium phosphate buffer (20 mL, 20 mM, pH 7.0 [*see* **Note 2**]) and spin at 700g in a centrifuge for 2 min. Mix macroporous polypropylene (100 mg) with ethanol (3 mL) and degas the mixture under vacuum. Add the supernatant from above to this mixture. Shake the suspension at room temperature for 15 h followed by filtration. Dry the filter cake (without washing) under vacuum for 1 h. If not used immediately in the esterification reaction, store the immobilized lipase at 4°C and under dry conditions (*see* **Note 3**).
2. Pre-equilibrate the immobilized enzyme from **step 1** to a known water activity (e.g., $a_w = 0.76$) over a saturated solution of the appropriate inorganic salt (e.g., NaCl) in a closed vessel for 24 h (*see* **Note 4**).

3. Stir (*see* **Note 5**) the cyclohexane solution (1 vol) (*see* **Note 6**) containing a methyl-branched carboxylic acid (0.15 M) (*see* **Note 7**), a long-chain alcohol (0.11 M)
   (*see* **Note 7**), an internal standard (6.1 mg/mL solution) (*see* **Note 8**), Na_2SO_4
   (1.33 mol/mol substrate), and $Na_2SO_4 \times 10\ H_2O$ (0.66 mol/mol substrate) (*see*
   Note 9) for 1 h in a sealed flask at room temperature.
4. Determine the starting ratio between the internal standard and the nucleophilic
   alcohol by gas chromatography (GC).
5. Add the pre-equilibrated immobilized enzyme.
6. Follow the progress of the esterification by withdrawal of samples from the stirred
   reaction and analyze by GC.
7. Terminate the reaction at the appropriate stage of conversion (*see* **Note 10**), by
   filtering off the enzyme and the salt mixture, wash with cyclohexane (0.5 vol)
   followed by pentane (0.5 vol) (*see* **Note 11**).

3.2. Isolation of the Remaining Enantiomerically Enriched Substrate from the Esterification

1. Extract the remaining substrate acid from the organic phase with aqueous sodium
   carbonate (10%, 5 × 0.2 vol).
2. Acidify the combined water phase containing the carboxylate salt and some alcohol to pH 1 with 6 M HCl and extract with diethyl ether (5 × 0.5 vol).
3. Wash the pooled organic phase with saturated sodium chloride solution (0.2 vol),
   dry ($MgSO_4$), and evaporate to dryness.
4. Distill at a suitable reduced pressure to give the methyl-branched carboxylic acid
   (*see* **Note 12**).
5. To enhance the enantiomeric excess of the acid (*see* **Note 13**), re-esterify
   enantioselectively (in a modified manner to that described in **Subheading
   3.1.**) to a conversion of about 25% (*see* **Note 10**) and use 6 molar equivalent
   of 1-dodecanol instead (*see* **Note 14**).

3.3. Isolation of the Enantiomerically Enriched Product from the Esterification

1. Dry ($MgSO_4$) the organic phase from above containing the produced ester, alcohol, and internal standard and evaporate to dryness.
2. Purify the ester by removing the internal standard and the remaining alcohol via
   liquid chromatography.
3. If needed, analyze the ee of the ester (*see* **Note 15**) and calculate an E value for
   the resolution reaction (*see* **Note 16**).
4. Hydrolyze the ester (0.75 M) in 95% EtOH containing 10% KOH (1 vol) under
   stirring at room temperature for 3 h (or as long as more than 5% of the ester is
   left) (*see* **Note 17**).
5. Interrupt the reaction by dilution with H_2O (0.5 vol) and acidify to pH 1 with 6 M
   HCl and then extract the water phase with diethyl ether (5 × 0.5 vol).

6. Extract the acid from the pooled organic phase with aqueous sodium carbonate (10%, 5 × 0.2 vol).
7. Acidify the combined water phase containing the carboxylate salt and some alcohol to pH 1 with 6 M HCl and extract with diethyl ether (5 × 0.5 vol).
8. Wash the pooled organic phase with saturated sodium chloride solution (0.2 vol), dry ($MgSO_4$), and evaporate to dryness.
9. Distill at a suitable pressure to give the methyl-branched carboxylic acid (*see* **Note 15**).
10. To obtain the acid with higher optical purity, re-esterify enantioselectively to a conversion of about 65% (*see* **Note 10**) using the standard procedure as in **Subheading 3.1.** This provides the ester of high ee (*see* **Note 15**).
11. Reduce the ester (0.5 M) (*see* **Note 18**) by dissolving it in diethyl ether (1 vol) and slowly add it to a solution of $LiAlH_4$ (0.5 M) in anhydrous diethyl ether (10 vol) under argon at room temperature.
12. Stir for 2 h (or at the point when no ester remains) and quench the reaction with H_2O (0.5 mL/g $LiAlH_4$), 15% NaOH (0.5 mL/g $LiAlH_4$) and then H_2O (0.5 mL/g $LiAlH_4$).
13. Reflux the mixture for 1 h, filter, dry ($MgSO_4$), evaporate to dryness, and distill at a suitable pressure to give the methyl-branched alcohol without any racemization. This three-step esterification procedure described so far has been used, for example, for the preparation of 2-methyloctanoic acid (*see* **Note 19**).
14. Dissolve the methyl-branched alcohol (0.05 M) in acetone (1 vol) and oxidize it to the corresponding acid by adding Jones reagent (0.025 vol of 2.67 M H_2CrO_4 in H_2SO_4/H_2O [*see* **Note 20**]).
15. Stir the reaction at room temperature for 2 h or until the starting alcohol has disappeared, then carefully add saturated sodium bisulfite (0.5 vol) to the reaction mixture and extract with pentane (3 × 0.5 vol).
16. Extract the acid from the pooled organic phase with aqueous sodium carbonate (10%, 5 × 0.2 vol).
17. Acidify the combined water phase containing the carboxylate salt to pH 1.0 with 6 M HCl, extract with diethyl ether (5 × 0.3 vol) and wash the pooled organic phase with saturated sodium chloride solution (0.2 vol), dry ($MgSO_4$), and evaporate to dryness.
18. Distill at a suitable pressure to give the methyl-branched carboxylic acid with no loss in ee.

4. Notes

1. The effect of immobilization is somewhat ambiguous in the esterification of 2-methylalkanoic acids. Compared with the crude enzyme, immobilization of CRL on polypropylene EP100 caused a dramatic improvement of the enantioselectivity for 2-methyldecanoic acid using long-chain alcohols like 1-octadecanol, whereas the E value was identical to the value of the crude enzyme in esterification of the same acid with 1-heptanol (*4*).
2. The pH should be close to the optimum for the enzyme used (*11*). The pH of a buffer from which an enzyme is going to be lyophilized has been shown to be of

    great importance and is sometimes referred to as the enzyme's "pH memory" *(12)*.

3. It is important to dry the immobilized enzyme carefully for lasting storage.

4. The rates of lipase reactions are dependent on the water activity, and different lipases require different levels of water for optimal performance. In the case of *C. rugosa* lipase, the water activity should be kept high and the pre-equilibration can be performed over saturated NaCl *(13)*.

5. Vigorous stirring using a magnetic bar can be applied, but the suspension is preferably mixed in an end-over-end rotation device to avoid destroying the enzyme particles *(14)*.

6. Other solvents than cyclohexane can be used. However, nonpolar solvents with a high log *P* are most suitable for lipase reactions *(15)*. Changing the solvent might influence both the enzyme activity and enantioselectivity. No simple correlation appears to exist between enzyme enantioselectivity and one or several physical parameters of the solvent *(16)*.

7. The alcohol can act as an enantioselective inhibitor that will be detrimental to the enantiomeric excess of the product. This has been reported for *C. rugosa* lipase-catalyzed resolution of 2-methylalkanoic acids *(4,17)*. The concentration of the alcohol should therefore be low. The *E* values of the resolution of some carboxylic acids are presented in **Table 1**.

8. The internal standard should be a hydrophobic aliphatic hydrocarbon, which will not participate in the enzyme reaction or change the nature of the hydrophobic solvent to any significant amount. Hexadecane, octadecane, or icosane has been used successfully in the reactions in **Table 1**.

9. A nonsoluble solid salt present in the reaction mixture in equilibrium with its hydrated form acts as a water buffer. This technique is especially useful in reactions like esterifications where water is produced during the course of the reaction. Sodium sulfate and sodium sulfate decahydrate will maintain a water activity of 0.8 *(22)*.

10. The enantiomeric excess of both substrate and product vary with the extent of conversion *(23)*. In order to know the ee at a certain stage of conversion, the *E* value of the reaction has to be determined (*see* **Note 16**).

11. If the enzyme powder is going to be reused, it can be recovered if washing of the filtered enzyme with aqueous solution is avoided. This has led to a fivefold loss of activity in some cases *(19)*. Although immobilization did not affect the *E* value, it facilitated the practical procedure and made recovery of the used enzyme-coated particles possible. The salt and salt hydrate present can be easily separated from the immobilized enzyme using chloroform. In this solvent, the enzyme particles are floating, whereas the salts sink. The surface layer is then collected by filtration and is dried as described in the immobilization method (*see* **Subheading 3.1.**). This procedure gave both retained activity and enantioselectivity *(19)*.

12. In the cases of 2-methyloctanoic acid, 2-methylnonanoic acid, 2-methyldecanoic acid, and other 2-methyl-branched acids, the isolated unreacted substrate acids all possess an *R*-configuration *(4,8,17,19,24)*. This is also the case when a 4-methyl-branched alkanoic acid is used *(20)*. However, when a 3-methyl-branched carboxylic acid is esterified, the remaining substrate is of the *S*-configuration *(21)*. *See* **Table 1**.

Table 1
***E* Values in Esterification of Selected Methyl-Branched Carboxylic Acids with Various Alcohols Catalyzed by *C. rugosa* Lipase in Cyclohexane**

Reacting enantiomer	1-Heptanol	1-Hexa-decanol	1-Octadecanol	1-Icosanol	2,2-Dimethyl-1-propanol
(structure) CO_2H (*S*)	57[a]	91[b]	84[b]	115[b]	—
(structure) CO_2H (*S*)	30[c]	85[c]	—	—	—
(structure) CO_2H (*S*)	83[d]	28[b]	120–150[b]	36[b]	98[e]
(structure, thiophene) CO_2H (*S*)	1.8[e]	—	—	—	—
(structure) CO_2H (*S*)	—	—	—	19[f]	—
(structure) CO_2H (*R*)	—	24[g]	—	—	37[e]
(structure) CO_2H (*R*)	—	17[g]	—	—	—

[a]This reaction was run using nonimmobilized lipase and with a molar excess of the alcohol, (*18*).
[b]Data from **ref. 4**.
[c]These reactions were run using a molar excess of the alcohol, which might lower the E value (*19*).
[d]Data from **ref. 17**.
[e]Data from Berglund, et al. (unpublished data).
[f]Data from **ref. 20**. [g]Data from **ref. 21**.

13. If the enantiomeric excess of the acid cannot be measured on a chiral GC or high-performance liquid chromatography (HPLC) column, it can be determined after derivatization to diastereomeric phenylethyl amides (*see* **Note 16**).

14. To obtain the remaining acid substrate in high enantiomeric excess, the equilibrium position is the most important parameter influencing the result. Therefore, using 6 eq. of the alcohol will displace the reaction equilibrium toward esterification. With this high concentration of alcohol, dodecanol simplifies the workup procedure compared to using a longer alcohol, which is precipitating. However, to obtain the ester product in a high enantiomeric excess, the *E* value rather than the equilibrium should be maximized (*4*).

15. The configuration of the isolated ester products varies with the substrate structure (cf. **Note 12**). The esters possess an *S*-configuration of 2-methyl-branched acids as well as for a 4-methyl-branched alkanoic acid. However, this is not the case when a 3-methyl-branched carboxylic acid is esterified.

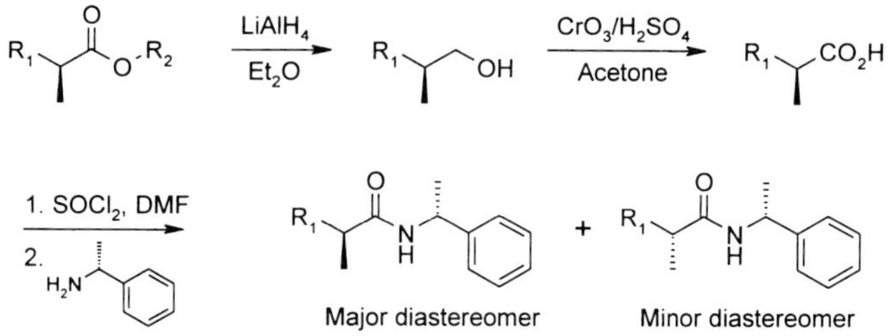

Fig. 2. Derivatization to diastereomeric amides for GC analysis of the enantiomeric excess of the product esters.

16. In order to obtain accurate E values in reversible reactions, the ee of the product should be determined at a low conversion, preferably less than 40%. The E value is then calculated from either **Eq. 1** or **2** *(23,25)*.

$$E = \frac{\ln\left[1 - \xi\,(1 + ee_p)\right]}{\ln\left[1 - \xi\,(1 - ee_p)\right]} \tag{1}$$

$$E = \frac{\ln\left[\dfrac{1 - ee_S}{1 + ee_S/ee_p}\right]}{\ln\left[\dfrac{1 - ee_S}{1 + ee_S/ee_p}\right]} \tag{2}$$

where ξ is the extent of conversion, ee_P is the ee of the product ester, and ee_S is the ee of the remaining nonreacted acid. If possible, **Eq. 2** should be used because ee is usually determined more accurately than the conversion. If determination of the ee is difficult by chiral gas chromatographic or HPLC methods, a reaction sequence starting with a reduction of the ester followed by reoxidation using Jones' reagent in acetone (*see* **Note 20**) and then derivatization to diastereomeric phenylethyl amides prior to GC analysis can be used **(Fig. 2)**. Reduction by LiAlH$_4$ *(26)*, as well as oxidation by Jones' reagent *(27–29)* are processes known to proceed without any racemization.

17. In order to minimize the racemization when 2-methyl-branched carboxylic esters are hydrolyzed, the reaction should be interrupted before 100% conversion. Following our protocol causes a decrease in ee of about 2% when 2-methyl-branched acids are used, but for 3- and 4-methyl-branched acids no racemization arises.

18. When using chemical hydrolysis, racemization is always a risk. Therefore, LiAlH$_4$ reduction of the 2-methyl-branched ester is preferred. This provides the corresponding methyl-branched alcohol, which can then be reoxidized to the acid. Thus, a 2-methyl-branched carboxylic acid is produced without any racemization *(24)*.

19. The complete three-step procedure for preparing the two enantiomers of 2-methyloctanoic acid is outlined in **Fig. 3**.

20. Preparation of 100 mL of Jones' reagent *(30)*: Dissolve 26.7 g of CrO$_3$ in 40 mL of H$_2$O. Then, carefully add 23 mL of H$_2$SO$_4$ and add water up to 100 mL.

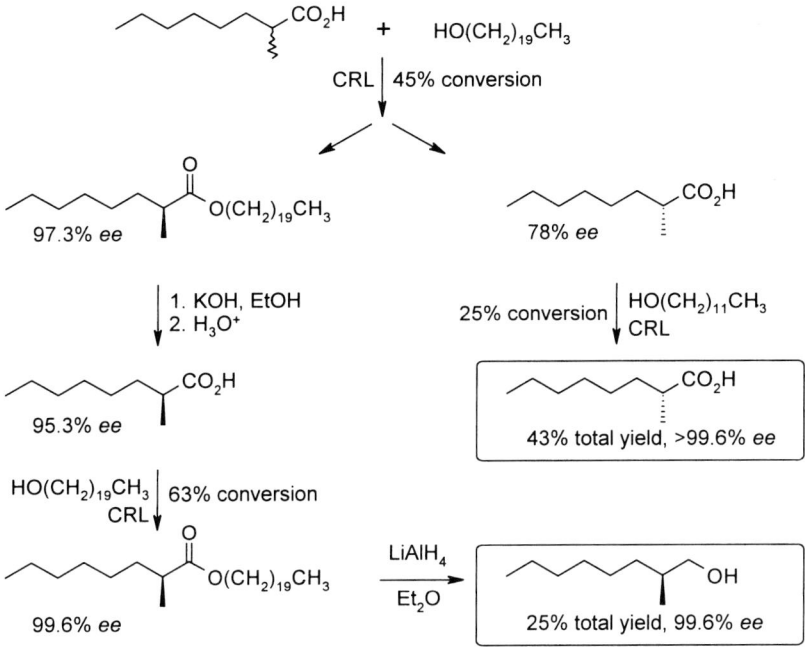

Fig. 3. Resolution of 2-methyloctanoic acid by a three-step esterification procedure with icosanol (0.75 eq.) and 1-dodecanol (6 eq.), respectively, catalyzed by CRL in cyclohexane. (From **ref. 4**.)

Acknowledgments

Financial support from the Swedish Council for Forestry and Agricultural Research and from the Swedish Natural Science Research Council is gratefully acknowledged.

References

1. Högberg, H.-E. (1995) Alkylation of amide enolates, in *Houben–Weiy Methods of Organic Chemistry*, Vol. E21a, *Stereoselective Synthesis* (Helmchen G., Hoffmann, R. W., Mulzer, J., Schaumann, E., eds.),Thieme, Stuttgart, pp. 791–915.
2. Myers, A. G., Yang, B. H., Chen, H., and Gleason, J. L. (1994) Use of pseudoephedrine as a practical chiral auxiliary for asymmetric synthesis. *J. Am. Chem. Soc.* **116,** 9361,9362.
3. Myers, A. G., Yang, B. H., Chen. H., McKinstry, L., Kopecky, D. J., and Gleason, J. L. (1997) Pseudoephedrine as a practical chiral auxiliary for the synthesis of highly enantiomerically enriched carboxylic acids, alcohols, aldehydes, and ketones. *J. Am. Chem. Soc.* **119,** 6496–6511.
4. Edlund, H., Berglund, P., Jensen, M., Hedenström, E., and Högberg, H.-E. (1996) Resolution of 2-methylalkanoic acids. Enantioselective esterification with long chain alcohols catalysed by *Candida rugosa* lipase. *Acta Chem. Scand.* **50,** 666–671.

5. Allen, C. F. and Kalm, M. J. (1963) 2-Methylenedodecanoic acid, in *Organic Syntheses, coll. vol. IV* (Rabjohn, R., ed.), Wiley, New York, p. 618, Note. 2.
6. Sonnet, P. E. and Baillargeon, M. W. (1989) Synthesis and lipase catalyzed hydrolysis of thiolesters of 2-, 3-, and 4-methyl octanoic acids. *Lipids* **24,** 434–437.
7. Dinh, P. M., Williams, J. M. J., and Harris, W. (1999) Selective racemisation of esters: relevance to enzymatic hydrolysis reactions. *Tetrahedron Lett.* **40,** 749–752.
8. Vörde, C., Högberg, H. E., and Hedenström, E. (1996) Resolution of 2-methyl-alkanoic esters: enantioselective aminolysis by (R)-1-phenylethylamine of ethyl 2-methyloctanoate catalysed by lipase B from *Candida antarctica. Tetrahedron: Asymmetry* **7,** 1507–1513.
9. Stecher, H. and Faber, K. (1997) Biocatalytic deracemization techniques: dynamic resolutions and stereoinversions. *Synthesis* 1–16.
10. Ebbers, E. J., Ariaans, G. J. A., Houbiers, J. P. M., Bruggnink, A., and Zwanenburg, B. (1997) Controlled racemization of optically active organic compunds: prospects for asymmetric transformation. *Tetrahedron* **53,** 9417–9476.
11. Montero, S., Blanco, A., Virto, M. D., Landeta, L. C., Agud, I., Solozabal, R., et al. (1993) Immobilization of *Candida rugosa* lipase and some properties of the immobilized enzyme. *Enzyme Microb. Technol.* **15,** 239–247.
12. Zaks, A. and Klibanov, A. M. (1985) Enzyme-catalyzed processes in organic solvents. *Proc. Natl. Acad. Sci. USA* **82,** 3192–3196.
13. Valivety, R. H., Halling, P. J., Peilow, A. D., and Macrae, A. R. (1992) Lipases from different sources vary widely in dependence of catalytic activity on water activity. *Biochim. Biophys. Acta* **1122,** 143–146.
14. Kvittingen, L., Sjursnes, B., Halling, P., and Anthonsen, T. (1992) Mixing condi-tions for enzyme catalysis in organic solvents. *Tetrahedron* **48,** 5259–5264.
15. Laane, C., Boeren, S., Vos, K., and Veeger, C. (1987) Rules for optimization of biocatalysis in organic solvents. *Biotechnol. Bioeng.* **30,** 81–87.
16. Carrea, G., Ottolina, G., and Riva, S. (1995) Role of solvents in the control of enzyme selectivity in organic media. *Trends Biotechnol.* **13,** 63–70; erratum: (1995) *Trends Biotechnol.* **13,** 122.
17. Berglund, P., Holmquist, M., Hult, K., and Högberg, H.-E. (1995) Alcohols as enantioselective inhibitors in a lipase in a lipase catalysed esterification of a chiral acyl donor. *Biotechnol. Lett.* **17,** 55–60.
18. Berglund, P., Holmquist, M., Hedenström, E., Hult, K., and Högberg, H.-E. (1993) 2-Methylalkanoic acids resolved by esterification catalysed by lipase from *Candida rugosa:* alcohol chain length and enantioselectivity. *Tetrahedron: Asymmetry* **4,** 1869–1878.
19. Berglund, P., Vörde, C., and Högberg, H.-E. (1994) Esterification of 2-methyl-alkanoic acids catalysed by lipase from *Candida rugosa:* enantioselectivity as a function of water activity and alcohol chain length. *Biocatalysis* **9,** 123–130.
20. Lundh, M., Smitt, O., and Hedenström, E., (1996) Sex pheromone of pine sawflies: enantioselective lipase catalysed transesterification of erythro-3, -dimethyl-pentadecan-2-ol, diprionol. *Tetrahedron: Asymmetry* **7,** 3277–3284.
21. Nguyen, B.-V. and Hedenström, E. (1999) Candida rugosa lipase as an enantio-selective catalyst in the esterification of methyl branched carboxylic acids: Reso-

lution of *rac*-3,7-dimethyl-6-octenic acid (citronellic acid). *Tetrahedron: Asymmetry* **10,** 1821–1826.

22. Halling, P. J. (1992) Salt hydrates for water activity control with biocatalysis in organic media. *Biotechnol. Tech.* **6,** 271–276.

23. Chen, C.-S., Fujimoto, Y., Girdaukas, G., Sih, C. J. (1982) Quantitative analysis of biochemical kinetic resolutions of enantiomers. *J. Am. Chem. Soc.* **104,** 7294–7299.

24. Bergström, G., Wassgren, A.-B., Anderbrant, O., Fägerhag, J., Edlund, H., Hedenström, E., et al. (1995) Sex pheromone of the pine sawfly Diprion pini (Hymenoptera: Diprionidae): chemical identification, synthesis and biological activity. *Experientia* **51,** 370–380.

25. Rakels, J. L. L., Straathof, A. J. J., and Heijnen, J. J. (1993) A simple method to determine the enantiomeric ration in enantioselective biocatalysis. *Enzyme Microb. Technol.* **15,** 1051–1056.

26. Noyce, D. S. and Denney, D. B. (1950) Steric effects and stereochemistry of lithium aluminium hydride reduction. *J. Am. Chem. Soc.* **72,** 5743–5745.

27. Sonnet, P. E. (1982) Synthesis of the stereoisomers of the sex pheromone of the southern corn rootworm and lesser tea tortrix. *J. Org. Chem.* **47,** 3793–3796.

28. Sonnet, P. E. (1987) Kinetic resolutions of aliphatic alcohols with a fungal lipase from *Mucor miehei. J. Org. Chem.* **52,** 3477–3479.

29. Guanti, G., Narisano, E., Podgorski, T., Thea, S., and Williams, A. (1990) Enzyme catalyzed monohydrolysis of 2-aryl-1, 3-propanediol diacetates. A study of structural effects of the aryl moiety on the enantioselectivity. *Tetrahedron* **46,** 7081–7092.

30. Bowers, A., Halsall, T. G., Jones, E. R. H., and Lemin, A. J. (1953) The chemistry of the triterpenes and related compounds. Part XVIII. Elucidation of the structure of polyporenic acid C. *J. Chem. Soc.* 2548–2560.

26

Optimization of Enzymatic Enantiomeric Resolutions Through Solvent Selection

Gianluca Ottlina, Francesco Secundo, Giorgio Colombo, and Giacomo Carrea

1. Introduction

One of the most challenging aspects in the synthesis of chiral molecules is the achievement of products with high enantiopurity. This can be sought through asymmetric synthesis or resolution of a racemic form into its components *(1)*. The resolution of racemic mixtures via enzyme catalysis is a highly studied and practiced methodology, and alcohols, acids and amines are routinely resolved with hydrolases in aqueous, aqueous/cosolvent, or organic media. The hydrolysis of esters in aqueous media is a procedure that has been utilized for a long time and changing the pH, temperature, or ionic strength of the medium can modify enzyme enantioselectivity. Another way to modify enzyme enantioselectivity is to add substantial proportions of organic cosolvents such as dimethyl sulfoxide (DMSO), dimethyl formamide (DMF), *t*-BuOH, acetone, and acetonitrile *(2)* to the aqueous buffer.

Since the early eighties, it has been demonstrated that several enzymes are active in pure organic media and that changing the nature of solvents can modify the enantioselectivity of the enzymes *(3–5)*. The opportunity to modify enzyme enantioselectivity by changing the nature of the medium has been mainly exploited for hydrolytic enzymes such as lipases and proteases and, in the literature, there is a wealth of examples referring to these enzymes (for review articles, *see* **refs.** *4* and *5*). Sometimes, the influence of the solvent nature is so remarkable that not only can it induce dramatic increases of enantioselectivity *(6)* but even reverse enzyme enantiopreference *(7)*.

Many attempts have been made trying to correlate enzyme enantioselectivity and solvent physico-chemical properties (polarity, hydrophobicity, dielectric

From: *Methods in Biotechnology, Vol. 15: Enzymes in Nonaqueous Solvents: Methods and Protocols*
Edited by: E. N. Vulfson, P. J. Halling, and H. L. Holland © Humana Press Inc., Totowa, NJ

constant, and so forth) *(4)*, but it seems that in most cases, these macroscopic characteristics cannot address the scope *(5,8)*. More recently, computational approaches have been proposed to rationalize the phenomenon, but, in addition to being demanding and time-consuming, they do not provide results of general applicability *(9,10)*.

Therefore, in our opinion the "trial- and-error" approach is still that of choice because, in principle, any solvent could be successfully used. Experience suggests the avoidance of solvents such as DMSO, DMF, ethanol, and methanol because they tend to denature enzymes. One remarkable exception is subtilisin, which is active in DMF. Other water-miscible solvents such as dioxane, acetonitrile, pyridine, tetrahydrofuran (THF), triethylamine, and acetone are widely used, but, with them, the control of water activity can be difficult. Instead, with water-immiscible solvents, it is relatively simple to set the water activity at the desired value, and the solvents most commonly used are benzene, CCl_4, methyl *t*-butyl ether, CH_2Cl_2, $CHCl_3$, nitrobenzene, 3-pentanone, *t*-amyl alcohol, hexane, and dodecane. The reaction medium can also be made of vinyl acetate or other vinyl esters that act both as solvent and acylating agent.

In addition to solvent properties, the water content in the reaction medium is another important factor to be considered. The parameter that better describes the effects of water on such enzyme properties as K_M, V_{max}, stability, and conformation is the thermodynamic water activity *(11,12)*. The effects of water activity on enzyme enantioselectivity are, however, controversial because an increase, decrease, or no variation in selectivity as a function of the water present in the reaction medium have been reported *(13–15)*. Therefore, when facing a resolution process, it is advisable to carry it out at different water activity values (*see* **Note 1**).

The resolution of a secondary alcohol via a transesterification reaction is a kinetic resolution where both enantiomers are transformed, but at different rates (*see* **Fig. 1**). For sufficiently long reaction times, all of the substrate will be converted into the product.

This means that the enantiomeric excess of the substrate increases as the conversion increases, whereas the enantiomeric excess of the product decreases as the conversion increases. The fact that the enantiomeric excess is related to the degree of conversion makes this parameter unsuitable for measurement of enzyme enantioselectivity in different solvents. A parameter that better describes enzyme enantioselectivity is the enantiomeric ratio (*E*), which is the ratio between the specificity constants of the two competing enantiomers. The enantiomeric ratio is commonly calculated from the conversion (*c*) and the enantiomeric excess of either the substrate (ee$_S$) or the product (ee$_P$) *(16)*:

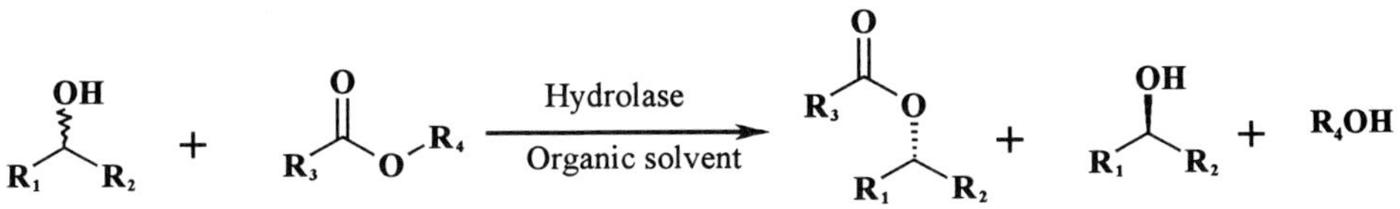

Fig. 1. Schematic representation of a generic secondary alcohol resolution through a transesterification reaction.

$$E = \frac{\ln[(1 - c)(1 - ee_S)]}{\ln[(1 - c)(1 + ee_S)]} \qquad E = \frac{\ln[(1 - c)(1 + ee_P)]}{\ln[(1 - c)(1 - ee_P)]} \qquad (1)$$

where $0 < c < 1$, $0 < ee_S < 1$, and $0 < ee_P < 1$.

To exemplify the influence of organic solvents on enzyme enantioselectivity, the resolution of (±) *trans*–sobrerol [(±) *trans*–5-(1-hydroxyl-1-methylethyl)-2-methyl-2-cyclohexen-1-ol] by transesterification with vinyl acetate catalyzed by lipase from *Pseudomonas cepacia* (Lipase PS) (*see* **Fig. 2**) will be described *(6)*.

2. Materials
2.1. Enzyme Immobilization

1. Lipase PS (Amano Pharmaceutical Co., Japan).
2. Hyflo Super Cel (Fluka, Switzerland).
3. 0.1 *M* potassium phosphate buffer, pH 7.0.
4. Watch glass or Petri dish.
5. Vacuum apparatus.

2.2. Solvent Preparation

1. Dioxane, acetone, THF, 3-pentanone, *t*-amyl alcohol of the highest grade of purity, >99%.
2. Activated 3 Å molecular sieves.

2.3. Reactants

1. (±) *trans*–Sobrerol (Aldrich).
2. Vinyl acetate (Aldrich).

2.4. Enzymatic Reaction

1. Thermostatted shaker.

2.5. Conversion and Enantiomeric Excess Determination

Gas chromatography (GC) apparatus equipped with a flame ionization detector and a chiral CP-cyclodextrin β 2,3,6-M-19 column (50 m, 0.25 inner diameter; Chrompack; The Netherlands) and H$_2$ as carrier gas.

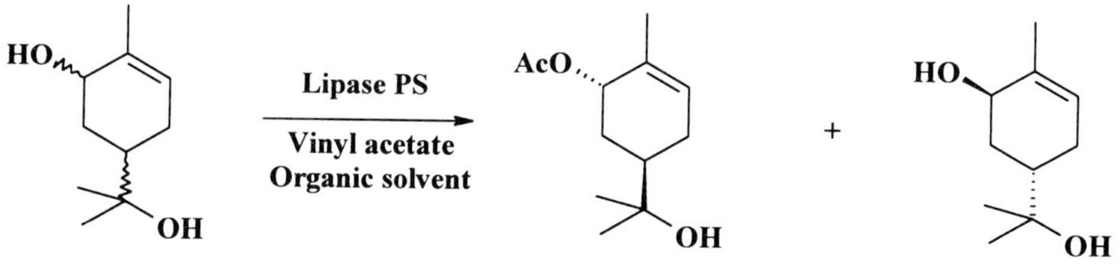

Fig. 2. Enzymatic resolution of (±) *trans*-sobrerol in organic solvents.

3. Methods

3.1. Enzyme Immobilization

1. 3 g of Lipase PS was mixed accurately with 10 g of Hyflo Super Cel on a watch glass.
2. 10 mL of buffer was added dropwise, and the mixture was then stirred with a spatula.
3. The mixture was dried over a vacuum pump (0.02 mbar) at room temperature for 24 h.
4. The powder was divided in several sealed vials and kept in a refrigerator.

3.2. Solvents Preparation

1. The 3-Å molecular sieves were new and activated. Activation was carried out in a vacuum apparatus at 200°C for 48 h. Activated molecular sieves were immediately used or kept in an oven at 150°C for subsequent employment.
2. Erlenmeyer flasks were filled to one-third volume with activated molecular sieves and two-thirds volume with solvent and then let stand for 2 d at room temperature to dry the solvents.

3.3. Reactants

For vinyl acetate, the same procedure as described in **Subheading 3.2.** was used.

3.4. Enzymatic Reactions

1. Five hundred milligrams of molecular sieves, 5 mL of solvent, 100 mg of *trans*-sobrerol, 100 µL of vinyl acetate, and 400 mg of immobilized Lipase PS were added to every 10 mL screwcapped vial. Vials were tightly capped and vigorously shaken by hand for a few seconds.
2. The vials were put in a thermostatted shaker at 45°C (*see* **Note 2**) and shaken at 250 rpm.

3.5. Conversion and Enantiomeric Excess Determination

1. Ten microliters of reactants were withdrawn at scheduled times (*see* **Note 3**), diluted in a 1-mL tube with 100 µL of acetone and centrifuged for a few seconds to clear the solution from enzyme powder and molecular sieves.

2. Ten microliters of the acetone solution were analyzed by gas chromatography (GC) under the following conditions: oven temperature 130°C for 10 min, then to 160°C with a heating rate of 0.5°C/min. The running time was approx 30 min (*see* **Note 4**).

3. The enantiomeric ratio was calculated with the second formula in **Eq. 1**, using the enantiomeric excess of the product (ee_P) and the degree of conversion (c) determined by GC (*see* **Notes 5–7**).

4. Notes

1. The technique described in this chapter is suitable for a water activity (a_w) value of ≤ 0.11. To study the effects of solvents at higher water-activity values, each reaction has to be pre-equilibrated at a precise a_w value using the saturated salt solution method *(12)*. In short, two open vials, one with reactants and the other with a saturated solution of salt, are equilibrated for about 2 d in a wide-mouth, tightly capped bottle. It is not advisable to pre-equilibrate different components in the same bottle because some reaction could occur.

2. For this type of application (45°C/250 rpm) thermostation by air is more convenient than by water bath.

3. The reaction time could last in certain circumstances up to 1 wk at 45°C. In this case, special care must be taken to ensure a tight capping of vials to avoid solvent evaporation.

4. If the GC apparatus and the chiral column are in good condition, substrate and product enantiomers are baseline separated. The retention times of (–) *trans*-sobrerol, (+) *trans*-sobrerol, (+) *trans*-sobrerol acetate and (–) *trans*-sobrerol acetate were 26.5, 27.6, 28.9, and 30.0 min, respectively.

5. Spontaneous reactions sometimes could occur, which would affect the accuracy of enantioselectivity determination. Setting up control reactions is advisable.

6. To achieve a good accuracy in the determination of the enantiomeric ratio, it is advisable to collect and analyze at least five samples in a range between 10% and 40% conversion and then average them. In the determination of the conversion degree, differences in the combustion factor for substrates and products have to be taken into account. If it is possible to determine both ee_S and ee_P, the conversion can be simply calculated with the following relation:

$$c = \frac{ee_S}{ee_P + ee_S}$$

7. The enantioselectivity, expressed as E, of Lipase PS in the resolution of (±) *trans*-sobrerol (**Fig. 2**) was as follows in the various solvents: THF, $E = 69$; vinyl acetate, $E = 89$; acetone, $E = 142$; dioxane, $E = 178$; 3-pentanone, $E = 212$; *t*-amyl alcohol, $E = 518$.

Acknowledgments

We thank the CNR Target Project on Biochemistry for financial support.

References

1. Collins, A. N., Sheldrake, G. N., and Cosby, J. (eds.) *Chirality in Industry II.* Wiley, New York.
2. Faber, K., Ottlina, G., and Riva, S. (1993) Selectivity-enhancements of hydrolases reactions. *Biocatalysis* **8,** 91–123.
3. Klibanov, A. M. (1989) Enzymatic catalysis in anhydrous organic solvents. *Trends Biochem. Sci.* **14,** 141–144.
4. Wescott, C. R. and Klibanov, A. M. (1994) The solvent dependence of enzyme specificity. *Biochim. Biophys. Acta* **1206,** 1–9.
5. Carrea, G., Ottlina, G., and Riva, S. (1995) Role of solvents in the control of enzyme selectivity in organic media. *Trends Biotechnol.* **13,** 63–70.
6. Bovara, R., Carrea, G., Ferrara, L., and Riva, S. (1991) Resolution of (±)-*trans*-sobrerol by lipase PS-catalyzed transesterification and effects of organic solvents on enantioselectivity. *Tetrahedron: Asymmetry* **2,** 931–938.
7. Hirose, Y., Kariya, K., Ssaki, I., Kurono, Y., Ebiike, H., and Achiwa, K. (1992) Drastic solvent effect on lipase-catalyzed enantioselectivity hydrolasis of prochiral 1,4-dihydropyridines. *Tetrahedron Lett.* **33,** 7157–7160.
8. Carrea, G., Ottlina, G., Riva, S., and Secundo, F. (1992) Effects of reaction conditions on the activity and enantioselectivity of lipase in organic solvents, in *Biocatalysis in Non-Conventional Media: Proceedings of an International Symposium* (Tramper, J., Vermue, M. H., Beeftink, H. H. and Stockar, U., eds.), Elsevier Science, New York, pp. 111–119.
9. Wescott, C. R. and Klibanov, A. M. (1997) Thermodynamic analysis of solvent effect on substrate specificity of lyophylized enzymes suspended in organic media. *Biotechnol. Bioeng.* **56,** 340–344.
10. Colombo, G., Ottlina, G., Carrea, G., Bernardi, A., and Scolastico, C. (1998) Application of structure-based thermodynamic calculations to the rationalization of the enantioselectivity of subtilisin in organic solvents. *Tetrahedron: Asymmetry* **9,** 1205–1214.
11. Greenspan, L. (1977) Humidity fixed points of binary satured aqueous solutions. *J. Res. Natl. Bur. Stand. A Phys. Chem.* **81A,** 89–96.
12. Valivety, R. H., Halling, P. J., and Macrea, A. R. (1992) Reaction rate with suspended lipase catalyst shows similar dependence on water activity in different organic solvents. *Biochim. Biophys. Acta* **1118,** 218–222.
13. Bovara, R., Carrea, G., Ottlina, G., and Riva, S. (1993) Water activity does not influence the enantioselectivity of lipase PS and lipoprotein lipase in organic solvents. *Biotechnol. Lett.* **15,** 169–174.
14. Wickli, A., Schmidt, E., and Bourne, J. R. (1992) Engineering aspects of the lipase-catalyzed production of (*R*)-1-ferrocenylethylacetate in organic media, in *Biocatalysis in Non-Conventional Media: Proceedings of an International Symposium* (Tramper, J., Vermue, M. H., Beeftink, H. H. and Stockar, U., eds.), Elsevier Science, New York, pp. 577–584.
15. Högberg, H. E., Edlund, H., Berglund, P., and Hedenström, E. (1993) Water activity influences enantioselectivity in lipase-catalysed resolution by esterification in an organic solvent. *Tetrahedron: Asymmetry* **4,** 2123–2126.
16. Chen, C. S. and Sih, C. J. (1989) General aspects and optimization of enantioselective biocatalysis in organic solvents: the use of lipase. *Angew. Chem. Int. Ed. Engl.* **28,** 695–707.

27

Chemoselective Amidification of Amino-Polyols Catalyzed with Lipases in Organic Solvents

Thierry Maugard, Magali Remaud-Simeon, and Pierre Monsan

1. Introduction

Glycamide surfactants are nonionic surfactants in which the hydrophilic moiety (an amino-alditol derivative) and the hydrophobic moiety (a fatty acid) are linked via an amide bond (*1*). Such sugar fatty amide surfactants can be obtained by chemical synthesis, using the Schotten–Baumann reaction between an amino-alditol and a fatty acid chloride in aqueous alkaline medium. An important drawback of this approach is the formation of salts by neutralization.

An alternative approach consists in using an enzymatic synthesis route, which avoids the formation of by-products and salts. The synthesis of amides can be catalyzed by proteases, using reverse hydrolysis. However, proteases are generally highly specific for the given amino acid and sensitive to organic solvents (*2*). In addition to proteases, lipases have also been shown to catalyze the synthesis of amides in nonconventional media (*3–4*) and are involved in the obtention of peptides (*2,5,6*), fatty amides (*7–9*), *N*-acyl-amino acids (*10,11*), and acyl-amino-propanol (*12*). However, the yields reported were insufficient to allow any industrial development.

In nonconventional media, numerous parameters affect the biocatalyst, such as water activity (*13–17*), solvent effects (*4,18*), and temperature, but also the acid–base condition of the media. Indeed, the activity, the specificity, and the stability of enzymes depend on the correct state of protonation of various ionizable groups in the protein molecule. The protonation state of ionizable groups of the catalytic site can be modified subsequently by exposure to acid or basic species, resulting in the alteration of enzyme activity. For example, an acid molecule could transfer a proton to the protein, its negative product remaining as a counterion (*19*).

From: *Methods in Biotechnology, Vol. 15: Enzymes in Nonaqueous Solvents: Methods and Protocols*
Edited by: E. N. Vulfson, P. J. Halling, and H. L. Holland © Humana Press Inc., Totowa, NJ

This chapter is dedicated to the operational procedure used to catalyze the amidification of a widely available amino-alditol derivative, by fatty acids or fatty acid methyl esters, using commercially available immobilized lipase preparations as catalysts. It provides the protocol for the *N*-oleoyl-*N*-methyl-glucamine synthesis and gives the analytical methods used to characterize the various products.

2. Materials

1. Lipozyme® (lipase from *Rhizomucor miehei* immobilized on an anionic macroporous resin, Duolite 568N) was a gift from Novo Industries (Denmark).
2. Novozyme® SP 435 (lipase from *Candida antarctica* immobilized on an acrylic resin) was a gift from Novo Industries (Denmark).
3. The solvents (*t*-amyl-alcohol, hexane, methanol, and so forth), all pure, were from Fluka and were predried over 4-Å molecular sieves (Merck).
4. *N*-Methyl-glucamine (Sigma Chemical Co., St. Louis, MO), *N*-methyl-galactamine (Aldrich), and *N*-octyl-glucamine (Aldrich) were more than 99% pure.
5. With the exception of Diester® (Colza fatty acid methyl esters), which was from Sidobre Sinova, all the fatty acids methyl and fatty acids were supplied by Sigma Chemical Co..
6. Dimethyladipate and ethyl lactate were from Fluka.
7. Reactor: The reactor is a flask mechanically stirred on a rotary evaporator (Büchi) that was used under atmospheric or reduced pressure at controlled temperature.
8. High-performance liquid chromatography (HPLC) analysis: reverse-phase chromatography (Ultrasep C18, 250 × 4 mm, 6 µm; ICS, France) was carried out using a Hewlett-Packard 1050 series system consisting of a pump, an ultraviolet (UV) detector, an injector, and a differential refractometer (RI) model 1047A. The recommended injection volume is 25 µL.
9. Carbon-13 nuclear magnetic resonance (^{13}C-NMR) spectra were recorded using an AC 250-MHz spectrometer from Brüker, with an internal reference of tetramethylsilane.
10. Mass spectra were obtained by chemical ionization (DCI/NH$_3$), using a NERMAG R10-10 spectrometer.
11. Infrared (IR) spectra were recorded using a Perkin-Elmer IRFT 1760-x spectrometer for KBr pellets.

3. Methods
3.1. General Procedure for the Enzymatic Acylation of N-Methyl-Glucamine

1. *N*-Methyl-glucamine (0.6 mmol), 0.6 mmol of oleic acid, and 100 mg of *Rhizomucor miehei* lipase (Lipozyme®) were mixed in 10 mL of hexane at 200 rpm (*see* **Notes 1** and **2**). The reactions were carried out in mechanically stirred flasks.

2. *N*-Methyl-glucamine (1.75 mmol), 1.75 mmol of oleic acid, and 100 mg of *Candida antarctica* lipase (Novozyme®) were mixed in 10 mL of *t*-amyl-alcohol (*see* **Notes 3** and **4**).
3. Reactions were run at 55°C or 90°C under atmospheric pressure or under reduced pressure.
4. The reaction can be realized with fatty acid methyl ester (*see* **Note 5**), but also with other substrates (*see* **Note 6**).

3.2. HPLC Analysis

1. Twenty-five microliters of the suitably diluted reaction mixture were injected. For reactions with long-chain fatty acids (more than 12 carbon atoms), a mixture of methanol/water/acetic acid, 90/10/0.3 (v/v/v) was used as the eluent at 40°C and a flow rate of 1 mL/min. For reactions with short-chain fatty acids (fewer than 12 carbon atoms), a mixture of methanol/water/acetic acid, 80/20/0.3 (v/v/v) was used as the fluent at 40°C and a flow rate of 1 mL/min. Products were detected using a UV detector at 210 nm and a differential refractometer.
2. The relative concentration of each species can be evaluated by calibration of the refractometer with pure material as references.
3. Yields and the selectivity of synthesis can be calculated by the following equations:

$$\frac{100 \text{ (Ester concentration)}}{\text{Oleic acid Initial concentration}} = \text{Ester yield} \qquad (\%)$$

$$\frac{100 \text{ (Amide concentration)}}{\text{Oleic acid Initial concentration}} = \text{Amide yield} \qquad (\%)$$

$$\frac{100 \text{ (Amide yield)}}{\text{Amide yield} + \text{Ester yield}} = \text{Selectivity for amide synthesis} \qquad (\%)$$

3.3. Lipase Activity Measurements

1. Aliquots of the reaction mixture were taken at intervals, filtered, and analyzed with an HPLC system.
2. The initial rates of amide and ester synthesis were determined by measuring the increase of corresponding products up to 5% conversion. Products were detected using a UV detector at 210 nm and a differential refractometer.

3.4. Purification of the Products Reaction

1. At the end of the reaction, the biocatalyst was removed by filtration and the solvent was evaporated under reduced pressure.
2. The remaining oil was separated into amide (*N*-acyl), monoesters of *N*-alkyl-glycamine (*O*-acyl), and amide esters of *N*-alkyl-glycamine (*N,O*-diacyl) by chromatography using a silica gel (60 H, Merck) column (30 cm × 20 mm).

3. The concentrated oil sample was diluted in a minimum volume of chloroform/ methanol (9/1, v/v) and was deposited at the top of the column previously equilibrated with chloroform/methanol (9/1, v/v).
4. The column was eluted with chloroform/methanol mixtures from 9/1 to 7/3 (v/v) for elution.
5. All the fractions obtained were analyzed by HPLC before structural analysis.
6. The solvent was evaporated by rotary evaporation.
7. The oily residue was rapidly triturated in ethyl ether and waxy, white solids were obtained.

3.5. Infrared Analysis of the Reaction Medium

Five hundred microliters of the hexane mixture at 55°C were injected into the analytical cell. Hexane was used as the reference (*see* **Note 1**).

4. Notes

1. *N*-methyl-glucamine is not soluble in apolar solvent such as hexane. It can be solubilised by ion-pair formation, in the presence of a fatty acid (e.g., oleic acid). This ion pair, identified by infrared spectroscopy, is essential for amide or ester synthesis. Its stability in hexane was also found to be the limiting factor of the reaction yield, which never exceeded 50% of acid conversion *(20)*.
2. The chemoselectivity of the reaction between oleic acid and *N*-methylglucamine toward amide or ester synthesis was under the control of the acid/ amine ratio. An excess of acid favored the ester synthesis. For an acid/amine ratio of 8, 100% of *N*-methyl-glucamine transformation was observed, yielding exclusively the C6 monoester derivative 6-*O*-oleoyl-*N*-methyl-glucamine. If the ratio was lower than 1 (excess amine), oleoyl-*N*-methyl-glucamide was the only product.
3. A more polar reaction solvent, *t*-amyl-alcohol, can be used to solubilize the *N*-methyl-glucamine substrate more efficiently. At 90°C and with a *N*-methylglucamine/acid ratio of 1, a 100% conversion yield with 97% of amide synthesis was obtained in <50 h. These results can be modified by changing the reaction conditions, especially the acid/*N*-methyl-glucamine concentration ratio, temperature and vacuum pressure *(21)*.
4. The influence of acid/*N*-methyl-glucamine molar ratio determines the protonation state of substrates and the protonation state of ionizable groups of the catalytic site, on which the enzyme activity is dependent. For additional detail, *see* **ref. 22**.
5. The transacylation reaction can be realized at 90°C and 500 mbar with an equimolar mixture containing fatty acid methyl esters (Colza fatty acid methyl esters) and *N*-methyl-glucamine as substrates. A 100% fatty acid methyl ester conversion was reached after 10 h reaction, with an amide yield of 80% *(23)*.
6. The process can be extended to various sources of amine (*N*-methylgalactamine, *N*-octyl-glucamine), and acyl donors (decanoic acid, lauric acid, palmitic acid, stearic acid, dimethyl adipate, and ethyl lactate) *(21)*.

References

1. Hildreth, J. (1982) *N*-D-Gluco-*N*-methylalkanamide compounds, a new class of non-ionic detergents for membrane biochemistry. *Biochem. J.* **207**, 363–366.
2. Matos, J. R., Blair West, J., and Wong, C. H. (1987) Lipase catalysed synthesis of peptides: preparation of a Penicillin G precursor and other peptides. *Biotechnol. Lett.,* **9**, 233–236.
3. Inada, Y., Nishimura, H., Takahashi, K., Yoshimoto, T., Ranjan Saha, A., and Saito, Y. (1984) Ester synthesis catalysed by polyethylene glycol modified lipase in benzene. *Biochem. Biophys. Res. Commun.* **122**, 845–850.
4. Zacks, A. and Klibanov, A. M. (1985) Enzyme catalysed processes in organic solvents. *Proc. Natl. Acad. Sci. USA* **82**, 3192–3196.
5. Margolin, A. L. and Klibanov, A. M. (1987) Peptide synthesis catalysed by lipases in anhydrous organic solvents. *J. Am. Chem. Soc.* **109**, 3802–3804.
6. West, J. B. and Wong, C. H. (1987) Use of nonproteases in peptide synthesis. *Tetrahedron Lett.* **28**, 1629–1632.
7. Montet, D., Pina, M., Graille, J., Renard, G., and Grimaud, J. (1989) Synthesis of *N*-lauryloleyl-amide by the *Mucor miehei* lipase in an organic medium. *Fatty Sci. Technol.* **1**, 14–18.
8. Bistline, G., Bilik, A., and Fearheller, S. H. (1991) Lipase catalysed formation of fatty amides. *J. Am. Oil Chem. Soc.* **68**, 95–98.
9. Tuccio, B. and Comeau, L. (1991) Lipase-catalysed synthesis of *N*-octyl-alkylamides in organic media. *Tetrahedron Lett.* **32**, 2763–2764.
10. Montet, D., Servat, F., Graille, J., Pina, M., Grimaud, J., Galzy, P., et al. (1990) Enzymatic synthesis of N-epsilon-acyllysines. *J. Am. Oil Chem. Soc.* **67**, 771–774.
11. Godtfredsen, S. and Björkling, F. (1990) An enzyme catalysed process for preparing *N*-acyl amino acids and *N*-acyl amino acid amides. International Patent Application No WO 90/14429.
12. Montet, D., Graille, J., Servat, F., Renard, G., and Marcou, I. (1989) Study of the acylation of aminopropanols catalysed by acyltransferases. *Rev. Fr. Corps Gras* **2**, 79–83.
13. Goldberg, M., Thomas, D., and Legoy, M. D. (1990) Water activity as a key parameter of synthesis reactions: the example of lipase in biphasic (liquid/solid) media. *Enzyme Microb. Technol.* **12**, 976–981.
14. Valivety, R. H., Halling, P., and Macrae, A. (1993) Water as a competitive inhibitor of lipase catalysed esterification in organic media. *Biotechnol. Lett.* **15**, 1133–1138.
15. Svensson, I., Wehtje, E., Adlercreutz, P., and Mattiasson, B. (1994) Effect of water activity on reaction rates and aquilibrium positions in enzymatic esterifications. *Biotechnol. Bioeng.* **44**, 549–556.
16. Halling, P. (1984) Effects of water on equilibria catalysed by hydrolytic enzymes in biphasic reaction systems. *Enzyme Microb. Technol.* **6**, 513–516.
17. Halling, P. (1994) Thermodynamic predictions for biocatalysis in nonconventional media: theory, tests and recommendations for experimental design and analysis. *Enzyme Microb. Technol.* **16**, 178–205.
18. Dordick, J. (1989) Enzymatic catalysis in monophasic organic solvents. *Enzyme Microb. Technol.* **11**, 194–211.

19. Valivety, R. H., Rakels, J. L. L., Blanco, R. M., Johnston, G. A., Brown, L., Suckling, C. J., et al. (1990) Measurement of pH changes in an inaccessible aqueous phase during biocatalysis in organic media. *Biotechnol. Lett.* **12,** 475–480.
20. Maugard, T., Remaud-Simeon, M., Petre, D., and Monsan, P. (1997) Lipase-catalysed chemoselective *N*-acylation of amino-sugar derivatives in hydrophobic solvent: acid-amine ion-pair effects. *Tetrahedron* **53,** 7587–7594.
21. Maugard, T., Remaud-Simeon, M., Petre, D., and Monsan, P. (1997) Enzymatic synthesis of glycamide surfactants by amidification reaction. *Tetrahedron* **53,** 5185–5194.
22. Maugard, T., Remaud-Simeon, M., and Monsan, P. (1998) Kinetic study of chemoselective acylation of amino-alditol by immobilised lipase in organic solvent: effect of substrate ionisation. *Biochim. Biophys. Acta* **1387,** 177–183.
23. Maugard, T., Remaud-Simeon, M., Petre, D., and Monsan, P. (1997) Lipase-catalysed synthesis of biosurfactants by transacylation of *N*-methyl-glucamine and fatty-acid methyl esters. *Tetrahedron* **53,** 7629–7634.

28

Synthesis of Esters Catalyzed by Lipases in Water-in-Oil Microemulsions

Haralambos Stamatis, Aristotelis Xenakis, and Fragiskos N. Kolisis

1. Introduction

There are two basic advantages in using enzymes as catalysts in organic media instead of aqueous solutions. First, organic solvents favor the solubility of hydrophobic substrates and, second, the presence of such solvents shifts the thermodynamic equilibrium of condensation/hydrolysis reactions in favor of the desired product. Different approaches have been proposed to facilitate the reversal of the normal hydrolytic action of enzymes. These include various macroheterogeneous biphasic systems such as liquid–liquid systems composed of a water-immiscible organic solvent and water, nearly anhydrous systems in which the enzyme is usually suspended as a powder or in an immobilized form adsorbed onto a suitable carrier in organic solvents or gases in a supercritical state, and various homogeneous and microheterogeneous media such as mixtures of water-miscible organic solvent and water as well as different types of microemulsion system (reverse micelles). The subject of enzyme catalysis in media with low water content has been reviewed by several authors (*1–6*).

Water-in-oil (w/o) microemulsions are thermodynamically stable, optically isotropic colloidal dispersions of water in oil (any apolar solvent) stabilized by surfactant molecules (*7*). The surfactant molecules are adsorbed spontaneously at interfaces and separate the nonpolar and aqueous phases, thus decreasing the interfacial tension down to very low values (10^{-5} dyn/cm). The optical transparency of the system is the result of one liquid being finely dispersed into the other, forming microdroplets (reverse

From: *Methods in Biotechnology, Vol. 15: Enzymes in Nonaqueous Solvents: Methods and Protocols*
Edited by: E. N. Vulfson, P. J. Halling, and H. L. Holland © Humana Press Inc., Totowa, NJ

micelles) with diameters typically of the order of 100 Å (*see* **Fig. 1**). The surfactant molecules used in the formation of microemulsions in apolar solvents include both natural membrane lipids and synthetic surfactants. **Table 1** provides some commonly used surfactants for enzymic studies.

The enzyme molecules can be entrapped in the water pools of the reverse micelles, avoiding direct contact with the organic solvent, potentially limiting their denaturation. The use of reverse micelles to solubilize enzymes (such as lipases) in organic solvents has attracted considerable interest in the past decade *(8,9)*.

The diameters of reverse micelles strongly depend on the molar ratio of water to surfactant. This ratio is generally expressed in terms of the parameter w_0 (also mentioned as R) as the ratio of the molarity of water to the molarity of surfactant in the system $w_0 = [H_2O]/[surfactant]$. For a wide variety of enzymes, a bell-shaped dependence of activity on w_0 has been observed *(8–10)*. It has been proposed that an optimum of enzyme activity occurs around a value of w_0 at which the size of the droplet is equal to the size of the enzyme molecule *(9)*. However, there is an ongoing discussion in the literature as to whether the water-content dependence of the enzyme activity is related to the size of the enzyme molecule *(8,11–13)*.

Microemulsion systems provide an enormous interfacial area (approx 100 m²/mL) through which the conversion of hydrophobic substrates can be catalyzed. Increasing the interfacial area is of great technological interest because this results in the increase of the number of substrate molecules available to react. Microemulsion systems offer several advantages as reaction media for bio-organic synthesis:

1. Both hydrophilic and hydrophobic substrates can be dissolved in high concentrations. The microemulsions represent a universal (i.e., an all-purpose microheterogeneous medium suitable for enzymatic reactions) *(9)*.
2. The substrates can be enzymatically converted with high yields because the thermodynamic equilibrium of condensation/hydrolysis reactions can be easily shifted by adjusting the water content. It is interesting to note that the use of reverse micelles succeeded in changing the equilibrium constants of various enzymatic reactions by a factor of 10^6 *(11)*.
3. The water content can be varied within a fairly broad range and the water content can be used as a tool to manipulate enzyme activity.
4. Multienzymatic reactions are feasible in reverse micelles.

The biotechnological relevance of enzymes in reverse micelles for the transformation of various substrates has been demonstrated by several authors. Some examples of enzymatic reactions in microemulsions are reverse hydrolytic reactions, such as peptide synthesis *(14)*, esterifications and transesterifications *(6,15)*, and oxidation and reduction of steroids *(16,17)*. A particular case of enzymatic studies in microemulsions is that of lipases, which act almost exclusively near interfaces in a classical heterogeneous procedure.

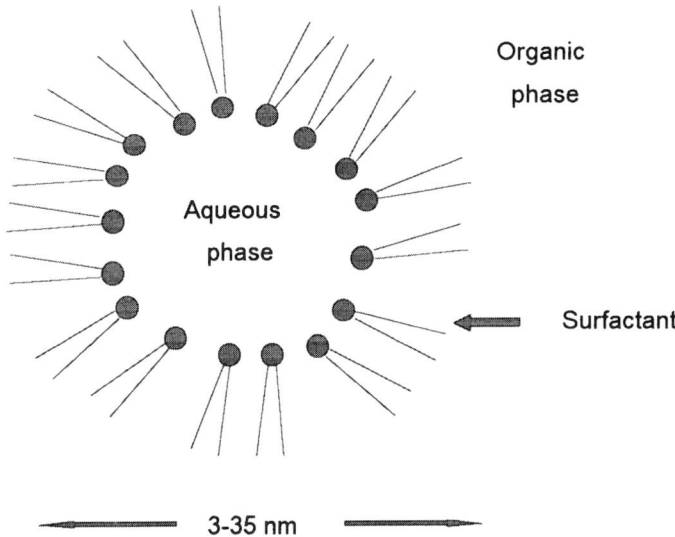

Fig. 1. Schematic presentation of reverse micelle.

Table 1
Commonly Used Surfactants for the Formation of w/o Microemulsions

Surfactants	Solvent systems
Bis-(2-ethylhexyl) sulfosuccinate sodium salt (AOT)	n-Hydrocarbons (C_6–C_{16}) Isooctane Cyclohexane
Cetyltrimethyl ammonium bromide (CTAB)	Heptane:chloroform n-Hydrocarbon:primary alcohols
Polyethylene glycol monododecylethers, $C_{12}E_n$ ($n = 3$–5)	n-Hydrocarbons (C_6–C_{12})
Phoshatidylcholine	n-Hydrocarbons (C_6–C_{12})
Phosphatidylethanolamine	

One of the most important problems that must be solved for the employment of a microemulsion system in industrial processes is the recovery of the products from the reaction mixture and the regeneration of the enzyme. The presence of surfactant leads to a poor separation by normal techniques, such as extraction and distillation, because of the problems of emulsion-forming and foaming caused by the surfactant. Larsson et al. *(18)* proposed a simple technique for enzyme reuse and product recovery, taking advantage of the phase behavior of the surfactant system to separate the microemulsion constituents into an oil-rich phase and a water-rich phase that contain almost all of the surfactant. By changing the temperature, a biphasic system was formed, where an oil-rich phase containing the product coexists with a water-rich phase containing the surfactant and the enzyme. The oil-rich phase may be replaced by a new solution of substrate in oil; then. the temperature is brought back to the level where the monophasic microemulsion is stable, and the reaction is repeated. A similar separation procedure (*see* **Fig. 2**), have been proposed by our group consisted of (1) a liquid–liquid extraction of the enzyme from the micellar phase into a new aqueous phase and (2) the equilibration of the remaining organic phase with an additional aqueous volume in order to separate the product from the surfactant. By this procedure, a significant amount of active lipase was extracted and the separation of the product from the surfactant system was almost complete *(19)*.

We have successfully used water-in-oil microemulsion system as a reaction media for the lipase-catalyzed esterification of both hydrophilic *(20,21)* and hydrophobic substrates *(6,21)*. The results of these experiments demonstrate that the present method should have quite general applicability for lipase-catalyzed synthetic reactions.

2. Materials

1. Lipase from *Penicillium simplicissimum* (from GBF, Germany), with a specific activity of 142 units/mg of protein (determined using triolein as the substrate) in 50 mM acetate buffer, pH 5.5.
2. Lipase from *Rhizopus delemar* (Fluka, Basel, Switzerland). The crude enzyme preparation was solubilized in a pH 5.5 buffer containing 20 mM acetate, 20 mM NaCl, 20 mM CaCl$_2$, and 1 mM NaN$_3$. The enzyme was precipitated by 70% (NH$_4$)$_2$SO$_4$ solution. After centrifugation, it was resolubilized in the same buffer and applied on a Sephadex G-100 gel filtration column and equilibrated with the above-mentioned buffer. The partially purified enzyme had a specific activity of 930 units/mg of protein, using tributyrin as the substrate.
3. Lipolase™ (Novo Nordisk A/S, Denmark), in the form of a straw-colored aqueous solution and was dialyzed before used at 4°C, against 25 mM Tris-HCl, pH 8.0. The enzyme exhibited a specific activity of 1350 units/mg of protein using *p*-nitrophenyl palmitate as the substrate.
4. Bis-(2-ethylhexyl)sulfosuccinate sodium salt (AOT) 99% pure (Sigma).
5. Isooctane, aliphatic alcohols, diols, glycerol, and fatty acids (Sigma).

3. Method
3.1. Esterification of Aliphatic Alcohols with Fatty Acids

1. Prepare a stock isooctane solution containing 100 mM AOT, 100 mM of alcoholic substrate (e.g., butanol, hexanol, octanol, geraniol, cholesterol), and 50 mM of fatty acid (e.g., lauric, myristic, palmitic, etc.).

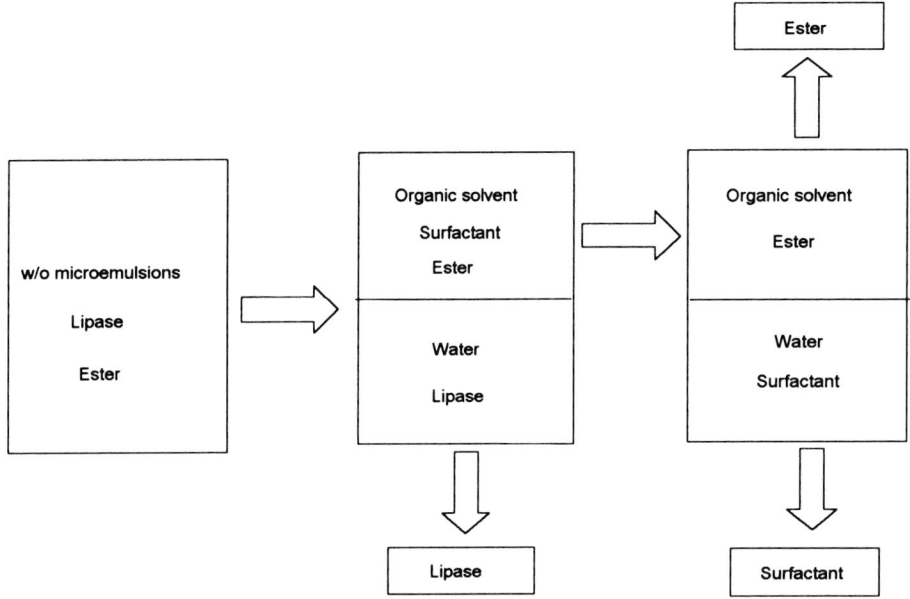

Fig. 2. Separation procedure of the enzyme and product recovery after completing an esterification reaction in a microemulsion system.

2. In a screw-cup reaction vessel, put 2 mL of the iso-octane solution and add 25 μL of 20 m*M* acetate (pH 5.5) buffer solution (*see* **Note 1**).
3. Vortex the mixture for 30–60 s until a clear solution (w/o microemulsion) formed.
4. Place the tube in a thermostable water bath, at 30°C for 10 min.
5. Add 10 μL of aqueous lipase solution (*Penicillium simplicissimum* or *Rhizopus delemar*) and mix the reaction mixture in vortex for 30 s (*see* **Notes 2** and **3**).
6. At various time intervals (total period 1–2 h), take aliquots of 1–2 μL and analyze them by gas chromatography (GC).

3.2. Esterification of Polar Diols with Fatty Acids

1. Prepare a stock iso-octane solution containing 100 m*M* AOT, 100 m*M* of a diol substrate (ethylene glycol, propane-1,3-diol, butane-1,4-diol, or pentane-1,4-diol), and 50 m*M* fatty acid (e.g., lauric, myristic, palmitic, and so forth) (*see* **Note 4**).
2. In a screw-cup reaction vessel, put 2 mL of the iso-octane solution and add 25 μL of a 25-m*M* Tris-HCl buffer, pH 8.0.
3. Vortex the mixture until a clear solution (w/o microemulsion) forms.
4. Place the tube in a thermostable water bath, at 25°C for 10 min.
5. Add 10 μL of aqueous lipase solution (Lipolase) and mix the reaction mixture in vortex for 30–60 s.
6. At various time intervals (total period 1–2 h), take aliquots of 1–2 μL and analyze them by GC (*see* **Note 5**).

3.3. Ester Recovery

1. After completing the esterification reaction, add on the micellar solution (microemulsion) an equivalent volume of a 25-mM Tris-HCl (pH 9.0) buffer solution containing different 100 mM NaCl. (*See* **Note 6**.)
2. Mix the two-phase system on a rotary shaker for 5 min at 350 rpm.
3. Separate the two phases by centrifugation for 10 min at 5000g.
4. Add on the upper organic phase (approx 5 mL) containing the product, the surfactant as well as some unreacted substrates), 3.5 mL fresh aqueous solution containing 5 mM Tris-HCl at pH 7.3. (*see* **Note 7**).
6. Incubate the mixture at 35°C for 30–60 min in order to separate the two phases (aqueous and organic).
7. Determine the ester and AOT concentrations in the oil top phase with GC.

3.4. Gas Chromatographic Analysis

Analysis of products is achieved by employing gas chromatography. Aliquots of the reaction mixture are withdrawn at selected time intervals and diluted four times with chloroform before analyzed. For the determination of the concentration of the fatty acid esters as well as AOT, a Perkin-Elmer 8500 chromatographer is used, equipped with a 2-m glass packed column, loaded with GP 5%-DEGS-PS (Supelco). The carrier gas is nitrogen at a flow rate of 15 mL/min, the oven temperature is kept at 180°C, and a flame-ionization detector (FID) detector is used. For the determination of diols monoesters and diesters a 3-ft glass packed column, loaded with 1%-Dexsil 300 (Supelco) and an FID detector are used. Nitrogen is used as the carrier gas at a flow rate of 20 mL/min, with detector port temperature at 300°C. The oven temperature is kept constant for 2 min at 185°C, linearly increased (10°C/min) up to 300°C.

4. Notes

1. The final w_o value ($w_o = $ [H$_2$O]/[AOT]), which affects the lipase catalytic behavior, depends on the water content of the system. The water content was adjusted by the addition of the required amount of buffer. The water content of the system as indicated in method (25 μL of buffer + 10 μL of lipase solution per 2 mL of microemulsion) corresponds to a w_0 =10.
2. The bioconversions were started by adding the lipase solution to the reaction mixture.
3. *Penicillium simplicissimum* lipase shows higher reaction rates in the esterification of long chain alcohols as well as secondary alcohols. *Rhizopus delemar* lipase shows a preference for the esterification of short-chain primary alcohols, whereas the secondary alcohols had a low rate of esterification and the tertiary ones could not be converted (*6*).
4. The esterification of various hydrophilic diols with fatty acids catalyzed by Lipolase in w/o microemulsions leads to synthesis of a monoester and a diester,

which are highly valuable products for industrial purposes *(20)*. The desired reaction was:

$$RCOOH + OH\text{-}(CH_2)_n\text{-}OH \longrightarrow RCOO\text{-}(CH_2)_n\text{-}OH + RCOO\text{-}(CH_2)_n\text{-}OOCR$$
$$(n=2, 3, 4, 5)$$

5. The maximum monoester formation is observed after 1–2 h incubation, when the overall conversion of fatty acid was 40–45%. Moreover, the product distribution can be composed almost quantitatively of the corresponding monoester if the reaction is stopped when the fatty acid conversion is lower than 25% *(20)*.
6. It is observed that when the pH value of aqueous phase is increased, an increase in lipase recovery yield is obtained. In the case of *R. delamar* lipase, the recovery yield is about 90% when the pH of the aqueous phase is 8.0–9.0 and 100 mM salt is used. This observation can be attributed to the electrostatic repulsive interactions between the negatively charged protein and surfactant molecules, at pH values of the aqueous phase beyond the isoelectric point of the protein. As a result of such interactions, the protein exhibits a decreased binding tendency to the surfactant molecules under alkaline conditions *(22)*.
7. The separation of a hydrophobic ester from the surfactant system was based on the phase behavior of such systems (i.e., the ability of the surfactant molecules to provoke, under certain conditions, the separation of the homogeneous system to an oil-rich phase and an aqueous phase). This ability depends on the incubation temperature of the system and also on the buffer-to-oil ratio *(19)*.

References

1. Ballesteros, A., Bornscheuer, U., Capewell, A., Combes, D., Condoret, J.-S., Koening, K., et al. (1995) Review article: enzymes in nonconventional phases. *Biocatal. Biotransform.* **13,** 1–42.
2. Dodrick, J. S. (1989) Enzymatic catalysis in monophasic organic solvent. *Enzyme Microb. Technol.* **11,** 194–211.
3. Khmelnitsky, Y. L., Levashov, A. V., Klyachko, N. L., and Martinek K. (1988) Engineering biocatalytic systems in organic media with low water content. *Enzyme Microb. Technol.* **10,** 710–724.
4. Klibanov, A. M. (1989) Enzymatic catalysis in anhydrous organic solvents. *Trends Biochem. Sci.* **14,** 141–144.
5. Stamatis, H., Xenakis, A. and Kolisis, F. N. (1999) Bioorganic reactions in microemulsions: the case of lipase. *Biotechnol. Adv.,* in press.
6. Stamatis, H., Xenakis, A., Provelegiou M., and Kolisis, F. N. (1993) Esterification reactions catalyzed by lipases in microemulsions. The role of enzyme localization in relation to its selectivity. *Biotechnol. Bioeng.* **42,** 103–110.
7. Eicke, H. F. and Rehak, J. (1976) On the formation of water/oil microemulsions. *Helv. Chim. Acta* **59,** 2883–2891.
8. Luisi, P. L. and Magid, L. (1986) Solubilization of enzymes and nucleic acids in hydrocarbon micellar solutions. *CRC Crit. Rev. Biochem.* **20,** 409–474.

9. Martinek, K., Levashov, A. V., Klyachko, N. L., Khmelnitsky, Y. L., and Berezin, I. V. (1986) Micellar enzymology. *Eur. J. Biochem.* **155,** 453–468.

10. Maestro, M. (1989) Enzymatic activity in reverse micelles—some modellistic considerations on bell-shaped curves. *J. Mol. Liquids* **42,** 71–82.

11. Verhaert, R. M. D. and Hilhorst, R. (1991) Enzymes in reverse micelles: 4. Theoretical analysis of a one-substrate/one-product conversion and suggestions for efficient application. *Recl. Trav. Chim. Pays-Bas* **110,** 236–246.

12. Sanchez-Ferrer, A., Perez-Gilabert, M., and Garcia-Carmona, F. (1992) Protein-interface interactions in reverse micelles, *Biocatalysis in Non-Conventional Media*, Progress in Biotechnology vol. 8 (Tramper, J., Vermue, M. H., Beeftink, H. H., and von Stockar, U., eds.), Elsevier, Amsterdam, pp. 181–188.

13. Otero, C., Rua, M. L., and Robledo, L. (1995) Influence of the hydrophobicity of lipase isoenzymes from *Candida rugosa* on its hydrolytic activity in reverse micelles. *FEBS Lett.* **360,** 202–206.

14. Luithi, P. and Luisi, P. L. (1984) Enzymatic synthesis of hydrocarbons—soluble peptides with reverse micelles. *J. Am. Chem. Soc.* **106,** 7285,7286.

15. Hayes, D. G. and Gulari, E. (1990) Esterification reactions of lipase in reverse micelles. *Biotechnol. Bioeng.* **35,** 793–801.

16. Smolders, A. J. J., Pinheiro, H. M., Noronha, P., and Cabral, J. M. S. (1991) Steroid bioconversion in a microemulsion system. *Biotechnol. Bioeng.* **38,** 1210–1217.

17. Hilhorst, R., Spruijt, R., Laane, C., and Verger, C. (1984) Rules for the regulation of enzyme activity in reversed micelles as illustrated by the conversion of apolar steroids by 20-b-hydroxysteroid dehydrogenase. *Eur. J. Biochem.* **144,** 459–462.

18. Larsson, K. M., Adlercreutz, P., and Mattiasson, B. (1990) Ezymatic catalysis in microemulsions: enzyme reuse and product recovery. *Biotechnol. Bioeng.* **36,** 135–141.

19. Stamatis, H., Xenakis, A., and Kolisis, F. N. (1995) Studies on enzyme reuse and product recovery in lipase catalyzed-reactions in microemulsions. *Ann. NY Acad. Sci.* **750,** 237–241.

20. Stamatis, H., Xenakis, A., Dimitriadis, E., and Kolisis, F. N. (1995) Catalytic behavior of *Pseudomonas cepacia* lipase in w/o microemulsions. *Biotechnol. Bioeng.* **45,** 33–41.

21. Stamatis, H., Macris, J., and Kolisis, F. N. (1996) Esterification of hydrophilic diols catalysed by lipases in microemulsions. *Biotechnol. Lett.* **18,** 541–546.

22. Stamatis, H. (1996) Enzymatic modification of amphiphilic substrates in microemulsions. Ph.D. thesis, University of Patras, Greece.

29

Enzymatic Conversion of Organosilicon Compounds in Organic Solvents

Takuo Kawamoto and Atsuo Tanaka

1. Introduction

Recently, the bioconversion of non-natural organic compounds has become increasingly important in order to deepen the knowledge of biocatalysts and to expand their application. In this sense, organosilicon compounds are very interesting targets as non-natural compounds.

Silicon, which is an abundant element in the Earth's crust, belongs to the same group as carbon, which is one of the most fundamental elements to organisms, but is different from carbon in its characteristics *(1)*. Silicon has a lower electronegativity and a larger covalent radius than carbon (**Table 1**). The bonds of silicon to oxygen and fluorine are stronger than the corresponding bonds between carbon and these elements, whereas its bonds to carbon and hydrogen are weaker. Although both silicon and carbon participate in sp^3 bonding, silicon also has vacant *d*-orbitals in the outer electronic configuration. Organosilicon compounds, which are compounds having Si-C bonds, are very useful in synthetic organic chemistry, and the usefulness of organosilicon compounds is mainly attributable to these features of silicon.

Silicon also plays an important role in the biosphere. Silicate bacteria, the simplest algae, and spore plants contain very large amount of silicon *(2,3)*. It has been also indicated that silicon is important for higher plants, animals, and man *(4)*, although its actual biochemical function is, at present, not unambiguously defined. It seems, however, that natural biochemical processes mainly involve silicon bound to oxygen, whereas organosilicon compounds have not been detected *(1–3)*. Therefore, the effects of organosilicon compounds on organisms and/or enzymes are of great interest.

From: *Methods in Biotechnology, Vol. 15: Enzymes in Nonaqueous Solvents: Methods and Protocols*
Edited by: E. N. Vulfson, P. J. Halling, and H. L. Holland © Humana Press Inc., Totowa, NJ

Table 1
Physical Properties of Silicon and Carbon Atoms

Atom	Electronegativity	Bond	Bond length
C	2.55	C–C	0.153
Si	1.90	C–Si	0.189

The progress of organosilicon chemistry has created a growing awareness of its considerable utility to organic chemists and has shown that wide spectra of organosilicon compounds often exhibit interesting biological properties *(2,3)*. Pharmaceutical application of organosilicon chemistry has also been gaining attention *(2,3,5–7)*.

Therefore, a combination of organosilicon chemistry with biochemistry is a very attractive new field. In particular, the production of organosilicon compounds with desired chemical and physical properties by the use of effective catalysts such as biocatalysts is of great importance. Because many organosilicon compounds are hydrophobic and/or water-insoluble, the use of organic solvents is desirable. In this chapter, the conversion of organosilicon compounds by such enzymes as hydrolases in organic solvents is described. Hydrolases do not require coenzymes and, consequently, can be easily handled. Furthermore, hydrolases are readily available from commercial sources and constitute a broadly applicable class of enzymes for reactions in organic solvent systems.

1.1. Organosilicon Compounds as Novel Substrates

Comparative studies were made by the use of organosilicon compounds having a different chain length between hydroxyl group and the silicon atom [$Me_3Si(CH_2)_nOH$, $n = 0, 1, 2$, and 3] and the corresponding carbon compounds [$Me_3C(CH_2)_nOH$, $n = 0, 1$, and 2) as another substrate (acyl acceptor) in the enantioselective esterification of 2-(4-chlorophenoxy)propanoic acid, whose (*R*)-enantiomer is useful as a herbicide, with lipase OF 360 of *Candida cylindracea* (**Fig. 1**) in benzene. The effects of the silicon atom on the enzymatic reaction in connection with the distance between the hydroxyl group and the silicon atom are discussed *(8)*. Organosilicon compounds and the carbon counterparts examined in this study, except for the case of $n = 0$ (trimethylsilylmethanol and 1,1-dimethylethanol), served as the substrates. In particular, trimethylsilylmethanol ($n = 1$) was found to be a particularly superior substrate, that is, the reaction rate was much higher than that with the corresponding carbon compound, and the enantiomeric excess of the acid remaining was also higher at about 50% conversion. In the case of conventional substrates such as the carbon analogs and linear-chain alcohols *(9)*, a

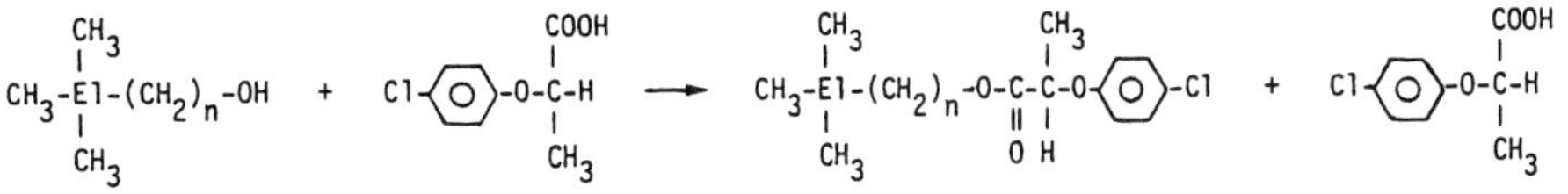

Fig. 1. Lipase-catalyzed enantioselective esterification of 2-(4-chlorophenoxy) propanoic acid with organosilicon compounds and the corresponding carbon analogs in benzene *(8)*. El = Si or C.

high enantioselectivity was not consistent with a high reaction rate. These results indicate that organosilicon compounds as novel substrates may solve some of the unavoidable and probably inherent problems of enantioselective reactions with conventional substrates. On the other hand, a difference was not observed between 2-trimethylsilylethanol ($n = 2$) and its carbon analog, 3,3-dimethylbutanol, in the enzymatic activity and enantioselectivity. Thus, the silicon atom mimicked the carbon atom for lipase in the case of 2-trimethylsilylethanol ($n = 2$) but made trimethylsilylmethanol ($n = 1$) an excellent substrate for enantioselective esterification with lipase in an organic solvent (*see* **Notes 1** and **2**). These phenomena were explained on the basis of the properties of the silicon atom, such as its lower electronegativity and larger atomic radius compared with carbon. A lower electronegativity of the silicon atom gives rise to a higher nucleophilicity of the oxygen atom of trimethylsilylmethanol ($n = 1$) compared with that of the corresponding carbon analog, and the hydroxyl group of trimethylsilylmethanol ($n = 1$) is less sterically hindered because of the longer bond of Si-C than that of C-C in the corresponding carbon analog (**Fig. 2**). Trimethylsilylmethanol ($n = 1$) is, therefore, more easily accessible to the acyl-enzyme intermediate and reacts as an much better acyl acceptor than the carbon analog. In the case of trimethylsilylethanol ($n = 2$), the favorable effect of the silicon atom previously mentioned was negligible because of the presence of a long ethylene group between the silicon atom and the hydroxyl group. Trimethylsilylethanol ($n = 2$) was, consequently, regarded as a substrate similar to the corresponding carbon compound by lipase. In spite of the favorable characteristics of the silicon atom, trimethylsilanol ($n = 0$) did not serve as a substrate of lipase, similar to the corresponding tertiary alcohol.

As far as we know, this work is the first study where the effects of the silicon atom in substrates for enzymatic reactions have been systematically discussed. Furthermore, this study demonstrated for the first time the ability of organosilicon compounds to break the limit of conventional substrates owing

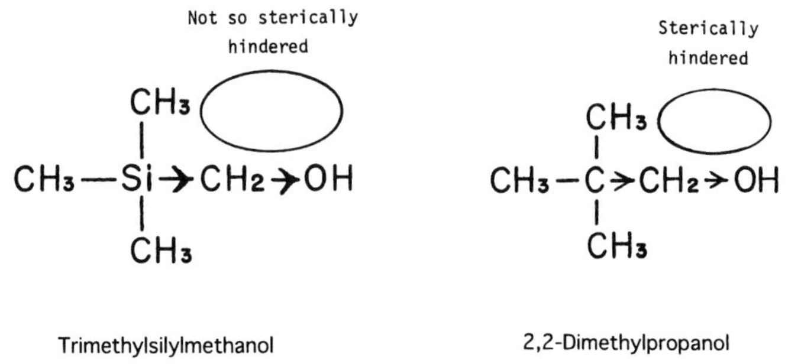

Not so sterically
hindered

$$CH_3—Si→CH_2→OH$$

Trimethylsilylmethanol

Sterically
hindered

$$CH_3—C→CH_2→OH$$

2,2-Dimethylpropanol

Fig. 2. Effects of the silicon atom in trimethylsilylmethanol ($n = 1$) on lipase-catalyzed esterification *(8)*.

to their unique characteristics. According to the above explanation, (aminomethyl)trimethylsilane ($Me_3SiCH_2NH_2$) can also be expected to be a better acyl acceptor than the corresponding carbon compound, 2,2-dimethylpropylamine ($Me_3CCH_2NH_2$), for hydrolase-catalyzed amide synthesis in organic solvents. In fact, (aminomethyl)trimethylsilane was a better substrate for such hydrolases as lipase OF 360 *(C. cylindracea)*, lipase KLIP-100 (*Pseudomonas* sp.), lipoprotein lipase Type A (*Pseudomonas* sp.) (*see* **Note 3**), and cholesterol esterase Type A (*Pseudomonas* sp.) than the corresponding carbon compound in amide synthesis with octanoic acid in isooctane *(10)*.

1.2. Optically Active Organosilicon Compounds

Preparation of optically active organosilicon compounds by biocatalysts is also of interest and importance in synthetic organic chemistry (*see* **Notes 4–7**). Three isomers of trimethylsilylpropanols (**Fig. 3**), 1-trimethylsilyl-2-propanol (**1a**), 1-trimethylsilyl-1-propanol (**2a**), and 2-trimethylsilyl-1-propanol (**3a**), were used as model compounds for studies of the enantioselective esterification of organosilicon compounds by hydrolases with 5-phenylpentanoic acid in isooctane *(11)*. Compared with the corresponding carbon analogs (**1b**, **2b**, and **3b**) (**Fig. 3**), it was found that the silicon atom in the substrates generally enhanced the enzyme enantioselectivity. Although it is generally difficult to perform the highly enantioselective conversion of primary alcohols, the efficient optical resolution of a primary alcohol, 2-trimethylsilyl-1-propanol (**3a**), was achieved with 95% enantiomeric excess (ee) at 50 % conversion by the enantioselective esterification with lipase Saiken 100 from *Rhizopus japonicus*. It is also worth noting that the β-hydroxyalkylsilanes, **1a** and **3a**, were effectively esterified by hydrolases (*see* **Note 8**), because they were unstable and easily converted to alkenes via β-elimination under both acidic and basic conditions (Peterson olefination) *(1)*; consequently, it is difficult to esterify them by chemical catalysts such as acids. This is a good

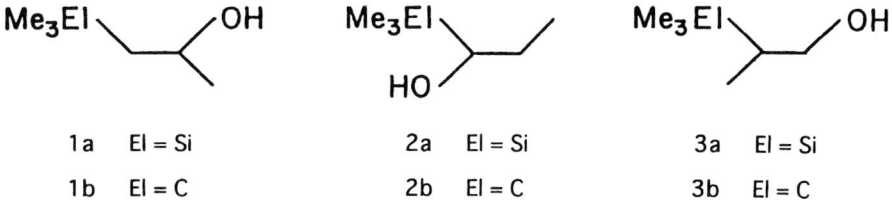

1a   El = Si

1b   El = C

2a   El = Si

2b   El = C

3a   El = Si

3b   El = C

Fig. 3. Three isomers of trimethylsilylpropanols and their carbon analogs used in hydrolase-catalyzed enantioselective esterification in isooctane.

example showing the effectiveness of introduction of biochemical methods into organosilicon chemistry.

Optical resolution of organosilicon compounds having a stereogenic silicon atom is also an important target for the application of enzymes. Furthermore, it is interesting to find out whether enzymes can recognize chirality at a silicon atom. Such chiral silanes are expected to be new materials, synthetic reagents *(12)* or biologically active compounds *(2)*. When $PhSi(Me)(Et)CH_2OH$ *(4)* **(Fig. 4)** was esterified with 5-phenylpentanoic acid in isooctane, hydrolases exhibited a good esterification activity but low enantioselectivity. Among these hydrolases examined, a commercial preparation of crude papain gave the best results (i.e., a moderate reaction rate and a high enantioselectivity [*see* **Note 9**]) *(13)*. Consequently, the enantioselective esterification of several primary alcohols, hydroxyalkylsilanes, having a stereogenic silicon atom (*see* **Notes 4–12**) was attempted **(Fig. 4)** with 5-phenylpentanoic acid by crude papain in water-saturated isooctane, and the effects of chain length between the silicon atom and the hydroxyl group (*see* **Notes 4–6**), and the substitution groups on the silicon atom (*see* **Notes 4,7–12**) were studied *(13)*. A short methylene chain between the silicon atom and the hydroxyl group, and a phenyl substitution on the silicon atom were found to be essential for high activity and high enantioselectivity in the crude papain-catalyzed optical resolution of these organosilicon compounds. The highly optically enriched $(+)-PhSi(Me)(Et)CH_2OH$ (*see* **Note 4**) (92% ee), $(+)-PhSi(Me)(Pr)CH_2OH$ (*see* **Note 7**) (93% ee), $(+)-p-Me-PhSi(Me)(Et)CH_2OH$ (*see* **Note 9**) (96% ee), and $(+)-p-F-PhSi(Me)(Et)CH_2OH$ (*see* **Note 10**) (99% ee) were obtained in 42%, 42%, 41%, and 41% yield from the racemic substrates, respectively.

1.3. Silicon-Containing Peptides

Peptides are very important compounds for organisms, and silicon-containing peptides are novel organosilicon compounds and are expected to have novel biological activity. For peptide synthesis, proteases are recognized as powerful catalysts. Therefore, synthesis of silicon-containing peptides by proteases is very interesting.

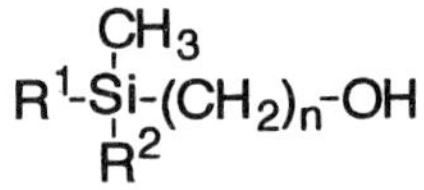

$$R^1\text{-}\underset{R^2}{\overset{CH_3}{\underset{|}{\overset{|}{Si}}}}\text{-}(CH_2)_n\text{-}OH$$

	R^1	R^2	n
4	Ph	Et	1
5	Ph	Et	2
6	Ph	Et	3
7	Ph	Pr	1
8	Ph	$n\text{-}C_6H_{13}$	1
9	p-Me-Ph	Et	1
10	p-F-Ph	Et	1
11	$n\text{-}C_6H_{13}$	Et	1
12	Pr	Et	1

Fig. 4. Organosilicon compounds having stereogenic silicon atoms employed in hydrolase-catalyzed enantioselective esterification in isooctane.

Thermolysin-catalyzed synthesis of silicon-containing peptides with 3-trimethylsilylalanine (TMS-Ala) (*see* **Note 10**), which is thought to be a useful silicon-containing amino acid as an analog of Leu, was attempted in ethyl acetate by Ishikawa et al. *(14)*. Considering the structure similarity of TMS-Ala to Leu, Leu-Leu analogs, (TMS-Ala)-Leu, Leu-(TMS-Ala), and (TMS-Ala)-(TMS-Ala), were selected as targets, and the effects of the introduction of TMS-Ala instead of Leu on enzymatic peptide synthesis were investigated. Leu-Leu is known to be an inhibitor of several zinc proteases *(15)* and can be used as the precursor of a bioactive peptide, leupeptin (Ac-Leu-Leu-Arg-al) *(16)*. It was found in this study that benzyloxycarbonyl (Z)-TMS-Ala was recognized as a better substrate by thermolysin than Z-Leu, and Z-(TMS-Ala)-Leu-OMe could be efficiently synthesized by thermolysin in ethyl acetate (*see* **Notes 11** and **12**). The acceleration of the reaction rate in the synthesis of Z-(TMS-Ala)-Leu-OMe compared with the case of Z-Leu-Leu-OMe was explained by the higher hydrophobicity of the trimethylsilyl group. On the other hand, TMS-Ala-OMe was not accepted as the amino component because of the bulkiness of the trimethylsilyl group.

As for the inhibitory activity of the silicon-containing dipeptides toward thermolysin, Leu-(TMS-Ala) was found to be a more potent inhibitor than Leu-

Leu, probably because of the higher hydrophobicity of trimethylsilyl group *(17)*, demonstrating that organosilicon compounds such as silicon-containing amino acids and/or silicon-containing peptides have great possibilities for application as bioactive compounds.

2. Materials

2.1. Organosilicon Compounds as Novel Substrates of Hydrolases in Organic Solvents

2.1.1. Esterification

1. Organosilicon compounds: trimethylsilanol (Shin-Etsu Kagaku Kogyo, Tokyo, Japan), trimethylsilylmethanol (Shin-Etsu Kagaku Kogyo), 2-trimethylsilylethanol (Aldrich, Milwaukee, WI), and 3-trimethylsilylpropanol (Aldrich).
2. Carbon analogs (for comparison): 1,1-dimethylethanol (Wako Pure Chemical Industries, Osaka, Japan), 2,2-dimethylpropanol (Nacalai Tesque, Kyoto, Japan), and 3,3-dimethylbutanol (Aldrich).
3. (*R,S*)-2-(4-Chlorophenoxy) propanoic acid (Aldrich).
4. Benzene saturated with deionized water.
5. Celite No. 535 (Wako Pure Chemical Industries).
6. Lipase OF 360 from *Candida cylindracea* (Meito Sangyo, Tokyo, Japan) *(see* **Note 1**).
7. Silica gel 60 (Merck, Darmstadt, Germany).

2.1.2. Amide Synthesis

1. Organosilicon compound: (aminomethyl)trimethylsilane (Shin-Etsu Kagaku Kogyo).
2. Carbon analog (for comparison): 2,2-dimethylpropylamine (Nacalai Tesque.
3. Octanoic acid.
4. Water-saturated isooctane.
5. Celite No. 535 (Wako Pure Chemical Industries).
6. Hydrolases: lipase OF 360 from *Candida cylindracea* (Meito Sangyo), lipase KLIP-100 from *Pseudomonas* sp. (Kurita Kogyo, Tokyo, Japan), lipoprotein lipase Type A from *Pseudomonas* sp. (Toyobo, Osaka, Japan), and cholesterol esterase Type A from *Pseudomonas* sp. (Toyobo).

2.2. Enzymatic Preparation of Optically Active Organosilicon Compounds in Organic Solvents

2.2.1. Preparation of Optically Active Trimethylsilylpropanols

1. Trimethylsilylpropanols (donated by Nitto Denko, Osaka, Japan): (*R,S*)-1-trimethylsilyl-2-propanol (**1a**), (*R,S*)-1-trimethylsilyl-1-propanol (**2a**), and (*R,S*)-2-trimethylsilyl-1-propanol (**3a**).
2. Carbon analogs (for comparison): 4,4-dimethyl-2-pentanol (**1b**) (Aldrich), 2,2-dimethyl-3-pentanol (**2b**) (Aldrich), and 2,3,3-trimethyl-1-butanol (**3b**) *(see* **Note 13**).
3. 5-Phenylpentanoic acid.

4. Water-saturated isooctane.
5. Celite No. 535 (Wako Pure Chemical Industries).
6. Hydrolases: lipase OF 360 from *Candida cylindracea* (Meito Sangyo), lipase Saiken 100 from *Rhizopus japonicus* (Osaka Saikin Kenkyusho, Osaka, Japan), lipase (Steapsin) from hog pancreas (Tokyo Kasei, Tokyo, Japan), lipoprotein lipase Type A from *Pseudomonas* sp. (Toyobo), and cholesterol esterase Type A from *Pseudomonas* sp. (Toyobo).
7. Sumichiral OA-4600 column (4.0 × 250 mm; Sumika Chemical Analysis Service, Osaka, Japan)
8. 3,5-Dinitro isocyanate (Sumika Chemical Analysis Service).
9. Mobil phase for high-performance liquid chromatography (HPLC): *n*-hexane/2-propanol (100:1 v/v) or *n*-hexane/2-propanol (98:2 v/v).
10. Silica gel 60 (Merck).
11. (*R*)-Methoxytrifluoromethylphenylacetic acid [(*R*)-MTPA] (Nacalai Tesque).

2.2.2. Preparation of Optically Active Silylmethanol Derivatives Having a Stereogenic Silicon Atom

1. Racemic alcohols having a stereogenic silicon atom (*see* **Note 14**): ethylmethylphenylsilylmethanol (*see* **Note 4**), 2-(ethylmethylphenylsilyl)ethanol (*see* **Note 5**), 3(ethylmethylphenylsilyl)propanol (*see* **Note 6**), methylphenyl-*n*-propylsilylmethanol (*see* **Note 7**), *n*-hexylmethylphenyl-silylmethanol (*see* **Note 8**), ethylmethyl (*p*-methylphenyl)silylmethanol (*see* **Note 9**), ethyl(*p*-fluorophenyl)methylsilylmethanol (*see* **Note 10**), ethyl-*n*-hexylmethylsilylmethanol (*see* **Note 11**), and ethylmethyl-*n*-propylsilylmethanol (*see* **Note 12**).
2. 5-Phenylpentanoic acid.
3. Water-saturated isooctane.
4. Celite No. 535 (Wako Pure Chemical Industries).
5. Papain (crude powder) from *Papaya latex* (Sigma, St. Louis, MO).
6. Silica gel 60 (Merck)
7. Sumichiral OA-4600 column (4.0 mm × 250 mm, Sumika Chemical Analysis Service).
8. 3,5-Dinitrophenyl isocyanate (Sumika Chemical Analysis Service).
9. Mobil phase for HPLC: *n*-hexane/2-propanol (98:2 v/v).
10. (*S*)-Cyanofluorophenylacetic acid [(*S*)-CFPA] (*see* **Note 15**).

2.3. Enzymatic Synthesis of Silicon-Containing Peptides in Organic Solvents

Amino acids (AAs) and their derivatives used are L-form unless otherwise indicated.

1. Benzyloxycarbonyl (Z)-3-trimethylsilylalanine (Z-TMS-Ala) (*see* **Note 10**) or Z-Leu (Kokusan Chemical Works, Osaka, Japan).
2. TMS-Ala-OMe - HCl (*see* **Note 10**) or Leu-OMe - HCl (Kokusan Chemical Works).
3. Chloroform.
4. Ethyl acetate.
5. 50 mM 2-(*N*-Morpholino)ethanesulfonic acid (MES)/NaOH buffer (pH 6.5) containing 5 mM CaCl$_2$.

6. Celite No.535 (Wako Pure Chemical Industries).
7. Thermolysin from *Bacillus thermoproteolyticus* Rokko (Daiwa Kasei, Osaka, Japan).
8. Mobil phase for HPLC: acetonitrile/water (pH 2.5 with phosphoric acid) 55:45 v/v.
9. Chemicals for isolation: ethyl acetate, hexane, $1N$ HCl, 2.5% Na_2CO_3 solution, and saturated NaCl solution.

3. Methods

3.1. Organosilicon Compounds as Novel Substrates of Hydrolases in Organic Solvents

3.1.1. Esterification

1. Prepare the Celite-adsorbed lipases as follows: Lipase (100 mg) suspended in 100 µL of deionized water is mixed thoroughly with 250 mg of Celite No. 535.
2. Carry out the reactions at 30°C with shaking (120 strokes/min). The reaction mixtures are composed of Celite-adsorbed lipase (corresponding to 100 mg lipase) and 10 mL of water-saturated benzene containing 100 mM (R,S)-2-(4-chlorophenoxy)propanoic acid and a 100-mM organosilicon compound described in **Subheading 2.1.1., item 1**. For comparison, the same reactions are carried out with the carbon analogs described in **Subheading 2.1.1., item 2.**
3. Measure the amount of esters produced by gas chromatography (GLC) using a glass column packed with PEG-HT supported on Uniport R (GL Science, Tokyo, Japan) (carrier gas, N_2; flame ionization detector [FID]).
4. Isolate the 2-(4-chlorophenoxy)propanoic acid remaining after the reactions from reaction mixtures by column chromatography (Silica gel 60).
5. Determine the optical purities of 2-(4-chlorophenoxy)propanoic acid from optical rotations measured with a JASCO DIP-140 type polarimeter (Japan Spectroscopic Co., Tokyo, Japan).

3.1.2. Amide Synthesis

1. Prepare the Celite-adsorbed hydrolases as follows: Hydrolase (30 mg) suspended in deionized water (30 µL) is mixed thoroughly with Celite No. 535 (75 mg).
2. Amide synthesis from (aminomethyl)trimethylsilane (100 mM) with 100 mM octanoic acid is carried out by Celite-adsorbed hydrolase (corresponding to 30 mg hydrolase) in water-saturated isooctane at 30°C with shaking (120 strokes/min). For comparison, the same reaction is performed with the carbon analog, 2,2-dimethylpropylamine.
3. To calculate the conversion rates, aliquots (20 µL) of reaction mixtures are subjected to GLC analysis using a glass column packed with silicon OX-17 supported on Chromosorb W A DMCS (Nash Cage, Tokyo, Japan) (carrier gas, N_2; detector, FID).

3.2. Enzymatic Preparation of Optically Active Organosilicon Compounds in Organic Solvents

3.2.1. Preparation of Optically Active Trimethylsilylpropanols

1. Prepare the Celite-adsorbed hydrolases as follows: Hydrolase suspended in 100 µL deionized water is mixed thoroughly with 250 mg Celite No. 535.

2. Carry out the enzymatic esterification at 30°C with shaking (120 strokes/min). The reaction mixtures are composed of Celite-adsorbed hydrolase (corresponding to 100 mg hydrolase) (*see* **Note 8**) and 10 mL water-saturated isooctane containing 100 m*M* racemic trimethylsilylpropanol (**Subheading 2.2.1., item 1**) and 100 m*M* 5-phenylpentanoic acid. For comparison, the same reactions are carried out with the carbon analogs (**Subheading 2.2.1., item 2**).

3. Measure the amount of the esters formed by GLC using a glass column packed with PEG-HT supported on Uniport R (GL Science) (carrier gas, N$_2$; detector, FID).

4. Determine the optical purity of remaining alcohols after esterification with hydrolases following derivatization with 3,5-dinitrophenyl isocyanate by HPLC using a Sumichiral OA-4600 column (4.0 × 250 nm). The mobile phase is *n*-hexane/2-propanol (100:1 v/v) for analysis of 1-trimethylsilyl-2-propanol (**1a**), 1-trimethylsilyl-1-propanol (**2a**), 4,4-dimethyl-2-pentanol (**1b**), and 2,2-dimethyl-3-pentanol (**2b**). For 2-trimethylsilyl-1-propanol (**3a**) and 2,3,3-trimethyl-1-butanol (**3b**), two columns of Sumichiral OA-4600 in series are used and mobile phase is *n*-hexane/2-propanol (98:2 v/v). The flow rate is 1.0 mL/min and the eluent is monitored at 254 nm in both cases. The enantiomeric excess (% ee) is calculated from the peak areas of the both enantiomers in the remaining alcohol.

5. Isolate the remaining alcohols by column chromatography on Silica gel 60.

6. To determine absolute configurations, the correlation method with ^{1}H-NMR (nuclear magnetic resonance) is used. The diastereomeric esters to use for analysis are prepared with (*R*)-MTPA.

3.2.2. Preparation of Optically Active Silylmethanol Derivatives Having a Stereogenic Silicon Atom

1. Adsorption of enzyme on Celite: enzyme preparation (100 mg) suspended in 100 μL deionized water is mixed thoroughly with 250 mg Celite No. 535.

2. Enzymatic reaction: The reaction mixtures composed of Celite-adsorbed enzyme (corresponding to 100 mg enzyme) and 10 mL water-saturated isooctane containing 100 m*M* racemic alcohol (**Subheading 2.2.2., item 1**) and 5-phenylpentanoic acid are shaken (120 strokes/min) at 30°C.

3. Determine the esters formed by GLC using a glass column packed with silicon SE-30 supported on Chromosorb W A AW-DMCS (Nishio Kogyo) (carrier gas, N$_2$; detector, FID).

4. The esters and the alcohols are isolated by column chromatography on Silica gel 60 after the reaction mixtures are filtered off to stop the reaction at more than 50% conversion.

5. Determine the optical purity of the remaining alcohols *(4,7–10)* after derivatization with 3,5-dinitrophenyl isocyanate with HPLC using two columns of Sumichiral OA-4600 (4.0 × 250 mm, Sumika Chemical Analysis Service) in series. The mobile phase is *n*-hexane/2-propanol (98:2 v/v) and the flow rate is 1.0 mL/min. The eluent is monitored at 254 nm. The enantiomeric excess is calculated from the peak areas of both the enantiomers. In the cases of **6** and **12**, the

enantiomeric excess is determined with ^{19}F-NMR after derivatization with (*S*)-CFPA *(see* **Note 15**).

6. Specific rotation of the alcohols isolated is measured with a JASCO DIP-140 polarimeter.

3.3. Enzymatic Synthesis of Silicon-Containing Peptides in Organic Solvents

1. Celite adsorption of thermolysin: thermolysin (54 mg) is suspended in 54 µL MES/NaOH buffer (50 m*M*, pH 6.5) containing 5 m*M* CaCl$_2$, and then mixed with 135 mg Celite No. 535.

2. AA-OME-HCl is converted to AA-OME by extraction with chloroform.

3. To carry out the thermolysin-catalyzed peptide synthesis at 30°C with shaking (120 strokes/min), the carboxyl component, Z-AA (40 m*M*), the amino component, AA-OME (40 m*M*), and thermolysin (54 mg) adsorbed on Celite are added to 10 mL ethyl acetate saturated with 50 m*M* MES/NaOH buffer (pH 6.5) containing 5 m*M* CaCl$_2$ *(see* **Notes 11** and **12**).

4. To calculate the conversion, aliquots (50 µL) of reaction mixtures are subjected to HPLC analysis using a Wakosil 5C18 column (4.6 × 150 mm, Wako Chemical Pure Industries), and the quantity of the residual Z-AA is determined. The mobile phase is acetonitrile/water (pH 2.5 with phosphoric acid) (55:45 v/v). The flow rate is 0.8 mL/min and the eluent is monitored at 254 nm.

5. Isolation of silicon-containing dipeptides: The enzyme preparation is removed from the reaction mixture by filtration and washed with ethyl acetate. The filtrate and washing are combined and washed successively with 1*N* HCl, 2.5% Na$_2$CO$_3$ solution, and saturated NaCl solution, and then evaporated *in vacuo*. Triturating the oily residue with *n*-hexane gives the dipeptides as a solid.

4. Notes

1. Trimethylsilylmethanol is a better substrate not only for lipase OF 360 but also for other lipases from different sources such as lipase Saiken 100 *(Rhizopus japonicus,* Osaka Saikin Kenkyusho Co., Osaka, Japan) and lipase (Steapsin) (hog pancreas, Tokyo Kasei Kogyo, Tokyo, Japan).

2. Trimethylsilylmethanol is also a very useful substrate for the enantioselective esterification of racemic naproxen [2-(6-methoxy-2-naphthyl)propionic acid], a nonsteroidal anti-inflammatory drug, by *C. cylindracea* lipase, as reported by Tsai and Wei *(18–20)*.

3. (Aminomethyl)trimethylsilane gives a homotropic effect under 100 m*M* in the amide synthesis by lipoprotein lipase Type A, whereas the carbon analog does not show such an effect.

4. Optical resolution of 2-methyl-1,3-propanediol monosilyl ethers (**13a** and **13b**) can be carried out by *Pseudomonas fluorescens* lipase (PFL)-catalyzed transesterification with vinyl acetate in chloroform **(Fig. 5)** *(21)*.

5. Various 1-trimethylsilyl-l-alkyn-3-ols (**14a** and **14b**), chiral building units useful for the synthesis of biologically active compounds, are resolved by

enantioselective acetylation with vinyl acetate mediated by immobilized lipase PS from *Pseudomonas cepacia* in diisopropyl ether (**Fig. 6**) *(22)*.

6. Enantioselective transesterification of racemic 1,1-dimethyl-l-sila-cyclohexan-2-ol (**15**) with triacetin in isooctane is successfully carried out by the use of crude lipase preparation of *Candida cylindracea* (**Fig. 7**) *(23)*.

7. The prochiral 2-sila-1,3-propanediol derivatives (**16a** and **16b**) are asymmetrized through transesterification with oxime ester or methyl isobutyrate by lipase from *Candida cylindracea* (LCC) and lipase from *Chromobacterium viscosum* (LCV) in diisopropyl ether or tetrahydrofuran (**Fig. 8**), although the enantiomeric excess (% ee) of the resulting silyl-chiral esters does not exceed 70–76% *(24)*.

8. Among the hydrolases in **Subheading 2.2.1.**, only lipase Saiken 100 cannot esterify 1-trimethylsilyl-l-propanol (**2a**).

9. Transesterification of ethylmethylphenylsilylmethanol (**4**) with 5-phenyl-pentanoic acid in water-saturated isooctane can be also carried out by various types of hydrolases, but high optical purity of the remaining alcohol is not obtained; 49% ee, shown by crude papain, is the highest value in the series of experiments.

10. Optically active silicon-containing amino acids, L- and D-3-trimethylsilylalanine (L- and D-TMS-Ala), can be obtained through enantioselective deacetylation of chemically synthesized *N*-acetyl-DL-TMS-Ala by acylase I from porcine kidney (PKA) (Sigma), as shown in **Fig. 9** *(25)*. These amino acids are used after benzyloxycarbonylation or esterification with methanol by the conventional methods *(26,27)*. *N*-Acetyl-DL-TMS-Ala is synthesized by acetylation of DL-TMS-Ala, which is synthesized from ethyl acetamidocyanoacetate and bromomethyltrimethylsilane according to the method of Porter and Shive *(28)*. The enantioselective deacetylation of *N*-acetyl-DL-TMS-Ala is carried out with PKA for 3.5 h. The reaction mixture consisted of 60 mL of 100 mM potassium phosphate buffer (pH 7.5) containing 12 mmol *N*-acetyl-DL-TMS-Ala, 12 mmol KOH, 30 μmol $CoCl_2$, and 30 mg PKA. D-TMS-Ala can be also obtained by chemical deacetylation of the residual substrate.

11. Thermolysin has a very high enantioselectivity not only for Leu but also for the silicon-containing amino acid, TMS-Ala, accepting only L-isomers. When Z-DL-TMS-Ala is used as the carboxyl component with L-Leu-OMe, only Z-L-(TMS-Ala)-Leu-OMe is found. This high enantioselectivity is a valuable merit of enzymatic peptide synthesis compared to the chemical method.

12. Various silicon-containing dipeptides, Z-(TMS-Ala)-X-OMe (X; Leu, Ile, Phe, and so forth), can be obtained when Z-TMS-Ala and X-OMe are used as the carboxyl component and the amino component in the thermolysin-catalyzed peptide synthesis, respectively.

13. 2,3,3-Trimethyl-l-butanol (**3b**) can be synthesized as follows: 2,3,3-Trimethyl-l-butene (3.5 g , 36 mmol) (Tokyo Kasei Kogyo) is dropped into a 0.3-L flask containing 80 mL of 0.5 M 9-borabicyclo[3,3,l]nonane in tetrahydrofuran (Aldrich) under an N_2 atmosphere, followed by stirring at room temperature for an additional 1 h. Then, water (5 mL), 3N NaOH (17 mL), and 30% H_2O_2 (17 mL) are added succes-

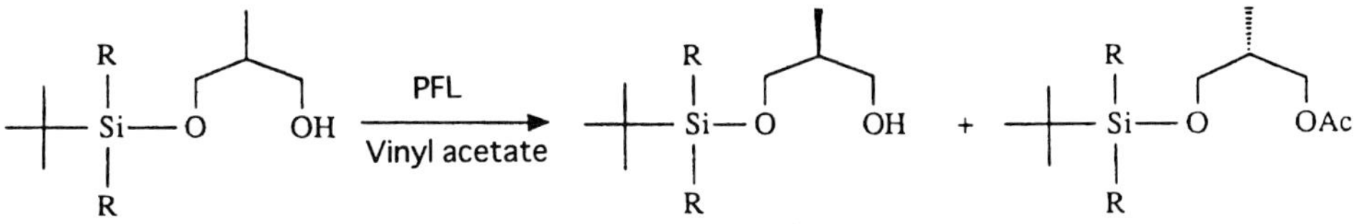

(RS)-13 (S)-13

a: R = Ph b: R = CH_3

Fig. 5. Lipase-catalyzed resolution of 2-methyl-1,3-propanediol monosilyl ethers in chloroform *(21)*.

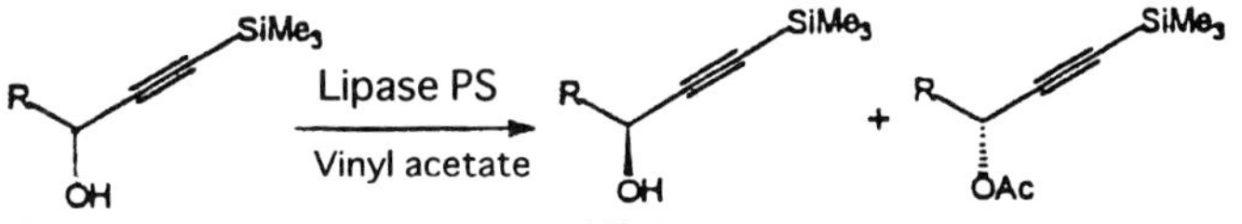

(RS)-14 (S)-14

a: R = C_2H_5 e: R = C_8H_{17}
b: R = Me_2CHCH_2 f: R = $C_{11}H_{23}$
c: R = C_5H_{11} g: R = $C_{13}H_{27}$
d: R = $C_3H_7CH{=}CH$ h: R = C_6H_5

Fig. 6. Lipase-catalyzed resolution of 1-trimethylsilyl-l-alkyn-3-ols in diisopropyl ether *(22)*.

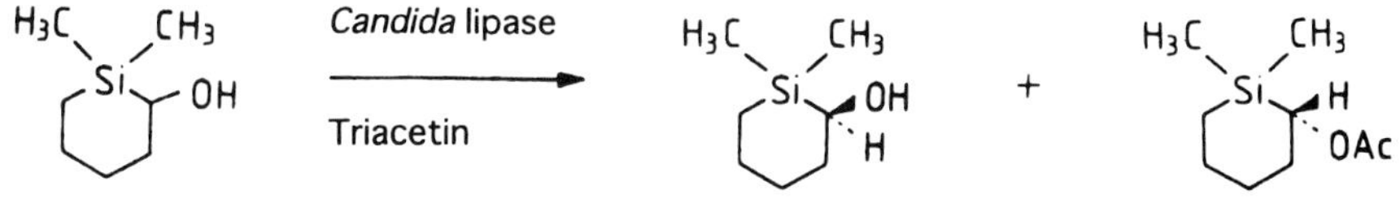

(RS)-15 (R)-15

Fig. 7. Lipase-catalyzed enantioselective transesterification of 1,1-dimethyl-l-sila-cyclohexan-2-ol with triacetin in isooctane *(23)*.

sively at a controlled temperature (less than 50°C) and the aqueous phase is saturated with K_2CO_3 after refluxing for 1 h. The organic phase, together with the ethyl ether extract of the aqueous phase, is dried over Na_2SO_4, concentrated, and distilled under atmospheric pressure. The boiling point 130–140°C fraction is collected to give 2,3, 3-trimethyl-l-butanol **(3b)** as colorless oil (3.0 g, 72% yield).

14. Ethylmethylphenylsilylmethanol *(4)* is synthesized as follows: To a solution of phenylmagnesium bromide in 150 mL dry tetrahydrofuran (THF) prepared from magnesium (2.4 g, 100 mmol) and phenyl bromide (17.3 g, 110 mmol) under an

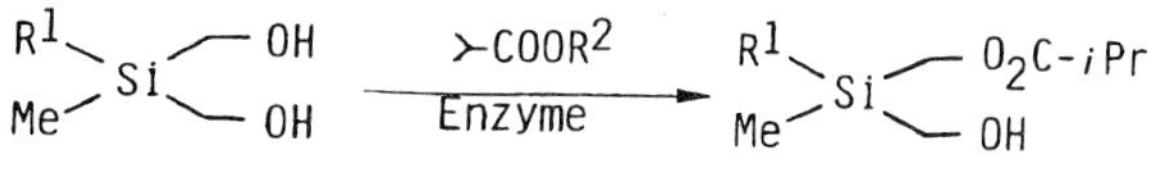

16a-b

	R^1	R^2	Enzyme
16a	Ph	Me	LCC
		ˋN=<	LCV
16b	n-C_8H_{17}	Me	LCC
		ˋN=<	LCV

Fig. 8. Asymmetrization of prochiral organosilicon compounds by lipase in organic solvents *(24)*.

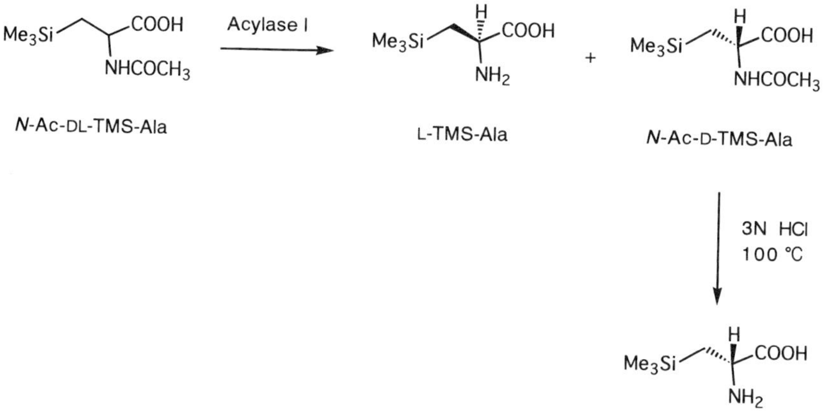

N-Ac-DL-TMS-Ala   L-TMS-Ala   *N*-Ac-D-TMS-Ala

D-TMS-Ala

Fig. 9. Preparation of optically active 3-trimethylsilylalanine (TMS-Ala) by acylase 1 *(25)*

N_2 atmosphere, (chloromethyl)diethoxymethylsilane (Petrarch Systems, Levittown, PA) (15.5 g, 85 mmol) is added dropwise at 0°C and stirred for 6 h. 10% NH_4Cl (100 mM) is added slowly and the mixture is warmed to room temperature. The organic layer is separated and the aqueous layer is extracted with diethyl ether (50 mL × 3). The combined organic layer is washed with saturated NaCl solution, dried over Na_2SO_4, and evaporated *in vacuo*. The residue is distilled under reduced pressure (3 mm Hg) followed by fractionation at a boiling point of 78–80°C to give crude (chloromethyl)ethoxymethylphenylsilane as a colorless oil. (Chloromethyl)ethoxymethylphenylsilane is added dropwise to a solution of ethylmagnesium bromide in 150 mL dry THF prepared from magnesium (3.8 g, 160 mmol) and ethyl bromide (18.5 g, 170 mmol) under an N_2 atmosphere, and the mixture is refluxed for 8 h. After cooling to room temperature, the mixture is worked up using the procedure just described, and then distilled. The fraction at 78–81°C (3 mm Hg) gives crude (chloromethyl)-ethylmethylphenylsilane. To a stirred mixture of magnesium (1.94 g, 80 mmol) and 100 mL dry THF, (chloromethyl)ethylmethylphenylsilane is added dropwise under an N_2 atmosphere. Ethyl bromide (0.22 g, 2 mmol) is added as an initiator, and the mixture is refluxed for 2 h and cooled to 0°C. Dry oxygen gas is slowly introduced into the mixture, followed by the addition of 10% NH_4Cl (80 mL). After stirring for 1 h at room temperature, the organic layer is separated and the aqueous layer is extracted with diethyl ether (50 mL × 3). The combined organic layer is washed with saturated NaCl solution, dried over Na_2SO_4, and evaporated. The residue is chromatographed on Silica gel 60 and distilled under reduced pressure to give compound **4**. The other racemic alcohols having a stereogenic silicon atom (*see* **Notes 5** and **12**) are prepared by the same procedure. Compounds **4** and **7–12** are synthesized from (chloromethyl)diethoxymethylsilane, **5** from (2-chloroethyl) dimehoxymethylsilane (Petrarch Systems), and **6** from (3-chloro-propyl) dimethoxymethylsilane (Petrarch Systems).

15. CFPA can be used in a similar manner as MTPA. CFPA was developed by Takeuch et al. *(29,30)*.

References

1. Colvin, E. W. (1981) *Silicon in Organic Synthesis,* Butterworths, London, UK.
2. Tacke, R. and Zilch, H. (1986) Sila-substitution—a useful strategy for drug design? *Endeavour New Series* **10,** 191–196.
3. Rioci, A., Seconi, G., and Taddei, M. (1989) Bioorganosilicon chemistry: trends and perspectives. *Chimica Oggi* **7,** 15–21.
4. Fessenden, R. J. and Fessenden, J. S. (1980) Trends in organosilicon biological research. *Adv. Organometal Chem.* **18,** 275–299.
5. Garson, L. R. and Kirchner, L. K. (1971) Organosilicon entities as prophylactic and therapeutic agents. *J. Pharm. Sci.* **60,** 1113–1127.
6. Creamer, C. E. (1982) Organosilicon chemistry and its application in the manufacture of pharmaceuticals. *Pharm. Technol.* March, 79–86.
7. Tacke, R. and Becker, B. (1987) Sila-substitution of drugs and biotransformation of organosilicon compounds. *Main Group Metal Chem.* **10,** 169–197.
8. Kawamoto, T., Sonomoto, K., and Tanaka, A. (1991) Efficient optical resolution of 2-(4-chlorophenoxy)propanoic acid with lipase by the use of organosilicon compounds as substrate: the role of silicon atom in enzymatic recognition. *J. Biotechnol.* **18,** 85–92.
9. Pan, S.-H., Kawamoto, T., Fukui, T., Sonomoto, K., and Tanaka, A. (1990) Stereoselective esterification of halogen-containing carboxylic acids by lipase in

organic solvent: effects of alcohol chain length. *Appl. Microbiol. Biotechnol.* **34**, 47–51.

10. Kawamoto, T., So, R. S., Masuda, Y., and Tanaka, A. (1999) Efficient enzymatic synthesis of amide with (aminomethyl)trimethylsilane. *J. Biosci. Bioeng.* **87**, 607–610.

11. Uejima, A., Fukui, T., Fukusaki, E., Omata, T., Kawamoto, T., Sonomoto, K., and Tanaka, A. (1993) Efficient kinetic resolution of organosilicon compounds by stereoselective esterification with hydrolases in organic solvent. *Appl. Microbiol. Biotechnol.* **38**, 482–486.

12. Larson, G. L. and Torres, E. (1985) Asymmetric induction by chiral silicon groups. *J. Organometal Chem.* **239**, 19–27.

13. Fukui, T., Kawamoto, T., and Tanaka, A. (1994) Enzymatic preparation of optically active silylmethanol derivatives having stereogenic silicon atom by hydrolase-catalyzed enantioselective esterification. *Tetrahedron: Asymmetry* **5**, 73–82.

14. Ishikawa, H., Yamanaka, H., Kawamoto, T., and Tanaka, A. (1999) Enzymatic synthesis of silicon-containing dipeptides with 3-trimethylsilylalanine. *Appl. Microbiol. Biotechnol.* **51**, 470–473.

15. Feder, J., Brougham, L. R., and Wildi, B. S. (1974) Inhibition of thermolysin by dipeptides. *Biochemistry* **13**, 1186–1189.

16. McConnel, R. M., York, J. L., Frizzell, D., and Ezell, C. (1993) Inhibition studies of some serine and thiol proteinases by new leupeptin analogues. *J. Med. Chem.* **36**, 1084–1089.

17. Ishikawa, H., Yamanaka, H., Kawamoto, T., and Tanaka, A. (1999) Inhibition of thermolysin by 3-trimethylsilylalanine derivatives. *Appl. Microbiol. Biotechnol.* **53**, 19–22.

18. Tsai, S.-W. and Wei, H.-J. (1994) Enantioselective esterification of racemic naproxen by lipases in organic solvent. *Enzyme Microb. Technol.* **16**, 328–333.

19. Tsai, S.-W. and Wei, H.-J. (1994) Effect of solvent on enantioselective esterification of naproxen by lipase with trimethylsilyl methanol. *Biotechnol. Bioeng.* **43**, 64–68.

20. Tsai, S.-W. and Wei, H.-J. (1994) Kinetics of enantioselective esterification of naproxen by lipase in organic solvents. *Biocatalysis* **11**, 33–45.

21. Grisenti, P., Ferraboschi, P., Manzocchi, A., and Santaniello, E. (1992) Enantioselective transesterification of 2-methyl-1,3-propenediol derivatives catalyzed by *Pseudomonas fluorescens* lipase in an organic solvent. *Tetrahedron* **48**, 3827–3834.

22. Allevi, P., Ciuffreda, P., and Anastasia, M. (1997) Lipase catalysed resolution of (*R*)- and (*S*)-1-trimethylsilyl-1-alkyn-3-ols: useful intermediates for the synthesis of optically active γ-lactones. *Tetrahedron: Asymmetry* **8**, 93–99.

23. Fritsche, K., Syldatk, C., Wagner, F., and Hengelsberg, H. (1989) Enzymatic resolution of *rac*-1,1-dimethyl-1-sila-cyclohexan-2-ol by ester hydrolysis or transesterification using a crude lipase preparation of *Candida cylindracea*. *Appl. Microbiol. Biotechnol.* **31**, 107–111.

24. Djerourou, A.-H. and Blanco, L. (1991) Synthesis of optically active 2-sila-1,3-propanediol derivatives by enzymatic transesterification. *Tetrahedron Lett.* **32**, 6325,6326.

25. Yamanaka, H., Fukui, T., Kawamoto, T., and Tanaka, A. (1996) Enzymatic preparation of optically active 3-trimethysilylalanine. *Appl. Microbiol. Biotechnol.* **45,** 51–55.
26. Farthing, A. C. (1950) Synthetic polypeptides. Part 1. Synthesis of oxazolid-2: 5-diones and a new reaction of glycine. *J. Chem. Soc.* **1950,** 3213–3217.
27. Brenner, M. and Huber, W. (1953) Herstellung von α-Aminosaureestern durch Alkoholyse der Methylester. *Helv. Chim. Acta* **36,** 1109–1115.
28. Porter, T. H. and Shive, W. (1968) DL-2-Indaneglycine and DL-β-trimethyl-silylalanine. *J. Med. Chem.* **11,** 402,403.
29. Takeuchi, Y., Itoh, N., Note, H., Koizumi, T., and Yamaguchi, K. (1991) α-Cyano-α-fluorophenylacetic acid (CFPA): a new reagent for determining enantiomeric excess that gives very large ^{19}F NMR $\Delta\delta$ values. *J. Am. Chem. Soc.* **113,** 6318.
30. Takeuchi, Y., Itoh, N., Satoh, T., Koizumi, T., and Yamaguchi, K. (1993) Chemistry of novel compounds with multifunctional carbon structure. 9. Molecular design, synthetic studies, and NMR investigation of several efficient chiral derivatizing reagents which give very large ^{19}F NMR $\Delta\delta$ values in enantiomeric excess determination. *J. Org. Chem.* **38,** 1812.

30

Synthetic Applications of Enzymes in Nonaqueous Media

Valérie Rolland and René Lazaro

1. Introduction
1.1. Peptide Synthesis

Instead of the conventional aqueous media, the use of organic solvents for enzymatic catalysis has become more and more popular over the past few years, allowing the preparative synthesis of asymmetric products *(1)*. Not only lipase, which naturally works at the lipid–water interfaces, but other enzymes (cytoplasmic proteases) have been shown to be stable and active enough to be used as efficient catalysts in nearly anhydrous organic media.

Enzymatic methods based on lipases in acyl transfer reactions and kinetic resolution of acids and alcohols in organic solvent have been widely described in **ref.** *2*; also *see* **refs.** *3* and *4*.

Enzymatic methods based on proteases *(5–7)* combined with chemical synthetic methods are attractive medium-sized alternatives for the practical syntheses of small and condensed peptides. Peptide synthesis catalyzed by protease is a smooth regioselective and stereoselective method devoid of racemization risk and is often carried out under mild conditions in aqueous or organic solvents, or in a mixture of both with a minimum side-chain protection. In spite of these advantages the protease approach suffers some drawbacks related to the inherent amidase activity of proteases that may cause secondary hydrolysis of the growing peptide chain, and to the fact that the substrates of proteases are generally limited to natural L-amino acids and the corresponding peptides, consistent with the native recognition properties of proteases.

From: *Methods in Biotechnology, Vol. 15: Enzymes in Nonaqueous Solvents: Methods and Protocols*
Edited by: E. N. Vulfson, P. J. Halling, and H. L. Holland © Humana Press Inc., Totowa, NJ

Compared with the well-established chemical stepwise peptide synthesis, enzymatic methods are very competitive in some particular cases such as the following:

1. The synthesis of short peptides having free side-chain functions [i.e., aspartame *(8,9)*]
2. The semisynthesis of peptides and proteins [i.e., insulin *(10)*]
3. The condensation of unprotected peptide fragments obtained through solid-phase peptide synthesis
4. the postmodification (i.e., amide formation) or the labeling of the C-terminal part of peptides obtained by recombinant DNA technology.

Thermodynamically controlled synthesis (i.e., a direct reversal of the catalytic hydrolysis of peptides) and kinetically controlled synthesis (i.e., aminolysis of N-protected amino esters or peptide esters) were the two enzymatic strategies applied *(5)*. In both cases, a large amount of water has been used, allowing the native protease catalysis to proceed efficiently in the reverse sense (synthesis vs hydrolysis). However, especially during the fragment coupling, some deleterious side effects may occur in the presence of water: simultaneously, the biocatalyst may work as expected (as a ligase) and as a protease cleaving either the newly formed bond or any other sensitive site already present in the peptide component. To minimize this undesired hydrolysis, the use of a high content of organic solvent or other "unfree" water-containing medium (ice or hydrated salt) has been considered and is reported hereafter.

In order to improve the amphiphilic character of the enzyme in an organic solvent *(11)*, α-chymotrypsin was modified on its ϵ-NH$_2$ lysyl groups with polyethylene glycol (PEG) via an amide bond *(12)*, resulting from the conversion of the terminal alcohol of monomethyl ether PEG molecule into a carboxylic group. This very efficiently soluble PEG-enzyme derivative catalyzed peptide synthesis in a mixture of *tert*-amyl alcohol/benzene containing 0.5% (v/v) water with N-protected (Z or Boc) esters of aromatic amino acid residues. Slightly activated esters [-OCarboxamidomethyl: OCam *(13)*] are better acyl donors than ethyl and Boc(Leu)^5 enkephalinamide has been prepared under these conditions in very good yield.

The process has been further improved by immobilization of α-chymotrypsin. Gels of copolymerized acrylated derivatives of PEG and α-chymotrypsin have been prepared and optimized *(14)*. α-Chymotrypsin has been modified on its ϵ-NH$_2$ lysyl groups (with 33% level of derivatization) by acryloyl chloride and with total retention of esterase activity. α-Chymotrypsin has been immobilized at the 90–95% level by crosslinking with acrylated derivatives of PEG 3400 and 5000 and a final loading around 105 mg α-chymotrypsin per gram of dry polymer has been obtained. Thus, by simple filtration the biocatalyst can be separated as a gel from the aqueous solution. Reaction conditions, in terms of enzyme substitution degree, nature of acyl donor and nucleophile, minimum water amount related to the medium hydrophobicity (1%) of *tert*-amyl alcohol and some peptide syntheses (i.e., enkephalinamide), have been optimized, leading to some peptide syntheses *(15)*. This amount of water is necessary for a good swelling of the acrylic gel (*see* **Subheading 3.1.**).

However, as the gel reticulation may be a limitation for the enzymatic site access to long peptide fragments, ultrasonic irradiation, a well-known method in organic chemistry for increasing the velocity of heterogenous reactions, has been applied. With this process, the gel is expanded and then the accessibility of the α-chymotrypsin site is improved. The water amount is now less important for swelling of the gel; only the enzyme-adsorbed water remains necessary (0.2%). The peptide maximum size is determined and this procedure could be applied to the preparation of some active peptides *(16)*. As expected under ultrasonic irradiations, the velocity of the peptide synthesis is increased sixfold without decreasing the final coupling yield.

Under the same conditions, it is noticeable that native α-chymotrypsin totally loses its activity by complete and irreversible denaturation. The enzyme must be immobilized to keep active. Sonication has to be continuously applied for good peptide coupling activation: the slope of the curve decreases as soon as the ultrasonic treatment stops. Ultrasonic treatment improves the substrate's and product's diffusion through the acrylic gel and also increases their solubilization.

This procedure has been applied to the preparation of polymers of dipeptides, Ac-Tyr-(Leu-Tyr-)$_n$-OEt and Z-Phe-(Leu-Tyr-)$_n$-OEt, in order to determine the maximum capacity of the enzymatic gel under ultrasonic treatment (*see* **Subheading 3.1.**). The maximum length of accepted peptide fragments was 27 aminoacid residues ($n = 13$), evaluated by mass spectroscopy. Following this result, it is easy to imagine long peptide synthesis with the help of ultrasonic treatment. This procedure was generalized for other proteases such as trypsin, papain, or pepsin, according to their specificity *(17)*.

In addition to V8 *Staphylococcus aureus* protease, which shows a good specificity toward acidic amino acid residues in the P1 position, a readily available glutamic-specific enzyme, called BL-GSE, isolated from *Bacillus licheniformis (18,19)* was used in peptide synthesis. It has been shown that this endopeptidase is 100-fold as active as the previous one and able to catalyze condensation as well as transesterification and transpeptidation reactions *(20,21)*. BL-GSE is also tolerant of high concentration of organic solvent; the enzyme has been incubated in Bicine buffer, containing 2 mM calcium chloride necessary for its activity *(18)* and different amounts of dimethyl formamide (DMF), acetonitrile, or methanol (*see* **Subheading 3.2.**): BL-GSE is quite stable in 60% DMF, 80 % CH$_3$CN, and 60 % MeOH. After 10 min, the activity level remains constant, but at higher proportions of organic solvent, the enzyme activity drastically decreases.

Moreover, specific replacements of C-terminal amino acids and the conversion of a peptide bond to an ester bond by enzymatically catalyzed alcoholysis, never described before, have been achieved. In 60% MeOH, BL-GSE has been used for catalyzing the direct esterification of the N-blocked dipeptide Z-Ala-Glu-OH; in 10 min at pH 8.5, the fraction of acylation reached 70%. After this time, the intermediate product, Z-Ala-Glu-OMe, is back hydrolyzed. This procedure could be applied to the direct esterification of peptides terminated by a nonprotected Glu (*see* **Subheading 3.3.**).

The semisynthesis of C-terminal peptides of gastrin, calcitonin, and cholecystokinin, or human neuropeptide Y was achieved by the introduction of Phe-NH$_2$ or Tyr-NH$_2$ at their carboxyl termini, respectively. Serine carboxypeptidase-catalyzed transpeptidations using peptide substrates and amino acid amides as nucleophiles have resulted in high yields of peptide amides. Carboxypeptidase Y serves as a very efficient catalyst for transpeptidation, especially for C-terminal alanine-containing peptides *(22)*. BL-GSE can be used for this purpose. The coupling is usually performed in the presence of 20% DMF, in which the secondary hydrolysis is negligibly small.

Because one of the most efficient substrates for BL-GSE is O-aminobenzoyl-Ala-Phe-Ala-Phe-Glu-Val-Phe-Tyr(NO$_2$)-AspOH, this peptide has been chosen as a model for transpeptidation of the Glu-Val bond with various nucleophiles (*see* **Subheading 3.4.**).

The yield of transpeptidation is generally improved by an increase in pH resulting from the deprotonation of the amino group of the entering nucleophile, although, simultaneously, the peptidase activity decreases, leading to lower rates of conversion.

The syntheses of monoesters of Glu or Asp is chemically not an easy task, whereas it is well known that amino esters of trifluoroethanol and benzyl alcohol are very good starting materials for kinetically controlled enzymatic peptide synthesis *(15,20)*.

Thus, it has been very tempting to perform peptide couplings in a one-pot reaction starting from free-carboxylic amino acid residues in two successive BL-GSE-catalyzed steps. The intermediate ester can be further subjected to BL-GSE catalytic aminolysis. With this new protocol in hand, it was not necessary to start from an amino ester, as is generally the case in kinetically controlled peptide synthesis. This would be of great advantage in the coupling of deprotected fragments obtained by solid-phase synthesis after cleavage from the resin (*see* **Subheading 3.2.**).

Aspartame (H-LAsp-LPhe-OMe) is one example of industrial enzymatic peptide preparation *(8)*. The coupling between Z-Asp-OH and H-Phe-OMe is performed using a metalloprotease (thermolysin) in water–glycerol mixtures under thermodynamic control conditions.

H-Asp-Phe-OMe can be also prepared using either α-chymotypsin or papain, belonging respectively to the serine and cysteine endopeptidase group, but only if the lateral chain is protected by methyl or ethyl ester as in the chemical strategy *(23)*. Masking the lateral carboxylic group is necessary for fitting the P1 requirement of these enzymes.

Recently, Yoshpe-Besançon et al. have directly prepared aspartame (although in low yields) from L-Asp-OH and L-Phe-OMe using a new aminopeptidase from *S. chromogenes* *(24)*. Even if BL-GSE is 1000 times less active when the P1 position is occupied by an aspartic acid instead of a glutamic acid, preparing Aspartame at low price could be an interesting challenge.

H-Asp-Phe-OMe is a poor substrate for BL-GSE:aspartame hydrolysis (10 m*M*) is carried out in water–bicine buffer with a very high amount of enzyme (1000-fold more than in a standard reaction) and only 5% hydrolysis was obtained. H-Phe-OMe is not a good nucleophile, as it is gradually transformed into the diketopiperazine derivative c(Phe-Phe). Other dipeptides were prepared with either more stable [H-Phe-NH-$(CH_2)_2$-OH] or more active [H-Met-OMe] entering nucleophiles. Hopefully, the resulting dipeptides H-Asp-Phe-NH-$(CH_2)_2$-OH and H-Asp-Met-OMe are sweet. To avoid starting ester hydrolysis and secondary hydrolysis of the new peptide bond, syntheses are carried out in water-miscible organic solvents, depending on the solubility of the nucleophile (70% or 80% of acetonitrile).

In conclusion, BL-GSE activity is higher for Glu-X substrates than for Asp-X substrates, so a lower amount of enzyme is needed for the Glu coupling. However the acyl-enzyme intermediate with Glu is probably less "stable" than that formed with Asp and is hydrolyzed faster.

It has been shown in the literature that a certain amount of structural water must be present for preserving the molecular flexibility of the enzyme molecules and, consequently, their activity, for example, in transesterification reaction *(25)*. To reduce the free water concentration in the enzyme surroundings, Jakubke et al. recently used serine or cysteine proteases in the frozen state for peptide synthesis in kinetically controlled reactions *(26,27)*. It is suprising that under frozen-state conditions, the endoproteinases are capable of coupling even free amino acids, acting as reverse carboxypeptidases. The explanation given for this freezing effect is that in this macroscopically frozen system, the acyl transfer and the following peptide synthesis would take place in the remaining liquid phase (*see* **Subheading 3.6.**).

Another way to reduce the content of free water is to use a hydrated salt suspended in a water-immiscible organic solvent: Thus, $Na_2CO_3 \cdot 10H_2O$ suspended in hexane is a water-scavenging inorganic salt. Surprisingly, although the enzyme (α-chymotrypsin) and most of the reactants are insoluble, the synthesis of some tripeptides proceeds quite well and it is believed that the salt hydrate buffers the thermodynamic water activity *(28,29)*.

1.2. Regioselective Carbohydrate Acylation

In one of our research programs aimed at preparing new trimodular nonionic surfactants made of a sugar with a lipophilic tail linked by an aminoacid residue *(30)*, several esterifications of suitable sugar derivatives have been attempted using different lipases.

The potential of lipases as catalysts for the acylation of carbohydrates is currently of considerable interest, as the reaction occurs selectively at the primary hydroxyl group of the sugar. The need for protection and deprotection steps of the nonfavorable secondary hydroxyl groups is avoided, allowing a reduction in the number of steps. In recent years, various reports have been published on the enzymatic transesterifications of carbohydrates using activated donors, such as trihaloethyl, vinyl, and oxime esters or acid anhydrides *(31)*. For example, 6-*O*-acyl derivatives of alkyl glucopyranosides, useful as biodegradable nonionic surfactants, were synthesized from fatty acids and the corresponding 1-*O*-ethyl-glucopyranosides in the presence of a thermostable lipase from *Candida antarctica* (CAL) *(32,33)* (*see* **Subheading 3.7.**). This lipase also catalyzed the regioselective acylation of 1-*O*-octyl-α-D-glucopyranoside by ethyl acrylate using zeolite CaA for selective adsorption of the produced ethanol or water; a 99% conversion and a 99% selectivity were observed in 4 h *(34)*.

Native enzymes such as *Porcine pancreatic* lipase (PPL), *C. antarctica* lipase (CAL), and proteinase N (crude subtilisin) or immobilized enzymes such as Lipozyme® (lipase from *Mucor miehei* adsorbed on anionic resin) or Novozyme SP 435 (immobilized *C. antarctica* lipase) were tested for regioselective acylation of the primary hydroxy group of a sugar (present in excess). The biocatalysis was carried out in organic solvents or in solvent-free systems under nearly anhydrous conditions, which favor lipase-catalyzed process. Nevertheless, the problem of carbohydrate solubility in these media created difficulties for choosing the appropriate conditions to perform the reaction. Glucose and others alkyl glucopyrannosides such methyl, ethyl, or 1,2-isopropylidene glucopyrannosides were not soluble. Consequently, the sugar used was butyl α-D-glucopyranoside, prepared enzymatically using almond meal *(35)*. As the melting point of butyl α-D-glucopyrannoside was low (65°C), experiments in the solvent-free system were performed on melted compounds for the efficient preparation of 6-*O*-lauroyl derivatives of butyl α-D-glucopyranoside (*see* **Subheading 3.8.**).

For our purpose of producing new surfactants, spacers used for the sugar acylation were bromoacetic acid and its active esters (*see* **Subheading 3.9.**), succinic acid esters (*see* **Subheading 3.10.**), and adipic acid, dodecanedioïc acid and hexadodecanedioïc acid (*see* **Subheading 3.11.**); increasing the length of the chain (2, 4, 6, 12, or 16 carbons, respectively) increased the lipophilicity of the compounds, giving better substrates for lipases. Proteinase N, catalyzing regioselective esterification of sugars in anhydrous *N,N*-dimethylformamide *(36)*, was not an efficient biocatalyst (data not shown).

Lipozyme in the presence of excess of bromoacetate catalyzed the acylation of butyl α-D-glucopyranoside in a solvent-free system, at 70°C in just 8 h (yield = 60%). With short diacids, this acylation was difficult. It was reported in 1993 that sugar acylation with succinic acid anhydride can be achieved, but with low regioselectivity, yielding only 7% for the C_6 monoester *(37)*. Unfortunately, the high melting point of this acid anhydride (120°C) prevented working in a solvent-free system. The active ester 2,2,2-trichloroethylsuccinate has a lower melting point (86°C). Two products were obtained resulting from an esterification followed by a transesterification: the butyl 6-*O*-succinoyl-α-D-glucopyranoside and the butyl 6-*O*-(2,2,2-trichloroethylsuccinate)-α-D-glucopyranoside.

With adipic acid (*n* = 4), dodecanedioic acid (*n* = 10) and hexadecanedioic acid (*n* = 14), sugar acylation using Lipozyme or PPL were performed (respectively 0%, 35%, and 60% yields), resulting in the formation of mixture of diesterified and triesterified sugars. There is a literature precedent showing that the length of the fatty chain plays a major role in the lipase-catalyzed reactions.

2. Materials

1. High-performance liquid chromatography (HPLC) analysis was done with a Waters Chromatograph equipped with a Nucleosil C18 analytical column operating at 1 mL/mm in an isocratic system for dipeptide analyses. Ultraviolet (UV) detection was achieved at 214 nm. For longer peptides, linear gradients were applied with 0.1% trifluoroacetic acid (TFA) in MeOH as buffer A and 0.1% TFA in water as buffer B and UV detection at 254 nm. Other HPLC analysis used Waters Associates equipment, with a Vydac C-18 reverse-phase column. Various linear and concave gradients were employed with 0.1% trifluoroacetic acid in water as buffer A and 0,1% trifluoroacetic acid in acetonitrile as buffer B. All separations were carried out at room temperature and monitored at 254 nm for the Z substrates. For components bearing the *o*-aminobenzoyl group as the dominant chromophore, the analysis was monitored at 320 nm.
2. Ultrasonic irradiations were performed by means of a commercial ultrasonic washing machine (Bandelin SONOREX TK52), using a thermstatted system.
3. Thin-layer chromatography (TLC) was performed on Merck precoated silica gel glass plates, and spots were revealed by 1% ninhydrin solution in ethanol, visualised by ultraviolet light or revealed by 5% ethanolic sulfuric acid after heating at 90°C or on precoated silica gel 60F254 plates, where spots were visualized by ultraviolet light or/and iodine vapor. The spots were revealed also by 0.2% ninhydrin solution.
4. Chiral stationary-phase HPLC was performed on the Waters HPLC system with a 486-nm detector apparatus on a Crown Pack CR+ column.
5. Optical rotations were recorded with a Perkin-Elmer 141 polarimeter at the sodium D line.
6. Column chromatography was performed using Geduran silica gel Merck 7734.

7. Melting points were obtained on a Büchi 510 apparatus and were not corrected.
8. Mass spectra were measured on Jeol JMS DX 300 apparatus with FAB ionization and on Platform II by electrospray (ESI).
9. ^{1}H-NMR (nuclear magnetic resonance) spectra were recorded on a Brucker spectrometer AC 250 or 400 MHz.
10. ^{13}C-NMR spectra were recorded on a Brucker spectrometer at 400 MHz.
11. Glu/Asp Specific Endopeptidase (BL-GSE) was isolated from *B. licheniformis* as previously described by Carlsberg Laboratory, Denmark *(18,19)*. Z-Phe-OH, Z-Asp-OBzl, Z-Glu-OBzl, H-Phe-OMe, and H-Met-OMe were from Bachem, Switzerland.
12. All other reagents were of analytical grade.
13. Aminolysis and hydrolysis products were collected and identified by amino acid analysis after hydrolysis with 6N HCl at 110°C for 24 h.

3. Methods

3.1. Peptide Coupling by Copolymerized Enzyme–Sonication

1. The immobilized biocatalyst is prepared as followed: *Bis*-acryloyl (bisac.) PEG 3400 acting as a crosslinking agent and the monomethylether of monoacrylated PEG 5000 serving as a matrix agent (the ratio was fixed at 4 bisac./1 monoac *[14]*) were prepared by acryloyl chloride esterification.
2. The acrylated derivatives (enzyme with 33% of derivatization and the two PEGs) were dissolved in borate buffer 0.05 M, pH 8.75, saturated by N$_2$.
3. Ammonium persulfate and N,N,N',N'-tetramethylethylenediamine (TEMED) were added in catalytic amounts to initiate the radical polymerization.
4. The mixture was stirred at 4°C under nitrogen until polymerization took place.
5. The resulting gel was broken and washed several times with distilled water (*see* **Note 1**).
6. General procedure for peptide coupling: *N*-Ac-Tyr-OEt (45 × 10^{-6} M) and H-Leu-NH$_2$ (45 × 10^{-6} M) were dissolved in 0.6 mL *t*-amyl alcohol containing 1% water (w/w). The gel containing 1 mg α-chymotrypsin (40 nmol) was added and the mixture shaken at 20°C for the required time (24 h) without ultrasonic treatment. *N*-Ac-Tyr -Leu-NH$_2$ was obtained in 98% yield (1.2% of the hydrolyzed product *N*-Ac-Tyr-OH).
7. Ultrasonic treatment was carried out at 20°C using a thermostatted bath. The different coupled products were submitted to HPLC analysis with different eluents.
8. After shaking the vessel for 10 min, the washing solvent was discarded by filtration. The whole cycle was repeated three times and the clean catalyst was then ready for the next coupling reaction (*see* **Note 2**).
9. The water content was determined by the Karl Fischer test on both the lyophilized copolymer and the organic solvent. For example, *N*-Ac-Tyr-Leu-NH$_2$ was obtained in *t*-amyl alcohol+ 0.2% water in 99.5% yield (0.5% of the hydrolyzed product *N*-Ac-Tyr-OH) after 4 h.

3.2. BL-GSE Stability in Organic Solvent

1. The BL-GSE stability at 25°C was determined by incubating the enzyme at concentrations around 2 mg/mL in 50 mM bicine, 2 mM CaCl$_2$ at pH 8.0, and various proportions of organic solvents (CH$_3$CN, DMF, MeOH).
2. The enzymatic activity toward Z-Glu-NH-Ph-NO$_2$ was determined as a function of time in the following way: 25 µL of a 8 mM substrate solution in methanol was added to 965 µL of 50 mM bicine, 2 mM CaCl$_2$, pH 8.0, followed by the addition of 10 µL of diluted solution of incubated enzyme.
3. The hydrolysis was determined spectrophotometrically at 410 nm using a Cary 219 or a Perkin-Elmer Lambda 7 spectrophotometer thermostatted at 25°C.

3.3. Coupling and Condensation Reactions in Organic Solvents

1. The assays were carried out at 20°C with 10 mM Z-Glu-OBzl (for coupling reactions) or Z-Ala-Glu-OH (for condensation reactions) as the acyl donor, in 50 mM Bicine, 2 mM CaCl$_2$ buffer at pH 8.5 and organic solvents, 35 nM BL-GSE, and different concentration of various nucleophiles (H-Val-NH$_2$, and so forth).
2. The reaction was followed by removing periodically 10-µL aliquots from the reaction mixture and the reactant composition was determined by HPLC. (HPLC running with linear gradient: 25–75% B in 20 min) (*see* **Notes 3** and **4**).
3. In condensation reactions, some assays in various experimental conditions in Bicine buffer containing MeOH, CF$_3$-CH$_2$OH (TFE) or Ph-CH$_2$OH with H-Val-NH$_2$ as the nucleophile were performed. The best results were obtained with methanol (60%): the coupling yield between Z-Glu-OH (10 mM) and H-Val-NH$_2$ (1 M) was 70% in 5 d without enzyme inactivation; with the dipeptide Z-Ala-Glu-OH, the same condensation yield was obtained in 2 d.

3.4. Transpeptidation Reactions

1. The nucleophiles (H-Val-NH$_2$, H-Val-Gly-NH$_2$ and H-Ala-Phe-NH$_2$) were dissolved in a final mixture of 20% DMF, 50 mM Bicine, and 2 mM CaCl$_2$ containing the subtrate ABz-Ala-Phe-Ala-Phe-Glu-Val-Phe-Tyr(NO$_2$)-AspOH dissolved in DMF (5 mM) at a final concentration of 0.2 mM; the pH was adjusted to 8.5 and the enzyme was then added from the aqueous solution to a final concentration of 10 nM. (HPLC running with linear gradients e: 30–80% B in 20 min).
2. With 2 M H-Val-NH$_2$, the maximum fraction of aminolysis (FA) obtained in 20% DMF was 0.65, but with the both dipeptides in the same experimental conditions, it was greatly increased, reaching FA 0.94 and 0.93 and remained constant (*see* **Note 5**). As BL-GSE is an endopeptidase, dipeptide amides as nucleophiles are more efficient for transpeptidation and only a small concentration of nonapeptide (0.2 mM) is necessary compared to the 10 mM of Z-Glu-OBzl used. Moreover, and for the same reasons, a shorter peptide like ABz-Ala-Ala-Glu-Gly-Tyr(NO$_2$)-Asp-OH and ABz-Ala-Ala-Glu-Ser-Tyr(NO$_2$)-Asp-OH gives a maximum FA 0.15 with 2 M H-Ala-Phe-NH$_2$ as the nucleophile.

3.5. Aspartame Analog Synthesis

1. To a mixture of 10 mM Z-Asp-OBzl and 1 M HBr, H-Phe-NH(CH$_2$)$_2$-OH in 70% CH$_3$CN, and 50 mM bicine/2 mM CaCl$_2$ buffer was added 2.5 µL BL-GSE solution (c = 1.2 mg/mL) after having adjusted the pH (8.0–9.0). This mixture was stirred at room temperature.
2. Periodically, an aliquot of 10 µL was withdrawn from the reaction mixture and analyzed by HPLC: The final yield was 44% in 30 min.
3. After enzymatic coupling, the aspartame analog H-Asp-Phe-NH(CH$_2$)$_2$-OH was obtained by catalytic hydrogenation on Pd/C in MeOH in a quantitative yield.

3.6. Peptide Synthesis in a Frozen System

1. Peptide synthesis reactions at –25°C were performed in 1.5-mL polypropylene tubes at a total sample volume of 0.1 mL.
2. The nucleophile (5 mmol) dissolved in 80 mL of water was adjusted with 1 N NaOH to give appropriate final pH's.
3. After the sequential addition of the amino components, the enzyme (10 µL), and the acyl donor (0.2 mL of the acyl donor dissolved in 10 mL dimethyl sulfoxide [DMSO]), the tubes were shaken and placed in liquid nitrogen for 20 s to achieve shock-freezing.
4. Then, they were transferred for the time of reaction into a constant-temperature cryostat (HAAKE, Germany). The reactions were stopped by adding 0.8 mL of 0.7N acetic acid in water. The reactions at room temperature were performed in the same manner as described without freezing using a total volume of 1 mL, containing 0.1 M phosphate buffer or carbonate buffer.
5. The coupling between Mal-His-ONb (2 mM) as the acyl donor and H-Lys-NH$_2$ (50 mM) catalyzed by α-chymotrypsin (15.2 µM) in carbonate buffer containing 10% DMSO (v/v) at pH 9.0 yielded 97% product in 3 h at –25°C.

3.7. Large Scale Synthesis of Alkyl 6-O-Alkanoyl-ᴅ-Glucopyranosides

1. Glucose (8 kg, 44 mol) was suspended in absolute ethanol (25 L, 418 mol). Ion-exchange resin amberlyst 15 (1.6 kg) was added and the reaction mixture was refluxed with efficient mechanical stirring for 18 h.
2. The ion-exchange resin was removed by filtration and excess EtOH was distilled off under reduce pressure, yielding crude ethyl-ᴅ-glucopyranoside as a variable syrup (about a 1/1 mixture of the α- and the β-anomers, containing about 4% unconverted glucose).
3. Melted fatty acids (lauric or stearic acid) (57 mol) were added and the mixture was heated under stirring to 70°C.
4. Immobilized *C. antarctica* lipase (400 g) was added and the stirring was continued at 70°C under reduced pressure (0.01 bar) until a >90% conversion was determined by HPLC (reached in 28 h).

5. The enzyme was removed by filtration (at 70°C under pressure), yielding a crude product (21.3 kg) consisting of 70 wt% of 6-*O*-monoester, 3% diester, 21% fatty acid, and 7% ethyl-D-glucopyranoside.
6. The excess of fatty acid was removed by short-path distillation (at 130°C and 0.004 mbar), resulting in a product of >85 wt% monoester, consisting of >96% 6-*O*-ester of ethyl D-glucopyranoside, the rest being the 6-*O*-monoester of D-glucose.

3.8. Butyl 6-O-Lauroyl-α-D-Glucopyranoside Synthesis by Lipozyme

1. Lauric acid (1.7 g, 8.5 mmol) and butyl α-D-glucopyranoside (1 g, 4.23 mmol) were added to boiling hexane in a Dean Stark apparatus (*see* **Note 6**).
2. Lipozyme (0.15 g) was added and the suspension was stirred at 70°C for 3 d, after which the formation of the diester was detected.
3. The mixture was filtered and the solvent was evaporated under reduce pressure.
4. Flash chromatography (94:6 DCM-MeOH) of the residue (2.6 g) gave the 6-*O*-lauroyl derivative (1.4 g, 80%).

3.9. Butyl 6-O-(Bromoacetyl)-α-D-Glucopyranoside Synthesis by Lipozyme

1. Butyl α-D-glucopyranoside (0.1 g, 0.42 mmol) and bromoacetic acid (0.6 g, 4.2 mmol, 10 equiv) were stirred at 70°C.
2. When the compounds melted, Lipozyme (0.1 g) was added and the mixture was mechanically stirred for 8 h at 70°C.
3. Then, the mixture was cooled to room temperature, and acetone (3 mL) was added.
4. The immobilized enzyme was filtered and washed with acetone (10 mL) and diethylether (3 mL).
5. The solution was concentrated under reduced pressure and the crude product was purified by a silica gel column chromatography, first with 300 mL diethyl ether and then a mixture of 95% diethyl ether/5% methanol as the eluent. The butyl 6-*O*-(bromoacetyl)-α-D-glucopyranoside was obtained in 60% yield as a yellow oil (*see* **Note 7**).

3.10. Synthesis of Butyl 6-O-Succinoyl α-D-Glucopyranoside and Butyl 6-O-(2,2,2-Trichloroethylsuccinate)-α-D-Glucopyranoside by Lipozyme

1. Butyl α-D-glucopyranoside (0.5 g, 2.12 mmol) and trichloroethyl succinate (4.22 g, 16.9 mmol, 8 equiv) were mechanically stirred at 90°C.
2. Lipozyme (0.5 g) was added to the melted compounds and the mixture was mechanically stirred for 24 h.
3. The pasty mixture was cooled to room temperature and acetone (10 mL) was added.

4. The immobilized enzyme was filtered and washed with acetone (50 mL) and diethyl ether (15 mL).

5. Butyl 6-*O*-succinoyl α-D-glucopyranoside was purified by chromatography on a silica gel column with a mixture of 99% diethylether/1% methanol as the eluent and was obtained as a yellow oil (*see* **Note 8**).

3.11. Synthesis of Butyl 6-O-Dodecanedioyl-α-D-Glucopyranoside and Butyl 6-O-Hexadodecanedioyl-α-D-Glucopyranoside by PPL

1. Butyl α-D-glucopyranoside (0.3 g, 1.27 mmol) and dodecanedioic acid (0.29 g, 1.27 mmol) or hexadodecandioic acid were dissolved in acetone (4.5 mL).

2. The mixture was stirred at 45°C and PPL was added (0.25 g).

3. The suspension was mechanically stirred for 3 d.

4. The mixture is then cooled to room temperature; the enzyme was filtered and washed with acetone (20 mL) and diethyl ether (5 mL).

5. The solution was concentrated under reduced pressure.

6. The compounds are purified by chromatography on a silica gel column with 90% diethylether/10% acetone as the eluent. Both products were obtained as yellow oils, yielding 35% in the case of butyl 6-*O*-dodecanedioyl-α-D-glucopyranoside and 60% for butyl 6-*O*-hexadodecanedioyl-α-D-glucopyranoside.

4. Notes

1. The enzyme retention could be estimated by a Bradford test applied to the collected water fractions and remained very high (90%) (enzyme/PEG ratio: 12.5). After lyophilization and weighing, the calculated loading degree was $(100–110) \times 10^{-6}$ g CT/mg dried gel.

2. After each experiment, the biocatalyst was easily recovered by filtration, and resuspended in fresh solvent.

3. The fraction of aminolysis, FA, is expressed as the ratio between the formed aminolysis product and the sum of all products being formed (i.e., the unconsumed substrate was disregarded in the calculations). In these conditions, the FA remained constant without back hydrolysis. The concentration of nucleophile from which the fraction of aminolysis was half the maximum value is designated KN(app) and was a measure for the dissociation constant of the complex between the acyl-enzyme intermediate and the nucleophile *(18,19)*.

4. In 60% CH$_3$CN or 80% DMF, the fraction of aminolysis is high and the BL-GSE stays stable in those conditions for more than 5 d. Moreover, the high value of organic solvent concentration avoids secondary peptide hydrolysis; indeed, the fraction of aminolysis remains constant.

5. Without DMF, back hydrolysis has been observed, thus decreasing the final yield of transpeptidation.

6. A gummy mixture was obtained when the reactants were mixed at room temperature.

7. With Novozyme® SP 435 *(37)*, total conversion was observed in 3 h but with a very low regioselectivity; all secondary hydroxyl groups were acylated in different yields and acylated glucopyranosides were not purified.

8. The respective yields were 36% for butyl 6-*O*-succinoyl α-D-glucopyranoside and 17% for butyl 6-*O*-(2,2,2-trichloroethylsuccinate)-α-D-glucopyranoside, resulting from competition between esterification and transesterification reactions.

References

1. Fitzpatrick, P. A. and Klibanov, A. M. (1991) How can the solvent affect enzyme enantioselectivity. *J. Am. Chem. Soc.* **113,** 3166–3171.
2. Rubin, Byron, ed. (1997) *Part B: Lipases—Enzyme Characterization and Utilization—Section III: Biocatalytic Utility*, Academic Press, London, p. 286.
3. Faber, K. (1997), *Biotransformation in Organic Chemistry*, 3rd ed., Springer-Verlag, New York.
4. Roberts, S. (1998) Preparative Biotransformations; the employment of enzymes and whole-cells in synthetic organic chemistry. *J. Chem. Soc. Perkin Trans.* **1,** 157–169.
5. Schellenberger, V. and Jakubke, H. D. (1991) Protease-catalysed kinetically controlled peptide synthesis. *Angew. Chem. Int. Ed. Engl.* **30,** 1437–1449.
6. Morihara, K. (1987) New developments in enzymatic peptide synthesis. *Trends Biotechnol.* **5,** 164–169.
7. Wong, C. H. and Wang, K. T. (1991) *Experientia* **47,** 1123–1129.
8. Isowa, Y., Ohmori, M., Ichikawa, T., Mori, K., Kihara, K., and Oyama, K. (1979) The thermolysin-catalysed reactions of N-substituted aspartic and glutamic acids with phenylalanine alkyl esters. *Tetrahedron Lett.* **28,** 2611,2612.
9. Oyama, K., Nishimura, S., Nonaka, Y., Kihara, K., and Hashimoto, T. (1981) Synthesis of an aspartame precursor by immobilized thermolysine in an organic solvent. *J. Org. Chem.* **46,** 5242–5244.
10. Morihara, K., Oka, T., and Tsuzuka, H. (1979) Semisynthesis of human insulin by trypsin-catalysed replacement of alaB30 by Thr in porcine insulin. *Nature* **280,** 412,413.
11. Inada, Y., Nishimura, H., Takahashi, K., Yoshimito, T., Saha, A. R., and Saito, Y. (1984) Ester synthesis catalysed by polyethylene glycol modified lipase in benzene. *Biochem. Biophys. Res. Commun.* **122(2),** 845–850.
12. Babonneau, M. T., Jacquier, R., Lazaro, R., and Viallefont, P. (1989) Enzymatic peptide synthesis in organic solvent mediated by modified α-chymotrypsin. *Tetrahedron Lett.* **30,** 2787.
13. Kuhl, P., Zacharias,V., Burckhardt, H., and Jakubke, H. D. (1986) On the use of carboxamidomethyl esters in the protease-catalysed peptide synthesis. *Monatsch. Chem.* **117,** 1195–1204.
14. Fulcrand, V., Jacquier, R., Lazaro, R., and Viallefont, P. (1990) New biocatalysts for peptide synthesis: gels of copolymerized acrylic derivatives of α-chymotrypsin and polyoxyethylene. *Tetrahedron* **46,** 3909.
15. Fulcrand, V., Jacquier, R., Lazaro, R., and Viallefont, P. (1991) Enzymatic peptide synthesis in organic solvent mediated by gels of copolymerized acrylic—derivatives of α-chymotrypsin and polyoxyethylene. *Int. J. Peptide Protein Res.* **38,** 273–277.

16. Rolland-Fulcrand, V., Hua, T. D., Lazaro, R., and Viallefont, P. (1991) Sono-enzymatic peptide synthesis. *Biomed. Biochim. Acta* **50,** 213–216.
17. Rolland-Fulcrand, V., Mai, N., Lazaro, R., and Viallefont, P. (1994) Enzymatic peptide synthesis by new supported biocatalysts. *Amino Acids* **6,** 311–314.
18. Svendsen, I. and Breddam, K. (1992) Isolation and aminoacid sequence of a glutamic acid specific endopeptidase from *Bacillus licheniformis. Eur. J. Biochem.* **204,** 165–171.
19. Breddam, K. and Meldal, M. (1992) Substrate preferences of glutamic acid specific endopeptidases assessed by synthetic peptide substrates based on intramolecular fluorescence quenching. *Eur. J. Biochem.* **206,** 103–107.
20. Rolland-Fulcrand, V. and Breddam, K. (1993) The use of a glutamic specific endopeptidase in peptide synthesis. *Biocatalysis* **7,** 75–82.
21. Rolland-Fulcrand, V., Lazaro, R., and Viallefont, P. (1994) Enzymatic esterification of N-protected Glu or Asp and synthesis of sweetener aspartame analogs by a new glutamic acid endopeptidase. *Catal. Lett.* **27,** 235–240.
22. Breddam, K., Widmer, F., and Meldal, M. (1991) Amidation of growth hormone releasing factor (1-29) by serine carboxypeptidase catalysed transpeptidation. *Int. J. Peptide Protein Res.* **37,** 153–160.
23. Adisson, L., Bolte, J., Demuynck, C., and Mani, J. C. (1988) Enzymatic synthesis of aspartyl-containing dipeptides. *Tetrahedron* **44,** 2185.
24. Yoshpe-Besançon, I., Auriol, D., Paul, F., Monsan, P., Gripon, J. C., and Ribadeau-Dumas, B. (1993) Purification and characterization of an aminopeptidase A from *Staphylococcus chromogenes* and its use for the synthesis of aminoacid derivatives and dipeptides. *Eur. J. Biochem.* **211,** 105–110.
25. Klibanov, A. M. (1989) Enzymatic catalysis in anhydrous organic solvents. *Trends Biotechnol. Sci.* **14,** 141–144.
26. Hänsler, M. and Jakubke, H. D. (1996) Reverse action of hydrolyses in frozen aqueous solutions. *Amino Acids* **11,** 379–395.
27. Beck-Pietraschke, K. and Jakubke, H. D. (1998) Protease-catalysed synthesis of peptides containing histidine and lysine. *Tetrahedron: Asymmetry* **9,** 1505–1518.
28. Kuhl, P. and Halling, P. (1991) Salt hydrates buffer water activity during chymotrypsin-catalysed peptide synthesis. *Biochim. Biophys. Acta* **1078,** 326–328.
29. Kvittingen, L., Sjurnes, B., Anthonsen, T., and Halling, P. J. (1992) Use of salt hydrates to buffer optimal water level during lipase catalysed synthesis in organic media: a practical procedure for organic chemists. *Tetrahedron* **48,** 2793–2802.
30. Boyat, C., Rolland, V., Roumestant, M. L., Viallefont, Ph., and Martinez, J. (2000) Chemoenzymatic synthesis of new non ionic surfactants. *Carbohydr. Res.* in press.
31. Watanabe, T., Matsue, R., Hoda, Y., and Kuwahara, M. (1995) Differential activities of a lipase and a protease toward straight- and branched-chain acyl donors in transesterification to carbohydrates in an organic medium. *Carbohydr. Res.* **275,** 215–220.
32. Adelhorst, K., Björkling, F., Godtfredsen, S. E., and Kirk, O. (1990) Enzyme catalysed preparation of 6-*O*-acylglucopyranoside. *Synthesis* **2,** 112–115.

33. Kirk, O., Björkling, F., and Godtfredsen, S. E. (1989), World Patent WO 90/09451 (to Novo-Nordisk).
34. de Goede, A. T. J. W., van Oosterom, M., van Deurzen, M. P. J., Sheldon, R. A., Van Bekkum, H., and Van Rantwijk, F. (1994) Selective lipase-catalysed esterification of alkyl glycosides. *Biocatalysis* **9,** 145–155.
35. Fabre, J., Paul, F., Monsan, P., Blonski, C., and Périé, J. (1994) Enzymatic synthesis of amino acid ester of butyl α-D-glucopyranoside. *Tetrahedron Lett.* **35(21),** 3535,3536.
36. Riva, S., Chopineau, J., Kieboom, A. P. G., and Klibanov, A. M. (1988) Protease-catalysed regioselective esterification of sugars and related compounds in anhydrous dimethylformamide. *J. Am. Chem. Soc.* **109,** 584–589.
37. Fabre, J., Betbeder, D., Paul, F., Monsan, P., and Périé, J. (1993) Versatile enzymatic diacid ester synthesis of butyl α-D-glucopyranoside. *Tetrahedron* **49,** 10,877.

31

Enzymes in Nonaqueous Solvents

Applications in Carbohydrate and Peptide Preparation

Shui-Tein Chen, Boonyaras Sookkheo, Suree Phutrahul, and Kung-Tsung Wang

1. Introduction

In nature, the vast majority of enzyme-catalyzed reactions are carried out in water. Traditionally, enzymic catalysis in organic synthesis is carried out in aqueous solution. In some cases, water-miscible organic cosolvents are added to improve the solubility of the hydrophobic substrate in the reaction solution. With the addition of an organic cosolvent, the equilibrium constant of the hydrolytic enzyme-catalyzed reaction can change and subsequently affect the yields of the product *(1)*. Recent studies on protease-catalyzed kinetic-controlled peptide bond formation showed that the addition of an organic cosolvent will drastically decrease the protease activity, with a lesser effect on esterase activity *(2)*. This novel enzymatic property favors the peptide bond formation in aqueous cosolvent, and the decrease of amidase activity prevents peptide bond hydrolysis in the synthesis of large peptides. Further mechanism-based studies aim to design an efficient ligase that favors peptide bond formation, and they use a genetic engineering approach to produce new enzymes that have been successfully used in peptide or glycosyl bond formation *(3,4)*.

Enzymatic reactions in aqueous solution often encounter problems of low substrate solubility and product hydrolysis. To avoid these problems, the search for enzymes for organic synthesis that are stable in organic solvents has been extensively pursued *(5–13)*. Traditionally, organic solvents such as acetone and ethanol are used in protein purification. Because most of the enzymes are insoluble in organic solvents, inconsistent results may occur when performing

From: *Methods in Biotechnology, Vol. 15: Enzymes in Nonaqueous Solvents: Methods and Protocols*
Edited by: E. N. Vulfson, P. J. Halling, and H. L. Holland © Humana Press Inc., Totowa, NJ

reactions in these conditions. This inconsistency is dependent on the reaction conditions or on the procedures for preparation of the enzyme.

As early as 1967, heterogeneous enzyme systems of chymotrypsin suspended in methylene chloride *(14)* and xanthine oxidase in nonpolar solvents *(15)* had been studied. Frank et al. found that a suspension of crystalline chymotrypsin in methylene chloride containing 0.25% water resulted in no loss of catalytic activity over 90 h, and the activity of the suspended enzyme can be progressively lost by the addition of the irreversible inhibitors diphenyl carbamyl chloride and *m*-nitrobenzene sulfonyl fluoride. Kinetic studies of the hydrolysis of *N*-acetyl-L-tyrosine ethyl ester (ATEE) by chymotrypsin crystals in methylene chloride were attempted and failed. The main factors were the turbidity of the suspensions, which obviated the use of ultraviolet (UV) absorbance as an analytical method, and the low conductivity of the solvent. These instrumental limitations stemmed further study in this field until the important observations of Zaks et al. that lipase dispersed in organic solvents exhibited catalytic activity *(16)*.

Over the past two decades, intense interest has been shown in studying enzyme reactions in organic solvents and in developing them for the synthesis of pharmaceuticals, chiral intermediates, specialty polymers, and biochemicals *(17)*. Investigations were no longer restricted to the conventional aqueous media. Many new enzymatic process have been developed such as the following: enzymes in organic solvents from non-polar to polar in nature; in eutectic mixtures; on a support suspended in organic solvents; covalently modified enzyme dissolved in organic solvents, or solubilized in solvents with a surfactant; enzymes enclosed in reversed micelles and so forth *(18–20)*. All these reactions avoid the uncertainty of the heterogeneous biocatalyst reaction. Although progress has been rapid, the technology remains in its infancy and choosing the right condition or process is still difficult. The technology still lacks a general method that will allow a competent scientist who is unfamiliar with the enzyme system to carry out the technique successfully at the first attempt by simply following detailed practical procedures.

Enzymes are proteins that act as catalysts in biological systems. The characteristics of an enzyme include substrate specificity and purification, molecular mass, pH optimum, assay, stability, and inhibition properties are all of interest to biochemists. From a synthetic chemist's point of view, enzymes are employed as reagents in organic synthesis, so the main concerns are the reactivity, selectivity, and durability of the enzymes used. About 90% of enzymatic reactions in organic solvents involve hydrolases (lipase, esterase, protease, phosphatases, glycosidase, etc.). These enzymes share the same catalytic mechanism of covalent-catalysis reaction *(21)*. **Figure 1** shows a typical kinetically controlled protease-catalyzed reaction. The enzyme initially hydrolyzes the substrate to form an acyl-enzyme intermediate (R-C(=O)-Ez) that

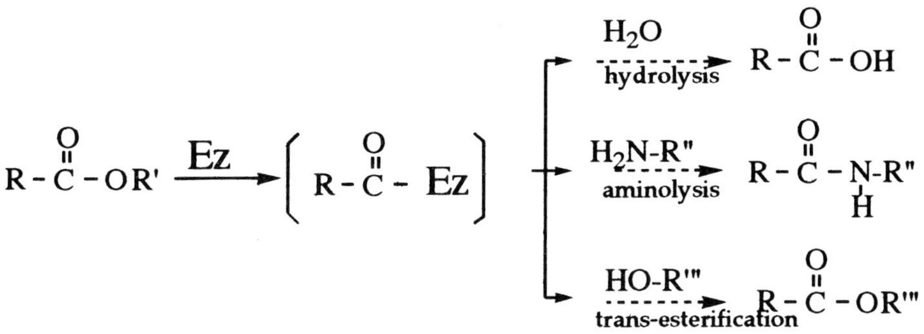

$$R - \overset{\overset{\textstyle O}{\|}}{C} - OR' \xrightarrow{\text{ Ez }} \left(R - \overset{\overset{\textstyle O}{\|}}{C} - Ez \right)$$

$$\begin{array}{l} \underset{\text{hydrolysis}}{\overset{H_2O}{\dashrightarrow}} \quad R - \overset{\overset{\textstyle O}{\|}}{C} - OH \\[2em] \underset{\text{aminolysis}}{\overset{H_2N\text{-}R''}{\dashrightarrow}} \quad R - \overset{\overset{\textstyle O}{\|}}{\underset{\underset{\textstyle H}{|}}{C}} - N\text{-}R'' \\[2em] \underset{\text{trans-esterification}}{\overset{HO\text{-}R'''}{\dashrightarrow}} \quad R - \overset{\overset{\textstyle O}{\|}}{C} - OR''' \end{array}$$

Fig. 1. Typical kinetically controlled protease-catalyzed reaction.

can be deacylated by water (hydrolysis reaction) or by an amine nucleophile (:NH$_2$-R") (peptide bond formation) *(22,23)*. If the procedure is carried out in an alcohol solution, it may lead to a transesterification reaction.

In aqueous solution, the yield of the peptide bond formation depends on two factors: (1) the relative rate of hydrolysis and aminolysis (i.e., on the nucleophilicity of water vs that of the amine nucleophile) and (2) the molar ratio of the nucleophiles (i.e., the concentration of water and that of the amine). When the reaction is carried out in low-water conditions, the hydrolysis of the acyl intermediate by water is decreased and the yield of product increased. However, the difficulties associated with detecting pH, conductivity, and enzyme activity in a nonaqueous solution are increased. Methods for measuring the enzyme activity in low-water systems have been documented *(24–27)*. When the reaction is carried out in anhydrous organic solvents, hydrolysis is minimized and the yield of a reaction depends on the reactivity or the durability of the enzyme. The first priority for a stable biocatalyst is to convert all the substrate to product. For enzymes with poor reactivity, the addition of more enzyme to complete the synthesis is necessary. Substrate selectivity of the biocatalyst is also important for reaction in organic solvents, and the effects of organic solvents on the enantioselectivity or regioselectivity have been documented. The geometry of the catalytic site of the enzyme may be different from that in aqueous solution because of binding of the solvent in the catalytic site, and this binding may alter the enantioselectivity or regioselectivity of the enzyme catalysis *(28–35)*.

The objectives of this chapter are to provide by the use of illustrative examples practical procedures for (1) the preliminary observation of enzyme activity in organic solvents, (2) a simple procedure for the screening of enzyme reactions, (3) the determination of enzymatic activity in organic solvents, (4) the determination of residual enzyme activity after a reaction in organic solvents, (5) the inversion of enantioselectivity by solvents, (6) the regioselective acylation of sugars derivatives, (7) the resolution of amino acids in mixture of 95%

t-butanol/5% water, and (8) enhancement of the rate of enzyme reactions by microwave irradiation.

1.1. The Preliminary Observation of Enzyme Activity in Organic Solvents

The use of modified enzymes or the addition of additives to solubilize enzymes in organic solvents have been well documented. Very limited studies have been performed on the solubility of enzyme in organic solvents *(36,37)*. It is expected that a soluble or partially soluble enzyme will achieve good reactivity in organic solvents, but, in general, enzymes in organic solvents are used as heterogeneous biocatalysts. The state of the suspended enzyme in the organic solvents can determine the reactivity of the biocatalyst. The same enzyme may have different appearances when suspended in different organic solvents, and, conversely, different enzymes suspended in the same organic solvent may also appear to have a different state. The easiest way to judge an active enzyme in an organic solvent is by the appearance of the suspended enzyme.

1.2. A Simple Procedure for the Screening of Enzyme Reactions

For synthetic purposes, thin-layer chromatography (TLC) is one of the most convenient tools for monitoring the course of enzyme-catalyzed reactions, especially in screening many enzyme libraries for one target compound. Currently, enzyme libraries that have 20 or 30 examples of the same types of enzyme such as proteases, lipases, transferases, and glycosidases are commercially available. Using TLC for screening enzymatic reactions in organic solvents is simpler than screening the reaction in aqueous solution. In the latter situation, when samples are loaded onto the TLC plate, the aqueous solution is diffused and the dissolved enzymes anchored onto the plate. In organic media, the biocatalyst is insoluble and does not adhere onto the TLC plate. The solvents evaporate faster when applying the sample to the TLC plate. A simple procedure for screening the lipase-catalyzed irreversible acylation of a diol is described in **Fig. 2** *(38,39)*. Twelve commercially available lipases suspended in vinyl acetate were investigated for catalysis of the irreversible acylation of a diol substrate, 3-phenoxy-1,2-propandiol (**1**), to form the monoesters 1-acetyl 3-phenoxy-1,2-propandiol or 2-acetyl 3-phenoxy-1,2-propandiol (**2**), or the diester 1,2-diacetyl 3-phenoxy-1,2-propandiol (**3**).

1.3. The Determination of Enzymatic Activity in Organic Solvents

Using a spectrophotometer with chromogenic substrates is one of the most convenient methods for the determination of enzymatic activity. Assays of enzymatic activity in organic solvents may encounter the following difficulties: (1) knowledge of the extinction coefficient of the chromogenic compound

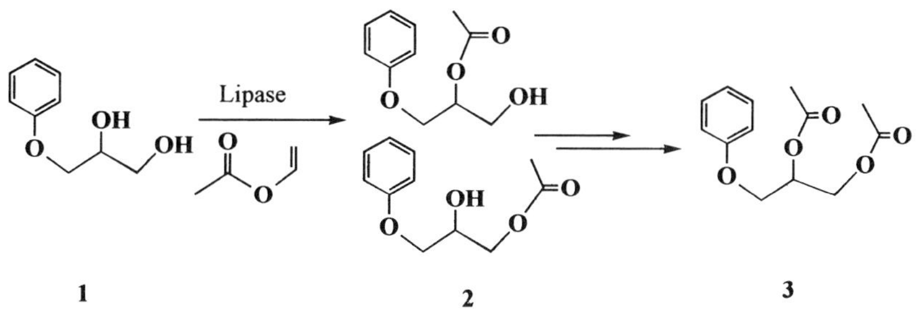

Fig. 2. Lipase-catalyszed irreversible acylation of 3-phenoxy-1,2-propandiol in vinyl acetate.

in organic solvents, (2) inconsistancy in activity of the hetereogeous biocatalyst in organic solvents, (3) the amount of the suspended enzyme taken from the organic solvents may not show consistency. For these and other reasons, other instrumental methods may be more reliable. thin-layer chromatography (TLC), gas chromatography (GC), and high-pressure liquid chromatography (HPLC) are popularly used in determining enzymatic activity in organic solvents. These methods are not substrate dependent as long as they have chromatographic properties. A densitometer may be used to quantify the colored spots on a TLC plate to determine the enzymatic activity in organic solvents. A GC equipped with a flame ionization detector is used frequently in lipase-catalyzed reactions, as in most cases, the substrates and reaction product of lipase are volatile. HPLC is also useful in the determination of enzymatic activity in organic solvents. The following example describes using HPLC to monitor a serine protease, alcalase, catalyzed transesterification of Cbz-Phe-OMe and Cbz-Phe-OBzl in ethanol solution to produce Cbz-Phe-OEt *(33,34)*. The rate of transesterification for the alcalase catalysis can be calculated from integrated peaks in the chromatogram.

1.4. The Determination of Residual Enzyme Activity After a Reaction in Organic Solvents

The remaining activity of an enzyme after use in an organic solvent provides useful information. The remaining activity of enzymes in aqueous solution is normally assayed by taking small aliquots of the incubated enzyme and adding a testing solution for measurement of enzymatic activity. However, because the enzyme suspended in organic solvents is in a heterogeneous form, a method based on aliquots cannot be used, but a more reliable procedure uses several fixed amounts of enzyme in organic solvent and incubates the enzyme solution at a desired temperature to measure its activity at specific time intervals. The following procedure describes the measurement of remaining activity of subtilisin Carlsberg in 95% alcohol using Moz-Trp-OBzl as substrate for the synthesis of Moz-Trp-OH (hydrolysis product) and Moz-Trp-OEt (transesterification product).

1.5. The Inversion of Enantioselectivity by Solvents

Enzymes are very efficient catalysts with high enantioselectivity, stereoselectivity, regioselectivity, and chemoselectivity. Crude enzymes may contain many isoforms when purchased. For the biochemical study of enzyme kinetics, specificity, and inhibition, a pure enzyme is a requirement. However, for synthetic purposes, the stability and reactivity of the biocatalyst may be the only concerns. Synthetic chemists may not be overly concerned with purity or the way in which an enzyme is purified; their interest lies in the enantioselectivity and regioselectivity of enzyme catalysis.

The enzyme may bind with water molecules to form fine crystals. In an aqueous solution, an enzyme molecule is usually surrounded by an aqueous core. When an enzyme is exposed to organic solvents, it may associated with that organic solvent. This association depends on the polarity of the solvent and the native structure of the enzyme. Different solvents may associate with the same enzyme at different regions. This association with the enzyme may alter the geometry surrounding the active site and result in changes in the enantiomeric outcome of the enzyme's catalysis. This section describes an example of the inversion of enantioselectivity by solvents in the Lipase MY-catalyzed acylation of 3-phenoxy-1,2-propanediol by suspending the enzyme in acetone, chloroform, and methylene chloride respectively *(40–43)*.

1.6. The Regioselective Acylation of Sugars Derivatives

Intensive research has been done on the selective protection and deprotection of the hydroxyl group of carbohydrates *(44–48)*. In general, chemical preparation of protected sugars with free primary hydroxyl groups may require several steps such as tritylation, esterification, and acid-catalyzed detritilation. Using enzyme-catalyzed selective acylation or deacylation, one or two steps may complete the synthesis. Some proteases possess both esterase and protease activities. Alcalase has both activities, with preferred esterase activity resulting in ester formation from a primary alcohol. The following procedure describes the alcalase-catalyzed regioselective acylation of (ethyl 2-acetamido-2-deoxy-1-thio-D-glucose) at the C_6 hydroxyl position using vinyl benzoate in 2-methyl-2-propanol as an irreversible acylation agent. The reaction is shown in **Fig. 3**.

1.7. The Resolution of Amino Acids in the Mixture of 95% 2-Methyl-2-Propanol/5% Water

Some amino acids having unusual side chains play an important role in many biologically and pharmaceutically active peptides. Most of the unnatural amino

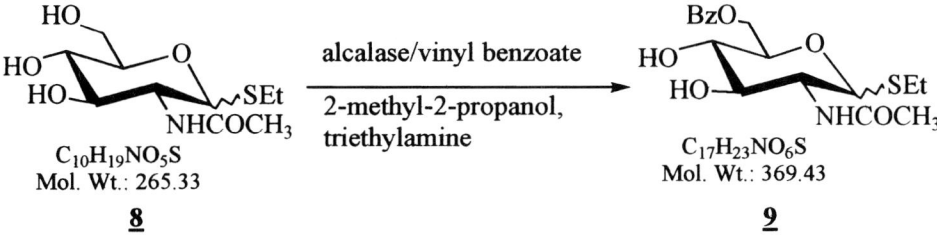

Fig. 3. Alcalase-catalyzed regioselective acylation at the C-6 hydroxyl position of ethyl 2-acetomido-2-deoxy-1-thio-D-glucose in 2-methyl-2-propanol using vinyl benzoate as an irreversible agent.

acids cannot be obtained by fermentation or recombinant DNA technology, and resolution is one of the best ways to produce optically pure unnatural amino acids. Kinetic resolution of amino acids in aqueous solution has been extensively studied, but resolution in water-limited conditions is rare. The following procedure describes the use of alcalase as a catalyst for resolution of amino acids in a mixture of 95% 2-methyl-2-propanol/5% water (**Fig. 4**) *(49–51)*.

1.8. Enhancement of the Rate of Enzyme Reactions by Microwave Irradiation

According to the report of Technology Vision 2020 of the US Chemical Industry, microwave technology will be widely used in the future to replace traditional heating methods in chemical synthesis *(52–55)*. Microwave heating involves direct absorption of energy by functional groups that bear ionic conductivity or a dipole rotation effect, which subsequently releases energy to the surrounding solution. This absorption of energy causes the functional groups involved to have a higher reactivity with surrounding reactants, different from that present when simply incubated with the reactants at a similar temperature. Microwave irradiation has been applied to continuous-flow process for preparative-scale synthesis *(56)*, and recently applied to the acceleration of enzyme catalysis *(57–62)*. Successful use of microwave irradiation to enhance the rate of lipase- and alcalase-catalyzed resolution and regioselective esterification has been documented.

We describe here a novel application of microwave technology to enhance the efficiency of alcalase-catalysis peptide bond formation (a rate increase of at least twofold to 10-fold). **Figure 5** shows the reactions of peptide bond formation using proline amide as a nucleophile. The reactions were carried out in 2-methyl-2-propanol and the products isolated by simple separation using flash column chromatography in yields of 85%.

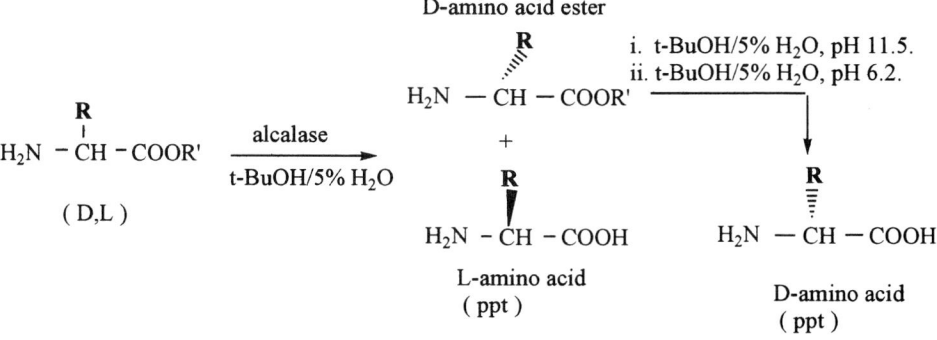

Fig. 4. Use of alcalase as a catalyst for resolution of amino acids in a mixture of 95% *t*-butanol/5% water.

Z-Ala-Phe-OMe + Pro-NH₂ $\xrightarrow{\text{i}}$ Z-Ala-Phe-Pro-NH₂ (85%)

i. alcalase/2-methyl-2-propanol; microwave irradiation 20 min.

Fig. 5. Enhancement of peptide bond formation by microwave irradiation.

2. Materials

2.1. The Preliminary Observation of Enzyme Activity in Organic Solvents

1. Lipase N (*Rhizopus niveus*, Amano).
2. Lipase p-1 (*Bacillus stereothermophilus*, isolated from the hot-spring area of ChiangMai, Thailand).
3. Alcalase 2.5 L (*Bacillus* subtilisin Carlsberg, serine protease, from Novo, Denmark).
4. Vinyl acetate (Merck, A.G., Germany)
5. 2-Methyl-2-propanol (G.R., Wako, Japan).

2.2. A Simple Procedure for the Screening of Enzyme Reactions

1. The following enzymes (100mg of each): (1) Lipase AP6 (*Aspergillus niger*, Amano), (2) lipase AY-30 (*Candida cylindracea*, Amano), (3) *Candida cylindracea* lipase (Meito-Sangyo), (4) lipase GC4 (*Geotrichum candidum*, Amano), (5) lipase MAP (*Mucor meihei*, Amano), (6) lipase MAP (*Mucor meihei*, Amano), (7) lipase N (*Rhizopus niveus*, Amano), (8) Lipase OF (*Candida cylindracea*, Meito-Sangyo), (9) Lipase R (*Humicola sp.*, Amano), (10) Lipase R-10 (*Humicola lanuginosa*, Amano), (11) *Candida rugosa* lipase (Sigma, L1754), (12) PPL type II (Sigma, L-3126).
2. Substrate, 3-phenoxy-1,2-propandiol (100 mg).

3. Solvent: vinyl acetate (3.0 mL) + water 30 (μL).
4. TLC plate: silica gel 60 F254, 20 × 20 cm aluminum sheets (Merck, A.G., Germany).
5. Developing solvent system: *n*-hexane:methylene chloride:ethyl acetate (1:1:1; v/v/v).
6. Visualization solution: 5% phosphomolybdic acid in alcohol.

2.3. The Determination of Enzymatic Activity in Organic Solvents

1. Enzyme solution: alcalase 2.5 L (2.0 mL).
2. Substrate solution: Cbz-Phe-OBzl, Cbz-Phe-OEt.
3. Internal standard: phenol (Merck AG Germany).
4. Reagents: absolute ethanol (Riedel-de Haen. AG, Germany).
5. Instrument: the HPLC apparatus consisted of an Hitachi 6200 intelligent-HPLC pump and a Hitachi 4200 UV detector. The data were collected on a Macintosh LCII using Rainin Chrompic software. Column: vidyc RP-18, 4.6 × 150 mm. Eluent: 80% CH_3CN in 0.1% trifluoracetic acid (TFA). Flow rate: 1 mL/min. UV wavelength: 214 nm.

2.4. The Determination of Residual Enzyme Activity After a Reaction in Organic Solvents

1. Enzyme solution: subtilisin Carlsberg (20.0 mg).
2. Substrate solution: Moz-Trp-OBzl (10 mg in 95% ethanol 2 mL).
3. Moz-Trp-OH and Moz-Trp-OEt (reference standard).
4. Instrument: the HPLC apparatus consisted of a Hitachi 6200 intelligent-HPLC pump and a Hitachi 4200 UV detector. The data were collected on a Macintosh LCII using Rainin Chrompic software. Column: RP-18, 4.6 × 150 mm. Eluent: 59% CH_3CN in 0.1% TFA. Flow rate: 1 mL/min. UV wavelength: 280 nm.

2.5. The Inversion of Enantioselectivity by Solvents

1. Enzyme: lipase MY(50 mg).
2. Substrates: racemic 3-phenoxy-1,2-propandiol, vinyl acetate.
3. Solvents: acetone, methylene chloride, chloroform.
4. Chiral column: Chiral OD-RH (Daicel Chemical Inc., Japan).
5. Instrument: the HPLC apparatus consisted of a Hitachi 6200 intelligent-HPLC pump and a Hitachi 4200 UV detector. The data were collected on a Macintosh LCII using Rainin Chrompic software. Eluent: $HClO_4$ (pH 2.0)/ACN = 70/30. Flow rate: 1 mL/min. UV wavelength: 254 nm.

2.6. The Regioselective Acylation of Sugars' Derivatives

1. Enzyme solution: alcalase, 2.5 L (1.0 mL).
2. Substrate solution: ethyl 2-acetamido-2-deoxy-1-thio-D-glucose.
3. Reagents: 2-methyl-2-propanol, triethylamine (Merck, A.G., Germany), vinyl benzoate.
4. TLC plate: Merck 20 × 20 cm silica gel 60 F254 TLC aluminum sheets.

5. Developing solvent system: EtOH:ethyl acetate, (1:9; v/v).
6. Visualization solution: 5% phosphomolybdic acid in alcohol.

2.7. The Resolution of Amino Acids in the Mixture of 95% t-Butanol/ 5% Water

1. Enzyme: alcalase 2.5 L.
2. Reagents: D,L-tyrosine-OMe (Bachem, Switzerland), methanol, ethyl acetate, methylene chloride, acetonitrile, dioxane, ether, and acetone (HPLC grade and reagent grade) were obtained from a local supplier, the ALPS Chemical Co. (Taiwan).
3. TLC was performed on silica gel G precoated plates (Merck A.G., Germany).
4. Instruments: Optical rotation was measured on a universal polarimeter (Schmidt & Haensch, Germany).
5. HPLC system consisted of two Waters model 6000 pumps, a Waters model 450 UV detector, and an M-660 solvent programmer. The Chiral CR-(+) HPLC column was purchased from the Daicel Chemical Company (USA).
6. A Suntex P.C. 303 Auto-pH Controller (Suntex Instruments Co., Taiwan).

2.8. Enhancement of the Rate of Enzyme Reactions by Microwave Irradiation

1. Enzyme: alcalase 2.5 L.
2. Reagents: 2-methyl-2-propanol, ethanol.
3. Substrate: Cbz-Ala-Phe-OMe, Pro-NH$_2$.
4. Instrument: kitchen microwave oven (Tatum microwave oven TMO-110,Taiwan), total power of the microwave was 650 W with nine power settings, the lowest of which was 72 W. In this study, 20% full power was used for preliminary tests. A Synthewave Reactor (Prolabo 402, France) was used for preparative scale synthesis.
5. The HPLC apparatus consisted of a Hitachi 6200 intelligent-HPLC pump and a Hitachi 4200 UV detector. The data were collected on a Macintosh LCII using Rainin Chrompic software. Column: RP-18, 4.6 × 150 mm. Eluent: 68% CH$_3$CN in 0.1% TFA. Flow rate: 1 mL/min. UV wavelength: 254 nm.

3. Methods

3.1. The Preliminary Observation of Enzyme Activity in Organic Solvents

1. To vials 1, 2, and 3 was added vinyl acetate (3 mL) and to vial 4 was added 2-methyl-2-propanol (3 mL).
2. Four enzymes, Lipase N (100 mg), lipase p-1 (100 mg), alcalase 2.5 L (0.5 mL), and alcalase 2.5L (0.5 mL) were added to the four corresponding vials and the resulting mixtures were stirred on a stirrer.
3. The results were recorded photographically after the enzyme mixtures were stirred for 10 min. **Figure 6** shows the photographs of three enzymes suspended in vinyl acetate and 2-methyl-2-propanol: vial 1: lipase N; vial 2: lipase P-1, vial 3: alcalase 2.5 L in vinyl acetate; and vial 4: alcalase 2.5 L in 2-methyl-2-pro-

Fig. 6. Appearance of enzymes (vial 1: lipase N; vial 2: lipase p-1; vial 3: alcalase 2.5 L; vial 4: alcalase) suspended in vinyl acetate (vials 1, 2, and 3) and in 2-methyl-2-propanol (vial 4).

panol. When the active enzyme is suspended in organic solvents, the enzyme is dispersed and the organic solvent solvate the enzyme molecules to make the reaction solution transparent, with the appearance of an insoluble compound forming a gel dissolved in the solvent (vials 3 and 4). Moderately active enzymes suspended in organic solvents form a turbid, aggregated solution. The degree of aggregation (which affects the amount of active site exposed to the substrate) determines the reactivity of the enzyme (vial 1, some enzyme coagulated and deposited on the bottom of the vial [*see* **Notes 1** and **2**]). When most of the enzyme is aggregated, the active site of the enzyme is buried in each enzyme molecule, resulting in a low enzymatic activity (vial 2) (*see* **Note 3**).

3.2. A Simple Procedure for the Screening of Enzyme Reactions

1. Twelve reaction vials are placed on the stirrer (**Fig. 7**).
2. To each of the reaction vials, lipase (100 mg) and 3-phenoxy-1,2-propandiol (**1**, 100 mg) are added, followed by the reaction solvent (3 mL).
3. At each time interval (2, 15, 60, and 120 min, respectively), small amounts of sample solution (20 µL) were taken and applied onto a TLC plate.
4. The TLC plate was developed, and after the solvent front reached a height of about 10 cm, the plate was removed, the solvent was dried, and then the TLC plate was quickly dipped into and taken out of the visualization solution. The TLC plate was heated on a hot plate for color to develop.

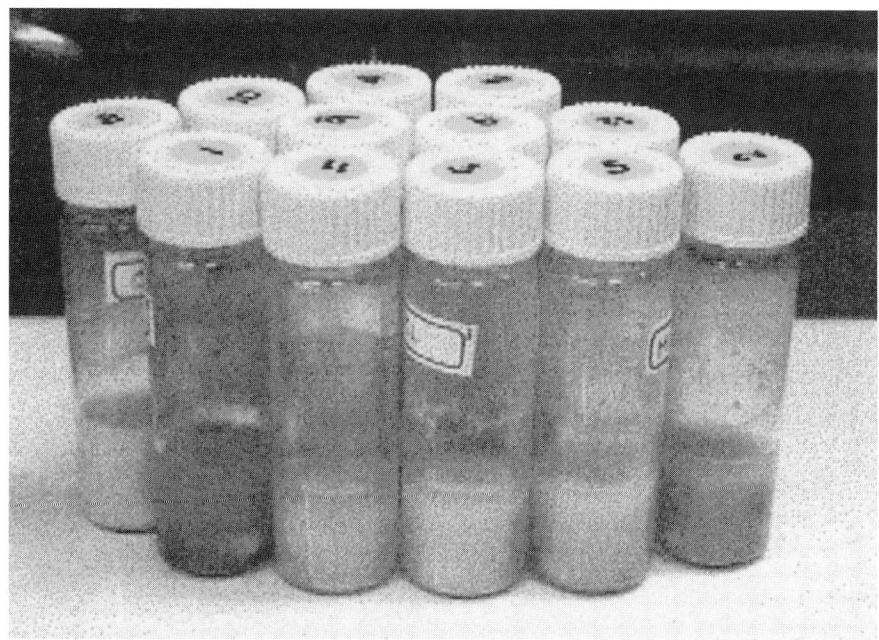

Fig. 7. The 12 vials on the stirrer. Each vial has a stirring bar to mix the enzyme solution.

5. **Figure 8** shows the results of the lipase-catalyzed acylation after the biocatalyzed reaction for 2 min **(Fig. 8A)**, 15 min **(Fig. 8B)**, 60 min **(Fig. 8C,)**, and 120 min **(Fig. 8D)**. TLC Rf: 0.20 for 3-phenoxy-1,2-propandiol **1**; 0.57 for 1-acetyl 3-phenoxy-1,2-propandiol **2**; 0.85 for 1,2-diacetyl 3-phenoxy-1,2-propandiol **3** (*see* **Note 4**). The lipase-catalyzed hydrolysis of vinyl acetate forms an acyl-enzyme intermediate, which, in turn, has been attacked by the nucleophilic oxygens of diol **1** to form, acetates **2** and **3**. The initial acetylation product will be a mixture of 1-acetate and 2-acetate. Most of the tested enzymes favor the C-1 alcohol as the nucleophile to form C-1 ester. After substantial amounts of the monoacetate **2** were formed, the acetylation continued on the second hydroxyl group to produce diacetate **3**.

3.3. Determining Enzymatic Activity in Organic Solvents

3.3.1. Removing Water from the Alcalase Solution

1. Alcalase 2.5 L (0.5 mL) and 2-methyl-2-propanol (5 mL) were added to a centrifuge tube (10 mL) and the mixture was agitated on a supermixer for 10 min.
2. The resulting mixture was centrifuged (1700g) for 10 min to spin down the enzyme, and the supernatant was decanted.
3. A fresh sample of 2-methyl-2-propanol (5 mL) was added, and the procedure was repeated three times until the water was removed completely.

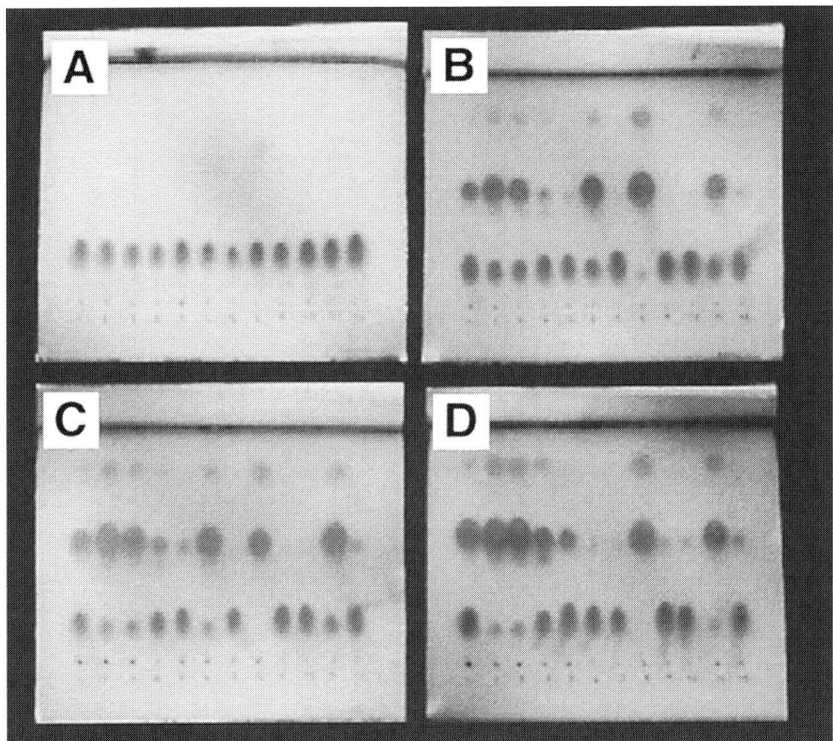

Fig. 8. The results of the developed TLC plate after enzyme catalysis for 2, 15, 60, and 120 min.

3.3.2. Enzymatic Activity

1. Cbz-Phe-OBzl (25 μmol, 9.74 mg) was dissolved in absolute ethanol (5 mL) and a freshly prepared alcalase (0.5 mL, prewashed with 2-methyl-2-propanol three times) solution was added.
2. The solution was stirred at room temperature.
3. Aliquots (20 μL) were taken and quenched by the addition of enough HCl (0.1 N, 4.8–4.9 mL) to make the final volume 5 mL.
4. The solution was centrifuged for 5 min at 3000 rpm, and 20-μL aliquots of supernantant were analyzed by HPLC using a RP-18 column.
5. The peak intensity corresponding to same retention time as Cbz-Phe-OEt and Cbz-Phe-OBzl were determined, and the reaction rates of transesterification were calculated against a calibration curve for Cbz-Phe-OEt (*see* **Note 5**).
6. **Figure 9** shows a typical HPLC of the transesterification of Cbz-Phe-OBzl (retention time 7.2 min) to Cbz-Phe-OEt (retention time 5.3 min) after the reaction solution was incubated for 0, 5, 30, and 60 min. Phenol (retention time 2.9 min) was added to the reaction solution as an internal standard, and Cbz-Phe-OH

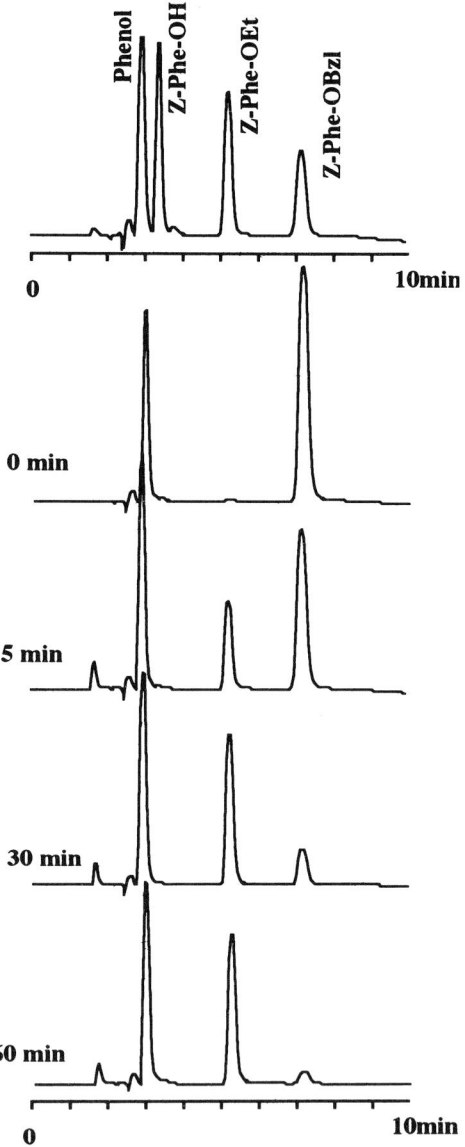

Fig. 9. HPLC analysis of alcalase-catalyzed transesterification of Cbz-Phe-OMe and Cbz-Phe-OBzl in ethanol solution to produce Cbz-Phe-OEt.

(retention time 3.3 min) was used to check for hydrolysis product present during the reaction. From the decreased area of the Cbz-Phe-OBzl or the increased area of Cbz-Phe-OEt, the activity of transesterification is calculated using the following equation:

Rate of transesterification of alcalase in ethanol = aS/Et

where: a is the percentage increase of Cbz-Phe-OEt or the percentage decrease of Cbz-Phe-OBzl, S is the molar number of the substrate, E is the enzyme unit used, and t is the reaction period.

3.4 The Determination of Residual Enzyme Activity After a Reaction in Organic Solvents

1. Subtilisin Carlsberg (20.0 mg) was dissolved in a cold phosphate buffer (50 mM, pH 8.0, 10 mL) in an ice bath.
2. The resulting solution was divided into 10 vials with 1 mL solution in each, and lyophilized to dryness (*see* **Note 6**).
3. To each vial, 95% ethanol (3 mL) was added and the mixture stirred at 35°C.
4. At time intervals of 0, 15, 30, 60, and 180 min. a substrate solution (Moz-Trp-OBzl, 100 µL) was added and stirring continued at 30°C for further 15 min before the reaction was quenched by the addition of 0.1N HCl (1 mL).
5. An aliquot of the solution was taken for HPLC analysis of the extent of hydrolysis and transesterification reactions. The results were analyzed using the initial time as 100% enzymatic activity. The remaining enzymatic activity of each aliquot was measured against initial activity.
6. **Figure 10** shows an HPLC analysis profile of the remaining activity of subtilisin Carlsberg incubated in 95% ethanol at 35°C. **Figure 10A**, shows the separation of the three peaks: Moz-Trp-OBzl (substrate, retention time 9.95 min), Moz-Trp-OEt (transesterification product, retention time 6.56 min), Moz-Trp-OH (hydrolysis product, retention time 3.87 min) that resulted from reacting the testing solution for 15 min at 35°C (*see* **Note 7**). The enzymatic activity at time zero was used as the 100% level for calculcation of remaining activity of enzymes incubated for longer periods. **Fig. 10B–D** show the HPLC of the reaction from enzyme incubated for 30, 60, and 180 min, respectively. Using the peak area of Moz-Trp-OEt and Moz-Trp-OH in **Fig. 10A–D**, the remaining activity of transesterification and hydrolysis are thus calculated. **Figure 11** shows the remaining enzymatic activity of subtilisin Carlsberg in 95% ethanol. The enzyme has a half life of about 60 minutes.

3.5. The Inversion of Enantioselectivity by Solvents

1. To each of the reaction vials were added lipase MY (50 mg), 3-phenoxy-1, 2-propanediol (168.4 mg), vinyl acetate (0.2 mL), and reaction solvents (3 mL of acetone, chloroform and methyl chloride, respectively).
2. The resulting mixture was stirred at 35°C for 24 h.
3. Aliquots (0.1 mL) were taken and transferred into a microcentrifuge tube. The solvent was evaporated and the residue was dissolved in the eluents for chiral HPLC analysis to determine the enantiomeric excess of unreacted 3-phenoxy-1,2-propanediol (*see* **Note 8**).
4. **Figure 12** shows the analysis of unreacted starting material by a chiral OD-RH column. R-(+)-3-phenoxy-1,2-propanediol eluted with a retention time of 8.2 min and the S-(–)-3-phenoxy-1,2-propanediol eluted at 10.45 min (**Fig. 12A**). **Figure**

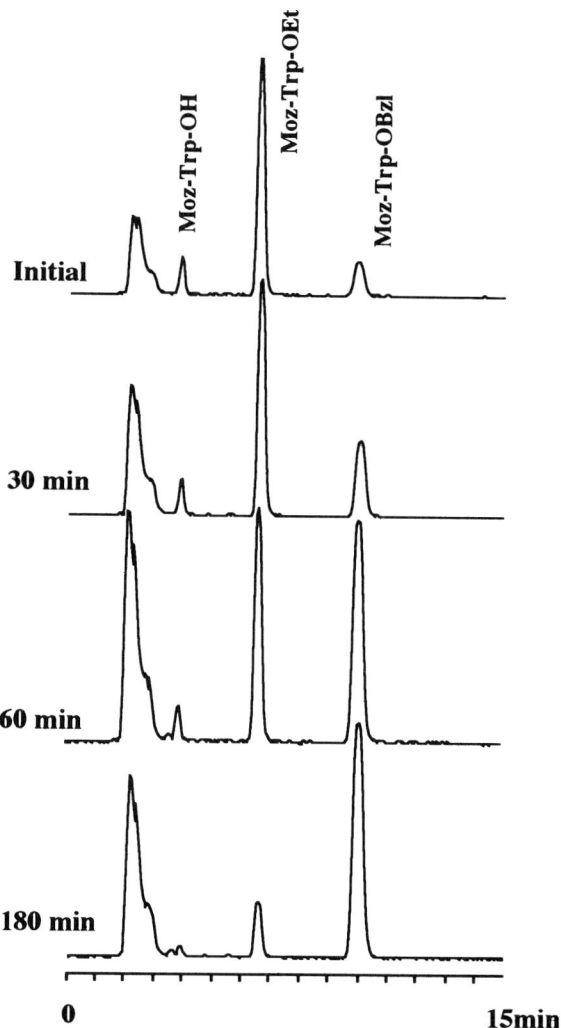

Fig. 10. HPLC analysis for measuring remaining activity of subtilisin Carlsberg incubated in 95% ethanol at 35°C.

12B–D. shows the analysis of the enzyme catalysis carried out in chloroform, methylene chloride, and acetone.

5. The lipase MY-catalyzed acetylation has an enantiomeric preference for R-(+)-3-phenoxy-1,2-propandiol when the reaction was performed in chloroform and in methylene chloride, with the ratio of unreacted substrate R/S substrate being 1:3. However, when the reaction was carried out in acetone, the enzyme had an enantiomeric preference for S-(–)-3-phenoxy-1,2-propandiol and gave a ratio of the unreacted substrate R/S of 45/42.

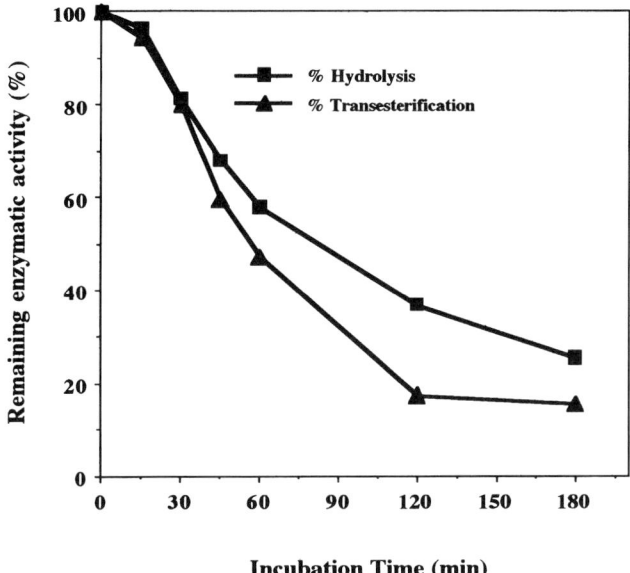

Fig. 11. The remaining enzymatic activity of subtilisin Carlsberg at 95% ethanol. The enzyme has a half-time of about 60 min.

3.6. The Regioselective Acylation of Sugar Derivatives

3.6.1. Removing Water from Alcalase

1. Alcalase (1.0 mL) and 2-methyl-2-propanol (10 mL) were added to a centrifuge tube (15 mL) and the mixture was mixed for 10 min.
2. The mixture was centrifuged at 1700*g* for 10 min and the supernatant was decanted.
3. Fresh 2-methyl-2-propanol (10 mL) was added, and the procedure repeated three times until the water was completely removed.

3.6.2. Regioselective Acylation

1. Pretreated alcalase is suspended in 2-methyl-2-propanol (10 mL), and compound **8** (ethyl 2-acetamido-2-deoxy-1-thio-D-glucose) (0.65 g, 2.44 mmol), vinyl benzoate (2 mL, 3 mmol), and triethylamine (1 mL) are added.
2. Mixtures were stirred for 0, 30, 60, 90, and 120 min.
3. After 120 min, TLC showed a complete conversion of **8** into **9** (ethyl 2-acetamido-2-deoxy-1-thio-D-glucose benzyl ester) (R_f = 0.65).
4. The reaction was stopped by filtration of the enzyme and evaporation of the solvent.
5. The residue was crystallized from *n*-hexane (20 mL) and dichloromethane (20 mL) and filtered to give product **9** (0.586 g, 1.71 mmol, 70%). (R_f = 0.33 for **8**; R_f = 0.65 for **9** in EtOAc EtOH = 90:10).

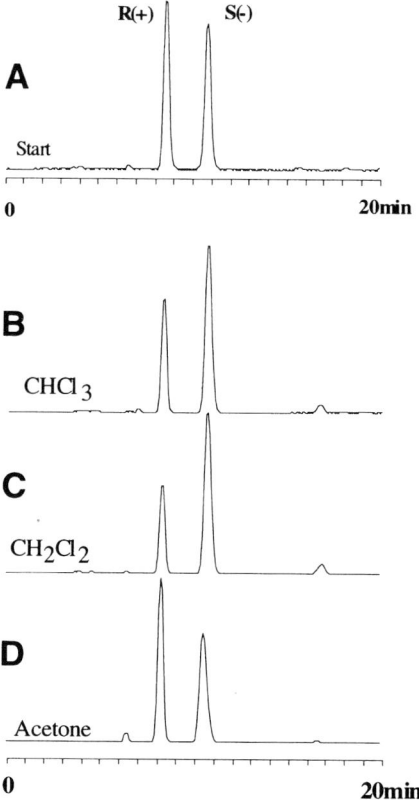

Fig. 12. The analysis of unreacted starting material by a chiral OD-RH column. R-(+)-3-phenoxy-1,2-propandiol eluted with retention time 8.2 min, and S-(–)-3-phenoxy-1,2-propandiol eluted at 10.45 min.

6. **Figure 13** shows the results of TLC monitoring of the reaction: disappearance of starting material ($R_f = 0.37$) and formation of the product ($R_f = 0.65$) within 120 min.

3.7. The Resolution of Amino Acids in the Mixture of 95% 2-Methyl-2-Propanol/5% Water

1. A solution of D,L-tyrosine methyl ester (0.10 mol, 19.5 g) and the alcalase 2.5L (5 mL) dissolved in a mixture of water (10 mL) and 2-methyl-2-propanol (190 mL) was incubated at 35°C (*see* **Note 9**).
2. The pH of the reaction solution was maintained continuously at 8.2 (*see* **Note 10**).
3. During the hydrolysis, L-amino acid precipitated.
4. The reaction was continued until 25 mL of 2 M NaOH was consumed (about 4 h).

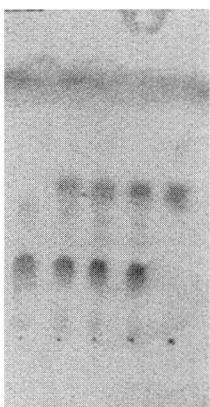

Fig. 13. TLC of regioselective benzoylation of sugar.

5. The resulting solution was filtered to yield L-tyrosine (8.68 g, yield 96%), melting point >300°C, $[\alpha]^{25}_D$ +21.7 ($c = 5$, 5 M HCl).
6. The pH of the filtrate was adjusted to 11.5 with NaOH (6 M) and left overnight (*see* **Note 11**).
7. After the amino acid was saponified (monitored by TLC), the pH of the resulting solution was adjusted to 6.2 with citric acid (10%).
8. The D-amino acid was precipitated after chilling for several hours.
9. The precipitation was filtered and washed with cold water (2×5 mL) to yield D-tyrosine (7.96 g, yield 88%) melting point >300°C, $(\alpha)^{25}_D$ −19.20 ($c = 5$, 6 M HCl).
10. The enantiomeric excesses of both antipodes were measured by Chiral CR-(+) column using aqueous perchloric acid (pH 1.5–2.0) as the eluent. The retention time depended on acidity, with a lower pH leading to a longer HPLC retention time. For hydrophobic amino acids, 5–15% methanol was added to the eluent to adjust the retention time of each amino acid. Peaks were detected at 200 nm. Peak identification was made by comparison of the retention times with those of authentic compounds prepared chemically (*see* **Note 12**).
11. **Figure 14** shows typical chromatographic results for the chiral stationary-phase HPLC analysis. **Figure 14A** shows the baseline separation of authentic D-tyrosine, L-tyrosine, D-tyrosine methyl ester, and L-tyrosine methyl ester. **Figure 14B** shows the analysis of D-tyrosine methyl ester in the solution, after L-tyrosine was separated by filtration. It shows that before the saponification, the filtrate contained D-tyrosine methyl ester (94.5%), unhydrolyzed L-tyrosine methyl ester (3%) and L-tyrosine (2.5%). **Figure 14C** shows the analysis of D-amino acid after saponification. Both the D-tyrosine methyl ester and unhydrolyzed L-Tyrosine methyl ester were totally converted to the free amino acid. Under the conditions employed, no racemization occurred in the base-catalyzed saponification of unhydrolyzed residue of either methyl ester or benzyl ester.

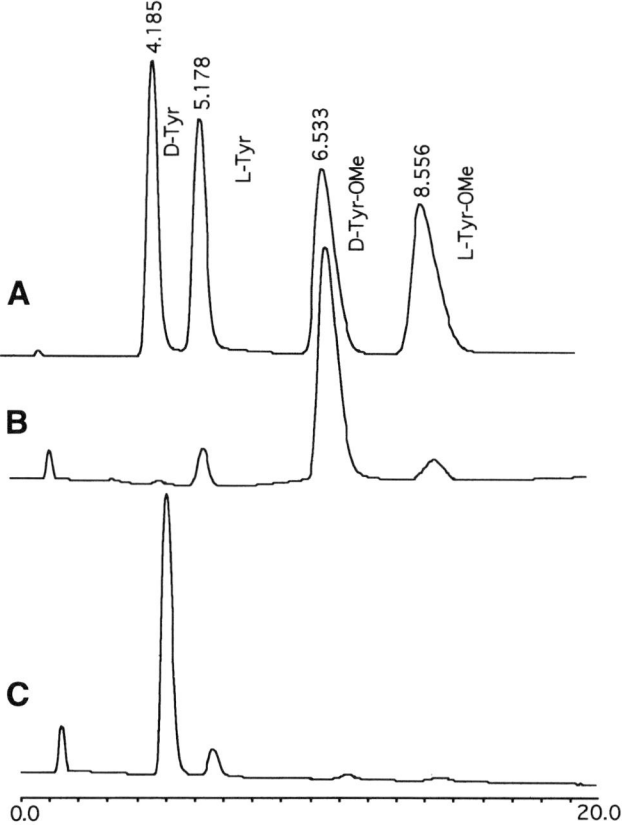

Fig. 14. Determination of enantiomeric excess of each product in resolution solution before isolation via the chiral separation of four components. (**A**) Standard of D, L-tyrosine and D, L-tyrosine methyl ester; (**B**) each of components in reaction solution after hydrolysis; (**C**) enantiomeric excess of D-tyrosine after saponification.

3.8. Enhancement of the Rate of Enzyme Reactions by Microwave Irradiation

3.8.1. Remaining Activity of Alcalase Following Microwave Irradiation (see **Note 13**)

The remaining activity of alcalase in 2-methyl-2-propanol was assayed by the procedure described in **Subheading 3.3.**

1. To each of eight reaction vials was added alcalase 2.5 L (0.1 mL) and 2-methyl-2-propanol (3 mL).
2. The resulting solution was put in a microwave oven and irradiated with 10% of full power for specified time (0, 1, 2, 5, 10, 20, 30, and 40 min, respectively).

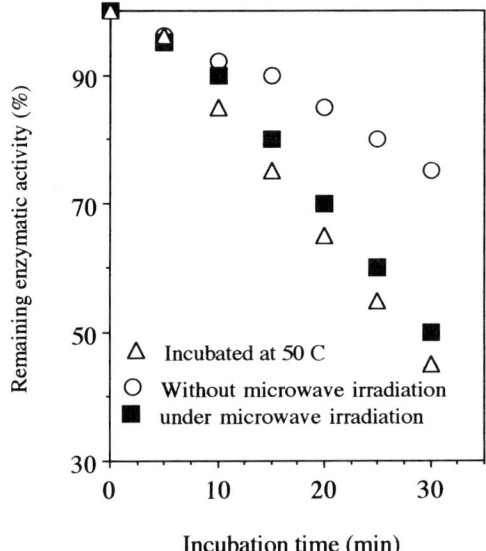

Fig. 15. Time-course of alcalase inactivation under the reaction conditions.

3. To the reaction vial was then added a substrate solution of Moz-Leu-OBzl, (35.5 mg, 0.1 mmol), dissolved in absolute ethanol (1.0 mL).

4. The resulting solution was allowed to stir at 25°C for 15 min and quenched by the addition of 0.1 *N* HCl (1 mL).

5. An aliquot of the solution was taken and analyzed by HPLC for the extent of transesterification.

6. Using the activity of enzyme without microwave irradiation as 100%, the remaining activity of alcalase for other irradiation times can be calculated.

7. **Figure 15** shows the time-course of alcalase inactivation under the reaction conditions. The temperature of the solution rose from 25°C to 50°C after the solution was irradiated for 30 min. After 30 min incubation, the alcalase maintained 50% of its original activity following irradiation, and maintained 75% of the activity without the irradiation and 44% of the activity after incubation at 50°C. The results showed that the thermo-inactivation of enzymatic activity might be the main reason for the inactivation of the alcalase under microwave conditions.

3.8.2. Peptide Bond Formation

1. Cbz-Ala-Phe-OMe (3.84 g, 10 mmol), Pro-NH$_2$HCl (4.50 g, 30 mmol), triethylamine (4.3 mL, 31 mmol) in 2-methyl-2-propanol (30 mL) and alcalase 2.5 L (5.0 mL), pretreated with anhydrous 2-methyl-2-propanol) were put in a Synthewave microwave oven and irradiated.

2. After all the Cbz-Ala-Phe-OMe disappeared (about 20 min, monitored by HPLC), the mixture was evaporated and residue was dissolved in ethyl acetate (200 mL).

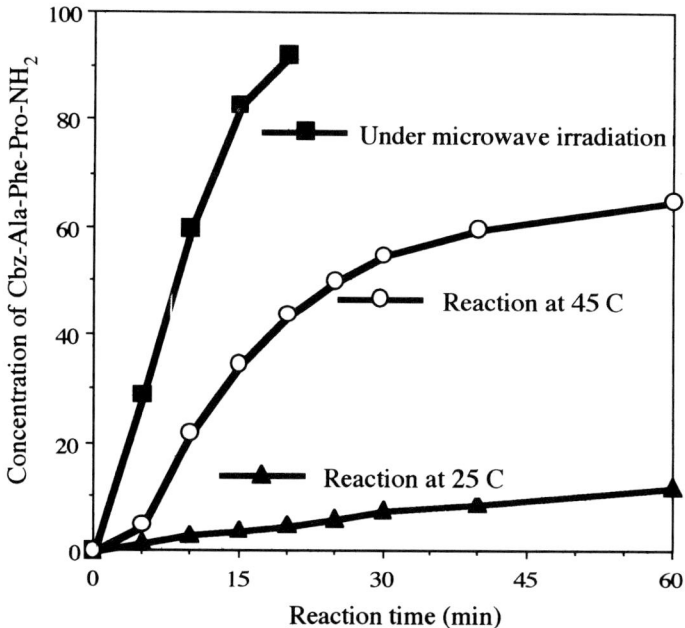

Fig. 16. Time-course of the increasing concentration of Cbz-Ala-Phe-Pro-NH$_2$ as the reaction proceeds.

3. The resulting solution was washed with 5% citric acid (3 × 25 mL), water (3 × 25 mL), 5% sodium bicarbonate (3 × 25 mL), dried over anhydrous sodium sulfate, and evaporated to give crude Cbz-Ala-Phe-Pro-NH$_2$.
4. This was purified via silica gel flash column chromatography eluted with MeOH:CH$_2$Cl$_2$ (4:1, v/v) to yield pure Cbz-Ala-Phe-Pro-NH$_2$ (3.62 g, 85% yield); melting point 138–142°C.
5. To further characterize the effect of microwave irradiation, HPLC was used to follow the course of the alcalase-catalyzed reaction with and without microwave irradiation.
6. **Figure 16** shows the time-course of the increasing concentration of Cbz-Ala-Phe-Pro-NH$_2$ (*see* **Note 14**).

4. Notes

1. The aggregation of enzyme suspended in organic solvents can be counteracted by dissolving the enzyme in an aqueous solution and lyophilization. Resuspension of the lyophilized powder in organic solvents improves the enzymatic activity significantly.
2. Poor enzymatic activity in organic solvents is apparent when the enzyme sticks together and is attached to the wall of the reaction vessel. In this situation, the enzymatic activity is decreased to near zero. However, when the enzyme is taken out and redissolved in aqueous solution, the enzyme recovers its activity. This means that an enzyme with poor activity in an organic solvent can be returned to

having good activity in aqueous solution. Enzymes with poor enzymatic activity in organic solvents will aggregate in order to stabilize the enzyme structure.

3. There are many procedures for the preparation of active enzyme suspended or dissolved in organic solvents, such as the following: (1) the use of lyophilized enzyme powder, (2) the addition of solid enzyme to a eutectic mixture, (3) the use of additives to solubilize the enzyme, and (4) precipitation of the enzyme by adding organic solvent followed by resuspension. The first three procedures have been described in detail (*17*). The latter procedure is worth mentioning for its simplicity in generating highly active enzymes. The procedure contains three steps: (1) dissolve the enzyme in aqueous solution and adjust the pH of the solution for optimal enzymatic activity; (2) add organic cosolvent to precipitate the enzyme (2-methyl-2-propanol, isopropanol, ethanol, or *t*-amyl alcohol can be used); (3) remove the water by repeated suspension and precipitation of the enzyme solution with the cosolvent. An example for the preparation of alcalase in 2-methyl-2-propanol involving removal of water from the alcalase solution is described in detail in **Subheading 3.3.**).

4. There are two inconsistency on the TLC plate (*see* **Fig. 8**).
   a. All the spots were applied using the same amounts of sample, but the visualized spots were not equal. This may be caused by the inhomogeneous heating of the TLC plate on the hot plate to visualize color.
   b. Some black color appeared on the plate, which was the result of the overheating of the visualization solution of 5% phosphomolybdic acid in alcohol on the hot plate.
   c. An unusual result was observed on the TLC plate for testing enzyme 6 at a reaction time of 120 min. This may have arisen from misapplication of sample solution 7 in place of solution 6.

5. Calibration curves for Cbz-Phe-OBzl and Cbz-Phe-OEt are prepared by injection of (20, 15, 10, 5, and 2.5 µL) of the same concentration of the Cbz-Phe-OBzl or Cbz-Phe-OEt solution to the HPLC and calculated from the area of the corresponding peak to give a linear line for the standard calibration curve.

6. Subtilisin Carlsberg is a protease, Therefore, the complete procedure should be done as quickly as possible at near 0°C to prevent autodigestion. Incubation of enzyme in pH 8.0 or in ethanol for different times may affect the remaining activity.

7. For measurement of remaining activity at initial time, a mixture of substrate solution (100 µL) and 95% ethanol (3 mL) at 35°C was added to the lyophilized powder of subtilisin Carlsberg on a stirrer. The resulting solution was reacted for 15 min and the following procedures were the same as those for measurement of the remaining activity at other incubation times.

8. 3-Phenoxy-1,2-propanediol has a chiral center at C-2. The acetylation will occur at C-1 and C-2 to form a complex product mixture. Determination of the enantiomeric excess of unreacted starting material is a simple way to determine the catalytic efficiency of this enzyme catalysis.

9. Methyl esters of racemic amino acids were used in the resolution. The alcalase had a higher esterase activity at pH 8.2 than at pH 7.0. Under these conditions,

the solubility of amino acid esters was also very high. Using a high concentration of the substrate (>10%) in resolution is potentially useful for a large-scale process. The free amino acid was less soluble in organic solvent containing 5% water. The precipitation of product occurred during the course of hydrolysis. The procedure illustrated a practical application of alcalase in amino acid resolution, particularly in the resolution of hydrophobic amino acid derivatives, which are insoluble in aqueous solutions. The results demonstrate that a small amount of water in the reaction solution is enough for the purpose of resolution.

10. The hydrolysis catalyzed by alkaline protease (alcalase) was conducted in a mixture of 95% 2-methyl-2-propanol/5% water at 25°C at pH 8.5 with a pH controller. The hydrolyzed L-amino acid, which was insoluble under these conditions, precipitated during the course of the hydrolysis and was separated by filtration.

11. The pH of the filtrate was adjusted to 11.5 to saponify the unreacted ester of the D-amino acid.

12. In order to make sure the racemization would not occur during the course of saponification, a chiral separation procedure to determine the ee of the D-amino acids before and after the saponification was applied. Baseline separation of the two enantiomeric pairs of the amino acid and the esters of amino acid was achieved by using the chiral column.

13. The remaining activity of alcalase after various irradiation times was determined by measuring the increased concentration of Moz-Leu-OEt via the trans-esterification of Moz-Leu-OBzl to Moz-Leu-OEt in absolute ethanol/2-methyl-2-propanol (1:3 v/v). The stability of the enzyme in ethanol was displayed as the basal activity of the enzyme remaining after incubation.

14. The temperatures of the reaction solution after microwave irradiation for 5, 10, 15, and 20 min were 35°C, 38°C, 42°C, and 45°C, respectively. When the reaction was carried out with microwave irradiation, the peak corresponding to the acyl donor gradually decreased and disappeared after the reaction solution had been irradiated for 20 min at which time the product appeared to have reached maximum concentration (91%). With a negative control, only a small quantity (7.5%) of product formed when the reaction was carried out at 25°C for 20 min with no irradiation. When reacted at 45°C, the product appeared rapidly for the first 30 min, reaching approx 44% of total at 20 min and 55% of total at 30 min, and more slowly after the first 30 min. The reaction continued for 5 h with approx 78% yield. At this point, there was hardly any activity of alcalase. This result indicates that microwave irradiation can enhance the rate of alcalase-catalyzed peptide bond formation between Cbz-Ala-Phe-OMe and Pro-NH_2 by approx 2 to 10 times that without microwave irradiation.

Acknowledgment

Support for this research provided by the National Science Council, Taiwan is gratefully acknowledged.

References

1. Homandberg, G. A., Mattis, J. A., and Laskowski, M., Jr. (1978) Synthesis of peptide bonds by proteinases. Addition of organic cosolvents shifts peptide bond equilibria toward synthesis. *Biochemistry* **17,** 5220–5227.
2. Barbas, C. F., III Matos, J. R., West, J. B., and Wong, C. H. (1988) A search for peptide ligase: cosolvent-mediated conversion of proteases to esterases for irreversible synthesis of peptides. *J. Am. Chem. Soc.* **110,** 5162–5166.
3. Jackson, D. Y., Burnier, J., Quan, C., Stanley, M., Tom, J., and Wells, J. A. (1994) A designed peptide ligase for total synthesis of ribonuclease a with unnatural catalytic residues. *Science* **266,** 243–247.
4. Abrahmsen, L., Tom, J., Burnier, J., Butcher, K. A., Kossiakoff, A., and Wells, J. A. (1991) Engineering subtilisin and its substrates for efficient ligation of peptide bonds in aqueous solution. *Biochemistry* **30,** 4151–4159.
5. Gomez-Puyou, M. T. D. and Gomez-Puyou, A. (1998) Enzymes in low water systems. *Crit. Rev. Biochem. Mol. Biol.* **33,** 53–89.
6. Klibanov, A. M. (1997) Why are enzymes less active in organic solvents than in water? *TIBTECH* **15,** 97–101.
7. Klibanov, A. M. (1989) Enzymatic catalysis in anhydrous organic solvents. *Trends Biochem. Sci.* **14,** 141–144.
8. Gupta, M. N. (1992) Enzyme function in organic solvents. *Eur. J. Biochem.* **203,** 25–32.
9. Wong, C. H. and Wang, K. T. (1991) New developments in enzymatic peptide synthesis. *Experientia* **47,** 1123–1129.
10. Wong, C. H. (1989) Enzymatic catalysis in organic synthesis. *Science* **244,** 1145–1152.
11. Chen, C. S. and Sih, C. J. (1989) General aspects and optimization of enantioselective biocatalysis in organic solvents: the use of lipases. *Angew. Chem. Int. Ed. Engl.* **28,** 695–707.
12. Dordick, J. S. (1992) Designing enzymes for use in organic solvents. *Biotechnol. Prog.* **8,** 259–267.
13. Klibanov, A. M. (1986) Enzymes that work in organic solvents. *Chemtech* **16,** 354–359.
14. Dastoli, F. R., Musto, N. A., and Price, S. (1966) Reactivity of active sites of chymotrypsin suspended in an organic medium. *Arch. Biochem. Biophys.* **115,** 44–47.
15. Dastoli, F. R. and Price, S. (1967) Further studies on xanthine oxidase in nonpolar media. *Arch. Biochem. Biophys.* **122,** 289–291.
16. Zaks, A. and Klibanov, A. M. (1985) Enzyme catalyzed processes in organic solvents. *Proc. Nat. Acad. Sci. USA* **82,** 3192–3196.
17. Koskinen, A. M. P. and Klibanov, A. M. (1996) *Enzymatic Reactions in Organic Media.* Blackie Academic & Professional, New York.
18. Dickinson, M. and Fletcher, P. D. I. (1989) Views and comment: enzymes in organic solvents. *Enzyme Microb. Technol.* **11,** 55,56.
19. Bell, G., Halling, P. J., Moore, B. D., and Partridge J. (1995) Biocatalyst behaviour in low-water system. *Trends Biotechnol.* **13,** 468–473.

20. Gill, L. and Vulfson, E. (1994) Enzymic catalysis in heterogeneous eutectic mixtures of substrates. *TIBTECH* **12,** 118–122.

21. Walsh, C. (1977) *Enzymatic Reaction Mechanisms*, W. H. Freeman and Co., San Francisco, pp. 9–48.

22. Faber, K. and Riva, S. (1992) Enzyme-catalyzed irreversible acyl transfer. *Synthesis* 895–910.

23. Jakubke, H. D., Kuhl, P., and Konnecke, A. (1985) Basic principles of protease-catalyzed peptide bond formation. *Angew. Chem. Int. Ed. Engl.* **24,** 85–93.

24. Wedcott, C. R. and Klibanov, A. M. (1994) The solvent dependence of enzyme specificity. *Biochim. Biophys. Acta* **1206,** 1–9.

25. Halling P. J. (1989) Lipase-catalysed reactions in low-water organic-media — effects of water activity and chemical modification. *Biochem. Soc. Trans.* **17,** 1142–1145.

26. Ingalls, R. G., Squires, R. G., and Butler, L. G. (1975) Reversal of enzymatic hydrolysis: rate and extent of ester synthesis as catalysed by chymotrypsin and subtilisin carlsberg at low water concentrations. *Biotechnol. Bioeng.* **17,** 1627–1637.

27. Brink, L. E. S., Tramper, J., Luyben, K. Ch. A. M., and Riet, K. V. (1988) Biocatalysis in organic media. *Enzyme Microb. Technol.* **10,** 736–743.

28. Kasche, V., Michaelis, G., and Galunsky, B. (1991) Binding of organic solvent molecules influences the p1'-p2' stereo- and sequence-specificity of α-chymotrypsin in kinetically controlled peptide synthesis. *Biotech. Lett.* **13,** 75–80.

29. Halling P. J. (1990) Solvent selection for biocatalyst in mainly organic system—prediction of effects on equilibrium position. *Biotechnol. Bioeng.* **35,** 691–701.

30. Nagashima, T., Watanabe, A., and Kise, H. (1992) Peptide synthesis by proteases in organic solvents: medium effect on substrate specificity. *Enzyme Microb. Technol.* **14,** 842–847.

31. Partridge, J., Halling, P. J., and Moore, B. D. (1998) Practical route to high activity enzyme preparations for synthesis in organic media. *Chem. Commun.* 841,842.

32. Wu, S. H., Lo, L. C., Chen, S. T., and Wang, K. T. (1989) Manipulation of enzymatic regioselectivity by structural modification of substrates. *J. Org. Chem.* **54,** 4220–4222.

33. Chen, S. T. and Wang, K. T. (1988) Papain catalysed esterification of N-protected amino acids. *Chem. Commun.* 327,328.

34. Chen, S. T., Hsiao, S. C., and Wang. K. T. (1991) Stable industrial protease catalyzed peptide bond formation in organic solvent. *Bioorg. Med. Chem. Lett.* **1,** 445–450.

35. Chen, S. T., Chen, S. Y., and Wang, K. Y. (1992) Kinetically controlled peptide bond formation in anhydrous alcohol catalyzed by an industrial protease "Alcalase". *J. Org. Chem.* **57,** 6960–6965.

36. Chen, S. T., Chen, S. Y., Tu, C. C., and Wang. K. T. (1993) Investigation of the conformation, solubility, structure and selectivity of alkaline proteases in anhydrous alcohols. *Bioorg. Med. Chem. Lett.* **3,** 1643–1648

37. Chen, S. T., Chen, S. Y., Tu, C. C., and Wang. K. T. (1995) Physicochemical properties of alkaline serine proteases in alcohol. *J. Protein Chem.* **14,** 205–216.
38. Wang, Y. F., Lalonde, J. J., Momongan, M., Bergbreiter, D. E., and Wong, C. H. (1988) Lipase-catalyzed irreversible trans-esterifications using enol esters as acylating reagents: preparative enantio- and regioselective synthesis of alcohols, glycerol derivatives, sugars, and organometallics. *J. Am. Chem. Soc.* **110,** 7200–7205.
39. Wang, Y. F., Chen, S. T., Liu, K. C., and Wong, C. H. (1989) Lipased-catalysed irreversible trans-esterification using enol esters: resolution of cyanohydrins andsynthesis of ethyl (*R*)-2-hydroxy-4-phenylbutyrate and (*S*)-propranolol. *Tetrahedron Lett.* **30,** 1917–1920.
40. Ueji, S., Fujino, R., Okubo, N., Miyazawa, T., Kurita, S., Kitadani, M., et al. (1992) Solvent-induced inversion of enantio-selectivity in lipase-catalysed esterification of 2-phenoxypropionic acids. *Biotechnol. Lett.* **14,** 163–168.
41. Janssen, A. E., Vaidya, A. M., and Halling, P. J. (1996) Substrate specificity and kinetics of *Candida rugosa* lipase in organic media. *Ann. NY Acad. Sci.* **799,** 257–261.
42. Janssen, A. E., Vaidya, A. M., and Halling, P. J. (1996) Substrate specificity and kinetics of *Candida rugosa* lipase in organic media. *Enzyme Microb. Technol.* **18,** 340–346.
43. Parida, S. and Dordick, J. S. (1993) Tailoring lipase specificity by solvent and substrate chemistries. *J. Org. Chem.* **58,** 3238–3244.
44. Sweers, H. M. and Wong, C. H. (1986) Enzyme-catalyzed regioselective deacylation of protected sugar in carbohydrate synthesis. *J. Am. Chem. Soc.* **108,** 6421,6422.
45. Riva, S., Chopineau, J., Kieboom, A. P. G., and Klibanov. A. M. (1988) Protease-catalyzed regioselective esterification of sugars and related compounds in anhydrous dimethylformamide. *J. Am. Chem. Soc.* **110,** 584–589.
46. Nicotra, F., Riva, S., Secundo, F., and Zucchelli, L. (1989) An interesting example of complementary regioselective acylation of secondary hydroxyl groups by different lipases. *Tetrahedron Lett.* **30,** 1703,1704.
47. Uemura, A., Nozaki, K., Yamashita, J. I., and Yasumoto, M. (1989) Lipase-catalyzed regioselective acylation of sugar moieties of nucleosides. *Tetrahedron Lett.* **30,** 3817,3818.
48. Faber, K. and Riva, S. (1992) Enzyme-catalyzed irreversible acyl transfer. *Synthesis* 895–910.
49. Houng, J. Y. and Chen, S. T. (1996) Kinetic resolution of amino acid esters catalyzed by lipases. *Chirality* **8,** 418–422.
50. Chen, S. T., Huang, W. H., and Wang, K. T. (1994) Resolution of amino acids in a mixture of 2-methyl-2-propanol (95%)/water (5%) catalyzed by alcalase via in situ racemization of one antipode mediated by pyridoxal 5-phosphate. *J. Org. Chem.* **59,** 7580,7581.
51. Chen, S. T., Huang, W. H., and Wang, K. T. (1994) Kinetic resolution of esters of amino acid in *t*-butanol containing 5% water catalyzed by a stable industrial alkaline protease. *Chirality* **6,** 572–576.

52. Technology vision 2020. The U. S. Chemical Industry. Copyright@ Dec. 1996 by The Amer. Chem. Soc., Amer. Inst. of Chem. Engin., The Chem. Manuf. Assoc. The Council of Chem. Res., The Synthetic Org. Chem. Manuf. Assoc.

53. Dagani, R. (1997) Molecular magic with microwaves: scientists are discovering new ways to use microwaves in organic synthesis, materials processing, waste remediation. *Chem. Eng. News* 26–33 (February 10).

54. Bose, A. K., Banik, B. K., Lavlinskaia, N., Jayaraman, M., and Manhas, M. S. (1997) More chemistry in a microwave: nontraditional techniques using domestic microwave ovens have been developed for conducing a wide variety of organic reactions that are fast, safe, inexpensive, and friendly to the enviroment. *CHEMTECH* 18–23 (September).

55. Caddick, S. (1995) Microwave assisted organic reactions. *Tetrahedron* **51,** 10,403–10,432.

56. Chen, S. T., Chiou, S. H., and Wang, K. T. (1990) Preparative scale organic synthesis using a kitchen microwave oven. *Chem. Commun.* 807.

57. Kabza, K. G., Gestwicki, J. E., McGrath, J. L., and Petrassi, H. M. (1996) Effect of microwave radiation on copper(ii) 2,2'-bipyridyl-mediated hydrolysis of bis(*p*-nitrophenyl) phospho-diester and enzymatic hydrolysis of carbohydrates. *J. Org. Chem.* **61,** 9599–9602.

58. Carrillo-Munoz, J. R., Bouvet, D., Guibe-Jampel, E., Loupy, A., and Petit, A. (1996) Microwave-promoted lipase-catalyzed reactions. resolution of (+)-1-phenylethanol. *J. Org. Chem.* **61,** 7746–7749.

59. Parker, M. C., Besson, T., Lamare, S., and Legoy, M. D. (1996) Microwave radiation can increase the rate of enzyme-catalysed reaction in organic media. *Tetrahedron Lett.* **37,** 8383–8386.

60. Gelo-Pujic, M., Guibe-Jampel, E., Loupy, A., Galema, S. A., and Mathe, D. (1996) Lipase-catalyzed esterification of some α-D-glucopyranosides in dry media under focused microwave irradiation. I. *Chem. Soc. Perkin Trans.* **1,** 2777–2780.

61. Lin, G. and Lin, W. Y. (1998) Microwave-promoted lipase-catalyzed reactions. *Tetrahedron Lett.* **39,** 4333–4336.

62. Chen, S. T., Tseng, P. H., Yu, H. M., W. C. Y., Hsiao, K. F., Wu, S. H., Wang, K. T. (1997) The studies of the microwave effects on the chemical reactions. *J. Chin. Chem. Soc.* **44,** 169.

32

Interface Bioreactor

Microbial Transformation Device on an Interface Between a Hydrophilic Carrier and a Hydrophobic Organic Solvent

Shinobu Oda, Takeshi Sugai, and Hiromichi Ohta

1. Introduction

Using this device referred to interface bioreactor, various microbial transformations of lipophilic substrate with a growing and/or living microorganism can be efficiently performed in a hydrophobic organic solvent (*see* **Fig. 1** and reviews in **refs.** *1* and *2*). In the device, the microorganism can grow using nutrients and water in the hydrophilic carrier and oxygen in the organic phase and spontaneously forms a thick microbial film on the surface of the carrier. Various lipophilic organic compounds added to the organic phase can be efficiently converted to the desired lipophilic products via microbial hydrolysis, esterification, oxidation, and reduction (*3*). Thus, we have applied this interface bioreactor in the syntheses of useful compounds as follows (**Fig. 2**):

1. Aliphatic carboxylic acids were synthesized via oxidation of 1-alkanols *(4)*.
2. *(R)*-2-Benzylcyclohexanone was synthesized via enantioface-differentiating hydrolysis of its enol ester *(5)*.
3. *(R)*-Sulcatol was synthesized via asymmetric reduction of 6-methyl- 5-hepten-2-one *(6)*.
4. Optically active citronellol and its derivatives were synthesized via enantioselective oxidation of racemic citronellol *(7,8)*.
5. Ethyl *(R)*-2-hydroxy-4-phenylbutanoate was synthesized via asymmetric reduction of ethyl 2-oxo-4-phenylbutanoate *(9)*.

In reactions 1–5, the toxicity of the substrate and the product on microbial cells grown on the carrier surface is efficiently alleviated, which we call as

From: *Methods in Biotechnology, Vol. 15: Enzymes in Nonaqueous Solvents: Methods and Protocols*
Edited by: E. N. Vulfson, P. J. Halling, and H. L. Holland © Humana Press Inc., Totowa, NJ

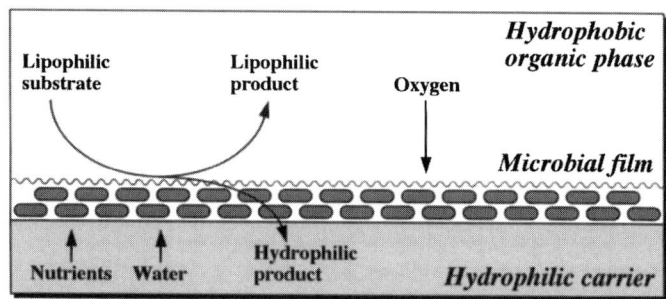

Fig. 1. Principle of interface bioreactor. Microorganism located on an interface between a hydrophilic carrier and an organic phase grows and forms a thick microbial film by using nutrients and water in the carrier. Oxygen is supplied from the organic phase.

toxicity alleviation on a solid–liquid interface (*10*; *see* **Note 1**). Consequently, concentration of the substrate and accumulation of the product in the organic phase drastically increased compared with those of an aqueous system (emulsion system). For example, although decanoic acid has a strong biotoxicity and kills many yeasts at only 0.2 g/L *(11,12)*, *Issatchenkia scutulata* var. *scutulata* IFO 10070 can produce 32.5 g/L of decanoic acid via oxidation of 1-decanol in the device *(4)*. We have proposed three types of interface bioreactor, an agar plate interface bioreactor (**Fig. 3**), a pad-packed interface bioreactor (**Fig. 4**), and a multistory interface bioreactor (**Fig. 5**). The agar plate interface bioreactor is the simplest one, consisting of a nutrient agar plate and an organic phase in a tray such as a glass Petri dish. In the pad-packed interface bioreactor, hydrophilic pads are axially located as carriers in a stainless-steel tank. In the multistory interface bioreactor, the agar plate interface bioreactors are stacked, and the organic phase in each reactor unit is connected with overflow lines and circulated from a bottom unit to head one with a diaphragm pump (*see* **Note 2**). The handmade reactor is easily and inexpensively constructed.

   It is expected that the interface bioreactor has some advantages compared with an emulsion and an organic–aqueous two-liquid-phase systems as follows:

1. A water-insoluble substrate and product can be solubilized into the organic phase (reaction solvent).
2. The substrate concentration and product accumulation in the organic phase can be drastically increased because the micro organism located on the solid–liquid interface can alleviate the toxicities of a substrate and a product presenting in the organic phase (*see* **Note 3**).
3. The recycling of coenzymes such as $NAD(P)^+$ and $NAD(P)H$ proceeds spontaneously because the microorganism on the solid–liquid interface is living.

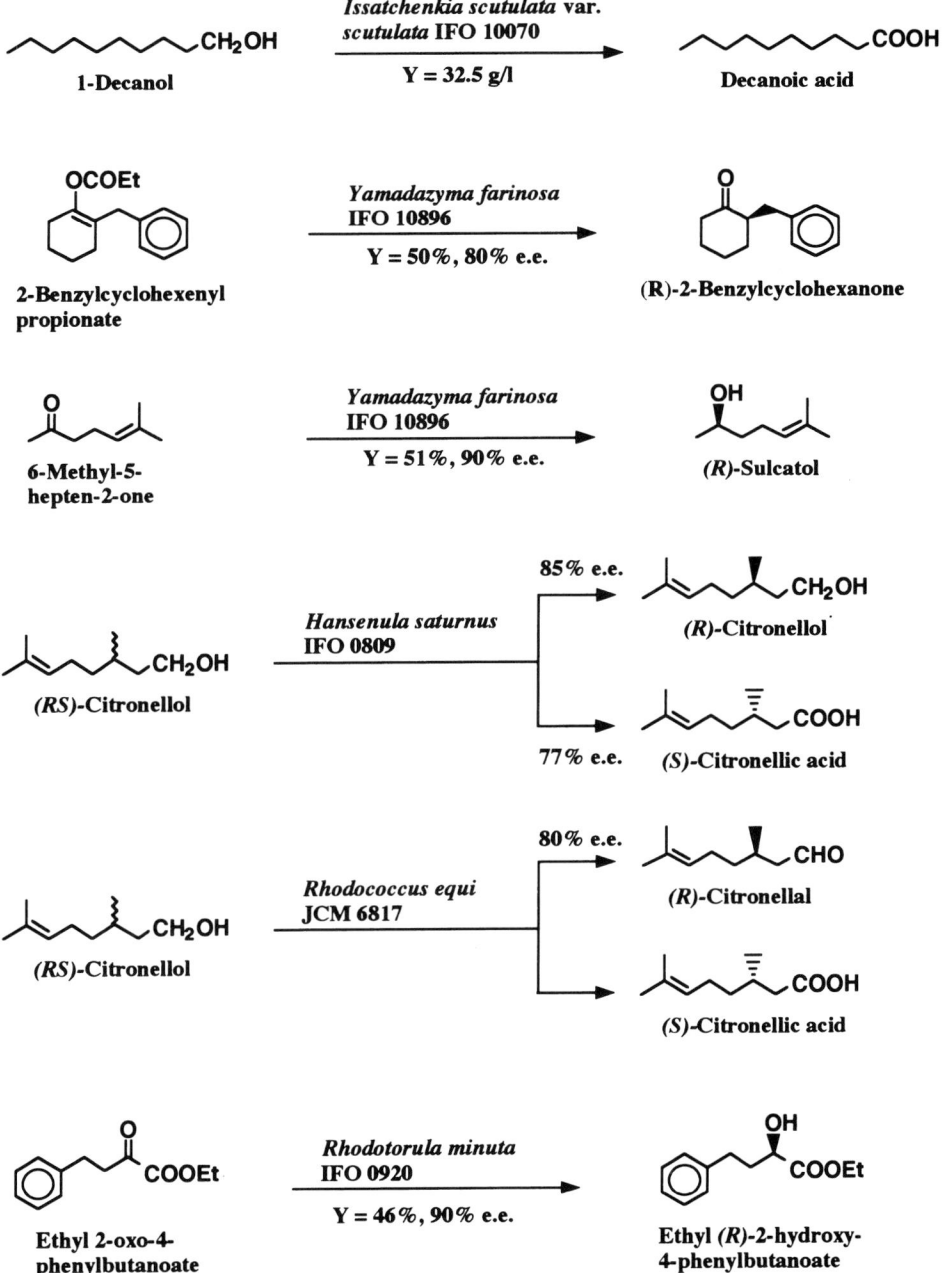

Fig. 2. Syntheses of some useful compounds with interface bioreactors. The interface bioreactor is applicable to various microorganisms and microbial transformations. Addition of the substrate and accumulation of the product are drastically increased because of a phenomenon, toxicity alleviation on the solid–liquid interface.

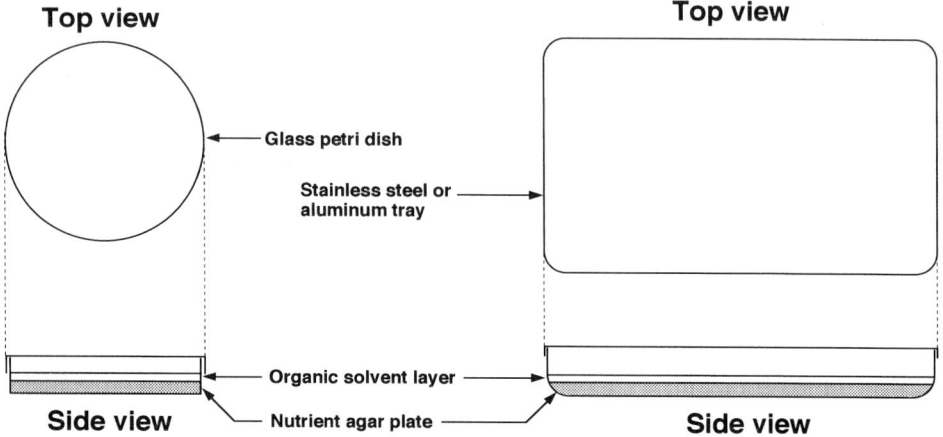

Fig. 3. Schematic diagrams of an agar plate interface bioreactor. The reactor is the simplest among three types of interface bioreactor. A microorganism growing on a nutrient agar plate can efficiently convert a substrate in an organic phase. A lipophilic product is highly accumulated in the organic phase.

Thus, metabolism, microbial oxidation, and reduction efficiently proceed in the device.

4. The cost of product recovery can be lowered because product accumulation in the organic phase is high, as mentioned earlier, and the separation of microbial cells adhering on the carrier surface from the reaction solvent is very easy.

5. The growth of the microorganism and microbial oxidation are effective because enough oxygen is spontaneously supplied from the organic phase (*see* **Note 4**).

6. Scale-up of the device is easy compared with the organic–aqueous two-liquid-phase system because the hydrophilic carriers (pads) can be located to three dimensions (*see* **Fig. 4**). Furthermore, a multistory interface bioreactor is also favorable to the scale-up because the reactor can be horizontally connected.

7. The application of the interface bioreactor is very wide because the device is applicable to many kinds of microorganism and microbial transformation. In this system, solvent-tolerant microorganisms *(13,14)* are not exactly essential.

8. The agar plate interface bioreactor can be efficiently used for the screening of microorganisms that have an activity of conversion of a lipophilic substrate because the conversion smoothly proceeds even by allowing the device to stand and the activity can be directly measured with gas chromatography (GLC) or high-performance liquid chromatography (HPLC) without any pretreatment such as extraction of the product.

Furthermore, we have reported a double- and a triple-coupling system as new types of acetylation system using the interface bioreactor *(15,16)*. In the double-coupling system, primary alcohols such as aliphatic, aromatic, and

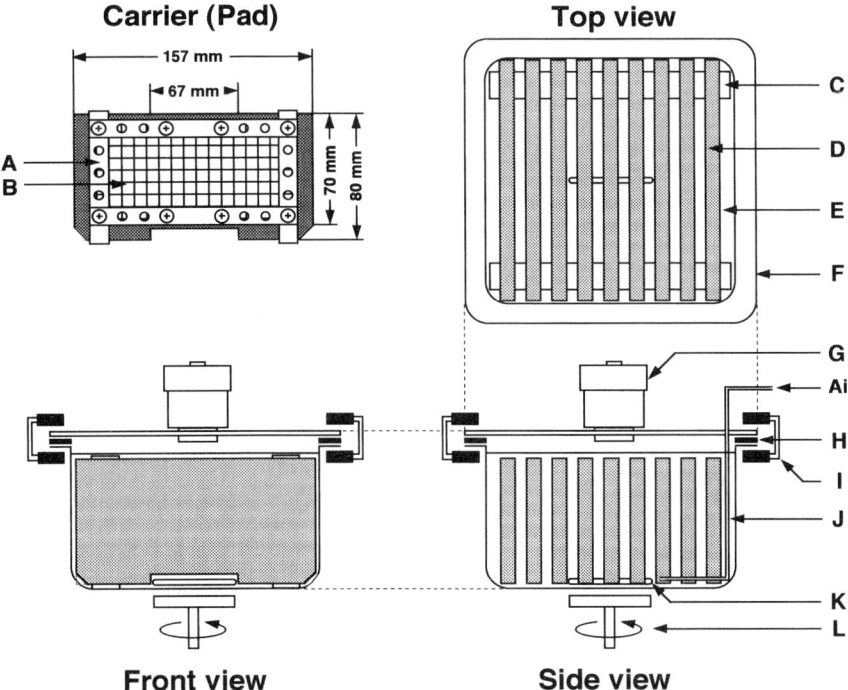

Fig. 4. Schematic diagram of a pad-packed interface bioreactor. A: Stainless-steel frame; B: stainless-steel net; C: Teflon spacer; D: hydrophilic carrier; E: organic phase; F: stainless-steel tank; G: oil mist trap; H: Viton packing; I: cramp; J: air line; K: magnet; L: magnetic stirrer. The reactor provides for a high volume of an organic layer containing a substrate.

terpene alcohols in the organic phase are efficiently acetylated with acetyl coenzyme A (acetyl-CoA) produced via metabolism of glucose by the aid of alcohol acetyltransferase [AATFase] (*see* **Note 5** and **Fig. 6**). On the other hand, the triple-coupling system consists of the microbial reduction of aldehydes to primary alcohols, the metabolism of glucose (formation of acetyl-CoA), and the acetylation of resulted primary alcohols with AATFase (**Fig. 6**). Moreover, if *Pichia kluyveri* IFO 1165 (corresponds to CBS 188) is used in the double-coupling system, racemic citronellol can be optically resolved via (*S*)-preferential acetylation as shown in **Fig. 7** *(17)*.

In this chapter, the interface bioreactor as a nonaqueous microbial transformation device and the double-coupling system, a new acetylation system without any acetyl donor are described. Furthermore, the application of two practical interface bioreactor, a pad-packed *(7,9,15,18)* and a multistory interface bioreactors *(17)* will be also explained.

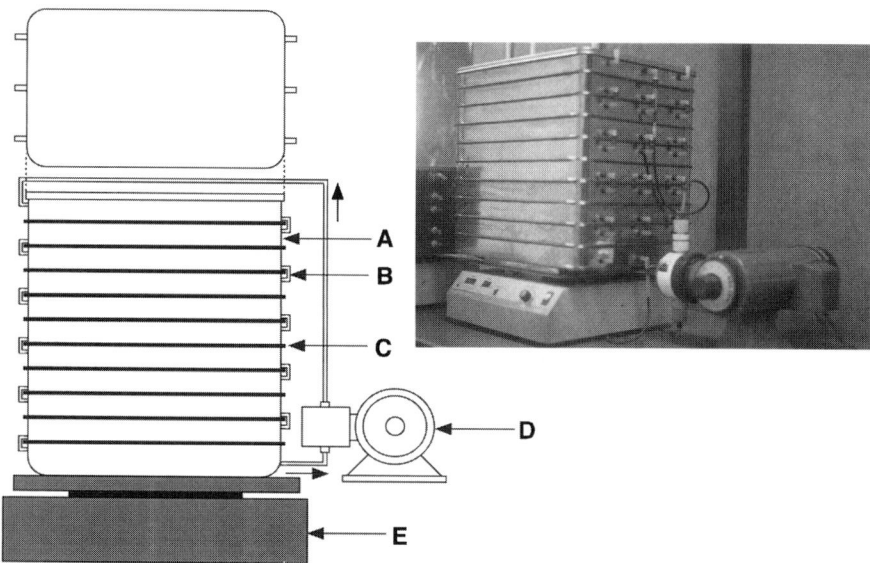

Fig. 5. Schematic diagram and a photograph of a multistory interface bioreactor. A: Reactor unit (agar plate interface bioreactor); B: overflow line; C: Viton packing; D: diaphragm pump; E: shaker. The reactor is amenable to simple and inexpensive scale-up. An organic phase overflowed from a bottom unit is circulated into a top one with a diaphragm pump. The organic phase in each reactor unit is connected.

2. Materials
2.1. Media

1. YMg agar medium: 5.0 g/L Bacto peptone, 3.0 g/L yeast extract, 3.0 g/L Bacto malt extract, 1.0 g/L $MgSO_4 \cdot 7H_2O$, 10.0 g/L glucose, and 15.0–30.0 g/L agar, pH 6.0 (*see* **Note 6**).
2. YP1 agar medium: 7.0 g/L Bacto peptone, 5.0 g/L yeast extract, 5.0 g/L K_2HPO_4, 10.0 g/L glucose, and 20.0 g/L agar, pH 7.5.
3. YP2 agar medium: 7.0 g/L Bacto peptone, 5.0 g/L yeast extract, 5.0 g/L K_2HPO_4, 50.0 g/L glucose, and 20.0 g/L agar, pH 6.5.
4. YMg/PB medium: 10.0 g/L Bacto peptone, 6.0 g/L yeast extract, 6.0 g/L Bacto malt extract, 2.0 g/L $MgSO_4 \cdot 7H_2O$, and 20.0 g/L glucose in 1 L of M/15 phosphate buffer, pH 6.0.

2.2. Reagents

1. 1-Decanol in decane (5% w/v).
2. 2-Benzylcyclohexenyl propionate in isooctane (1% w/v).
3. 6-Methyl-5-hepten-2-one in isooctane (1% w/v).
4. Ethyl 2-oxo-4-phenylbutanoate in decane (1.5% w/v).

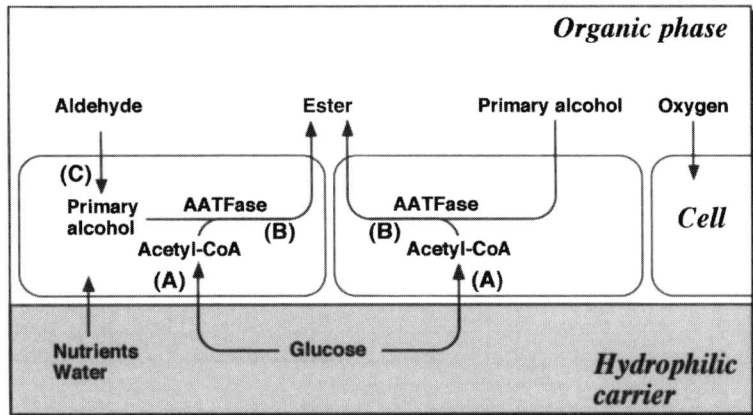

Fig. 6. Principle of double- and triple-coupling systems. A: metabolism of glucose; B: acetylation of a primary alcohol; C: reduction of an aldehyde. The primary alcohol is acetylated with acetyl-CoA by the aid of AATFase. Microbial oxidation of the primary alcohol is repressed by a high concentration of glucose in a carrier.

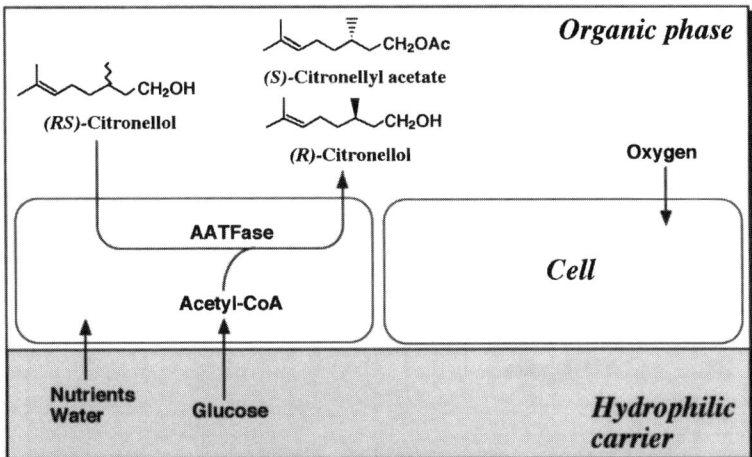

Fig. 7. Optical resolution of (*RS*)-citronellol in the double coupling system with *Pichia kluyveri* IFO 1165. (*S*)-Citronellol is preferentially acetylated with acetyl-CoA by the aid of AATFase in *Pichia kluyveri* IFO 1165.

5. Sodium alginate in deionized water (5% w/v).
6. Polyvinyl alcohol (PVA) (polymerization, 1500) in deionized water (10% w/v).
7. Sulfuric acid in deionized water (20% w/v).
8. Glutaraldehyde in deionized water (10% w/v).

9. Calcium chloride in deionized water (0.1 *M*).
10. *(RS)*-Citronellol in dodecane (2% w/v).
11. A solvent system of hexane–ethyl acetate (5:1).
12. A solvent system of hexane–2-propanol (975:25).
13. A solvent system of hexane–ethyl acetate (9:1).
14. A solvent system of hexane–ethyl acetate (1:1).
15. Sodium hydroxide in deionized water (10% w/v).

2.3. Apparatuses

2.3.1. Gas Chromatography Columns

1. A glass column (diameter, 2.6 mm; length, 3 m) containing 5% Thermon-3000/ Chromosorb W (Chromato Packing Center, Kyoto).
2. A glass column (diameter, 2.6 mm; length, 3 m) containing 5% silicon OV-1.
3. A capillary column (diameter, 0.25 mm; length, 6 m) containing silicon TC-1.
4. A capillary column (diameter, 0.25 mm; length, 30 m) containing β-DEX™ 225 (Supelco Co., Ltd.).

2.3.2. High-Performance Liquid Chromatography Columns

1. A stainless-steel column (diameter, 4.6 mm; length, 250 mm) containing Chiralcel OJ (Daisel Chemical Industries Co., Ltd., Tokyo).
2. A stainless-steel column (diameter, 4.6 mm; length, 250 mm) containing TSK-Gel silica 60 (Tosoh Co. Ltd., Tokyo).
3. A stainless-steel column (diameter, 4.6 mm; length, 250 mm) containing Chiralcel OD (Daisel Chemical Industries Co., Ltd.).

3. Methods

3.1. Microbial Transformations in an Agar Plate Interface Bioreactor

The procedure of microbial transformations in the reactor is simplest among the three types of interface bioreactor as shown in **Fig. 8**.

3.1.1. Synthesis of Decanoic Acid via Oxidation of 1-Decanol

1. Prepare a YMg agar plate (volume, 25 mL) in a glass Petri dish (diameter, 70 mm; height, 20 mm). The agar plate is used as a hydrophilic carrier in the agar plate interface bioreactor.
2. Spread 300 μL of the cell suspension of *Issatchenkia scutulata* var. *scutulata* IFO 10070 (1 loopful in 1 mL YMg liquid medium) on the agar plate (*see* **Note 7**).
3. Remove excess moisture by allowing the plate to stand; then seal the glass Petri dish with parafilm (*see* **Note 8**).
4. Precultivate the cells at 30°C by allowing the plate to stand for 1 d (*see* **Note 9**).
5. Add 8 mL of a 5 wt% solution of 1-decanol in decane to the plate (overlay the agar plate with the decane solution; *see* **Note 10**).

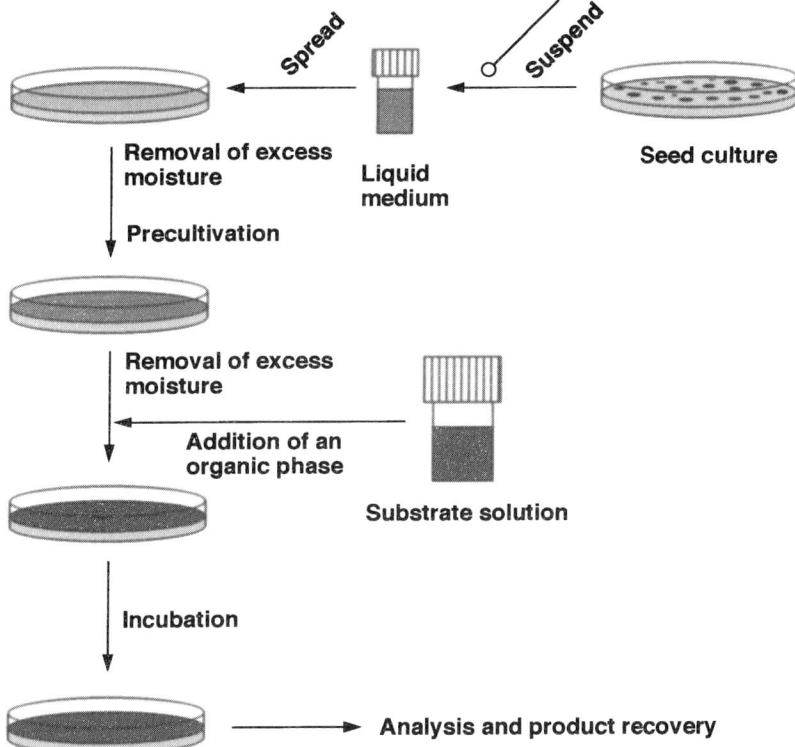

Fig. 8. Flowchart of microbial transformation with an agar plate interface bioreactor.

6. Incubate the agar plate interface bioreactor at 30°C by allowing the device to stand (*see* **Note 11**).
7. Determine produced decanoic acid and residual 1-decanol with GLC: the column (diameter, 2.6 mm; length, 3 m) contains 5% Thermon-3000/Chromosorb W, the column temperature is increased from 100°C to 240°C at a rate of 7°C/min, and the carrier gas is N_2 (flow rate, 60 mL/min). The retention times of 1-decanol and decanoic acid are 10.1 and 19.7 min, respectively.

3.1.2. Synthesis of (R)-2-Benzylcyclohexanone via Enantiofacially Selective Hydrolysis of Its Enol Ester

1. Prepare a YP1 agar plate (volume, 500 mL) in an aluminum tray (width, 290 mm; depth, 220 mm; height, 30 mm).
2. Inoculate starkly cells of *Yamadazyma farinosa* IFO 10896 on the agar plate with a loop (*see* **Note 12**).
3. Precultivate the cells at 30°C by allowing the plate to stand for 2 d.

4. Add 50 mL of a 1 wt% solution of 2-benzylcyclohexenyl propionate in isooctane to the plate (overlay the agar plate with the isooctane solution) and seal the tray with a glass plate.

5. Incubate the agar plate interface bioreactor at 25°C by allowing the device to stand for 24 h.

6. Determine produced *(R)*-2-benzylcyclohexanone and residual 2-benzylcyclohexenyl propionate with GLC: the column (diameter, 2.6 mm; length, 2 m) contains 5% silicon OV-1, the column temperature is increased from 100°C to 225°C at a rate of 2°C/min, and the carrier gas is N_2 (0.6 kg/cm^2). The retention times of *(R)*-2-benzylcyclohexanone and 2-benzylcyclohexenyl propionate are 2.3 and 5.7 min, respectively.

7. Recover the organic layer from the agar plate interface bioreactor.

8. Wash the microbial film by vigorous stir with ethyl acetate.

9. Wash the combined organic layer with brine, and concentrate *in vacuo.*

10. Dissolve the residue in hexane–ethyl acetate (20:1).

11. Pass the solvent mixture through a pad of SiO_2 (10 g).

12. Distill the elute at 170–175°C/1.5 torr (bulb-to-bulb distillation).

13. Determine the enantiomeric excess of recovered *(R)*-2-benzylcyclo-hexanone with HPLC: The column is Chiralcel OJ (diameter, 4.6 mm; length, 250 mm), the eluent is hexane–2-propanol (180:1), the flow rate is 1.0 mL/min, and ultraviolet (UV) detection is at 254 nm.

3.1.3. Synthesis of (R)-Sulcatol via Asymmetric Reduction of 6-Methyl-5-Hepten-2-One

1. Prepare a YP2 agar plate (volume, 50 mL) in a glass petri dish (diameter, 84 mm).

2. Inoculate starkly *Yamadazyma farinosa* IFO 10896 on the agar plate with a loop (*see* **Note 12**).

3. Precultivate the cells at 30°C by allowing the plate to stand for 2 d.

4. Add 6 mL of a 1 wt% solution of 6-methyl-5-hepten-2-one in isooctane onto the plate (overlay the agar plate with the isooctane solution).

5. Incubate the agar plate interface bioreactor at 25°C by allowing the plate to stand for 24 h (*see* **Note 11**).

6. Determine produced *(R)*-6-methyl-5-hepten-2-ol [sulcatol] and residual 6-methyl-5-hepten-2-one with capillary GLC: the column (diameter, 0.25 mm; length, 6 m) contains silicon TC-1, the column temperature is 120°C (isothermal), and the carrier gas is N_2 (0.6 kg/cm^2). The retention times of 6-methyl-5-hepten-2-one and *(R)*-sulcatol are 7.5 and 7.7 min, respectively.

7. Recover the organic layer from the agar plate interface bioreactor.

8. Wash the microbial film by vigorous stir with ethyl acetate.

9. Concentrate the combined organic layer *in vacuo.*

10. Dissolve the residue in hexane–ethyl acetate (5:1).

11. Pass the solvent mixture through a pad of SiO_2 (10 g).

12. Distill the eluent at 105–120°C/35 torr (bulb-to-bulb distillation).

13. Determine the enantiomeric excess of *(R)*-sulcatol with nuclear magnetic resonance (NMR) after *(R)*-α-methoxy-α-(trifluoromethyl)phenylacetyl chloride [*(R)*-MTPA] treatment.

3.2. Microbial Transformations with a Pad-Packed Interface Bioreactor

The reactor is favorable for treatment of the organic layer containing a substrate.

3.2.1. Synthesis of (R)-2-Hydroxy-4-Phenylbutanoate via Asymmetric Reduction of Ethyl 2-Oxo-4-Phenylbutanoate

1. Pour a YMg agar medium (agar content, 3.0 wt%; volume, 2.0 L) into a stainless-steel cast (width, 470 mm; depth, 350 mm; height, 20 mm) set six stainless-steel frames attached each a stainless-steel net (*see* **Note 13**).
2. After solidified the agar medium, cut out six pieces of nutrient agar pad as shown in **Fig. 4** (*see* **Note 14**).
3. Inoculate a condensed 1-d broth (200 mL from 1 L broth) of *Rhodotorula minuta* IFO 0920 on the surface of the pads by dipping.
4. After removal of excess moisture from the pad surface, set the pads with Teflon spacers in a stainless-steel tray (width, 350 mm; depth, 280 mm; height, 110 mm) (*see* **Notes 13** and **15**).
5. Precultivate the cells at 30°C by allowing the pads for 2 d.
6. Set the pads with Teflon spacers in a stainless-steel tank (width, 160 mm; depth, 160 mm; height, 110 mm) (*see* **Note 16**).
7. Add 600 mL of a 1.5 wt% solution of ethyl 2-oxo-4-phenylbutanoate (EOPB) in decane into the tank (*see* **Note 10**).
8. Run the bioreactor at 30°C with agitation (700 rpm) with a magnetic stirrer.
9. Determine the concentration of ethyl *(R)*-2-hydroxy-4-phenylbutanoate [(R)-EHPB] produced and residual EOPB with HPLC: the column is TSK–gel silica 60 (diameter, 4.6 mm; length, 250 mm), the eluent is hexane–2-propanol (975:25), and the flow rate is 0.8 mL/min. The retention times of *(R)*-EHPB and EOPB are 5.5 and 8.5 min, respectively.
10. After the determination of the concentration of *(R)*-EHPB and EOPB, put 10 mL of the decane layer on a silica gel column (Wakogel C-200, 5 g; Wako Pure Chemicals Co. Ltd. Tokyo).
11. After removal of decane by elution with 30 mL of hexane, produced *(R)*-EHPB and EOPB are eluted with 20 mL of ethyl acetate.
12. Remove ethyl acetate *in vacuo*.
13. Purify *(R)*-EHPB with thin-layer chromatography (TLC) (Silica 60F$_{254}$, 200 × 200 mm) developed with a solvent system of benzene–ethyl acetate (9:1). The R_f values (mobility of a compound on TLC relative to the eluent as 1.00) of EOPB and *(R)*-EHPB are 0.44 and 0.26, respectively.
14. Determine the absolute configuration and enantiomeric excess of *(R)*-EHPB with HPLC: the column is Chiralcel OD (diameter, 4.6 mm; length, 250 mm), the eluent is hexane–2-propanol (95:5), and the flow rate is 1.0 mL/min. The retention times of *(S)*- and *(R)*-EHPB are 7.2 and 9.7 min, respectively. Calculate the enantiomeric excess of *(R)*-EHPB based on the peak areas.

15. After the running, collect the decane layer by decanting.
16. Extract *(R)*-EHPB five times with a half-volume of methanol.
17. Remove the methanol *in vacuo*.
18. Put the recovered yellow oil on a silica gel column (Wakogel C-200; diameter, 55 mm; length, 350 mm).
19. Eluted *(R)*-EHPB with benzene–ethyl acetate (9:1).
20. After removal of the eluent *in vacuo*, determine the yield and the chemical and optical purities of *(R)*-EHPB by the same manner as mentioned earlier.

3.2.2. Hydrolysis of Neat 2-Ethylhexyl Acetate

1. Add 10 portions of an aqueous solution of Na–alginate (ALG; viscosity, 100–150 cP; 5% w/v) to 100 portions of an aqueous solution of PVA (polymerization, 1500; 10% w/v) (*see* **Note 17**).
2. Adjust the mixture to pH 2.6–2.8 with 20% H_2SO_4, and thoroughly mix with a blender until the disappearance of alginate aggregates (*see* **Note 17**).
3. Add one portion of an aqueous solution of glutaraldehyde (GA; 10% w/v) to the mixture, and sufficiently mix (*see* **Notes 17** and **18**).
4. Pour the mixture into a stainless-steel cast (width, 470 mm; depth, 350 mm; height, 20 mm), set a stainless steel frame attached to a stainless-steel net (*see* **Fig. 4**) into the resin mixture, and lid with a glass plate.
5. Crosslink PVA with GA by storing at room temperature for 2 d.
6. Immerse the PVA–GA crosslinking plate in a 0.1 *M* $CaCl_2$ solution at room temperature for 2 d (*see* **Note 19**).
7. Cut out the carriers holding the stainless-steel frame (thickness, 19 mm; volume, 226 mL) and wash the plates in flowing water for 2 d (*see* **Note 20**).
8. Immerse the carriers in an equal volume of YMg/PB liquid medium and sterilize the carriers by autoclaving (121°C, 20 min) (*see* **Note 21**).
9. Inoculate a condensed 1-d broth (200 mL from 1 L broth) of *Candida cylindracea* ATCC 14830 on the surface of six pieces of the PVA-based pad by dipping.
10. Set the pads with Teflon spacers in a stainless-steel tray (width, 350 mm; depth, 280 mm; height, 110 mm) (*see* **Fig. 9** and **Notes 13** and **15**).
11. Precultivate the cells at 30°C by allowing the pads to stand for 1 d.
12. Set the pads with Teflon spacers in a stainless-steel tank (width, 160 mm; depth, 160 mm, height, 110 mm) (*see* **Note 16**).
13. Add neat 2-ethylhexyl acetate into the tank, and set up the pad-packed interface bioreactor as shown in **Fig. 4** (*see* **Note 10**).
14. Run the bioreactor at 30°C with agitation (700 rpm) with a magnetic stirrer.
15. Determine produced 2-ethyl-1-hexanol with GLC: the column is Thermon-3000/Chromosorb W (diameter, 2.6 mm; length, 3 m), the column temperature is 120°C, and the carrier gas is N_2 (flow rate, 60 mL/min). The retention times of 2-ethylhexyl acetate and 2-ethyl-1-hexanol are 3.7 and 5.3 min, respectively.

3.3. Syntheses of (R)-Citronellol and (S)-Citronellyl Acetate via Double-Coupling System in a Multistory Interface Bioreactor

The reactor is suitable for convenient and inexpensive scale-up.

1. Prepare 10 pieces of 2.1 L of YMg agar plate with agar and glucose contents of 2 and 20 wt % (*see* **Note 22**) respectively, in 10 stackable stainless-steel trays (reactor units; each width, 362 mm; depth, 470 mm; height, 50 mm) having 6 holes (diameter, 10 mm) (*see* **Fig. 5**).
2. Inoculate starkly 30 mL of a 1-d broth of *Pichia kluyveri* IFO 1165 cultured in YMg medium on the agar plate in each tray.
3. Precultivate the cells at 30°C by allowing the plate to stand for 1 d.
4. Construct the multistory interface bioreactor as shown in **Fig. 5**.
5. Introduce a solution of *(RS)*-citronellol in dodecane (2% w/v) into a head unit of the multistory interface bioreactor through a inlet line with a diaphragm pump (*see* **Note 10**).
6. Run the reactor at 30°C with reciprocal shaking (60 strokes/min) and organic layer circulation (below 200 mL/min).
7. Determine the concentration of *(S)*-citronellyl acetate and *(R)*-citronellol with GLC every 24 h. The column is Thermon-3000/Chromosorb W (diameter, 2.6 mm; length, 3 m), the column temperature is increased from 100°C to 240°C at a rate of 7°C/min, and the carrier gas is N_2 (flow rate, 60 mL/min). The retention times of *(S)*-citronellyl acetate and *(R)*-citronellol are 9.1 and 10.4 min, respectively.
8. Check the enantiomeric excess of *(R)*-citronellol with GLC every 24 h. The capillary column is β-DEX™ 225 (diameter, 0.25 mm; length, 30 m); the column temperature is 85°C (isothermal); the injector and detector temperatures are 210 and 220°C, respectively; The split ratio is 1:00; and the carrier gas is He (linear velocity, 54 cm/s). The retention times of *(S)*- and *(R)*-citronellol are 30.1 and 30.8 min, respectively. The enantiomeric excess of citronellol is determined based on the peak areas.
9. After the run, collect the organic phase by decantation.
10. Introduce the organic phase in a silica gel column (Wakogel C-200, total 2.1 kg) (*see* **Note 23**).
11. Remove decane by elution with hexane (*see* **Note 24**).
12. Separate *(S)*-citronellyl acetate and *(R)*-citronellol by elution with hexane–ethyl acetate (9:1 and 5:5, respectively).
13. Determine the enantiomeric excess of *(S)*-citronellyl acetate by the same manner as in the case of *(R)*-citronellol after alkaline hydrolysis with ethanol–10% NaOH solution.

4. Notes

1. The mechanism for the alleviation of toxicity on the solid–liquid interface is unclear at present. The organic phase may function as a reservoir of the toxic compound. Furthermore, we think that the formation of a polysaccharide layer (capsule) is accelerated by solid cultivation, and the hydrophilic layer may be function as a barrier against lipophilic poisons (*see* reviews in **refs. *1* and *2***).

2. In the multistory interface bioreactor, the diaphragm pump is favorable because air is also circulated with the organic phase. Overflow and circulation lines must be constructed with a solvent resistant tube, such as Viton tube.

3. For example, the oxidation of toxic 2-octanol with *Rhodococcus equi* IFO 3730 in the agar plate interface bioreactor results in 730 times the accumulation of 2-octanone compared with that in an emulsion system *(3)*.

4. In general, hydrophobic organic solvents, such as decane, exhibit higher oxygen solubility compared with that of water.

5. Acetyl-CoA is very expensive. Thus, this system cannot be practically performed with a purified enzyme (alcohol acetyltransferase, AATFase). Using a living microorganism, acetyl-CoA is produced via the metabolism of glucose and thus is available to the coupling system.

6. If the microbial transformation extends over a long time, it is favorable that 0.2 wt% of AgarMate (Toho Co., Ltd., Tokyo; a polysaccharide preventing dryness of the agar plate) is added to the agar plate.

7. The seed cultivation of the strain is performed on a nutrient agar plate (YMg agar medium). The cell suspension is easily spread without a stirring stick and a turntable.

8. Care should be taken to avoid contamination during the removal of excess moisture.

9. A thick microbial film is spontaneously formed during the precultivation.

10. It is essential that the excess moisture must be removed before adding the organic layer because the presence of excess moisture on the immobilized carrier surface is a fatal problem, as the microbial film on the carrier surface is washed out by water droplets in the organic phase.

11. Excess airflow into the device must be avoided because the organic layer is reduced by the airflow. Therefore, it is favorable to stand the device in an adequate box or seal the device.

12. The strain is favorably inoculated on the agar plate with a loop because the spreading of the cell suspension leads to low growth.

13. Stainless-steel cast must be sterilized by irradiation of UV.

14. Avoid contamination. A blade for cutting the pads must be sterilized by flame.

15. The strain can not grow on the carrier surface contacting a Teflon pad. The pads must not contact each other.

16. The stainless-steel tank must be sterilized by irradiation of UV. Pay attention to peel off a microbial film by contact during the setup.

17. Avoid air bubble contamination.

18. The total weight of the resin is 2220 g.

19. The stickiness of the pad surface is reduced according to the alginate–Ca^{2+} crosslinkage.

20. Cutting of the PVA-based pads must be carefully done without cracking.

21. In this operation, an aqueous phase in a pad is substituted by a liquid medium.

22. In the double-coupling system with *Pichia kluyveri* IFO 1165, the favorable glucose content in the carrier is ranging from 15 to 30 wt% *(17)*. The high concentrations of glucose repress the production of a citronellol-oxidizing enzyme. Production of alcohol dehydrogenases are generally repressed by high concentrations of glucose *(19,20)*.

23. For the separation of 1 g of *(S)*-citronellyl acetate and *(R)*-citronellol, 25 g of Wakogel C-200 is necessary.
24. For the separation of 1 g of *(S)*-citronellyl acetate and *(R)*-citronellol, the required amount of hexane-ethyl acetate (9:1 and 5:5, respectively) are 100 and 80 mL, respectively.

References

1. Oda, S. and Ohta, H. (1997) Bioconversion on an interface between an organic and an aqueous phase. *Recent Res. Dev. Microbiol.*, **1**, 85–99.
2. Oda, S. and Ohta, H. (1997) Interface bioreactor: bioconversion process on an interface between an organic solvent and a polymer gel. *J. Jpn. Soc. Colour Mater.* **70**, 538–546.
3. Oda, S. and Ohta, H. (1992) Microbial transformation on interface between hydrophilic carriers and hydrophobic organic solvents. *Biosci. Biotechnol. Biochem.* **56**, 2041–2045.
4. Oda, S., Kato, A., Matsudomi, N., and Ohta, H. (1994) Production of aliphatic carboxylic acids *via* microbial oxidation of 1-alkanols with interface bioreactor. *J. Ferment. Bioeng.* **78**, 149–154.
5. Katoh, O., Sugai, T., and Ohta, H. (1994) Application of microbial enantiofacially selective hydrolysis in natural product synthesis. *Tetrahedron: Asymmetry* **5**, 1935–1944.
6. Sugai, T., Katoh, O., and Ohta, H. (1995) Chemo-enzymatic synthesis of *(R,R)*-(−)-pyrenophorin. *Tetrahedron* **51**, 11,987–11,998.
7. Oda, S., Inada, Y., Kato, A., Matsudomi, N., and Ohta, H. (1995) Production of *(S)*-citronellic acid and *(R)*-citronellol with an interface bioreactor. *J. Ferment. Bioeng.* **80**, 559–564.
8. Oda, S., Kato, A., Matsudomi, N., and Ohta, H. (1996) Enantioselective oxidation of racemic citronellol with an interface bioreactor. *Biosci. Biotechnol. Biochem.* **60**, 83–87.
9. Oda, S., Inada, Y., Kobayashi, A., and Ohta, H. (1998) Production of ethyl *(R)*-2-hydroxy-4-phenylbutanoate *via* reduction of ethyl 2-oxo-4-phenylbutanoate in an interface bioreactor. *Biosci. Biotechnol. Biochem.* **62**, 1762–1767.
10. Oda, S. and Ohta, H. (1992) Alleviation of toxicity of poisonous organic compounds on hydrophilic carrier/hydrophobic organic solvent interface. *Biosci. Biotechnol. Biochem.* **56**, 1515–1517.
11. Kato, N. and Shibasaki, I. (1975) Comparison of antimicrobial activities of fatty acids and their esters. *J. Ferment. Technol.* **53**, 793–801.
12. Kabara, J. J., Swieczkowski, D. M., Conley, A. J., and Truant, J. P. (1972) Fatty acids and derivatives as antimicrobial agents. *Antimicrob. Agents Chemother.* **2**, 23–28.
13. Inoue, A. and Horikoshi, K. (1989) A *Pseudomonas* thrives in high concentrations of toluene. *Nature* **338**, 264–266.
14. Inoue, A. and Horikoshi, K. (1991) Estimation of solvent-tolerance of bacteria by the solvent parameter Log *P*. *J. Ferment. Bioeng.* **71**, 194–196.
15. Oda, S., Inada, Y., Kobayashi, A., Kato, A., Matsudomi, N., and Ohta, H. (1996) Coupling of metabolism and bioconversion: microbial esterification of citronellol

with acetyl coenzyme A produced via metabolism of glucose in an interface bioreactor. *Appl. Environ. Microbiol.* **62,** 2216–2220.

16. Oda, S. and Ohta, H. (1997) Double coupling of acetyl coenzyme A production and microbial esterification with alcohol acetyltransferase in an interface bioreactor. *J. Ferment. Bioeng.* **83,** 423–428.

17. Oda, S., Sugai, T., and Ohta, H. (1999) Optical resolution of racemic citronellol via double coupling system in an interface bioreactor. *J. Ferment. Bioeng.* **87,** 473–480.

18. Oda, S., Tanaka, J., and Ohta, H. (1998) Interface bioreactor packed with synthetic polymer pad: application to hydrolysis of neat 2-ethylhexyl acetate. *J. Ferment. Bioeng.* **86,** 84–89.

19. Schimpfessel, L. (1968) Presence and regulation of the synthesis of two alcohol dehydrogenases from *Saccharomyces cerevisiae. Biochim. Biophys. Acta* **151,** 317–329.

20. Lutstorf, U. and Megnet, R. (1968) Multiple forms of alcohol dehydrogenase in *Saccharomyces cerevisiae.* 1. Physiological control of ADH-2 and properties of ADH-2 and ADH-4. *Arch. Biochem. Biophys.* **126,** 933–944.

33

Yeast-Mediated Reactions in Organic Solvents

Andrew J. Smallridge and Maurie A. Trewhella

1. Introduction

The use of baker's yeast as a reagent in organic synthesis has been well documented *(1,2)*. Yeast is capable of catalyzing a wide range of different types of reactions although it is most commonly used for the asymmetric reduction of carbonyl groups and alkenes. By far, the majority of yeast-mediated reduction reactions have been conducted using fermenting yeast in an aqueous reaction system. Reactions of this type involve the addition of yeast and nutrient, glucose or sucrose, to a buffered solution followed by the addition of the substrate. Under these conditions, the yeast is actively growing, all of the biochemical pathways are operating, and, as part of the fermentation process, the substrate is converted into the desired product, usually with a high degree of stereoselectivity. Although this methodology is well established and in many cases highly successful, it does have a number of inherent experimental problems. The use of an aqueous environment for transformations involving organic substrates is undesirable, because of the limited water solubility of many substrates. The greatest problem, by far, associated with the use of yeast in aqueous systems is the isolation of the product. At the end of the reaction, the yeast mass is finely dispersed throughout the water and this makes filtration of the yeast and extraction of the product a difficult procedure. In addition, the fermenting yeast has produced a range of metabolic products that will need to be separated from the desired product.

Recently, a much simpler reaction system for yeast reactions involving the use of an organic solvent system has been reported *(3–6)*. Using this new methodology, the reaction is carried out using dried baker's yeast in an organic solvent. The yeast is not actively fermenting or growing, so no metabolic

From: *Methods in Biotechnology, Vol. 15: Enzymes in Nonaqueous Solvents: Methods and Protocols*
Edited by: E. N. Vulfson, P. J. Halling, and H. L. Holland © Humana Press Inc., Totowa, NJ

products are produced and there is no need for a food source. More importantly, at the end of the reaction, the yeast remains at the bottom of the reactor vessel and the solvent can be readily decanted from the yeast and evaporated to give the desired product, in high yield.

This chapter describes the procedures and methodology required to carry out the yeast-mediated reduction of a keto group or an alkene in an organic solvent system.

2. Materials
2.1. Yeast

For the yeast-mediated reduction reactions, we use Mauripan active dried yeast, but most dried baker's yeasts are suitable for the reactions (*see* **Note 1**). Any dry baker's yeast that is available from a supermarket is appropriate. No special precautions are required for the handling or storage of the yeast. Although the yeast is packaged in airtight containers, we have found that exposure to air for prolonged periods of time, >1 yr, in no way diminishes the activity of the yeast. In our laboratory, the yeast is stored in jars or plastic bags at ambient temperature. No effort is made to limit exposure to moisture or air.

2.2. Organic Solvent

The solvents used for the reactions are standard LR solvents and it is not necessary to dry or purify them prior to use. For the majority of our reactions, we use petroleum spirit, with a boiling range of 40–60°C. In this case, we use AR solvent, which is slightly more expensive than the LR grade; however, the absence of higher boiling fractions simplifies purification of the product, justifying the additional cost.

2.3. Water

All of the yeast reaction mixtures contain a small amount of water; it is not necessary to use distilled water, tap water is routinely used.

3. Method
3.1. General Comments

The general method for the yeast-mediated reduction reactions involves placing yeast, the solvent, the substrate, and a small amount of water in a reaction vessel (*see* **Note 2**) and stirring or shaking (*see* **Note 3**) the mixture at room temperature (approx 20°C) for 24 h. A number of factors have been shown to influence the reaction.

1. Water. The water is needed to activate the yeast, no reaction occurs in the absence of the water. We have found for all of the reactions we have conducted

so far that 0.8 mL water/g yeast is the minimum amount required to obtain maximum yeast activity. Other researchers have reported that as little as 0.2 mL/g yeast is sufficient *(7)*.

2. Time. All of the reactions are conducted for 24 h. Leaving a reaction for a longer period of time does not result in an increase in the extent of reduction because the yeast enzymes involved in the reaction are slowly deactivated in the reaction system. After a period of 24 h in contact with the organic solvent–water mixture very little reductase activity is left in the yeast. The yeast retains its activity in neat petroleum spirit indefinitely; it is only after the water is added that deactivation is initiated. The exact nature of the deactivation is unknown.

3. Amount of Yeast. Because the yeast activity virtually ceases after 24 h, it is important that the reaction goes to completion in that period of time in order to maximize the yield. The yeast/substrate ratio is therefore crucial. A minimum of 1 g yeast/mmol substrate is required for these reactions; however, ratios of up to 15 g yeast/mmol substrate have been found to be necessary in some cases. As a general rule, if the reaction does not go to completion, it should be repeated using more yeast. The addition of extra yeast to an incomplete reaction does not appear to be as effective. It is better to isolate the unreacted starting material and repeat the reaction. Yeast-reduction reactions involve the coenzyme NAD(P)H, which, in a system involving fermenting yeast, is continually recycled by the various metabolic pathways. In the nonfermenting organic system, the coenzyme cannot be regenerated and, therefore, somewhat larger amounts of yeast are required in the organic solvent system. The ease with which the product can be isolated from the yeast ensures that high isolated yields are obtained even with apparently excessive quantities of yeast. We have obtained a 68% isolated yield from a reaction utilizing 11 g yeast/mmol substrate.

4. Solvent. For the yeast reactions, it is important to use a nonpolar solvent. We have found that the most suitable solvent for all of the yeast reactions we have conducted so far is hexane or petroleum spirit *(8)*. Reactions have also been reported to proceed in benzene *(7)*, toluene, carbon tetrachloride, and diethyl ether *(8)*. The yeast is slightly less active in these solvents and more yeast may be required to obtain complete reduction. Mixtures of polar and nonpolar solvents can be used for the reaction with only a small decrease in reductase activity. For example, reactions conducted in petroleum ether containing 5% (v/v) ethyl acetate, tetrahydrofuran (THF), or dichloromethane and up to 30% (v/v) chloroform give yields and stereoselectivity similar to a reaction conducted in petroleum spirit. An apparent drawback to the use of petroleum spirit as a solvent is the lack of solubility of a large number of organic substrates in this solvent. In some cases, larger amounts of yeast may be required. We have found that compounds that are largely insoluble in the petroleum-spirit-based solvents are readily reduced in this system. Presumably, over the 24 h reaction period, the equilibration is such that the substrate is solubilized and reacts with the yeast. Consequently, the use of more polar solvents to aid substrate solubility has not been necessary.

3.2. Reduction of Carbonyl Groups (9)

1. Isopropyl acetoacetate (1 g, 6.3 mmol) was added to a 500-mL round-bottom flask with yeast (12.6 g) and water (10 mL) (0.8 mL/g yeast) and petroleum spirit (300 mL) and stirred at room temperature for 24 h.
2. The reaction mixture was filtered and the yeast washed with ethyl acetate (100 mL) (*see* **Note 4**).
3. The solvent was then removed under reduced pressure (*see* **Note 5**), and bulb-to-bulb distillation (100°C/20 mm) gave isopropyl (*S*)-(+)-3-hydroxybutyrate (0.95 g, 96%). literature boiling point 78–79°/21 mm (*10*). ^{1}H-NMR δ(CDCl$_3$) 1.23, d, *J*-6.3 Hz, H4; 1.27, d, *J*-6.3 Hz, (CH$_3$)$_2$; 2.40, dd, *J*-16.2, 8.4 Hz, H2; 2.48, dd, *J*-16.2, 3.6, H2; 2.66, s(br), OH; 4.21, m, H3; 5.07, sept, *J*-6.3 Hz, CH. ^{13}C-NMR δ21.1, C4; 21.8 (CH$_3$)$_2$; 42.5, C2; 63.7, C3; 67.5, CH; 171.8, C1. [α]$_D$ +38.89° (CHCl$_3$, *c* = 1). Chiral gas chromatography showed a ratio of 98.4 : 1.6 (97% ee) (*see* **Note 6**).

3.3. Reduction of Carbon–Carbon Double Bond (11)

1. (*E*)-2-Nitro-1-phenylethene (0.3 g, 2.0 mmol), yeast (22.13 g, 11g/mmol), water (17.7 mL) (0.8 mL/g yeast), and petroleum spirit (200 mL) was stirred at room temperature for 24 h.
2. The solvent was removed by filtration and the yeast was washed with ethyl acetate (3X, 50mL).
3. The filtrates were combined and evaporated *in vacuo* to give a residue that was bulb-to-bulb distilled (150°C/0.5 mm) to give 2-nitro-1-phenylethane (0.216 g, 72%). ^{1}H-NMR δ(CDCl$_3$) 3.34, 2H, t, *J*-7.5, 1-H; 4.63, 2H, t, *J*-7.5, 2-H; 7.3, 5H, m, Ph.

4. Notes

1. We have found that there is a difference in the activity of the various preparations of dried yeast sold. For example, Sigma Type II yeast has about half the activity of the Mauripan dried yeast. This does not affect its suitability for the reduction reactions, because if the quantity of yeast added to the reaction is doubled, a similar yield of product is obtained. In general, if the reaction does not go to completion in 24 h using a particular preparation of the yeast, the reaction should be conducted with a larger amount of yeast.

2. Although the order of addition of reagents is not crucial, for simplicity we generally add the water last.
3. In our initial work, the reactions were conducted in standard round-bottom flasks containing a magnetic stirring bead. Stirring the reaction mixture gradually becomes more difficult as the reaction progresses because the yeast absorbs the water, becomes sticky, and ends up as a thick mass at the bottom of the flask. The magnetic stirrer bead then stops stirring because of the consistency of the yeast. In an attempt to overcome the problems associated with stirring the yeast reactions, we now use Schott bottles for the reactions and these are placed in an orbital shaker and shaken for 24 h. This technique is especially useful when rather large quantities of yeast are being used. It does not appear to matter whether the reaction is stirred or shaken, as we have not detected any appreciable differences in reaction rate between the two techniques. As a preference, it is easier to shake the reaction when large amounts of yeast (>10 g) are used.
4. We have found that ethyl acetate is the best solvent for washing the yeast. It removes the greatest amount of product and the least amount of water and lipid material from the yeast.
5. It is not necessary to dry the organic layer. The yeast acts as a drying agent and removes all of the water from the solvent.
6. Chiral gas chromatography is carried out on the trifluoroacetate derivative using a Chiraldex G-TA (30 m × 0.25 mm) column with a cyclodextrin phase thickness of 0.125 μm. The trifluoroacetate is prepared by adding 0.2 mL of dichloromethane and 0.1 mL of trifluoroacetic anhydride to approx 1 mg of isopropyl 3-hydroxybutyrate in a capped vial that was kept at room temperature for 30 min and then evaporated to near dryness. The residue was dissolved in 0.2 mL ethanol and injected into the gas chromatograph.

References

1. Servi, S. (1990) Baker's yeast as a reagent in organic synthesis. *Synthesis* 1.
2. Csuk, R. and Glanzer, B. I. (1991) Baker's yeast mediated transformations in organic chemistry. *Chem. Rev.* **91,** 49.
3. Jayasinghe, L. Y., Smallridge, A. J., and Trewhella, M. A. (1993) The yeast mediated reduction of ethyl acetoacetate in petroleum ether. *Tetrahedron Lett.* **34,** 3949.
4. Nakamura, K., Kondo, S., Kawai, Y., and Ohno, A. (1993) Effect of organic solvent on reduction of α-keto esters mediated by baker's yeast. *Bull. Chem. Soc. Jpn.* **66,** 2738.
5. North, M. (1996) Baker's yeast reduction of β-keto esters in petrol. *Tetrahedron Lett.* **37,** 1699.
6. Rotthaus, O., Kruger, D., Dermuth, M., and Schaffner, K. (1997) Reduction of keto esters with baker's yeast in organic solvents—a comparison with the results in water. *Tetrahedron* **53,** 935.
7. Nakamura, K., Kondo, S., Kawai, Y., and Ohno, A. (1991) Reduction by baker's yeast in benzene. *Tetrahedron Lett.* **32,** 7075.

8. Jayasinghe, L. Y., Smallridge, A. J., and Trewhella, M. A. (1994) The use of organic solvents in the yeast mediated reduction of ethyl acetoacetate. *Bull Chem. Soc. Jpn.* **67,** 2528.

9. Medson, C., Smallridge, A. J., and Trewhella, M. A. (1994) The stereoselective preparation of β-hydroxy esters using a yeast reduction in an organic solvent. *Tetrahedron: Asymmetry* 1049.

10. Tai, A., Harada, T., Hiraki, Y., and Murakami, S. (1983) Stereochemical investigation on asymmetrically modified Raney nickel catalyst. Mode of interaction between modifying reagent and substrate in the enantioface-differentiating process. *Bull. Chem. Soc. Jpn.* **56,** 1414.

11. Bak, R., McAnda, A., Smallridge, A. J., and Trewhella, M. A. (1996) The yeast mediated reduction of nitrostyrenes in organic solvent systems. *Aust. J. Chem.* **49,** 1257.

34

Biocatalysis in Pharmaceutical Process Development

SCH56592, A Case Study

Brian Morgan, David R. Dodds, Michael J. Homann, Aleksey Zaks, and Robert Vail

1. Introduction

The literature is now replete with articles covering the use of biocatalysis, with two commercially available databases (*1,2*), each listing over 20,000 reports of reactions using biologically derived catalysts, and several texts which extensively cover the subject (*3–7*). Even the more specific subtopic, the utility of biocatalysis in the pharmaceutical industry, is the subject of recent reviews listing hundreds of references (*8,9*). With this presence in the literature, it is timely to describe how biocatalysis has been applied by a Process Development Group to the synthesis of a single pharmaceutical entity. Using the antifungal compound SCH56592 (**Fig. 1**, compound **1**) as an example, this chapter will describe several biocatalytic approaches that were explored for the synthesis of key intermediates of this molecule. In keeping with the subject of this volume, the description will concentrate on the use of commercially available enzymes in organic solvents, although hydrolytic reactions and microbial reductions will also be briefly addressed for comparison and to complete the narrative.

Fungal infections are a particularly serious class of disease, and the number of antifungal drugs available is considerably less than antibacterial drugs. At this time, the most widely used class of antifungal drugs are the "azoles," so named because of the five-member nitrogen-containing ring, usually a triazole ring, common in all of their structures (**Fig. 1**). The triazole ring binds the P450 enzyme responsible for the 14α-demethylation of lanosterol and inhibits this enzyme's activity. This prevents the synthesis of ergosterol, which performs

From: *Methods in Biotechnology, Vol. 15: Enzymes in Nonaqueous Solvents: Methods and Protocols*
Edited by: E. N. Vulfson, P. J. Halling, and H. L. Holland © Humana Press Inc., Totowa, NJ

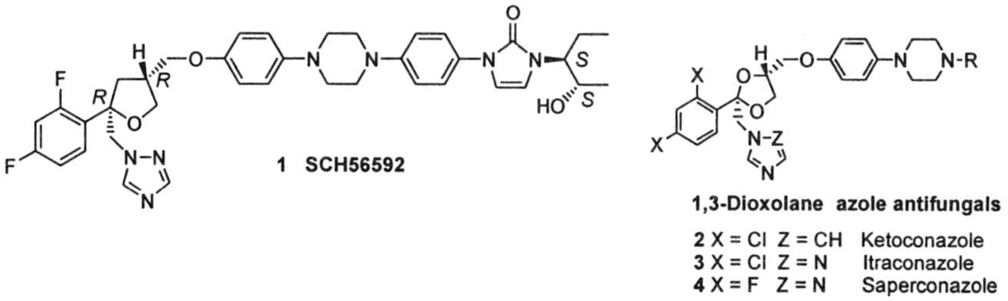

Fig. 1. Antifungal compound SCH56592 and 1,3-dioxolane azole antifungals.

the same essential structural function in fungal cells as cholesterol does in mammalian cells. These drugs also decrease the ratio of saturated to unsaturated fatty acids in the fungal cell, in addition to their blockade of ergosterol synthesis. The result is growth inhibition, membrane dysfunction, and ultimately death of the fungal cell, with low toxicity to the cells of the infected mammalian host *(10)*.

SCH56592 (**1**) is an azole antifungal, currently in Phase II clinical trials, with superior activity against systemic *Candida* infections and pulmonary *Aspergillus* infections, and is useful for treating both normal and immunocompromised patients. Like other azole antifungals, SCH56592 possesses the triazole ring, a dihalophenyl ring, and a rigid side chain, all distributed around a central five-member ring. Unlike the other azole antifungals, the central ring in SCH56592 is a tetrahydrofuran rather than a 1,3-dioxolane ring (**Fig. 1**). This hydrolytically stable tetrahydrofuran ring is believed to be responsible for the increased antifungal activity *(11)*, the absolute configuration of the two stereochemical centers in this ring being essential. In addition, another pair of stereochemical centers exists at the other end of the molecule at the extreme end of the side chain. The configuration of this pair of stereocenters is also essential to the antifungal activity of SCH56592.

Construction of each of the pairs of stereochemical centers was a significant synthetic challenge. A convergent synthesis of the drug was planned. The left-hand 2,2,5-trisubstituted tetrahydrofuran (THF) portion of the molecule would be condensed with the linear tetracyclic fragment, which would already contain the vicinal stereocenters on the five-carbon side chain. The two chiral key intermediates **5** and **6**, which are the result of this strategy, are shown in **Fig. 2**.

2. Enzymatic Approaches to the Chiral Tetrahydrofuran Sulfonate (5)
2.1. Enzymatic Desymmetrization of (R)-Triol 8

The initial synthesis of the key 2,2,4-trisubstituted tetrahydrofuran **5** was somewhat lengthy (**Fig. 3**) *(12)*. The *(R)*-tertiary hydroxyl chiral center was introduced

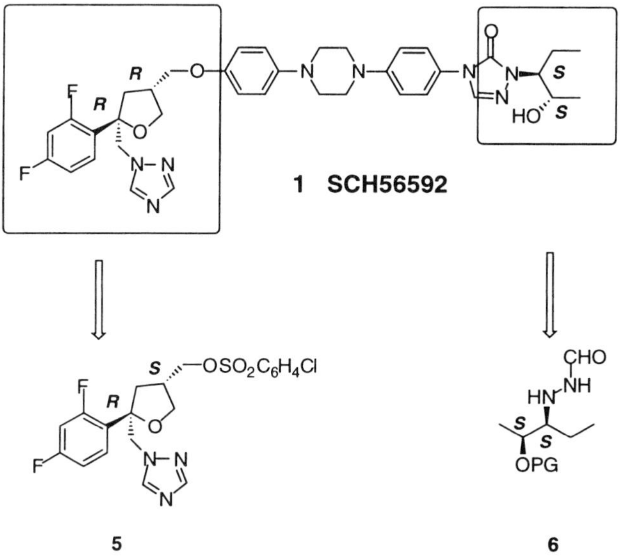

Fig. 2. Two chiral key intermediates

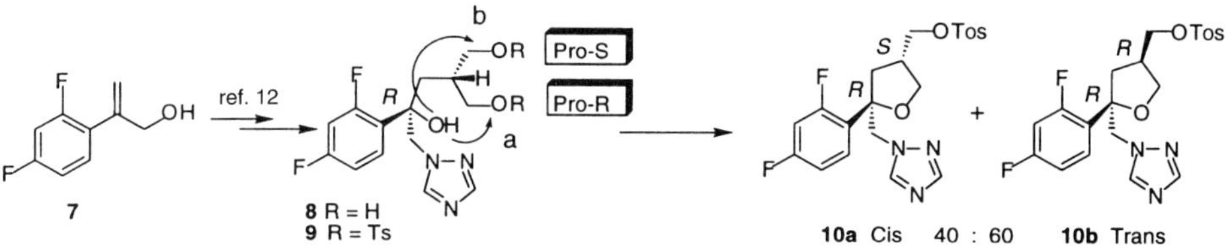

Fig. 3. Differentiating between primary alcohols.

via a Sharpless–Katsuki epoxidation of the allylic alcohol **7**, followed by introduction of the triazole and malonate alkylation/reduction to form the *R*-triol **8**. The crucial reaction involved activation of both primary hydroxyls as tosylates followed by nucleophilic displacement. Unfortunately, the existing chirality at the tertiary hydroxyl did not control the stereochemistry of the cyclization and a mixture of isomers was obtained under all the conditions examined. The best result obtained by this cyclization route was a 2:3 mixture of (2*R*,4*S*)-**10a** and (2*R*,4*R*)-**10b**, in which the undesired (2*R*,4*R*)-isomer predominated. Careful chromatography was required to isolate the desired (2*R*,4*S*)-**10a** (25–30% from the *R*-triol **8** and <10% from **7**).

Clearly, what was required was some way to differentiate between the two primary alcohols. If the pro-*S* alcohol* could be blocked, then tosylation and cyclization would furnish the desired (2*R*,4*S*)-**10a** exclusively (path a in **Fig. 3**).

Because *(R)*-triol **8** is a 2-substituted-1,3-propanediol, enzymatic desymmetrization suggested itself as a convenient method to differentiate between the two primary alcohols. Therefore, the selective enzymatic formation of a monoester of **8** was investigated *(14–17)*.

The enzymatic acetylation of *(R)*-triol **8** in neat methyl acetate using porcine pancreatic lipase (PPL Sigma Type II) had been previously reported to display pro-*R* selectivity, giving the monoacetate *(R,R)*-**11** (**Fig. 4**) *(18)*. Six subsequent steps (including inversion of the newly formed *R*-chiral center) were required to arrive at the desired (2*R*,4*S*)-*cis* tosylate **10a**. However, in our hands, the reported conditions for this acylation resulted in a mixture of triol, diacetate, and monoacetate, the latter having only marginally useful diastereomeric excess (85–90% de).

Therefore, a screen of our enzyme collection was undertaken to identify enzymes showing high pro-*S* selectivity for the acylation of *(R)*-triol **8** *(19)*. Of the 86 commercially available enzyme preparations that were initially tested, 16 showed significant pro-*R* selectivity, using vinyl acetate (10 equiv.) as the acylating agent in EtOAc. These gave the less desired monoacetate *(R,R)*-**11**, in some cases with moderate to high diastereomeric excess. The 12 enzymes that showed the desired pro-*S* selectivity all suffered from low selectivity and/or low rate of acylation. From these 12, Amano Lipase R (*Penicillium roquefortii*) [yielding the *(R,S)*-**11** isomer in 78% de at 20% conversion] was chosen for a study of different solvents and different acetylating agents, but no conditions were found that improved the diastereoselectivity.

Lipases generally display the same prochiral selectivity for the acylation of 2-substituted-1,3-propanediols as for the hydrolysis of the corresponding diesters. As a result, products of opposite absolute stereochemistry may be obtained using the same enzyme under either acylating or hydrolytic conditions (**Fig. 5**) *(15)*.

Of the 16 enzymes that displayed pro-*R* selectivity in the enzyme screen under acylation conditions, the best results under hydrolytic conditions were observed with Amano Lipase AK (*Pseudomonas* sp.) [*(R,S)*-**11**, 98% de]. Its pro-*R* selectivity could be exploited to prepare the desired monoester *(R,S)*-**11** by hydrolysis of a diester *(R)*-**12** (**Fig. 5**). Although hydrolysis of various diesters did provide the desired *(R,S)*-monoester with high de, the selectivity of lipase AK under aqueous conditions was not as good as in organic solvent, and

---

*A prochiral carbon center is defined as one containing a pair of identical substituents. If one of the paired substituents is arbitrarily chosen and elevated in priority, while keeping the priorities selective to the other groups unchanged, a chiral center results, which can be designated according to the Cahn–Prelog–Ingold rules. If the derived chiral center possesses the *S*-configuration, then the elevated substituent is defined as the pro-*S* group, and vice versa *(13)*.

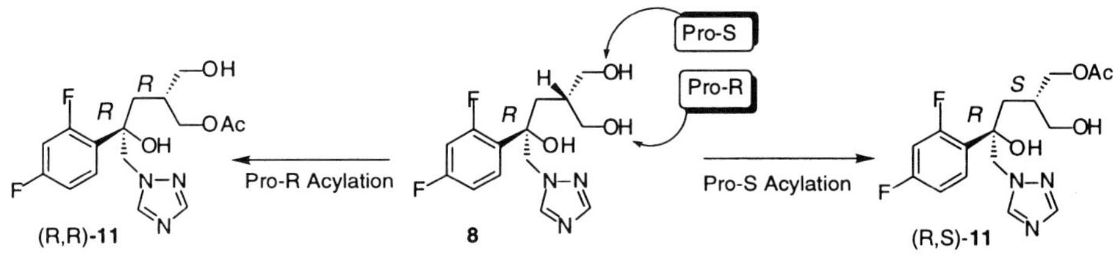

Fig. 4. Enzymatic acylation of (R)-triol **8**.

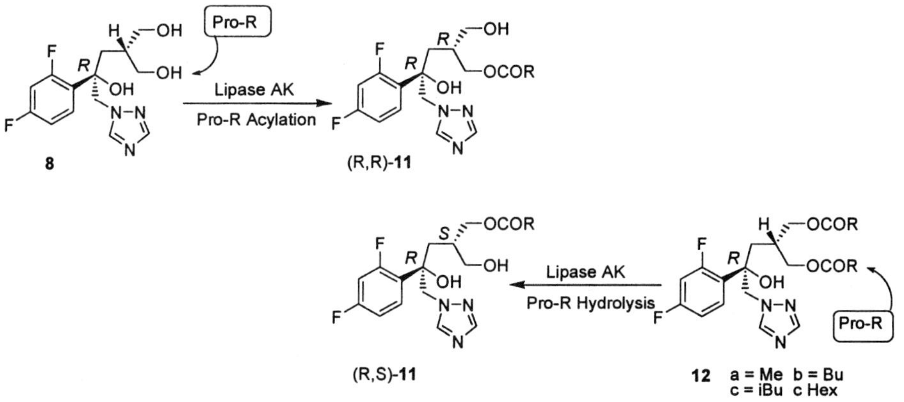

Fig. 5. Products of absolute stereochemistry.

substantial amounts of overhydrolysis to the triol occurred before complete consumption of the diester starting material. Hydrolysis of the dibutyrate **12b** appeared convenient because, (1) it reacted almost 10 times faster than the diacetate **12a** or diisobutyrate **12c** and three times faster than the hexanoate **12d** and (2) it was easily separated from the more water-soluble *(R)*-triol **8**. Furthermore, the chemical acylation and enzymatic hydrolysis reactions could be carried out in one pot. When butyrylation of *(R)*-triol **8** in THF was complete, the entire reaction mixture was diluted 10-fold with 50 m*M* KCl, the pH adjusted to 7.0, lipase AK added (1 g/g diester), and the pH maintained at 7.0 using a pHstat. After complete consumption of diester **12b**, the monobutyrate *(R,S)*-**11b** was obtained in good yield (64%) with 99% de.

As the hydrolysis of the dibutyrate **12b** was being developed, continued screening under nonaqueous conditions finally identified an enzyme displaying useful levels of the desired pro-*S* selectivity. Treatment of *(R)*-triol **8** with Novo SP435 (*Candida antarctica* lipase B) in EtOAc or MeCN with vinyl acetate as the acylating agent formed the monoacetate *(R,S)*-**11** with 94–97% de, the selectivity being better in MeCN (**Table 1**).

Table 1
Acetylation of (R)-Triol 8 with Novo SP435

Run	Solvent	Time min	% Triol 8	% MonoOAc 11	% DiOAc 12a	% de
1	EtOAc	75	5	83	12	94.0
2	MeCN	55	4	95	1	96.6

Conditions: Run 1: *(R)*-Triol **8**, 64 mg; Novo SP435, 15 mg; EtOAc, 1 mL; VinylOAc, 10 equiv; room temperature (RT). Run 2: *(R)*-Triol **8**, 0.5 g; Novo SP435, 51 mg; MeCN, 5 mL; VinylOAc, 2 equiv; RT.

Introduction of the enzymatic desymmetrization of *(R)*-triol **8** approximately doubled the throughput of the key intermediate (2*R*,4*S*)-tosylate **10a** and avoided a difficult chromatographic purification. Nevertheless, the synthesis was lengthy, with the two chiral centers being introduced separately at different steps. As so often happens in pharmaceutical process development, before either the enzymatic hydrolysis with lipase AK or the acylation with Novo SP435 was run on a large scale, a more efficient route to tosylate **10a** was identified.

2.1.1. Synthesis of (2S,2'R)-2[2'-(2",4"-Difluorophenyl)-2'-Hydroxy-3'-(1",2",4"-Triazolyl)]Propyl-1-Acetoxy-3-Hydroxypropane (R,S)-11

A mixture of triol **8** (2.0 g, 6.38 mmol), vinyl acetate (1.4 mL, 13.06 mmol) and Novo SP435 (0.20 g) in MeCN (20 mL) was stirred at room temperature and monoitored periodically by high-performance liquid chromatography (HPLC). After 85 min the reaction mixture was filtered, the enzyme beads washed with EtOAc (10 mL) and the filtrate evaporated (<30°C). The crude product was purified by column chromatography, eluting with 0–6% MeOH/CH$_2$Cl$_2$. The required fractions were pooled and evaporated to obtain a foam (2.04 g, 89.9%; 96.6% de): $[\alpha]_D^{25}$ = – 49.19 (*c* 1.118, MeOH).

2.2. Enzymatic Desymmetrization of Diol 13

In the improved synthesis *(20)*, shown in **Fig. 6**, a diol olefin **13** could be quickly assembled from bromodifluorobenzene. This underwent an iodocyclization reaction to form a 2,2,4-trisubstituted tetrahydrofuran in which the desired *cis*-isomer

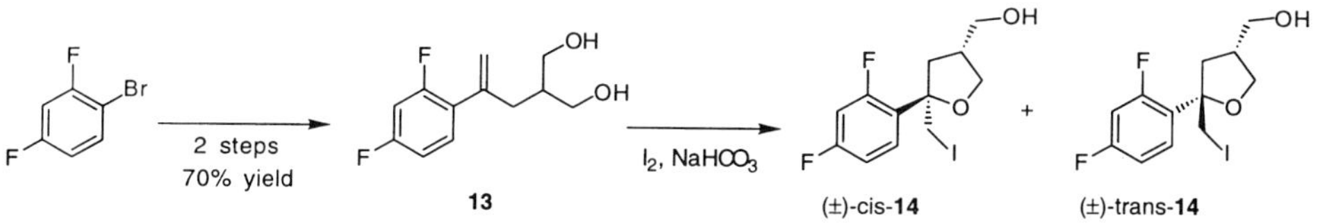

Fig. 6. Improved synthesis.

14 predominated (85:15 *cis:trans*) *(21,22)*. Introduction of the triazole followed by formation of the chlorosulfonate provided the key intermediate **5**, which could be condensed with the side chain to form SCH56592.

Like the *(R)*-triol **8** in the previous route, diol **13** is also a 2-substituted-1,3-propanediol, providing a similar opportunity for biocatalysis. If a suitable enzyme could be found to desymmetrize diol **13**, then the stereoselectivity of the iodocyclization would transfer the chirality generated by the enzyme to the newly formed benzylic center. Acylation of the pro-*S* hydroxyl of **13** would result in a five-step sequence to (*R,S*)-sulfonate **5** (acylation, iodocyclization, triazole introduction, deacylation, and sulfonation), whereas pro-*R* acylation would require two extra steps to invert the (2*R*) chiral center. Pro-*R* hydrolysis, requiring one extra step to generate diester **15,** would result in a six-step sequence to **5 (Fig. 7)**.

Approximately 205 commercially available enzyme preparations were screened for selective acetylation of diol **13** in toluene using vinyl acetate as the acyl donor. The best results are collected in **Table 2**; *C. antarctica* lipase B showed the best pro-*S* selectivity, whereas a lipase from *Humicola lanuginosa* displayed excellent pro-*R* selectivity. As earlier, the major effort was to obtain not only the highest possible enantiomeric excess (ee) of the monoester product, but to maximize the yield of monoester **16** and minimize the yield of both unreacted diol **13** and the diester **15** formed by overreaction.

The availability of two enzymes with opposite prochiral selectivity allowed three approaches to the preparation of the desired (2*S*)-monoester precursor; pro-*R* acylation of **13** or pro-*R* hydrolysis of **15** using Amano lipase CE, or pro-*S* acylation of **13** using Novo SP435 **(Fig. 8)**.

2.2.1. Pro-R Acylation

Pro-*R* acetylation of diol **13** using lipase CE gave the monoacetate (*R*)-**16a** with high chemical and optical yield. As the reaction profile illustrates **(Fig. 9)**, the prochiral selectivity was excellent with monoacetate being formed in high enantiomeric excess and very little diacetate being formed; $v^R_1 \gg v^S_1$ or v^S_2 **(Fig. 8)**. However, protection/deacylation was required to invert the chiral center, resulting in a seven-step sequence from diol **13** to sulfonate **5**. Because attempts to introduce

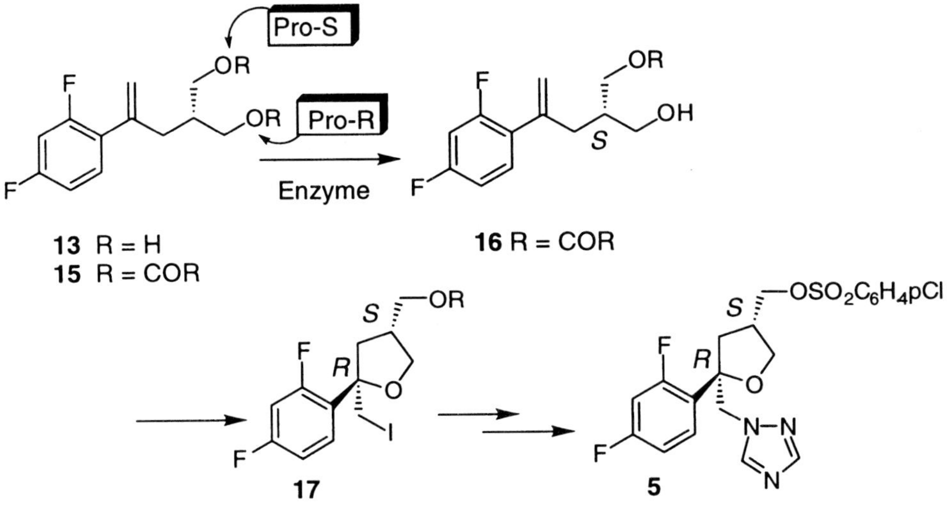

Fig. 7. Six-step sequence.

Table 2
Acetylation of Diol 13: Initial Screen Results

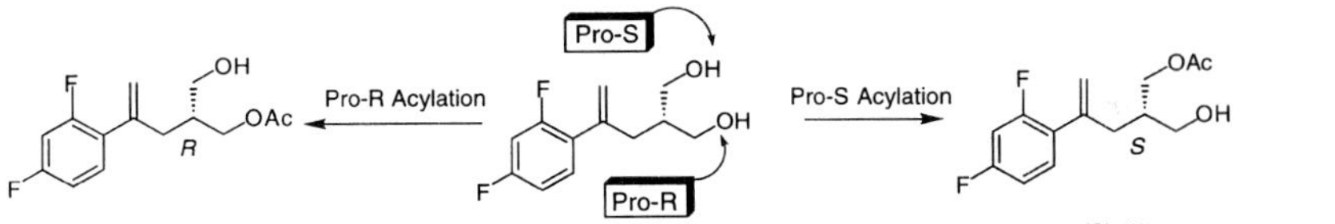

Enzyme (Source)(Vendor)	Time	Diol	MonoOAc	DiOAc	ee	Preference
	min	13	16	15	%	
		%	%	%		
Lipase SP435 (*C. antarctica*) (Novo)	60	0	83	17	97	Pro-S
Chirazyme L-2 (*C. antarctica*) (Boehringer-Mannheim)	1440	0	60	40	97	Pro-S
Lipase CE (*H. lanuginosa*) (Amano)	95	0	97	3	99	Pro-R
Lipase *H. lanuginosa* (Biocatalysts Ltd.)	220	1	98	1	98	Pro-R

Conditions: Diol **13**, 50 mg; enzyme, 10–200 mg; VinylOAc, 5–10 equiv; toluene, 1.0 mL; room temperature.

a tetrahydropyranyl protecting group resulted in significant racemization, presumably via 1,3 acyl migration, this route was not pursued.

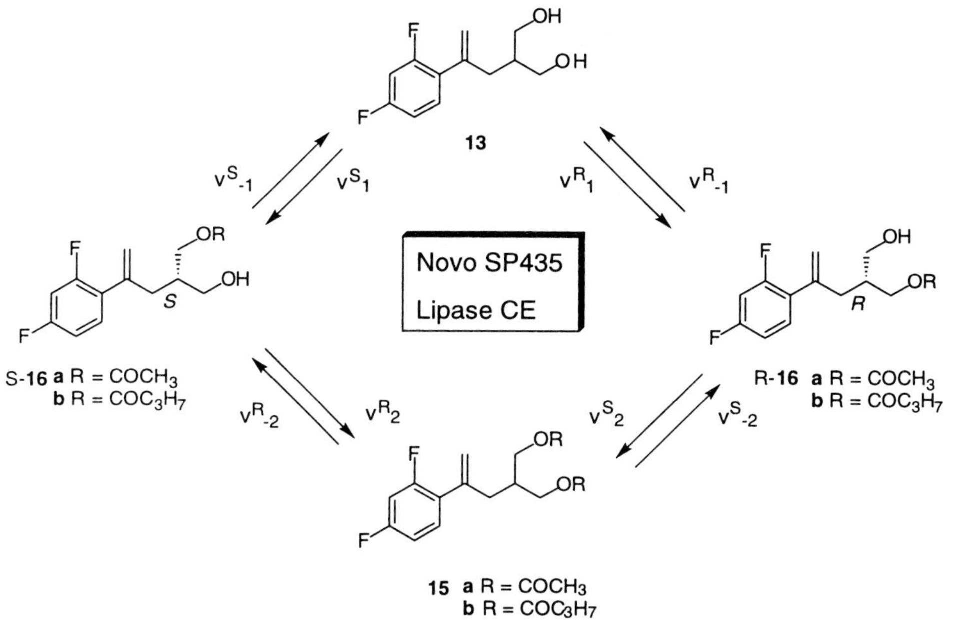

Fig. 8. Approach to the preparation of the desired 2*S*-monoester precursor.

2.2.1.1. SYNTHESIS OF (2R)-2[2'-(2",4"-DIFLUOROPHENYL)]-2'-PROPENYL-1-ACETOXY-3-HYDROXYPROPANE (R)-**16B**

Lipase CE (2.0 g) was added to a solution of diol **13** (10.0 g, 43.8 mmol) and vinyl acetate (8.0 mL, 86.8 mmol) in HPLC-grade toluene (200 mL). The mixture was stirred at room temperature with periodic HPLC monitoring. After 26 h the reaction mixture was filtered through a Celite pad that was washed with tert-butyl metyl ether (TBME) (20 mL). The filtrate was evaporated (<30°C) and purified by column chromatography, eluting with 10–35% EtOAc/hexane. Pooling the relevant fractions yielded the diacetate **15b** (0.66 g), a mixture of monoacetate and diacetate (0.89 g), and pure monoacetate (*R*)-**14b** as a viscous oil (10.37 g, 87.6%; 97.0% ee): $[\alpha]_D^{24} = +13.9$ (*c* 1.085, EtOH).

2.2.2. Pro-R Hydrolysis

Lipase CE-catalyzed pro-*R* hydrolysis of the diester **15** would yield the *(S)*-monoester, for a six-step total sequence from diol **13** to the key sulfonate **5**. Although hydrolysis of the dibutyrate **15b** produced the monobutyrate (*S*)-**16b** with high enantiomeric excess **(Fig. 10)**, the chemoselectivity was poor and the reaction mixture consisted of a mixture of diol **13**, dibutyrate **15b**, and the desired monoester (*S*)-**16c**.

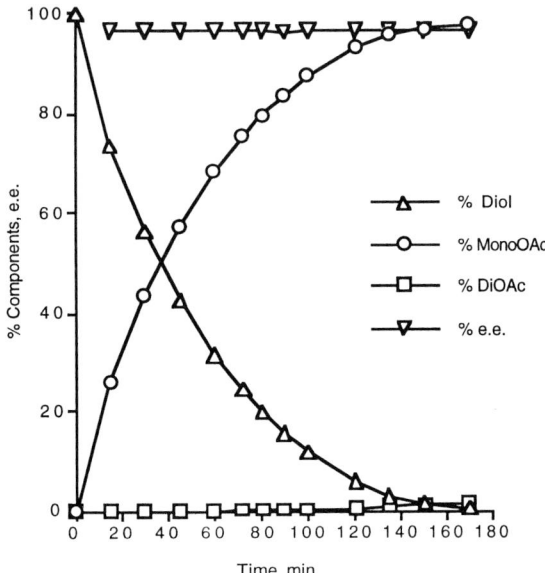

Fig. 9. Acylation of **13** with Amano lipase CE: Diol **13**, 5 g; lipase CE, 5 g; vinyl acetate, 5 equivalents.; toluene, 100 mL; room temperature.

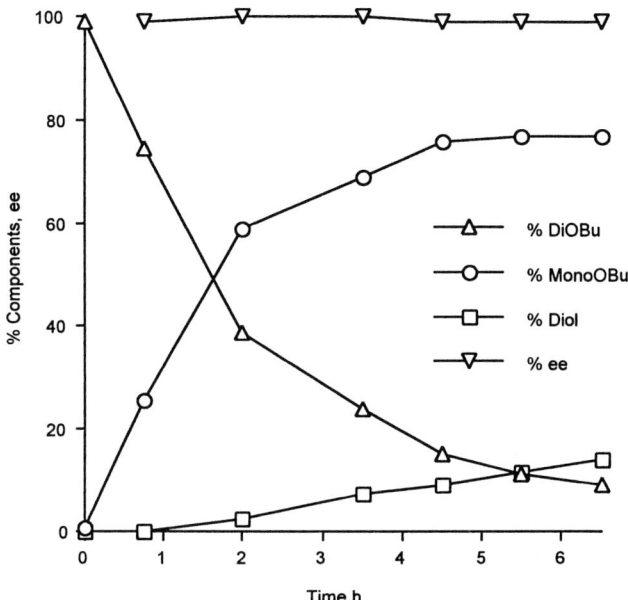

Fig. 10. Amano lipase CE hydrolysis: DiOBu **15b**, 7 g; Amano CE, 5 g; 50 m*M* KCl, 63 mL; pH 7.5; 22°C.

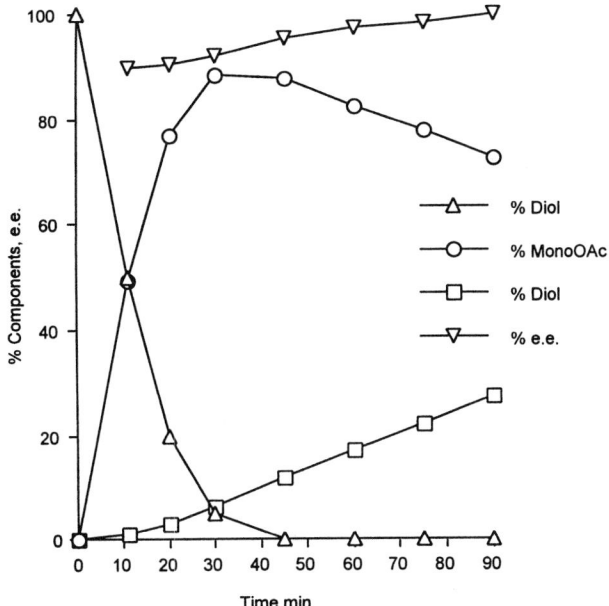

Fig. 11. Acylation of diol **13** with Novo SP435: Diol **13**, 50 mg; Novo SP435, 10 mg; VinylOAc, 10 equivs; Toluene, 1.0 mL; room temperature.

2.2.3. Pro-S Acylation

Novo SP435-catalyzed pro-*S* acetylation of diol **13** provided (*S*)-monoacetate **16a** directly, for an overall five-step route to the key sulfonate **5**. Because the prochiral selectivity is lower than for Lipase CE ($v^S_1/v^R_1 \sim 20$), optical purity was purchased at the cost of chemical yield, and a significant amount of the diacetate **15a** had to be formed before the ee of the remaining (*S*)-monoacetate **16a** reached useful levels (**Figs. 8** and **11**). In other words, the moderate prochiral selectivity of the initial desymmetrization reaction is compensated for in a subsequent kinetic resolution, in which the initially formed monoester **16** reacts further to form the diester **15** *(23)*. Because the enzyme displays the same preference in both the desymmetrization and the kinetic resolution steps ($v^S_2 > v^R_2$), the undesired (*R*)-monoester is acylated faster than the desired *(S)*-isomer and the ee of the monoester increases with conversion. However, this comes at the cost of reduced yield of monoester **16** and increased levels of the diester byproduct **15**.

Because it provided the desired (*S*)-monoester directly under operationally simple conditions, the Novo SP435-catalyzed acetylation of diol **13** was selected for optimization. Because the prochiral selectivity of the enzyme was not absolute, the enantiomeric excess of the (*S*)-monoacetate **16a** increased over the course of the reaction at the expense of chemical yield. The question

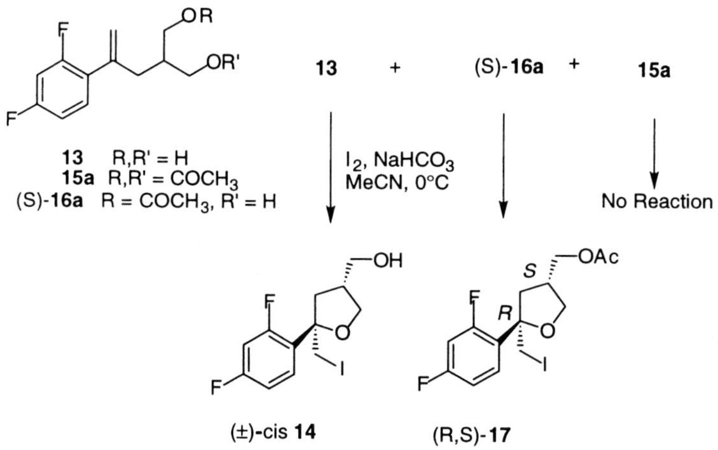

Fig. 12. Iodocyclization of unreacted diol **13**, yielding ravemic material **14**.

was: What compromise should be struck between optical and chemical yield? Like the (S)-monoacetate **16a**, unreacted diol **13** also undergoes iodocyclization to yield racemic material **14** (**Fig. 12**), which, if carried forward, would ultimately degrade the optical purity of the key intermediate sulfonate **5**. The diester **15a**, lacking a hydroxyl function, does not participate in the iodocyclization reaction and, in fact, may be recovered as the starting diol **13** at a subsequent stage.

Although minimization of both diol and diester would be desirable and would give the maximum yield, it was more important that the amount of unreacted diol be minimized. Thus, it was preferable to run the enzymatic acetylation reaction until no diol **13** remained and to accept the yield loss as a result of the formation of the diacetate **15a**. This ensured that the enantiomeric excess of the (S)-monoacetate **16a** was sufficiently high. In practice, the reaction was carried out until <1–2% of diol **13** remained, at which point the optical purity of the monoacetate was typically >95% ee.

The optimization of this process has been described in detail elsewhere *(19)*. Examination of a number of solvents, acetylating agents, and temperatures indicated that the reaction worked best using vinyl acetate in either toluene or acetonitrile at 0–15°C. As the iodocyclization reaction that followed this enzyme step was expected to be run in acetonitrile, this was chosen to avoid changing solvents between steps. Efforts to reduce the quantity of vinyl acetate showed that the reaction could be run with as little as 1.3 equivalents.

It was hoped that the use of a bulky acylating agent would decrease diester formation and increase the selectivity of the initial desymmetrization reaction. Using trifluoroethyl isobutyrate in toluene did suppress diester formation, but the monoester was formed with low ee; even after extended reaction and formation of diisobutyrate, the ee of the monoisobutyrate remained low. With trifluoroethyl 2-methylbutyrate, diester formation was completely suppressed, but the monoester was again formed in very low ee. Thus, vinyl acetate remained the acylating agent of choice.

Good volumetric productivity was attained by increasing the concentration of substrate to 20% and reducing the loading of the enzyme to 5% by weight of diol. To demonstrate enzyme recovery and reuse, a 50-mg sample of Novo SP435 was carried through multiple reactions over a period of 14 d. After 10 cycles, the enzyme sample still produced monoacetate of acceptable quality, but the rate of the reaction had slowed down considerably. This decrease in rate was attributed more to mechanical losses of enzyme during recycling than to inactivation of the enzyme. The conditions initially transferred to the pilot plant were the following:

- 20% Diol **13** solution
- 5% Novo SP435 (w/w)
- 2.0 Equivalents vinyl acetate
- Industrial-grade MeCN
- 0°C Reaction temperature
- Monoacetate specification: (*S*)-**16a** > 95% ee; <2% diol **13** remaining

A representative result is shown in **Table 3** (run 1). The specifications for the enzymatic acylation were subsequently tightened to produce monoacetate (*S*)-**16a** of 98–99% ee with <1% unreacted diol remaining. This could be achieved using the above conditions but running the acetylation out to approx 30% diacetate formation (**Table 3**, run 2). The latter result also showed that there was little or no loss of optical purity during the iodocyclization reaction.

2.2.3.1. SYNTHESIS OF (2S)-2[2'-(2",4"-DIFLUOROPHENYL)]-2'-PROPENYL-1-ACETOXY-3-HYDROXYPROPANE (S)-**16A**

A mixture of diol **13** (5.01 g, 21.95 mmol) and Novo SP435 (0.26 g) in industrial-grade MeCN (25 mL) was cooled in an ice bath for 15 min. Vinyl acetate (4.0 mL, 43.4 mmol) was added, and the mixture stirred at 0°C. After 6 h, the reaction mixture was filtered, the beads washed with EtOAc (30 mL), and the combined organics evaporated (< 30°C). The residue was purified by column chromatography, eluting with 10–50% EtOAc/hexanes. Pooling the relevant fractions yielded the monoacetate (*S*)-**16a** (4.23 g, 71.3%; 98.2% ee): $[\alpha]_D^{23} = -13.9$ (*c* 1.688, EtOH).

2.2.4. Diol Desymmetrization: The Role of Solvent and Acylating Agent

The Novo SP435-catalyzed acetylation of diol **13** was used routinely for the preparation of hundreds of kilograms of the monoacetate (*S*)-**16a**. However,

Table 3
Scale-Up of Enzymatic Acetylation

Run	Diol 13	% Diol 13	% MonoOAc (S)-16a	% DiOAc 15a	% ee
1	9 kg	2	81	17	97[a]
2	33 kg	<1	74	26	99[b]

[a] ee of S-16a.
[b] ee of R,S-17.

the continued loss of up to 30% of product as the diacetate, even though it was potentially recoverable and reusable, was a major weakness in the process. The eventual solution of this problem illustrates the effect that a subtle interplay of solvent and acylating agent can have on the development of a commercial enzymatic process and the hazards of making assumptions while screening in an empirical field.

The enzymatic acetylation of diol **13** had been examined in a series of common solvents **(Table 4)**, and the best results, in terms of product distribution and selectivity, were observed in MeCN and toluene, the reaction being faster and the diol more soluble in MeCN.

The enzymatic acylation was also investigated using 11 different acylating agents in either MeCN or toluene. Vinyl acetate and acetic anhydride showed similar selectivities when compared in MeCN **(Table 5)**, as would be expected if the prochiral discrimination was dictated solely by approach of the diol substrate to the covalent acyl-enzyme intermediate at the active site of the enzyme. It was hoped that the use of bulky acylating agents would improve the product distribution of the reaction by (1) increasing the prochiral selectivity of the initial desymmetrization and (2) decreasing diester formation. Although the use of trifluoroethyl isobutyrate (TFEOiBu) in toluene did decrease formation of the diester, the enantiomeric excess of the monoester remained low; extending the reaction time did not improve the ee of the monoester even though substantial amounts of the diester were formed. Because MeCN and toluene performed equally well as media in the enzymatic reaction with vinyl acetate **(Table 4**, runs 7 and 8), and because the selectivity of the desymmetrization was presumed to depend *only* on the approach of the diol to

Table 4
Product Distribution and Enantiomeric Excess at >94% Diol Conversion in Various Solvents

Run	Solvent	Vinyl acetate equiv.	Diol/enzyme ratio (w/w)	Time (min)	Product distribution (%)			% ee (S)
					Diol 13	Monoacetate (S)-16a	Diacetate 15a	
1	i-Pr₂O	10.0	4.0	90	6	84	10	91
2	THF	10.0	4.0	120	2	77	22	90
3	Dioxane	10.0	4.0	90	1	75	24	93
4	Acetone	10.0	4.0	90	1	83	16	94
5	i-PrOAc	2.0	9.0	260	1	78	21	97
6	TBME	2.0	9.0	226	3	75	22	94
7	MeCN	5.0	4.0	60	1	85	14	96
8	Toluene	5.0	4.0	120	1	85	14	96

Conditions: Diol **13**, 50 mg; solvent, 1.0 mL; 0°C (except run 3, 20°C).

Table 5
Influence of Acylating Agent

	Acyl agent	Equiv	Solvent	Diol 13 (M)	Diol/enzyme ratio (w/w)	Temp (°C)	Time (min)	Product distribution (%)			
								Diol **13**	Monoester (S)-**16**	Diester **15**	%ee (S)
1	VinylOAc	1.4	MeCN	0.93	5.0	0	115	1	78	21	96
2	VinylOAc	Neat	Neat	0.90	10.0	0	165	1	84	15	96
3	VinylOAc	2.0	MeCN	0.88	19.0	0	270	0	84	16	95
4	TFEOAc	2.0	MeCN	0.85	15.0	0	385	12	86	2	91
5	TFEOiBu	10.0	Toluene	0.22	5.0	RT	420	0	96	4	89
							1250	0	87	13	91
6	Ac$_2$O	2.0	MeCN	0.85	4.6	RT	120	1	81	18	97
7	VinylOAc	2.0	MeCN	0.85	20.0	0	200	2	80	19	97
8	VinylOAc	2.0	MeCN	0.85	94.0	0	1150	1	82	17	97

the acyl-enzyme intermediate, no other combinations of solvent and isobutyrate reagents were examined.

However, when the acylation of diol **13** was later examined in MeCN with isobutyric anhydride as the acylating agent, the reaction was found to be more selective than with vinyl acetate *(24)*. At first, this result seemed strange, given that TFEOiBu had been a poor acylating agent and that the identical acyl-enzyme intermediate should be formed from either TFEOiBu or isobutyric anhydride. Subsequent investigation showed that the role of the solvent was crucial. When run under identical conditions, the Novo SP435-catalyzed acylation with TFEOiBu was found to be faster and significantly more selective in MeCN than in toluene (compare runs in **Table 6** and **Fig. 13**).

When isobutyric anhydride was used as the acylating agent, a similar solvent difference was also observed. In this case, although the reaction was faster in toluene, it was more selective in MeCN. After 2 h in toluene, the monoisobutyrate **16c** was formed in only 95% ee even though 22% of the diester had been formed (**Table 7**, run 1). In the same period in MeCN (run 2), only 9% diester had been formed, but the monoester **16c** had already reached 99% ee. This difference in reactivity and selectivity is illustrated in **Fig. 14**.

Two other changes were made to the reaction conditions. The temperature was lowered to avoid any possible nonenzymatic acylation by the anhydride, resulting in a small improvement in selectivity. Solid $NaHCO_3$ was also added to the reaction mixture to prevent any racemization by acid-catalyzed 1,3-acyl migration. These conditions have been used to prepare monoester **16c** consistently with >98% ee. Based on these results, the plant process was modified as follows:

- 20% Diol **13** solution
- 5% Novo SP435 (w/w)
- 1.2 Equivalents isobutyric anhydride
- Industrial-grade MeCN
- –10 to –15°C Reaction temperature
- 0.7 Weight equiv. $NaHCO_3$
- Monoacetate specification: (*S*)-**16c** < 98% ee; <1% diol **13** remaining

In developing the desymmetrization of diol **13**, 205 enzymes, 8 solvents and 11 acylating agents were examined. However, every one of the 18,040 possible combinations could not be individually examined. The linear approach to identifying the best enzyme followed by the optimum solvent followed by the optimum acylating agent did lead to a local maximum on the reaction yield surface: Novo SP435-catalyzed acetylation with vinyl acetate in MeCN. This local maximum was achieved after making some reasonable assumptions:

1. Because the reaction showed the same selectivity with vinyl acetate in toluene and MeCN, a similar lack of solvent effect would be seen with isobutyrate acylating agents *(25)*.

Table 6
Acylation of Diol 13 with Trifluoroethyl Isobutyrate

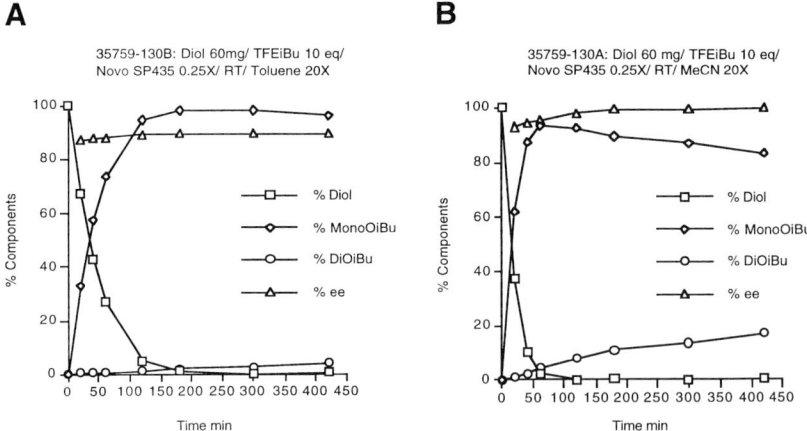

Run	Solvent	Time min	% Diol 13	% MonoOiBu (S)-16c	% DiOiBu 15c	% ee
1	Toluene	435	0.4	97.2	2.4	89
2	Toluene	420	0.3	96.0	3.7	89
		1250	0	87.2	12.8	91
3	MeCN	114	0.2	93.0	6.8	97
4	MeCN	120	0	92.0	8.0	98

Conditions: Runs 1, 2, and 4: Diol **13**, 60 mg; Novo SP435, 13–14 mg; TFEOiBu, 10 equiv.; solvent, 1.0 mL; RT. Run 3: Diol **13**, 1.0 g, Novo SP435, 0.1 g; TFEOiBu 5 equiv.; solvent, 5.0 mL; RT.

A

35759-130B: Diol 60mg/ TFEiBu 10 eq/
Novo SP435 0.25X/ RT/ Toluene 20X

□— % Diol
◇— % MonoOiBu
○— % DiOiBu
△— % ee

Time min

B

35759-130A: Diol 60 mg/ TFEiBu 10 eq/
Novo SP435 0.25X/ RT/ MeCN 20X

□— % Diol
◇— % MonoOiBu
○— % DiOiBu
△— % ee

Time min

Fig. 13. The enzymatic isobutyrylation of diol **13** with TFEOiBu in Toluene (Graph A, Run 2, **Table 6**), or in MeCN (Graph B, Run 4, **Table 6**).

Table 7
Acylation of Diol 13 with Isobutyric Anhydride

Run	Solvent	Time min	% Diol 13	% MonoOiBu (S)-16c	%DiOiBu 15c	% ee
1	Toluene	120	0.0	77.5	22.4	95
2	MeCN	120	0.0	90.5	9.5	99

Conditions: Diol **13**, 1.0 g; solvent, 5.0 mL; Novo SP435, 50 mg; (*i*-PrCO)$_2$O, 1.3 equiv.; RT.

A

35759-142B:Toluene 5X/ Diol 1.0g/ (iPrCO)₂O 1.3 eq./ SP435 5%/ RT

B

35759-142A: MeCN 5X/ Diol 1.0g/ (iPrCO)₂O 1.3 eq./ SP435 5%/ RT

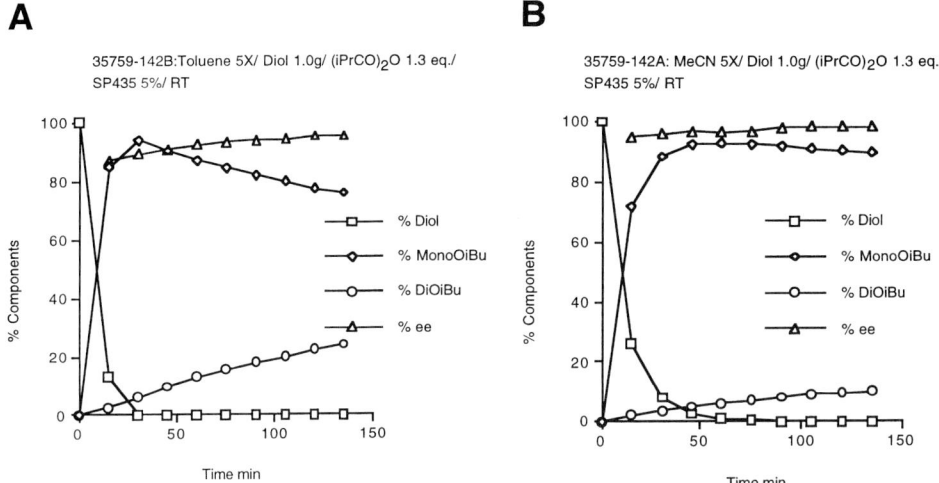

Fig. 14. The enzymatic isobutyrylation of diol **13** in Toluene (Graph A, Run 1, **Table 7**), or in MeCN (Graph B, Run 2, **Table 7**).

2. The reaction mechanism suggested that prochiral selectivity would be determined solely by approach of the substrate diol to the acyl-enzyme intermediate. Reactions in MeCN using vinyl acetate and acetic anhydride as acylating agents and proceeding through the same acyl-enzyme intermediate showed similar selectivities. Thus, it was assumed that because trifluoroethyl isobutyrate showed poor selectivity in toluene, isobutyric anhydride would also show poor selectivity in MeCN.

These assumptions proved to be incorrect. By failing to recognize the interplay of enzyme/solvent/acylating agent in determining selectivity, the initial process subjected to scale-up resulted in significant loss of yield. However, the proper choice of solvent and acylating agent has improved the yield of the desired product by up to 20%. Although seductive in their rational appearance and in their ability to simplify the task of reaction optimization, even simple assumptions concerning interrelated variables can be misleading. An automated approach, testing every possible variable, or a multivariate investigation of reaction parameters *(26,27)*, might have led to a more global maximum sooner. This reaction is a specific example of the need for process chemists to utilize these tools.

2.2.4.1. SYNTHESIS OF (2S)-2[2'-(2",4"-DIFLUOROPHENYL)]-2'-PROPENYL-1-ISOBUTYROXY-3-HYDROXYPROPANE (S)-**16c**

A mixture of diol **13** (1.0 g, 4.4 mmol) and Novo SP435 (52 mg) was stirred in industrial-grade MeCN (5 mL) under N_2 at 25°C. Isobutyric anhydride

(0.84 mL, 5.1 mM) was added and the reaction monitored by chiral HPLC every 15 min. After 100 min, the reaction mixture was filtered through a Celite pad that was washed with TBME (25 mL). The filtrate was washed with water (20 mL), saturated NaHCO$_3$ (20 mL) water (20 mL), saturated NaCl (10 mL), and dried (MgSO$_4$), filtered and evaporated. The residue was purified by column chromatography, eluting with 20–30% EtOAc/hexanes. Pooling the relevant fractions yielded the monoisobutyrate (S)-**16c** (1.2 g, 90.2%; 98.2% ee): $[\alpha]_D^{25} = -13.95$ (c 1.584, EtOH).

3. Enzymatic Approaches to the Formylhydrazine 6

The other key chiral intermediate for SCH56592 is the N-formylhydrazine **6**, possessing two adjacent chiral centers fixed in a *syn* configuration. This is condensed with a suitable linear tricycle, installing the triazole with the essential hydroxylated side chain; condensation with sulfonate **5** completes the synthesis of SCH56592. Some of the biocatalytic approaches that were explored for the production of **6** are collected in **Fig. 15** and include both the microbial reduction of ketones and the acylation of alcohols by isolated enzymes in organic solvent. Again, it will be the latter reactions that will be discussed more fully in this section.

3.1. Microbial Reduction of Formylhydrazone 18

Sodium borohydride reduction of (2S)-2-benzyloxy-N-formylhydrazone **18** results in the predominant formation of the less desired diastereomer (2S,3R)-**6**. Although we knew of no literature precedent for the microbial reduction of hydrazones, we decided to screen our culture collection for such activity. Because **18** decomposed in both culture media and buffer solutions at pH < 7.0, the screen was carried out at pH 8.0. Cells from 340 cultures were collected, washed, and then assayed against hydrazone **18** at 2 g/L substrate in pH 8.0 buffer. In no case was any sign of **6** detected; hydrolysis to the ketone **19** and reduction of this to 2-benzyloxy-3-pentanol (**20**), sometimes with high diastereoselectivity, were the only identified products (**Fig. 16**).

3.2. 2-Nitro-3-pentanol (21) as a Precursor to 6

Another approach that was studied in some detail is outlined in **Fig. 17**. In this scenario, the formylhydrazine **6** would be prepared from *syn* (2S,3S)-3-amino-2-pentanol (**22**) via the diaziridine **23** *(28)*. The aminopentanol **22**, in turn, would be derived from reduction of (2S,3S)-3-nitro-2-pentanol (**21**), which would be accessed by enzymatic kinetic resolution of the racemate or by microbial reduction of the corresponding ketone **24** (**Fig. 18**).

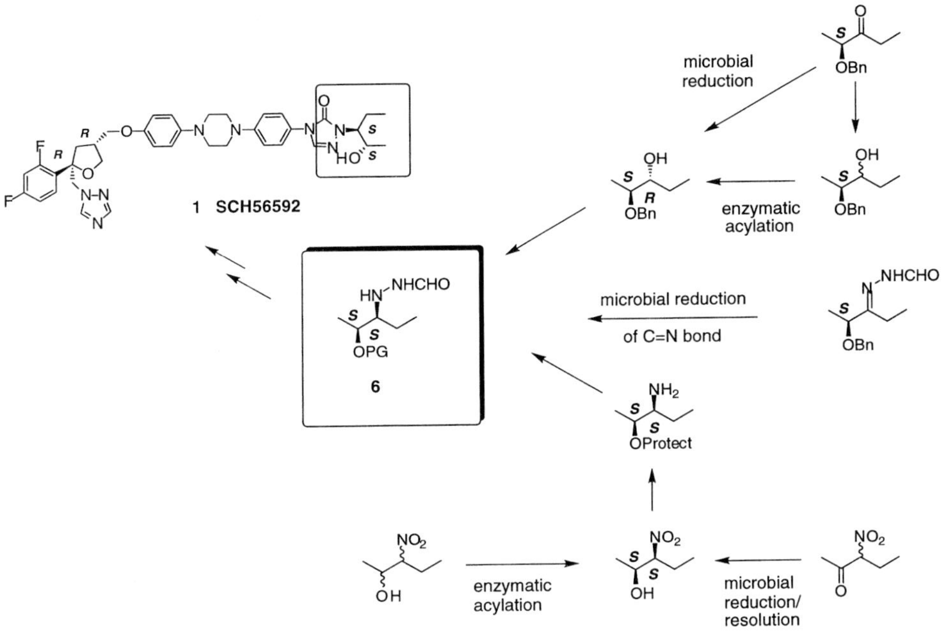

Fig. 15. Biocatalytic approaches to produce diol **6**.

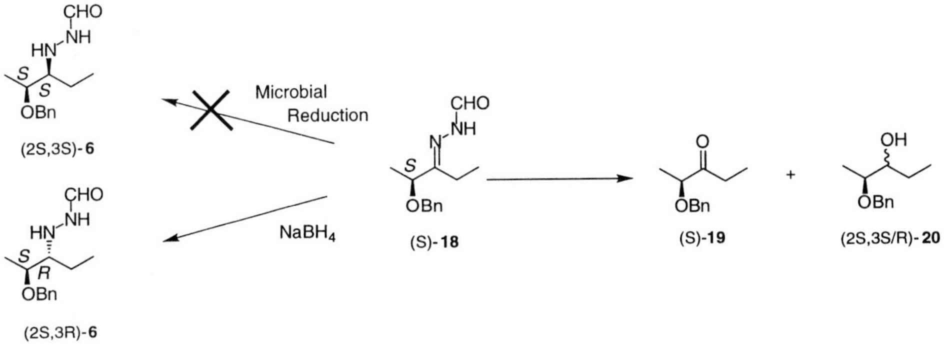

Fig. 16. Identified products.

3.2.1. Microbial Reduction of 3-Nitro-2-Pentanone (24)

A major attraction for investigating the enantioselective microbial reduction of (±)-3-nitro-2-pentanone (**24**) was the possibility that the unreacted enantiomer might undergo racemization under the reaction conditions, resulting in a high-yield dynamic resolution. Therefore, a screen for the reduction of (±)-**24** (0.5–1.0 g/L) to (2S,3S)-3-nitro-2-pentanol (**21**) was conducted. Among the

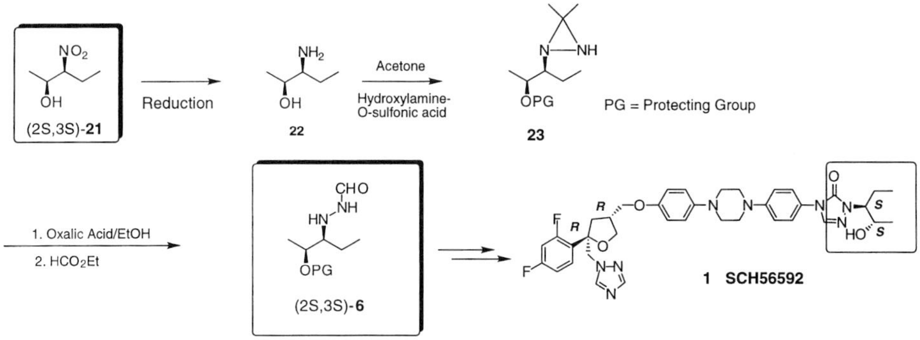

Fig. 17. Another biocatalytic approach.

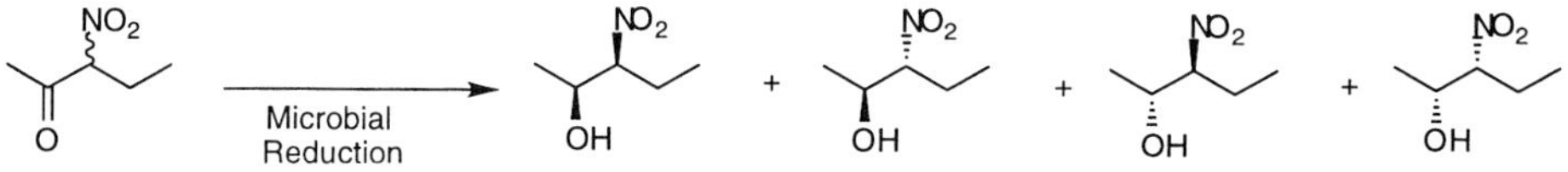

Fig. 18. Derivitization of aminopentanol **22**.

1082 cultures of yeast, filamentous fungi, and bacteria surveyed, over half appeared to either completely degrade the substrate or accumulate unknown products. Most of the remaining cultures demonstrated poor diastereo-selectivity and many produced all four possible isomers. However, 50 cultures produced **21** with high ee but with variable de. Of these, *Hansenula subpelliculosa* (ATCC 16766) predominantly produced the desired (2*S*,3*S*)-**21** and was chosen for process optimization *(29,30)*.

Bioconversion of **24** (2 g/L) using *H. subpelliculosa* revealed that most of the substrate was consumed following 6 h of incubation to give nitropentanol **21** along with nitropropane. Retro-Henry reversion of **21** to nitropropane occurred in the presence or absence of culture at pH > 3.0, but it was signifi-cantly reduced by the addition of glucose. Therefore, bioconversions were con-ducted employing resting cells resuspended in phosphate–citrate buffer (pH 3.0) supplemented with glucose (3%). Conversion yields as high as 80–90% at 5 g/L of **24** were achieved within 2 h using 10X cell concentrates. However, reaction selectivity decreased with increased substrate concentration employ-ing both low or high density cell cultures. Elevated substrate concentrations, especially in the presence of low cell density resulted in increasing yields of the undesired 2*R* product diastereomers, suggesting that 3-nitro-2-pentanone (**24**) was likely reduced by more than one enzyme. In order to minimize this activity, low substrate concentrations (<1 g/L) and high cell densities (20X concentrate) were necessary but were not economical. Consequently, no further development of this reaction was pursued.

3.2.2. Enzymatic Kinetic Resolution of (±)-3-Nitro-2-pentanol (21)

Commercially available (±)-3-nitro-2-pentanol *(21)* exists as an approx 2:1 *syn/anti* mixture *(31)*. A screen of 210 enzyme preparations was undertaken to identify catalysts capable of the diastereoselective and enantioselective acylation of **21**. Twenty enzyme preparations were reactive under the conditions of the screen (vinyl acetate in TBME), all of them showing a slight preference for acylation of the *syn* isomer at low conversion. With one exception, all the enzymes preferentially acetylated the 2*R* chiral center. Two enzymes, Novozyme 435 and ChiroCLEC™ BL, showing opposite enantioselectivity (*R* and *S* selective, respectively*, were chosen for further investigation **(Fig. 19)**.

3.2.2.1.ENZYMATIC KINETIC RESOLUTION OF (±)-3-NITRO-2-PENTANOL (21) WITH CHIROCLEC™ BL

Because ChiroCLEC BL catalyzed the acylation of the isomers with the desired *S* configuration at C-2, the reaction is a direct resolution. In general, direct resolutions are less convenient because the enantiomeric excess of the product (ee$_p$) decreases with increasing conversion, and for reactions with moderate selectivity, ee$p$ may have declined below a useful level well before the 50% maximum theoretical yield is reached [e.g., for an Enantiomer ratio E *(33)* of 50, the ee$_p$ will have decreased to 92.7% by 40% conversion]. Furthermore, an extra step is required to convert the product ester back to the desired alcohol. Deacylation of (2*S*,3*S/R*)-**25** under basic conditions was expected to be a problem because of a facile retro-Henry reaction of the nitropentanol **21**.

Because ChiroCLEC BL showed unique selectivity, 21 other hydrolytic enzyme preparations (derived from *B. licheniformis* or *Bacillus* sp.) were examined for their ability to acetylate (±)-**21** using vinyl acetate in TBME. Again, only ChiroCLEC BL showed any activity under these conditions, providing (2*S*,3*S*)-**25** with high ee but with poor diastereoselectivity (*syn*-**25**, 94.3% ee; conversion, 49%; *E* >100). Interestingly, in a companion screen of 24 *Bacillus* hydrolases, ChiroCLEC BL showed no reactivity for the hydrolysis of (±)-nitroacetate **25** in phosphate buffer (pH 7) (both the dry form, for

---

*The absolute configurations were established as follows. Pure nitroacetate **25a** (*syn/anti* mixture), from the Novozyme 435-catalyzed acetylation of **21**, was reduced (Bu^{3}SnH/AIBN) and compared on chiral GC with the acetate prepared from commercial *(R)*-2-pentanol, establishing the reactive *syn*-isomer as 2*R*. Similarly, pure *anti*-nitroacetate **25a** was reduced to *(R)*-pentyl acetate. This established the reactive isomers in the enzymatic acetylation as (2*R*,3*R*)-*syn* and (2*R*,3*S*)-*anti*. In addition, Novozyme 435-catalyzed deacylation of pure (±)-*syn*-**25a** provided the (2*R*,3*R*)-**21**, whose structure was confirmed by X-ray crystallography of the brosylate. The assignments were also confirmed by 1H NMR examination of the corresponding Mosher esters *(32)*.

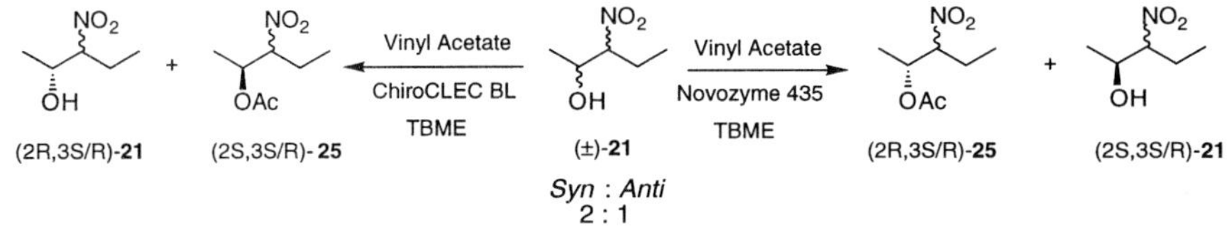

Fig. 19. Enzymes Novozyme 435 and ChiroCLEC BL, showing opposite enantioselectivity.

reaction in organic solvents, and the wet form, for hydrolysis reactions, were essayed). The only two enzymes that did show activity, Interspex Protease/ Esterase (*B. licheniformis*) and Fluka Esterase (*Bacillus* sp.), showed a preference for hydrolysis of the (2*R*,3*S/R*)-isomers, albeit with poor enantioselectivity and diastereoselectivity.

The CLEC-catalyzed acetylation worked well in a number of common solvents, with the best results being observed in *t*-amyl alcohol, TBME, and THF. In all cases, the enantioselectivity was good, with the *syn*-(2*S*,3*S*)-**25** being formed in 93–98% ee; the *anti*-(2*S*,3*R*)-25 was formed less selectively (89–97% ee). No attempt was made to adjust the water content of the reactions even though the performance of CLECs are reported to be susceptible to water activity *(34)*.

The CLEC-catalyzed acylation showed a marked dependence on the nature of the acetylating agent, with the best results being observed with vinyl acetate. Poor results were observed in neat alkyl acetates and also, surprisingly, with acetic anhydride. When tested again under similar conditions, acetic anhydride showed poor reactivity and selectivity, whereas propionic, butyric and isobutyric anhydrides showed good reactivity; isobutyric anhydride also displayed good enantioselectivity in the acylation of *syn*-nitroalcohol (ee$_S$ >95%, ee$_P$ 91.6%, $E = 105$ at 52% conversion). Reaction with chloroacetic anhydride occurred rapidly and, presumably, nonenzymatically to produce racemic product.

Because enzymes show the same enantiomeric preference for acylation and deacylation reactions, CLEC-catalyzed alcoholysis of nitropentanol esters appeared attractive because (1) it would produce the alcohol with the desired 2*S* configuration and (2) it was possible to prepare *syn* esters chemically from *syn/anti* mixtures (*vide infra*).

In a comparison of the CLEC-catalyzed deacylation/hydrolysis of three nitroesters **25a–c (Fig. 20)**, the best results were observed for the deacylation of the propionate **25b** and the chloroacetate **25c** with *n*-PrOH and *n*-BuOH; MeOH performed poorly (approx 16% conversion for **25c**) and no reaction was observed with the acetate **25a** under all conditions. One major problem with the ChiroCLEC BL-catalyzed kinetic resolution of nitropentanol, under both acylation and deacylation conditions, was a lack of reproducibility, particularly on scale-up. The deacylation of the chloroacetate **25c** with propanol

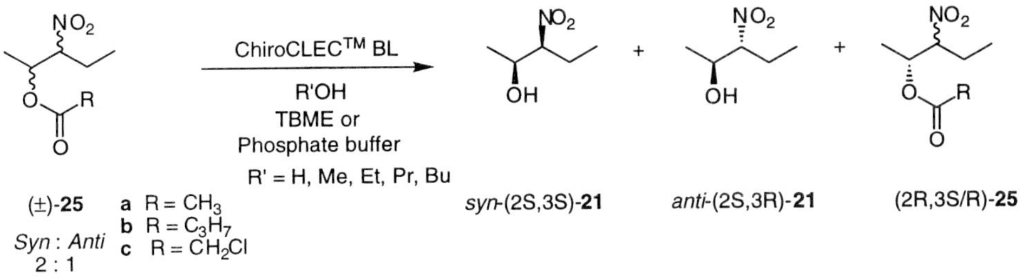

Fig. 20. Nitroesters **25a–c**.

in heptane gave good conversion on a 100-mg scale (50% conversion after 24 h using 25% enzyme loading). When scaled to 1 g (using 10% enzyme loading), the reaction reached only 34% conversion after 45 h. Because of the apparent lack of reproducibility and the relative expense of the catalyst, the alternative Novozyme 435 catalyzed resolution was favored.

*3.2.2.1.1. Synthesis of Syn-(2R,3R)-3-Nitro-2-Pentyl Chloroacetate (**25c**).* A mixture of *syn*-(±)-**25c** (1.0 g, 4.8 mmol), *n*-PrOH (3.6 mL, 47.7 mmol), and ChiroCLEC BL (0.1 g) in heptane (100 mL) was shaken at 40°C. After a fast initial rate (20% conversion in 1.5 h), the reaction slowed considerably (34% conversion after 41 h). At this stage, the reaction mixture showed ee$_S$ 45.4%, ee$_P$ 95.0%, $E = 61$ at 32.3% conversion. A further portion of enzyme was added (150 mg) and shaking continued for a total of 64 h (ee$_S$ 98.2%, ee$_P$ 88.0%, $E = 73$ at 52.7% conversion). The reaction mixture was filtered and the solvent evaporated. The residue was dissolved in toluene (10 mL) and extracted with 50% MeOH/water (5X 10 mL). The toluene layer was dried (MgSO$_4$), filtered, and evaporated to obtain (2R,3R)-**25c** (0.40 g, 40.3%; 98.1% ee); $[\alpha]_D^{25} = +13.87$ (*c* 1.810, EtOH). The MeOH/water extracts were concentrated, extracted with EtOAc, and the EtOAc extract dried (MgSO$_4$), filtered, and evaporated to yield *syn*-(2S,3S)-3-nitro-2-pentanol **21** (0.18 g, 27.8%; 88.0% ee).

3.2.2.2. ENZYMATIC KINETIC RESOLUTION OF (±)-3-NITRO-2-PENTANOL (**21**) USING NOVOZYME 435

Novozyme 435 catalyzed the acylation of the undesired (2R,3S/R)-isomers, thus permitting a more convenient subtractive resolution. Subtractive resolutions are generally preferred because the desired enantiomer is recovered unchanged from the reaction mixture. Furthermore, because the enantiomeric excess of the unreacted starting material (ee$_S$) increases with increasing conversion, subtractive resolutions can provide the desired enantiomer in high ee even for reactions with only moderate selectivity, if the reaction proceeds to high enough conversion. For example, for an E value of 20, unreacted starting material with 96.0% ee can be obtained if the conversion proceeds to 58% conversion. This was not a

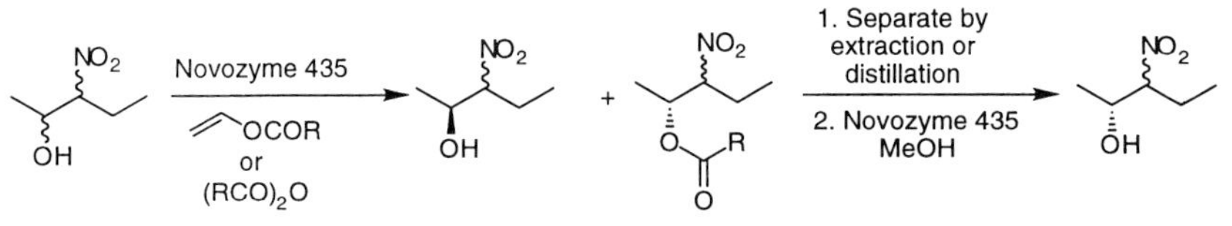

(±)-**21**
Syn : *Anti*
2 : 1

(2S,3S/R)-**21**

(2R,3S/R)-**25** **a** R = CH_3
b R = C_3H_7
c R = CH_2Cl
d R = iC_3H_7

(2R,3S/R)-**21**

Fig. 21. Enzymatic acylation/separation/enzymatic deacylation sequence.

consideration for Novozyme 435 because it showed high enantioselectivity for the acylation of *syn*-**21** ($E > 100$), but poor diastereoselectivity.

At first, we had mistakenly assigned the 2*S* configuration to the product of the Novozyme 435 acylation. Consequently, initial experiments concentrated on an enzymatic acylation/separation/enzymatic deacylation sequence, which allowed access to the undesired (2*R*,3*S*/*R*)-**21** (**Fig. 21**).

Enantioselectivity was high using a number of vinyl esters as acylating agents. Chloroacetate formation was particularly convenient because (2*R*,3*S*/*R*)-**25c** might be easily isolated by distillation from the initial enzymatic resolution, and the subsequent enzymatic deacylation was rapid. Enzymatic deacylation of the acetate (2*R*,3*S*/*R*)-**25a** and particularly the isobutyrate (2*R*,3*S*/*R*)-**25d** was often sluggish when using the enzyme recovered from the initial resolution step.

*3.2.2.2.1. Synthesis of (2R,3R/S)-3-Nitro-2-Pentanol (**21**) (Enzymatic Acylation/Enzymatic Deacylation).* A mixture of (±)-3-nitro-2-pentanol (**21**) (30 mL, 0.24 mol), vinyl acetate (13 mL, 0.12 mol), and Novozyme 435 (1.0 g) was shaken in toluene (200 mL) at room temperature at 175 rpm. After 23 h, the reaction was filtered and the enzyme beads washed with toluene. The filtrate was extracted with 50% MeOH/water (8X 200 mL), by which time gas chromatography (GC) indicated that <1% of the unreacted *syn*-alcohol was present. The toluene solution was then washed with saturated NaCl (100 mL), dried ($MgSO_4$), and filtered.

The filtrate, containing (2*R*,3*R*/*S*)-**25**, was treated with MeOH (25 mL, 0.62 mol) and Novozyme 435 (1 g fresh enzyme and 1 g recovered enzyme) and shaken at room temperature and 175 rpm. After 120 h, the *syn*-acetate was completely deacylated (approx 7% *anti*-acetate remained). The reaction mixture was filtered, the enzyme beads washed with toluene, and the combined filtrate evaporated. The residue (9.8 g) was distilled under reduced pressure to yield (2*R*,3*R*/*S*)-**21** (7.5 g, 23.3%) (the product contained approx 10% *anti*-**25**) (*syn*-**21**, 99.1% ee; *anti*-**21**, 92.2% ee; 79.5% de): $[\alpha]_D^{25} = 2.54$ (*c* 1.024, EtOH).

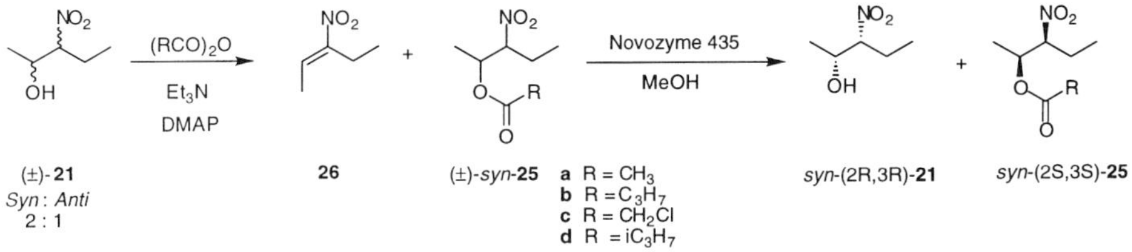

Fig. 22. Alternative chemical acylation/enzymatic deacylation procedure.

It has now been established that, as for other secondary alcohols *(35)*, Novozyme 435 in fact acylates the 2*R* enantiomer. When investigating the alternative chemical acylation/enzymatic deacylation procedure (**Fig. 22**), a significant observation was made. During chemical acylation of (±)-**21** (anhydride/Et₃N) a major by-product was formed, identified as *(E)*-3-nitro-2-pentene (**26**). More importantly, the nitropentene was formed more rapidly from the undesired *anti* diastereomer. Although the preparation of nitroalkenes from acylosing nitro compounds has been reported *(37,38)*, we are unaware of any reports of preferential elimination. Consequently, controlling the acylation conditions would result in complete elimination of the undesired *anti* nitropentanol isomer to nitropentene **26**, which could be easily removed by distillation, leaving disatereomerically pure *syn* nitroalcohol to be enzymatically resolved. Depending on the order of the reactions (chemical acylation/elimination/enzymatic acylation/enzymatic deacylation), the choice of enzyme and the workup procedure, access to optically enriched (2*S*,3*S*)- or (2*R*,3*R*)-**21** or the corresponding esters **25** would be possible.

3.2.2.2.2. Synthesis of (2R,3R)-3-Nitro-2-Pentanol (21) (Chemical Acylation/Elimination followed by Enzymatic Deacylation). Acetic anhydride (50 mL, 0.53 mol) was added dropwise to a solution of 3-nitro-2-pentanol (**21**) (50 mL, 0.40 mol) and triethylamine (85 mL, 0.61 mol) in TBME (200 mL) at 0°C, followed by dimethylaminopyridine (0.25 g, 0.2 mmol). After 45 min at 0°C, the reaction was warmed to room temperature and monitored closely by GC. After 4 h, the *anti*-nitroacetate **25a** was completely consumed. The reaction mixture was quenched with water (100 mL) and washed with 1.2 *M* HCl (4X 100 mL), water (100 mL), saturated NaHCO₃ (2X100 mL), water (100 mL), and saturated NaCl (100 mL), then dried (MgSO₄), filtered, and evaporated. The residue was distilled under reduced pressure collecting the *syn*-nitroacetate **25a** (100–107°C/24 mm Hg) (27.5 g, 38.9%).

A mixture of *syn*-**25a** (27 g, 0.15 mol), MeOH (30 mL, 0.8 mol), and Novozyme 435 (5.4 g) was shaken in toluene (200 mL) at 40°C at 200 rpm. After 39 h, chiral GC indicated the following: ee$_S$ 93.9%; ee$_P$ >99%; conversion, 48.6%. The reaction mixture was filtered after 111 h. The filtrate was extracted with 50% MeOH/water (10X100 mL). The toluene layer was dried (MgSO₄), filtered, and evaporated to obtain (2*S*,3*S*)-**25a** (10.8 g), which was purified by column chromatography [8.54 g, 31.6% yield based on (±)-acetate; 96.4% ee]: $[\alpha]_D^{25}$ = –22.0 (*c* 1.391, EtOH).

The MeOH/water extracts were concentrated to remove MeOH, saturated with NaCl (50 g) and extracted with EtOAc (100 mL). The organic layer was dried (MgSO₄), filtered, and evaporated, and the residue purified by column chromatography to obtain (2*R*,3*R*)-**21** [7.41 g, 33.7% yield based on (±)-acetate; > 99% ee]: $[\alpha]_D^{25}$ = +3.24 (*c* 1.111, EtOH).

Enzymatic methanolysis of *syn* nitroacetate (±)-**25a**, isolated after elimination of the *anti*-isomer, was sometimes slow and the results erratic. The possibility that nitropentene **26** was inhibiting the enzyme was suggested by several reactions using crude reaction mixtures. For example, after chemical acetylation, elimination, and workup, a mixture of *syn*-(±)-**25a** and **26** showed no reaction when treated with Novozyme 435 and MeOH. Furthermore, the enzyme recovered from this failed methanolysis reaction showed no further activity for the methanolysis of purified (±)-**25a** nor in the acetylation of (±)-nitroalcohol **21**. Nitropentene is a powerful Michael acceptor and might be expected to irreversibly inactivate hydrolytic enzymes by reaction with the serine hydroxyl at the active site. The propanolysis of (±)-*syn*-**25c** using ChiroCLEC BL in the presence and absence of nitropentene **26** showed essentially no reaction (<5%), whereas the deacylation of (2*S*,3*S*)-**25c** and *syn/anti*-(±)-**25c** ran as expected. This suggested that the enzyme might still be inactivated by traces of nitropentene remaining after purification of the *syn* ester or even by nitropentene adventitiously formed during the deacylation reaction.

Accordingly, the best course of action was to carry out the enzymatic acylation first and isolate the *syn/anti* (2*S*,3*S/R*)-**21**, then chemically acylate and eliminate using a second labile acylating agent. Following this, the pure *syn*-alcohol (2S,3S)-**21** could be separated cleanly from nitropentene. For this purpose the use of trifluoroacetic anhydride was especially convenient (**Fig. 23**).

Using this enzymatic acylation/chemical acylation/elimination method, (2*S*,3*S*)-**21** could be obtained in >99% ee and 98% de. The approx 23% isolated yield was quite respectable because the theoretical yield was only 30% (kinetic resolution of a 2:1 *syn:anti* mixture). However, the process was not very attractive because (1) it was not robust (the elimination reaction required careful monitoring and temperature control) and (2) a commercial source of pure *syn*-nitroalcohol was not identified.

*3.2.2.2.3. Synthesis of (2R,3R)-3-Nitro-2-Pentanol (**21**) (Enzymatic Resolution Followed by Chemical Acylation/Elimination).* Novozyme 435 (3 g) was added to a solution of (±)-**21** (30.1 g, 0.23 mol) and vinyl butyrate (25.2 g, 0.22 mol) and the mixture stirred at room temperature for 17 h. The reaction was filtered and the filtrate extracted with 50% MeOH/water (7X100 mL). The

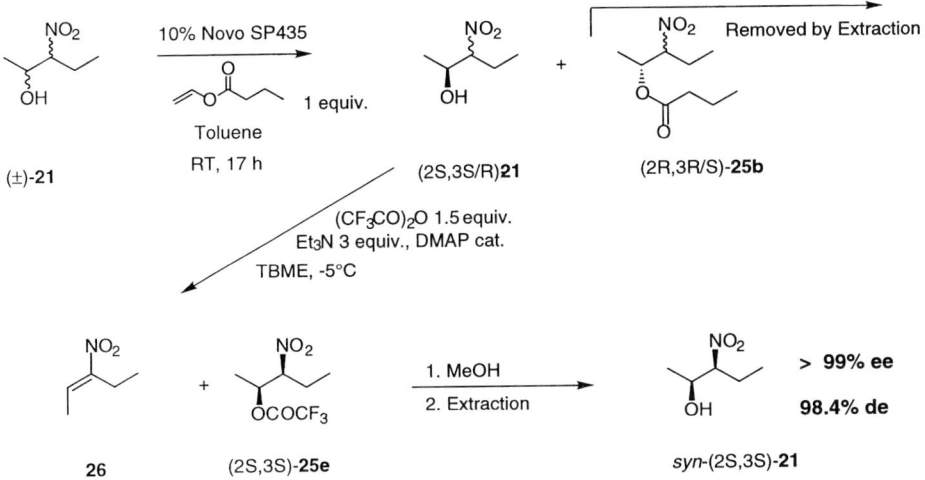

Fig. 23. Enzymatic acylation using trifluoroacetic anhydride.

aqueous MeOH extracts were back-extracted with heptane (100 mL), then concentrated to remove the MeOH. NaCl (60 g) was added to the residue, which was then extracted with EtOAc (2X100 mL). The organic layer was dried, filtered, and evaporated to obtain nitroalcohol **21** (14.9 g; 49.4%) (>99% ee, 37.1% de; approx 2% butyrate **25b** present).

A solution of the nitroalcohol **21** (14.9 g; 0.11 mol), triethylamine (46.7 mL, 0.34 mol), and dimethylaminopyridine (74 mg, 0.5%) in TBME (200 mL) was stirred under N_2 and cooled to –10°C. Trifluoroacetic anhydride (23.8 mL, 0.17 mol) was added slowly dropwise, maintaining the internal temperature of the reaction at <–5°C. After 3 h, GC indicated complete consumption of the *anti*-isomer. The reaction mixture was quenched with water (100 mL), washed with 1.2 *M* HCl (2X100 mL), water (100 mL), saturated $NaHCO_3$ (100 mL), water (100 mL) and saturated NaCl (50 mL), dried ($MgSO_4$), filtered, and evaporated. The residue, containing *syn*-trifluoroacetate **25e**, *syn*-nitroalcohol **21**, and nitropentene **26**, was dissolved in MeOH (50 mL) and stirred overnight. The mixture was then diluted with MeOH (50 mL) and water (100 mL) and extracted with heptane (3X100 mL) to remove nitropentene. The aqueous layer was then concentrated and saturated with NaCl, and extracted with EtOAc (2X50 mL). The EtOAc layer was dried ($MgSO_4$), filtered, and evaporated to give (2S,3S)-**21** (8.2 g, crude 27%; >99% ee and 98.4% de).

3.3. 2-Benzyloxy-3-Pentanol (29) as a Precursor to 6

The required (2S,3S)-formylhydrazine (**6**) was initially prepared from ethyl lactate, which provided the C2 chiral center (**Fig. 24**). Subsequent chain elongation

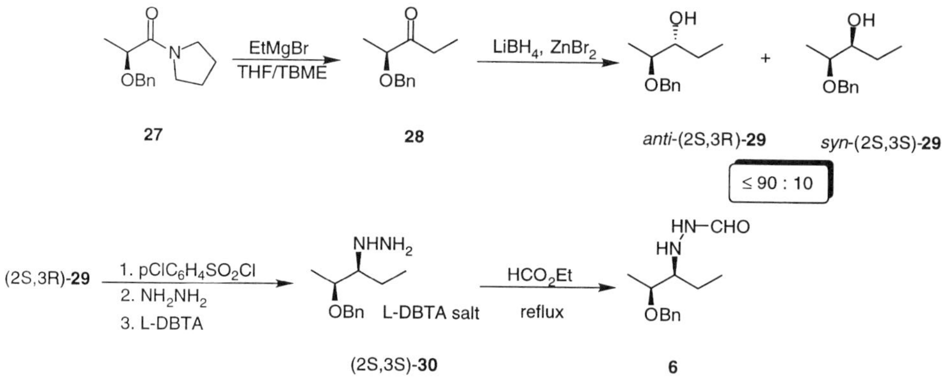

Fig. 24. Preparation of (2*S*,3*S*)-formylhydrazine fom ethyl lactate.

via the pyrrolidine amide **27** provided ketone **28**. However, under all conditions tried, reduction of **28** resulted in a mixture (90:10) of the (2*S*,3*R*)- and (2*S*,3*S*)-2-benzyloxy-3-pentanol (**29**), with separation of the minor (2*S*,3*S*)-diastereomer at this or subsequent steps proving difficult. Displacement of the chlorobenzene-sulfonate with hydrazine inverted the C3 chiral center to the required *S* configuration, and formylation completed the synthesis of **6**.

Because the chemical reduction of ketone **28** to the alcohol **29** showed poor diastereoselectivity, various chemoenzymatic routes to pure (2*S*,3*R*)-2-benzyloxy-3-pentanol (**29**) were considered. These approaches could include a microbial reduction of the ketone **28**, a hydroxylation of 2-benzyloxypentane at the C3 position, or a selective enzymatic acylation 2-benzyloxy-3-pentanol (**29**). The following sections will mainly describe our effort to effect a diastereoselective enzymatic acylation of *syn/anti*-**29**, with a brief description of the microbial reduction of 2-benzyloxy-3-pentanone (**28**). Attempts to produce the desired **29** by microbial hydroxylation of 2-benzyloxypentane were unsuccessful.

3.3.1. Microbial Reduction of 2-Benzyloxy-3-Pentanone (**28**)

Based on our previous experience in the microbial reduction of alkyl aryl ketones, only a truncated panel of our culture collection was examined for the reduction of 2-benzyloxy-3-pentanone (**28**) (**Fig. 25**). Of the 122 cultures that were screened, 94 showed <5% product formation at 1 g/L of **28**. Most cultures produced (2*S*,3*S*)-**29**, but seven cultures provided the desired (2*S*,3*R*)-isomer in high ee. Of these seven cultures, only *Rhodococcus* sp. (ATCC 19070) consistently provided the desired SR alcohol in high ee (approx 98%). Altering fermentation conditions enabled 70–80% conversion at 1 g/L of **28**; however, this decreased to 45% at 2 g/L and increased cell concentration (5X) afforded

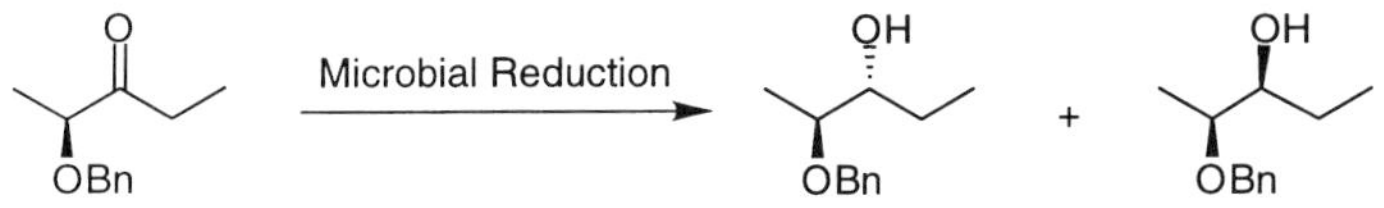

Fig. 25. Reduction of 2-benzyloxy-3-pentanone **28**.

no increase in conversion yield. Unfortunately, accumulation of product **29** at a level >0.4 g/L was detrimental to the ongoing conversion, and early exposure of cells to elevated levels of substrate was toxic. Cell extract specific activity was similar in cultures grown with and without substrate, indicating that reductase activity was not substrate inducible. Under both conditions, maximum activity occurred early in culture growth (8–12 h) when available carbohydrate levels were high, indicating that catabolite repression of enzyme activity was highly unlikely. Peak enzyme specific activity was short lived and decreased to <50% prior to the onset of stationary phase. Isolated enzyme activity was also inhibited by product at concentrations similar to that previously observed with intact cells. Decreasing enzyme levels as well as product accumulation limited the ability of *Rhodococcus* sp. #19070 to reduce ketone **28** to concentrations <1g/L. The toxicity of both the substrate and the product suggests delivery of the substrate via desorption from a resin with reabsorption of the product *(40)*. However, this strategy would not resolve the problem of decreasing enzyme levels, hence, our effort was concentrated on an enzymatic diastereoselective acylation.

3.3.2. Diastereoselective Enzymatic Acylation of (2S,3R/S)-2-Benzyloxy-3-Pentanol *(29)*

Because the chemical reduction of benzyloxyketone **28** showed poor diastereoselectivity, the enzymatic acylation of (2*S*,3*R/S*)-**29** was explored. If the minor (2*S*,3*S*)-isomer from the chemical reduction could be selectively acylated with a suitable acylating agent containing an acidic *(41–46)* or basic functional group, then isolation of the desired (2*S*,3*R*)-**29** would become a simple acid/base extraction.

From a screen of 148 enzyme preparations in THF, only 8 enzymes showed significant activity and only 4 displayed significant selectivity for the acylation of (2*S*,3*R/S*)-**29** (49% de) with succinic anhydride (**Fig. 26**).

The diastereomeric excesses of the hemisuccinate product **31** were poor (67–80% de), with the best results obtained with Toyobo Lipoprotein Lipase 701 (*Pseudomonas* sp.), which acylated the (2*S*,3*R*)-diastereomer. Using LPL-701, 14 solvents, other anhydrides, and a series of nitrogen-containing acylating

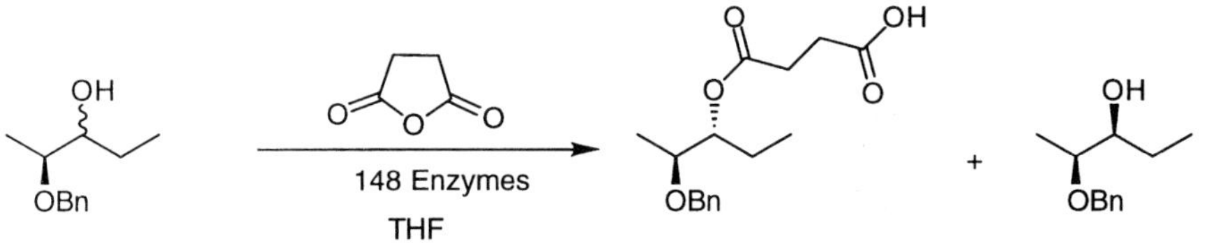

(2S,3S/R)-**29** *anti*-(2S,3R)-**31** *syn*-(2S,3S)-**29**

Fig. 26. Acylation of (2*S*,3*R*/*S*)-29 with succinic anhydride.

Table 8
Enzymatic Butyrylation of 2-Benzyloxy-3-Pentanol (29)

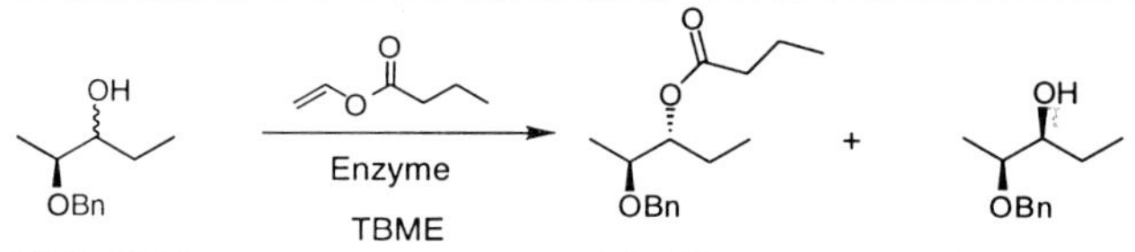

(2S,3R/S)-**29** (2S,3R)-**32** (2S,3S)-**29**

	Enzyme	mg	(2S,3R)-**32a** % de	% Conversion
1	Meito Sangyo Lipase OF	65	97.8	71.9
2	Novo Lipozyme IM-20	65	99.0	60.0
3	Amano PS-30	57	98.0	39.5
4	Novo SP 524	52	99.1	50.7
5	Sigma Pork Pancreatic Lipase	96	96.9	24.2
6	Sigma Acylase I	94	98.4	29.3
7	Novo SP 526	50	98.2	22.9

Conditions: (2*S*,3*R*/*S*)-**29**, 50 mg, VinylOBu 5 equiv., TBME 2 mL, RT, 35 h except 7 (5 h).

agents were also examined without any improvement in either reactivity or selectivity. Hydrolysis of the hemisuccinate **31** (49% de) using LPL-701 in 10 m*M* phosphate buffer/THF (10/2 mL) (pH 7.0) was also slow; after 44 h, the reaction showed approx 50% conversion with 50% de in the product.

These results suggested that functionalized esters would not be obtained enzymatically with high diastereoselectivity, so the enzyme catalyzed acylation of 2-benzyloxy-3-pentanol (**29**) was examined using vinyl butyrate as an alternate acylating agent. From a screen of 147 hydrolytic enzymes a number of enzymes showed promising reactivity (**Table 8**). However, in all cases, the major diastereomer (2*S*,3*R*)-**29** reacted preferentially, resulting in a less convenient direct purification and requiring a subsequent hydrolysis to restore the desired alcohol function.

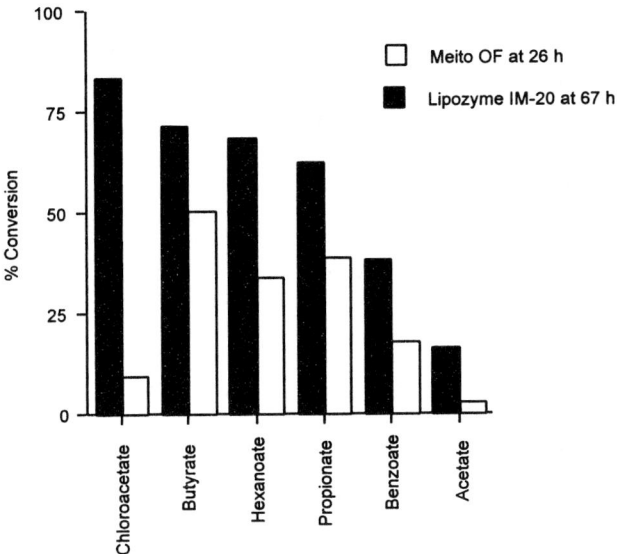

Fig. 27. The effect of ester chain length on the acylation of (2*S*,3*R/S*)-**29** in TBME. Conditions: **29**, 0.2 g, 50% de; enzyme, 0.2 g; TBME, 2 mL; acylating agent, 2 equivs.; room temperature; % Conversion was measured at 26 h for Meito OF and at 67 h for Lipozyme IM-20.

The 14 best enzymes from the vinyl butyrate screen were also screened for their ability to use trifluoroethyl hemisuccinate as an acylating agent. Although three enzymes showed some selectivity [(2*S*,3*R*)-**31**, 80–90% de], the reactivity was low (3–14% after 65 h).

Because of their high diastereoselectivity, reasonable reactivity, cost, and commercial availability, Meito Sangyo Lipase OF (Meito OF) (*C. rugosa*)* and Novo Lipozyme IM-20 (*Mucor miehei*) were chosen for further development. (Novo IM is Novo Lipozyme immobilized on a macroporous resin; IM-20 and IM-60 are different loadings.)

Optimization of this reaction has been reported previously *(47)*. Both enzymes showed a similar dependence on the nature of the acylating agent except in the case of vinyl chloroacetate, which performed best for Lipozyme IM-20 while reacting poorly with Meito OF (**Fig. 27**). For both enzymes, the enzymatic butyrylation was examined in 20 common solvents. The best results were obtained in ether and hydrocarbon solvents, and TBME was chosen as the

---

*Formerly called *C. cylindracea*. *C. rugosa* will be used throughout the text except when referring to manufacturers' labeling.

solvent of choice for all further reactions. On a weight basis, the reactions were generally faster with Meito OF than with Novo Lipozyme IM-20.

In the case of Meito OF, the presence of molecular sieves (g/g) resulted in a 10–20% higher conversion compared to identical reactions lacking sieves. The presence of sieves has been reported to both enhance *(48,49)* and retard *(50)* enzymatic acylations. Although no significant difference in moisture content could be determined by Karl Fischer titration of reaction mixtures in the presence or absence of sieves, it is possible that water in the reaction mixture, or the butyric acid or acetaldehyde resulting from hydrolysis of the acylating agent, adversely affects the enzymes' performance *(51)*. Variability in the performance of some batches of substrate was observed and the possibility that some samples of **29** contained an enzyme inhibitor was briefly examined. Pyrrolidine, dimethoxyethane, and residual pyrrolidine amide **27** were considered as potential contaminants. Of these three, only pyrrolidine showed a significant inhibitory effect on the enzymatic acylation; in the presence of 5% (v/v) (1.25 equiv.) pyrrolidine, only approx 10% reaction was observed after 50 h. Guo and Sih have shown that some chiral amines are effective inhibitors of *C. rugosa* lipase *(52)*. However, the level of pyrrolidine in the crude samples was not established.

It was still hoped that after the enzyme catalyzed butyrylation, the crude reaction mixture might be acylated chemically with succinic anhydride, and the (2S,3S)-hemisuccinate **31** so formed from the unwanted (2S,3S)-**29** diastereomer could then be removed by base extraction. However, it proved difficult to cleanly and completely acylate all of the unreacted alcohol, so a column chromatographic separation of the (2S,3R)-butyrate **32** from the unreacted (2S,3S)-alcohol **29** was performed.

The reaction sequence was successfully scaled up in the pilot plant as shown by the results in **Table 9** for a 7-kg reaction. After complete acylation [10 d for >95% conversion of the (2S,3R)-isomer], the butyrate **32** was isolated by column chromatography (5X SiO$_2$), then hydrolyzed to yield enantiomerically and diastereomerically pure (2S,3R)-**29**.

There were some process problems on scaling up the reaction: The high enzyme loading and the presence of sieves caused problems in filtration; the reaction required extended periods to approach completion; the enzyme was relatively costly; chromatography was required for isolation. Because an immobilized enzyme would greatly facilitate filtration, Meito Sangyo Lipase OFG (an immobilized form of *C. rugosa*, kindly provided by Meito Sangyo, Tokyo) and Lipozyme IM-60 were compared with Meito OF (acetone powder) and Lipozyme IM-20. The best results were observed using three weight equivalents of Meito OFG (which has a listed activity one-third that of Meito OF) or Lipozyme IM-60 (activity three times that of Lipozyme IM-20).

Table 9
Scale-up of Enzymatic Butyrylation of 2-Benzyloxy-3-Pentanol (29)

	kg	Yield		% de	% ee
		kg	%		
(2S,3R/S)-**29**	7.00			74.8	96.7
(2S,3R)-**32**		6.67	69.1	99.0	99.4
(2S,3R)-**29**		4.88	63.7	98.8	99.4

Conditions: (2S,3R/S)-**29**, 7 kg (74.8% de); Meito OF, 7 kg; VinylOBu, 8.2 kg; 4-Å Sieves, 7 kg; TBME, 77 L; RT, 10 d.

The use of Lipozyme IM-60 with vinyl chloroacetate as the acylating agent looked promising; the faster reaction, simplified and less odorous workup, separation of products by distillation, and the possibility of enzyme reuse potentially outweighing the increased costs of enzyme and acylating agent. However, before these improvements could be reduced to practice, a more concise synthesis of the key hydrazone intermediate was developed.

3.3.2.1. SYNTHESIS OF (2S,3R)-2-BENZYLOXY-3-PENTANOL (29)

A mixture of 4-Å molecular sieves (8 × 12 mesh) (1.0 g), vinyl butyrate (1.3 mL, 10.3 mmol), and Meito OF (1.0 g) was suspended in TBME (7.4 mL) and stirred at room temperature for 95 min. A solution of (2S,3R/S)-**29** (1.0 g, 5.15 mmol) (70% de) in TBME (2.6 mL) was added and the mixture stirred for 163 h. The reaction mixture was then filtered through a Celite pad, which was washed with TBME (10 mL), and the combined filtrate evaporated. The residue was placed on a silica gel column (15 g) and eluted with hexane (50 mL), 2% (50 mL) and 4% (50 mL) TBME/hexane, collecting fractions of 15–20 mL. The desired fractions were combined and evaporated to obtain a colorless liquid [0.97 g, 98.2% de; 60.4% based on total alcohol, 71.3% based on (2S,3R)-**29** in starting mixture]: $[\alpha]_D^{23} = +16.7$ (c 1.323, EtOH).

Four Lipozyme IM-20 catalyzed acylations [1.2 g of (2S,3R)-alcohol per reaction] were combined and the butyrate (2S,3R)-(**32a**) (5.36 g, 20.3 mmol) obtained by column chromatography was dissolved in a mixture of *i*-PrOH (30 mL) and 2 *M* NaOH (100 mL). The mixture was heated at reflux for 17.5 h, then cooled to room temperature and the *i*-PrOH removed. The residue was extracted twice with EtOAc (100 mL and 50 mL), and the combined organic extracts were washed with saturated aqueous NaCl (50 mL), dried (MgSO$_4$), filtered, and evaporated. The crude product was distilled in a Kugelrohr oven (oven temperature 100–105°C, 1.0 mm Hg) to obtain a colorless viscous liquid [3.70 g, 93.9%; 76.6% overall from starting (2*S*,3*R*)-alcohol] (97.4% de): [α]$_D^{21}$ = +35.7 (*c* 1.181, EtOH).

3.3.3. A Comparison of Commercially Available C. rugosa and Mucor miehei Enzymes Under Acylation Conditions

Because of the low rate of the enzymatic butyrylation of (2*S*,3*R/S*)-**29**, *Candida* and *Mucor* enzymes from different commercial sources were compared. Although different reactivities were to be expected given the different activities of the enzyme preparations, a surprisingly wide variation in selectivity was also observed.

Of the seven *Mucor miehei* preparations examined, three showed no reactivity, whereas the other four showed similar activity and selectivity (**Table 10**, entries 31–37).

The 17 *C. rugosa* enzyme preparations that were surveyed fell into three groups. In general, the purified enzyme preparations showed poor activity (**Table 10**, entries 1–4). Altus Analytical Grade Lipase CR, a purified form of Meito OF, is known to be inactive under acylation conditions. Similarly, purified Sigma lipase (L-8525) showed no activity in TBME.

In contrast, the crude Sigma lipase preparation (Type VII), which contains 38% added lactose, showed 40% conversion after 89 h (entry 15). The addition of carbohydrates to previously dialyzed *C. rugosa* has been shown to affect the enzyme's behavior under both hydrolytic and acylation conditions *(53)*. The highly purified crosslinked enzyme crystal formulation [Altus ChiroCLEC-CR(dry)] also showed poor reactivity at 5% w/w loading. At higher loading (10–50% w/w) with vinyl butyrate as acylating agent, the CLEC showed lower selectivity, and the reactions seemed to stall at <20% conversion in TBME and iso-octane and when using pure (2*S*,3*R*)-**29** (data not shown). Similar behavior in the presence of vinyl acetate has recently been reported *(54)*. The CLEC-CR also showed poor reactivity when using trifluoroethyl butyrate as acylating agent (entry 21), the reaction terminating at approx 11% conversion. The presence of surfactants and optimal water activity have been shown to be important determinants of the reactivity/selectivity of crosslinked enzyme crystals *(55)*, and purified preparations of *C. rugosa* lipase *(56)*, but were not addressed in the present study.

Candida rugosa Meito OF and its immobilized forms, OFG and OFC, showed 40–60% conversion after 89 h, as did Biocatalysts' *C. cylindracea* and EDC Lipase XX Concentrate (**Table 10**, entries 5–9). This set of enzymes also showed good diastereoselectivity, forming (2*S*,3*R*)-**32** with 89–92% de. (The de was lower than previously recorded above because an impurity coeluted on HPLC with the product, depressing the observed de.) The chiral preference

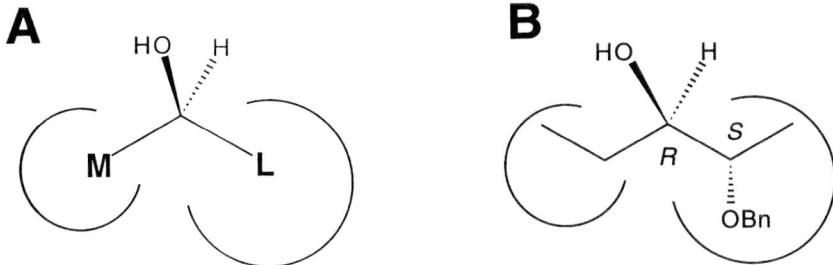

Fig. 28. (**A**) Empirical rule which predicts the faster reacting enantiomer of secondary alcohols with hydrolases, e.g., *C. rugosa* (adapted from **ref. 57**). (**B**) The faster-reacting (2*S*,3*R*)-**29** obeys the model.

shown by this group of enzymes agrees with the model proposed by Kazlauskas *(57,58)*, which is supported by X-ray crystal structures of the purified enzyme *(59)*, if the medium and large groups are assigned to the ethyl and benzyloxyethyl groups, respectively (**Fig. 28**).

A third group of enzymes, comprising Sigma Type VII, Meito Sangyo MY, Amano AY-30, Boehringer-Mannheim Chirazyme L-3 and Lipase 305 (*Candida* sp.), and Genzyme *C. cylindracea* all showed poor selectivity (**Table 10**, entries 10–16). In fact, the de of the product butyrate **32** was less than that of the starting material (47% de), indicating a preference for acylation of the minor (2*S*,3*S*)-diastereomer. Although Sigma Type VII has been shown to be heterogeneous *(60–63)*, it is also reported to be free of contaminating enzymes displaying opposite stereochemical preference for the enantioselective hydrolysis of racemic aryl- and (aryloxy)propionic esters *(64)*. However the present study suggests that some *C. rugosa* preparations contain fractions with opposing diastereoselectivity under acylating conditions.

A similar grouping of enzymes was observed when trifluoroethyl butyrate was used as acylating agent (**Table 10**, entries 19–30). The reactions were slower than with vinyl butyrate, but Meito OF and Biocatalysts' *C. cylindracea* acylated the (2*S*,3*R*)-isomer with 90–91% de, whereas Amano AY-30, Meito MY, Chirazyme L-3, and Genzyme *C. cylindracea* showed a preference for the minor (2*S*,3*S*)-isomer (the de of the product [14–21% de] was less than that of the starting material [47% de]). Although crude *C. rugosa* (Meito OF) reportedly consists of several hydrolases and at least one protease *(65)*, the lack of reactivity of the esterases from Boehringer-Mannheim and Altus might suggest that a contaminating esterase is not responsible for Meito OF's reactivity.

In contrast to their behavior in organic solvent, all the *Candida* enzymes were hydrolytically competent but showed uniformly poor selectivity. As earlier, the increase in de of the unreacted starting material with increasing

Table 10
Butyrylation of (2S,3R/S)-29 with Selected *Candida rugosa* and *Mucor miehei* Enzymes

Entry	Vendor	Enzyme	Wt. (mg)	Time (h)	29 (% de)	32 (% de)	Conversion (%)
Candida rugosa **Enzymes**/Vinyl Butyrate							
1	Altus	Lipase CR Analytical Grade 001	36	89	43.8	n/d*	< 1
2	Sigma	Lipase L8525 (purified)	20	93	45.6	n/d*	1.2
3	Altus	ChiroCLEC CR (dry)	5	89	42.1	72.8	9.2
4	Altus	ChiroCLEC CR (dry)	81	93	42.6	68.6	13.1
5	Biocatalysts	Lipase *Candida cylindracea*	94	89	11.8	91.0	41.0
6	Enzyme Development Corp.	Enzeco Lipase XX Concentrate	89	89	2.2	91.6	45.9
7	Meito Sangyo	Lipase OF	107	89	5.4	90.2	44.7
8	Meito Sangyo	Lipase OFG	141	89	25.8	89.1	57.3
9	Meito Sangyo	Lipase OFC	110	89	13.4	90.7	40.9
10	Amano	Lipase AY-30	104	89	61.1	10.6	33.0
11	Boehringer-Mannheim	Lipase 305	46	89	51.2	1.3	14.8
12	Boehringer-Mannheim	Chirazyme L3	65	89	51.9	29.9	38.0
13	Genzyme	Lipase *Candida cylindracea*	33	89	49.2	13.1	15.5
14	Meito Sangyo	Lipase MY	100	89	50.8	21.0	26.2
15	Sigma	Lipase Type VII	122	89	57.6	23.2	40.3
16	Sigma	Lipase Type VII-A (immobilized)	145	89	54.4	6.7	20.3
17	Altus	*C. rugosa* Esterase (lyophilate)	22	89	42.5	69.0	7.4
18	Boehringer-Mannheim	Cholesterol esterase	50	89	43.6	42.6	9.4

Candida rugosa Enzymes/Trifluoroethyl Butyrate

19	Altus	Lipase CR Analytical Grade 001	64	93	44.6	48.1	0.3
20	Sigma	Lipase L8525 (purified)	16	93	45.0	20.0	0.3
21	Altus	ChiroCLEC CR (dry)	18	93	40.7	69.3	11.2
22	Meito Sangyo	Lipase OF	88	93	27.6	90.5	29.2
23	Biocatalysts	Lipase *Candida cylindracea*	90	93	32.1	90.3	23.9
24	Amano	Lipase AY-30	135	93	54.9	6.4	23.0
25	Boehringer-Mannheim	Chirazyme L3	107	93	55.7	14.5	28.3
26	Genzyme	Lipase *Candida cylindracea*	27	93	46.6	2.7	7.1
27	Meito Sangyo	Lipase MY	98	93	50.1	12.0	16.1
28	Sigma	Lipase Type VII	113	93	54.8	14.0	25.9
29	Sigma	Lipase Type VII-A (immobilized)	96	93	47.4	10.9	6.5
30	Altus	*C. rugosa* Esterase (lyophilate)	8	89	45.2	n/d*	1.6

Mucor miehei Enzymes/Vinyl Butyrate

31	Novo	Lipozyme IM-20	96	89	21.3	92.8	35.0
32	Novo	Lipozyme IM-60	91	89	27.4	92.5	57.6
33	Novo	SP 524 Lipase (nonimmobilized IM-20/60)	72	89	25.0	92.5	30.9
34	Fluka	Esterase	37	89	40.1	93.0	10.3
35	Enzyme Development Corp.	S-4827 Enzeco Esterase–Lipase	104	89	42.3	n/d*	<1
36	Gist-Brocades	Piccantase A	106	89	43.0	n/d*	<1
37	Solvay	Lipase G-1000	174	89	44.0	n/d*	<1

Conditions: (2S,3R/S)-**29**, 0.1 g, 47.0% de; vinyl butyrate, 3 equiv., or trifluoroethyl butyrate, 5 equiv.; TBME, 2.0 mL; RT, 250 rpm. *n/d, not determined.

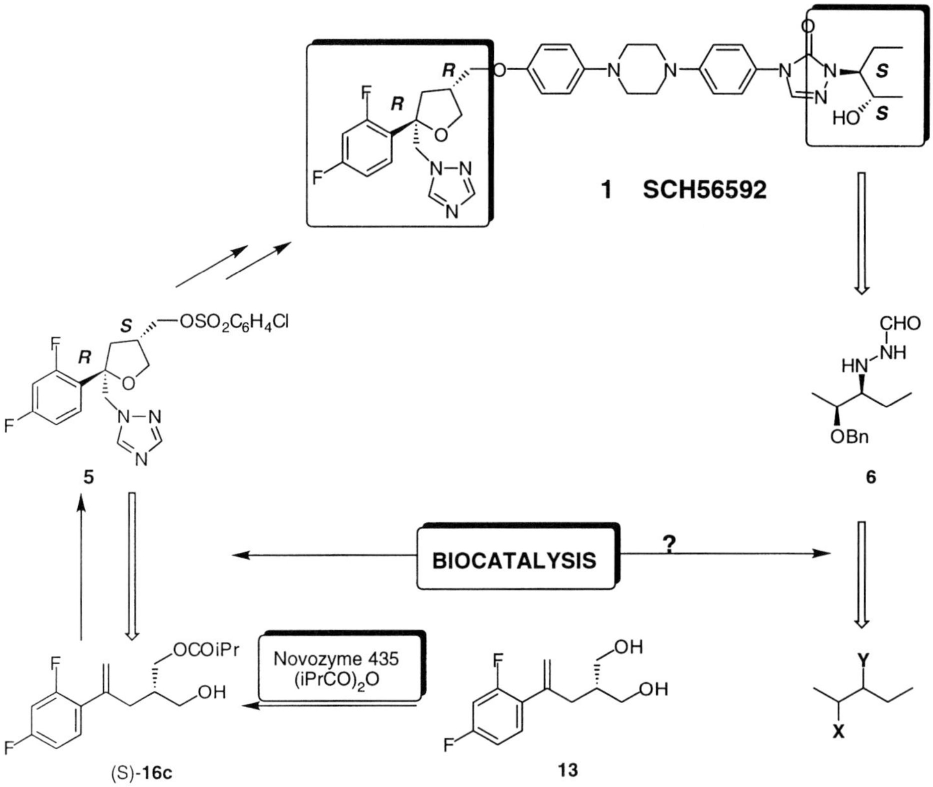

Fig. 29. Final commercial synthesis.

conversion indicates that Amano AY-30, Meito MY, and Sigma Type VII prefer for the minor (2*S*,3*S*)-isomer. Of the *Mucor* enzymes tested, only Novo Lipozyme 10,000L (a liquid preparation) showed activity; the immobilized IM-60 showed no activity under similar conditions.

The disadvantages of using crude commercial enzyme preparations in synthetic reactions have long been recognized, and many attempts to improve the enantioselectivity of *C. rugosa* lipases have been reported. (References *31*, *65*, and *66* provide useful reviews.) Our results reinforce the observation that an enzyme's performance may critically depend on commercial source and purity.

4. Conclusion

Regulatory pressure has encouraged the development and marketing of chiral drugs as single enantiomers *(67)*, and sales of synthetic single enantiomer drugs as a percentage of drug sales are expected to reach 34% by the year 2000 *(68)*. This has also been driven to some extent by advances in the synthesis of optically pure compounds, which has allowed the synthesis of increasingly complex single enantiomer drug candidates to become more economical. Biocatalysis is now recognized as a powerful technology for the contruction of optically pure pharaceutical intermediates.

In this chapter, we have described our experience in integrating biocatalysis into the process development of a new azole antifungal. Our account has concentrated on the use of commercially available isolated enzymes; microbial reduction efforts were briefly described, but microbial hydroxylation was beyond the scope of this volume. The enzymatic desymmetrization of the prochiral 2-substituted-1,3-prochiral **13** has been fully integrated into the synthesis and will be a key step in the final commercial synthesis **(Fig. 29)**. Attempts to develop a chemoenzymatic synthesis of formylhydrazine **6** were less successful. Nevertheless, a diastereoselective enzymatic acylation was temporarily used to prepare material until an alternative chemical asymmetric synthesis could be developed. As a case study in biocatalysis, SCH56592 highlights the transition of biocatalysis from a potentially useful technology to a well-accepted technology for the production of pharmaceuticals.

References

1. *Biotransformations* CD-ROM. Chapman and Hall, London.
2. *BioCatalysis,* Synopsis Scientific Systems, Ltd., Leeds, UK.
3. Faber, K. (1997) *Biotransformations in Organic Chemistry*, 3rd edition, Springer-Verlag, Berlin.
4. Drauz, K. and Waldmann, H., eds. (1995) *Enzyme Catalysis in Organic Synthesis: A Comprehensive Handbook*, VCH Publishers, New York.
5. Wong , C.-H. and Whitesides, G. (1994) *Enzymes in Synthetic Organic Chemistry*, Pergamon, New York.
6. Poppe L. and Novak, L. (1992) *Selective Biocatalysis*, VCH Publishers, New York.
7. Roberts, S. M., Wiggins, K., Casy, G., and Phythian, S., eds. (1992) *Preparative Biotransformations: Whole Cell and Isolated Enzymes in Organic Synthesis*; Wiley, Chichester, UK.
8. Zaks, A. and Dodds, D. R. (1997) Application of biocatalysis and biotransformation to the synthesis of pharmaceuticals. *Drug Discovery Today* **2,** 513–531.
9. Patel, R. N. (1997) Stereoselective biotransformations in synthesis of some pharmaceutical intermediates. *Adv. Appl. Microbiol.* **43,** 91–140.
10. Georgopapaddakou, N. H., Dix, B. A., Smith, S. A., Freudenberger, J., and Funke, P. T. (1987) Effect of antifungal agents on lipid biosynthesis and membrane integrity in *Candida albicans. Antimicrob. Agents and Chemother.* **31,** 46–51.
11. Saksena, A. K., Girijavallabhan, V. M., Lovey, R. G., Desai, J. A., Pike, R. E., Jao, E., et al. (1995) Sch 51048, a novel broad-spectrum orally active antifungal agent: synthesis and preliminary structure-activity profile. *Bioorg. Med. Chem. Lett.* **7,** 127–132.

12. Saksena, A. K., Girijavallabhan, V. M., Lovey, R. G., Pike, R. E., Desai, J. A., Ganguly, A. K., et al. (1994) Enantioselective synthesis of the optical isomers of broad-spectrum orally active antifungal azoles, SCH 42538 and SCH 45012. *Biorg. Med. Chem. Lett.* **4,** 2023–2028.

13. Allworth, W. L. (1972) *Stereochemistry and Its Application in Biochemistry* Wiley, New York.

14. Ramos Tombo, G. M., Schar, H.-P., Busquets, X. F., and Ghisalba, O. (1986) Synthesis of both enantiomeric forms of 2-substituted 1,3 propanediol monoacetates starting from a common prochiral precursor, using enzymic transformations in aqueous and in organic media. *Tetrahedron Lett.* **27,** 5707–5710.

15. Boland, W., Fröbl, C., and Lorenz, M. (1991) Esterolytic and lipolytic enzymes in organic synthesis. *Synthesis* 1049–1072.

16. Danieli, B., Lesma, G., Passerella, D., and Riva, S. (1993) Chiral synthons via enzyme-mediated asymmetrization of meso-compounds. *Adv. Use Synthons Org. Chem.* **1,** 143–219.

17. Schoffers, E., Golebiowski, A., and Johnson, C. R. (1996) Enantioselective synthesis through enzymic asymmetrization. *Tetrahedron* **52,** 3769–3826.

18. Lovey, R. G., Saksena, A. K., and Girijavallabhan, V. M. (1994) PPL-catalyzed enzymic asymmetrization of a 2-substituted prochiral 1,3-diol with remote chiral functionality: improvements toward synthesis of the eutomers of SCH 45012. *Tetrahedron Lett.* **35,** 6047–6050.

19. Morgan, B., Dodds, D. R., Zaks, A., Andrews, D. R., and Klesse, R. (1997) Enzymic desymmetrization of prochiral 2-substituted-1,3-propanediols: a practical chemoenzymatic synthesis of a key precursor of SCH 51048, a broad-spectrum orally active antifungal agent. *J. Org. Chem.* **62,** 7736–7743.

20. Sudhakar, A. R. (1995) Process and catalysts for preparing dialkyl malonate intermediates for the synthesis of antifungal agents. *US Patent* 5,442,093.

21. Saksena, A. K., Girijavallabhan, V. M., Lovey, R. E., Pike, R. E., Wang, H., Ganguly, A. K., et al. (1995) Highly stereoselective access to novel 2, 2,4-trisubstituted tetrahydrofurans by halocyclization: practical chemoenzymic synthesis of SCH 51048, a broad-spectrum orally active antifungal agent. *Tetrahedron Lett.* **36,** 1787–1790.

22. Saksena, A. K., Girijavallabhan, V. M., Pike, R. E., Wang, H., Lovey, R. G., Liu, Y.-T., et al. (1995) Preparation of chiral 2-azolylmethyl-2-phenyl 4-sulfonyloxymethyltetrahydrofurans as antifungal intermediates. *US Patent* 5,403,937.

23. Wang, Y.-F., Chen, C.-S., Girdaukas, G., and Sih, C. J. (1984) Bifunctional chiral synthons via biochemical methods. III. Optical purity enhancement in enzymic asymmetric catalysis. *J. Am. Chem. Soc.* **106,** 3695,3696.

24. Nielsen, C. M. and Sudhakar, A. (1998) Process for preparing intermediates for the synthesis of antifungal agents. *US Patent* 5,756,830.

25. Ke, T., Wescott, C. R., and Klibanov, A. M. (1996) Prediction for the solvent dependence of enzymatic prochiral selectivity by means of structure-based thermodynamic calculations. *J. Am. Chem. Soc.* **118,** 3366–3374.

26. Ebert, C., Ferluga, G., Gardossi, L., Gianferra, T., and Linda, P. (1992) Improved lipase-mediated resolution of mandelic acid esters by multivariate investigation of experimental factors. *Tetrahedron: Asymmetry* **3**, 903–912.

27. Shieh, C.-J., Akoh, C. C., and Yee, L. N. (1996) Optimized enzymic synthesis of geranyl butytate with lipase AY from *Candida rugosa. Biotechnol. Bioeng.* **51**, 371–374

28. Kuznetsov, V. V., Makhova, N. N., Strelenko, Y. A., Khel'nitskii, L. I. (1991) Role of pH in the synthesis of diaziridines. *Izv. Akad. Nauk. SSSR, Ser. Khim.* **12**, 2861–2871.

29. Occhiato, E. G., Guarna, A., DeSarlo, F., and Scarpi, D. (1995) Baker's yeast reduction of a γ-nitro ketones. II. Straightforward enantioselective synthesis of 2,7-dimethyl-1, 6-dioxaspiro [4.4] nonanes. *Tetrahedron: Asymmetry* **6**, 2971–2976.

30. Molinari, F., Occhiato, E. G., Aragozzini, F., and Guarna, A. (1998) Microbial biotransformations in water/organic solvent system. Enantioselective reduction of aromatic β- and γ-nitro ketones. *Tetrahedron: Asymmetry* **9**, 1389–1394.

31. Seebach, D., Beck, A. K., Mukhopadhyay, T., and Thomas, E. (1982) Diastereoselective synthesis of nitroaldol derivatives. *Helv. Chim. Acta* **65**, 1101–1133.

32. Ohtani, I., Kusumi, T., Kashman, Y., and Kakisawa, H. (1991) High-field FT NMR application of Mosher's method. The absolute configurations of marine terpenoids. *J. Am. Chem. Soc.* **113**, 4092–4096.

33. Chen, C.-S., Fujimoto, Y., Girdaukas, G., and Sih, C. J. (1982) Quantitative analyses of biochemical kinetic resolutions of enantiomers. *J. Am. Chem. Soc.* **104**, 7294–7299.

34. Wang, Y.-F., Yakovlevsky, K., Zhang, B., and Margolin, A. L. (1997) Crosslinked crystals of subtilisin: versatile catalyst for organic synthesis. *J. Org. Chem.* **62**, 3488–3495.

35. Anderson, E. M., Larsson, K. M., and Kirk, O. (1998) One biocatalyst—many applications: the use of *Candida antarctica* B-lipase in organic synthesis. *Biocatal. Biotransform.* **16**, 181–204.

36. Kitayama, T. (1996) Asymmetric synthesis of pheromones for *Bactrocera nigrotibialis, Andrena wikella,* and *Andrena haemorrhoa* F from a chiral nitro alcohol. *Tetrahedron* **52**, 6139–6148.

37. Barton, D. H. R., Kervagoret, J., and Zard, S. Z. (1996) A useful synthesis of pyrroles from nitroolefins. *Tetrahedron* **46**, 7587–7598.

38. Denmark, S. E. and Senanayake, C. B. W. (1996) Tandem inter [4+2]/intra [3+2] cycloadditions. 8. Cycloadditions with unactivated dipolarophiles. *Tetrahedron* **52**, 11,579–11,600.

39. Sasai, H., Tokunaga, T., Watanabe, S., Suzuki, T., Itoh, N., and Shibasaki, M. (1995) Efficient diastereoslective and enantioselective nitroaldol reactions from prochiral starting materials: utilization of La-Li-6,6'-disubstituted BINOL complexes as asymmetric catalysts. *J. Org. Chem.* **60**, 7388,7389.

40. Roddick, F. A. and Britz, M. L. (1997) Production of hexanoic acid by free and immobilized cells of *Megasphaera elsdenii*: influence of in-situ product removal using ion exchange resin. *J. Chem. Technol. Biotechnol.* **69**, 383–391.

41. Yamamoto, K., Nishioka, T., and Oda, J. (1988) Asymmetric ring opening of cyclic acid anhydrides with lipase in organic solvents. *Tetrahedron Lett.* **29,** 1717–1720.
42. Terao, Y., Tsuji, K., Murata, M., Achiwa, K., Nishio, T., Watanabe, N., et al. (1989) Facile process for enzymic resolution of racemic alcohols. *Chem. Pharm. Bull.* **37,** 1653–1655.
43. Fiaud, J.-C., Gil, R., Legros, J.-Y., Aribi-Zouioueche, L., and Konig, W. A. (1992) Kinetic resolution of 3-tert-butyl and 3-phenylcyclobutylidenethanols through lipase-catalyzed acylation with succinic anhydride. *Tetrahedron Lett.* **33,** 6967–6970.
44. Gutman, A., Brenner, D., and Boltanski, A. (1993) Convenient practical resolution of racemic alkyl-aryl alcohols via enzymic acylation with succinic anhydride in organic solvents. *Tetrahedron: Asymmetry* **4,** 839–844.
45. Ozegowski R., Kunath, A., and Schick, H. (1993) Lipase-catalyzed asymmetric alcoholysis of 3-substituted pentanedioic anhydrides. *Liebigs Ann. Chem.* 805–808.
46. Hyatt, J. A. and Skelton, C. (1997) A kinetic resolution route to the (S)-chromanmethanol intermediate for synthesis of the natural tocols. *Tetrahedron: Asymmetry* **8,** 523–526.
47. Morgan, B., Stockwell, B. R., Dodds, D. R., Andrews, D. R., Sudhakar, A. R., Nielsen, C. M., et al. (1997) Chemoenzymic approaches to SCH 56592, a new azole antifungal. *J. Am. Oil. Chem. Soc.* **74,** 1361–1370.
48. Degueil-Castaing, M., DeJeso, B., Drouillard, S., and Maillard, B. (1987) Enzymic reactions in organic synthesis: ester interchange of vinyl esters. *Tetrahedron Lett.* **28,** 953,954.
49. Izumi, I., Tamura, F., and Sasaki, K. (1992) Enzymic kinetic resolution of [4](1,2) ferrocenophane derivatives. *Bull. Chem. Soc. Jpn.* **65,** 2784–2788.
50. Herradon, B. and Valverde, S. (1994) Biocatalytic synthesis of chiral polyoxygenated compounds: modulation of the selectivity upon changes in the experimental conditions. *Tetrahedron: Asymmetry* **5,** 1479–1500.
51. Weber, H. K., Stecher, H., and Faber, K. (1995) Sensitivity of microbial lipases to acetaldehyde formed by acyl-transfer reactions from vinyl esters. *Biotechnol. Lett.* **17,** 803–808.
52. Guo, Z.-W. and Sih, C. J. (1989) Enantioselective inhibition: strategy for improving the enantioselectivity of biocatalytic systems. *J. Am. Chem. Soc.* **111,** 6836–6841.
53. Sanchez-Montero, J. M., Hamon, V., Thomas, D., and Legoy, M. D. (1991) Modulation of lipase hydrolysis and synthesis reactions using carbohydrates. *Biochim. Biophys. Acta* **1078,** 345–350.
54. Lundh, M., Smitt, O., and Hedenstrom, E. (1996) Sex pheromone of pine sawflies: enantioselective lipase catalyzed transesterification of erythro-3,7-dimethyl-pentadecan-2-ol, Diprionol. *Tetrahedron: Asymmetry* **7,** 3277–3284.
55. Perischetti, R. A., Lalonde, J. J., Govardhan, C. P., Khalaf, N. K., and Margolin, A. L. (1996) *Candida rugosa* lipase; enantioselectivity enhancements in organic solvents. *Tetrahedron Lett.* **37,** 6507–6510.
56. Tsai, S.-W. and Dordick, J. S. (1996) Extraordinary enantiospecificity of lipase catalysis in organic media induced by purification and catalyst engineering. *Biotechnol. Bioeng.* **52,** 296–300.

57. Kazlauskas, R. J., Weissfloch, A. N. E., Rappaport, A. V., and Cuccia, L. A. (1991) A rule to predict which enantiomer of a secondary alcohol reacts faster in reactions catalyzed by cholesterol esterase, lipase from *Pseudomonas cepacia,* and lipase from *Candida rugosa. J. Org. Chem.* **56,** 2656–2665.

58. Franssen, M. C. R., Jongejan, H., Kooijman, H., Spek, A. L., Mondril, N. L. F. L., Boavida dos Santos, P. M. A. C., and de Groot, A. (1996) Resolution of a tetrahydrofuran ester by *Candida rugosa* lipase (CRL) and an examination of CRL's stereochemical preference in organic media. *Tetrahedron: Asymmetry* **7,** 497–510.

59. Cygler, M., Grochulski, P., Kazlauskas, R. J., Schrag, J. P., Bouthillier, F., Rubin, F. B., et al. (1994) A structural basis for the chiral preferences of lipase. *J. Am. Chem. Soc.* **116,** 3180–3186.

60. Wu, S.-H., Guo, Z.-W., and Sih, C. J. (1990) Enhancing the enantioselectivity of *Candida* lipase-catalyzed ester hydrolysis via noncovalent enzyme modification. *J. Am. Chem. Soc.* **112,** 1990–1995.

61. Allenmark, S. and Ohlsson, A. (1992) Studies of the heterogeneity of a *Candida cylindracea (rugosa)* lipase: monitoring of esterolytic activity and enantioselectivity by chiral liquid chromatography. *Biocatalysis* **6,** 211–221.

62. Rua, M. L., Diaz-Maurino, T., Fernandez, V. M., Otero, C., and Ballesteros, A. (1993) Purification and characterization of two distinct lipases from *Candida cylindracea. Biochim. Biophys. Acta* **1156,** 181–189.

63. del Rio, J. L. and Faus, I. (1998) Resolution of (±)-trans-2-phenylcyclohexan-1-ol by lipases from *Candida rugosa:* effect of catalyst source and reaction conditions. *Biotechnol. Lett.* **20,** 1021–1025.

64. Guo, Z.-W. and Sih, C. J. (1989) Enantioselective inhibition: strategy for improving the enantioselectivity of biocatalytic systems. *J. Am. Chem. Soc.* **111,** 6836–6841.

65. Lalonde, J. J., Govardhan, C., Khalaf, N., Martinez, K. V., and Margolin, A. L. (1995) Cross-linked crystals of *Candida rugosa* lipase: highly efficient catalysts for the resolution of chiral esters. *J. Am. Chem. Soc.* **117,** 6845–6852.

66. Colton, I. J., Ahmed, S. N., and Kazlauskas, R. J. (1995) A 2-propanol treatment increases the enantioselectivity of *Candida rugosa* lipase toward esters of chiral carboxylic acids. *J. Org. Chem.* **60,** 212–217.

67. FDA (1992) Food and Drug Administration Policy Statement for the Development of New Stereoisomeric Drugs. *Chirality* **4,** 338–340.

68. Stinson, S. C. (1995) Chiral drugs. *Chem. Eng. News* Oct. 9, p. 52.

PART III

Reaction Systems and Bioreactor Design

Evgeny N. Vulfson

The majority of reactions in low-water systems are carried out using suspension of enzyme (free or immobilized) in organic solvents containing a certain amount of water. Optimization of water activity in such systems and the preparation of biocatalyst for use in organic solvents are discussed in detail in Part II (edited by Prof. Halling). However, as contributions combined in this part amply demonstrate, it is also possible to use alternative media to achieve full benefits of the low-water environment. For example, supercritical fluids, and especially CO_2, have been successfully used to carry out synthetic reactions with hydrolytic and other enzymes (*see* Chapters 39, 44, and 46). The main attraction of SC-SO_2 is that it can be easily and quantitatively removed from the product. In addition, the reaction can be run continuously with the recycle of SC-SO_2, if required (Chapter 46), and it is possible to combine the biotransformation and product recovery step in a single robust bioreactor.

Solvent-free systems, i.e., systems where the reaction takes place in a neat mixture of substrates in the absence of added solvent, are especially attractive for the synthesis and modification of lipids (Chapters 36 and 38). However, this approach can also be used successfully for the preparation of numerous other products (*see* Chapter 41 and references cited therein). In general, it is often the properties of substrates that define the best or most efficient media for their conversion. For example, compounds that are poorly soluble in those organic solvents that are generally considered suitable for biocatalysts can be efficiently transform in supersaturated substrate solutions (Chapter 42) or even in frozen aqueous solutions (Chapter 40). Chapter 37 deals with biotransformations in gas phase. This system may be particularly attractive for enzymatic syntheses involving highly volatile reactants.

From: *Methods in Biotechnology, Vol. 15: Enzymes in Nonaqueous Solvents: Methods and Protocols*
Edited by: E. N. Vulfson, P. J. Halling, and H. L. Holland © Humana Press Inc., Totowa, NJ

Biotransformations involving solid substrates are considered separately due to certain nonobvious peculiarities in kinetics and thermodynamics of such reactions. A general overview with some practical examples have been prepared by Kasche and Spiess (Chapter 43) and Straathof et al. (Chapter 47). Both groups of authors also discuss "bioengineering" parameters of transformations in suspensions such as, e.g., the effect of particle size. On the other hand, Erbeldinger et al. (Chapter 35) are focused specifically on solid-to-solid synthesis of dipeptides. This chapter also provides useful references to earlier, relevant work on enzymatic synthesis of short peptides in eutectic mixtures. Finally, no book on low-water enzymology would be complete without reverse micelles or microemulsions. Hence two contributions were included. Levashov and Klyachko (Chapter 45) prepared an excellent overview of the field with several examples of how reverse micelles can be used in synthetic applications, whereas chapter 48, by Carvalho et al., deals specifically with microemulsion-based bioreactors.

Finally, this part contains two contributions from industry. Morgan et al. have prepared an excellent case study that illustrates the use of enzymes in process development in the pharmaceutical industry, and Peilow and Misbah (Chapter 49) provided an insight into the preparation and use of biocatalysts in the food industry.

The editors emphasize that the separation of the articles collected in this chapter from those included in Part II is, to some extent, artificial because all the fundamental principles of low-water enzymology apply equally well to organic solvents, solvent-free substrate mixtures, and SC-SO$_2$. It was felt, however, that such a separation would help the reader to "navigate" through the volume more easily.

35

Enzymatic Solid-to-Solid Peptide Synthesis

Markus Erbeldinger, Uwe Eichhorn, Peter Kuhl, and Peter J. Halling

1. Introduction
1.1. Background

Solid-to-solid peptide synthesis is an enzyme-catalyzed reaction carried out in a mixture consisting of solid substrates and up to 20% (w/w) of enzyme solution in water. No organic solvents are necessary for the preparation of the initial reaction mixtures. Generally, solid-to-solid synthesis is considered to be a low-water reaction system because of the very high overall concentration of substrates used. However, from the enzyme's "viewpoint," the reaction mixture is just an aqueous solution saturated with substrates, as this is where the actual biotransformation takes place. Therefore, this approach combines advantages of both water- and solvent-based systems (i.e., high enzyme activity, high substrate concentration, and high degree of conversion to the final product). Another attraction of solid-to-solid synthesis is that it enables improved volumetric productivity in the reactor to be achieved. The avoidance of organic solvents is often advantageous too, especially for applications in the pharmaceutical and food industry.

Solid-to-solid synthesis is reported to work for a wide range of substrates and enzymes using both thermodynamically and kinetically controlled approaches *(1–8)*. The overall appearance of the reaction mixture can vary from visually all solid to a suspension of solid substrate(s) in liquid. Thus, the terms "enzymatic synthesis in suspension systems" or "suspension-to-suspension" synthesis have sometimes been used. In all cases, however, some starting material remains as an undissolved solid. For kinetically controlled reactions larger quantities of water and volumes of liquid phase are acceptable, whereas in thermodynamically controlled systems, this would lead to a substantial reduction in yields.

From: *Methods in Biotechnology, Vol. 15: Enzymes in Nonaqueous Solvents: Methods and Protocols*
Edited by: E. N. Vulfson, P. J. Halling, and H. L. Holland © Humana Press Inc., Totowa, NJ

The preparation of the reaction mixture and sampling are slightly more complicated for solid-to-solid reactions than for those which take place in a conventional solvent-based medium. In the latter case, reactants are normally well mixed, whereas in solid-to-solid systems, there will always be some sampling errors caused by the heterogeneity of the reaction mixture. Consequently, a slightly higher scatter is usually obtained in time-course analyses of solid-to-solid reactions. The control of pH is also more complicated as compared to conventional transformations in liquid media. Therefore, substrates and additional salts should be chosen with care.

1.2. Experimental Strategy: Analytical Scale Synthesis

As mentioned earlier, solid-to-solid reaction mixtures always contain undissolved reactants and hence are not homogenous. Consequently, it is difficult in practice to analyze samples containing less than 0.1 mmol of substrate with reasonable accuracy. This is because withdrawing an aliquot from a heterogeneous sample as small as 0.1 mmol is exceedingly difficult. Hence, it is advisable in the case of a time-course analysis to divide the reaction mixture into several small portions immediately after mixing and to quench individual samples at specified time intervals. It is also possible to run the reaction in a larger single batch so that the ratio of the sample's weight (or volume) to the weight (or volume) of the overall batch would remain within acceptable limits. Either way, a substantial amount of substrate (typically 1–2 mmol per experiment) is required for optimization or kinetic studies, which can be rather expensive.

1.3. Experimental Strategy: Preparative Scale Synthesis

Because of the specific constituency of solid-to-solid reaction mixtures, not all types of laboratory mixers are equally suitable. It is best to use a mixer where the agitator can reach every part of the reactor (e.g., a form of scraping). Eichhorn et al. *(9)* has successfully used a ploughshare mixer. Typically, this mixer has two ploughshares mounted on the rotating axis, which scrape the walls of the reactor. Stirred tank mixers where either the agitator or the mixer bowl rotates with planetary motion are also a good choice. This design is basically the same as in many conventional kitchen food processors. Thus, if an appropriate food processor is placed in a temperature-controlled cabinet, it can be used instead of more expensive laboratory mixers with water jackets.

The handling of samples is rather similar regardless of whether the synthesis is carried out on an analytical or preparative scale. For analysis, a sample of around 0.1 mmol has to be removed from the reactor using a spatula or a special device (*see* **Note 1**) and dissolved as described in **Subheading 2.5**.

2. Materials

2.1. Substrates

N-Components (nucleophiles) can be in the form of either free amines or salts (e.g., a hydrochloride), provided the pH in the reaction mixture is adequate (*see* **Subheading 2.4.**). Suitable amino acid derivatives can be purchased from all major suppliers of fine chemicals.

2.2. Enzymes

Enzyme loading of 3 g per mole of substrates is usually a good starting point for optimization, when using serine or cystein proteases. In the case of thermolysin (Sigma Chemical Co, UK), this loading is probably somewhat excessive. If crude industrial-grade enzymes are used, much higher quantities (by weight) may be required. Note that the loading of the enzyme can be higher than its solubility in the water added, but this does not seem to cause a problem.

2.3. Water

Water content and the resulting viscosity of the reaction mixture are the key parameters in solid-to-solid synthesis. A comparison of different systems suggests that a water level of 10% (w/w) is usually not too far away from the optimum (*see* **ref. *10***) and **Notes 2** and **3**).

2.4. Bases or Acids

These may be needed to obtain optimal pH. Hydrochloride salts of the amino component (commonly used as substrates) must be neutralized. It is convenient to use a basic salt such as $KHCO_3$, which, when adding, in an equimolar amount to the substrate is sufficient for neutralization. However, a higher $KHCO_3$/substrate ratio between 1.5 and 2 (mol/mol) has been shown to increase enzymatic activity, particularly in the case of acidic amino acid derivatives (*see* **Notes 4** and **5**). Alternatively a molar excess of one substrate over another can be used depending on the pK values (*see* **ref. *11*** and **Note 6**). In some cases it is necessary to use an acid *e.g.* 10 m*M* H_3PO_4 to stop the reaction.

2.5. Organic Solvent

A solvent can be used for the preparation of reaction mixtures *(2,3)* and is needed to dissolve the reactants prior to analysis. In the latter case, 50% (v/v) acetonitrile/water is often satisfactory.

3. Method

3.1. Analytical Scale Synthesis

Two main methods are used to prepare an initial reaction mixture for solid-to-solid synthesis on a laboratory scale. First (A), by mixing the solid substrates with an aqueous enzyme solution (solid lines in top part of **Fig. 1**). In the second method (B), substrates (including a hydrochloride form of nucleophiles) are neutralized by mixing with a basic salt and water (dotted line in top part of **Fig. 1**) and the enzyme is then added as a dry powder. In either case, the following apply:

1. Weigh the two solid substrates, the enzyme (*see* **Note 7**) and, for method B, the basic salt (*see* **Note 8**).
2. Mix the appropriate dry solids thoroughly with a spatula, vortex mixer, or just shake the reaction vessel by hand (methods A and B).
3. a. For method A, add the required amount of water into an Eppendorf test tube containing the enzyme to dissolve or suspend the catalyst.
   b. For method B, add the required amount of water and mix immediately with a spatula to ensure good distribution of the water and to prevent foaming when CO_2 is released on neutralization. Keep the vial under supervision for 15–30 min, depending on the salt used, and mix the content time to time because of foaming.
4. Start the reaction by either adding the enzyme solution (**step 3a**) or dry enzyme powder to the substrate mixture resulting from **step 3b** (refer to **Fig. 1**).
5. Mix the final reaction mixture very intensively with a spatula to ensure that the enzyme is evenly distribution throughout.
6. If required, separate the resulting viscous creamy mixture into several batches using a spatula. For the suggested 0.1-mmol scale, the required amount is enough to cover the tip of a 3-mm spatula. Incubate the Eppendorf tubes in a water bath. A temperature of 40°C should be suitable for most enzymes. Immerse the sample tube completely to avoid condensation inside (*see* **Notes 9–11**).
7. At the required time intervals, take one Eppendorf tube out of the water bath, remove the solid or highly viscous substrate/product mixture (*see* **Note 12**), terminate the reaction, and dissolve the sample for analysis. If high-performance liquid chromatography (HPLC) analysis is employed, it is preferable to use a solvent mixture of the same composition as the mobile phase. The volume depends on the solubility of substrates and products. Ten milliliters of 50% (v/v) acetonitrile/H_2O mixture is suitable for dissolving most dipeptides and is a good starting point. Prior to HPLC analysis, remove the precipitated enzyme particles from the samples by filtration.

3.2. Preparative Scale Synthesis

The method described in **Subheading 3.1.** can be easily adjusted to a preparative scale by increasing the reaction volume. The flowchart in **Fig. 1** is still valid except for the separation and sampling steps. The composition of the reaction mixture should be optimized prior to scaling up the synthesis. The reaction mixture can be prepared manually or with a laboratory mixer (*see* **step 5**). When carrying the synthesis, the reaction mixture should be mixed at regular time intervals (e.g., 1 h, and preferably continuously, to facilitate mass transfer.

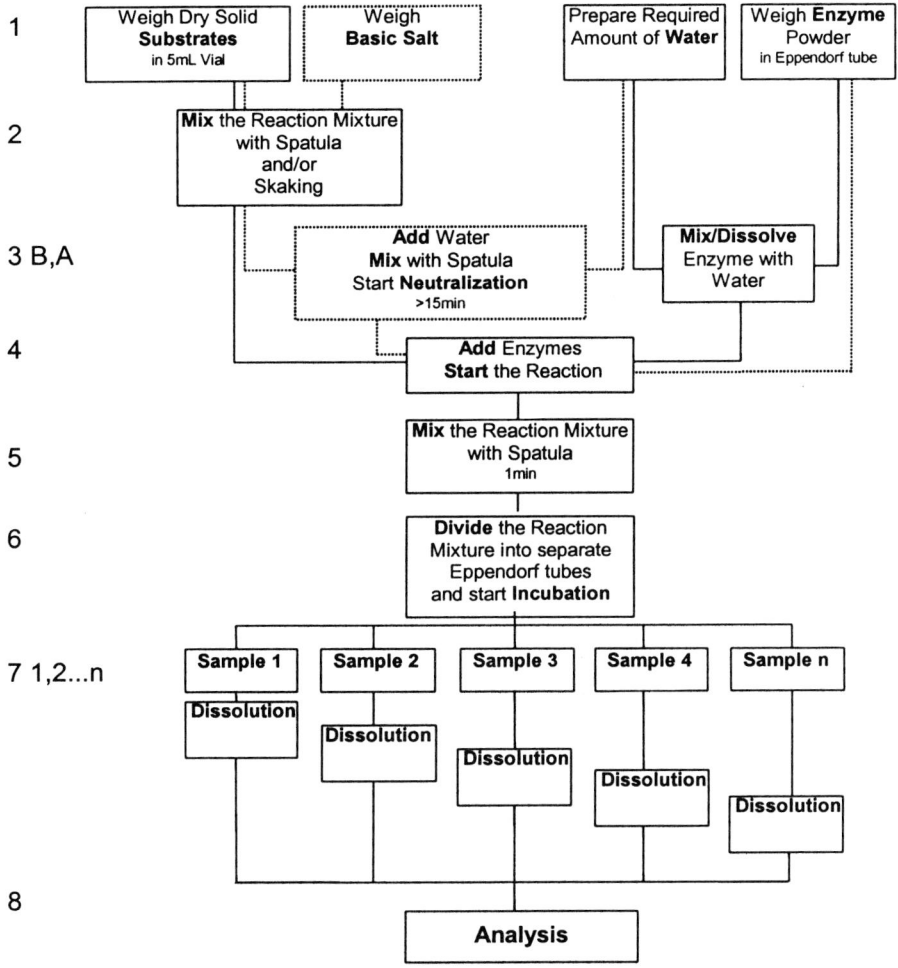

Fig. 1. Flowchart of solid-to-solid synthesis on a laboratory scale.

4. Notes

1. To withdraw a sample from the reactor, it is recommended to use a "sample thief." This simple device consists of a stainless-steel rod with horizontal holes right through and a stainless-steel tube fitted inside that has holes in the corresponding positions. By a 90° rotation of the tube, the holes in the rod can be either closed or opened. The closed sample thief is inserted into the solid reaction mixture; it is then opened to fill the holes with material, and closed prior to removing it.

2. A low level of water in the system often results in a low mass transfer of substrates and products to and from the enzyme. Enzymes also require some water to retain sufficient catalytic activity. However, if the water level is too high, the product yield

of thermodynamically controlled peptide syntheses decreases. Hence, an optimization of water content is advisable for each new reaction system.

3. A useful indication of a reaction mixture having the right viscosity is that it would move slowly down the wall of an Eppendorf tube at 40°C. Ideally, the reaction mixture should appear homogenous and have "creamy" consistence. If no liquid phase is visable, the water level is probably too low.

4. The use of a weakly basic salt like $KHCO_3$ helps to prevent an "overshooting" of pH. "Stronger" salts like K_2CO_3 are likely to make the pH too basic; this leads to the reduction in final yields (a conversion of 80% or more is usually considered to be satisfactory). Again, some optimization has to be done for each pair of substrates.

5. It is not advisable to control pH with organic buffers because of the very high (up to 10 M) substrate concentrations employed. The amount of buffering compound(s) needed to control the pH is likely to be too high in terms of their weight or volume. The use of organic buffers may also complicate the recovery of the final product.

6. A higher conversion of the limiting substrate is obtained when using one of the reactant in excess, whereas the use of equimolar quantities of substates helps to obtain a higher overall yield of the product per unit weight of the reaction mixture. This, together with possible difficulties associated with the product recovery, should be taken into consideration when deciding on how to buffer the system for preparative scale applications.

7. This can be easily done by using an Eppendorf tube placed in a 5-mL vial or in any other suitable holder.

8. A 5-mL vial is sufficient for working on a 2-mmol scale. For a larger-scale synthesis, the size of the reactor should be increased accordingly.

9. Because the tubes are fully immersed into the water bath, there is a risk of water penetrating into the sample. This is not a problem with Eppendorf tubes, which do not require any additional sealing.

10. To avoid errors in the initial rate determination, where accurate timing is crucial, all samples should be incubated in one batch for the first hour or so. In practical terms, this means that the Eppendorf tubes should be put aside until all the samples required for the experiment are prepared so that they can be incubated together.

11. This can be achieved by putting a stainless-steel weight on top of a commercially available Eppendorf tube rack.

12. With increasing product conversion and/or reaction time, the reaction mixture may appear to "dry out" (i.e., to change from a highly viscous liquid to a more solid structure). In this case, a sample containing a small amount of this reaction mixture may stick to the wall of the Eppendorf tube. If so, it is easy to scratch it off with a rod (1 mm∅) (e.g., an Allan key). Alternatively, one can add 1 mL solvent and stir the reaction mixture with the Allan key to obtain a suspension that is much easier to handle. If not all the material is removed from the Eppendorf tube with solvent, this "washing" out procedure can be repeated one more time.

References

1. Kuhl, P., Eichhorn, U., and Jakubke, H. D. (1992) Thermolysin- and chymotrypsin-catalysed peptide synthesis in the presence of salt hydrates, in *Biocatalysis in Non-Conventional Media* (Tramper, J., et al., eds.), Elsevier Science, Amsterdam, pp. 513–518.
2. Gill, I. and Vulfson, E. N. (1993) Enzymatic synthesis of short peptides in heterogeneous mixtures of substrates. *J. Am. Chem. Soc.* **115**, 3348,3349.
3. López-Fandiño, R., Gill, I., and Vulfson, E. N. (1994) Protease-catalysed synthesis of oligopeptides in heterogenous substrate mixtures. *Biotechnol. Bioeng.* **43**, 1024–1030.
4. Halling, P. J., Eichhorn, U., Kuhl, P., and Jakubke H. D. (1995) Thermodynamics of solid-to solid conversion and application to enzymic peptide synthesis. *Enzyme Microb. Technol.* **17**, 601–606.
5. Eichhorn, U., Beck-Piotraschke, K., Schaaf, R., and Jakubke, H. D. (1997) Solid-phase acyl donor as a substrate pool in kinetically controlled protease-catalysed peptide synthesis. *J. Peptide Sci.* **3**, 261–266.
6. Erbeldinger, M., Ni, X., and Halling, P. J. (1998) Effect of water and enzyme concentration on thermolysin catalysed solid-to-solid peptide synthesis. *Biotechnol. Bioeng.* **59**, 68–72.
7. Cerovsky, V. (1992) Protease-catalysed peptide synthesis in solvent-free system. *Biotech. Tech.* **6**, 155–160.
8. Klein, J. U. and Cerovsky, V. (1996) Protease-catalysed synthesis of Leu-enkephalin in asolvent-free system. *Int. J. Peptide Protein Res.* **47**, 348–352.
9. Eichhorn, U., Bommarius, A. S., Drauz, K., and Jakubke, H. D. (1997) Synthesis of dipeptides by suspension-to-suspension conversion via thermolysin catalysis: from analytical to preparative scale. *J. Peptide Sci.* **3**, 245–251.
10. Erbeldinger, M., Ni, X., and Halling, P. J. (1998) Enzymatic synthesis with mainly undissolved substrates at very high concentrations. *Enzyme Microb. Technol.* **23**, 141–148.
11. Erbeldinger, M., Ni, X., and Halling, P. J. (1999) Kinetics of enzymatic solid-to-solid peptide synthesis: intersubstrate compound, substrate ratio and mixing effects. *Biotechnol. Bioeng.* **63**, 316–321.

36

Enzymatic Synthesis and Hydrolysis Reactions of Acylglycerols in Solvent-Free Systems

Cristina Otero, Jose A. Arcos, Hugo S. Garcia, and Charles G. Hill, Jr.

1. Introduction

Lipases (E.C. 3.1.1.3) are conventionally defined as enzymes that catalyze the hydrolysis of acylglycerols to release a fatty acid and lower acylglycerols or glycerol itself. However, this definition is rather restrictive, because these enzymes also cleave a wide variety of other esters as well as amides. In addition, lipases catalyze the corresponding reverse reactions that are normally carried out in media containing only small amounts of water. Indeed, water formed as a by-product of these synthetic reactions should be removed from the reaction mixture in order to shift the position of thermodynamic equilibrium so as to favor the synthetic reactions. A wide variety of industrially important lipase-mediated transformations have been carried out to date *(1)* in numerous organic solvents *(2)*. Cernia et al. *(3)* have recently reviewed the role of solvent polarity and hydrophobicity in modulating lipase activity. The present work describes lipase-catalyzed biotransformations in mixtures of liquid substrates because this approach circumvents the need for organic solvents.

The generic classes of lipase-catalyzed reactions are shown in **Fig. 1**. Enzymatic hydrolysis (lipolysis) is particularly beneficial for applications where only partial hydrolysis of triglycerides is required. A variety of bioreactor configurations can be employed to effect the hydrolysis reaction, but in our experience, a hollow-fiber reactor (*see* **Fig. 2**) is particularly attractive for this application (see **refs.** *4–8* and **Notes 1** and **2**). This type of immobilized enzyme bioreactor has several advantages over conventional two-phase (emulsion) batch systems because it (1) enables one to obtain higher

From: *Methods in Biotechnology, Vol. 15: Enzymes in Nonaqueous Solvents: Methods and Protocols*
Edited by: E. N. Vulfson, P. J. Halling, and H. L. Holland © Humana Press Inc., Totowa, NJ

Hydrolysis

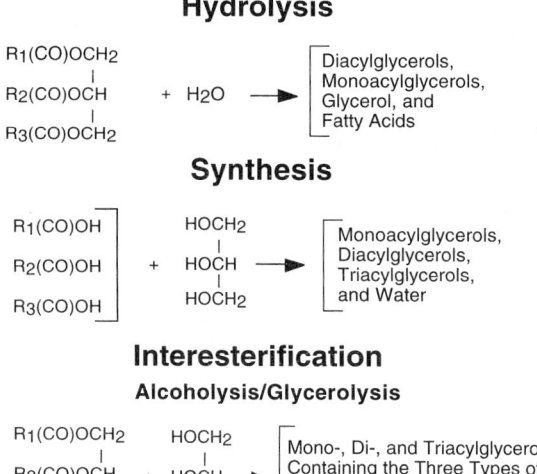

Synthesis

Interesterification
Alcoholysis/Glycerolysis

Acidolysis

Transesterification

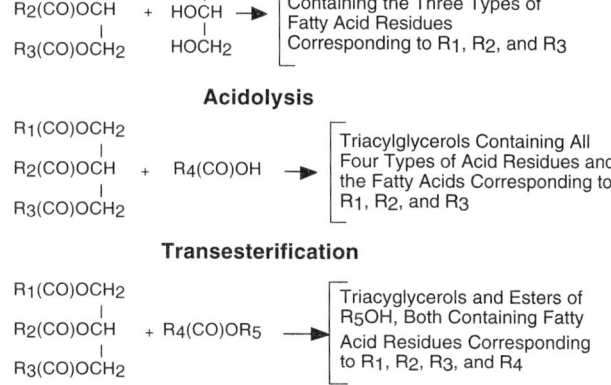

Fig. 1. Schematic representations of hydrolysis, synthesis, and interesterification reactions involving triacylglycerols. All of the reactions are reversible, but they are named according to the indicated direction of the arrow.

yields of product per unit of enzyme, (2) can be operated continuously, and (3) is easily controlled to compensate for enzyme deactivation by adjusting the residence time of the oil feedstock.

The ability of lipases to maintain high catalytic activity and selectivity in nonaqueous media when used either as free enzymes or in immobilized form can be exploited in a wide range of other applications. Thus, immobilized lipases can be utilized for the production of monoacylglycerols and diacylglycerols enriched in polyunsaturated fatty acids (PUFAs or omega-3 fatty acids) via hydrolysis of fish oil *(4)*, direct esterification of glycerol with conjugated linoleic acid (CLA) *(9)*, or acidolysis of an acylglycerol with CLA *(10)*. Enzymatic preparation of acylglycerols in solvent-free media (by esterification,

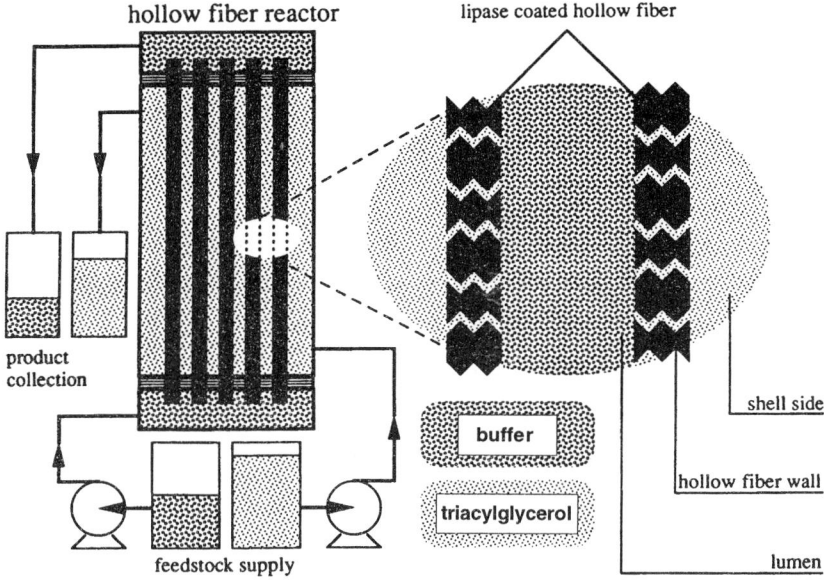

Fig. 2. Schematic diagram of the reactor with a magnified view of the vicinity of the hollow fiber. (Adapted from **ref. 6**.)

acidolysis, tranesterification, and so forth) containing minimal amounts of water has the following advantages over the use of organic solvent-based systems: (1) There is no necessity for purification/elimination of the solvent; this is especially important for the preparation of products intended for human consumption. (2) The reaction rate would normally be faster and the reactor volume smaller; hence, higher productivity can be obtained. The major challenge is to ensure sufficient miscibility of the reactants in a solvent-free mixture and/or good mixing to minimize mass transfer limitations on reaction rates. As an example of a solvent-free lipase-catalyzed esterification process, we shall consider the synthesis of acylglycerols from conjugated linoleic acid and glycerol.

It should be noted that acidolysis, alcoholysis, and transesterification **(Fig. 1)** constitute a class of reactions that are sometimes referred to as interesterification reactions. As the first step, these reactions involve hydrolysis of the substrate ester bond. The carboxylic acid and hydroxyl moieties generated by the hydrolysis subsequently react to form a new ester. The essential feature of all interesterifications is that they enable one to replace the acid or alcohol residue associated with the ester bond of interest. These reactions have long been used to modify the properties of lard *(11)*. More recently, there has been renewed interest in the use of these processes for the modification of edible oils *(12)*. For example, acidolysis has been successfully employed for the production of cocoa butter substitutes from inexpensive starting materials *(13)* and to incorporate omega-3 fatty acids in vegetable oils *(14,15)*.

In this chapter, interesterification is exemplified by the production of monooleoylglycerols and dioleoylglycerols from triolein and glycerol. Both interesterification reactions and direct esterification of glycerol with fatty acids can be used to produce monoesters and diesters in high yields at low temperatures to suppress undesirable migration of the secondary acyl group. In general terms, the former reaction is probably more selective and hence more economically attractive for the production of pure diacylglycerols, whereas the latter is of greater interest for the preparation of glycerol monoesters *(16–18)*. The experimental procedures to be described show how one can employ alcoholysis (specifically glycerolysis) of triacylglycerols to produce lower acylglycerols and acidolysis to exchange the acyl groups present in a triacylglycerol. Following the descriptions of the experimental techniques recommended for the aforementioned reactions, a general methodology for developing appropriate mathematical forms of the rate expressions for lipase-catalyzed transformations is presented.

2. Materials

2.1. Hydrolysis of Acylglycerols

1. Enzyme: a soluble form of the lipase of interest.
2. Substrate: oil to be subjected to lipolysis.
3. Solvents for cleaning bioreactors: hexane, diethyl ether, and ethanol.
4. McIlvane buffer (*see* **Note 3**): dibasic anhydrous sodium phosphate, citric acid monohydrate, glycerol, sodium hydroxide, and deionized water.
5. Reagents for titrimetric analyses: phenolphthalein, denatured alcohol, methanol, potassium hydroxide, and potassium hydrogen phthalate.
6. Standards for high-performance liquid chromatography (HPLC) analyses: carboxylic acids, which are expected to be released in the course of hydrolysis.
7. Derivatisation reagents: methanol, chloroform, 2,4'-dibromophen-acetophenone (PBPB), 18-crown-6 ether, acetonitrile, and potassium carbonate. Stabilizer: butylated hydroxyanisole.
8. HPLC-grade solvents: methanol, acetonitrile, and deionized water.
9. Equipment: A schematic diagram of a hollow-fiber reactor used for hydrolysis is shown in **Fig. 2** (*see* **Note 2**).

2.2. Esterification

1. Enzyme: immmobilized *Mucor miehei* lipase (Lipozyme IM, Novo-Nordisk, Denmark or Chirazyme L-9, Bohringer Mannheim, Germany).
2. Substrates: glycerol and conjugated linoleic acid produced by chemical isomerization of linoleic acid according to Chin et al. *(19)*.
3. Standards: monolinoleylglycerol, 1,2-dilinoleylglycerol, and trilinoleylglycerol for use as standards (Sigma Chemical Co., Spain).

4. Molecular sieves with an effective pore diameter of 0.4 nm (Fisher Scientific, USA).
5. Equipment: Esterification reactions can be carried out in a conventional well-stirred batch reactor or in shaker flasks with sufficient agitation to maintain the immobilized enzyme and molecular sieves in suspension (*see* **Notes 4** and **5**).

2.3. Interesterification

2.3.1. Glycerolysis

1. Enzyme: *Pseudomonas* sp. lipase (Amano Pharmaceuticals, Japan).
2. Substrates: triolein (99%) and glycerol.
3. Celite for chromatographic analyses.
4. Equipment: see **Subheading 2.2.**, **item 5**.

2.3.2. Acidolysis

1. Enzyme: a soluble or immobilized lipase (e. g., lipases PS and F [free], PS-C and PS-D [immobilized] from Amano [Japan] or immobilized *Mucor miehei* [Lipozyme IM-60] and *Candida antarctica* [Novozyme 435] lipases from Novo (*see* **Note 6**). These immobilized enzymes are also currently marketed by Bohringer Mannheim (Germany).
2. Substrate: anhydrous butter fat and high-purity CLA (approx 98–99%).
3. Solvents and standards (tri-, di-, and monoacylglycerols and free fatty acids) for gas chromatography and HPLC (*see* **Subheading 2.1.**, **item 8**).
4. Equipment: Acidolysis reactions can be carried out in a conventional stirred batch reactor or tubes or flasks provided that sufficient agitation is maintained and the immobilized enzyme stays in suspension. For example, the reaction of butter oil and CLA can be carried out in Erlenmeyer flasks in an orbital shaker or water bath (*see* **Note 7**).

3. Methods
3.1. Hydrolysis of Acylglycerols
3.1.1. Immobilization of the Enzyme in the Hollow-Fiber Bioreactor

1. Allow the temperature to stabilize at the desired value.
2. Clean the reactor prior to immobilization of the enzyme by washing sequentially with a 1:1 (v/v) mixture of hexane and diethyl ether (1 h), 1:1 (v/v) mixture of ethanol and diethyl ether (1 h), and ethanol (1 h).
3. Immediately prior to the immobilization process, rinse both the tube side and the shell side of the reactor for three consecutive 30-min periods with ethanol, deionized water, and McIlvane buffer at flow rates of 500 mL/h.
4. Recirculate lipase solution containing 3.0 g of lipase per 100 mL of buffer through both sides of the hollow-fiber reactor for 3 h.
5. Pump 100 mL of the McIlvane buffer solution through both sides of the reactor to remove any soluble enzyme that has not adsorbed on the fibers.

6. Prior to the measurements, flush the shell side of the reactor with the contents of two oil syringes at a total flow rate of 104 mL/h in order to displace the aqueous solution with which the shell side had been filled.

3.1.2. Continuous Operation of the Bioreactor

1. Pump the acylglycerol upward through the shell side of the vertically oriented reactor and the buffer upward through the tube side (*see* **Note 8**).
2. Conduct experiments at a series of space times (*see* **Note 9**) in order to obtain the data necessary to determine a rate expression. If the rate expression is known, employ a flow rate or sequence of flow rates that is appropriate to obtain the desired degree of conversion.
3. Once the reactor has reached quasi-steady-state operating conditions, collect samples of the effluent stream(s) for titrimetric and HPLC analyses (*see* **Note 10**).

3.1.3. Analyses

1. Titrimetric: To determine the total concentration of free fatty acids released in the effluent stream, dissolve 0.2-mL samples of the effluent oil stream in 8 mL of a 1:1 (v/v) ethanol/diethyl ether mixture and titrate it with a methanolic (0.1 *N*) solution of potassium hydroxide using phenolphthalein as the indicator. (It is necessary to previously standardize the titrant solution with potassium biphthalate using phenolphthalein as the indicator.)
2. HPLC: Analyze samples of both effluent streams by HPLC to determine the rates of release of individual fatty acids. A reliable procedure is described in **Note 11**.
3. Regression: A variety of statistical techniques can be employed to determine the rate expression that provides the best fit of the experimental data (*see* **Subheading 3.4.**).

3.2. Esterification

3.2.1. Reaction Conditions

1. Place appropriate amounts of glycerol, CLA, and molecular sieves in a flask. Typically, 2.0 g CLA, 0.8 g molecular sieves, and sufficient glycerol to provide the desired molar ratio of CLA to glycerol are added to a 25-mL flask (*see* **Notes 12–15**).
2. Preheat the mixture to the desired temperature.
3. Place the flask containing the reactants in an orbital shaker maintained at the desired (constant) temperature and add the biocatalyst (0.3 g Lipozyme IM, for the quantities of material specified in **step 1** (*see* **Note 16**). Depending on the stoichiometric ratio of reactants, the reaction conditions employed, and the time for reaction, one can produce primarily mono-, di-, or triacylglycerols or mixtures thereof (cf. **Fig. 3**).
4. Analyze the reaction mixture by HPLC (*see* **Note 17**).

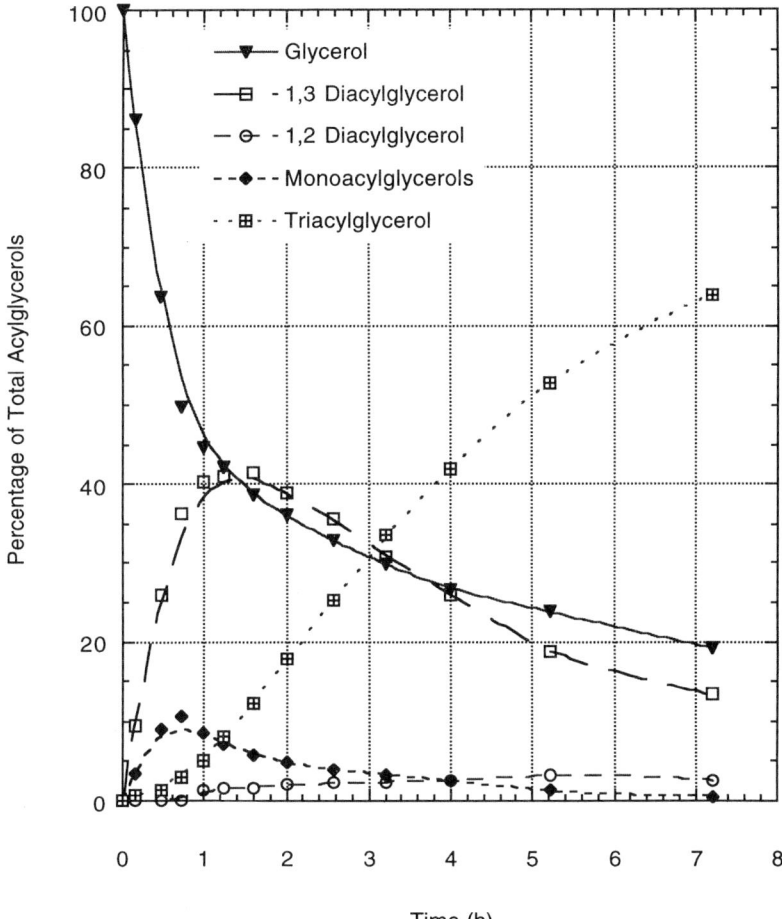

Fig. 3. Time-course of the distribution of reactant and product species in the esterification of glycerol with CLA. Conditions: 2000 mg CLA, 217 mg glycerol, 300 mg Lipozyme IM, 800 molecular sieves, 50°C.

3.3. Interesterification

3.3.1. Glycerolysis

3.3.1.1. ADSORPTION OF LIPASE ON THE SUPPORT

1. Dissolve 5 g of lipase powder in 20 mL of 0.1 M phosphate buffer, pH 7.5, and stir for 20 min.
2. Centrifuge the solution at 0°C for 10 min at 11,950g.
3. Add 5 g of Celite to the enzyme solution (<20% [w/w]) and continue stirring for 1 h.
4. Slowly add 5 mL of cold (–20°C) acetone, with stirring.

5. Filter the suspension of the immobilized enzyme, wash twice with 20 mL of acetone, dry for 30 min *in vacuo*, and store in a closed vial at 4°C.

3.3.1.2. Reaction Conditions

1. Weigh out the required amounts of triolein, glycerol, and biocatalyst into a flask: 145 mg triolein, 50 mg biocatalyst, and sufficient glycerol to provide the desired molar ratio of reactants (*see* **Notes 18** and **19**).
2. Place the reactants in a controlled-temperature environment. Stir the contents of the flask with a magnetic stirrer. Once the contents of the flask solidify, cease stirring (*see* **Note 20**).
3. Prior to analyzing the reaction mixture by HPLC, dissolve the product mixture in chloroform and remove the biocatalyst by filtration (*see* **Note 21**).

3.3.2. Acidolysis

3.3.2.1. Reaction Conditions

1. Prepare mixtures containing butter oil and CLA in a 10:1 (w/w) ratio (1.1 g total) in 13×100 mm tubes (*see* **Note 22**).
2. Add 50 mg of the immobilized enzyme, seal the tube under nitrogen, and incubate at 50°C in an orbital shaker set at 200 rpm (*see* **Note 23**).
3. To prepare the sample for analysis, dilute with 5 mL hexane and filter through a 0.45 μm membrane filter.

3.3.2.2. Analyses

1. Take two 100 μL aliquots from the diluted samples and dry under nitrogen. Then, add 500 μL chloroform/methanol (2:1, v/v) to each aliquot.
2. Add 1.0 mL of 0.1 *M* methanolic NaOH to one of the tubes and allow it to stand at room temperature for 30 min. Then, add 200 μL deionized water and extract the mixture with 2 mL of hexane.
3. To the second tube, add 1 mL of 0.2 *M* methanolic HCl, flush with nitrogen, and leave overnight on a heating block at 80°C.
4. After methylation is completed, add 200 μL water to each of the tubes and extract the mixtures twice with 2 mL of hexane. Dilute the extracts with hexane to obtain solutions containing approximately 1.2 mg lipid per milliliter.
5. Inject 1 μL of each solution into a gas chromatograph equipped with a 60-m Suplecowax-10 column, flame ionization detection (FID), and split injector.
6. Take 20-μL aliquots of the samples diluted with hexane, dry under nitrogen, and suspend in 4 mL chloroform:methanol (2:1, v/v). Inject 20 μL into an HPLC apparatus consisting of a quaternary pump, a 4.6×250-mm Alltech Econosil normal-phase column, and a laser light-scattering mass detector (Alltech, ELSD 500). Elute with a mobile phase composed of hexane, 2-propanol, ethyl acetate, and acetic acid. Lipids are eluted from the column in the following order: free fatty acids, triacylglycerols, 1,3-diacylglycerols, 1,2-diacylglycerols, 1-monoacylglycerols, and 2-monoacylglycerols.

3.4. Rate Expressions for Lipase-Catalyzed Reactions

The development of rate expressions to characterize various reactions catalyzed by lipases is a challenging task. The main problems are the large variety of molecular species that can be involved in these reactions and the concomitant mathematical complexity of the associated equations. Water can be consumed (hydrolysis) or generated (esterification), or sequentially consumed and released with no net consumption or production of water (interesterification). In addition, a very large number of regioisomers can be present when natural fats and oils are used as substrates. Furthermore, nonenzymatic side reactions (e.g., deactivation of the enzyme or intramolecular migration of an acyl group) may occur concurrently with the enzyme-mediated transformations.

To analyze the kinetics of interesterification, one usually assumes that this reaction can be described in terms of a two-step mechanism: initial hydrolysis of an acylglycerol followed by subsequent esterification. This working hypothesis requires multiple entrances and exits of substrate and product species at the active site of the enzyme in such a manner that the overall mechanism can be considered as a generalized ping-pong mechanism. The mathematics required to derive the corresponding rate expressions is frequently tedious, but can be greatly simplified using the schematic approach of King and Altman *(20,21)*. (*See* **Note 24** for an example in which this technique is employed.)

Although the King–Altman method permits one to readily derive rate equations for very complex mechanisms, the number of parameters involved in the rate expression associated with a ping-pong bi-bi mechanism in its generic form is rather large. Hence, overparameterization frequently occurs if one tries to fit experimental data with this type of rate expression. There are several methods that can be employed to circumvent this problem:

1. Cha's method (rapid equilibrium) *(22)*: One selects as the rate-limiting step a reaction involving either the free enzyme or an enzyme–substrate complex. All other steps are assumed to be in rapid equilibrium.
2. Segel's method (product release) *(21)*: The rates of certain reactions are assumed to be much faster than the net rate of product formation. The composite system is then treated as if the net rate of the product formation is equal to the rate of the limiting step.
3. Net rate constant *(23)*: This approach permits one to simplify the derivation of rate expressions for several consecutive enzymatic reactions that do not involve branched pathways. In this method, the true rate constants for the forward and reverse directions are replaced by a "net rate constant" that has the effect of producing the same flux through the step in question by treating it as if it was irreversible.

4. Notes

1. Other types of reactors may also be employed to effect hydrolysis of triacylglycerols. For example, Garcia et al. *(24)* have utilized *C. rugosa* lipase

immobilized on polypropylene sheets to effect the hydrolysis of butter oil. These sheets are separated by spacers, wound in a spiral configuration, and placed in a tube so that the axis of the spiral is parallel to the longitudinal axis of the tube. A pump is employed to force an oil–water emulsion to flow perpendicular to the plane of the spiral. One could also employ a conventional batch reactor or shaken flask in which the hydrolysis is mediated by an immobilized enzyme suspended in an oil–water emulsion. However, the hollow-fiber configuration offers advantages over these types of reactor because it eliminates the necessity of using an emulsified feedstock with the attendant requirements for employing an emulsifying agent and for breaking the emulsion once the reaction is completed. Furthermore, the products of the reaction from the hollow-fiber reactor are not adulterated by the presence of a residual emulsifier.

2. The hollow-fiber reactor module (**Fig. 2**) was purchased from Hoechst Celanese Corp. (Liqui-Cel Laboratory Module, Charlotte, NC). The reactor shell is 30 cm long and 2.5 cm in diameter. The effective surface area of the fibers on the tube side is 0.4 m². The shell side can hold up to 26 mL of fluid and the tube side holds 27 mL. Each polypropylene hollow fiber has a nominal internal diameter of 400 μm, a nominal wall thickness of 30 μm, and contains micropores with nominal dimensions of 0.075 × 0.15 μm. A multichannel syringe pump (KD Scientific, Boston, MA) is used to pump the oil and the buffer solution from 60-mL disposable syringes (Becton Dickson, Rutherford, NJ) through the system. The effluent streams are collected in separate bottles. The rubber stoppers used to seal the bottles contain two holes. A 1/8-in.-diameter plastic tube is inserted through one hole to permit the effluent stream (either buffer or menhaden oil) to enter the bottle. The other hole permits air to escape as the liquid accumulates. Inside the bottles, vials are used for collecting the samples of the oil and buffer streams. The bath is maintained at a constant temperature using an Isotemp Immersion Circulator (Fisher, Pittsburgh, PA). A peristaltic pump (Cassettes pump, Manostat, New York, NY) is used for loading the enzyme and for cleaning the reactor. The pump can provide flow rates from 10 to 6000 mL/h.

3. The McIlvane buffer is 0.1 *M* dibasic anhydrous sodium phosphate, 0.05 *M* citric acid monohydrate, and 15% (w/w) glycerol, with the pH adjusted to 7.0 with 0.1 *M* NaOH.

4. High levels of esterification of glycerol with fatty acids can also be achieved by using various flow reactor configurations. Either continuous-flow stirred tank reactors or packed-bed tubular reactors could be employed. We have achieved good yields in the preparation of acylglycerols by pumping a mixture of glycerol and fatty acids (specifically CLA) through a packed-bed reactor containing not only an immobilized lipase but also molecular sieves. The molecular sieves can be either mixed with the enzyme to get a uniform distribution throughout the packed bed or one can employ a "segmented" reactor in which the packed bed consists of alternating regions of immobilized enzyme and desiccant.

5. In any type of reactor, the support for the biocatalyst might restrict access of the substrate to the active site of the enzyme. Therefore, when selecting an appropriate matrix, one must consider whether the solid is likely to impose diffusional

limitations on the reaction rate by restricting the access of substrates through the pore structure to the site of the enzyme. In some cases, it may be advantageous to select a support with a high affinity for the substrate to enhance its local concentration in the vicinity of the enzyme. High local concentrations of polar substrates can be obtained inside supports of similar polarity (*see*, e.g., **ref. 17**).

6. Evaluation of several commercially available lipases showed that Novozyme 435 and Lipozyme IM-60 had the highest catalytic activity for acidolysis of butter oil with CLA. Immobilized *Pseudomonas* sp. enzyme (Lipase PS-D from Amano) showed similar activity but is not a food-grade enzyme and is therefore not suitable for the preparation of products intended for human consumption. Because Lipozyme IM-60 is a 1,3-specific lipase and Novozyme 435 is nonspecific, the choice between these two enzymes may be dictated by the desirability of exchanging some acyl groups at the *sn*-2 position. Additional studies of the kinetics of the acidolysis reactions were carried out with Novozyme 435 in order to assess the rate at which acyl groups at the *sn*-2 position were exchanged. Equilibrium with respect to the exchange reaction is approached within 24 h.

7. This equipment enables one to simultaneously carry out a number of batch reactions at the same temperature. The kinetics of the reaction can be followed by either withdrawing a sample from a sufficiently large single flask or tube or by preparing a sufficient number of smaller tubes to be used as individual samples. Flasks or tubes containing the reaction mixture are capped under nitrogen to avoid oxidation of substrates and products.

8. The volumetric flow rate of the aqueous buffer stream is relatively unimportant because virtually all of the fatty acids released by hydrolysis exit in the organic effluent stream. However, when using a syringe pump with three syringes, it is convenient to employ a flow rate of the acylglycerol that is twice that of the buffer (i.e., two of the syringes are normally filled with oil and one with buffer solution).

9. The space time for a reactor is defined as the ratio of the volume of the reactor occupied by the reacting fluid to the volumetric flow rate of that fluid *(25)*. For liquid phases, the space time is normally equivalent to the residence time of the fluid in the reactor.

10. To determine whether the reactor has reached quasi-steady-state conditions, samples of the effluent organic stream are periodically titrated until no change in the total acidity is observed. The titrimetric data enable one to model the total release of fatty acids in terms of a single response kinetic model. The results of the HPLC analyses permit one to determine the distribution of free fatty acids produced by hydrolysis. These data form the basis for the development of a multiresponse model of the reaction kinetics.

11. To conduct the HPLC analysis of the effluent streams, add 500 µL samples to 2.5 mL of an internal standard solution (0.5–1.0 g/L of fatty acid in a 1:1 [v/v] mixture of chloroform/methanol stabilized with 0.05% butylated hydroxlyanisole). Then, add 1 mL of the chloroform/methanol mixture to bring the total volume to 4.0 mL and shake vigorously. Withdraw 88 µL and add this aliquot to 3.0 mL of a 1-g/L solution of 2-4'-dibromoacetophenone in acetonitrile. To the resultant solution,

add 80 µL of a 5-g/L solution of 18-crown-6-ether in acetonitrile. Then, add 0.2 g of potassium carbonate. Incubate the resulting mixture at 80°C for 30 min and then cool it to room temperature. During the incubation step, the free fatty acids are derivatized as bromophenacyl esters. Quench the derivatization reaction by addition of 40 µL of a 40-g/L solution of formic acid in acetonitrile. Incubate the resultant solution at 80°C for another 5 min. Refrigerate the samples for 90 min and then cold filter through 0.2-µm nylon membranes. Place the cold-filtered samples in 4 mL HPLC vials and seal them with a rubber septum and plastic cap. Inject 20 µL into a Waters Novo Pack 4 µm spherical C18 stainless column held at 33°C. An appropriate mobile phase is water, methanol, and acetonitrile *(26)*. Bromophenacyl esters are detected spectrophotometrically at 254 nm.

12. These solvent-free reaction mixtures are heterogeneous and often contain two liquid phases as well as one or more solid phase. Glycerol is not completely miscible with fatty acids and a precipitate may form during the reaction. Hence, it is important to provide adequate agitation in order to facilitate good contact of the immiscible liquids and the suspended solid phases (the molecular sieves, the immobilized lipase, and any precipitates that form). On a laboratory scale, orbital shakers and rollers provide a reasonable balance between the degree of agitation needed to provide adequate mixing and the requirement for little or no attrition of brittle solids. In large-scale continuous-flow packed-bed reactors, a portion of the monoacylglycerol and diacylglycerol products can be recycled to the reactor inlet to aid with mixing/emulsification of the feedstocks and with wetting the solid packing.

13. Some enzyme supports (e.g., Celite) or other solids (e.g., molecular sieves, silica gel), when added to the reaction mixture, may adsorb polar substrates such as glycerol and fatty acids, thus affecting the reaction kinetics *(16)* and the regiospecificity of the biotransformation *(27)*. These effects are attributed to either an increase in the local concentration of substrate in the vicinity of the immobilized enzyme (*see* **Note 5**) or an increase in the interfacial area between two poorly miscible liquid substrates. The support must also provide both sufficient mechanical strength and good enzyme retention as well as have an appropriate distribution of pore sizes and a chemical composition that facilitates diffusion (and perhaps adsorption) of substrates with different polarities (e.g., glycerol, fatty acids, and esters thereof). For esterification reactions, it is desirable to employ a hydrophobic support to facilitate the removal of water from the system. For the choice of immobilization materials, see **ref. 28**.

14. The stoichiometry of the synthesis of the glycerol triester of CLA requires a molar ratio of fatty acid to glycerol of 3 (*see* **Fig. 3**). However, higher ratios of CLA to glycerol generally give greater percentages of triacylglycerol in the final mixture of products, provided that the reaction is allowed to proceed for sufficient time. In studies of reaction kinetics, it is desirable to forego addition of the biocatalyst until the suspension is at the temperature of interest. This approach facilitates studies of the first and second stages of the polyesterification reactions.

  Note that formation of the monoesters and diesters proceeds rapidly at a temperature of 40°C, but for the synthesis of triesters using 1,3-specific lipases, migration of an acyl group from either the *sn*-1 or *sn*-3 position to the *sn*-2 position is rate limiting. Thus, selective selective esterification at the *sn*-1 and *sn*-3 positions is possible because of differences in the reactivities of the hydroxyl groups at these positions relative to that at the *sn*-2 position because the hydrolysis at this position is hindered.

When an initial molar ratio (CLA/glycerol) of 2 is employed at 50°C in the presence of Lipozyme IM and molecular sieves, approx 94% of the fatty acid is esterified in 7 h to give a product mixture consisting of 6% (w/w) fatty acid, 8% monoester, 35% diester, 49% triester, and 2% glycerol.

15. The use of a drying agent (e.g., molecular sieves) increases both the reaction rate and the product yield. Make sure that the capacity of the drying agent is adequate. For molecular sieves that absorb about 20% (w/w) water, use of a 3:1 weight ratio of fatty acid to sieves is recommended. It is advisable to activate the molecular sieves by heating in an oven at 200°C for 3 h prior to use. Alternatively, use sieves with a capacity indicator or add a larger excess.

16. An orbital shaker is preferred to a mechanical stirrer or magnetic stirring bar in order to minimize mechanical attrition of either the support for the enzyme or the molecular sieves. An alternative for situations where more extensive agitation is required is bubbling dry nitrogen through the suspensions.

17. The use of a normal-phase column for the HPLC analysis enables one to separate the 1,2-diacylglycerol from the 1,3-diacylglycerol even when a mixture of fatty acids is used as substrate. If standards for the glycerol esters are not commercially available, it is helpful to employ a light-scattering or refractive index detector.

18. To avoid excessive formation of hydrolytic side products, the water content of the reaction medium must be kept at a low level; 2.5–5.0% (w/w) water relative to the glycerol is often an appropriate level

19. A good enzyme for this application is *Pseudomonas* sp. lipase adsorbed on Celite. The efficacy of Celite as a catalyst support is attributed to its relatively large pores, minimal propensity for adsorption of water, and significant capacity for adsorbing glycerol *(17)*. This support also shows good mechanical strength. An alternative approach employing a continuous-flow reactor has been utilized by Garcia et al. *(29)* to effect the glycerolysis of butteroil in the absence of solvent.

20. To arrest or promote migration of acyl groups in the glyceride product, one may vary the temperature using the considerations presented in **Note 14**. Formation of monoacylglycerols and diacylglycerols is encouraged if the reaction is carried out at temperatures close to their melting point to achieve partial precipitation of the product *(30)*. Vigorous agitation should be provided until the reaction mixture solidifies. In this case, use of a magnetic stirrer is preferred to orbital shakers because it gives better agitation of viscous media.

21. Chloroform is usually used to dissolve solid or semiliquid samples for analysis. To avoid undesirable alcoholysis, use ethanol-free chloroform (ethanol is often added as a stabilizing agent) or remove ethanol by aqueous extraction followed by drying with molecular sieves.

22. The initial water content of the reaction mixture plays a key role in determining the kinetics of acidolysis. Commercially available enzyme preparations, as well as the oil and fatty acid feedstocks, contain small amounts of water. In our experiments, the immobilized enzyme, butter oil, and CLA contained 1.6, 0.06, and 0.3% (w/w) moisture, respectively. At higher water contents, the reaction rate declines. Because this reaction is carried in a relatively dry medium, temperatures of up to 70°C can be used. In the acidolysis of butter oil with CLA, the optimal temperature was 50°C.

23. Acidolysis reactions were carried out using different ratios of butter oil to CLA. At high ratios of butter oil to CLA and in the presence of sufficient catalyst, the reaction approaches equilibrium after about 6 h, although longer reaction times (over 24 h) may be necessary at very low water contents. The level of incorporation of CLA in triglycerides can be readily controlled by changing the ratio of reactants, the quantity and type of enzyme employed, and the reaction conditions (e.g., water content and temperature).

24. As an example, consider the derivation of a rate expression for the enzymatic synthesis of triacylglycerols from glycerol and a fatty acid as mediated by a 1,3-specific lipase. In particular, consider the reaction of CLA with glycerol for which the synthesis procedure was described in **Subheading 3.2**. For a 1,3-specific lipase, isomerization of the 1,3-diester to the 1,2-diester must occur prior to esterification of the third hydroxyl group. Consequently, after the 1,3-diester is formed in an enzymatic step, the nonenzymatic isomerization must occur to obtain the triester. This reversible isomerization step is assumed to obey first-order kinetics in both the forward and reverse directions. Now, consider a simplified ping pong bi-bi mechanism in which only two enzymatic species are involved: the free lipase [E] and the covalently bound acyl intermediate [E-FA]. According to this mechanism, the lipase first binds fatty acid to form [E-FA] and release water. The acylated form of the enzyme then reacts with glycerol, a monoacylglycerol, or an appropriate diacylglycerol, to give a higher acylglycerol [HAG], thus regenerating the free lipase.

To simplify the kinetic analysis, the rates of all reverse enzymatic reactions will be assumed to be negligible. Then, the acylated form of the enzyme will be produced by only one reaction, but there will be three possible reactions by which the free enzyme can be regenerated. The four steps involved in the King–Altman method can now be employed to derive the rate expression corresponding to the situation described above.

> **Step 1.** Arrange the different enzyme species into a polygon with each enzyme species at a vertex. Connect the vertices with lines that represent the steps in the mechanism by which the two species are interconnected. Label the lines associated with each step in terms of either a unimolecular rate constant or the product of the rate constant and the concentration of reactant (or product) involved in a bimolecular reaction of interest. Do this for both forward and reverse reactions. (*See* **Fig. 4.** In this case, the polygon is a straight line.)
>
> **Step 2.** Write all possible patterns containing one less line than the number of enzyme species.

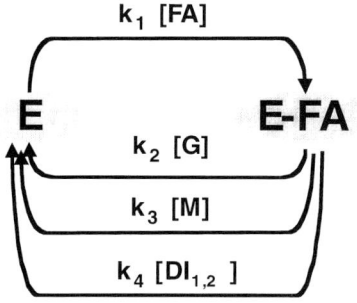

Fig. 4. Working diagram for King–Altman analysis.

Step 3. Express the concentration of each enzyme species as a fraction of the total enzyme present $[E]_t$. This fraction is given by the sum of the labels of all the indicated patterns by which the particular species of interest is produced, divided by the sum of the labels of all reactions of the various enzyme forms:

$$\frac{[E]}{[E]_t} = \frac{k_2[G] + k_3[M] + k_4[DI_{1,2}]}{k_1[FA] + k_2[G] + k_3[M] + k_4[DI_{1,2}]}$$

$$\frac{[E-FA]}{[E]_t} = \frac{k_1[FA]}{k_1[FA] + k_2[G] + k_3[M] + k_4[DI_{1,2}]}$$

Step 4. Write the differential equations in terms of the rate expressions for each mechanistic step:

$$\frac{d[M]}{dt} = k_2[E-FA][G] - k_3[E-FA][M]$$

$$\frac{d[DI_{1,3}]}{dt} = k_3[E-FA][M] + k_6[DI_{1,2}] - k_5[DI_{1,3}]$$

$$\frac{d[DI_{1,2}]}{dt} = k_5[DI_{1,3}] - k_6[DI_{1,2}] - k_4[E-FA][DI_{1,2}]$$

$$\frac{d[\text{T}]}{dt} = k_4[\text{E} - \text{FA}] \, [\text{DI}_{1,2}]$$

Combination of the equations developed in steps 3 and 4 gives:

$$\frac{d[\text{M}]}{dt} = \frac{[k_1 k_2[\text{G}] \, [\text{FA}] - k_1 k_3 \, [\text{M}] \, [\text{FA}]][\text{E}]_t}{k_1 \, [\text{FA}] + k_2 \, [\text{G}] + k_3 \, [\text{M}] + k_4 \, [\text{DI}_{1,2}]}$$

$$\frac{d[\text{DI}_{1,3}]}{dt} = \frac{k_1 k_3 \, [\text{M}] \, [\text{FA}][\text{E}]_t}{k_1[\text{FA}] + k_2 \, [\text{G}] + k_3 \, [\text{M}] + k_4 \, [\text{DI}_{1,2}]} + k_6 \, [\text{DI}_{1,2}] - k_5 \, [\text{DI}_{1,3}]$$

$$\frac{d[\text{DI}_{1,2}]}{dt} = k_5 \, [\text{DI}_{1,3}] - k_6 \, [\text{DI}_{1,2}] - \frac{k_1 k_4[\text{DI}_{1,2}] \, [\text{FA}][\text{E}]_t}{k_1 \, [\text{FA}] + k_2 \, [\text{G}] + k_3 \, [\text{M}] + k_4 \, [\text{DI}_{1,2}]}$$

$$\frac{d[\text{T}]}{dt} = \frac{k_1 k_4[\text{DI}_{1,2}] \, [\text{FA}][\text{E}]_t}{k_1 \, [\text{FA}] + k_2 \, [\text{G}] + k_3 \, [\text{M}] + k_4 \, [\text{DI}_{1,2}]}$$

In addition to the differential equations presented above, the concentrations of the various species as functions of time must also satisfy the following stoichiometric constraints resulting from material balances on the fatty acid and the glycerol backbone:

$$[\text{FA}] = [\text{FA}]_0 - [\text{M}] - 2[\text{DI}] - 3[\text{T}]$$
$$[\text{G}] = [\text{G}]_0 - [\text{M}] - [\text{DI}] - [\text{T}]$$

where [M], [DI], and [T] represent the instantaneous total concentrations of monoacylglycerols, diacylglycerols, and triacylglycerols, respectively. The subscript zero refers to the initial concentration of the species of interest.

Acknowledgments

This work was supported in part by the Colleges of Engineering and Agriculture and Life Sciences and the Center for Dairy Research of the University of Wisconsin-Madison through funding from Dairy Management Incorporated and by a postdoctoral fellowship to Jose A. Arcos provided by the Fundacion Ramon Areces. Additional support was provided by the United States National Science Foundation Grant BES-9320536 and the Spanish CICYT Grant Number BIO96-0837.

References

1. Zaks, A. and Klibanov, A. M. (1985) Enzyme-catalyzed processes in organic solvents. *Proc. Natl. Acad. Sci. USA* **82**, 3192–3196.
2. Gandhi, N. N. (1997) Applications of lipase. *J. Am. Oil Chem. Soc.* **74**, 621–634.
3. Cernia, E., Palocci, C., and Soro, S. (1998) The role of the reaction medium in lipase-catalyzed esterifications and transesterifications. *Chem. Phys. Lipids* **93**, 157–168.

4. Rice, K. E., Watkins, J., and Hill, C. G., Jr. (1999) Hydrolysis of menhaden oil by a *Candida cylindracea* lipase immobilized in a hollow fiber reactor. *Biotechnol. Bioeng.* **63,** 33–45.

5. Malcata, F. X., Hill, C. G., Jr., and Amundson, C. H. (1992) Hydrolysis of butteroil by immobilized lipase using a hollow-fiber reactor: Part II. Uniresponse kinetic studies. *Biotechnol. Bioeng.* **39,** 984–1001.

6. Malcata, F. X., Hill, C. G., Jr., and Amundson, C. H. (1992) Hydrolysis of butteroil by immobilized lipase using a hollow-fiber reactor: Part III. Multiresponse kinetic studies. *Biotechnol. Bioeng.* **39,** 1002–1012.

7. Malcata, F. X, Hill, C. G., Jr., and Amundson, C. H. (1993) Hydrolysis of butteroil by immobilized lipase using a hollow fiber reactor, Part V: Effects of pH. *Biocatalysis* **7,** 177–219.

8. Garcia, H. S., Qureshi, A., Lessard, L., Ghannouchi, S., and Hill, C. G., Jr. (1995) Immobilization of pregastric esterases in a hollow fiber reactor for continuous production of lipolyzed butteroil. *Lebensm. Wiss. Technol.* **28,** 253–258.

9. Arcos, J. A., Otero, C., and Hill, C. G., Jr. (1998) Rapid enzymatic production of acylglycerols from conjugated linoleic acid and glycerol in a solvent-free system. *Biotechnol. Lett.* **20,** 617–621.

10. Garcia, H. S., Storkson, J. M., Pariza, M. W., and Hill, C. G., Jr. (1998) Enrichment of butteroil with conjugated linoleic acid via enzymatic interesterification (acidolysis) reactions. *Biotechnol. Lett.* **20,** 393–395.

11. Hoerr, C. W. and Waugh, D. F. (1955) Some physical characteristics of rearranged lard. *J. Am. Oil Chem. Soc.* **32,** 37–41.

12. Marangoni, G. and Rosseau, R. (1995) Engineering triacylglycerols: the role of the interesterification. *Trends Food Sci. Technol.* **6,** 329–335.

13. Macrae, A. R. (1983) Lipase-catalyzed interesterification of fats and oils. *J. Am. Oil Chem. Soc.* **60,** 243A–246A.

14. Huang, K. H., Akoh, C. C., and Erickson, M. C. (1994) Enzymatic modification of melon seed oil: incorporation of eicosapentaenoic acid. *J. Agri. Food Chem.* **42,** 2646–2648.

15. Yamane, T., Suzuki, T., Sahashi, Y., Vikersveen, L., and Hoshino, T. (1992) Production of *n*-3 polyunsaturated fatty acid-enriched fish oil by lipase-catalyzed acidolysis without solvent. *J. Am. Oil Chem. Soc.* **69,** 1104–1107.

16. Arcos, J. A. and Otero C. (1996) Enzyme, medium and reaction engineering to design a low-cost, selective production method for mono- and dioleoylglycerols. *J. Am. Oil Chem. Soc.* **73,** 673–683.

17. Otero, C., Pastor, E., Fernández, V. M., and Ballesteros, A. (1990) Influence of the support on the reaction course of tributyrin hydrolysis catalyzed by soluble and immobilized lipases. *Appl. Biochem. Biotechnol.* **23,** 237–247.

18. Otero, C., Pastor, E., Fernandez, V., and Ballesteros, A. (1990) Synthesis of monobutyrylglycerol by transesterification with soluble and immobilized lipases. *Appl. Biochem. Biotechnol.* **26,** 35–44.

19. Chin, S. F., Liu, W., Storkson, J. M., Ha, Y. L., and Pariza, M. W. (1992) Dietary sources of conjugated dienoic isomers of linoleic acid, a newly recognized class of anticarcinogens. *J. Food Compos. Anal.* **5,** 185–197.

20. King, E. L. and Altman, C., (1956) Schematic method of deriving the rate laws for enzyme-catalyzed reactions. *J. Phys. Chem.* **60,** 1375–1378.
21. Segel, I. H. (1975) *Enzyme Kinetics.* Wiley, New York.
22. Cha, S. (1968) A simple method for derivation of rate equations for enzyme-catalyzed reactions under the rapid equilibrium assumption or combined assumptions of equilibrium and steady state. *J. Biol. Chem.* **243,** 820–825.
23. Cleland, W. W. (1975) Partition analysis and the concept of net rate constant as tools in enzyme kinetics. *Biochemistry* **14,** 3220–3224.
24. Garcia, H. S., Malcata, F. X., Hill, C. G., Jr., and Amundson, C. H. (1992) Use of *Candida rugosa* lipase immobilized in a spiral wound membrane reactor for the hydrolysis of milkfat. *Enzyme Microb. Technol.* **14,** 535–545.
25. Hill, C. G., Jr. (1977) *An Introduction to Chemical Engineering Kinetics and Reactor Design.* Wiley, New York, p. 255.
26. Garcia, H. S., Reyes, H. R., Malcata, F. X., Hill, C. G., Jr., and Amundson, C. H. (1990) Determination of the major free fatty acids in milkfat using a three-component mobile phase for HPLC analysis. *Milchwissenschaft* **45,** 747–759.
27. Selmi, B., Gontier, E., Ergan, F., and Thomas, D. (1997) Enzymatic synthesis of tricaprylin in a solvent-free system: lipase regiospecificity as controlled by glycerol adsoption on silica gel. *Biotechnol. Tech.* **11,** 543–547.
28. Malcata, F. X., Reyes, H. R., Garcia, H. S., Hill, C. G., Jr., and Amundson, C. H. (1990) Immobilized lipase reactors for modification of fats and oils-A review. *J. Am. Oil Chem. Soc.* **67,** 890–910.
29. Garcia, H. S., Yang, B. K., and Parkin, K. L. (1996) Continuous reactor for enzymic glycerolysis of butteroil in the absence of solvent. *Food Res. Int.* **28,** 605–609.
30. McNeill, G. P., Shimizu S., and Yamane T. (1990) Solid phase enzymatic glycerolysis of beef tallow resulting in a high yield of monogliceride. *J. Am. Oil Chem. Soc.* **67,** 779–783.

37

Solid–Gas Catalysis at Controlled Water Activity

Reactions at the Gas–Solid Interface Using Lipolytic Enzymes

Sylvain Lamare and Marie Dominique Legoy

1. Introduction

The development of nonaqueous biocatalysis in recent years has expanded the field of applications of enzymes and cells in new bioprocesses. Systems involving biocatalysts suspended in organic solvents, supercritical fluids, or gas phases have overcome numerous problems associated with the use of aqueous media such as low solubility of hydrophobic substrates, unfavorable thermodynamic equilibrium for many reactions, and thermal instability of the biocatalyst itself. Enzymes or cells, when used in low-water systems, were found to be efficient catalysts with the level of their activity being highly dependent on the degree of hydration. However, low mass transfer efficiency between liquid and solid phases in these nonconventional multiphase systems often remains a serious limitation. This chapter aims to discuss the utility of enzymes in bioreactors containing a mixture of substrates and water vapor as these reactors are relatively unexplored in comparison to those where enzymes are placed directly in aqueous or nonaqueous solvents.

Some years ago, Yagi and co-workers *(1)* have demonstrated that purified hydrogenase from *Desulfovibrio desulfuricans* in the dry state can bind hydrogen and render it activated. This results in parahydrogen–orthohydrogen conversion with no participation of aqueous protons in the reaction. Clearly, hydrogenase is a unique enzyme whose substrate is gaseous hydrogen. Nevertheless, several examples of utilization of the gas–solid system using either whole cells or isolated enzymes can be found in the literature *(2–4)* and have recently been reviewed *(5)*. Furthermore, gas–solid bioreactors have already been developed for the use of enzymes whose natural substrates are

From: *Methods in Biotechnology, Vol. 15: Enzymes in Nonaqueous Solvents: Methods and Protocols*
Edited by: E. N. Vulfson, P. J. Halling, and H. L. Holland © Humana Press Inc., Totowa, NJ

nongaseous (e.g., horse liver alcohol dehydrogenase *[6]*, *Sulfolobus solfataricus* alcohol dehydrogenase *[7]*, *Pischia pastoris* alcohol oxydase *[8–10]*, and, finally, lipolytic enzymes *[11–15]*). Numerous applications can be envisaged for this technology. The production of volatile compounds (aldehydes, ketones, and esters) for the food and especially the aroma and fragrances industry is one of them. However, the scope for potential applications of solid–gas biocatalysis is much wider. For example, biocatalytic treatment/detoxication of industrial waste gases and the development of new specific biosensors are also of great interest. In this chapter, a transesterification (alcoholysis) reaction between propionic acid methyl ester and *n*-propanol at controlled water activity (a_w) in a continuous gas–solid reactor catalyzed by a *Fusarium solani pisii* cutinase will be described in some detail and the extension of this approach to reactions of hydrolysis and esterification will be briefly discussed.

2. Materials

1. The experimental setup for performing solid–gas biocatalysis is depicted in **Fig. 1**. The bioreactor is composed of two ovens, one for the formation of the gas and the other for enzymatic reaction. Nitrogen is used as a carrier gas. The gas is dried over silica gel and molecular sieves before it is divided into four lines, regulated by electronic mass flow meters (Brooks, model 5850 E). All the lines are made using 3.18 × 2.10 mm (outer diameter [O.D.] × inner diameter [I.D.]) stainless steel tubing and the connections are made with Swagelock connectors. Each line is equipped with a 1.5-m-long coil in order to heat the gas to the desired temperature prior to it entering the saturation flasks. Three of the lines are used for the saturation of gas with substrates and water that are contained in different flasks and the fourth line is used for adjusting partial pressures of the components in the mixing chamber prior to entering the bioreactor. The temperature in each saturation flask is measured via thermocouples K (0.1°C resolution). At the bioreactor level, the entering gas is heated or cooled using a 4-m-long coil in order to adjust the thermodynamic activity of each component. One thermocouple K and a pressure gauge are placed in the gaseous flow just upstream of the bioreactor in order to have a better control of the activities. The bioreactor itself is composed of a 9-cm-long glass tube (6.5 mm O.D, 3.5 mm I.D) in which biocatalyst is packed between two layers of glass wool. The gas leaving the reactor is then filtered using an on-line 7-µm gas filter and directed (passing through a six-way valve for injection in a gas chromatograph) into a condenser composed of a cold trap.
2. Five hundred to 1000 hydrolytic units of free, adsorbed, or covalently immobilized *Fusarium solani pisii* cutinase (*see* **Notes 1–4**).
3. Analytical-grade propionic acid methyl ester and *n*-propyl alcohol (gas chromatograph [GC] purity >99%) dehydrated over 5 Å molecular sieves, and pure water (MilliQ grade).
4. Compressed nitrogen of at least 99.99% purity for use as a carrier gas (*see* **Note 5**).

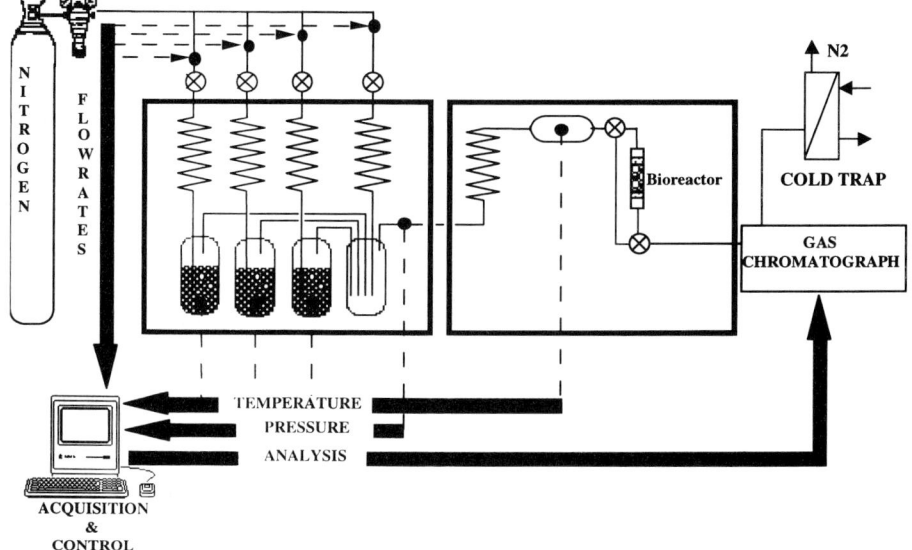

Fig. 1. Experimental setup.

3. Methods

1. Enzyme preparation should be loaded in the reactor just before starting the experiment in order to avoid moisturizing of the dehydrated biocatalyst. Alternatively, it can be stored over silica gel or P_2O_5. The charge of enzyme is placed between two deactivated glass wool layers. Care must be taken during the loading of the reactor to ensure the homogeneity of the catalytic bed. To this end, realize the first stopper in the reactor glass tube when packing uniformly 25 mg of dimethylchlorosilane (DMCS)-treated glass wool (Chrompack, France). Add the catalyst and pack the catalytic bed by patting the reactor gently to avoid inhomogeneity. Then, realize the second glass wool stopper and place the Swagelock connectors on the glass tubing using either graphite or polytetrafluoroethylene (PTFE) ferules. Connect the reactor to its oven.

2. Fill three saturation flasks with the ester and alcohol (previously dehydrated) and water, respectively (*see* **Note 6** for details). Connect them to the substrate oven. Activate the three-way valves in order to connect the inlet and outlet ports of the reactor.

3. Flush all the parts of the bioreactor setup with dried nitrogen using the dilution line for 30 min at 20 mLn/min (mLn: normal milliliter at 0°C and 1 atm) and set the two ovens at appropriate temperatures (*see* below).

4. Determination of operational parameters. As molar concentrations are not particularly useful when working at the gas–solid interface, operational parameters are expressed in terms of thermodynamic activities (a_X) for any compound X and as a_w for water activity. Note that a_X can have any value between 0 and 1 and implies the "availability" of the compound *(16)*.

Transesterification (alcoholysis) between *n*-propanol and propionic acid methyl ester leading to the formation of propionic acid propyl ester and methanol is taken as a representative biotransformation (for more details and explanations, refer to **refs.** *(11–14)*. The reaction is carried out at 70°C, with a water activity of 0.6 (*see* **Note 17**) and activities for the ester (a_{ester}) and the alcohol ($a_{alcohol}$) being equal to 0.2. The total flow rate is set at 1000 μmol/min of gas and the absolute working pressure is 1 atm. Resolution of relevant equations (*see* **Note 8**) leads to the following operating parameters: temperature of saturation flasks, 72.3°C; carrier gas flows through the alcohol, ester and water saturation flasks, 2.64, 0.98 and 9.48 mLn/min, respectively; carrier gas flow in the dilution line, 0.08 mLn/min; partial pressures for the alcohol, ester and water, 0.064, 0.143, and 0.183 atm, respectively. Thus, molar flows of each compound for the alcohol, ester, water, and nitrogen are 64, 143, 183, and 610 μmol/min, respectively. The total volumetric flow is then equal to about 28.2 mL/min.

5. Isolate the reactor using the two three-way valves at the inlet and at the outlet ports of the packed bed and set the different nitrogen flow meters at their appropriate level (*see* **Note 9**). Wait for 30 min and check that the composition of the gas phase is stable using gas chromatography (GC) (*see* **Note 10**).

6. Once the composition of the gas phase is stabilized, the two three-way valves of the reactor are activated, thus allowing the gas to pass through the catalytic bed. The reaction rate, as well as productivity of the bioreactor (*see* **Note 11**) can be determined by analysis of the gas at the outlet of the reactor (*see* **Note 10**).

7. Products are recovered by using a cold trap at the end of the line. The recovered products can then be analyzed and, if necessary, purified at the end of the experiment.

8. In order to reuse the biocatalyst in consecutive runs, the reactor must be thoroughly rinsed. To this end, shut down all the carrier flows, except the one used for the dilution in the mixing chamber. Isolate the saturation flasks. Set the flow rate of pure nitrogen at 5–10 mL/min. Analyze the outlet gas by GC. Once no residual matter is detected, continue flushing the reactor with dried gas for 2 h in order to ensure sufficient dehydration of the catalyst. Dismantle the reactor, adapt two caps to the glass tube, and store at 4°C over silica gel or P_2O_5.

9. Other reactions can also be carried out at the gas–solid interface using lipolytic enzymes. However, operational conditions have to be tested and optimized in every case and, especially when using free carboxylic acids as substrates or when an acid is one of the reaction products. A high water activity (>0.3) combined with high acidity often result in a rapid denaturation of biocatalysts. This denaturation was found to be irreversible. Nevertheless, esterification reactions using acetic, propionic, butyric, iso-butyric, and valeric acids and different C1–C5 alcohols have been carried out successfully with the operational stability of enzyme being sufficient for the development of an industrial process. Hydrolysis can also be performed, but because of high-a_w requirement for this reaction, the temperature in the bioreactor must be lowered to minimize thermal denaturation of the enzyme. Note that the denaturation can also be the result of the release of acid into the medium.

4. Notes

1. *Fusarium solani pisii* cutinase cloned and expressed in *Escherichia coli* was a gift from Corvas (Ghent, Belgium). Fungal cutinases are monomeric hydrolytic enzymes of low molecular weight (25 kDa). They display high hydrolytic activity toward both long- and short-chain fatty acid esters. The catalytic triad comprises Ser 120, His 188, and Asp 175. The optimal pH of *Fusarium* cutinase is slightly basic (8.5–9.5) and it loses half of its catalytic activity in 2.5 h at 80°C at pH 7.5 in phosphate buffer. A highly purified enzyme preparation (>98% as judged by sodium dodecyl sulfate electrophoresis [Pharmacia PhastSystem, France]) was used in this work. Other enzymes sources can also be used successfully e.g., commercial immobilized lipases, Novozyme 435 or Lipozyme IM20 (both Novo-Nordisk, Denmark), or *Candida rugosa* lipase (Sigma Chemical Co, France). However, for use in the solid–gas system, the latter enzyme must be purified and immobilized by adsorption as described in **Note 2**.

2. Chromosorb P Acid Washed (Prolabo, Rhône Poulenc, France), mesh 30/60 can be used for enzyme adsorption. Typically, an enzyme (50 mg) is dissolved in 20 mM phosphate buffer pH 7.5 (3 mL), and dry Chromosorb P (1.5 g) is added to the solution. After vigorous shaking for 5 min, the preparation is placed in a vacuum dessicator and dried overnight. The resulting enzyme-loaded Chromosorb is stored at 4°C over silica gel. Hydrolytic activity of the preparation is measured as described in **Note 4**.

3. The covalent attachment of the cutinase to a cation-exchanger resin (Ion Exchanger IV [mesh 20/50], Merck, France) is carried out using 1-cyclohexyl-3-(2-morpholinoethyl)carbodiimide metho-*p*-toluenesulfonate as the coupling agent. Previously dried resin (1 g) is added to a 5-mM carbodiimide solution prepared in 25 mM phosphate buffer, pH 8.0 (20 mL). The preparation is shaken at room temperature for 15 min and the resin is filtered and washed with MilliQ water (1 L). The resin is then packed into a chromatographic column (LKB, France, 20 cm length, 2.5 cm I.D) and percolated in a closed loop with 25 mM phosphate buffer pH 8.0 (25 mL), containing purified cutinase (10 mg). The flow rate of 5 mL/min is applied using a peristaltic pump (Minipulse 2 ,Gilson, France). The covalent binding of the protein to the resin is monitored as a decrease in absorbance at 280 nm by circulating the enzyme solution through a cuvet placed in a spectrophotometer (Uvikon 930, Kontron, France). When the absorbance is stabilized (typically a plateau is reached after passing the enzyme solution through the column for 1 h), the column is rinsed with water (500 mL) and the excess of carbodiimide is neutralized by passing a 200 mM glycine solution for 20 min. The resin is then rinsed with water (500 mL), collected and dried under vacuum overnight. The dry material (the protein loading is 5–6 mg of enzyme per gram of resin) is placed in a sealed vessel and stored at 4°C over silica gel. Hydrolytic activity of the preparation is measured as described in **Note 4**.

4. The hydrolytic activity of the different enzyme preparations is determined by a pH-stat method (Metrohm, Switzerland), using a 150-mM solution of triacetin in water as substrate. Experiments are carried out at 30°C and pH 8.0. A substrate

solution (3.0 mL) is placed in the pH-stat vessel, and after setting the initial pH at 8.0, different amounts of solid immobilized cutinase or, in the case of free enzyme, 10 μL of a 1-mg/mL enzyme solution are added. The acetic acid produced is titrated with a freshly prepared 25 mM NaOH solution.

Typical hydrolytic activities of the cutinase preparations are as follows:

Free enzyme: 880 U/mg of protein

Adsorbed enzyme: 36 U/mg of protein

Covalently bound enzyme: 78 U/mg of protein

(1 U is defined as 1 μmol of acetic acid liberated per min)

5. Nitrogen is preferred to air for two main reasons. First, substrates and products can be flammable, and the use of nitrogen minimizes the risks of explosion or fire. Second, solid–gas catalysis is always carried out at relatively high temperatures and the absence of oxygen greatly minimizes the inactivation of the biocatalyst through the combined deleterious effects of oxygen and temperature. Before use, all the tubing in the reactor must be passivated by flushing it with pure nitrogen.

6. For satisfactory saturation of the carrier gas, it is recommended to use flasks such as the one depicted in **Fig. 2**. This vessel allows good saturation at high flow rates of the carrier gas to be achieved. Typically, these flasks contain about 150 mL of pure substrate (50 mL in the first chamber and 100 mL in the second one) and can be operated at flow rates of up to 20 mL/min. The other advantage of this design is that most of the saturation occurs at the first stage. Because each compound has a different enthalpy of vaporization, it is easier to obtain the same temperature in the first oven when the energy consuming vaporization is minimal in the second stage.

7. As the equilibrium binding of water molecules to proteins is determined by water activity, it is one of the key parameters in the system. Thus, one has to be aware of the effect of alterations to the hydration level. Numerous studies in liquid systems showed that the amount of water associated with the enzyme correlates well with the retention of catalytic activity. Furthermore, enzymes in dehydrated (or partially hydrated) form are more resistant to thermal inactivation. Thus, water plays a dual (and antagonist) role. On the one hand, it helps to maintain the catalytically competent conformation of enzymes by giving some flexibility to the protein molecules (i.e., enhances the catalytic rate). On the other hand, water is involved in the reactions leading to thermal inactivation of the biocatalyst. By controlling the total amount of water available in the microenvironment of the enzyme, an optimal degree of hydration can be established and applied with the view of achieving good catalytic rate and thermal stability *(11–14)*. Further information on the effects of a_w on the properties of enzyme in nonaqueous systems can be found in a recent review *(16)*.

8. One of the key parameters in the solid–gas biocatalysis is the "availability" of different chemical species in the gas phase to the enzyme. The thermodynamic parameter corresponding to each "availability" is the activity of the compound. Because of the lack of sensors for on-line measurements, a calculation method is needed in order to obtain numerical values. For each compound, determination of its thermodynamic activity requires the knowledge of its partial pressure in the

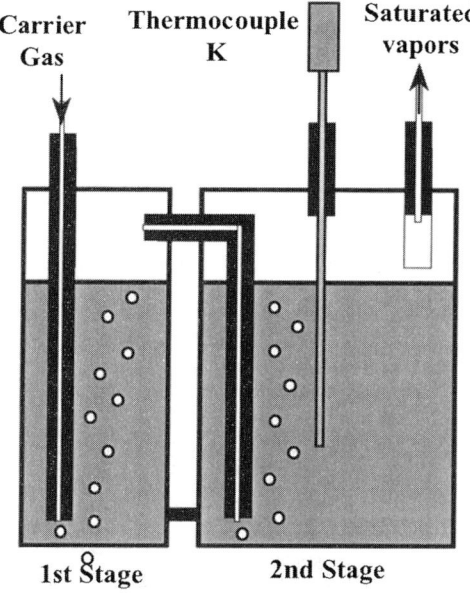

Carrier Gas   Thermocouple K   Saturated vapors

1st Stage   2nd Stage

Fig. 2. Saturation flask.

reaction gas phase and access to the vapor pressure curve as a function of temperature. One can assume that a carrier gas, after bubbling through a substrate solution, is in equilibrium with the liquid phase and, therefore, the partial pressure of the substrate in the gas leaving the saturation apparatus is equal to the vapor pressure corresponding to the saturation pressure above the pure compound. In order to calculate the composition of the gas, different molar flows for each compound (carrier + substrates + water) have to be known.

The carrier flow in each line is calculated using:

$$Q_{N2}^n = \frac{Q_V^n \, N_2 \text{ normalized}}{RT} \quad \text{(mol/h for } T = 273.15 \text{ K)} \tag{1}$$

Then, with the knowledge of the molar flow of the carrier gas, it is possible to calculate the different molar flows in the outlet of the saturation flasks using the equation that includes saturation pressure:

$$Q_X^n = \frac{P_X^n \text{sat}}{Q_{N2}^n \, (\text{Pa} - P_X^n \text{sat})} \quad \text{(mol/h)} \tag{2}$$

From these, the partial pressure of each compound in the gas entering the bioreactor is determined by:

$$P^n_{PX} = \frac{Q^n_X}{\overset{n}{\Sigma} (Q^n_{N2} + Q^n_X)} \text{ Pa (atm)} \qquad (3)$$

and the activity of each compound in the reactor stage is calculated as follows:

$$a_x = \frac{P^n_{Px}}{P_{P^n_X}\text{sat at the temperature of the bioreactor}} \qquad (4)$$

For the foregoing calculations, the following parameters are used:

R	Ideal gases constant (8.314 J/mol.K)
T	Temperature (K)
Pa	Absolute pressure in the system (atm)
$Q^n_{N_2}$normalized	Normalized volumetric flow of carrier gas in line n (L/h at 273 K and 1 atm)
$Q^n_{N_2}$	Molar flow of carrier gas in line n (mol/h)
Q^n_X	Molar flow of substrate X in line n (mol/h)
Pp^n_Xsat	Saturation pressure of compound X in line n (atm)
Pp^n_X	Partial pressure of compound X in the gas entering the bioreactor (atm)
a_X	Activity of compound X in the bioreactor (dimensionless)
Q_{Vtotal}	Total volumetric flow in the bioreactor (L/h)
Q_{Vtotal}	Total molar flow in the bioreactor (mol/h)

To determine Pp^n_X sat, see **Note 12**.

Assuming that the applied conditions are far from the critical temperature and the critical pressure for each compound, and using the ideal gas law $PV = nRT$, one can have a good estimate of the total volumetric flow and the residence time in the bioreactor:

$$Q_{Vtotal} = \frac{RT \overset{n}{\underset{i=1}{\Sigma}}(Q^n_{N2} + Q^n_X)}{Pa} \text{ (L/h at } T = \text{temperature of the bioreactor in K)} \qquad (5)$$

As a consequence of these calculations, special care should be taken to ensure accurate control of the physical parameters, such as temperatures in all stages of the reactor, absolute pressure, and the molar flows of the carrier gas.

9. When no carrier gas is passed through the lines, the saturation flasks will be equilibrated at the temperature of the saturation oven. However after applying the flow, the temperature in the saturation flasks may decrease because of insufficient heat transfer to counteract the loss of energy resulting from vaporization of substrates. If so, the oven temperature must be corrected and monitored until the desired value in the final saturation chamber is obtained. In some cases, when the three flasks give three different stable temperatures, flow rates of the carrier gas must be recalculated.

10. For analyses, the vapor phase leaving the reactor is sampled using a 0.25-mL loop on a six-way valve (Valco) maintained at 150°C. Samples are automatically injected

in the split injector of a gas chromatograph (Hewlett Packard, model 5890 A) equipped with a thermal conductivity detector (TCD) for the detection of water and a flame ionization detector (FID) for the analysis of all other products. An CP-Sil 19-CB fused-silica capillary column (25 m × 0.32 mm I.D. × 1.2 μm film thickness, Chrompack, France) is used. The split ratio is 25:1. The injector is kept at 200°C and detectors are kept at 250°C. The column temperature is held at 50°C for 1 min, then programmed to increase to 110°C at 10°C/min and kept for 2 min at this temperature. The carrier gas is nitrogen and the flow rate is 2 mL/min. Hydrogen and air are supplied to the FID at 30 and 300 mL/min, respectively. Quantitative data are obtained after integration on a HP 3396A integrator.

11. Note that thermodynamic activity is an equilibrium parameter. Thus, for meaningful measurements, a certain equilibration time prior to the analysis is necessary for the solid phase to equilibrate with the gas phase. This time depends on the quantity of the solid phase packed in the reactor and the molar flows of different compounds. For example, in the case of enzyme hydration with water, the equilibration time can be assessed from the isotherm sorption curve for the enzyme preparation (*see* **Note 13**) and the molar flow of water through the reactor. For a transesterification, where water does not participate in the reaction, the amount of water in the gas phase entering and leaving the reactor at equilibrium should be the same. This can be easily monitored by GC and, once this is achieved, it is assumed that the system is equilibrated. When all the compounds present in the gas phase participate in the reaction, calibration curves for all of them have to be constructed as a function of partial pressure (*see* **Note 14**). Then, a correct mass balance between the inlet and the outlet gas is a good indicator that the solid and the gas phases are in equilibrium.

12. If saturation pressure curves for pure compounds are not available in the literature, determination of vapor pressure curves as a function of temperature can be carried out using a MiniVap VP apparatus (Grabner, Austria). In a typical experiment, vacuum is applied over a closed cell with an accurate temperature regulation. Once the zero absolute pressure is reached in the vessel and the temperature is at the low starting point value, a sample of pure compound (1.0 mL) is injected into the cell, using a three-way valve. Then, the temperature gradient is programmed to reach the end point selected and continuous measurements of temperature and pressure in the cell are performed. The data (Psat and T) are recorded and the curve is fitted using the exponential formula for determining saturation pressure at a given temperature in the range tested.

13. Isotherm sorption curves can be determined using an Autosorb system (Biosystemes, France). A dried preparation of supported (100 mg) or free (20 mg) enzyme is deposited into a glass container that is placed in a temperature/water activity controlled chamber. The temperature can be varied in the range from 10° to 80°C. The enzyme sample should be previously incubated at $a_w = 0$ for at least 48 h under dried nitrogen. The water activity is programmed to reach different plateaus using increments of 0.1, and once a new plateau is reached, the corresponding water activity is maintained stable for 48 h. Every hour, the water

content is determined by weighing on an electronic balance in the chamber until the weight is stabilized for all the samples loaded in the carousel. The water content for each a_w is then calculated. When working with dehydrated enzymes, the isotherm sorption curve is only constructed by increasing the water activity. Isotherm sorption curves can also be determined manually by gravimetric methods and equilibration of the sample over saturated salt solutions *(17)*.

14. The calibration procedure requires special attention. As a fixed volume is injected in the gas chromatograph and because gaseous samples are subject to variations of temperature or pressure, the calibration must be performed under conditions close to those used in the experiment. The total flow, the resulting back-pressure, and the temperature of the sampling loop must be the same under all the conditions tested. Then, different gas compositions can be created by mixing the flows from the three saturated nitrogen lines in various proportions. Calculations presented in Note 8 allow the determination of the respective partial pressures in the gas phase. If the back-pressure is significantly different in different experiments, a correction and normalization on the absolute pressure must be applied. Knowing that,

$$P_{PX}^n = \frac{Q_X^n}{\sum^n (Q_{N2}^n + Q_X^n)} \text{ Pa (atm)} \tag{6}$$

response factors can be determined from the curve $\text{Area}_X = f[Pp_X]$. The response factors for products of the reactions can be obtained by injecting liquid standards made to contain known amounts of nontransformed substrates and products.

References

1. Yagi, T., Tsuda, M., Mori, Y., and Inokuchi, H. (1969) Hydrogenase activity in the dry state. *J. Am. Chem. Soc.* **91**, 2801.
2. Kim, C. and Rhee, S.-K. (1992) Enzymatic conversion of ethanol into acetaldehyde in a gas–solid bioreactor. *Biotechnol. Lett.* **14**, 1059–1064.
3. Uchiyama, H., Oguri, K., Yagi, O. and Kokufuta, E. (1992) Trichloroethylene degradation by immobilized resting-cells of *Methylocystis sp.* M in a gas–solid bioreactor. *Biotechnol. Lett.* **14**, 619–622.
4. Zilli, M., Converti, A., Lodi, A., Del Borghi, M., and Ferraiolo, G. (1992) Phenol removal from waste gases with a biological filter by *Pseudomonas putida*. *Biotechnol. Bioeng.* **41**, 693–699.
5. Lamare, S. and Legoy, M. D. (1993) Biocatalysis in the gas phase. *TIBTECH* **11**, 413–418.
6. Pulvin, S., Legoy, M. D., Lortie, R., Pensa, M., and Thomas, D. (1986) Enzyme technology and gas phase catalysis: alcohol dehydrogenase example. *Biotechnol. Lett.* **8(11)**, 783,784.

7. Pulvin, S., Parvaresh, F., Thomas, D., and Legoy, M. D. (1988) Solid–gas reactors. A comparison between horse liver and the thermostable *sulfolobus solfataricus* ADH. *Ann. NY Acad. Sci.* **545,** 434–439.

8. Barzana, E., Klibanov, A., and Karel, M. (1987) Enzyme-catalysed gas-phase reactions. *Appl. Biochem. Biotechnol.* **15,** 25–34.

9. Barzana, E., Klibanov, A., and Karel, M. (1989) Enzymatic oxidation of ethanol in the gaseous phase. *Biotechnol. Bioeng.* **34,** 1178–1185.

10. Barzana, E., Klibanov, A., and Karel, M. (1989) A colorimetric method for the enzymatic analysis of gases: the determination of ethanol and formaldehyde vapors using solid alcohol oxydase. *Anal. Biochem.* **182,** 109–115.

11. Lamare, S. and Legoy, M. D. (1995) Working at controlled water activity in a continuous process: the gas/solid system as a solution. *Biotechnol. Bioeng.* **45,** 387–397.

12. Lamare, S. and Legoy, M. D. (1995) Solid–gas biocatalysis: how to fully define the system? *Biotechnol. Tech.* **9,** 127–132.

13. Lamare, S. (1996) Engineering of/with lipases: importance of water in new biocatalytic processes: the solid/gas catalysis example, in *Engineering of/with Lipases* (Malcata, F. X., ed.), NATO ASI Series, **317,** 357–390.

14. Lamare, S. and Legoy, M. D. (1997) Kinetic studies of a *Fusarium solani pisii* cutinase used in a gas/solid system. Hydrolysis and transesterification reactions. *Biotechnol. Bioeng.* **57,** 1–8.

15. Lamare, S. and Legoy, M. D. (1998) Procédé réactionnel par catalyse solide/gaz en milieu non conventionnel, réacteur correspondant et utilisation de ce réacteur. French Patent PCT/FR98/01592.

16. Halling, P. (1994) Thermodynamic predictions for biocatalysis in nonconventionnal media: theory, tests and recommendations for experimental design and analysis. *Enzyme Microb. Technol.* **16,** 178–206.

17. Halling, P. (1992) Salt hydrates for water activity control with biocatalysts in organic media. *Biotechnol. Tech.* **6,** 271–276.

38

Solvent-Free Biotransformations of Lipids

Tsuneo Yamane

1. Introduction

Although it has not been explicitly recognized in the literature, biotransformations in organic media can be classified as either solvent-based or solvent-free reactions. In the former systems, one or more substrates are dissolved in an inert organic solvent that does not participate in the reaction in any respect, but to provide an environment for the enzyme to exert its action on the dissolved substrate(s). In a solvent-free system, no other compounds but substrate(s) and enzyme are present in a reactor. In principle, one substrate can be used in a large excess over another and, if so, it may also act as a solvent for other reactants. Examples of such "neat" biotransformations can be found even in the early literature *(1–3)*.

One of the big attractions of enzymatic solvent-free synthesis is potentially very high volumetric productivity. This, however, does not apply to all reactions, and in many instances, it may actually take longer to achieve the desired degree of substrate conversion in the absence of added solvent. In this case, volumetric productivity in the reactor [(mass of product formed)(reactor volume)$^{-1}$ h^{-1}] should be calculated for both solvent-based and solvent-free systems using the same volume of the reaction mixture and the same amount of the enzyme in order to make an economically justified choice between the two. Similarly, there is no risk of solvent-induced inactivation of the biocatalyst in a solvent-free system, but the overall loss of enzyme activity can still be significant if the reaction time is too long. It should be added that the avoidance of organic solvents is particularly advantageous to the food industry where stringent regulations related to the use of organic solvents are in force. Also, no fireproof and explosion-proof equipment/procedures are necessary for the solvent-free processing and the environment at the factory is less hazardous to the health of the workers.

From: *Methods in Biotechnology, Vol. 15: Enzymes in Nonaqueous Solvents: Methods and Protocols*
Edited by: E. N. Vulfson, P. J. Halling, and H. L. Holland © Humana Press Inc., Totowa, NJ

As enzymatic reactions in organic solvents, solvent-free systems can be classified in several groups depending on particular features of the biotransformation in question (**Table 1**). Thus, the reaction can be carried out with either free or immobilized enzymes with one or more substrates that are either used in nearly stoichiometric quantities or one is present in large excess. Further, solvent-free biotransformations can be implemented in a monophasic reaction mixture as well as in biphasic liquid–liquid and even biphasic liquid–solid systems. The latter, "solid-phase glycerolysis of fats," was extensively studied in the author's laboratory for many years *(4–9)*.

Currently, a number of biotransformations of low- to mid-value symmetrical fats and oils (e.g., 1,3-distearoyl-2-oleoyl-glycerol, 1,3-dioleoyl-2-palmitoyl-glycerol and 1,3-dibehenoyl-2-oleoyl-glycerol) are being produced in solvent-free systems. Similar solvent-free transesterifications are used for the manufacture of low-volume high-value food additives and pharmaceuticals, including simple alkyl and terpenyl esters such as ethyl isovalerate, heptyl oleate, geranyl acetate, and citronellyl acetate (flavors and fragrances), and glycerides enriched in polyunsaturated fatty acids (PUFAs) for biomedical applications *(10,11)*. Esters of C2–C8 alcohols (e.g., isopropyl myristate, isopropyl palmitate, and 2-ethylhexyl palmitate are currently produced commercially for applications in cosmetics and personal care products [see the review article by Vulfson et al. *(12)*]. Structured lipids containing mono-, di-, or tri-PUFAs are also expected to be manufactured in a solvent-free enzymatic process *(13,14)*, as triglycerides enriched with *cis*-5,8,11,14,17-eicosahexaenoic acid (EPA) or *cis*-4,7,10,13,16,19-docosapentaenoic acid (DHA) have numerous health benefits and are better adsorbed in the intestine *(15)*.

In addition to "conventional" solvent-free systems, where substrates and products are completely miscible liquids at the reaction temperature, two other more sophisticated approaches to increasing mutual miscibility or solubility of substrate(s) in the reaction mixture have been developed (*see* review by Vulfson et al. *[12]*). In these systems, one of the substrates is a solid at ambient temperature and is only sparingly soluble in another substrate (e.g., a sugar and fatty acids). To overcome the problem of solubility, a prior "hydrophobization" of the sugar moiety either by 1-*O*-alkylation (ethyl-, *n*- and *iso*-propyl or butyl-) or by acetonization (2,3-isopropylidene sugar, sugar acetal) is carried out and the product is esterifiied with molten fatty acids in a solvent-free process *(16–20)*. Alternatively Vulfson and coworkers used molten substrate(s), the crystallization of which under the reaction conditions was deterred for sufficient time to complete the biotransformation.

The enzyme performance in solvent-free substrate mixtures (as well as in conventional solvent-based biotransformation) is crucially dependent on the water content (or thermodynamic water activity, a_w) of the medium. Thus, a_w

Table 1
Classification of Solvent-Free Biotransformations

Number of phases:
      Monophasic (homogeneous, liquid)
      Biphasic (heterogeneous, liquid)
      Biphasic (heterogeneous, solid)
Number of substrates:
      One substrate
      Two substrates:
            Nearly stoichiometric quantities
            Excess amount of one substrate
Extended systems:
      Molten substrate/eutectic mixtures
      Substrates derivatized to increase mutual miscibility
Enzyme
      Free (suspended enzyme powder)
      Immobilized (on or within solid support)

in the system must be carefully controlled, as some water is essential for the maintenance of high catalytic activity of the bicatalyst (*see 21–23*) and other contributions to this volume). However, an excess of water in the system is undesirable, as it leads to decreased yields of product resulting from hydrolytic side reactions and/or an unfavorable position of equilibrium. Thus, in esterification and transesterification reactions, the removal of water and volatile alcohols, respectively, is necessary to shift the equilibrium in favor of the product formation. Strategies to achieve this include application of vacuum, bubbling of dry inert gas, and the use of selective adsorbents such as silica gel, molecular sieves, and so forth. Other methods to increase the product yield in reactions where (side) products are not volatile are available too: fractional crystallization (wintering) of saturated fatty acid liberated from acidolysis *(11)* and tautomerization of liberated vinyl alcohol when using vinyl esters of carboxylic acids as substrates for transesterifications *(24)*.

In the following sections, lipase-catalyzed solvent-free biotransformations of lipids are exemplified by two reactions. One is transesterification between EPA ethyl ester (EPAEE) and tricaprylin to yield 1(3)-EPA-2,3(1,2)-dicaprylin *(25)*. This is a typical monophasic solvent-free system:

$$\text{EPAEE} + C_8 \left[\begin{array}{c} C_8 \\ \\ C_8 \end{array} \right. \longrightarrow C_8 \left[\begin{array}{c} \text{EPA} \\ \\ C_8 \end{array} \right. + \text{C8-EE} \qquad (1)$$

The other is the biphasic solvent-free esterification of glycerol with free fatty acid (FFA) to give to 1,3-diacylglycerol *(26)*:

$$\text{HO} \underset{\text{OH}}{\overset{\text{OH}}{\Big[}} + 2\text{FFA} \quad \longrightarrow \quad \text{HO} \underset{\text{FA}}{\overset{\text{FA}}{\Big[}} + 2\text{H}_2\text{O} \qquad (2)$$

2. Materials

1. Lipozyme IM60 (*Rhizomucor miehei* lipase immobilized on an anion-exchange resin) was provided by Novo-Nordisk Bioindustry Ltd. (Chiba, Japan) (*see* **Note 1**).
2. EPAEE was kindly donated by Nippon Suisan Co., Ltd. (Tokyo, Japan) (*see* **Note 2**).
3. All other substrates (i.e., glycerol, triglycerides and fatty acids), organic solvents (methanol, chloroform), and salts were reagent grade and can be obtained from traditional suppliers.

3. Methods

3.1. Adjustment of Initial a_w of Lipase and Substrate

To obtain the desired initial a_w, both the enzyme and substrate(s) are equilibrated over a saturated salt solution in a closed desiccator for 2 d at room temperature (i.e., about 25°C) (*see* **Note 3**).

3.2. Reaction Vessel

The reaction is carried out in a flat-bottomed glass bottle or a vial with a magnetic stirrer bar. The bottle is tightly stoppered to prevent the absorption of moisture from the atmosphere and placed into a water-jacketed round vessel made of glass. This setup is positioned on a magnetic stirrer. The temperature is maintained by circulating thermostatted water through the jacket.

3.3. Control of Vacuum

Application of vacuum is a convenient method to remove water and/or volatile by-products on a laboratory scale. This is achieved by connecting a conventional vacuum pump to the reaction vessel. To monitor the vacuum, a Pirani gage sensor probe (AVP 202N13, Okano Works Ltd., Osaka, Japan) is inserted in the vacuum tube between the pump and the reaction vessel. A digital meter connected to the sensor gives pressure reading in a range of 2.0×10^3 to 2.0×10^{-1} Pa (15 – 0.001 torr). A high-precision, manually operated needle valve is used to control the vacuum. The reading/adjustments are made every several hours, as pressure in the reaction vessel tends to decrease gradually during batch reactions (*see* **Note 4**).

3.4. Analysis of the Reaction Mixture

In the transesterification between EPAEE and tricaprylin, 9–10 different chemical species, including positional isomers, are formed in the reaction mixture. They can be successfully separated and analyzed by silver-ion high-performance liquid chromatography (HPLC) *(25)*. The incorporation of EPA in glycerides can be monitored quantitatively by high-temperature gas chromatography using on-column injection *(25)*. For quantitative analysis of various acylglycerols formed in the esterification reaction between glycerol and FFA, a thin-layer chromatography/flame-ionization detector (TLC/FID) (Iatroscan MK-5, Iatron Laboratories, Tokyo, Japan) is most convenient *(27)*.

3.5. Determination of Water Content

The water content of the reaction mixture is determined by Karl Fischer titration (MKS-1, Kyoto Electronics Ltd., Kyoto, Japan).

3.6. Preparation of Structured Triglycerides Containing EPA by Transesterification (25)

Values of a_w for both enzyme and tricaprylin are adjusted to 0.11 over saturated LiBr solution prior to the reaction (*see* **Note 5**). The transesterification reaction is performed with 1.41 g (3 mol) of tricaprylin and 0.33 g (1 mol) of EPAEE using 50 mg of Lipozyme. The substrates are first mixed in a closed vial with a magnetic stirrer. The reaction is initiated by adding the lipase, and the vial is purged with N_2 gas. The incubation is carried out at 40°C with stirring at 400 rpm. After 2 h incubation at atmospheric pressure, a vacuum (3 mm Hg) is applied. Typically, approx 95 mol% of EPAEE is incorporated into the triglyceride in 24 h. The final reaction mixture contains the desired product (1-EPA-2,3-C8; *see* **Note 6**), ethyl caprylate (by-product), and tricaprylin (remaining substrate). The final product is purified by column chromatography.

3.7. Preparation of Symmetrical 1,3-Diacylglycerol by Esterification of Free Fatty Acids with Glycerol (26)

The esterification is performed in a closed vial with 25 mmol free fatty acid (FFA), 12.5 mmol glycerol, and 4% (w/w) Lipozyme. The reaction is carried out at atmospheric pressure for 1 h, prior to applying vacuum (3 mm Hg) or bubbling dry N_2 gas (0.7 L/min) through the reaction mixture.

When caprylic acid is used as the substrate at 30°C (*see* **Note 7**), the molar yield of 1,3-dicapryloylglycerol reaches maximum (84.6%) after 8 h and decreases thereafter, owing to the formation of tricaprylin (98% conversion after 12 h). 1,2-Dicapryloylglycerol is present at low levels throughout the reaction (<1%) and 2-monocapryloylglycerol is not detected. The water content of the reaction mixture is reduced from 4.5% at the beginning to 0.45% at the end. When the molar ratio of caprylic acid to glycerol is increased from 2:1 to 4:1, up to 98% of 1,3-dicapryloylglycerol with less than 1% of 1-monocapryloylglycerol and tricaprylin is obtained after extraction (*see* **Notes 8** and **9**). The pure product is recovered by evaporating the extraction solvent.

4. Notes

1. In Europe, the immobilized enzyme is available from Novo-Nordisk A/S, Bagsvaerd, Denmark.
2. EPAEE must be stored at −80°C in small (0.5–1.0 g each) sealed ampoules filled with N_2 gas because PUFAs are very prompt to oxidation.
3. Different a_w values can be obtained by using different saturated salt solutions: 0.11 (LiBr), 0.12 (LiCl), 0.20 (CH$_3$COOK), 0.31 (MgCl$_2$·6H$_2$O), 0.44 (K$_2$CO$_3$), 0.55 [Mg(NO$_3$)$_2$·6H$_2$O], 0.73 (NaCl), 0.76 (Na$_2$SO$_4$·10H$_2$O+Na$_2$SO$_4$), 0.86 (KCl), and 0.90 (ZnSO$_4$·7H$_2$O).
4. An automatic pressure controller is commercially available from the same manufacturer, but it is relatively expensive.
5. Strictly speaking, the a_w of the enzyme and both substrates should be adjusted to the same values prior to the reaction. However, to avoid oxidation of EPAEE, only the enzyme and tricaprylin are incubated over the saturated LiBr solution.
6. The absolute steric configuration has not yet been elucidated; the product may be racemic.
7. The reaction temperature depends on the melting point of FFA used, typically ranging from 25°C to 45°C.
8. The reaction mixture is first dissolved in chloroform/methanol (2/1 v/v). The immobilized enzyme is removed by centrifugation and glycerol is extracted in water. For effective recovery of the product, the water layer is twice re-extracted with chloroform, as partial glycerides are, to some extent, water soluble. The two chloroform extracts are combined with the initial chloroform/methanol solution-containing products.
9. Almost the same molar yield and purity of the product are obtained when using capric instead of caprylic acid. With other FFAs, the corresponding symmetrical 1,3-diacylglycerols are obtained in lower (60–85%) yields.

References

1. Zaks, A. and Klibanov, A. M. (1984) Enzymatic catalysis in organic media at 100°C. *Science* **224**, 1249–1251.
2. Yamane, T., Hoq, M. M., Itoh, S., and Shimizu, S. (1986) Glycerolysis of fat by lipase. *J. Jpn. Oil Chem. Soc.* **35**, 625–631.
3. Hirata, H., Higuchi, K., and Yamashina, T. (1989) Enzyme reaction in organic solvent. III. Effect of water content and inhibition of alcohol for the lipase catalyzed transesterification in tributyrin and 1-octanol. *J. Jpn. Oil Chem. Soc.* **38**, 48–52.
4. McNeill, G. P., Shimizu, S., and Yamane, T. (1990) Solid phase enzymatic glycerolysis of beef tallow resulting in a high yield of monoglycerides. *J. Am. Oil Chem. Soc.* **67**, 779–783.
5. McNeill, G. P., Shimizu, S., and Yamane, T. (1991) High yield enzymatic glycerolysis of fats and oils. *J. Am. Oil Chem. Soc.* **68**, 1–5.
6. McNeill, G. P. and Yamane, T. (1991) Further improvements in the yield of monoglycerides during enzymatic glycerolysis of fats and oils. *J. Am. Oil Chem. Soc.* **68**, 6–10.

7. Yamane, T., Kang, S.-T., Kawahara, K., and Koizumi, Y. (1994) High-yield diacylglycerol formation by solid-phase enzymatic glycerolysis of hydrogenated beef tallow. *J. Am. Oil Chem. Soc.* **71,** 339–342.
8. Bornsheuer, U. T. and Yamane, T. (1994) Activity and stability of lipase in the solid-phase glycerolysis of triolein. *Enzyme Microb. Technol.* **16,** 864–869.
9. Rosu, R., Uozaki, Y., Iwasaki, Y., and Yamane, T. (1997) Repeated use of immobilized lipase for monoacylglycerol production by solid phase glycerolysis of olive oil. *J. Am. Oil Chem. Soc.* **74,** 445–450.
10. Yamane, T., Suzuki, T., Sahashi, Y., Vikersveen, L., and Hoshino, T. (1992) production of *n*-3 polyunsaturated fatty acid enriched fish oil by lipase-catalyzed acidolysis without solvent. *J. Am. Oil Chem. Soc.* **69,** 1104–1107.
11. Yamane, T., Suzuki, T., and Hoshino, T. (1993) Increasing *n*-3 polyunsaturated fatty acid content of fish oil by temperature control of lipase-catalyzed acidolysis. *J. Am. Oil Chem. Soc.* **70,** 1285–1287.
12. Vulfson, E. N., Gill, I., and Sarney, D. B. (1996) Productivity of enzymatic catalysis in non-aqueous media: New concepts. Section 9.2. Enzymatic solvent-free synthesis, in *Enzymatic Reactions in Organic Media* (Koskinen, A. M. P. and Klibanov, A. M., eds.), Blackie Academic & Professional, Glasgow, pp. 245–254.
13. Lee, K.-T. and Akoh, C. C. (1998) Solvent-free enzymatic synthesis of structured lipids from peanut oil and caprylic acid in a stirred tank batch reactor. *J. Am. Oil Chem. Soc.* **75,** 1533–1537.
14. Iwasaki, Y., Han, J.-J., Narita, M., Rosu, R., and Yamane, T. (1999) Enzymatic synthesis of structured lipids from single cell oil of high docosahexaenoic acid content. *J. Am. Oil Chem. Soc.* **76,** 563–569.
15. Christensen, M. S., Høy, C.-E., Becker, C. C., and Redgrave, T. G. (1995) Intestinal absorption and lymphatic transport of eicosapentaenoic (EPA), docosahexaenoic (DHA), and decanoic acids: dependence on intramolecular triacylglycerol structure. *Am. J. Clin. Nutr.* **61,** 56–61.
16. Adelhorst, K., Björkling, F., Godtfredsen, S. E., and Kirk, O. (1990) Enzyme-catalyzed preparation of 6-*O*-acylglucopyrasides. *Synthesis* 112–115.
17. Fregapane, G., Sarney, D. B., and Vulfson, E. N. (1991) Enzymatic solvent-free synthesis of sugar acetal fatty acid esters. *Enzyme Microb. Technol.* **13,** 796–800.
18. Fregapane, G., Sarney, D. B., Greenberg, S. G., Knight, D. J., and Vulfson, E. N. (1994) Enzymatic synthesis of monosaccharide fatty acid esters and their comparison with conventional products. *J. Am. Oil Chem. Soc.* **71,** 87–91.
19. Fregapane, G., Sarney, D. B., and Vulfson, E. N. (1994) Facile chemo-enzymatic synthesis of monosaccharide fatty acid esters. *Biocatalysis* **11,** 9–18.
20. Sarney, D. B., Kapeller, H., Fregapane, G., and Vulfson, E. N. (1994) Chemo-enzymatic synthesis of disaccharide fatty acid esters. *J. Am. Oil Chem. Soc.* **71,** 711–714.
21. Yamane, T. (1987) Enzyme technology for the lipids industry: an engineering overview. *J. Am. Oil Chem. Soc.* **64,** 1657–1662.
22. Yamane, T., Kozima, Y., Ichiryu, T., and Shimizu, S. (1988) Biocatalysis in microaqueous organic solvent. *Ann. NY Acad. Sci.* **542,** 282–293.
23. Yamane, T. (1988) Importance of moisture content control for enzymatic reactions in organic solvents: a novel concept of "microaqueous." *Biocatalysis* **2,** 1–9.

24. Bornscheuer, U. T. and Yamane, T. (1995) Fatty acid vinylesters as acylating agents: a new method for the enzymatic synthesis of monoacylglycerols. *J. Am. Oil Chem. Soc.* **72,** 193–197.
25. Han, J.-J., Iwasaki, Y., and Yamane, T. (1999) monitoring of lipase-catalyzed transesterification between eicosapentaenoic acid ethyl ester and tricaprylin by silver ion high-performance liquid chromatography and high-temperature gas chromatography. *J. Am. Oil Chem. Soc.* **76,** 31–39.
26. Rosu, R., Yasui, M., Iwasaki, Y., and Yamane, T. (1999) Enzymatic synthesis of symmetrical 1,3-diacylglycerols by direct esterification of glycerol in solvent-free system. *J. Am. Oil Chem. Soc.* **76,** 839–844.
27. Tatara, T., Fujii, T., Kawase, T., and Minagawa, M. (1983) Quantitative determination of tri-,di-, monooleins and free oleic acid by the thin layer chromatography/flame ionization detector system using internal standards and boric acid impregnated chromarod. *Lipids* **18,** 732–736.

39

Lipase-Catalyzed Synthesis of Sugar Fatty Acid Esters in Supercritical Carbon Dioxide

Haralambos Stamatis, Vasiliki Sereti, and Fragiskos N. Kolisis

1. Introduction

Fatty acid monoesters and diesters of sugars are widely used as emulsifiers in a great variety of food and cosmetics formulations (1). Traditionally, these surfactants are produced by transesterification at high temperatures in the presence of an alkaline catalyst. Among drawbacks of the conventional chemical process, significant coloration of the final product and the formation of undesirable side products are especially worth mentioning. Consequently, the enzyme-catalyzed regioselective acylation of sugars in nonpolar organic solvents and supercritical fluids (SCFs) at ambient temperatures has received much attention in recent years (2–5).

Since the first reports on the use of SCFs (6,7), notably supercritical carbon dioxide ($SCCO_2$), the feasibility of enzymatic transformations in this medium has been established in numerous studies of oxidation and (trans)esterification reactions (8,9). The use of supercritical CO_2 has certain advantages over conventional organic solvents, as it is nontoxic, nonflammable, and inexpensive. As compared to many other solvents, $SCCO_2$ has superior mass transfer characteristics because of the high diffusivity of reactants, low surface tension, and relatively low viscosity (8,9). The critical temperature (31.1°C) and pressure (7.34 MPa) are relatively low, which facilitate the use of biocatalysts. It is also easily removed from the final product. An additional benefit of using $SCCO_2$ is that it can be used as a common solvent in an integrated continuous biotransformation/separation process (10).

One of the limitations of $SCCO_2$ as a medium for biotransformations is that it is a rather nonpolar solvent, more suitable for dissolution of hydrophobic

From: *Methods in Biotechnology, Vol. 15: Enzymes in Nonaqueous Solvents: Methods and Protocols*
Edited by: E. N. Vulfson, P. J. Halling, and H. L. Holland © Humana Press Inc., Totowa, NJ

compounds. Several methods have been proposed in the literature to overcome the problem of poor solubility of polar substrates (e.g., carbohydrates) in nonpolar organic solvents and supercritical fluids. These include complexation of sugars and polyols with phenylboronic acid *(11)* or modification with acetone *(12)*, and successful lipase-catalyzed acylation of sugars in both systems have been reported. However, the implementation of these approaches necessitates additional step(s) to recover the modified substrates and/or products.

Another method proposed for the introduction of highly polar substrates in nonpolar media is based on the adsorption of polar compounds onto an inert material with a high internal surface such as silica gel *(13,14)*. It was found that the adsorbed glycerol and other polyols can be esterified with various fatty acids in *n*-hexane, solvent-free systems, as well as in $SCCO_2$ using lipase as the catalyst *(10,15,16)*. Our group has successfully applied this preadsorbtion technique to introduce various monosaccharides such as glucose or fructose, in $SCCO_2$ and their regioselective esterification with fatty acids has been described *(17,18)*. The use of silica gel and α-cellulose as inert supports for preadsorbtion of polar sugars **(Fig. 1)** enabled us to obtain the desired product in about 50% yields and they could be easily separated from the remaining substrate upon completion of the reaction.

2. Materials

1. Immobilized lipase from *Mucor miehei* (Lipozyme™), Novo-Nordisk (Denmark). The enzyme activity was 25 BIU/g.
2. Glucose, fructose, lauric, palmitic, and oleic acids, 99% pure (Sigma Chemical Co., St. Louis, MO).
3. Silica gel (mesh 70–230) and α-cellulose (Sigma Chemical Co.).
4. Chloroform, methanol, ethanol, acetic acid, hexane, and diethyl ether (Merck, Darmstadt, Germany).
5. Carbon dioxide 99.999% pure.
6. Equipment: A schematic diagram of the high-pressure device used for the lipase-catalyzed esterification of sugars in $SCCO_2$ is presented in **Fig. 2**. The system was placed in a temperature-controlled room. Carbon dioxide of 99.999% purity was used.

3. Methods
3.1. Adsorption of Sugars onto a Solid Support

1. Prepare methanolic solution (20 mL) containing 1 g of glucose or fructose.
2. Add 0.5 g of silica gel or α-cellulose, under stirring at room temperature. Stir the mixture for 30 min. (*See* **Note 1**.)
3. Evaporate the methanol under continuous agitation at reduced pressure and dry resulting homogenous powder over P_2O_5 for 72 h.

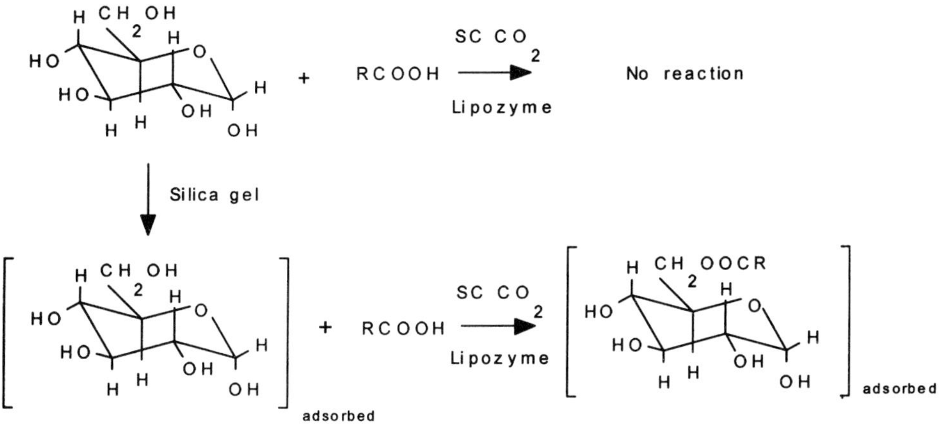

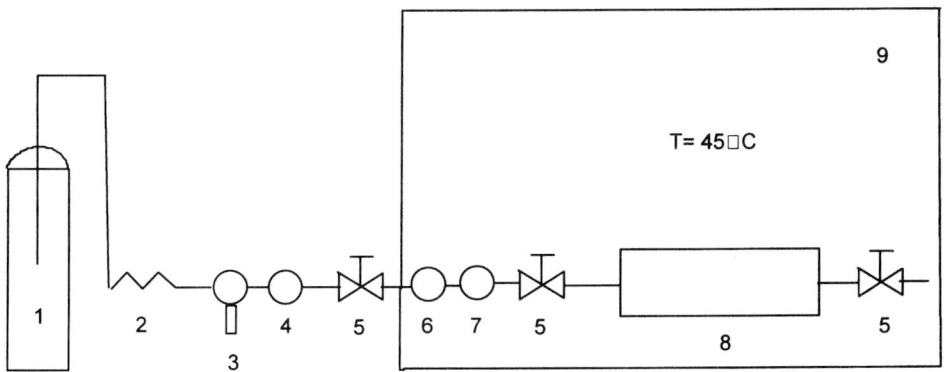

Fig. 1. Schematic representation of the enzymatic esterification of glucose preadsorbed on silica gel in supercritical CO_2.

Fig. 2. Schematic diagram of the $SCCO_2$ apparatus used. 1: Liquid CO_2 tank; 2: cooler; 3: high pressure pump; 4: 20.5-MPa security disk; 5: valve; 6: manometer; 7: thermocouple; 8: reaction vessel; 9: temperature-controlled room.

3.2. Bioconversion in SCCO₂

1. Put the preadsorbed dried substrate (75 mg; 0.28 mmol monosaccharide), fatty acid (0.56 mmol), and Lipozyme (75 mg) in a 7.0-mL metal tubular $SCCO_2$ reactor containing a magnetic stirrer bar (*see* **Note 1**).
2. Fill the reactor with cooled liquid carbon dioxide that was heated up to the reaction temperature of 45°C and pressurized to 15.4 MPa by a piston pump. Incubate under stirring for 48 h (*see* **Notes 2** and **3**).
3. At the end of the incubation period, cool the reactor down and slowly depressurize the content of the vessel in 10 mL of methanol. Dissolve the solid reactants in

the same solvent (10 mL) and remove the remaining silica gel (or α-cellulose) and the enzyme by filtration.

4. Analyze the filtrate (methanolic solution that contained fatty acid, monosaccharide, and monosaccharide fatty ester) by TLC or HPLC (*see* **Note 4**).

3.3. Analytical Methods

Qualitative analysis of products was done by TLC on silica gel 60 plates (Merck). A 6-cm plate was first developed in chloroform:methanol:water (64:12:1) up to 2 cm, then in a mixture of chloroform:methanol:acetic acid (96:3 :1) up to 4 cm and, finally, in a mixture of hexane:diethyl ether:acetic acid (70:30:1) up to 5.5 cm. Plates were sprayed with 5% (v/v) ethanolic solution of H_2SO_4 and incubated for 10 min at 150°C. Monoesters, diesters, and triesters are separated by this procedure.

Quantitative analysis of samples was carried out by HPLC using a C_{18} Nucleosil column and a refractive-index detector. Elution was performed with methanol at a flow rate of 1 mL/min. Standard solutions of sugars, fatty acids, and sugar monoesters solutions methanol were used for calibration.

3.4. Purification of Sugar Esters

The product was recovered from the reactor as described in **Subheading 3.2.**, but the solid was dissolved in chloroform (10 mL). In order to obtain sugar fatty esters of high purity, the filtrate was applied on a silica gel 60 (230–400 mesh) column (1 × 30 cm) equilibrated with chloroform. The column was washed with chloroform at 1.5 mL/min to remove free fatty acid, then with the chloroform/methanol mixture (90/10, v/v) to elute diesters, and, finally, with a mixture of chloroform/methanol/water (64/10/1, v/v/v) to obtain monosaccharide fatty acid esters. The yield of monosaccharide esters was 45 and 34% when silica gel and α-cellulose were used as substrate supports, respectively.

4. Notes

1. The enzyme and the substrates (fatty acid and adsorbed sugar) were separately preincubated in closed vessels containing saturated solutions of LiCl at 25°C for 72 h, to obtain a thermodynamic water activity (a_w) of 0.11. At this a_w value, the conversion of monosaccharides is about 50% (*17,18*).

2. The start of the reaction is assumed to be the point when both temperature and pressure of the system reached the desired values ($T = 45°C$ and $P = 15.4$ MPa).

3. Molar ratio of fatty acid to sugar affects the composition of the final product. At the molar ratio of 2, the monoesters are the predominant products (approx 70–80% of all the esters species present). Further excess of fatty acid leads to a significant reduction in the proportion of monoesters formed (*17*).

4. As fatty acids are very soluble in $SCCO_2$, the method described enables the reuse of this substrate when it is present in excess.

Acknowledgments

The authors would like to thank Novo-Nordisk A/S Denmark, for the generous gift of *Mucor miehei* lipase.

References

1. Akoh, C. C. (1994) Synthesis of carbohydrate fatty acid polyesters, in *Carbohydrate Polyesters as Fat Substitutes* (Akoh, C. C. and Swanson, B. G., eds.), Marcel Deckker, New York, pp. 9–35.
2. Therisod, M. and Klibanov, A. M. (1986) Facile enzymatic preparation of monoacylated sugars in pyridine. *J. Am. Chem. Soc.* **108,** 5638–5640.
3. Riva, S., Chopineau, J., Kieboom, A. P. G., and Klibanov, A. M. (1988) Protease-catalyzed refioselective esterification of sugars and related compounds in anhydrous dimethylformamide. *J. Am. Chem. Soc.* **110,** 584,585.
4. Klibanov, A. M. (1990) Assymetric transformations catalyzed by enzymes in organic solvents. *Acc. Chem. Res.* **23,** 114–120.
5. Vulfson, E. N. (1998) Enzymatic synthesis of surfactants, in *Novel Surfactants: Preparation, Application and Biodegrability* (Holmberg, K., Norberg, M., eds.), Marcel Dekker, New York, pp. 279–300.
6. Randolph, T. W., Blanch, H. W., Prausnitz, J. M., and Wilke, C. R. (1985) Enzymatic catalysis in a supercritical fluid. *Biotechnol. Lett.* **7,** 325–328.
7. Randolph, T. W., Clark, D. S., Blanch, H. W., and Prausnitz, J. M. (1988) Enzymatic oxidation of cholesterol aggregates in supercritical carbon dioxide. *Science* **239,** 387–390.
8. Ballesteros, A., Bornscheuer, U., Capewell, A., Combes, D., Condoret, J.-S., Koening, K., et al. (1995) Enzymes in non-conventional phases. *Biocatal. Biotransform.* **13,** 1–42.
9. Nakamura, K. (1990) Biochemical reactions in supercritical fluids. *Trends Biotechnol.* **8,** 288–292.
10. Marty, A., Combes, D., and Condoret, J. S. (1994) Continuous reaction-separation process for enzymatic esterification in supercritical carbon dioxide. *Biotechnol. Bioeng.* **43,** 497–504.
11. Castillo, E., Marty, A., Combes, D., and Condoret, J. S. (1994) Polar substrates for enzymatic reactions in supercritical CO_2: how to overcome the solubility limitation. *Biotechnol. Lett.* **16,** 169–174.
12. Fregapane, G., Sarney, D. B., Sydney, G. G., Knight, D. J., and Vulfson, E. N. (1994) Enzymatic synthesis of monosaccharide fatty acid esters and their comparison with conventional products. *J. Am. Oil Chem. Soc.* **71,** 87–91.
13. Berger, M., Laumen, K., and Schneider, M. P. (1992) Enzymatic esterification of glycerol I. Lipase-catalyzed synthesis of regioisomerically pure 1, 3-*sn*-diacylglycerols. *J. Am. Oil Chem. Soc.* **69,** 955–960.
14. Berger, M. and Schneider, M. P. (1992) Enzymatic esterification of glycerol I. Lipase-catalyzed synthesis of regioisomerically pure 1(3)-*rac*-monoacylglycerols. *J. Am. Oil Chem. Soc.* **69,** 961–965.

15. Castillo, E., Dossat, V., Marty, A., Condoret, J. S., and Combes, D. (1997) The role of silica gel in lipase-catalyzed esterification reactions of high-polar substrates. *J. Am. Oil Chem. Soc.* **74,** 77–85.
16. Charlemagne, D. and Legoy, M. D. (1995) Enzymatic synthesis of polyglycerol-fatty acid esters in a solvent free-system. *J. Am. Oil Chem. Soc.* **72,** 61–65.
17. Stamatis, H., Sereti, V., and Kolisis, F. N. (1998) Studies on the enzymatic synthesis of sugar esters in organic medium and supercritical carbon dioxide. *Chem. Biochem. Eng. Quart.* **12,** 151–156.
18. Tsitsimpikou, C., Stamatis, H., Sereti, V., Daflos, H., and Kolisis, F. N. (1998) Acylation of glucose catalysed by lipases in supercritical carbon dioxide. *J. Chem. Technol. Biotechnol.* **71,** 309–314.

40

Transformations in Frozen Aqueous Solutions Catalyzed by Hydrolytic Enzymes

Marion Haensler and Hans-Dieter Jakubke

1. Introduction

Because of the high specificity of hydrolytic enzymes, their synthetic applications require only minimal regioselective protection of substrates. Hence, hydrolases are an interesting alternative to the use of conventional catalysts in the synthesis of chiral and polyfunctional compounds. Numerous synthetic applications of lipases (see other chapters in this volume), proteases *(1–3)*, glycosidases *(4–7)*, and ribonucleases *(8–12)* have been reported in the literature.

Enzymatic formation of a peptide bond can be accomplished in an equilibrium-controlled or kinetically controlled process. The former is the direct reversal of the hydrolytic cleavage reaction, whereas the kinetic approach to enzymatic peptide synthesis relies on the protease acting as a "transferase." Proteolytic enzymes which form covalent acyl-enzyme intermediates (e.g., serine and cysteine proteases) readily catalyze the transfer of the acyl group from an activated donor to a nucleophile. In this process, the most important yield-limiting factors are competitive hydrolysis of the acyl-enzyme intermediate and secondary cleavage of the newly formed bond. These undesirable side reactions can be suppressed by manipulation of the reaction medium.

Traditionally, the hydrolytic side reactions have been minimized and even prevented by performing the synthesis in organic solvents under conditions of controlled water activity *(13)*. Unfortunately, enzymes are often inactivated in these systems and changes in the biocatalyst specificity may occur. To address these problems and for ecological reasons, an alternative approach to reduce the effective water content in an aqueous reaction mixture by freezing it has been developed *(14–17)*. In such a system, most but not all of the water in the mixture is frozen and the enzyme-catalyzed reaction takes place in a

From: *Methods in Biotechnology, Vol. 15: Enzymes in Nonaqueous Solvents: Methods and Protocols*
Edited by: E. N. Vulfson, P. J. Halling, and H. L. Holland © Humana Press Inc., Totowa, NJ

much diminished liquid phase where all other reactants are concentrated. Thus, the natural medium for enzymes, water, is used and, at the same time, the disadvantages of having it present at a very high concentration are avoided. As a result of freezing of the reaction mixture, significantly higher yields of peptides are obtained and even free amino acids considered to be inefficient nucleophiles at room temperature can be coupled in good yields.

Furthermore, glycosidases and ribonucleases are also catalytically active in frozen aqueous solutions *(18–21)*. It should be pointed out that there are similarities between the formation of peptide and glycosidic bonds because both include the condensation of an activated donor and a nucleophilic acceptor. However, because of the lower nucleophilicity of the hydroxyl as compared to the amino group, the enzyme-bound intermediate, in the case of glycosidases, is more prone to competitive hydrolysis than the covalently bound acyl enzyme in protease-catalyzed peptide synthesis. The use of frozen aqueous systems helps to overcome this problem as demonstrated by β-galactosidase-mediated synthesis of β-D-galactopyranosyl-*N*-acetyl-D-glucosamine *(18)*.

Ribonucleases are known to catalyze the synthesis of dinucleoside monophosphates from nucleoside 2′,3′-cyclic phosphates and free nucleosides because of the reversibility of the transesterification step in the hydrolysis of RNA. As described earlier, the predominant competitive hydrolysis of the 2′,3′-cyclic donor in RNase A-catalyzed synthesis of cytidylyl-(3′→5′)uridine can be suppressed to a large extent by performing the reaction in frozen aqueous solutions *(18)*.

Transformations in frozen reaction mixtures can be easily performed by shock-freezing of small reaction volumes followed by incubation at subzero temperatures in a cryostate. However, in preparative biotransformations (e.g., protease-catalyzed peptide synthesis), there is a considerable risk of undesired hydrolysis of the acyl donor during the "starting" period because shock-freezing of large reaction volumes can be slow and is often accompanied by temperature gradients. To overcome these limitations, a simple apparatus was constructed *(22)*.

2. Materials

2.1. Analytical-Scale Peptide Synthesis

1. Aqueous enzyme stock solution, 10-fold concentrated (*see* **Note 1**); make fresh daily.
2. Aqueous acyl-donor stock solution, 2.5-fold concentrated, make fresh daily (*see* **Note 2**).
3. Aqueous amino-component stock solution, twofold concentrated, pH adjusted to 0.2 U above the desired final pH (*see* **Note 2**) with 2 *N* NaOH or 1 *N* HCl; store at 4°C, make fresh after 3 d.
4. Polypropylene tubes (1.5 mL).
5. Liquid nitrogen.
6. Cryostate (Lauda or Haake, Germany), temperature adjusted to –15°C.
7. 2.5% (v/v) trifluoroacetic acid.

2.2. Preparative-Scale Peptide Synthesis

1. Apparatus for peptide synthesis (*see* **Fig. 1**).
2. Aqueous stock solution containing amino component and enzyme, 2.5-fold concentrated (*see* **Note 3**), pH adjusted to 0.2 units above the desired final pH (*see* **Note 2**); make fresh daily.
3. Aqueous stock solution containing acyl-donor ester, 1.67-fold concentrated (*see* **Note 3**); make fresh daily.
4. Liquid nitrogen.
5. Freezer (–18°C).
6. 2.5% (v/v) trifluoroacetic acid.

2.3. β-Galactosidase-Catalyzed Disaccharide Synthesis

1. 0.1 *M* Potassium phosphate buffer, pH 6.5, 1 m*M* $MgCl_2$, 5 m*M* dithioerythritol (DTE).
2. Stock solution of glycosidase (β-galactosidase from *Escherichia coli*) in the phosphate buffer (300 U/mL); make fresh daily.
3. Stock solution of glycosyl donor (*o*-nitrophenyl-β-D-galactopyranoside, 125 m*M*) in the phosphate buffer, make fresh daily.
4. Stock solution of glycosyl acceptor (*N*-acetyl-D-glucosamine, 400 m*M*) in the phosphate buffer; store at 4°C; make fresh after 3 d.
5. Polypropylene tubes (1.5 mL).
6. Liquid nitrogen.
7. Cryostate, adjusted to –5 °C.
8. Water bath with boiling water.

2.4. Ribonuclease A-Catalyzed Synthesis of Cytidylyl-(3'→5')Uridine

1. 0.1 *M* Tris-HCl buffer, pH 7.0.
2. Stock solution of ribonuclease A from bovine pancreas (1 mg/mL) in the Tris-HCl buffer, store at 4°C, make fresh after 3 d.
3. Stock solution of 2',3'-cytidine cyclophosphate (6.75 m*M*) in 0.1 *M* Tris-HCl buffer, pH 7.0; make fresh daily.
4. Stock solution of uridine (0.7 *M*) in 0.1 *M* Tris-HCl buffer, pH 7.0, store at 4°C; make fresh weekly.
5. Polypropylene tubes (1.5 mL).
6. Liquid nitrogen.
7. Cryostate, adjusted to –10°C.
8. Water bath with boiling water.

2.5. Ribonuclease T1-Catalyzed Synthesis of Guanylyl-(3'→5')Cytidine

1. 0.1 *M* Tris-HCl buffer, pH 7.0.
2. Stock solution of ribonuclease T1 from *Aspergillus oryzae* (9 µg/mL) in the Tris-HCl buffer, store at 4°C (*see* **Note 4**); make fresh after 3 d.

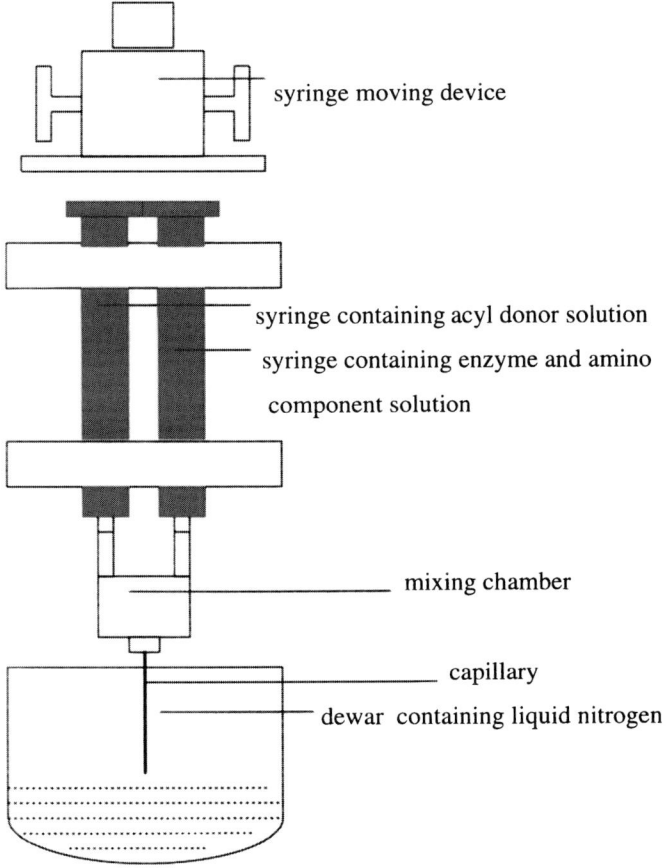

Fig. 1. Schematic diagram of the apparatus for preparative protease-catalyzed peptide synthesis. (Adapted from **ref. 22**.)

3. Stock solution of guanosine-2',3'-cyclophosphate (6.75 mM) in the Tris-HCl buffer; make fresh daily.

4. Stock solution of cytidine (0.125 M) in the Tris-HCl buffer, store at 4°C; make fresh weekly.

5. Polypropylene tubes (1.5 mL).

6. Liquid nitrogen.

7. Cryostate, adjusted to –10°C.

8. 0.4 M ZnSO$_4$.

3. Methods

3.1. Analytical-Scale Peptide Synthesis

Eight to ten identical samples are typically prepared for each experiment.

1. Precool the stock solutions of acyl donor, enzyme, and amino component to 0°C in an ice bath.
2. Add 50 µL of the amino-component stock solution to 40 µL of the stock solution of the acyl donor in a 1.5-mL polypropylene tube. Start the reaction by adding 10 µL of the enzyme stock solution. Shake the tube rapidly and allow shock-freezing of the sample in liquid nitrogen for 30 s.
3. Transfer the tube to a cryostate and incubate at −15°C.
4. After appropriate time intervals, stop the reaction by adding 0.1 mL of 2.5 % (v/v) trifluoroacetic acid to the frozen sample.
5. Analyze by high-performance liquid chromatography (HPLC) (RP 18 column, 5 µm, isocratic or gradient elution using acetonitrile/water mixtures containing 0.1% trifluoroacetic acid, v/v) (*see* **Note 5**).

3.2. Preparative-Scale Peptide Synthesis

A special peptide synthesis apparatus is used (**Fig. 1**). Two syringes (10 and 20 mL) are connected to a conical mixing chamber by flexible tubes. At the outflow of the chamber, a capillary (1×50 mm) is positioned through which drops of the mixed reaction solution are supplied into a dewar (2 L) containing liquid nitrogen. Ice spheres with a diameter of 3–6 mm are formed. To obtain single spheres, the outlet of the capillary is positioned 8–10 cm above the liquid-nitrogen surface. The ice spheres are collected in a plastic container adapted to fit into the dewar and incubated in a freezer at −18°C. To ensure a constant ratio of both reaction solutions in the mixture, injections are performed manually using a rack gear.

1. Fill the 10-mL syringe completely with the stock solution of amino component and enzyme. The acyl-donor stock solution (15 mL) is placed in the 20-mL syringe.
2. Inject the mixture of the two syringes into the dewar using the rack gear at approximately 5 mL/min.
3. Allow shock-freezing for 2 min.
4. Transfer the ice spheres collected into the plastic container to a freezer.
5. To monitor the course of the reaction, stop the reaction by picking single spheres after desired reaction times and analyze by HPLC.
6. After all the acyl donor is consumed, add 2.5% (v/v) trifluoroacetic acid to stop the reaction (trifluoroacetic acid should completely cover the reaction mixture) and isolate the peptide product.

3.3. β-Galactosidase-Catalyzed Disaccharide Synthesis

Eight to ten identical samples are typically prepared for each experiment.

1. Precool the stock solutions of enzyme, glycosyl donor, and glycosyl acceptor to 0°C in an ice bath.
2. Add 50 µL of the glycosyl-acceptor stock solution to 40 µL of the stock solution of the glycosyl donor in a 1.5-mL polypropylene tube. Start the reaction by adding 10 µL of the enzyme stock solution. Shake the tube rapidly and allow shock-freezing of the sample in liquid nitrogen for 30 s.

3. Transfer the test tube to a cryostate and incubate at –5°C (*see* **Note 6**).
4. After appropriate time intervals, stop the reaction by incubating the reaction mixture in a boiling water bath for 10 min.
5. Analyze by HPLC (Nucleosil amino column, 5 µm, isocratic elution using aceto-nitrile/water, 75/25, v/v, 235 nm) (*see* **Note 6**).

3.4. Ribonuclease-(A and T1)-Catalyzed Dinucleoside Monophosphate Synthesis

Eight to ten identical samples are typically prepared for each experiment.

1. Precool the stock solutions of enzyme, 2',3'-cyclic nucleoside monophosphate and free nucleoside to 0°C in an ice bath.
2. Add 40 µL of the stock solution of 2',3'-cyclic nucleoside monophosphate to 50 µL of the stock solution of the free nucleoside in a 1.5-mL polypropylene tube. Start the reaction by adding 10 µL of the enzyme stock solution. Shake the tube rap-idly and allow shock-freezing of the sample in liquid nitrogen for 30 s.
3. Transfer the test tube to a cryostate and incubate at –10°C (*see* **Note 7**).
4. After appropriate time intervals, inactivate the enzyme by incubating the reaction mixture in boiling water for 10 min (RNase A) or by adding 100 µL of 0.4 M $ZnSO_4$ (RNase T1).
5. Analyze by HPLC (RP 18 column, 5 µm, elution with 0.01 M potassium phosphate buffer, pH 4.15, for 15 min, followed by a linear gradient of methanol from 0 to 30% applied in further 20 min) (*see* **Note 7**).

4. Notes

1. The final enzyme concentration to be used depends on the enzyme and the acyl-donor ester. The amount of enzyme should be sufficient to complete the conver-sion of the ester in 24 h. In the case of α-chymotrypsin, this is about 10–25 µg/mL. For papain, the required enzyme concentration is typically 0.2–0.4 mg/mL *(23)*. Note that cysteine proteases should be dissolved in 5 mM dithiothreitol (DTT)/ 0.7 M EDTA to activate the enzyme.
2. A final concentration of the acyl donor should be in the range of 2–25 mM, depending on its solubility in water. The amino component (free base) should be present in a 10-fold excess over the acyl donor, calculated according to Henderson–Hasselbalch.
   For serine proteases, a final pH of 9.0, after mixing of all the reaction components, before freezing is recommended. This pH, however, is not necessarily suitable for cysteine proteases in frozen solutions *(23)*. For papain, pH 7.8 is recommended.
   Usually, it is sufficient to adjust pH of the amino-component stock solution to about 0.2 units above the desired final pH, but this must be checked for every particular reaction system.
3. The complete reaction mixture should contain 25–50 mM acyl-donor ester. The final enzyme concentration is varied according to the amount of the acyl donor used to achieve the complete conversion of the ester within 24 h at most.

4. Concentration of RNase T1 is determined spectrophotometrically using $\varepsilon_{278nm} = 17,300/M/cm$ *(24)*.
5. Depending on the system, the improvement in yields in peptide syntheses through freezing can be very impressive (i.e., the conversions reach 90%) *(15)*.
6. At reaction temperatures below –5°C, glycosidase-catalyzed reactions are very slow. The disaccharide yield of 53% is obtained in the frozen system as compared to 35% when the same reaction is carried out at room temperature *(18)*.
7. The temperature of –10°C is optimal for RNase-catalyzed reactions in frozen solutions. With RNase T1, the yield of dinucleoside monophosphate of up to 90% is obtained in frozen solutions *(19)*. Under similar conditions, RNase A gives yields of up to 30%, which still compare favorably to about 18% obtained in conventional aqueous mixtures *(18)*.

References

1. Bongers, J. and Heimers, E. P. (1994) Recent applications of enzymatic peptide synthesis. *Peptides* **15,** 183–193.
2. Wong, C. H. and Wang, K. T. (1991) New developments in enzymatic peptide synthesis. *Experentia* **47,** 1123–1129.
3. Jakubke, H.-D., Eichhorn, U., Haensler, M., and Ullmann, D. (1996) Non-conventional enzyme catalysis: Application of proteases and zymogens in biotransformations. *Biol. Chem.* **377,** 455–464.
4. Nilsson, K. G. I. (1988) Enzymatic synthesis of oligosaccharides. *TIBTECH* **6,** 256–264.
5. Ichikawa, Y, Look, G. C., and Wong, C. H. (1992) Enzyme-catalyzed oligosaccharide synthesis. *Anal. Biochem.* **202,** 215–238.
6. Halcomb, R. L. (1995) Hydrolysis and formation of C-O-bonds. Glycosidases, in *Enzyme Catalysis in Organic Synthesis* (Drauz, K. and Waldmann, H., eds.), VCH, Weinheim, Germany, pp. 303–306.
7. Monsan, P. and Paul, F. (1995) Enzymatic synthesis of oligosaccharides. *FEMS Microbiol. Rev.* **16,** 187–192.
8. Mohr, S. C. and Thach, R. E. (1969) Application of ribonuclease T1 to the synthesis of oligoribonucleotides of defined base sequence. *J. Biol. Chem.* **244,** 6566–6576.
9. Bauer, S., Lamed, R., and Lapidot, Y. (1972) Large scale synthesis of dinucleoside monophosphates catalysed by ribonuclease from *Aspergillus clavatus*. *Biotechnol. Bioeng.* **14,** 861–870.
10. Rowe, M. J. and Smith, M. A. (1972) Synthesis of oligoribonucleotides with ribonuclease T1. *Biochim. Biophys. Acta* **281,** 338–346.
11. Bratovanova, E. K., Kasche, V., and Petkov, D. D. (1993) Kinetically controlled enzymic synthesis of ribonucleotide bond in vitro. *Biotechnol. Lett.* **15,** 347–352.
12. Backmann, J., Doray, C. C., Grunert, H. P., Landt, O., and Hahn, U. (1994) Extended kinetic analysis of ribonuclease T1 variants leads to an improved scheme for the reaction mechanism. *Biochem. Biophys. Res. Commun.* **199,** 213–219.
13. Jakubke, H.-D. (1994) Protease-catalyzed peptide synthesis: basic principles, new synthesis strategies and medium engineering. *J. Chin. Chem. Soc.* **41,** 355–370.

14. Schuster, M., Aaviksaar, A., and Jakubke, H.-D. (1990) Enzyme-catalyzed peptide synthesis in ice. *Tetrahedron* **46,** 8093–8102.
15. Haensler, M. and Jakubke, H.-D. (1996) Nonconventional protease catalysis in frozen aqueous solutions. *J. Peptide Sci.* **2,** 279–289.
16. Haensler, M. and Jakubke, H.-D. (1998) Endoproteinase Pro-C-catalyzed peptide bond formation in frozen aqueous systems. *Enzyme Microb. Technol.* **22,** 617–620.
17. Haensler, M., Wehofsky, N., Gerisch, S., Wissmann, J.-D., and Jakubke, H.-D. (1998) Reverse catalysis of elastase from porcine pancreas in frozen aqueous systems. *Biol. Chem.* **379,** 71–74.
18. Haensler, M. and Jakubke, H.-D. (1996) Reverse action of hydrolases in frozen aqueous solutions. *Amino Acids* **11,** 379–395.
19. Haensler, M., Hahn, U., and Jakubke, H.-D. (1997) Reverse action of ribonuclease T1 in frozen aqueous systems. *Biol Chem.* **378,** 115–118.
20. Haensler, M., Schuerer, H., Hahn, U., and Jakubke, H.-D. (1998) Reverse action of ribonuclease T1 variants in ice. *Nucleosides Nucleotides* **17,** 1267–1274.
21. Hubner, B., Haensler, M., and Hahn, U. (1999) Modification of ribonuclease T1 specificity by random mutagenesis of the substrate binding segment. *Biochemistry* **38,** 1371–1376.
22. Haensler, M., Thust, S., Klossek, P., and Ullmann, G. (1999) Enzyme-catalyzed preparative peptide synthesis in frozen aqueous systems. *J. Mol. Catal. B* **6,** 95–98.
23. Haensler, M., Ullmann, G., and Jakubke, H.-D. (1995) The application of papain, ficin and clostripain in kinetically controlled peptide synthesis in frozen aqueous solutions. *J. Peptide Sci.* **1,** 283–287.
24. Georgalis, Y., Zouni, A., Hahn, U., and Saenger, W. (1991) Synthesis and kinetic study of transition state analogs for ribonuclease T1. *Biochim. Biophys. Acta* **1118,** 1–5.

41

Enzymatic Synthesis of Sugar Fatty Acid Esters in Solvent-Free Media

Douglas B. Sarney and Evgeny N. Vulfson

1. Introduction

Sugar esters are employed as surfactants in a wide range of food, pharmaceutical and agrichemical applications (*1*). The use of enzymes for carrying out esterification of sugars has been extensively studied over the last decade because of exquisite regioselectivity and mild reaction conditions associated with the use of biocatalysts. Another advantage of enzyme-based systems is the ability to prepare esters of reducing sugars with no side reaction (e.g., caramelization that often takes place under conditions conventionally used for [trans]esterification).

In general, two main approaches have been explored for the lipase-catalyzed preparation of sugar fatty acid esters. The syntheses have been performed either by employing lipases in an organic solvent suitable for solubilization of both substrates (*2–6*) or by conducting the reaction in a solvent-free mixture of substrates using sugars that have been modified to improve the miscibility of reactants (*7,8*). The use of highly polar solvents (e.g., dimethyl formamide [DMF], pyridine) is considered to be practically more facile, but the enzyme stability and reaction rates are poor, thus hindering applications on a preparative scale. However, prior acetalization of sugars improves their solubility in or miscibility with molten fatty acids, thus avoiding the use of highly polar solvents in the reaction medium. These solvent-free mixtures allow much higher conversions in a shorter time, even when the reaction is performed in a nearly equimolar mixture of substrates (*7–9*) and the high operational stability of the enzyme under these conditions facilitates biocatalyst recycling.

From: *Methods in Biotechnology, Vol. 15: Enzymes in Nonaqueous Solvents: Methods and Protocols*
Edited by: E. N. Vulfson, P. J. Halling, and H. L. Holland © Humana Press Inc., Totowa, NJ

This chapter describes the synthesis of a range of sugar-based surfactants, namely monosaccharide and disaccharide fatty acid monoesters, by the enzymatic esterification of sugar acetals with fatty acids. Commercially available, immobilized lipases were used as catalysts for the esterifications in both organic solvents and under solvent-free conditions, and the hydrolysis of the resulting sugar acetal monoesters was accomplished using aqueous acids as exemplified in **Fig. 1** with the 1,2-*O*-isopropylidene derivative of xylose.

Products synthesized using this methodology include 6-*O*-acyl-D-galactopyranoses, 6-*O*-acyl-D-glucopyranoses, 5-*O*-acyl-D-xylofuranoses, 6'-*O*-acyl-lactoses, and 6'-*O*-acyl-sucrose monoesters. Overall yields of 57–87% (monosaccharide esters), up to 70% (lactose esters), and 20–27% (sucrose esters) were obtained. The products were fully characterized by fast atom bombardment mass spectrometry (FAB-MS) and ^{13}C nuclear magnetic resonance (NMR). A range of medium- and long-chain fatty acids were used with similar efficiency.

2. Materials

1. Lipozyme IM-60™ (E.C. 3.1.1.3 lipase from *Mucor miehei*) and Novozyme™ 435 (*Candida antarctica* lipase) were supplied by Novo-Nordisk A/S (Bagsvërd, Denmark).
2. 1,2-*O*-Isopropylidene-α-D-xylofuranose (**1a, Fig. 2**), 1,2-*O*-isopropylidene-α-D-glucofuranose (**3a, Fig. 2**), 1,2:3,4-di-*O*-isopropylidene-α-D-galactopyranose (**5a, Fig. 2**), fatty acids, disaccharides, Amberlite IRA-400 OH anion-exchange resin, silica gel, molecular sieves, and all organic solvents used in this study were obtained from Aldrich Chemical Co. (Gillingham, UK) and were of the highest available purity, generally >99%.
3. Tetrafluoroboric acid (50% aqueous solution) and silica gel C60 TLC plates were supplied by Merck (Eastleigh, UK).
4. ^{1}H and ^{13}C spectra were recorded on a Bruker WH 400 Fourier transform spectrometer at 400.0 MHz and 100.62 MHz, respectively, using d_6-dimethyl sulfoxide (DMSO) as the solvent. FAB-MS spectra were obtained on a Kratos MS9/50TC spectrometer, using xenon at 5–8 keV. Accurate mass measurements were recorded at 1.0 milli-amu resolution, using PEG 600 ions as reference. Gas chromatography (GC) analysis was performed on a Hewlett-Packard series 5890A gas chromatograph equipped with a flame-ionization detector (FID). One microliter of trimethylsilyl derivatives, prepared according to Sweeley et al. *(10)*, was applied to a Hewlett-Packard Ultra 2, 25 m × 0.2 mm fused silica capillary column *(8,11)*.

3. Methods

3.1. Synthesis of Monosaccharide Fatty Acid Esters (see Note 1)

3.1.1. General Preparation of 5-O-Acyl-1,2-O-Isopropylidene-α-D-Xylofuranoses 1b–f and 5-O-Acyl-D-Xylofuranoses 2b–f

The general approach is exemplified by the enzymatic esterification of 1,2-*O*-isopropylidene-α-D-xylofuranose **1a** with tetradecanoic acid to form

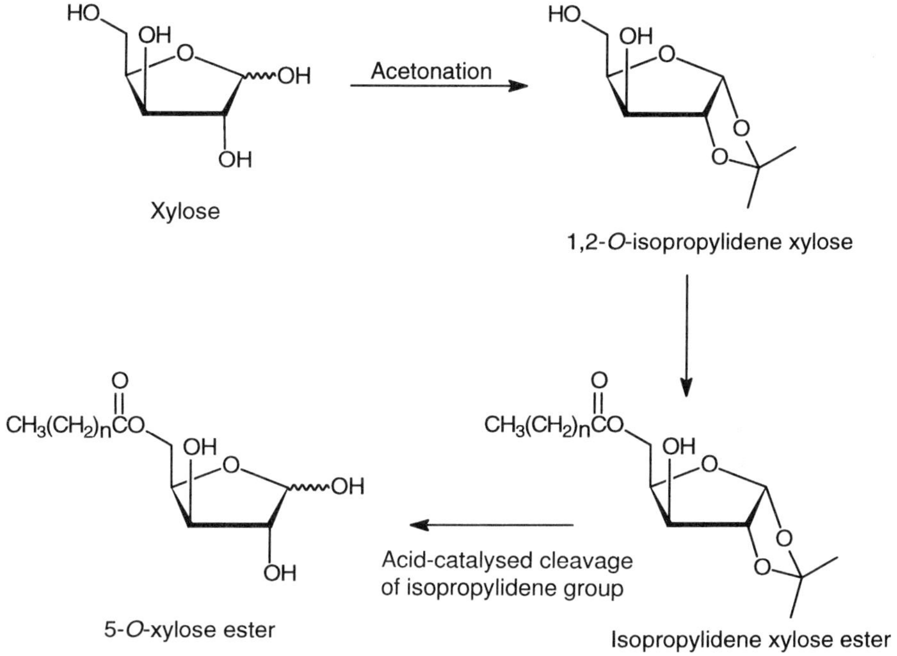

Fig. 1. The principle of sugar modification followed by enzymatic esterification and removal of modifying groups in synthesis of sugar esters (exemplified by xylose).

5-*O*-tetradecanoyl-1,2-*O*-isopropylidene-α-D-xylofuranose **1d** followed by cleavage of the isopropylidene group to yield 5-*O*-tetradecanoyl-α-D-xylofuranose **2d** (**Fig. 2**).

The product **1a** (10.0 g, 52.6 mmol) and tetradecanoic acid (12.0 g, 52.6 mmol) were mixed together and heated to 75°C. Lipozyme (10% w/w) was added, and the mixture was stirred (150 rpm) at 75°C under vacuum (<20 mbar) until approx 80% conversion had been reached (typically 18 h). The reaction mixture was then diluted with acetone (approx 100 mL) and filtered, and the filtrate was evaporated under vacuum at room temperature. The solid obtained was dissolved in ether (approx 400 mL) and washed with 0.3 M K$_2$CO$_3$/0.3M NaCl$_{(aq)}$ (approx 500 mL). The organic phase was evaporated under vacuum at room temperature, and the product was dried by repeated azeotropic distillation with absolute ethanol to yield a white solid **1d** (16.3 g, 78%, purity >98% by GC).

The product **1d** was dissolved in formic acid:water:ethyl acetate (50:20:30, v/v/v) (65 mL) in a stoppered flask and left to react for 1 h at 75°C. Acetonitrile (65 mL) was added and the crude product was precipitated by cooling to 0°C. The white precipitate was filtered off and dried, and traces of residual acid were removed by successive azeotropic distillations with toluene at 65°C under

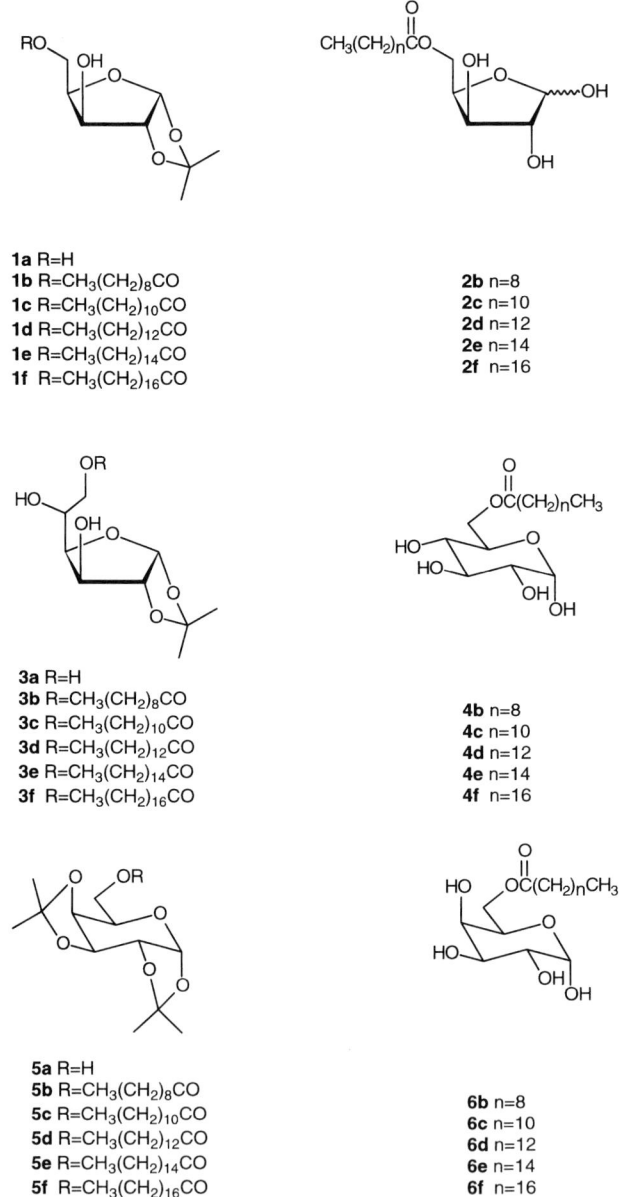

Fig. 2. Chemical structures of monosaccharide acetals and their esters.

reduced pressure. Minor quantities of **1d** and ester cleavage products (xylose and tetradecanoic acid) were removed by washing with ether and acetone to yield a white solid **2d** (>99% pure by GC).

3.1.2. General Preparation of 6-O-Acyl-1,2-O-Isopropylidene-α-D-Glucofuranoses **3b–f** and 6-O-Acyl-α-D-Glucopyranoses **4b–f**

The general approach is exemplified by the enzymatic esterification of 1,2-*O*-isopropylidene-α-D-glucofuranose **3a** with dodecanoic acid to form 6-*O*-dodecanoyl-1,2-*O*-isopropylidene-α-D-glucofuranose **3c** followed by cleavage of the isopropylidene group to yield 6-*O*-dodecanoyl-α-D-glucopyranose **4c** (Fig. 2).

The product **3a** (250 mg, 1.14 mmol) and dodecanoic acid (227 mg, 1.14 mmol) were mixed together and heated to 75°C in a sealed vial with 1 mL of cyclohexanone. Lipozyme (25% w/w) was added and then the mixture was shaken at 75°C (200 rpm) for 24 h. The reaction mixture was diluted with acetone (2 mL) and filtered, and the filtrate was evaporated to dryness under high vacuum. The white solid obtained was dissolved in ether (10 mL), and residual glucose acetal and fatty acid were removed by repeated extraction with 5 mL of 0.3 M K_2CO_3/0.3M $NaCl_{(aq)}$. The organic phase was dried over anhydrous Na_2SO_4, filtered, and evaporated to dryness under vacuum at room temperature to yield a white solid **3c** (375 mg, 82%, purity >98% by GC).

The glucose acetal ester **3c** was dissolved in 80% trifluoroacetic acid (TFA) (0.5 mL), shaken, and left to react for 1 h at room temperature. Acetonitrile was added at –20°C to precipitate the product, which was subsequently filtered and dried under high vacuum to leave a white, lustrous solid **4c** (>98% pure by GC). Because glucose acetal esters were very sensitive to accompanying ester-bond cleavage during the acetal hydrolysis reaction, the deprotection of esters **3b** and **3d–f** were optimized individually. The following protocols were used: **3b**—80:20 TFA:water (room temperature); **3d**—70:30 TFA:water (room temperature); **3e** and **3f**—50:20:30 formic acid:water:ethyl acetate (75°C).

3.1.3. General Preparation of 6-O-Acyl-1,2:3,4-di-O-Isopropylidene-α-D-Galactopyranoses **5b–f** and 6-O-Acyl-α-D-Galactopyranoses **6b–f**

The general approach is exemplified by the enzymatic esterification of 1,2:3,4-di-*O*-isopropylidene-α-D-galactopyranose **5a** with hexadecanoic acid to form 6-*O*-hexadecanoyl-1,2:3,4-di-*O*-isopropylidene-α-D-galactopyranose **5e** followed by cleavage of isopropylidene groups to yield 6-*O*-hexadecanoyl-α-D-galactopyranose **6e** (*see* Fig. 2).

The product **5a** (10.0 g, 38.4 mmol) and hexadecanoic acid (9.85 g, 38.4 mmol) were mixed together and heated to 75°C. Lipozyme (10% w/w) was added, and the mixture was stirred at 75°C (150 rpm) under vacuum (<20 mbar) until >90% conversion had been reached (typically 24–48 h). The reaction mixture was diluted with acetone (50 mL) and filtered, and the filtrate was evaporated under vacuum at room temperature. The solid obtained was dissolved in ether (approx 400 mL) and washed with 0.3 M K_2CO_3/0.3 M $NaCl_{(aq)}$ (approx 500 mL). The organic phase was evaporated under vacuum at room temperature, and the product was dried by repeated azeotropic distillation with absolute ethanol to yield a white solid **5e** (17.5 g, 92%, purity >98% by GC).

The product **5e** was dissolved in 90% TFA (70 mL) in a stoppered flask and shaken for 5 min. The mixture was then left for 20 min to react. Absolute ethanol (200 mL) was added, the solution was evaporated under vacuum at 35°C, and after drying by azeotropic distillation with absolute ethanol (4X 100 mL), a white solid was obtained. Minor quantities of **5e** and ester cleavage products (galactose and hexadecanoic acid) were removed by washing with ether and acetone to yield a white solid **6e** (>99% pure by GC).

3.2. Synthesis of Disaccharide Fatty Acid Esters

3.2.1. General Preparation of 6'-O-Acyl-4-O-(3',4'-O-Isopropylidene-β-D-Galactopyranosyl)-2,3:5,6-Di-O-Isopropylidene-1,1-di-O-Methyl-D-Glucose Monoesters (Lactose Tetra-Acetal Monoesters) 7b–g and 6'-O-Acyl-4-O-β-D-Galactopyranosyl-D-Glucose Monoesters (Lactose Monoesters) 8b–g (Fig. 3; also see Note 2).

4-*O*-(3',4'-*O*-Isopropylidene-β-D-galactopyranosyl)-2,3:5,6-di-*O*-isopropylidene-1,1-di-*O*-methyl-D-glucose (lactose tetra-acetal) **7a** was prepared in a 50% yield by refluxing a suspension of α-lactose monohydrate and *p*-toluene sulfonic acid (p-TSA) in 2,2-dimethoxypropane and purified as described by Thelwall et al. *(12)*. The crude preparation of **7a** was obtained by neutralization of *p*-TSA with ion-exchange resin (Amberlite IRA-400), filtration, and evaporation of the 2,2-dimethoxypropane. The resulting syrup consisted of 80% **7a**, partial lactose acetals, and traces of lactose and was used directly for enzymatic esterification.

Enzymatic esterification reactions were performed either on a 2-g scale in open-top glass vials in an incubator–shaker at 75°C (180 rpm) or on a 15 g scale in 100 mL round bottom flasks stirred at 180 rpm at 75°C and under vacuum (20 mbar). All reactions were conducted with an initial addition of toluene or *t*-butyl acetate (50% v/w), to aid miscibility of the substrates, which was allowed to evaporate after 6 h of reaction time by removing the caps from the vials or by applying a vacuum (20 mbar) with the larger-scale synthesis.

The general approach is exemplified by the enzymatic esterification of 4-*O*-(3',4'-*O*-isopropylidene-β-D-galactopyranosyl)-2,3:5,6-di-*O*-isopropylidene-1,1-di-*O*-methyl-D-glucose (lactose tetra-acetal) **7a** with octadecanoic acid to form 6'-*O*-octadecanoyl-4-*O*-(3',4'-*O*-isopropylidene-β-D-galactopyranosyl)-2,3:5,6-di-*O*-isopropylidene-1,1-di-*O*-methyl-D-glucose (6'-*O*-octadecanoyl-tetra-acetal-lactose) **7g** followed by cleavage of acetal groups to yield 6'-*O*-octadecanoyl-4-*O*-(β-D-galactopyranosyl)-D-glucose (6'-*O*-octadecanoyl-lactose) **8g**.

The product **7a** (1.0 g, 1.95 mmol) and octadecanoic acid (0.56 g, 1.95 mmol) were mixed together with 1 mL toluene (water saturated 25°C)

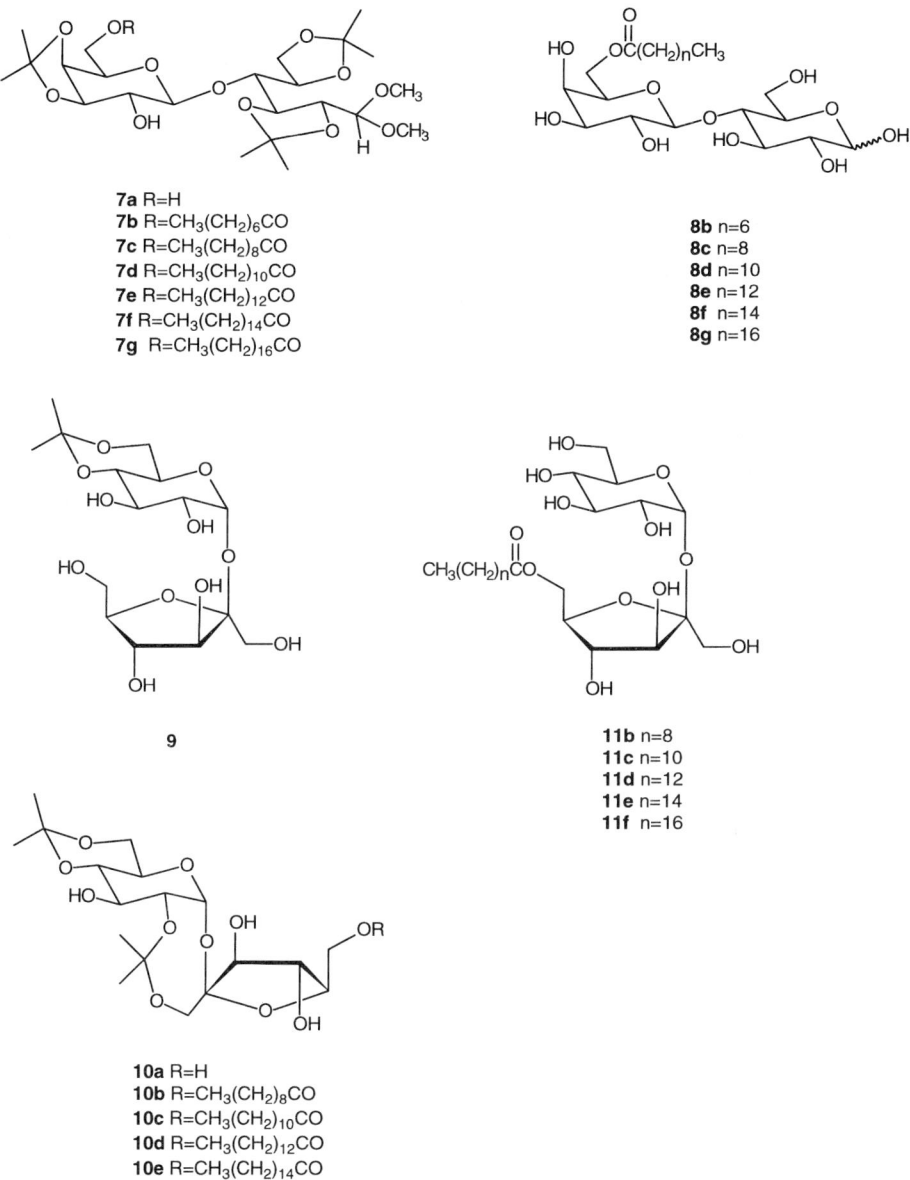

7a R=H
7b R=CH₃(CH₂)₆CO
7c R=CH₃(CH₂)₈CO
7d R=CH₃(CH₂)₁₀CO
7e R=CH₃(CH₂)₁₂CO
7f R=CH₃(CH₂)₁₄CO
7g R=CH₃(CH₂)₁₆CO

8b n=6
8c n=8
8d n=10
8e n=12
8f n=14
8g n=16

9

11b n=8
11c n=10
11d n=12
11e n=14
11f n=16

10a R=H
10b R=CH₃(CH₂)₈CO
10c R=CH₃(CH₂)₁₀CO
10d R=CH₃(CH₂)₁₂CO
10e R=CH₃(CH₂)₁₄CO
10f R=CH₃(CH₂)₁₆CO

Fig. 3. Chemical structures of disaccharide acetals and their esters.

and heated to 75°C to give a homogeneous liquid phase. Lipozyme (10% w/w, 0.156 g) was added, and the mixture was stirred at 75°C (180 rpm) for 6 h, after which the vial was opened to allow evaporation of the solvent. After 80% conversion had been reached (typically 24 h), the reaction mixture was diluted with acetone (5 mL) and filtered, and the filtrate was evaporated under vacuum at room temperature. The syrup obtained was dissolved in petroleum ether (approx 16 mL) and washed twice with distilled water to recover unreacted **7a**. Small amounts of **7g** were back-extracted from the combined water phases into petroleum ether (8 mL). Unreacted fatty acids were removed from the pooled ether phases by washing three times with 0.2 M K_2CO_3/0.3 M $NaCl_{(aq)}$ (approx 10 mL). The organic phase was dried with Na_2SO_4 and evaporated under vacuum at room temperature, and the product was dried by repeated azeotropic distillation with absolute ethanol to yield a viscous, colorless solid **7g** (1.18 g, 78%, purity >99% by GC).

The product **7g** was dissolved in 0.2:1:99 tetrafluoroboric acid:water:acetonitrile (11.8 mL) in a stoppered flask and shaken for 2 h at 30°C. The final product, lactose monostearate, was insoluble in this mixture and hence precipitated out. This white solid (>95% purity) was filtered and washed with acetonitrile to remove traces of residual acid and subsequently dried before recrystalization from methanol to yield white crystals **8g** (>99% pure).

This methodology could also be extended to the synthesis of maltose monomyristate via the enzymatic esterification of maltose triacetal [4-*O*-(4,6-*O*-isopropylidene-α-D-glucopyranosyl)-2,3:5,6-di-*O*-isopropylidene-1,1-di-*O*-methyl-D-glucose] with myristic acid followed by HBF_4-catalyzed cleavage of the acetal groups to yield maltose monomyristate in the same way.

3.2.2. Synthesis of Sucrose Esters (see **Note 3**)

3.2.2.1. PREPARATION OF SUCROSE ACETALS (**9** AND **10A**)

Two sucrose ketals, 4,6-*O*-isopropylidenesucrose (**9**) and 2,1':4,6-di-*O*-isopropylidenesucrose (**10a**) (**Fig. 3**), were prepared according to published procedures *(13–16)* with the minor modifications described below. A solution of sucrose (8.55 g, 25 mmol) in dry *N,N*-dimethylformamide (100 mL) containing molecular sieve pellets (Type 3Å) was stirred with 2-methoxypropene (3.2 mL, 32.5 mmol, **9** and 12 mL, 125 mmol, **10a**) in the presence of *p*-TSA. The reaction was allowed to proceed for 40 min at 70°C. The reaction mixture was then cooled, neutralized with sodium carbonate, and filtered and the solvent removed by rotary evaporation. The desired sucrose acetal was purified by column chromatography. The following yields were obtained: 3.2 g, 32% (**9**); 3.8 g, 36% (**10a**).

A crude mixture of isopropylidene sucrose was prepared as described for **10a**. This typically contained some unreacted sucrose, 60–70% of **10a**, 30–35% of **9**, and a small quantity of a sucrose triacetal [2,1':4.6-di-*O*-isopropylidene-6'-*O*-(1-methyl-1-methoxyethyl) sucrose] *(16)*. This preparation was used for enzymatic esterification as obtained, after neutralization of the *p*-TSA catalyst with Amberlite IRA-400 OH anion-exchange resin, filtration, and removal of the solvent.

3.2.2.2. GENERAL PREPARATION OF SUCROSE ACETAL 6'-*O*-MONOESTERS **10** B–F AND SUCROSE 6'-*O*-MONOESTERS **11**B–F*F)*

The approach is exemplified by the enzymatic esterification of 2,1':4,6-di-*O*-isopropylidene-sucrose (**10a**) with octadecanoic acid (stearic acid, C18:0) to form 6'-*O*-octadecanoyl-2,1':4,6-di-*O*-isopropylidenesucrose (**10f**) followed by cleavage of the acetal groups to yield 6'-*O*-octadecanoyl sucrose (**11f**, **Fig. 3**). Crude 2,1': 4,6-di-*O*-isopropylidenesucrose syrup (20.66 g) was incubated with stearic acid (13.80 g) at 75°C for 48 h with Novozyme™ 435 (1.70 g) with an initial addition of 21 mL of toluene to promote miscibility of the reagents. The solvent was allowed to evaporate during the first few hours of the reaction. After the completion of the reaction (typically 48 h), the resultant mixture was dissolved in ethanol and the enzyme was recovered by filtration. Unreacted fatty acid was then removed by passing the solution through a pad of basic alumina, and after evaporation of the ethanol *in vacuo*, the product was redissolved in ether (1:10 w/v). Unreacted sucrose acetals were then extracted with aqueous NaCl/K_2CO_3 (0.3 M each) and the organic solution was dried over $MgSO_4$, filtered, and evaporated to dryness.

Hydrolysis of the isopropylidene groups to obtain the sucrose ester was carried out by dissolving the product (12.24 g) in a solution of acetonitrile:water:tetrafluoroboric acid (500:5.5:0.5 v/v, 120 mL) and stirring at ambient temperature for 8 h. The solution was cooled in ice and the resultant precipitate (9.91 g, **11f**) was filtered off. The reaction conditions for the acid-catalyzed hydrolysis of acetal groups were tailored for different esters (**10b–e**). Sucrose acetal monoesters **10b–d** (10% w/v) were hydrolyzed in ethyl acetate:TFA:water 95:3:2 at 60°C for 8 h. TFA was then removed by azeotropic distillation with ether. Reaction conditions for **10e** were identical to those described for **10f**.

Final purification was achieved by column chromatography. The structure of the resultant monoester (>98% pure by GC) was determined to be 6'-*O*-stearoyl sucrose (**11f**) by ^{1}H- and ^{13}C-NMR spectroscopy and high resolution FAB-MS. This was in agreement with the structures assigned to the intermediate sucrose acetal ester (**10f**). Esters **11b–e** were prepared and characterized in an analogous manner.

4. Notes

1. Monosaccharide esters. The acylation of **1a** and **5a** using Lipozyme resulted in 76–92% yields of the monosaccharide acetal esters **1b–f** and **5b–f** when the reaction was carried out in equimolar mixtures of substrates in the absence of added solvents. These yields could be further improved by employing a slight excess

(typically 25–50 mol%) of one of the substrates. Lipase-catalyzed esterification of **1a** yielded practically no diester as detected by HPLC analysis. This was because the esterification rate of the secondary hydroxyl group in **1a** (as measured by HPLC analysis) was more than two orders of magnitude lower than that of the primary hydroxyl group. Hence, only traces of the 3,5-diester were formed during the incubation time used. However, under the same solvent-free conditions used for the acylation of **1a** and **5a**, acylation of 1,2-*O*-isopropylidene-α-D-glucofuranose **3a** led to much lower yields of the monosaccharide acetal esters **3b–f** (typically 20–40%). This was the result of the accompanying formation of diesters and triesters and the **3b–f** monoester content decreased even further on continuation of the incubation.

When comparing the esterification of the substrates **1a** and **3a**, we believe that the loss of regioselective control at the primary hydroxyl position in **3a** could be explained by the difference between the physico-chemical properties of the reaction mixtures formed by substrates **1a** and **3a**, respectively. Thus 1,2-*O*-isopropylidene-α-D-xylofuranose **1a** had a melting point of 70°C and was completely miscible with molten fatty acids at the reaction temperature employed (75°C). However, 1,2-*O*-isopropylidene-α-D-glucofuranose **3a** melted at 160°C and was only sparingly soluble in the lipophilic phase under the same conditions. The accumulation of monoesters **3b–f** (which were significantly more soluble in the liquid phase of fatty acid) at the initial stages of the reaction could have led to their subsequent esterification in preference to the starting material **3a.** Hence, the effective competition of the monoester products **3b–f** with **3a** results in the rapid formation of diester and triester products.

Therefore, the addition of a relatively small amount of organic solvent, in order to increase the solubility of **3a** in the lipophilic phase, was undertaken to improve regioselectivity. This modification of the reaction mixture allowed the monoesters **3b–f** to be obtained in the presence of 2 parts of cyclohexanone (w/w) in 72–92% yields after 24 h. Higher esters were not detected in the reaction mixture by HPLC analysis during the same period of time. The addition of other organic solvents such as 2-methoxyethyl ether or tertiary amyl alcohol resulted in a similar effect, thus excluding the specific effect of cyclohexanone on the enzyme regiospecificity. It should be noted, however, that although the solubility of **3a** was significantly improved, the reaction mixture remained heterogeneous at the initial stages of the reaction, even in the presence of organic solvents.

The monosaccharide esters **2b–f**, **4b–f**, and **6b–f** were isolated in overall yields of 57–87% after acid-catalyzed cleavage of the acetal group(s) of monosaccharide acetal esters **1b–f**, **3b–f**, and **5b–f**, respectively.

The ^{1}H and ^{13}C-NMR spectroscopic analyses revealed that galactose esters **6b–f** and glucose esters **4b–f** were isolated as pure α-anomers, whereas xylose esters **2b–f** were isolated as a mixture of α- and β-anomers in approximately a 1:1 ratio. Interconversion from furanose to pyranose ring configuration occurred during the acid-catalyzed acetal cleavage of isopropylidene glucose esters **3b–f.**

2. Lactose esters. Lactose tetra-acetal (**7a**) was found to be an appropriate substrate in spite of possible steric hindrances and up to 80% conversion was obtained with this

substrate in 24 h. The product **7a** appeared to be sufficiently soluble in molten fatty acid for the reaction to proceed even in the absence of added solvents. Reaction vials were left open to allow evaporation of the water produced during the reaction, thus driving the equilibrium in the desired direction of synthesis.

No purification of **7a** was required prior to enzymatic esterification and the reaction could be carried out under solvent-free conditions. Crude **7a** was obtained as an oil that contained a mixture of lactose tetra-acetal and numerous other partial acetals. Owing to the high regioselectivity of the lipases used, partial acetals either remain unreactive (e.g., 6'-protected species) or are esterified to yield the same final product after hydrolysis of the acetal groups.

HBF_4 was the most efficient catalyst for hydrolysis of the acetal groups after esterification of **7a**, but similar results could be obtained with acetic acid (1:4:4 $CH_3COOH:H_2O:CH_3CN$, 75°C) over a longer period (30 h). Final products precipitated directly from the reaction mixture to yield both α-and β-anomers of the lactose esters where the ratio was dependent on the deprotection conditions. Pure α-anomers of **8b–g** could be obtained by subsequent recrystalisation from methanol.

3. Sucrose esters. The regioselectivity of acylation with *M. miehei* (Lipozyme IM60) and *C. antarctica* (Novozyme 435) lipases was established by carrying out the reactions with the two sucrose acetals (**9** and **10a**) on a 0.65-mol scale using myristic acid and 3,3-dimethyl-2-butanone (pinacolone) or 2-hexanone as solvents. The products obtained were analyzed by ^{13}C-NMR for each of the enzyme/ solvent combinations. In all cases, one major monoester product could be isolated by column chromatography, although the formation of minor amounts of other regioisomers, particularly in the case of **9**, was noted. Significant quantities of diesters were formed only when the reaction was continued for a long period (7 d). Characterization of the major monoester products by NMR spectroscopy revealed that irrespective of the solvent or enzyme employed, the position of acylation was the same for both acetals (**9** and **10a**), namely the primary hydroxyl group of the fructose ring. Therefore, regardless of the nature of sucrose acetals used for enzymatic esterification, 6'-*O*-acyl sucrose was the same final product of the synthesis after cleavage of the acetal groups. Among the enzymes studied, the *M. miehei* lipase was generally the most effective at catalyzing the esterification of acetals **9** and **10a** with myristic acid (C14:0) and that between the sucrose acetals tested, 2,1':4,6-di-*O*-isopropylidenesucrose (**10a**) gave the highest yield of monoester. *C. antarctica* lipase (Novozyme 435) showed a similar preference for **10a** in solvent-free esterifications, but relatively poor yields were obtained with **9**.

Although **9** was a less efficient substrate than the di-isopropylidene-derivative (**10a**), the observation that **9** and **10a** can be acylated at the same 6'-position (thus, giving the same final product after removal of isopropylidene group[s]) prompted us to investigate the feasibility of enzymatic esterification of a mixture of **9** and **10a**. To this end, a crude preparation of 2,1':4,6-di-*O*-isopropylidenesucrose (**10a**), containing up to 30% of **9** and some unreacted sucrose and higher acetals, was used directly as obtained after neutralization of *p*-TSA and evaporation of the solvent. When this was attempted using the *M. miehei* enzyme as the catalyst, yields were significantly reduced (from >40% with pure **10a** to 13% with the crude product). However, *C.antarctica* lipase worked well in the crude mixture of substrates and gave even better yields of **10d**, as determined by GC, than the *M. miehei* enzyme and pure di-isopropylidene sucrose (**10a**) with all the fatty acids tested. Improved rates observed with the mixture of 2,1':4,6 di-*O*-isopropylidene-sucrose and *C. antarctica* lipase were attributed to a better miscibility of this crude isopropylidene sucrose syrup with molten fatty acid. In accordance with earlier experiments involving pure acetals, the same major product, 6'-*O*-monoester, was obtained. Consequently, *C. antarctica* lipase and the crude preparation of 2,1':4,6-di-*O*-isopropylidenesucrose (**10a**) are the best combination for the large-scale preparation of sucrose-6'-*O*-monoesters. Under these conditions, a high degree of regioselectivity was observed where the final product contained up to 98% of 6'-*O*-monoester (**11b–f**) by GC.

References

1. Colbert, J. C. (1974) *Sugar Esters: Preparation and Application.* Noyes Data Co., NJ.
2. Therisod, M. and Klibanov, A. M. (1986) Facile enzymatic preparation of monoacylated sugars in pyridine. *J. Am. Chem. Soc.* **108,** 5638–5640.
3. Chopineau, J., McCafferty, F. D., Therisod, M., and Klibanov, A. M. (1988) Production of biosurfactants from sugar alcohols and vegetable oils catalysed by lipases in a non-aqueous medium. *Biotechnol. Bioeng.* **31,** 208–214.
4. Khaled, N., Montet, D., Pina, M., and Graille, J. (1991) Fructose oleate synthesis in a fixed catalyst bed reactor. *Biotechnol. Lett.* **13,** 167–172.
5. Oguntimein, G. B., Erdmann, H., and Schmid, R. D. (1993) Lipase catalysed synthesis of sugar ester in organic solvents. *Biotechnol. Lett.* **15,** 175–180.
6. Schlotterbeck, A., Lang, S., Wray, V., and Wagner, F. (1993) Lipase-catalysed monoacylation of fructose. *Biotechnol. Lett.* **15,** 61–64.
7. Adelhorst, K., Bjorkling, F., Godtfredsen, S. E., and Kirk, O. (1990) Enzyme catalysed preparation of 6-*O*-acylglucopyranosides. *Synthesis* 112–115.
8. Fregapane, G., Sarney, D. B., Greenberg, S. G., Knight, D. J., and Vulfson, E. N. (1994) Enzymatic synthesis of monosaccharide fatty acid esters and their comparison with conventional products. *J. Am. Oil Chem. Soc.* **71,** 87–91.
9. Fregapane, G., Sarney, D. B., and Vulfson, E. N. (1991) Enzymic solvent-free synthesis of sugar acetal fatty acid esters. *Enzyme Microb. Technol.* **13,** 796–800.
10. Sweeley, C. C., Bentley, R., Marita, M., and Wells, W. W. (1963) Gas-liquid chromatography of trimethylsilyl derivatives of sugars and related substances. *J. Am. Chem. Soc.* **85,** 2497–2507.
11. Karrer, R. and Herberg, H. (1992) Analysis of sucrose fatty-acid esters by high-temperature gas-chromatography. *J. High Resolut. Chromatogr.* **15,** 585–588.
12. Thelwall, L. A. W., Hough, L., and Richardson, A. C. (1981) U.S. Patent 4,284,763.
13. Khan, R. (1974) Sucrochemistry, Part XIII. Synthesis of 4,6-*O*-benzylidene-sucrose. *Carbohydr. Res.* **32,** 375–379.

14. Khan, R. and Mufti, K. S. (1975) Synthesis and reactions of 1',2:4,6-di-*O*-isopropylidene-sucrose. *Carbohydr. Res.* **43,** 247–253.
15. Khan, R., Mufti, K. S. and Jenner, M. R. (1978) Synthesis and reactions of 4, 6-acetals of sucrose *Carbohydr. Res.* **65,** 109–113.
16. Fanton, E., Gelas, J., and Horton, D. (1981) Kinetic acetonation of sucrose—preparative access to a chirally substituted 1,3,6-trioxacyclooctane system. *J. Org. Chem.* **46,** 4057–4060.

42

Biotransformations in Supersaturated Solutions

David A. MacManus, Anna Millqvist-Fureby, and Evgeny N. Vulfson

1. Introduction

It has been shown in the last few years that enzymatic reactions can be carried out efficiently in nearly equimolar mixtures of substrates in the absence of added solvents *(1,2)*. However, for solvent-free reactions to be efficient, a reasonable degree of miscibility between substrates is usually required. This limits the practical utility of this approach to biotransformations of substances that are either liquid at the reaction temperature (e.g., lipids [see other chapters in this volume]) or can be chemically "engineered" to become soluble in the other liquid substrate *(3–9)*. In order to overcome this problem, we have developed a complementary bioreaction system where enzymatic transformations take place in the liquid phase of a low-melting-point eutectic mixture of substrates *(10–14)*. Such eutectics can be formed by mixing molten substrates together at elevated temperatures and then cooling the mixture down to an ambient temperature *(11)*. A small amount of organic solvent or water is often used to further suppress the melting point and/or to decrease the viscosity of the reaction medium *(12)*. Enzymatic transformations in eutectic mixtures have been successfully used for the preparation of numerous peptide derivatives *(10,13,14)*.

This latter development posed an interesting question of whether it would be possible to apply similar principles to the transformation of other complex natural products (e.g., carbohydrates). Enzymes are valuable tools in the synthesis and modification oligosaccharides, as they can be used to generate the required type and configuration of glycosidic linkage with no requirement for elaborate protection/deprotection strategies. Numerous examples of the preparative use of glycosidases and glycosyltransferases, either on their own or in conjunction with conventional chemistry, have been described in the

From: *Methods in Biotechnology, Vol. 15: Enzymes in Nonaqueous Solvents: Methods and Protocols*
Edited by: E. N. Vulfson, P. J. Halling, and H. L. Holland © Humana Press Inc., Totowa, NJ

literature (see reviews in **refs. *15–17***). In general, the use of biocatalysts leads to a significant simplification in synthetic protocols because of a reduction in the overall number of steps required. In this chapter, we describe the use of metastable supersaturated solutions of carbohydrates as an attractive low-water media for the preparation of glycosides and disaccharides *(18–21)* and give some examples of the synthetic utility of this methodology.

The physico-chemical characteristics of binary systems formed by a simple monosaccharide or disaccharide and water are well studied and understood *(22,23)*. Probably the most interesting feature of practically any phase diagram involving carbohydrates and water is the glass region, shown as a shaded area in **Fig. 1**, as it is a thermodynamically unstable phase. Glasses are readily formed on cooling saturated solutions of sugars and, generally, the lower the water content of the system, the higher the glass transition temperature (T_g). However, it is the region above the glass transition and below the solubility curve **(Fig. 1)** that is of particular relevance to this discussion. In this region, many carbohydrates form metastable supersaturated solutions. Further cooling of these solutions may lead to the formation of a glass, but, on standing for a sufficiently long time, they would inevitably revert to the thermodynamically more stable crystalline state. It is also well known that the addition of plasticizers, such as water, alcohols, and polyols, significantly depresses the temperature of glass transition and often retards the crystallization of sugar even at lower temperatures. It is this region of the phase diagram that we have exploited for carrying out numerous glycosidase-catalyzed reactions of practical utility *(18–21)*.

In general, the advantages of supersaturated solutions as a reaction medium include better volumetric productivity of the biotransformation and higher yields of the desired product *(19)*. It is also a means of redirecting the regioselectivity of glycosylation from primary to secondary hydroxyl groups by, for example, using a 6-*O*-modified glycoside acceptor. Thus, it has been shown that 6-*O*-acetyl glycosides serve as good substrates for many glycosidases that are, under the conditions of very high substrate concentrations, "forced" to catalyze the transfer of monosaccharide units to the secondary hydroxyl groups *(21)*. These substrates have not been widely used in enzymatic transglycosylation because of their often poor reactivity under conventional reaction conditions.

2. Materials

1. Enzymes: All glycosidases used in this work were obtained from Aldrich Chemical Co. (Dorset, UK). Lipases (Novozyme 435 and Lipozyme IM) were purchased from Novo-Nordisk, Bagvërd, Denmark. The enzymes (many of them crude) were used as obtained.
2. 6-*O*-Acyl glycosides were prepared using the lipase-catalyzed acylation of commercially available alkyl glycosides according to Millqvist-Fureby et al. *(19)*.

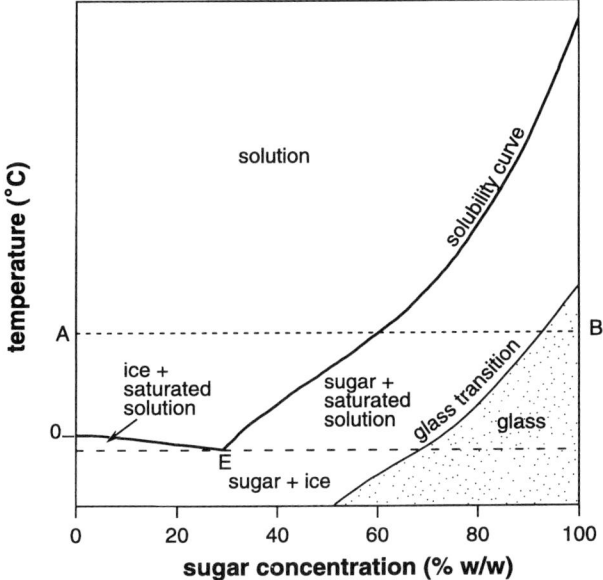

Fig. 1. A typical phase diagram for a binary system containing a monosaccharide or disaccharide and water. The shaded area represents the glass region. E is the eutectic point. The T_g's for some common monosaccharide and disaccharides, as determined by Orford et al. *(24)*, are 7°C (D-fructose), 13°C (D-xylose), 32°C (D-galactose), 38°C (D-glucose), 70°C (sucrose), and 95°C (maltose). (Adapted from **ref. *18*.**)

3. All other substrates, solvents, and buffer salts were obtained from Sigma-Aldrich Chemical Co. (Dorset, UK).
4. Only conventional laboratory equipment is required to perform the syntheses.

3. Methods

3.1. Preparation of Supersaturated Solutions

In a typical experiment two substrates (i.e., glycon donor [monosaccharide or glycoside] and a glycon acceptor [alcohol or glycoside]) are mixed together in a top-capped vial. A small amount of water or buffer is added and the resulting mixture is heated up to approx 95°C to obtain a homogenous solution. This solution is then cooled down to the reaction temperature (typically 37°C) in a temperature-controlled shaker–incubator (*see* **Notes 1** and **2**).

3.2. Synthesis of Glycosides

In a typical example, glucose (100 mg), ethylene glycol (100 mg), and 50 m*M* sodium citrate buffer, pH 5.2 (*see* **Table 1** for details) were mixed in a capped vial and heated up to approx 95°C to dissolve the reactants. The vials were then

Table 1
Almond β-Glucosidase-Catalyzed Synthesis of Glucosides
in Supersaturated Solutions

Alcohol	Alcohol:Glucose (mol:mol)	Initial rate (mmol/h)	Yield (%)	Yield (g/g)	Time (h)
Ethylene glycol[a]	2.9:1	15.6	58	0.28	54
Diethylene glycol	1.7:1	25.5	44	0.26	48
Tetraethylene glycol	0.9:1	5.8	30	0.25	72
Hexaethylene glycol	0.6:1	5.2	18[a]	0.18	24
1,3-Propanediol	2.3:1	35.2	55	0.29	24
2,3-Propanediol	2.3:1	2.6	41	0.22	48
1,4-Butanediol	2.0:1	25.6	47	0.27	24
2,3-Butanediol	2.0:1	4.7	21	0.12	48
1,5-Pentanediol	1.7:1	16.2	40	0.24	24
1,6-Hexanediol	1.5:1	11.5	41	0.26	72

Note: The reaction mixtures contained glucose (100 mg), alcohol (100 mg), 50 m*M* citrate buffer, pH 5.2 (50 µL), and β-glucosidase (20 U). No glucosylation of the secondary hydroxyl group in glycerol was observed.
[a]This sample contained 35 µL of buffer.
Source: Adapted from **ref. 18**.

transferred to an incubator at 37°C, and after 10 min, the reaction was started by the addition of β-glucosidase (20 U in 20 µL of 50 m*M* citrate buffer, pH 5.2, containing 5 m*M* dithiothreitol [DTT]). At specified time intervals, aliquots (approx 2 mg) were withdrawn, methanol (100 µL) was added, and the samples were dried in a heating block (60°C) under a stream of nitrogen. The progress of the reaction is monitored by gas chromatography (GC) as described in **Note 3**. Reaction conditions and yields obtained are summarized in **Table 1**.

3.2. Synthesis of Disaccharides

In a typical experiment, the *p*-nitrophenyl glycoside donor (0.2 g, 0 66 mmol), 6-*O*-acyl-protected acceptor glycoside (1.26 g, 5.3 mmol), and citrate buffer (50 m*M*, pH 5.2, 2.2 cm³) were placed in a screw-cap vial and heated on a hot plate to 90–95°C. When the substrates had dissolved, the vials were transferred to an incubator at 37°C and the reaction was started by the addition of enzyme. The following amounts of enzymes were used: β-amylase (2500 U), green coffee bean α-galactosidase (10 U), *Escherichia coli* β-galactosidase (500 U), jack bean α-mannosidase (20 U), rice α-glucosidase (50U), and almond β-glucosidase (500 U). The course of the reaction was followed by thin-layer chromatography (TLC) (propan-2-ol:nitromethane:water, 10:9:3 v/v/v) or high-performance liquid chromatography (HPLC) (**Note 4**), and when the donor glycoside had been consumed, the mixture was filtered. The solvent was evaporated *in vacuo* and the mixture was dissolved in acetic acid anhydride–pyridine (50:50 v/v, 50 cm^3). *N,N*-Dimethylaminopyridine (10 mg) was added and the mixture stirred overnight. The solvent was evaporated *in vacuo* and the resulting syrup applied to a column of silica gel and eluted with chloroform, followed by chloroform–methanol (100:1 v/v). In cases where mixtures of regioisomers were formed, it was often necessary to carry out further purification by silica gel chromatography. The product disaccharide acetates were obtained as foams on evaporation of the solvent *in vacuo*. Deacetylation, if required, was carried out using conventional methods. Yields of different disaccharides prepared by this method are given in **Table 2**. Nuclear magnetic resonance (NMR) spectra for all of the products mentioned can be found in **refs.** *19* and *21*.

4. Notes

1. The necessary amount of water should be determined experimentally for each particular pair of substrates. The minimum is the amount required to dissolve the reactants at about 95°C. In practice, a little more water is needed to ensure that the substrate(s) do not crystallize out too rapidly and/or that the reaction mixture does not turn into glass on cooling. Suitable for biotransformation solutions can be obtained at substrate concentrations from 1.5 to 8 times higher than their solubility in water at the reaction temperature, depending on the temperature and reactants used.

2. The solubility curves can be determined by preparing mixtures of substrates (with no heating!) and incubating them for up to a few weeks at the required temperature. Periodically, small aliquots are withdrawn and the concentration of the compound of interest in the liquid phase is determined by a suitable method (e.g., HPLC or GC using a standard calibration curve). It is assumed that equilibrium is established when the amount of material in solution reached a plateau. The solubilities of some of the substrates used in this work are given in **Table 3**. Differential scanning calorimetry (DSC) is particularly useful for determining the glass transition temperature (for details, *see* **ref.** *18*).

3. For the GC analysis, the dried samples were derivatized by adding 50% (v/v) *N,O-bis*-(trimethylsilyl)-trifluoroacetamide (BSTFA) in pyridine (100 µL) and dry pyridine (900 µL), followed by incubation at 70°C for 30 min. One microliter was then withdrawn for the GC analysis, which was carried out using a Hewlett-Packard GC series 5890 (Palo Alto, CA), fitted with a split injector (split ratio of 1:25), a fused-silica capillary column (HP Ultra 2, Hewlett Packard, 25 m × 0.22 mm, 0.33-mm film thickness), and a flame-ionization detector. The injector was held at 300°C, and the detector at 350°C. The temperature program was as follows: injection temperature 100°C, the temperature was then increased at 20°C/min up to 325°C and held at 325°C for 2–15 min depending on the compounds analyzed. When appropriate, calibration curves were constructed for the accurate determination of concentrations.

Table 2
Regioselectivity of Transglycosylation in Supersaturated Solutions

Enzyme preparation (%)	Donor	Acceptor	Products	Ratio	Yield[a]
E. coli	Gal-β-OPNP	Gal-β-OMe	1-6 and 1-3	2:3	14
E. coli	Gal-β-OPNP	Gal-β-OMe(6-OAc)	1-3	n.a.	15
β-Amylase	Gal-β-OPNP	Gal-β-OMe	1-6 and 1-3[b]	1:1	28
β-Amylase	Gal-β-OPNP	Gal-β-OMe(6-OAc)	1-3 and 1-2[b]	8:1	26
β-Amylase	Gal-β-OPNP	Gal-α-OMe(6-OAc)	1-3 and 1-4	1:1	6
β-Amylase	Gal-β-OPNP	Glc-β-OMe(6-OAc)	1-3 and other	14:1	16
β-Amylase	Gal-β-OPNP	Glc-α-OMe(6-OAc)	1-3 and 1-2	1:1	18
β-Amylase	Gal-β-OPNP	Man-α-OMe(6-OAc)	1-3 and 1-2	3:2	15
Green coffee bean	Gal-α-OPNP	Gal-β-OMe(6-OAc)	1-3[b]	n.a.	6
Almond	Glc-β-OPNP	Gal-β-OMe(6-OAc)	1-3	n.a.	8
Rice	Glc-α-OPNP	Gal-β-OMe(6-OAc)	1-3 and 1-2	1:1	12
Jack bean	Man-α-OPNP	Gal-β-OMe(6-OAc)	1-3 and 1-6[b,c]	4:1	9

[a]Overall yield of purified disaccharides.
[b]Evidence of some self-transfer products.
[c]The formation of small quantities of (1-6)-linked disaccharides in this case resulted from partial deacylation of the acceptor catalyzed by contaminant esterases in the crude enzyme preparation.
Source: Adapted from **refs. *19*** and ***21***.

Table 3
Solubility of Donor and Acceptor Glycosides at 37°C

Compound	Solubility (wt%)	Solubility (*M*)
Gal-α-D-OPNP	5.5 ± 0.6	0.18 ± 0.02
Gal-β-D-OPNP	3.4 ± 0.04	0.11 ± 0.001
Fuc-α-L-OPNP	0.44 ± 0.04	0.015 ± 0.0008
Fuc-β-D-OPNP	0.42 ± 0.04	0.015 ± 0.001
Fuc-α-L-OPNP[a]	0.67 ± 0.03	0.023 ± 0.001
Fuc-β-D-OPNP[a]	1.1 ± 0.1	0.039 ± 0.004
Xyl-β-D-OPNP	1.8 ± 0.03	0.068 ± 0.001
Gal-α-D-OMe	36 ± 1	2.1 ± 0.1
Gal-β-D-OMe	51 ± 2	3.2 ± 0.1
Fuc-α-L-OMe	65 ± 3	4.7 ± 0.2
6-*O*-AcGal-α-D-OMe	56 ± 1	3.0 ± 0.1

[a]The saturating solubility of these glycosides was determined at 50°C.

4. For HPLC analysis, aliquots of the reaction mixture (4 μL) were withdrawn at the required time intervals and diluted with 400 μL of acetonitrile:water (3:1 v/v) prior to injection. HPLC was carried out using a Gilson 305/306 pump system equipped

with a Gilson 234 autoinjector (Anachem, Luton, UK) and a Sedex 55 evaporative light-scattering detector. The samples were analyzed with a Hypersil Hypercarb column (5 µm, 100 mm × 4.6 mm) using a water/acetonitrile gradient (from 100 to 90% water over 10 min, to 80% water over 5 min, to 40% water over 8 min, to 100% water over 4 min) at a flow rate of 0.75 mL/min.

Acknowledgment

The financial contribution of the Biotechnology and Biological Sciences Research Council is gratefully acknowledged.

References

1. Sarney, D. B. and Vulfson, E. N. (1995) Application of enzymes to the synthesis of surfactants. *Trends Biotechnol.* **13,** 164–172.
2. Vulfson, E. N., Gill, I., and Sarney, D. B. (1995) Productivity of enzymatic catalysis in non-aqueous media: new developments, in *Enzymatic Reactions in Organic Media* (Klibanov, A. M., Koskinen, A. M. P., eds.), Blackie Academic and Professional, Glasgow, pp. 244–265.
3. Adelhorst, K., Bjorkling, F., Godtfredsen, S. E., and Kirk, O. (1990) Enzyme catalyzed preparation of 6-*O*-acyl-glucopyranosides. *Synthesis* **2,** 112–115.
4. Fregapane, G., Sarney, D. B., and Vulfson, E. N. (1991) Enzymic solvent-free synthesis of sugar acetal fatty acid esters. *Enzyme Microb. Technol.* **13,** 796–800.
5. Fregapane, G., Sarney, D. B., Greenberg, S., Knight, D. J., and Vulfson, E. N. (1994) Enzymatic synthesis of monosaccharide fatty acid esters and their comparison with conventional products. *J. Am. Oil Chem. Soc.* **71,** 87–91.
6. Sarney, D. B., Fregapane, G., and Vulfson, E. N. (1994) Facile chemo-enzymatic synthesis of monosaccharide fatty acid esters. *Biocatalysis* **11,** 9–18.
7. Sarney, D. B., Kappeler, H., Fregapane, G., and Vulfson, E. N. (1994) Chemo-enzymatic synthesis of disaccharide fatty acid esters. *J. Am. Oil Chem. Soc.* **71,** 711–714.
8. Sarney, D. B., Barnard, M. J., MacManus, D. A., and Vulfson, E. N. (1996) Application of lipases to the regioselective synthesis of sucrose fatty acid monoesters. *J. Am. Oil Chem. Soc.* **73,** 1481–1487.
9. Valivety, R., Gill, I., and Vulfson, E. N. (1997) Enzymatic synthesis of novel bola- and gemini-surfactants. *J. Surfact. Deterg.* **1,** 177–185.
10. Gill, I. and Vulfson, E. N. (1993) Enzymatic synthesis of short peptides in heterogeneous mixtures of substrates. *J. Am. Chem. Soc.* **115,** 3348,3349.
11. Gill, I. and Vulfson, E. N. (1994) Enzymatic catalysis in heterogeneous eutectic mixtures of substrates. *Trends Biotechnol.* **12,** 118–122.
12. Gill, I., Lopez-Fandino, R., and Vulfson, E. N. (1994) Protease-catalysed synthesis of oligopeptides in heterogeneous substrate mixtures. *Biotechnol. Bioeng.* **43,** 1024–1030.
13. Lopez-Fandino, R., Gill, I., and Vulfson, E. N. (1994) Enzymatic catalysis in heterogeneous mixtures of substrates: the role of the liquid phase and the effects of adjuvants. *Biotechnol. Bioeng.* **43,** 1016–1023.
14. Jorba, X., Gill, I., and Vulfson, E. N. (1995) Enzymatic synthesis of the "delicious peptide" fragments in eutectic mixtures *J. Agric. Food Chem.* **73,** 2536–2541.

15. Nilsson, K. G. I. (1988) Enzymatic synthesis of oligosaccharides. *Trends Biotechnol.* **6,** 256–264.
16. Wang, P. G., Fitz, W., and Wong, C. -H. (1995) Making complex carbohydrates via enzymatic routes. *CHEMTECH* **25,** 22–32.
17. Fernandez-Mayoralas, A. (1997) Synthesis and modification of carbohydrates using glycosidases and lipases. *Topics Curr. Chem.* **186,** 1–20.
18. Millqvist-Fureby, A., Gill, I. S., and Vulfson, E. N. (1998) Enzymatic transformations in supersaturated solutions: I. A general study with glycosidases. *Biotechnol. Bioeng.* **60,** 190–196.
19. Millqvist-Fureby, A., MacManus, D. A., Davies, S., and Vulfson, E. N. (1998) Enzymatic transformations in supersaturated solutions: II Synthesis of disaccharides via transglycosylation. *Biotechnol. Bioeng.* **60,** 197–203.
20. Millqvist-Fureby, A., Gao, C., and Vulfson, E. N. (1998) Enzymatic synthesis of ethoxylated glycoside esters using glycosidases in supersaturated solutions and lipases in organic solvents. *Biotechnol. Bioeng.* **59,** 747–753.
21. MacManus, D. A. and Vulfson, E. N. (2000) Regioselectivity of enzymatic glycosylation of 6-*O*-acyl-glycosides in supersaturated solutions. *Biotechnol. Bioeng.*, in press.
22. Franks, F. (1993) Solid aqueous solutions. *Pure Appl. Chem.* **65,** 2527–2537.
23. Slade, L. and Levine, H. (1988) Non-equilibrium behavior of small carbohydrates–water systems. *Pure Appl. Chem.* **12,** 1841–1864.
24. Orford, P. D., Parker, R., and Ring, S. G. (1990) Aspects of glass transition behaviour of mixtures of carbohydrates of low molecular weight. *Carbohydr. Res.* **196,** 11–18.

43

Enzymatic Transformations in Suspensions (I)

One Solid Substrate and Product

Volker Kasche and Antje Spieß

1. Introduction

1.1. Why Biotransformations in Suspensions?

The use of aqueous suspensions or emulsions, with pure solid or liquid substrate and/or product phases to increase the solubility of reactants, is an alternative to conventional biotransformations in organic solvents. The first suspension systems were studied qualitatively some 20–30 yr ago in connection with the development of biotransformations of steroids and mineralization of polycyclic aromatics, as these compounds have low solubility in water-based reaction media (1). In biological systems, many enzymatic transformations involve solid substrates such as aggregates of biopolymers (e.g., cellulose fibrils, starch granules, blood clots, collagen, bacterial cell walls, bone, and so forth) (2). However, these reactions differ from biotransformations involving suspensions of low-molecular-weight compounds. In the former case, the adsorption of enzyme on the surface of insoluble biopolymer aggregates or particles is required and the reaction is catalyzed by the adsorbed enzyme with the formation of soluble products, which can be further degraded in the liquid phase.

It is often crucial in biotransformations of practical importance to obtain high space-time and product yields at high substrate and product concentrations. At constant enzyme loading, the space-time yield can only be increased by a factor of <2 when the substrate concentration is larger than Michaelis–Menten constant, K_m. Typically, K_m's for natural substrates are below 1 mM, but for nonnatural substrates frequently used in biotransformations, these values are often much higher [i.e., in a 100 mM range (3)]. Also, a high degree

From: *Methods in Biotechnology, Vol. 15: Enzymes in Nonaqueous Solvents: Methods and Protocols*
Edited by: E. N. Vulfson, P. J. Halling, and H. L. Holland © Humana Press Inc., Totowa, NJ

of substrate conversion is desired in practical applications and the space-time yield is difficult to increase in the upper range of substrate concentrations of 0.1 to 1 *M* because of unavoidable product inhibition. In practice, however, the solubility of many nonpolar substrates of biotechnological interest in aqueous solutions is much lower than the concentration necessary for a viable industrial biotransformation. For such substrates, the reactions can be carried out with high space-time yields in aqueous suspensions. As the concentration of reactants at the "solid–liquid"-phase boundary exceeds the respective value at the phase boundary in an analogous aqueous organic two liquid-phase system, the higher space-time yields are expected in suspensions (**Fig. 1**). Clearly, in the case where the solubility is higher than K_m, the space-time yield is only marginally increased by increasing the solubility of the substrates.

Recently, biotransformations of low-molecular-weight compounds (amino acids and their derivatives, peptides, β-lactams) in suspensions have been investigated in some detail *(3–8)*. A number of different systems were studied, ranging from low-water eutectic mixtures of substrates, with and without inert liquid adjuvants, to suspensions in reaction solvent of varying water content. Lipase-catalyzed synthesis of sugar fatty acid esters in dry organic solvent suspensions have also been performed *(9,10)*. The biotransformations studied were either equilibrium- or kinetically controlled processes (**Fig. 2**), catalyzed by free and immobilized enzymes *(11)*. Typically these reactions involve the consumption and formation of acids and bases that may change pH in the liquid phase. These pH shifts may reduce the space-time and product yields of the biotransformations, especially when immobilized enzymes are used *(12,13)*.

Generally, in low-water systems, the rate of mass transfer from the solid to liquid phase is the factor that limits the space-time yield, and pH shifts can be difficult to control by conventional procedures used in (bio)chemical engineering. The aim of this chapter is to provide the reader with guidelines and tools to achieve optimal water content and buffer capacity, when performing biotransformations in suspensions involving low-molecular-weight substrates.

1.2. Classification of Biotransformations in Suspensions

Biotransformations can be carried out in suspensions in three ways:

1. From solid substrate to dissolved product
2. From solid substrate to solid product
3. From dissolved substrate to solid product

In the first case, the yield does not increase in suspensions as compared to analogous reaction with completely dissolved reactants, but in the latter two

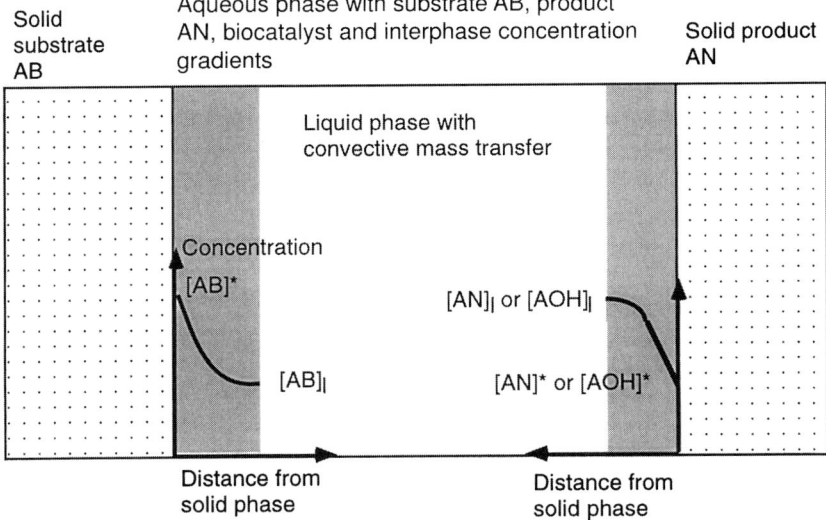

Fig. 1. Schematic representation of an aqueous suspension with solid substrate (AB) and product (AN) where the reaction takes place in the liquid phase (denoted by the subscript l) and is under equilibrium (AB + H$_2$O = AOH + BH) or kinetic (AB + NH = AN + BH) control. The asterisk denotes the concentration at the solid–liquid interface at equilibrium with the solid phase. The shaded areas denote an unstirred liquid layer on the surface of the solid particle with exclusively diffusive mass transfer. The volume of the unstirred layer can be reduced by stirring and it increases in proportion with decreasing water content in the system.

cases, the yield of the desired product will be higher in the suspension. For an equilibrium-controlled process, where the product precipitates (e.g., condensation of amino acids), the increase in yield follows from the mass action law (5). For kinetically controlled reactions, the yield increases too as a result of the reduction in the rate of product hydrolysis (**Fig. 2**).

1.3. Theory

1.3.1. Maximum Space-Time Yield in Well-Mixed Suspensions

The system analyzed here is schematically presented in **Fig. 1**. The maximum space-time yield in a system containing a suspension of one solid substrate is equal to the maximum rate of mass transfer from the solid to the liquid phase. The biocatalyst activity V_{max} required to convert this amount of substrate per unit time can be estimated from the following pseudo-steady-state condition, assuming simple Michaelis–Menten kinetics:

$$k_La\ ([AB]^* - [AB]_l) = V_{max}[AB]_l/(K_m + [AB]_l) \tag{1}$$

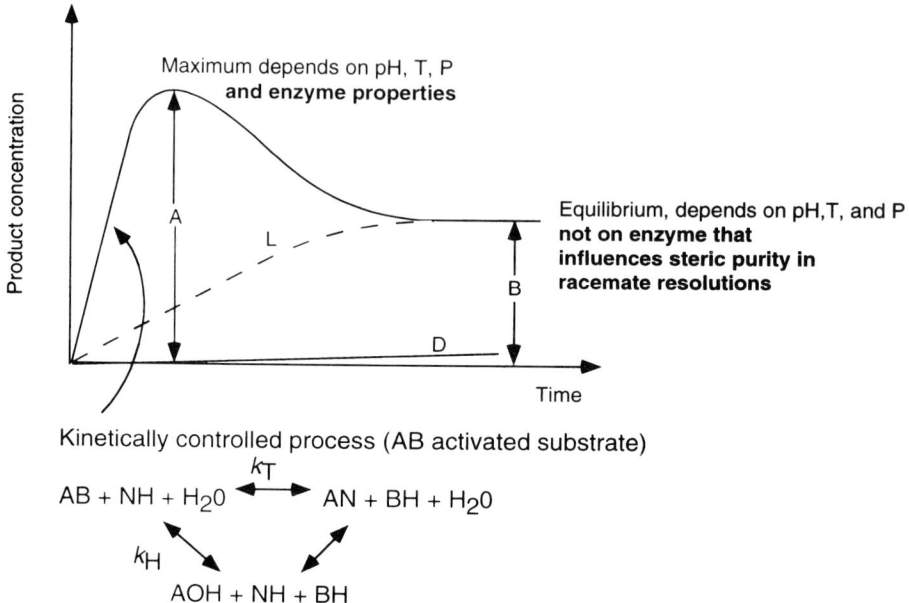

Fig. 2. Progress curves for equilibrium- or kinetically controlled biotransformations for dissolved substrates and products. A and B give the optimal end points of these processes. After the reaching the maximum, the yield in kinetically controlled reactions declines as the rate of product (AN) hydrolysis becomes larger than the rate of its formation. The loss of AN can be reduced if the reaction is carried out under conditions where the product precipitates (i.e., in suspensions).

where k_L is the mass transfer coefficient and a is the surface area per unit suspension volume. The maximum rate of mass transfer is obtained when $[AB]^* \gg [AB]_l$. When $[AB]_l > K_m$, the enzyme activity required to convert the dissolved substrate is $V_{max} > k_L a\ [AB]^*$. The estimation of V_{max} is done as follows. The solid substrate is considered to exist as spherical particles with radius $r_{AB,0}$ at the start of the reaction and occupy the volume fraction β of the suspension. Then the rate of dissolution is:

$$k_L a\ [AB]^* = \frac{3 \mathrm{Sh} D_{AB} \beta}{2 r_{AB}}\ [AB]^* \qquad (2)$$

where Sh is the Sherwood number (the ratio of the particle diameter to the thickness δ of the diffusion layer [Sh ≈ 10 in well-mixed suspensions *(14)*] and D_{AB} is the diffusion coefficient (approx 5×10^{-6} cm²/s for substrates with the molecular weight in the range 300–500 Daltons). With $\beta = 0.1$ and $[AB]^* = 10$ mM, the maximum dissolution rate or the minimum enzyme content required for the conversion is:

$$750 \times 10^{-6} M/s = 45000 \text{ U/L for } r_{AB} = 100 \text{ } \mu m$$
$$8300 \times 10^{-6} M/s = 500000 \text{ U/L for } r_{AB} = 30 \text{ } \mu m$$

It has been observed with different substrates that the particle size in suspensions is about 100 μm, reducing to 30 μm at the very latest stages of the reaction (approx 3% of the remaining substrate) because of the continuous reduction of the particle size during the biotransformation (*see* **Note 1**). Thus, the required enzyme activities can be easily obtained with commercially available enzymes in the free and immobilized forms.

To assess the potential of biotransformations in suspension for practical applications, some of the factors that may influence the space-time yield in this system should also be considered:

- The rate of mass transfer and control of pH, if required, and the dependence of these parameters on the suspension content β, which also reflects the water content of the system.
- The feasibility of using immobilized enzymes, especially when one of the products precipitates out. In this case, a method of separating the insoluble product and immobilized enzyme must be available.

1.3.2. Influence of Water Content and Control of pH

For this discussion, the solid substrate particles are assumed to be solid spheres of the same size. A constant amount of enzyme in different buffer volumes is added to a constant particle volume. To fill the empty space between the settled particles, about 30–40% (v/v) H_2O is required, depending on how dense the particles are packed. At a lower water content, some of the particles would have virtually no aqueous phase around them, unless a wetting agent is used. At such water contents, the mass transfer to the aqueous phase can hardly be influenced by stirring and the Sherwood number in **Eq. 2** will be near the value for unstirred system (= 2) *(14)*. This and the reduced interface area in the absence of the wetting agent decreases *a* (*see* **Note 2**). Particle agglomeration may reduce it further. This is expected to result in a space-time yield that increases with the water content until it becomes larger than the empty space between the particles and the volume of the unstirred diffusion layer. The volume of the unstirred diffusion layer is seven times the volume of the solid particles in an unstirred system *(14)*. The thickness of the unstirred diffusion layer ($\delta = 2$ r_{AB}/Sh) could be reduced by a factor of approx 10 as Sherwood numbers of 10–20 can be obtained by stirring or agitating the suspension *(12)*. Then, the mass transfer rate from the suspended particles into the liquid phase is increased (**Eq. 2**). Once the reaction mixture is well stirred, the shape of the curve of the space-time yield or rate as a function of water content depends on whether the system is "reaction" or mass transfer limited. In a reaction-limited system, the space-time yield increases with the enzyme activity V_{max}. In this case, the space-time yield will have a maximum, as the same enzyme activity is diluted in increasing amount of water. In a mass-transfer-limited system, the space-time yield does not depend on V_{max}, as the enzyme activity by far exceeds the rate of mass transfer from the solid to liquid phase. In this case, the space-time yield will become constant and independent of water content once the maximum rate of mass transfer has been achieved.

Indeed, this relationship has been observed experimentally in equilibrium-controlled peptide synthesis in buffered and unbuffered systems *(15)*. In such reactions, only small pH changes are expected. However, it does not apply to equilibrium-controlled hydrolysis of esters and amides, or kinetically controlled peptide synthesis *(3,16,17)*. One should also note that at low water contents, the pH is difficult to control. It is beneficial, therefore, to maintain pH near the optimum by using a buffer with a large buffering capacity (*see* **Notes 3** and **4**). It then follows from the above that the value of β selected for a biotransformation in suspension is determined by the following factors:

- β must not be too large to allow for an optimal rate of mass transfer resulting from stirring or agitating. The stirring must also be sufficient to minimize particle agglomeration that would reduce *a* and allow one to control the pH by the addition of acids or bases.
- When particles of immobilized enzymes are used as biocatalysts, their abrasion resulting from collisions with a suspended substrate (or product) must be minimal. In batch reactors, this limits the biocatalyst density to <10 % (v/v).

It follows from this that β should be selected in the range between 0.1 and 0.2.

1.3.3. Free Versus Immobilized Enzymes

For a process in which a solid substrate is converted to dissolved product (*see* **Subheading 1.2.**), immobilized enzymes can be used. In this case, the biocatalyst can be easily separated from the reaction mixture. The only additional problem to consider is the possible abrasion of the immobilized enzyme caused by collisions with substrate particles. This may limit the solid substrate content in the bioreactor.

In the other two cases in which a solid substrate is formed in the reaction (*see* **Subheading 1.2.**), the possible precipitation of products in the pores of the immobilized enzyme particles must also be considered, as it hampers the reuse of biocatalyst. The precipitation of product in the pores can be avoided by using an immobilization support with small pores (<50 nm), as the solubility increases rapidly with decreasing particle size *(18)*. This has been verified *(3)* for the equilibrium-controlled hydrolysis of D-phenylglycine-OMe (case 3 reaction in **Subheading 1.2.**).

Another problem that should be solved is the separation of the solid product from the particles of the immobilized enzyme (*see* **Note 5**).

2. Materials

1. *N*-Acetyl-L-tyrosine-ethyl ester (ATEE) and arginine amide (Arg-NH$_2$) were obtained from Serva (Heidelberg, Germany) and Bachem Biochemica (Heidelberg, Germany), respectively.
2. Bovine α-chymotrypsin (CT, E.C. 43.4.21.1) was purchased from Worthington (Freehold, NJ) and used as obtained.
3. All other chemicals and solvent (purchased from traditional suppliers) were of analytical grade.
4. Stock solutions: A 10-mg/mL stock solution of CT was prepared in 0.1 *M* formate buffer (pH 3.0) to avoid autolysis during the dissolution of the enzyme (*see* **Note 6**). It was then diluted with carbonate buffer (*I* = 0.2 *M*; pH 9.0) and kept at –20°C. A 1 *M* solution of Arg-NH$_2$ was prepared in carbonate buffers, pH 9.0, of different ionic strengths and buffer capacity (*I* = 0.2 or 1 *M*).

3. Methods

3.1. Experimental Procedures

3.1.1. Biotransformation

Different amounts of Arg-NH$_2$ stock solution were added to 250 mg ATEE in Poly-Prep columns (www.biorad.com), which were closed at the bottom. The suspension was vortexed to mix the solid and liquid phases and the reaction was started by adding 10 µL of a suitable diluted CT solution in the carbonate buffer. The suspension was mixed by placing the columns on a roller mixer (Mixer 820, Ortho Diagnostic Systems with 20 rpm) at 25°C. At different time intervals, aliquots of liquid were withdrawn from those sample that had liquid-phase volume larger than that of the solid phase and analyzed by high-performance liquid chromatography (HPLC). In the case of samples prepared at low water content, approx 5–10 mg of the suspension was taken, accurately weighed, and diluted to dissolve the solids prior to HPLC analysis. The pH of the suspension was measured using a small pH electrode (Inlab 0423) with a 3-mm tip diameter from Mettler-Toledo (www.Mettler-Toledo.com). If necessary, the pH was adjusted to 9.0 by adding 3.3 *M* NH$_4$OH.

3.1.2. HPLC Analysis

The samples were analyzed using a 10-cm RP-18 column packed with 5-µm particles (Merck, Darmstadt, Germany) at 30°C. The column was eluted isocratically with 70% (v/v) 3 m*M* KH$_2$PO$_4$ (pH 4.7) and 30% (v/v) MeOH. The peak areas at 220nm (Arg-NH$_2$) and 280 nm (ATEE, *N*-Acetyl-L-Tyr and *N*-Acetyl-L-Tyr-Arg-NH$_2$) were used to quantify the concentration of substrates and products, using the calibration curves constructed with respective standards.

3.2. Results and Discussion

The initial rate of the dipeptide formation (i.e., the initial space-time yield) as a function of water content at high buffer capacity ($I = 1\ M$) is given in **Fig. 3**. The reaction rate was found to be constant until 30% of the dipeptide product had been formed. The dependence of the rate on the water content is as predicted by the theoretical analysis for a reaction-limited system. This was found to apply as the initial rate (or space-time yield) increased with enzyme activity in the sample. The theoretical analysis assumes that the pH in the suspension does not change during the reaction (*see* **Subheading 1.3.**). Indeed, the pH in the suspension was found to decrease by less than 0.2 pH units.

This was not the case for the biotransformation carried out with a low-capacity buffer ($I = 0.2\ M$). The pH drop during the first 10 min of the reaction was found to range from 2 pH units at the highest water content to about 3 pH units at the lowest water content. It is well known that in kinetically controlled peptide synthesis, the ratio of the rate of amino acyl transfer onto the accepting amino acid (the formation of dipeptide) to the rate of hydrolysis decreases with decreasing pH *(11)*. **Figure 4** shows the dependence of this ratio on water content of the system at two different buffer capacities. The lower ratios at the lower buffer capacity are the result of the reduction in pH that is observed in the suspensions. This cannot, however, explain the reduction of the ratio at decreased water contents. One possible explanation can be offered which is based on the scheme in **Fig. 1**. When the water content of the system is reduced significantly, an increasing proportion of dipeptide molecules is formed in the unstirred diffusion layer. In an unstirred system of spherical particles the volume of the unstirred layer is seven times the volume of the suspended particles. In a stirred system with a Sherwood number of approx 10, this volume is approximately three times that of the particle *(14)*. The nucleophile Arg-NH$_2$ must diffuse into the unstirred layer where it is consumed, and, consequently, its effective concentration there will be lower than in the bulk phase. This should lead to the reduction in the ratio of the rate of transfer to hydrolysis at decreasing water content, as observed in the experiment.

3.3. Conclusions

1. Large space-time and product yields can be obtained in equilibrium- and kinetically controlled biotransformations of aqueous or organic suspensions with solid substrates and/or products *(3–10)*.
2. The maximum rate of mass transfer of reactants from solid particles into the aqueous phase is achieved when the volume of the latter is larger than seven times the volume of the solids in suspension. Only then can the volume of the unstirred diffusion layer outside the solid particles be reduced by stirring or agitating the system. Similar volume fractions are also required to control the pH in the aque-

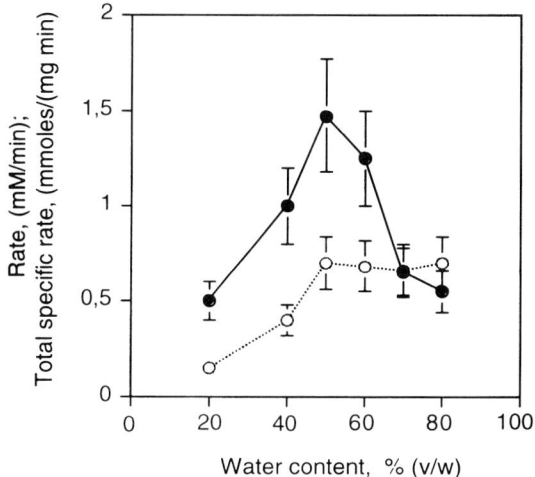

Fig. 3. Initial volumetric rate (closed symbols) and total specific activity (open symbols) for the formation of the dipeptide *N*-Acetyl-Tyr-Arg-NH$_2$ using 0.8 mmol *N*-acetyl-Tyr-*O*-Et (250 mg) suspended in different volumes of 800 m*M* ArgNH$_2$ dissolved in carbonate buffer ($I = 1$ M) in a kinetically controlled reaction catalyzed by CT (8 µg) at 25°C. The rates reflect changes in the liquid phase.

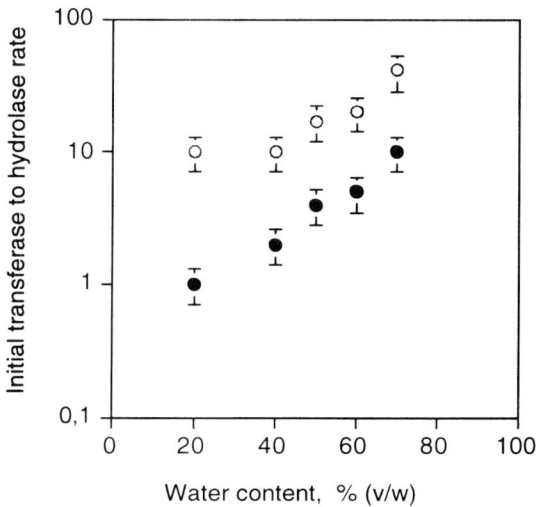

Fig. 4. The ratio of the initial rate of amino acyl transfer to the rate of ATEE hydrolysis in kinetically controlled synthesis of *N*-Acetyl-L-Tyr-Arg-NH$_2$ in suspensions of 250 mg ATEE with varying volumes of 1 M Arg-NH$_2$ at 25°C catalyzed by 8 µg CT in carbonate buffer of pH 9.0 of different buffer capacities (open symbols: $I = 1$ M; closed symbols: $I = 0.2$ M).

ous phase for reactions where acids and bases are formed or consumed. This limits the volume fraction of the suspended particles to about 0.2 or less.

3. In suspensions with immobilized enzymes the volume fraction of solids and biocatalyst should be limited to approx 10% of the total volume to minimize abrasion of the immobilized enzyme particles.

4. It follows from **steps 2** and **3** that a β of approx 0.1 should be selected. Then, the total substrate and product concentrations of low-molecular-weight compounds (<1000 Daltons) in excess of 0.1 M can be efficiently biotransformed in aqueous suspensions.

4. Notes

1. When possible, the particle size in suspension before and after the biotransformation should be checked using microscopy .

2. The use of wetting agents can only be recommended when they can be easily separated from the desired product of the biotransformation.

3. In systems in which acids or bases are produced or consumed in the aqueous phase, the pH must be kept at or near to the optimum for the biotransformation by adding buffer with a pK value around this pH. Buffers should be chemically inert (i.e., not react with intermediates such as acyl-enzymes). This excludes, for example, ammonia or Tris at pH values around 8.0, as the unprotonated amino groups can deacylate acyl-enzymes *(11,19)*. Carbonate buffers are recommended at pH > 8.0.

4. In preparative-scale and/or industrial biotransformations, the amount of buffer should be minimized, as it has to be removed from the product.

5. The sedimentation velocity of the immobilized enzyme particles generally differs from the corresponding velocities for substrates and products. In small-scale experiments, the immobilized enzyme can therefore be recovered after one or two sedimentations using a narrow filter (Bio-Rad) that can be closed at the bottom. After the sedimentation, the liquid phase is sucked off, and the layers with immobilized enzyme and solid reactants are separated. Prior to reuse, immobilized enzymes should be thoroughly washed to dissolve solid substrate and/or products associated with the particles. (Precipitated product within the particles tends to dissolve very slowly.) To check for possible abrasion, the immobilized enzyme particles should be visually inspected in a microscope or particle size distribution can be analyzed.

6. Many peptidases can autolyze. When preparing stock solutions of enzymes, high concentrations are often obtained locally during the dissolution and this may cause considerable loss of enzyme activity. This has been observed for numerous proteases but especially for trypsin where > 90% of the activity can be lost when stock solutions are prepared at neutral pH *(20)*. It is advisable therefore to prepare the stock solutions at a pH far from the enzyme's pH optimum, provided that the enzyme is stable at this pH. For trypsin and α-chymotrypsin stock solutions can be prepared in dilute HCl (pH 2.0–3.0), and kept at –20°C. Acetic acid buffers are sometimes not suitable because their use may lead to denaturation upon freezing and thawing, as acetate binds to the enzyme surface (proteins are precipitated by acetic acid derivatives such as trichloro-

acetic acid!). For preparative-scale biotransformations, a special filtration procedure has been patented *(21)*.

Acknowledgments

This work has been partly supported by Max-Buchner Forschungsstiftung and DFG (Graduiertenkolleg Biotechnologie GK 95-2).

References

1. Goodhue, C. T., Rosazza, J. P., and Peruzzotti, G. P. (1986) Methods for the transformation of organic compounds, in *Manual of Industrial Microbiology and Biotechnology* (Demain, A. L., Solomon, N. A., eds.), American Society of Microbiology, Washington, DC.
2. McLaren, A. D. and Packer, L. (1970) Some aspects of enzyme reactions in heterogeneous systems. *Adv. Enzymol.* **50,** 245–303.
3. Kasche, V. and Galunsky, B. (1995) Enzyme catalyzed biotransformations in aqueous two-phase systems with precipitated substrate and/or product. *Biotechnol. Bioeng.* **45,** 261–267.
4. Gill, I. and Vulfson, E. (1994) Enymatic ctalysis in heterogeneous eutectic mixtures of substrates, *TIBTECH* **12,** 118–122.
5. Halling, P. J., Eichhorn, U., Kuhl, P., and Jakubke, H. D. (1995) Thermodynamics of solid-to-solid conversion and application to enzymatic peptide synthesis. *Enzyme Microb. Technol.* **17,** 601–606.
6. Kuhl, P., Eichhorn, U., and Jakubke, H. D. (1995) Enzymatic peptide synthesis in microaqueous solvent-free systems. *Biotechnol. Bioeng.* **45,** 276–278.
7. Mincheva, Z., Stambolieva, N., and Petrova, K., and Galunsky, B. (1996) Penicillin amidase catalysed preparative synthesis of cephem-7-(2-benzoxazolon-3-yl-acetamido)-desacetoxycephalosporanic acid using a non-specific polyethylene-glycol-modified acyl donor. *Biotechnol. Tech.* **10,** 727–730.
8. Wolff, A., Zhu, L., Kielland, V., Straathof, A. J. J., Jongejan, J., and Heijnen, J. J. (1997) Simple dissolution-reaction model for the enzymatic conversion of suspension of solid substrate. *Biotechnol. Bioeng.* **56,** 433–440.
9. Cao, L., Fischer, A., Bornscheuer, U. T., and Schmid, R. D. (1997) Lipase-catalyzed solid phase synthesis of sugar fatty acid esters. *Biocatal. Biotransform.* **14,** 269–283.
10. Sarney, D. B., Vitro, M., Barnard, M., and Vulfson, E. N. (1997) Enzymatic synthesis of sorbitan esters using low boiling azeotropes as a reaction solvent. *Biotechnol. Bioeng.* **54,** 351–356.
11. Kasche, V. (1986) Mechanisms and yields in enzyme catalyzed equilibrium and kinetically controlled synthesis of β-lactam antibiotics, peptides and other condensation products. *Enzyme Microb. Technol.* **8,** 4–16.
12. Spieß, A., Schlothauer, R., Hinrichs, J., Scheidat, B., and Kasche, V. (1999) pH gradients in heterogeneous biocatalysts and their influence on rates and yields of the catalyzed processes. *Biotechnol. Bioeng.* **62,** 269–277.
13. Tischer, W. and Kasche, V. (1999) Immobilized enzymes: crystals or carriers? Trends Biotechnol. **17,** 326–335.

14. Kasche, V. and Kuhlmann, G. (1980) Direct measurements of the thickness of the unstirred diffusion layer outside immobilized biocatalysts. *Enzyme Microb. Technol.* **2,** 309–312.
15. Erbeldinger, M., Ni, X., and Halling, P. (1999) Kinetics of enzymatic solid-to-solid peptide synthesis: inter-substrate compound, substrate ratio and mixing effects. *Biotechnol. Bioeng.* **63,** 316–321.
16. Gill, I., Lopez-Fandino, R., Jorba, X., and Vulfson, E. N. (1996) Biologically active peptides and enzymatic approach to their production. *Enzyme Microb. Technol.* **18,** 163–183.
17. Eichhorn, U., Beck-Piotraschke, K., Schaaf, R., and Jakubke, H. D. (1997) *J. Pept. Sci.* **3,** 261–266.
18. Freundlich, H. (1909) *Kapillarchemie*, Akademische, Verlags-Gesellschaft, Leipzig, pp. 143–145.
19. Kasche, V. and Zöllner, R. (1982) Tris(hydroxymethyl)methylamine is acylated when it reacts with α-chymotrypsin. *Hoppe Seyler´s Z. Physiol. Chem.* **363,** 531–534.
20. Gabel, D. and Kasche, V. (1973) Autolysis of beta-trypsin. Influence of calcium ions and heat. *Acta Chem. Scand.* **27,** 1971–1981.
21. Clausen, K (1995) Method for the preparation of certain β-lactam antibiotics. US Patent 5, 470,717.

44

Biotransformations in Supercritical Fluids

Nuno Fontes, M. Conceição Almeida, and Susana Barreiros

1. Introduction

A substance is said to be supercritical (SC) above a singular point on the phase diagram, the so-called critical point. The use of SC fluids as solvents for enzymatic transformations is a relatively new area of research *(1,2)* that is expected to expand in the future. Close to the critical point, small changes in temperature or pressure can effect large changes in the density/solvation ability of SC-fluid. This property of SC-fluids is currently used in a wide range of extraction applications. It can also be fruitfully exploited for the integration of biotransformation and downstream processing steps in a single bioreactor. In addition, lower viscosity and higher diffusivity of SC-fluids as compared to most organic solvents enable better mass transfer, which is often a limiting factor in reaction systems where the enzyme and reactants are not contained in the same phase.

Water plays multiple roles in nonaqueous biocatalysis by enhancing the molecular mobility, and hence the catalytic activity of enzymes *(3)*, by stabilizing charged transition states *(4)* and by shifting the equilibrium between hydrolysis and synthesis. Thus, the properties of an enzyme are largely dependent on the amount of water directly associated with it; consequently, the degree of biocatalyst hydration must be carefully controlled. A convenient way to do this is by fixing the water activity (a_w) of the system *(5)*. This may be accomplished by, for example, pre-equilibrating separately enzyme, solvent, and reactants with saturated salt solutions *(3)*, although this approach is not practical when the solvent is a gas at normal temperature and pressure. However, if the system is equilibrated with water, the value of the a_w will be the same in all the phases present. Therefore, the a_w of the whole system can be set and measured in the most convenient phase *(5)*. When using supercritical

From: *Methods in Biotechnology, Vol. 15: Enzymes in Nonaqueous Solvents: Methods and Protocols*
Edited by: E. N. Vulfson, P. J. Halling, and H. L. Holland © Humana Press Inc., Totowa, NJ

fluids, the direct addition to the reaction mixture of pairs of salt hydrates that confer a certain a_w is particularly useful, and guidelines for selecting adequate salt hydrates are available *(6)*. Sodium phosphate salts have been used often in organic media, and we find that they are also effective in SC-CO_2. Using these salts, it is possible to establish the dependence of water concentration in a given solvent as a function of the a_w. The extrapolation of the data to $a_w = 1$ (water saturation) gives values that are in good agreement with those available in the literature *(7,8)*.

The protocol described here is a general method for biotransformations in supercritical fluids (e.g., SC-CO_2, SC-ethane) using biocatalyst suspensions in the form of freeze-dried powders, an immobilized enzyme, or enzyme microcrystals, and it is an update of an earlier version *(9)*.

2. Materials

1. Solid enzyme preparation.
2. Carbon dioxide, ethane, and nitrogen with purities of ≥99.95 mol%.
3. Molecular sieves, 0.3 nm pore diameter.
4. Teflon bar (for pushing the piston out of the cell at the end of an experiment).
5. Lyophilization equipment for preparing freeze-dried enzyme powders.
6. Safety glasses for using the high-pressure equipment, gloves for manipulating organic solvents and substrates.
7. Experimental apparatus. A schematic representation of the experimental apparatus used for enzymatic biotransformations in SC fluids at pressures up to 300 bar is shown in **Fig. 1**. The gas from the gas bottle (GB) at a pressure indicated by the manometer (M) is supplied through a line filter (LF). It is compressed at C and then enters a chamber with molecular sieves (MS) as a liquid. At this point, pressure is measured with the pressure transducer (PT1). Valve ADM is a chromatographic valve for admission of the substrates with syringe S1. The valve is connected to the variable volume high-pressure cell VVC via valve V6. Also connected to the cell is the two-way valve SAMP for sampling. Valve V8 is also a two-way valve that allows the release of samples directly into the titration chamber of a Karl Fischer water titrator, and collection of samples in a flask for gas chromatographic/high-performance liquid chromatography (GC/HPLC) analysis. Valve SAMP is connected via valve V7 either to the syringe S2, containing an appropriate solvent for diluting samples for analysis, or to a nitrogen line. Low-humidity nitrogen, which was further dried over molecular sieves, is used as a rinsing fluid when taking samples for Karl Fischer titrations. The back-pressure fluid used to move the piston of the cell is a liquid pressurized with the manual syringe pump (MP). The pressure in the reaction mixture can be determined indirectly from that of the back-pressure fluid, as indicated by the pressure transducer PT2. The cell assembly comprises the high-pressure cell and all the parts connected to it, up to but excluding valves V5 and V9, which are permanently connected to the solvent gas and the back-pressure fluid lines, respectively. The

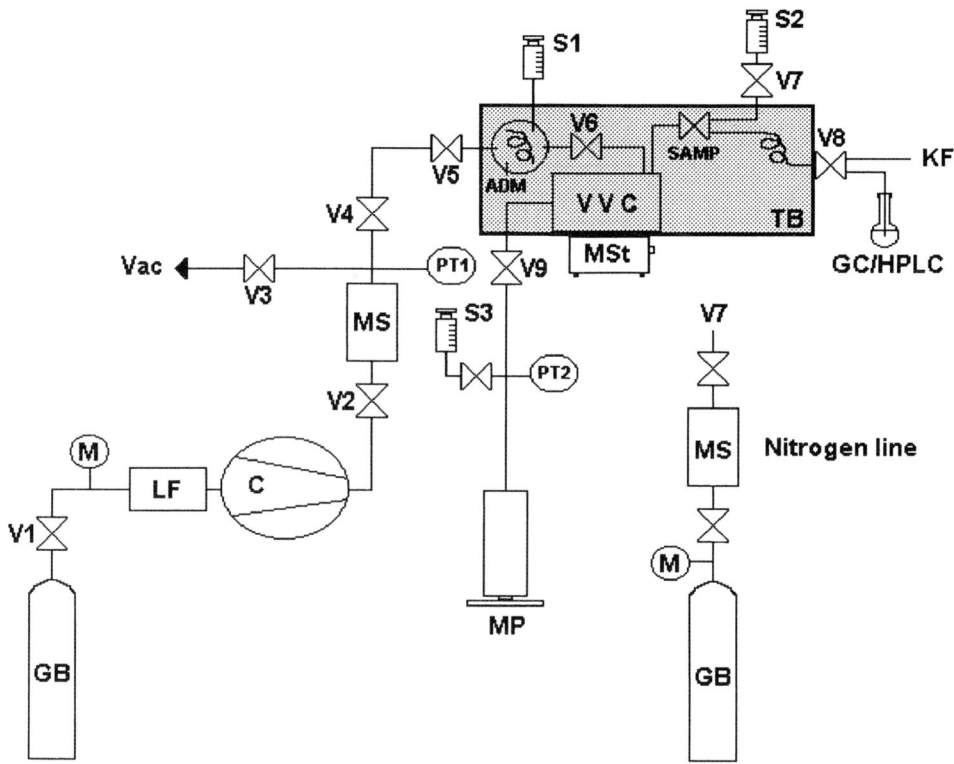

Fig. 1. Schematic representation of the experimental apparatus used for enzymatic biotransformations in SC fluids.

cell, valve V6, and the loops of valves SAMP and ADM are immersed in a thermostatted liquid bath (TB). The reaction mixture is stirred with a stirring bar (SB) operated by a magnetic stirrer (MSt).

Figure 2 is a schematic representation of the stainless steel cell VVC. The seal at both ends is provided by a Teflon O-ring (Tor). The polyacetal washer (PAWa) prevents damage to the sapphire window (SWin). The tubing connected to the back-pressure fluid line (BPF) is soldered to the rear end screw of the cell. The polyacetal piston with Buna M O-rings (BOr) separates the back-pressure fluid from the reaction mixture. A stainless steel rod (R) with marks corresponding to certain volumes of the cell (calibrated prior to experiments) is screwed onto the rear end of the piston, going through a nut with Teflon ferrules to allow for movement of the rod. The connection to valve SAMP is behind the plane of the drawing at a 45° angle with the connection to valve V6.

Figure 3 shows schematically how the chromatographic valve ADM is used. In position 1, the loop is filled with a mixture of substrates; in position 2, the content of the loop is admitted into the cell.

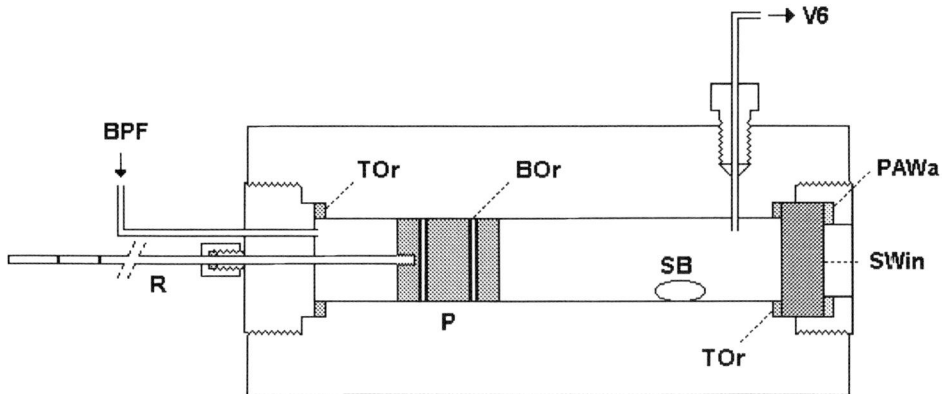

Fig. 2. Schematic representation of the stainless-stell cell VVC.

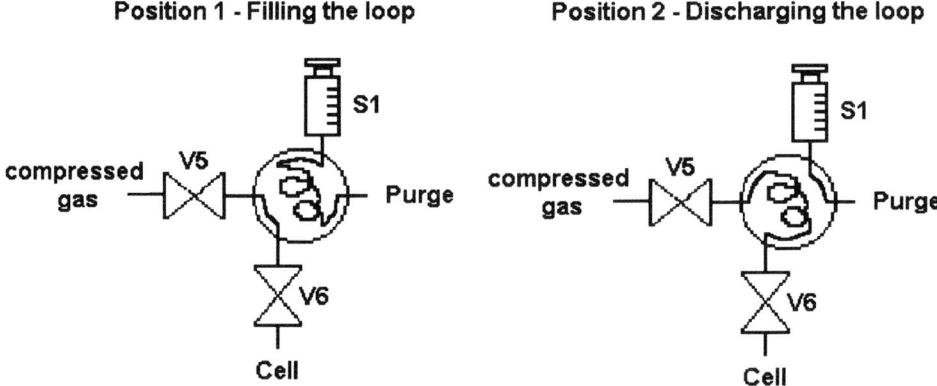

Fig. 3. Schematic representation of the use of chromatographic valve ADM.

3. Methods

1. Take the cell assembly and the loose parts referred to in **step 14** from the oven and allow them to cool down (this should take about 30–60 min). Close all the valves. Cover the mouth of the needle of syringe S1 and the orifice on valve V7 with parafilm. (The tubing soldered to the needle for syringe S2 will be connected later.)
2. Place the Buna M O-rings on the piston and screw the metal rod onto it. Mark the rod with a fine pen at points corresponding to the desired initial and final volumes of the cell, V_i and V_f (see also **steps 3,5**, and **9**). Push the piston into the cell, from the back, so that the rear end of the piston stays about 5 mm below the bed of the Teflon O-ring. Put the Teflon O-ring in place. Allow the rod to go through the small nut with Teflon O-rings and screw on the rear end of the cell, with the cell

held tightly on a bench vice (do not bend the tubing for connection to the back-pressure fluid too close to the soldered joint). Adjust the torque on the small nut to avoid leaks of the back-pressure fluid, while allowing the rod to move with the piston.

3. Connect the back of the cell to the back-pressure fluid line via valve V9. Open V9. Pressurize the back-pressure fluid with a manual pump to push the piston forward to a position corresponding to mark V_i on the rod. Keep valve V9 open.

4. Hold the cell in the vice so that its front end is slightly tilted upward. Introduce the enzyme, all substrates but one, any additional reaction mixture components (e.g., salt hydrate pairs), and a stirring bar. Put the Teflon O-ring in place. Put the polyacetal washer in position on the front end screw of the cell where it should have a tight fit and let the sapphire window rest on it. Screw on the front end screw. Do not connect syringe S2 to valve V7. Do not connect the KF line of valve V8 to the Karl Fischer apparatus. Do not connect syringe S1 to valve ADM. Place the cell assembly in a thermostatted bath. Connect the cell to the solvent admission line via valve V5.

5. Open valves V1, V2, and V4 and pressurize the solvent gas to a pressure about 50 bar higher than the required reaction pressure. Make sure valve ADM is in position 1 in **Fig. 3**. Open valve V5 gently, and admit the liquefied solvent up to valve V6. Close valve V5. Open valve V6 gently to admit a small amount of solvent into the cell. Keep valve V6 open. Open valve V5 and admit the solvent until the reading in the pressure transducer PT1 drops to the vapor pressure of the gas solvent. (The piston tends to go slightly backward.) Close valve V5. Immediately raise the pressure of the back-pressure fluid to keep the piston in its original position (*see* **Note 1**). Pressurize the solvent gas circuit, to a pressure of about 50 bar higher than the required reaction pressure. Open valve V5 to admit more solvent into the cell until PT1 gives a reading of about 20 bar higher than that of the back-pressure fluid on PT2. Once again, check the mark on the rod. Repeat the process until the cell is filled with the solvent at the desired pressure and volume, as indicated by the pressure transducer PT2 and the mark V_i on the metal rod. Close valve V6. Start stirring.

6. Check for leaks, in particular for gas bubbles from parts immersed in the thermostated bath and for a pressure drop on PT2 (care should be taken to avoid overtightening the nut through which the rod is allowed to move). Final adjustments should be made once the temperature of the cell has stabilized, which takes place about 1 h after starting the admission of the solvent. Valve V9 is kept open to monitor the pressure via PT2. The reaction mixture should be allowed to equilibrate with water before taking samples; in the case of a_w control via salt hydrate pairs, the rate of water transfer depends on the salts chosen.

7. To take a sample for Karl Fischer titration, connect valve V7 to the nitrogen line. Adjust the pressure regulator of the nitrogen bottle to a relative pressure of 0.5 bar. Open valves V7 and V8-to-KF. Allow the nitrogen to flow through the lines and into the atmosphere for about 10 min to eliminate any humidity in the system. With nitrogen still flowing, immerse the KF line in the solution

of the Karl Fischer apparatus and allow the nitrogen to bubble through the solution until a constant drift is reached (the drift is the amount of water (in μg) that enters the titration chamber per minute). Register the drift (the average drift if the values fluctuate slightly) — drift 1. Stop stirring the reaction mixture to allow solids to settle down. While this takes place, close valves V7, SAMP-to-V7, and V8-to-KF and wait until the drift reaches a constant value — drift 2. Also *see* **Notes 2–4**.

8. Taking a sample causes a pressure drop in the cell. The magnitude of the drop depends on the volume of the sampling loop and the compressibility of the reaction mixture. When operating sufficiently close to the critical point of the latter and for loops with a volume of 1–2% of the volume of the cell, the pressure drops should not exceed 20 bar. Thus, increase the pressure of the back-pressure fluid, P_1, by about 20 bar (ΔP) and open valve SAMP-to-V8. If ΔP was chosen correctly, the pressure should recover to the value of P_1; if not, do the necessary readjustment. Close valve SAMP-to-V8. Press the start button on the Karl Fischer titrator and slowly open valve V8-to-KF so that the content of the sampling loop is discharged directly into the titration chamber. Once the gaseous solvent stops bubbling through the solution, open valves SAMP-to-V7 and V7 to rinse the expansion zone with nitrogen. Register the time interval between the opening of valve V7 and the end point of the titration — Δt. The drift will eventually stabilize (drift 1). In the second titration, drift 1 is usually, but not necessarily, similar to that determined as described in **step 7**. Further water titrations are done in the same way. Use the adequate ΔP and keep stirring off.

9. To initiate the reaction, close valves V7 and V8 and resume stirring. Prepare an adequate mixture of substrates — the missing substrate plus appropriate amounts of other substrate(s) that have already been added to the solvent (*see* **Note 5**). About a fivefold excess of this mixture over the volume of the loop of valve ADM should be prepared. Make sure valves V6 and SAMP are closed and that the pressure reading in PT1 exceeds that in PT2 by at least 50 bar. Open and immediately close valve V5; the latter condition should remain. Connect syringe S1 to valve ADM. Introduce the mixture in syringe S1, wash, and then fill up the loop with it. Turn valve ADM to position 2. Open valve V6 slowly, thereby releasing the contents of the loop into the cell with solvent. This marks the start of reaction. As before, alternately admit solvent and back-pressure fluid so as to reach the desired pressure and volume — final volume of the cell, mark V_f on the rod — at the selected temperature (*see* **Note 1**). Keep valve ADM in position 2 (the solvent in the loop is released at the end of the experiment). Wash the substrate admission lines (this is particularly important in the case of solid substrates that may plug the lines once the liquid in which they were dissolved evaporates).

10. Close valves V6, V5, V4, V2, and V1 while keeping valve V9 open. Periodically, stop stirring, wait until all solids settle down, and take a sample to follow the reaction. To do this, connect syringe S2 to valve V7. Increase the back-pressure fluid pressure by ΔP as described in **step 8**, open valve SAMP-to-V8, immediately make any necessary pressure adjustments, and close valve SAMP-to-V8. Take a volumetric flask with some of the collecting solvent and immerse the

GC/HPLC line in it. By slowly opening valve V8-to-analysis, allow the gaseous solvent to bubble through the collecting solvent (it is important that the gas be released very gently). Once the gas stops bubbling, introduce an appropriate amount of the collecting solvent into the syringe S2, open valve V7, and wash the contents of the sampling loop into the volumetric flask. Disconnect syringe S2, fill it up with air, reconnect it, and push the collecting solvent that remained in the lines into the flask. Add the solvent directly to the flask to reach the marked volume. Close valves V7 and V8. Typically, six samples are taken for analysis. The volume of collecting solvent chosen must be sufficient for quantitative recovery of the content of the loop.

11. At the end of experiment (*see* **Note 6**), with valve V9 open, depressurize the back-pressure fluid down to atmospheric pressure. The piston will move backward and a meniscus that indicates the presence of vapor and liquid phases should form. Close valve V9. Disconnect the cell assembly from valves V5 and V9 and take it out of the thermostated bath.

12. To avoid frequent disassembling and reassembling of connections to the cell, the latter is emptied through valve V7. In a hood, disconnect syringe S2, open valve V7, and then slowly open valve SAMP-to-V7 to release the solvent. After that, unscrew the nut with Teflon ferrules that holds the metal rod so that it will be possible to unscrew the rear end screw of the cell. Remove this screw and the Teflon O-ring, which may or may not be reutilized, depending on the pressure at which the experiment was performed. (O-rings are often too deformed after being extruded at 300 bar and hence discarded.) Unscrew the front end-screw of the cell, remove the Teflon O-ring (the same consideration applies), the polyacetal washer, and the sapphire window. Use the Teflon bar referred to in **Subheading 2.4., item 4** to push the piston backward, out of the cell. Remove the Buna M O-rings carefully with a stylus and discard them.

13. Wash the interior of the cell, all the lines of the cell assembly, loops, piston, O-rings, washer, and sapphire window, first with water and then with acetone, using a syringe. Make sure no solids remain in the tubing. Blow out the remaining acetone, first with a syringe, then with nitrogen (compressed air usually has a higher level of humidity).

14. Place the cell assembly with all valves open and the loose parts — sapphire window, polyacetal washer, piston, Teflon O-rings, Buna M O-rings, stir bar, tubing, and needle for connecting syringe S2, syringe S1 — in an oven at 60°C and leave them there to dry overnight. Make sure the liquid substrates are stored over molecular sieves.

15. It is important that there are no leaks; hence, the apparatus should be pressure tested with nitrogen after replacing parts.

4. Notes

1. The reaction mixture must never be exposed to surfaces previously in contact with the back-pressure fluid. The allowed volume change will depend on the distance between the two O-rings of the piston. Include the volume of the stirring bar in all cell volume calculations.

2. The concentration of water in the reaction mixture inside the cell, $[H_2O^{solv}]$, is given by:

$$[H_2O^{solv}] = \text{Volume of sampling loop/Karl Fischer reading} - \Delta t \text{ (drift 1 – drift 2)}$$

Here, $[H_2O^{solv}]$ is in grams of water per cubic decimeter of solvent mixture, the Karl Fischer reading is in micrograms of water, Δt is in minutes, drift 1 and drift 2 are in micrograms of water per minute and the volume of the sampling loop is in microliters. The meaning of the parameters Δt, drift 1, and drift 2 is given in **steps 7 and 8**.

3. Taking a sample for Karl Fischer titration involves careful opening of valve V8 to release the gaseous solvent into the titration chamber of the apparatus, waiting until the gas stops bubbling (a), opening valve V7 and rinsing with nitrogen (b). Between (a) and (b), there is a time period when the water intake is low. In order to avoid this being considered as the end point of the titration, an adequate value for the delay function of the Karl Fischer apparatus should be selected. Karl Fischer readings will vary with the delay selected and will become constant once the delay is set at an appropriate value (e.g., 120 s).

4. Sometimes, the first of the three samples taken for Karl Fischer titration gives too high a value and has to be discarded. This most likely reflects insufficient drying of the sampling loop and/or small-diameter tubing.

5. The last substrate is washed into the cell with solvent. To ensure that this addition does not change the concentration of the substrate(s) added initially, the last substrate should be mixed with appropriate amounts of those already present. Knowledge of the cell volumes V_i and V_f of the number of samples taken for Karl Fisher titration and the volume of the sampling loop and of the desired concentrations of substrates (which should have a safety margin relative to the solubility limit under the conditions of the experiments) allows the calculation of the amounts that should be added through valve ADM and also the dimensioning of the loop of valve ADM.

6. Because the catalytic performance of enzymes in SC-CO$_2$ can be orders of magnitude slower than in, for example, SC-ethane *(7,8)*, experiments in the former solvent may take considerably longer under otherwise identical conditions.

References

1. Randolph, T. W., Blanch, H. W., Prausnitz, J. M., and Wilke, C. R. (1985) Enzymatic catalysis in a supercritical fluid. *Biotechnol. Lett.* **7,** 325–328.
2. Hammond, D. A., Karel, M., Klibanov, A. M., and Krukonis, V. J. (1985) Enzymatic reactions in supercritical gases. *Appl. Biochem. Biotechnol.* **11,** 393–400.
3. Partridge, J., Dennison, P. R., Moore, B. D., and Halling, P. J. (1998) Activity and mobility of subtilisin in low water organic media: hydration is more important than solvent dielectric. *Biochim. Biophys. Acta* **1386,** 79–89.
4. Xu, Z.-F., Affleck, R., Wangikar, P., Suzawa, V., Dordick, J. S., and Clark, D. S. (1994) Transition state stabilization of subtilisins in organic media. *Biotechnol. Bioeng.* **43,** 515–520.

5. Bell, G., Halling, P. J., Moore, B. D., Partridge, J., and Rees, D. G. (1995) Biocatalyst behaviour in low-water systems. *Trends Biotechnol.* **13,** 468–473.
6. Zacharis, E., Omar, I. C., Partridge, J., Robb, D. A., and Halling, P. J. (1997) Selection of salt hydrate pairs for use in water control in enzyme catalysis in organic solvents. *Biotechnol. Bioeng.* **55,** 367–374.
7. Almeida, M. C., Ruivo, R., Maia, C., Freire, L., Corrêa de Sampaio, T., and Barreiros, S. (1998) Novozym 435 activity in compressed gases. water activity and temperature effects. *Enzyme Microb. Technol.* **22,** 494–499.
8. Fontes, N., Almeida, M. C., Peres, C., Garcia, S., Grave, J., Aires-Barros, M. R., et al. (1998) Cutinase activity and enantioselectivity in supercritical fluids. *Ind. Eng. Chem. Res.* **37,** 3189–3194.
9. Corrêa de Sampaio, T. and Barreiros, S. (2000) Transesterification reactions catalyzed by subtilisin Carlsberg suspended in supercritical carbon dioxide and in supercritical ethane, in *Methods in Biotechnology, vol. 13: Supercritical Fluid Methods and Protocols* (Williams, J. R. and Clifford, A. A., eds.), Humana, Totowa, NJ, pp. 179–188.

45

Reverse Micellar Systems

General Methodology

Andrey V. Levashov and Natalia L. Klyachko

1. Introduction

Intensive investigations of enzymes in reverse micelles began more than 20 yr ago, and since then micellar enzymology has become a research area in its own right (*see* reviews in **refs. 1–9**). In early days, the systems of "surfactant-water–organic solvent" were considered as a model for biomembranes, and the entrappment of enzymes into surfactant aggregates (micelles) was seen as an approach to better understanding the role of membranes in biocatalysis (*4*). It should be stressed, however, that numerous applications of reverse micelles to biotechnology and particularly to bio-organic synthesis, including transformations of water-insoluble substrates, have also been suggested (*see*, for example, the review in **refs. 10, 11–16**, and **Note 1**). The rational for trying reverse micelles as a medium for biotransformations is that the position of equilibrium for numerous reactions of practical interest is unfavorable in aqueous solutions, and many conventional chemical reagent are poor, if at all soluble, in water.

There are three [and only three (*17*)] issues that one has to address when carrying out enzymatic reactions in nonaqueous systems: (1) solvent selection, (2) biocatalyst preparation (e.g., immobilization), and (3) spatial separation of the solvent and biocatalyst. The first two are "system specific" (i.e., require optimization for a particular combination of the enzyme and solvent), whereas the third is more general in the sense that the approach to spatial separation is often independent of the enzyme–solvent combination used. Perhaps the simplest illustration of this latter principle is the use of aqueous organic two-phase reaction systems, where the enzyme is contained in the aqueous phase and substrates/products are dissolved in a water-immiscible organic solvent

From: *Methods in Biotechnology, Vol. 15: Enzymes in Nonaqueous Solvents: Methods and Protocols*
Edited by: E. N. Vulfson, P. J. Halling, and H. L. Holland © Humana Press Inc., Totowa, NJ

(18). The variation in the nature of the solvent and the volume ratio between the two phases enables one to change the apparent equilibrium of the reaction and obtain a higher yield of the product *(19,20)*. Isolation of products from the organic solvent is not too difficult and this is a clear technological advantage of the system. However, the necessity of providing vigorous agitation to increase the mass transfer and inactivation of enzymes at the aqueous–organic interface are obvious drawbacks. These, to a large extent, can be overcome by producing a fine stabilized dispersion of one phase in another.

The high degree of "dispersion" of two-phase systems can be achieved by preparing a water-in-oil (w/o) type of microemulsion. These microemulsions (or hydrated reverse micelles) consist of water droplets, containing enzymes and other water-soluble compounds, which are surrounded by stabilizing layer of surfactant and dispersed in organic solvent (**Fig. 1**). In other words, the micellar system is an extreme case of the aqueous–organic two-phase system; consequently, their characteristics are rather similar. Both enable the compartmentalization of water- and organic-soluble reactants in the respective phases and allow control over the position of equilibrium to favor synthesis over hydrolysis.

As compared to two-phase systems, microemulsions contain an additional component, a surfactant, which stabilizes the interface and influences both the state of micellar matrix and the behavior of enzymes entrapped (*see* **Subheading 3.**). It should be mentioned that the surfactant can also have a dramatic effect on the thermodynamics of the reaction. **Figure 2** shows the position of equilibrium in oxidation of iso-butanol to iso-butyraldehyde, catalyzed by alcohol dehydrogenase, in the water–octane two phase system (curve 1) and in the microemulsion based on hydrated micelles of Aerosol OT in octane (curve 2). In this admittedly rather extreme case, the difference in equilibrium constants between the two reaction systems is about five orders of magnitude, and the effect observed is obviously caused by the presence of the surfactant.

2. Materials

Reagents for the preparation of reverse micelles (e.g., surfactants, solvents, buffers, and enzymes) can be obtained from traditional suppliers of fine chemicals. When using enzymes, purified preparations are preferred to obtain higher catalytic activity and to avoid possible artifacts.

3. Methods
3.1. Entrapment of Proteins into Reverse Micelles

Proteins (enzymes) can be entrapped into surfactant reverse micelles in organic solvent using one of the following procedures. The first procedure has been proposed by ourselves *(14)* and is now used most extensively. It is known as the "injection" method. According to this method, a small amount of aqueous

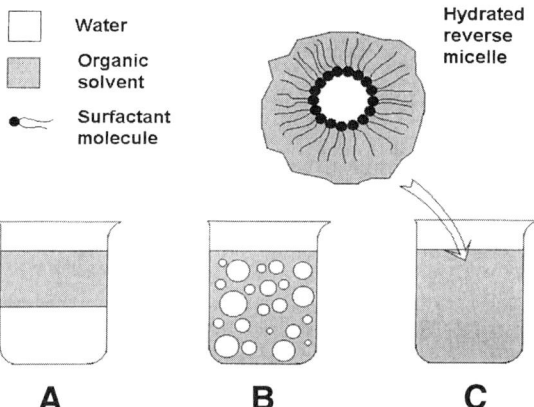

Fig. 1. Some special cases of aqueous organic two phase systems: (**A**) the simplest system consisting of water and water-immiscible organic solvent; (**B**) water-in-oil emulsion; and (**C**) surfactant-stabilized microemulsion (reverse micelles).

protein solution (typically, several percent by volume) is added to a solution of surfactant in organic solvent (dry or slightly hydrated). The actual volume ratio of the aqueous to organic solutions is determined by experimental conditions (e.g., the degree of hydration of the surfactant, its concentration, etc.). The resulting mixture is vigorously shaken until an optically transparent solution is obtained. This procedure is very simple and effective, although, in some instances, the question of time required to attain equilibrium in this system has been raised (*see*, for example, **ref. 22**). Generally, long equilibration times (as well as undesirable side reactions), if a problem, can be avoided by separately preparing micellar solutions of reagents with the required degree of hydration and then mixing them in different proportions. As a result, there is no change in the hydration of the surfactant molecules; hence, the size of micelles remains constant. (The change in the size of micelles is believed to be the main reason for slow equilibration and the occurrence of side reactions.)

The second procedure, proposed by Menger and Yamada *(23)*, consists of the initial introduction of the required amount of water into the surfactant solution in the solvent in order to attain the required degree of hydration (w_0). This is followed by dissolution of dry (lyophilized) protein in the micellar solution under vigorous shaking (*see* **Note 2**). Usually, an excess of the protein is used and the undissolved material is removed by centrifugation. A new portion of the protein is then added to the supernatant and the abovementioned procedure is repeated several times with careful control of the surfactant and water content of the supernatant (*see* **Note 3**). One of the drawbacks of this method is a prolonged contact between the enzyme molecules and the organic solvent/sur-

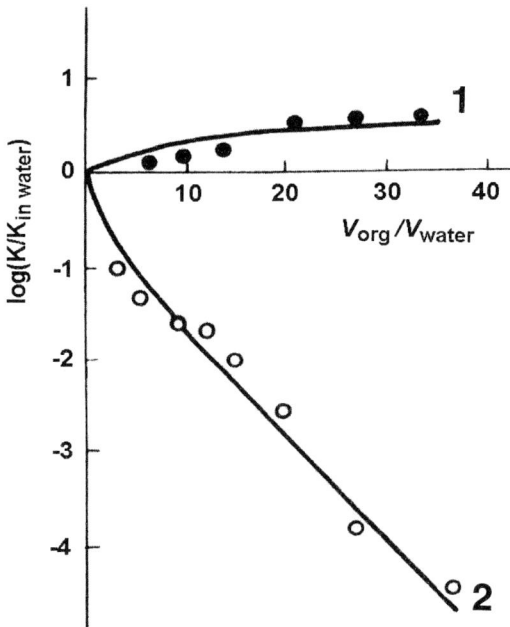

Fig. 2. The dependence of the equilibrium constant of the reaction of oxidation of iso-butanol into iso-butyraldehyde (catalyzed by horse liver alcohol dehydrogenase) on the volume ratio of organic to aqueous phase (V_{org}/V_{water}). Curve 1: Two-phase system (not containing surfactant) octane/water (0.02 M phosphate buffer, pH 8.8); Curve 2: the system of reverse micelles of AOT–water (0.02 M phosphate buffer, pH 8.8)–octane at initial AOT concentration of 0.7 M, 20°C. (From **ref. *1***.)

factant, as this may lead to partial denaturation of the former. By using this procedure, it is difficult to control the changes in the content of impurities (and contaminants in the proteins preparation), such as other proteins or salts. However, the method allows one to obtain a much higher final concentration of enzymes in the system as compared to the one described previously.

The third procedure, first described by Hanahan *(13)* and developed by Luisi and coworkers *(24,25)* (see also reviews in **refs. *2*** and **26**), is based on the phenomenon of spontaneous interfacial transfer of proteins in a two-phase system that consists of approximately equal volumes of an aqueous protein solution and an organic solvent containing surfactant (i.e., micellar system). However, it is relatively slow to equilibrate (*see* **Note 4**) and the interaction of proteins with the interface could lead to their inactivation. The degree of protein inclusion into the micellar phase can be regulated by, for example, pH and ionic strength. In our opinion, this method is more useful in the context of protein purification than for applications to bio-organic synthesis.

3.2. Separation of the Components of Micellar Systems: Product Isolation

Micellar systems are microheterogeneous (or pseudohomogeneous) and, consequently, the separation of one particular component from the system is a difficult task. However this has to be accomplished in preparative syntheses to recover the reaction product. Some general approaches to the problem are outlined as follows:

The simplest and the most elegant way of achieving this is a phase separation that can be caused by changing some physical parameters of the system (e.g., temperature *[27]*). Altering the temperature sometimes leads to the formation of two phases, the organic solvent phase enriched with the reaction product and containing some surfactant, and the aqueous phase where the surfactant is preferentially concentrated. The final product is then worked up after solvent evaporation. Another method of isolating low-molecular-weight products from micellar systems without destroying the micelles involves the use of semipermeable membranes. This approach is more suitable for the use in continuous bioreactors.

To recover water-soluble components (including water-soluble products), the addition of water or an aqueous salt solution to obtain a macrobiphasic system is probably the method of choice. The addition of water-miscible organic solvents, such as acetone or alcohol, is often used to "destroy" the micellar system irreversibly. The resulting organic solution, containing the surfactant and reaction products, is then subjected to chromatography. This method is also effective when it is necessary to isolate the protein or enzyme in an active state *(28)*. Varying the nature and the proportion of organic solvent added, it is also possible to form two separate phases and to concentrate the compound of interest in one of them *(29)*.

Finally, one may consider a reversible phase transfer as a general approach to the separation of components in micellar systems. In this case, a liquid-crystalline phase is formed with an enzyme entrapped into surfactant aggregates that are insoluble in the bulk organic solvent *(29)*. To avoid the necessity of dealing with residual soluble surfactants, one can also consider the use of immobilized micelles, polymeric micelles, micelle-containing organic gels, or polymerizable surfactants *(30)*.

3.3. Enzyme Activity in Reverse Micelles

In general, enzyme kinetics of many reactions in micellar systems obey the classical Michaelis–Menten equation. According to this equation, the reaction rate is determined by the Michaelis constant (K_M), catalytic constant (k_{cat}), and the concentration of substrate (*see* **Note 5**). The actual efficiency of an enzyme is characterized by V_{max} or k_{cat}, the latter being a first-order constant representing intrinsic properties of the enzyme. It should also be noted that the catalytic constant is independent of the effects of concentration and mass transfer (diffusion).

One of the most striking phenomena of micellar enzymology is the bell-shaped dependence of k_{cat} on the degree of surfactant hydration (w_0), which is defined as the ratio of molar concentrations of water to surfactant and determines the size of reverse micelles **(Fig. 3)**. The maximum activity is usually observed at those w_0 when the size of the "inner cavity" of reverse micelle is equal to the size of the entrapped protein *(32)* (see inset in **Fig. 3**, where the correlation between radii of different enzymes [r_p] and corresponding optimal aqueous micellar cavities [r_m] is shown). It should be noted that r_p represents an effective radius (i.e., the radius of the sphere of the same volume as that of the enzyme molecule). Then, it can be calculated from the molecular mass (M_p) of the protein (assuming that the protein has more or less spherical form) using

$$r_p = 0.7 \sqrt[3]{M_p}$$

For a widely used surfactant, Aerosol OT [*bis*-(2-ethylhexyl) sulfosuccinate sodium salt], the radius of the inner core of the micelle (r_m) is in a good correlation with w_0:

$$r_m = 4 + 1.5w_0$$

These two simple equations enable one to evaluate *a priori* the value of $w_{0,opt}$ at which an enzyme of known molecular mass is expected to display maximum activity:

$$w_{0,opt} = 0.5 \sqrt[3]{M_p} - 2.7$$

In the case of nonspheric proteins, deviations from the correlation between r_p and r_m are found. Typically, the highest catalytic activity is observed in reverse micelles with the diameter of the aqueous cavity equal to the longest dimension of the enzyme molecule *(32)* (see also the review in **ref. 6**). A possible reason for this is given in **Note 6**.

3.4. Synthetic Reactions in Reverse Micelles

In principle, reverse micelles (microemulsions) are as suitable for carrying out enzymatic reactions involving poorly water-soluble substances as aqueous–organic two-phase and other solvent-based systems. A classic example of transformation of water-insoluble substrates in reverse micelles is the elegant stereospecific reduction of progesterone with concurrent coenzyme regeneration *(33)*. The application of phospholipase *(13)* has already been mentioned, and, not surprisingly perhaps, there is a large body of published work dealing with lipolysis and lipase-catalyzed (trans)esterifications. However, there are only a few

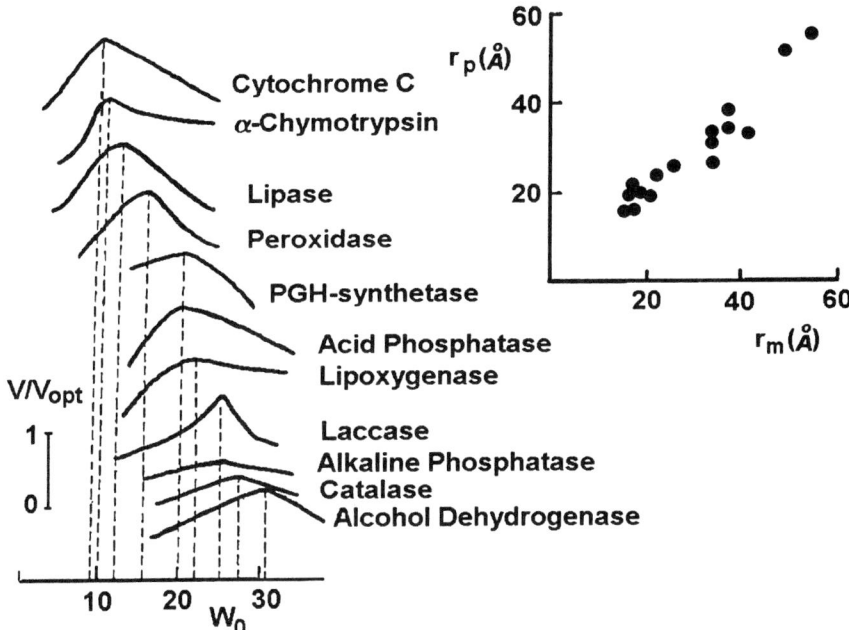

Fig. 3. Regulation of the relative catalytic activity (V/V_{opt}) of different enzymes solubilized in Aerosol OT reverse micelles, by variation of the surfactant hydration degree, $w_0 = [H_2O]/[Surf.]$. Inset: correlation between effective radii of entrapped enzyme (r_p) and optimal aqueous micellar cavities (r_m).

examples of other synthetic reactions carried out in reverse micelles (*see* the review in **ref. 7**). This is mainly the result of the product recovery from micellar systems being somewhat hampered by the presence of surfactants. However, reverse micelles offer a unique advantage of being individual "nanobioreactors" and this property may be fruitfully exploited in numerous other applications.

Thus, the use of reverse micelles enables one to introduce a limited number of hydrophobic groups into proteins to make them "membrane active" (*33a,34*), to carry out selective modification of proteins with water-insoluble reagents such as steroids (*35*) and organometallics (*36*), to synthesize protein–polymer conjugates (*see* **refs. 12,37,** and **38** and **Note 7**), and the perform polymerization reactions (*39,40*), including those involving proteins (*41,42*) (*see* also **Note 8**).

4. Notes

1. It should be mentioned that the first, in our opinion, research in micellar enzymology (published by Hanahan *[13]* in 1952) dealt with the phospholipase-catalyzed synthesis of lysolecithin from phosphatydilcholine, which was solubilized in diethyl ether with the help of a substrate in a micellar form.

2. The time required to dissolve dry proteins is usually much greater than that needed to prepare a protein solution in water. The former may take from several minutes to several hours.

3. The loss of both surfactant and water may occur when separating the undissolved material. This can be checked chromatographically (*see*, for example, **ref.** *43*). Also, the water content of reverse micelles can be determined by nuclear magnetic resonance (NMR) *(44)*. The surfactant concentration in micellar solution can be determined gravimetrically (i.e., by weighing out the dry residue after evaporation of the solvent). Note that in order to calculate the degree of hydration correctly, it is necessary to know the water content of the solvent and surfactant used. To this end we have successfully used infrared spectroscopy (water band has characteristic frequency 3420 cm^{-1}) *(45)*.

4. It may take from several minutes up to several days to reach equilibrium in this system. A gentle stirring, although it is not essential, tends to accelerate the process.

5. In reverse micelles, the substrate partition between phases may play a role in the enzyme regulation through *Km* when, for example, the local concentration of substrate is increasing because of changes in the ratio of the bulk organic to the aqueous micellar phase *(31,46)*. Such concentration effects can control parameters of the reaction, including equilibrium constant.

6. In our opinion, the main reason for observing the maximum catalytic activity under the conditions where the size of a protein molecule and the dimensions of the cavity are the same is a tight "fixation" of the protein via close contact with the inner micellar interface. This makes the protein more rigid and helps the enzyme to maintain its catalytically active conformation *(47,48)*. Further increase in rigidity (and the activity) can be achieved by partial replacement of water in reverse micelles with a water-miscible organic solvent *(49,50)*.

7. At a low degree of hydration (small micelle size), the reaction of intermolecular coupling between a protein and a polymer (polyelectrolyte) does not occur because the reagents are being held in separate micelles. As soon as w_0 reaches a critical value (threshold) (i.e., the size of micelles becomes large enough to accommodate the complex), the yield of the coupling reaction increases dramatically, often reaching 100% *(38)*. This approach can also be used to control the composition and catalytic activity of oligomeric enzymes in micellar systems *(51,52)*.

8. For example, enzymes and a modified copolymer of polyethylene- and polypropylene-glycole can be used as macromonomers to give granules of highly thermostable biocatalyst in the form of nanoparticles that are soluble both in water and organic solvents *(41,42)*. These stabilized enzyme nanoparticles can be used for carrying out synthetic reactions in organic solvents *(42)*.

References

1. Matinek, K., Levashov, A. V., Klyachko, N. L., Khmelnitsky, Yu. L., and Berezin, I. V. (1986) Micellar enzymology. *Eur. J. Biochem.* **155,** 453–468.
2. Luisi, P. L. and Magid, L. J. (1986) Solubilization of enzymes and nucleic acids in hydrocarbon micellar solutions. *CRC Crit. Rev. Biochem.* **20,** 409–474.

3. Luisi, P. L., Giomini, M., Pileni, M. P., and Robinson, B. H. (1988) Reverse micelles as hosts for proteins and small molecules. *Biochem. Biophys. Acta* **947,** 209–246.

4. Martinek, K., Klyachko, N. L., Kabanov, A. V., Khmelnitsky, Yu. L., and Levashov, A. V. (1989) Micellar enzymology: its relation to membranology. *Biochim. Biophys. Acta* **981,** 161–172.

5. Khmelnitsky, Yu. L., Kabanov, A. V., Klyachko, N. L., Levashov, A. V., and Martinek, K. (1989) Enzymatic catalysis in reversed micelles, in *Structure and Reactivity in Reverse Micelles* (Pileni, M. P., ed.), Elsevier, Amsterdam, pp. 230–261.

6. Klyachko, N. L., Levashov, A. V., Khmelnitsky, Yu. L., and Martinek, K. (1991) Catalysis by enzymes entrapped in hydrated surfactant aggregates having various structures in organic solvents, in *Kinetics and Catalysis in Microheterogeneous Systems* (Gratzel, M., Kalyanasundaram, K., eds.), Marcel Dekker, New York, pp. 135–181.

7. Oldfield, C. (1994) Enzymes in water-in-oil microemulsions ("reversed micelles"): principles and applications. *Biotechnol. Genet. Eng. Rev.* **12,** 255–327.

8. Nicot, C., and Waks, M. (1995) Proteins as invited guests of reverse micelles: conformational effects, significance, applications. *Biotechnol. Genet. Eng. Rev.* **13,** 268–314.

9. Tuena De Gomez-Puyou, M. and Gomez-Puyou, A. (1998) Enzymes in low water systems. *Crit. Rev. Biochem. Mol. Biol.* **33,** 53–89.

10. Martinek, K., Berezin, I. V., Khmelnitsky, Yu. L., Klyachko, N. L., and Levashov, A. V. (1987) Micellar enzymology: potentialities in applied areas (biotechnology). *Collect. Czech. Chem. Commun.* **52,** 2589–2602.

11. Martinek, K., Berezin, I. V., Khmelnitsky, Yu. L., Klyachko, N. L., and Levashov, A. V. (1987) Enzymes entrapped into reversed micelles of surfactants in organic solvents: key trends in applied enzymology (biotechnology). *Biocatalysis* **1,** 9–15.

12. Levashov, A. V. and Klyachko, N. L. (1995) Micellar enzymology for enzyme engineering. Ideas and realization. *Ann. NY Acad. Sci.* **750,** 80–84.

13. Hanahan, D. J. (1952) The enzymatic degradation of phosphatidylcholine in diethyl ether. *J. Biol. Chem.* **195,** 199–206.

14. Martinek, K., Levashov, A. V., Klyachko, N. L., and Berezin, I. V. (1977) Catalysis by water-soluble enzymes in organic solvents. Stabilization of enzymes against denaturation (inactivation) through their inclusion in reversed micelles. *Dokl. Akad. Nauk SSSR* **236,** 920–923; Engl. transl. **236,** 951 (1978).

15. Martinek, K., Levashov, A. V., Khmelnitsky, Yu. L., Klyachko, N. L., and Berezin, I. V. (1982) Colloidal solution of water in organic solvents: a microheterogeneous medium for enzymatic reactions. *Science* **218,** 889–891.

16. Martinek, K., Levashov, A. V., Berezin, I. V., Kurganov, B. I., Klyachko, N. L., Khmelnitsky, Yu. L., et al. (1983) Method of enzymatic transformation of water-insoluble substrates. USSR Patent SU 1113406, 15.09.84. Bul. N 34.

17. Khmelnitsky, Yu. L., Levashov, A. V., Klyachko, N. L., and Martinek, K. (1988) Engineering biocatalytic systems in organic media with low water content. *Enzyme Microb. Technol.* **10,** 710–724.

18. Antonini, E., Carrea, G., and Cremonesi, P. (1981) Enzyme catalyzed reactions in water-organic solvent two-phase systems. *Enzyme Microb. Technol.* **3,** 291–296.

19. Semenov, A. N., Khmelnitsky, Yu. L., Berezin, I. V., and Martinek, K. (1987) Water–organic solvent two-phase systems as media for biocatalytic reactions: the potential for shifting chemical equilibria towards higher yield of end products. *Biocatalysis* **1,** 3–8.

20. Semenov, A. N., Cerovsky, V., Titov, M. I., and Martinek, K. (1988) Enzymes in preparative organic synthesis: the problem of increasing the yield of product. *Collect. Czech. Chem. Commun.* **53,** 2963–2985.

22. Kabanov, A. V., Nametkin, S. N., Matveeva, E. G., Klyachko, N. L., Martinek, K., and Levashov, A. V. (1988) Relaxation effects in the systems of protein-containing reverse micelles of surfactants in organic solvents. *Mol. Biol.* **22,** 473–483.

23. Menger, F. M. and Yamada, K. (1979) Enzyme catalysis in water pools. *J. Am. Chem. Soc.* **101,** 6731–6734.

24. Luisi, P. L., Henningen, F., and Joppich, M. (1977) Solubilization and spectroscopic properties of α-chymotrypsin in cyclohexane. *Biochem. Biophys. Res. Commun.* **74,** 1384–1389.

25. Leser, M. E., Wei, G., Luisi, P. L., and Maestro, M. (1986) Application of reverse micelles for the extraction of proteins. *Biochem. Biophys. Res. Commun.* **135,** 629–635.

26. Luisi, P. L. and Laane, C. (1986) Solubilization of enzymes in apolar solvents via reverse micelles. *Trends Biotechnol.* **4,** 153–161.

27. Larsson, K. M., Adlercreutz, P., and Matiasson, B. (1990) Enzyme catalysis in microemulsions: enzyme reuse and product recovery. *Biotechnol. Bioeng.* **36,** 135–141.

28. Bogdanova, N. G., Klyachko, N. L., Kabakov, V. E., Martinek, K., and Levashov, A. V. (1989) Enzymatic synthesis of *N*-benzoyl-L-tyrosine *p*-nitroanilide in the system of Aerosol OT/water : 2,3-butanediol/octane. *Bioorg. Chem.* **15,** 634–637.

29. Miethe, P., Gruber, R., and Voss, H. (1989) Enzymes in lyotropic liquid crystal — a new method of bioconversion in non-aqueous media. *Biotechnol. Lett.* **11,** 449–454.

30. Khmelnitsky, Yu. L., Gladilin, A. K., Roubailo, V. L., Martinek, K., and Levashov, A. V. (1992) Reversed micelles of polymeric surfactants in nonpolar organic solvents. A new microheterogeneous medium for enzymatic reactions. *Eur. J. Biochem.* **206,** 737–745.

31. Levashov, A. V., Pantin, V. I., Martinek, K., and Berezin, I. V. (1980) Kinetic theory of reactions catalyzed by enzymes solubilized in organic solvents by means of surfactants. *Dokl. Akad. Nauk SSSR* **252,** 133–136.

32. Klyachko, N. L., Pshezhetsky, A. V., Kabanov, A. V., Vakula, S. V., Martinek, K., and Levashov, A. V. (1990) Enzymatic catalysis in surfactant agregates: optimal structure of surfactant matrix. *Biol. Membr.* **7,** 467–472. Engl. transl. *Biol. Membr.* **4(5),** 698–707.

33. Hilhorst, R., Laane, C., and Veeger, C. (1983) Enzymatic conversion of apolar compounds in organic media using an NADH regenerating system and dihydrogen as reductant. *FEBS Lett.* **159,** 225–228.

33a. Levashov, A. V., Kabanov, A. V., Khmelnitsky, Yu. L., Berezin, I. V., and Martinek, K. (1984) Chemical modification of proteins (enzymes) with water-insoluble reagents. *Dokl. Akad. Nauk SSSR* **278,** 246–248. Engl. transl. 295–297.

34. Kabanov, A. V., Levashov, A. V., and Alakhov, V. Ya. (1989) Lipid modification of proteins and their membrane transport. *Protein Eng.* **3,** 39–42.
35. Yatsimirskaya, E. A., Gavrilova, E. M., Egorov, A. M., and Levashov, A. V. (1993) Preparation of conjugates of progesterone with bovine serum albumine in the reversed micellar medium. *Steroids* **5,** 547–550.
36. Ryabov, A. D., Trushkin, A. M., Baksheeva, L. I., Gorbatova, R. K., Kubrakova, I. V., Mozhaev, V. V., et al. (1992) Chemical attachment of organometallics to proteins in reverse micelles. *Angew. Chem. Int. Ed. Engl.* **31,** 789,790.
37. Levashov, A. V. (1992) Microheterogeneous surfactant-based systems as the media for enzymatic reactions. *Pure Appl. Chem.* **64,** 125–128.
38. Kabanov, A. V., Levashov, A. V., Khrutskaya, M. M., and Kabanov, V. A. (1990) Tailoring of macromolecule conjugates using reversed micelles as matrix microreactors. *Makromol. Chem.* **191,** 2801–2814.
39. Speiser, P. (1984) Hardened reverse micelles as drug delivery systems, in *Reverse Micelles* (Luisi, P. L., Straub, B. E., eds.), Plenum, New York, pp. 339–346.
40. Candau, F. and Carrer, M. (1989) Inverse microlatexes: mechanism of formation and characterization, in *Structure and Reactivity of Reverse Micelles* (Pileni, M. P., ed.), Elsevier, Amsterdam, pp. 361–370.
41. Abakumova, E. G., Levashov, A. V., Berezin, I. V., and Martinek, K. (1985) Universal enzymatic catalyst for reaction both in water and organic solvent. Catalytic activity of α-chymotrypsin entrapped in nanoparticles of cross-linked acrylamide. *Dokl. Akad. Nauk SSSR* **283,** 136–139.
42. Khmelnitsky, Yu. L., Neverova, I. N., Momcheva, R., Yaropolov, A. V., Belova, A. B., Levashov, A. V., et al. (1989) Surface-modified polymeric nanogranulates containing entrapped enzymes: a novel biocatalyst for use in organic media. *Biotechnol. Tech.* **3,** 275–280.
43. Larsson, K. M., Janssen, A., Adlercreutz, P., and Matiasson, B. (1990) Three systems used for biocatalysis in organic solvents—a comparative study. *Biocatalysis* **4,** 163–175.
44. Levashov, A. V., Klyachko, N. L., Martinek, K., Polyakov, V. M., and Sergeev, G. B. (1987) NMR-control of water content in hydrated micellar systems of Aerosol OT in octane with solubilized protein. *Vestnik MGU Ser. 2, Chemistry* **28,** 287–290.
45. Levashov, A. V., Pantin, V. I., and Martinek, K. (1979) 2,4-Dinitrophenol as an acid-base indicator in reversed micelles of the surfactant AOT in octane. *Colloid Zh* **41,** 453–460. *Engl. transl.* 380–386.
46. Martinek, K., Levashov, A. V., Klyachko, N. L., Pantin, V. I., and Berezin, I. V. (1981) The principles of enzyme stabilization. VI. Catalysis by water-soluble enzymes entrapped into reversed micelles of surfactants in organic solvents. *Biochem. Biophys. Acta* **657,** 277–294.
47. Belonogova, O. V., Likhtenstein, G. I., Levashov, A. V., Khmelnitsky, Yu. L., Klyachko, N. L., and Martinek, K. (1983) Use of the spin label method to study the state of the active site and microsurroundings of α-chymotrypsin, solubilized in octane, using surfactant Aerosol OT. *Biochemistry* **48,** 379–386, *Engl. transl.* 329–335.

48. Levashov, A. V., Klyachko, N. L., Bogdanova, N. G., and Martinek, K. (1990) Fixation of highly reactive form of α-chymotrypsin by micellar matrix. *FEBS Lett.* **268,** 238–240.

49. Klyachko, N. L., Bogdanova, N. G., Koltover, V. K., Martinek, K., and Levashov, A. V. (1989) The relationship between the activity and conformational mobility of α-chymotrypsin in reverse micellar systems. *Biochemistry* **54,** 1224–1230, Engl. transl. 1004–1009.

50. Klyachko, N. L., Bogdanova, N. G., Levashov, A. V., and Martinek, K. (1992) Micellar enzymology: superactivity of enzymes in reversed micelles of surfactants solvated by water/organic cosolvent mixtures. *Collect. Czech. Chem. Commun.* **57,** 625–640.

51. Kabanov, A. V., Klyachko, N. L., Nametkin, S. N., Merker, S., Zaroza, A. V., Bunik, V. I., et al. (1991) Engineering of functional supramolecular complexes of proteins (enzymes) using reversed micelles as matrix microreactors. *Protein Eng.* **4,** 1009–1017.

52. Levashov, A. V., Ugolnikova, A. V., Ivanov, M. V., and Klyachko, N. L. (1997) Formation of homo- and heterooligomeric supramolecular structures by D-glyceraldehyde-3-phosphate dehydrogenase in reversed micelles of Aerosol OT in octane. *Biochem. Mol. Biol. Int.* **42,** 527–534.

46

Enzymatic Transformations in Supercritical Fluids

Alain Marty and Jean-Stéphane Condoret

1. Introduction

Water is the most common solvent for biochemical reactions both in vivo and in vitro in enzymological experiments. Unfortunately, synthetic reactions that can be carried out by reversing the hydrolytic action of certain enzymes as well as other biotransformations (e.g., oxidation) are difficult or even impossible to operate in water. Oxidation reactions are limited by the poor solubility of oxygen in water, and syntheses are impeded by high water activity. In order to shift the thermodynamic equilibrium in favor of the synthesis, it is necessary to use nonaqueous solvents. However, the use of solvents can be problematic because of toxicity, flammability, and increasing environmental concerns. As a result, supercritical fluids (SCFs) have attracted much attention in recent years as an alternative to organic solvents for carrying out enzymatic reactions. Up to now, SCFs have only been used in large-scale industrial processing to extract plant materials (e.g., coffee, hops). Nevertheless, interest in the use of SCF as a solvent for biocatalysis is growing rapidly (see reviews in **refs. *1–9***).

The phase diagram for one of the most attractive SC fluids, CO_2, is presented in **Fig. 1**. Liquid and gaseous phases may coexist below the critical point (73.8 bar and 31.8°C). When moving upward along the liquid–vapor equilibrium curve, the gaseous phase is compressed and its density increases until it reaches that of the liquid phase. Beyond the critical point, both phases are indistinguishable and the fluid is monophasic and occupies all the volume. It can be described as a dense gas or as an expanded liquid. Generally, SCFs exhibit liquidlike density and therefore good solvating power, but they retain gaslike compressibility. Consequently, it is possible to control their solvating power by changing the pressure and/or temperature, with a continuous transition from a good to poor solvent. Moreover,

From: *Methods in Biotechnology, Vol. 15: Enzymes in Nonaqueous Solvents: Methods and Protocols*
Edited by: E. N. Vulfson, P. J. Halling, and H. L. Holland © Humana Press Inc., Totowa, NJ

low viscosity and high diffusion coefficients of these fluids enhance mass transport and therefore reaction kinetics. These unique properties of SCFs enable one to design efficient integrated processes by coupling an enzymatic reaction with subsequent fractionation and product recovery steps.

In practical terms, supercritical carbon dioxide ($SCCO_2$) has considerable advantages over other SCFs and has been most frequently used as a medium for biotransformations (*see* **Note 1**). Its critical pressure (7.38 MPa) is "acceptable" and its critical temperature (31°C) is consistent with the use of enzymes and/or labile solutes. It is inexpensive, nonflammable, and nontoxic and it has the GRAS (Generally Regarded As Safe) status. In addition, its "naturalness" is greatly appreciated by the food and health-care-related industries. Lipase-catalyzed reactions are by far the best studied in SCFs so far, although many other enzymes (e.g., alkaline phosphatase, polyphenol oxidase, cholesterol oxidase, α-chymotrypsin, subtilisin, thermolysin, dehydrogenases, glucoamylase, and cellulase isomerase) have all been successfully used in SCFs too (*see* **Note 1**).

2. Materials

1. Carbon dioxide is usually sold in bottles of 37 kg (20 m^3, 49.5 bar at 15°C). Industrial CO_2 has a purity of 99.7%, whereas N45 CO_2 is 99.9 % pure. The main difference is the water content (6 ppm in the case of N45 CO_2) and the price (the N45 CO_2 being 10 times more expensive). When very low water content is crucial, an adsorbing device at the outlet of the CO_2 bottle is required.
2. Tubings, adapters, reducers, and valves are in stainless steel, with the 316 stainless steel being the most common type. They are usually designed to withstand pressures of up to 100 MPa.
3. Stainless-steel high-pressure hand valves (100 MPa) are used at the ends of the reactor to isolate a part of the circuit, at the bottom of a reactor or separator (Whitey, Hoke, Top-Industrie, Nova Swiss, Sitec).
4. Fine micrometering valves to control flow rate and pressure drops in a cascade of depressurization steps. Special attention must be given to the choice of valves for accurate setting and maintenance of given flow conditions using a minimum number of turns. It is also possible to use high-pressure-controlled valves with a PID control of pressure or flow rate (Nova Swiss).
5. A safety valve has to be located at the outlet of the pump if the latter is not equipped with an automatic pressure control. It is also obligatory to install a safety valve (Top-Industrie, Nupro, Nova Swiss) at the top of the reactor, and at the top of each separator in the case of continuously operating plants.
6. High-pressure gauges with a conventional Bourdon tube in stainless steel are used for pressure measurements.
7. In a continuous pilot plant, it is advisable to install a mass flow meter system to measure the flow rate and the density of the fluid at the reactor's entrance (Rheonik sensor, Bopp & Reuther, Mannheim, Germany; Micro Motion, Rosemount, Rungis, France). They are generally based on the Coriolis force measurement.

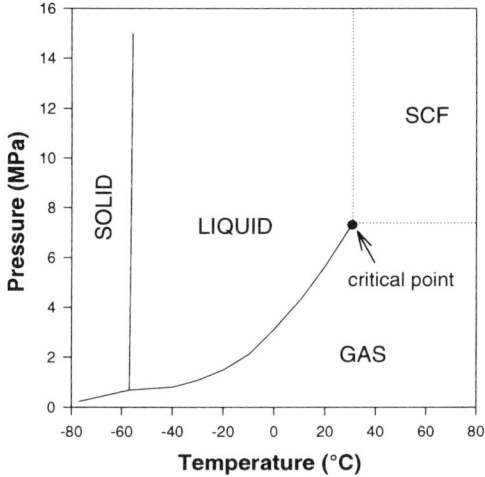

Fig. 1. Pressure–temperature diagram for pure CO_2.

8. Sapphire tube (length: 80 mm, external diameter: 25 mm, internal diameter: 15 mm) or sapphire windows enable the solubility of the reactants to be checked visually (RSA Le Rubis, Jarrie, France).
9. Carbon dioxide is usually pumped and pressurized as a liquid, and the bottle must be equipped with a plunger in order to pump the liquid phase. To avoid cavitations in the pump, it is necessary to cool the liquid CO_2 and also the pump head. A pump with a stainless-steel diaphragm is recommended. Double diaphragm pumps allow one to obtain pulse-free flows. These pumps are very robust and the only maintenance required is cleaning of the ball checks. Centrifugal pumps can also be used but they are usually more suited for large-scale applications. A very good survey of mechanical aspects and pump technologies can be found in **ref. *10***. It is recommended to flush all the installation with CO_2, then to equilibrate it at the pressure of the CO_2 bottle (around 5 MPa). When vessels for the reaction and fractionation are used sequentially, it is more effective to increase only the pressure in the reactor in order to rapidly reach high fluid density, where volumetric pumps are most efficient (one movement of the diaphragm displaces one fluid volume and the corresponding mass which rises with density). When 15 or 20 MPa pressure is reached, the needle valve separating the reactor and the first separator is opened (care should be taken not to decrease the pressure below 10 MPa). When operating a continuous process with solvent recycling, it is convenient, for the stability of the system, to leave the CO_2 bottle with an open connection to the last separation vessel. Therefore, its pressure is fixed and stable at the value of the bottle pressure.
10. In a setup in which the reaction is coupled with fractionation and recirculation of the fluid, it is necessary to locate a check valve at the outlet of the CO_2 bottle in order to control unidirectional flow.

11. It is necessary, upstream of the pump, to cool the CO_2 to a temperature around $-5°C$, but it is recommended not to decrease the temperature below $0°C$ if water is present, because icing and clogging of the check balls may occur.

12. Enzymes are not soluble in SCFs, hence, immobilization is not necessary as the recovery and reuse of the catalyst is relatively easy. Nevertheless, in order to achieve good dispersion of the catalyst in the fluid phase, it is better to use immobilized enzymes. In this case, physical adsorption is the simplest and the least expensive method of immobilization. If lipases are used, suitable immobilized preparations (e.g., Lipozyme® or Novozyme® [Novo-Nordisk]) can be obtained commercially.

3. Methods

3.1. Choice of SCF

As mentioned previously, among SCFs available, $SCCO_2$ is the first-choice solvent for enzymatic transformations. Although some light hydrocarbons also have low critical pressure and a critical temperature in the ambient range (*see* **Note 2**), they are not as convenient as CO_2 because of their flammability (*see* **Note 3**).

3.2. Choice of Pressure

Pressure is likely to affect the reaction rate by changing either the reactant's solubilities or the rate constant directly. Indeed, an increase in pressure leads to enhanced fluid density and, therefore, improved solvating power of the fluid. Furthermore, if the activation volume ΔV^* is negative, increased pressure should have a positive effect on the reaction rate constant. However, in the case of enzymatic catalysis, the latter effect has not yet been demonstrated. Another important consequence of pressure changes is the modification of the reactant's partition between SCF and the solid phase containing the enzyme *(11)*. Other relevant phenomena induced by increase in pressure are the acidification of the medium and a change in diffusion coefficients of solutes *(12)*.

As a first step, use a pressure around 200 bar as a good compromise between the solubility of reactants and the process cost. Indeed, the solvating power of $SCCO_2$ increases enormously in the vicinity of the critical point, but above 200 bar, the effect is significantly smaller (**Fig. 2**). Density values in **Fig. 2** were obtained using the equation of state of Lee–Kesler–Plocker. The modeling results correlate very well with experimental values (not shown on the curve) obtained by Ely et al. *(13)*.

3.3. Choice of Temperature

Temperature, in addition to its effect on the enzyme stability, is directly related to the SCF's solvating power, which is the sum of two competing parameters: SCF density and the vapor pressure of the solute (or solid sublimation pressure of the solute). The first parameter decreases as tempera-

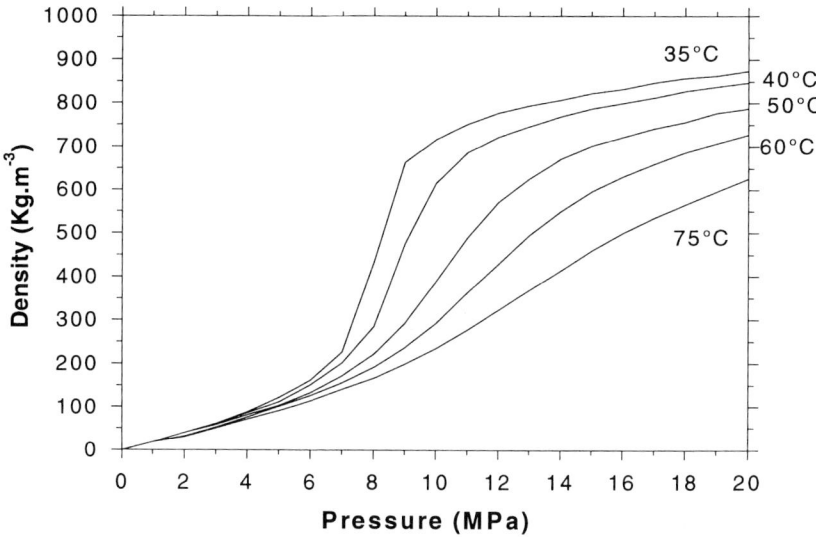

Fig. 2. Computed CO_2 density versus temperature and pressure.

ture increases **(Fig. 2)**, whereas the opposite is the case for the second one. At high pressure (from two to three times the critical pressure), the density is not significantly affected by temperature and, consequently, solubilities increase with temperature. At lower pressure, the negative effect on density can be predominant and the solubility may decrease with temperature. This effect, known as the retrograde solubility phenomenon, is exemplified for water in **Fig. 3**, through the crossing of the solubility curves in the range of 70–150 bar. Thus, it is necessary when operating in SCFs to find a good compromise among optimum solvating power, enzyme stability, and solute thermolability. Most of the time, a temperature range between 35°C and 45°C is a good initial choice.

3.4. Solubilisation Rules

3.4.1. General

The simplest method for evaluating the solubility of a compound was proposed by Chrastil *(15)*. It is based on the concept of complex formation between the solvent and solutes and is expressed by:

$$c = d^k \exp [(a/T) + b]$$

where c is solubility (g/L), d is the density of the fluid (g/L), T is the temperature (K), and a, b, and k are constants characteristic of the solvent–solute couple. k is the total number of molecules involved in the complex and should be independent of both pressure and temperature. a and b are dependent on the heat of

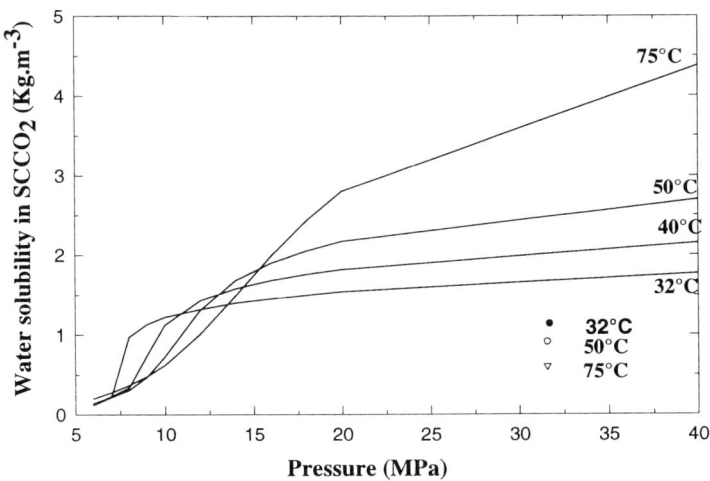

Fig. 3. Water solubility in SCCO$_2$; experimental points from **ref. *14*** and modeling from the equation of Chrastil *(15)* ($k = 1.549$, $a = -2826.4$, and $b = -0.807$) as described in the text.

solvation and the molecular weight of the two compounds. Density may be computed with an equation of state for pure SCF by assuming that the concentrations of solutes are low enough to have no effect on this parameter. Such modeling is shown in **Fig. 3** for solubilization of water in CO$_2$. Other methods were proposed to calculate solubilities in SCCO$_2$ *(16–19)*. When no data are available in the literature, experiments with a microextractor could provide the necessary information (*see* **Note 4**).

3.4.2. Enhancement of Solubility by Polar Cosolvents

The main drawback of SCCO$_2$ is that it has limited solvating power with respect to polar compounds. This is a serious limitation for biotechnological applications where many natural molecules of interest (e.g., alkaloids, carotenoids, phenols, proteins, and sugars) are only sparingly soluble in SCCO$_2$. In this case, a polar cosolvent or so-called "entrainer," such as acetone, ethanol, methanol, or water, is added in order to increase the polarity of the medium and to solubilize the target solute via the formation of hydrogen bonds. Typically, cosolvents are added to the SCF at moderate concentrations of less than 10 mol% (*see* **Note 5**). In a batch reactor, the cosolvent can be added directly into the reactor prior to pressurization, whereas in a continuous process, the addition should be made to the CO$_2$ inflow via an high-performance liquid chromatography (HPLC) pump to deliver a constant flow rate at the operating pressure. However, the use of another component in the system further increases the complexity and may also complicate downstream processing. In addition, the solubility enhancement effect of cosolvents is usually limited in the case of very polar compounds.

Two alternative methods have been developed for some specific cases. To solubilize polyols (e.g., glycerol and sugars), it has been proposed *(20)* to form a hydrophobic complex between the polyol and phenylboronic acid (PBAC), which is much more soluble in the supercritical phase **(Fig. 4)**. This method was used to perform the esterification of glycerol and sugar with oleic acid in $SCCO_2$ (*see* **Note 6**).

Another method involves the adsorption of polar substrates onto a solid hydrophilic support such as silica gel *(20)*. Compared with the former, this approach is more general because it is not necessary to have two vicinal hydroxyl groups in the substrate molecule (*see* **Note 7**).

3.5. Reaction Vessels

3.5.1. Batch Reactors

Most of the time, a batch reactor **(Fig. 5)** is the most convenient mode of operation: The substrate is introduced into the reaction vessel, the enzyme is added and the reactor is closed. Cold CO_2 (generally it is not necessary to decrease the temperature below $-5°C$) is pressurized using a volumetric pump (membrane or piston), heated to the working temperature and introduced into the reactor. When the desired pressure is obtained, the pump is shut off, the vessel is isolated, and the reaction proceeds. At a given time, the reactor is rapidly cooled in dry ice or in liquid nitrogen to stop the reaction. The solid CO_2 sublimates on opening and the products remaining in the reactor are assayed. However, it is difficult using this procedure to accurately define the start of the reaction and impossible to obtain a whole kinetic curve in a single run (*see* **Note 8**).

3.5.2. Semicontinuous Processes

These were proposed by several authors *(21–23)*. In these reactors, $SCCO_2$ flows through a saturation vessel containing substrates impregnated into glass–wool, upstream of the reaction vessel **(Fig. 6)**.

3.5.3. Continuous Process

The main value of $SCCO_2$ is the great variability of its density and, consequently, its solvating power as a function of temperature and pressure. This facilitates the fractionation of residual substrates and/or products after the reaction. The most convenient system is a fixed bed of immobilized enzyme connected to a separation unit where the target compound(s) are recovered by reducing the pressure through a valve into a separation vessel, thus lowering the fluid velocity and allowing the "desolubilized" solute to settle. High yields are

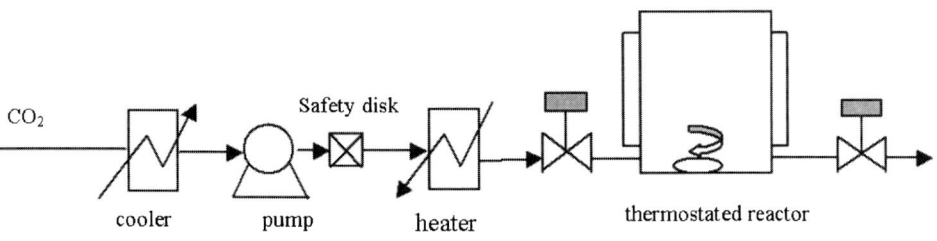

Fig. 4. Esterification of glycerol with oleic acid using the glycerol–PBAC complex.

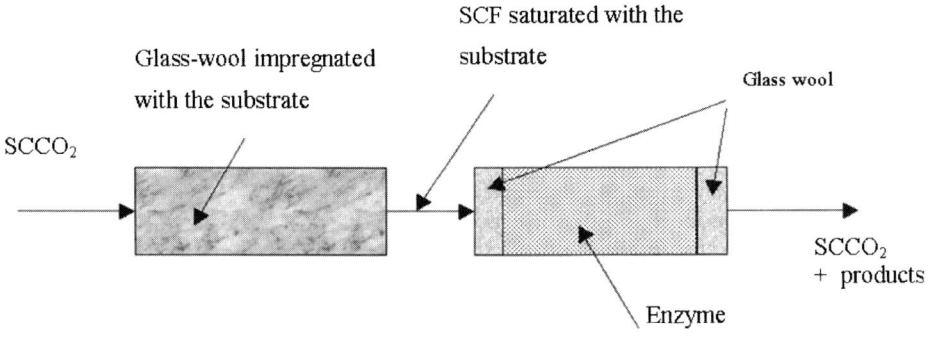

Fig. 5. Batch reaction process.

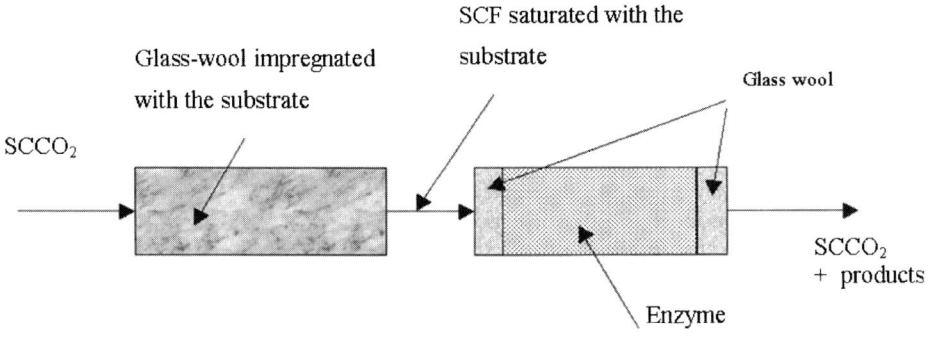

Fig. 6. Schematic representation of a semicontinuous reactor vessel.

usually difficult to obtain in the separation of the biphasic medium unless specific demisting devices are used (e.g., cyclonic separators [Separex patent] or packing loaded vessels). A cascade of depressurization vessels enables the release of the fractionated compounds. Such a system, including the recycling of the supercritical solvent, has been successfully operated by Marty et al. *(24)* for the lipase-catalyzed production of ethyl oleate from ethanol and oleic acid (*see* **Note 9**). The experimental setup, which was built by Separex Company (Nancy, France), is presented in **Fig. 7**.

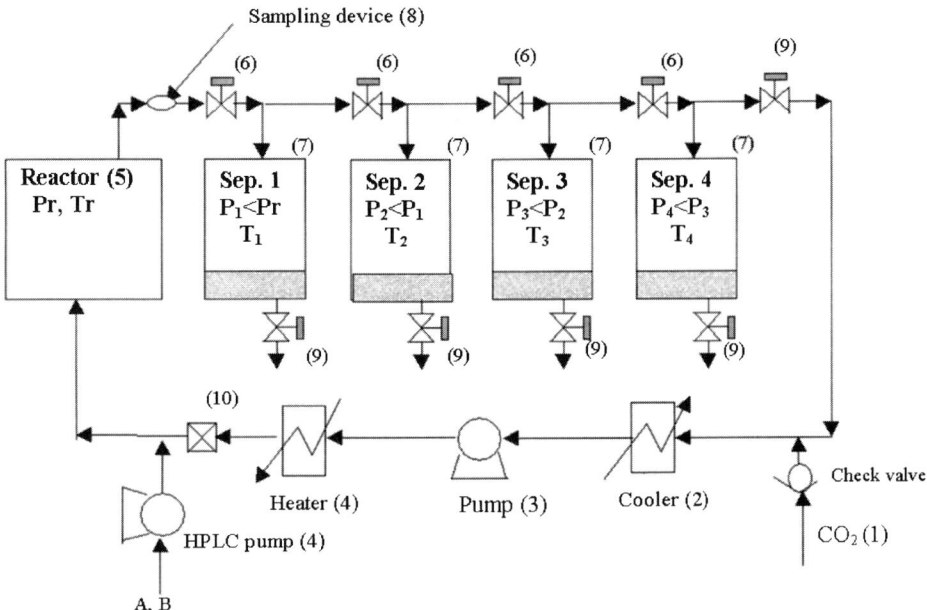

Fig. 7. Schematic representation of pilot plant for continuous reaction coupled with on-line fractionation of solutes and recycling of the solvent. (*See* **Note 10** for the detailed description.)

The same type of processing enables partial fractionation of the reaction mixture in order to avoid product inhibition or to shift thermodynamic equilibrium of the reaction by eliminating one of the product, as and when it is formed. In the case shown in **Fig. 8**, one product (D) is condensed in the separation vessel, whereas A, B, and C remain soluble at the pressure and temperature conditions used in the separation vessel (*see* **Note 11**).

3.6. Enzyme Stability, Activity, and the Role of Water

Enzyme stability in SCFs is similar to that in organic solvents and is much higher than in water (*see* **Note 12**), as pressure does not have a significant effect on the stability provided that compression and decompression are slow. As with organic solvents, irreversible denaturation of enzymes occurs at higher levels of hydration *(25,26)*. $SCCO_2$ as a reaction medium is also superior to water for enzyme-catalyzed oxidations resulting from the greater solubility of oxygen in the former *(23)*. The comparison of reaction rates in $SCCO_2$ and *n*-hexane for lipase-catalyzed reactions *(25,26)* produced similar results when the degree of hydration in the system was taken into account. However, for continuous lipase-catalyzed esterifications, $SCCO_2$ may well prove to be a better solvent

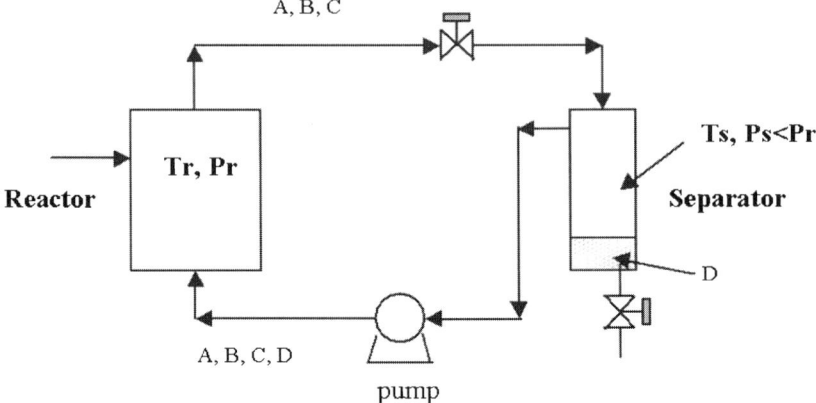

Fig. 8. Batch reactor designed for elimination of one of the products.

than *n*-hexane. The higher water solubility in CO_2 enables the water produced in the course of the reaction to be continuously evacuated, whereas in hexane, it is largely adsorbed onto the immobilization support that leads to a drastic decrease in enzyme activity *(27)*.

In general, it can be concluded that water plays a very similar role in enzyme-catalyzed reactions carried out in SCFs and organic solvents (for a detailed discussion of the role of water in nonaqueous biocatalysis *see* **refs. 28** and **29** and other chapters in this volume). Hence, accurate control of water activity in SCF-based reaction systems is crucial. To achieve this, the knowledge of water-adsorption isotherms that describe the water partition between the solid phase containing the enzyme and SCF is required. These isotherms may be obtained experimentally for every set of pressure and temperature conditions used (*see* **ref. 30** and other chapters in this volume). A predictive method using limited experimental data has also been developed *(31)*.

3.7. Conclusions

In conclusion, the authors wish to reiterate the answer to the key question: What is the value of using a supercritical fluid as a medium for biocatalysis? In the case of $SCCO_2$, the main reason is that it is a "natural" and "environmentally friendly" solvent. In addition, the variable solvating power of $SCCO_2$ and other SCFs facilitates the integration of biocatalytic and downstream processing steps in a single robust bioreactor. Finally, the control of water content in continuous operations is often easier in SCFs than in organic solvents. Thus, it has been shown that constant productivity can be maintained longer in $SCCO_2$ than in *n*-hexane.

4. Notes

1. Most of the work has been carried in $SCCO_2$ but other SCFs (e.g., ethane, etylene, fluroform, propane, sulfure, and hexafluoride) were also studied in some detail.
2. Critical pressure and temperature of selected fluids are shown in **Table 1**.
3. In this case, the installation has to be adapted for the use of flammable solvents and all safety valves and vents must be connected to a tank, whereas CO_2 can be vented into the atmosphere.
4. Operating the reaction in a sapphire vessel or in a vessel with a sapphire window enables one to at least check the number of phases in the system. Such devices are highly advisable, especially when operating with "exotic" components or when a cosolvent is present.
5. For example, Wong and Johnston *(32)* have improved the solubility of cholesterol (2.6×10^{-5} *M/M* in $SCCO_2$ at 15 MPa and 35°C) with an enhancement factor of 1.2–7.2 depending on the cosolvent used (3.5 mol%).
6. Whereas glycerol solubility in $SCCO_2$ is <1 m*M*, with this method 20 m*M* glycerol can be solubilized (15 MPa and 40°C) using an equimolar quantity of PBAC *(20)*. During the lipase-catalyzed esterification, 50% oleic acid was converted to the main product, mono-olein, as would be expected when two hydroxyl groups in the glycerol molecule are bloked with PBAC. Another advantage of this method is that the complex is rather stable in SCF but can be easily hydrolyzed by small amounts of water. After the reaction, recovery of PBAC can be accomplished by extraction. Using the same approach, 20 m*M* of a fructose–PBAC complex were solubilized in $SCCO_2$ (15 MPa, 40°C). Conversion of oleic acid was, in this case, about 30% *(20)*. Also, note that the use of sapphire reaction vessels allows visual and qualitative control of this pseudosolubilization.
7. The two procedures can be used for adsorption, depending on whether the solute is a liquid or a solid. For a liquid like glycerol, 1 g of glycerol and 1 g of silica gel are gently mixed until a homogeneous powder is obtained *(20)*. In the case of solids (e.g., fructose), 1 g is dissolved in 12 mL of ethanol and 1 g of silica gel is added to the solution. The solvent is evaporated under reduced pressure with continuous agitation to give a homogeneous powder. Enzymatic esterification of the adsorbed glycerol or fructose with oleic acid was carried out with 1.3 g of thus obtained powders, oleic acid (30 m*M*), and 200 mg of *Rhizomucor miehei* lipase in a 8-mL tubular sapphire reaction vessel.
8. Marty et al. *(30)* proposed a specific device to overcome these two limitations (*see* **Fig. 9**). It consists of a 16 mL sapphire tubular reactor (1) in which the enzyme is confined by liquid chromatographic filters (11 and 12) (cotton can also be used). Sapphire walls enable the solubilization to be checked visually. This tubular reactor is an element of a complete loop as shown in **Fig. 9**. The volume of the whole loop is 70 mL, mainly because of the dead volume of the pump (2). This system is equivalent to a well-stirred reactor because the high-pressure gear pump (2) is assumed to recirculate in a time that is short compared to the reaction time. The whole system is placed in a temperature-controlled room (3). Pressure and temperature sensors are located in this loop, and the sample

Table 1
Critical Pressure and Temperature
of Selected Fluids

	Critical point	
Solvent	T_c (°C)	P_c (MPa)
CO_2	31.1	7.38
Freon 13	28.9	3.87
Ammoniac	132.5	11.13
water	374.2	21.8
Ethylene	9.3	5.0
Ethane	32.3	4.83
Propane	96.7	4.2
Butane	152	3.8
Cyclohexane	280.3	4.07
Acetone	235.1	4.7
Ethanol	243.4	6.3
Isopropanol	235.3	4.76
Ethylamine	183	5.62
Pyridine	347	5.55

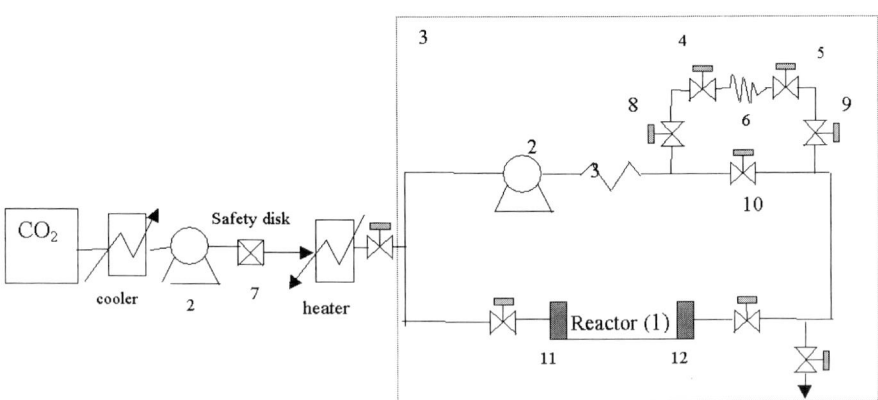

Fig. 9. Batch-circulated reaction vessel with sampling possibility.

loop (composed of two valves [4 and 5]) is separated by a tube (6). A small magnetic bar in the horizontal sapphire tube and a magnetic agitator placed below may not provide perfect stirring because of nonoptimal geometry, but it is sufficient to maintain the enzyme in suspension. A safety disk (7) is located between the pump and the entrance to the reactor. To take samples, a sample loop is installed between valves 8 and 9, which are in a closed position. Valves 8 and 9 are opened and valve 10 is then closed. The fluid is therefore recirculated through the sample loop. Then the circulation is restored by opening valve 10, and at the same time, valves 8 and 9 are closed. This results in the isolation of the sample loop. The sample loop is then disconnected from valves 8 and 9 and is slowly decompressed into methanol. This methanol is also used to rinse and recover all the solute condensed inside the loop. Before reusing, the sample loop is filled with 5-MPa CO_2 from the bottle. When one substrate is liquid and highly soluble in $SCCO_2$ (e.g., ethanol), the sample loop can also be used to introduce it into the system, giving the zero time of the reaction accurately.

Dumont et al. *(26)* used a well-designed equipment with a 100-mL jacketed stainless-steel reactor stirred by a turbine driven by an electric motor. This device enables the addition of small quantities of catalyst after pressurization and thermostatation. The sampling for analysis (without depressurization) is accomplished using a six-way HPLC valve (Rheodyne 7125).

The kinetics can be followed using on-line analysis. The concentration of ultraviolet (UV)-active compounds can be measured by circulating the reaction medium through a UV-vis detector equipped with a high pressure cell. It is also possible to couple the reactor to a supercritical fluid chromatograph with a packed or capillary column.

9. The influence of the cascade of depressurization units and the effect of temperature on the separation selectivity have also been studied by the same authors. An example of a "good" separation is given in **Table 2**. The greatest proportion of the ester produced was found in separators 2 and 3, leading to 84.6% combined recovery of the product (*see* **Table 2**).

10. Cooled liquid carbon dioxide (1 and 2) is pressurized by a metallic membrane pump (Dosapro Milton Roy, USA) (3) and then heated to the reaction temperature (4). The reactor (5) is connected downstream to a cascade of four separators (7), where the pressure in each separator is adjusted using needle valves (6). After the last separator, at the lowest pressure (5 MPa), CO_2 is liquefied by cooling (2), to be recycled. A HPLC pump (4) enables substrates to be injected into the $SCCO_2$ flow with a precise flow rate at whatever operating pressure. At the outlet of the reactor, a sample loop (8) enables the sampling of the effluent mixture. At the bottom of each separator, a discontinuous recovery of the condensed liquid phase is possible, using simple high-pressure valves (9). A mass flow meter (10), based on the Coriolis effect, is used to determine the mass flow rate. The recycling of the solvent can be accomplished in the gaseous state via a compressor, but this is uneconomical in small-scale reactors. At the compression stage, high output temperatures (>100°C) are generated.

11. This process has been proposed by Marty et al. *(33)* for transterifications. Doddema et al. *(34)* described a process involving two countercurrent columns to improve a continuous transesterification by extracting ethanol that, in this case, was the product of the reaction and inhibitor. Another attempt was made by Adshiri et al. *(35)* who improved the degree of incorporation of oleic acid into

Table 2
Results of the Separation Process in SCCO$_2$

	SEP1 10.2 MPa 18.5 mL	SEP2 8.1 MPa 9.9 mL	SEP3 7.2 MPa 17.5 mL	SEP4 5.2 MPa 2.8 mL
Oleic acid (g/L)	605	148	99	56
Ethyl oleate (g/L)	138	741	790	489
Ethanol (g/L)	2.1	11	15.5	91
Water (g/L)	2.8	5.9	4.4	327
Ester purity (mol%)	15.8	68.7	73.1	7.2
Ester recovery (%)	10.2	29.2	55.1	5.5
			yield:84.6 %; purity 71.5 mol%	

Note: Output of the reactor. Conversion: 58% based on oleic acid; 2.4 g/L ethyl oleate: 3.7 g/L of CO$_2$ (*P*: 15 MPa, *T*: 44°C, flow rate: 40mL/min, duration: 185 min).
Source: Adapted from **ref. 24**.

triglyceride (tricaprylin was used as the second substrate) by selective removal of the product using a batch reactor.

12. For example, immobilized *Mucor miehei* lipase (Lipozyme®) loses about 10% of its activity after 6 d at 40°C, when incubated in SCCO$_2$ at pressures of 13–18 MPa *(25)*. A very similar loss of the catalytic activity was observed in n-hexane under the same conditions. High stability of nine commercial enzymes in SCCO$_2$ (20 MPa, 35°C) was also demonstrated by Taniguchi et al. *(36)*, although a relatively short contact time (1 h) was used in these experiments.

Interestingly, Cernia et al. *(37)* reported that enantioselectivity and conversion in the lipase-catalyzed acylation of different alcohols is superior in SCCO$_2$ when compared to organic solvents.

References

1. Aaltonen O. and Rantakylä, M. (1991) Biocatalysis in supercritical CO$_2$. *CHEMTECH* 240–248.
2. Perrut, M. (1994) Enzymatic reactions and cell behaviour in supercritical fluids. *Chem. Biochem. Eng. Quart.* **8(1),** 25–30.
3. Russell, A. J., Beckman, E. J., and Chaudhary, A. K. (1994) Studying enzyme activity in supercritical fluids. *CHEMTECH* **24,** 33–37.
4. Jarzebski, A. B. and Malinowski, J. J. (1995) Potentials and prospects for application of supercritical fluid technology in bioprocessing. *Process Biochem.* **30,** 343–352.
5. Kamat, S., Beckman, E. J., and Russell, A. J. (1995) Enzyme activity in supercritical fluids, *Crit. Rev. Biotechnol.* **15,** 41–71.
6. Ballesteros, A., Bornscheuer, U., Capewell, A., Combes, D., Condoret, J. S., Koenig, K., et al. (1995) Enzymatic reaction in non conventional media. *Biocatal. Biotransform.* **13,** 1–42.

7. Nakamura, K. (1996) Enzymic synthesis in supercritical fluids. *Supercrit. Fluid Technol. Oil Lipid.*

8. Cernia, E. and Palocci, C. (1997) Lipases in supercritical fluids. *Methods Enzymol.* **286**(Lipases, Part B), 495–508.

9. Ikushima, Y. (1997) Supercritical fluids: an interesting medium for chemical and biochemical processes. *Adv. Colloid Interf. Sci.* **71–72,** 259–280.

10. King, M. B. and Bott, T. R. (1983)*Extraction of Natural Products Using Near Critical Solvents.* Blackie Academic and Professional/Chapman & Hall, London.

11. Erickson, J. C., Schyns, P., and Cooney, C. L. (1990) Effect of pressure on an enzymatic reaction in a supercritical fluid. *AIChE J.* **36,** 299–301.

12. van Eijs, A. M. M., de Jong, J. P. J., Doddema, H. J., and Lindeboom, D. R. (1988) Enzymatic transesterification in supercritical carbon dioxide. Proceedings of the International Symposium on Supercritical Fluids (Nice, France) (Perrut, M., ed.), pp. 933–942.

13. Ely, J. F., Haynes, W. M., and Bain, B. C. (1989) Isochoric measurements on CO2 and on (0.982 CO2 +0.018 N2) from 250 to 330 K at pressures to 35 MPA. *J. Chem. Thermodynam.* **21,** 879–894.

14. Wiebe, R. and Gaddy, V. L. (1941) Vapour phase composition of carbon dioxide–water mixtures at various temperature and at pressures to 700 atmosphere. *J. Am. Chem. Soc.* **63,** 475–477.

15. Chrastil, J. (1982) Solubility of solids and liquids in supercritical gases. *J. Phys. Chem.* **86,** 3016–3021.

16. Giddings, J. C., Meyers, M. N., McLaren, L., and Keller, R. A. (1968) High pressure gas chromatography of nonvolatile species. *Science* **162,** 67.

17. King, J. W. and Friedrich, J. P. (1990) Quantitive correlations between solute molecular-structure and solubility in supercritical fluids. *J. Chromatogr. Sci.* **517,** 449–458.

18. Peng, D. Y. and Robinson, D. B. (1976) A new two constant equation of state. *Eng. Chem. Fundam.* **15,** 59.

19. Valle, J. M. and Aguilera J. M. (1988) An improved equation for predicting the solubility of vegetable oils in supercritical CO_2. *Eng. Chem. Res.* **27,** 1551–1553.

20. Castillo, E., Marty, A., Combes, D., and Condoret, J. S. (1994) Polar substrates for enzymatic reactions in supercritical CO_2. How to overcome the solubility limitation. *Biotechnol. Lett.* **16,** 169–174.

21. Hammond, D. A., Karel, M., Klibanov, A. M., and Krukonis, V. J. (1985) Enzymatic reactions in supercritical gases. *Appl. Biochem. Biotechnol.* **11,** 393–400.

22. Miller, D. A., Blanch, H. W., and Prausnitz, J. M. (1991) Enzyme-catalyzed interesterification of triglycerides in supercritical carbon dioxide. *Ind. Eng. Chem. Res.* **30,** 939–946.

23. Randolph, T. W., Blanch, H. W., and Prausnitz, J. M. (1988) Enzyme-catalyzed oxidation of cholesterol in supercritical carbon dioxide. *AIChE J.* **34,** 1354–1360.

24. Marty, A., Combes, D., and Condoret, J. S. (1994) Continuous reaction-separation process for enzymatic esterification in supercritical carbon dioxide. *Biotechnol. Bioeng.* **43,** 497–504.

25. Marty, A., Chulalaksananukul, W., Condoret, J. S., Willemot, R. M., and Durand, G. (1990) Comparison of lipase-catalyzed esterification in supercritical carbon dioxide and in *n*-hexane. *Biotechnol. Lett.* **12,** 11–16.

26. Dumont, T., Barth, D., Corbier, C., Branlant, G., and Perrut, M. (1992) Enzymatic reaction kinetic: comparison in an organic solvent and in supercritical carbon dioxide. *Biotechnol. Bioeng.* **40,** 329–333.

27. Marty, A., Dossat, V., and Condoret, J. S. (1997) Continuous operation of lipase-catalyzed reactions in non-aqueous solvents: influence of the production of hydrophilic compounds. *Biotechnol. Bioeng.* **56,** 232–237.

28. Halling, P. (1994) Thermodynamic predictions for biocatalysis in non-conventional media: theory, tests and recommendations for experimental design and analysis. *Enzyme Microb. Technol.* **16,** 178–206.

29. Colombier, S., Tweddell, R., Condoret, J. S., and Marty, A. (1998) Water activity control: a way to improve the efficiency of a continuous lipase esterification. *Biotechnol. Bioeng.* **60,** 362–368.

30. Marty, A., Chulalaksananukul, W., Willemot, W., and Condoret, J. S. (1992) Kinetics of lipase-catalyzed esterification in supercritical CO_2. *Biotechnol. Bioeng.* **39,** 273–280.

31. Condoret, J. S., Vankan, S., Joulia, X., and Marty, A. (1997) Prediction of water adsorption curves for heterogeneous biocatalysis in organic and supercritical solvents. *Chem. Eng. Sci.* **52,** 213–220.

32. Wong, J. M. and Johnston, K. P. (1986) Solubilization of biomolecules in carbon dioxide based supercritical fluids. *Biotechnol. Process.* **2,** 29–38.

33. Marty, A., Manon, S., Ju, D. P., Combes, D., and Condoret, J. S. (1995) The enzymatic reaction–fractionation process in supercritical carbon dioxide. *Enzyme Engineering XII*, Vol. 750, (Dordick, J. S. and Russell, A. J., eds.), New York Academy of Sciences, New York, pp. 408–411.

34. Doddema, H. J., Janssens, R. J. J, de Jong, J. P. J., van der Lugt, J. P., and Oostrom, H. H. M. (1990) Enzymatic reactions in supercritical carbon dioxide and integrated product-recovery, 5th European Congress on Biotechnology, Copenhagen (Christiansen, et al., eds.), pp. 239–242.

35. Adshiri, T., Akiya, H., Chin, L. C., Arai, K., and Fujimoto, K. (1992) Lipase-catalyzed interesterification of triglycerides with supercritical carbon dioxide. *J. Chem. Eng. Jpn.* **25,** 104–105.

36. Taniguchi, M., Kamihira, M., and Kobayashi, T. (1987) Effect of treatment with supercritical carbon dioxide on enzymatic activity. *Agric. Biol. Chem.* **51(2),** 593–594.

37. Cernia, E., Palocci, C., Gasparrin, F., Misiti, D., and Fagano, N. (1994) Enantioselectivity and reactivity of immobilized lipase in supercritical carbon dioxide. *J. Mol. Catal.* **89,** L11–L18.

47

Enzymatic Transformations in Suspensions (II)

Adrie J. J. Straathof, Mike J. J. Litjens, and Joseph J. Heijnen

1. Introduction

Biotransformations in organic media with none of the components being in suspension are rare. As enzymes are virtually insoluble in organic solvents, homogeneous reactions are only feasible if the enzyme is solubilized by, for example, chemical modification *(1)* or coating it with a lipid *(2)*. In general, one or more components are present in the reaction mixture in solid state (i.e., the enzyme, substrate[s], product[s], and adsorbents [e.g., molecular sieves or a mixture of a salt and its hydrate]). Thus, the majority of enzymatic transformations in organic media are suspension reactions. Interestingly, Kuhl et al. *(3)* described a reaction where as many as five solids (chymotrypsin powder, two amino-acid-derived substrates, one peptide product, and partly hydrated Na_2CO_3) were suspended in hexane. In this chapter, we will deal with enzymatic reactions in suspension where at least one of the reactants (substrate or product) is only partially dissolved. In practice, this means that the actual biotransformation is preceded by the dissolution of substrate and/or followed by the precipitation of product.

Suspended reactants can be used both in aqueous and nonaqueous media. If one of the substrates is a liquid, another solid substrate may be suspended in it without any added solvent. If all substrates are solids, one might try to form a low-melting-point eutectic mixture where the excess of one of the substrates is suspended in a liquid eutectic phase *(4)*. Alternatively, suitable eutectic mixtures can be formed by adding a small amount of solvent or water, the so-called adjuvant. Typically, the amount of adjuvant added is insufficient to form a bulk liquid phase and its function is to promote the formation of a eutectic mixture of substrates with the melting point below that of the reaction temperature (and of the pure crystalline substrates) *(5,6)*.

From: *Methods in Biotechnology, Vol. 15: Enzymes in Nonaqueous Solvents: Methods and Protocols*
Edited by: E. N. Vulfson, P. J. Halling, and H. L. Holland © Humana Press Inc., Totowa, NJ

Three types of reaction with solid substrates or products can be discerned: (1) suspended substrate reactions, (2) suspended product reactions, and (3) suspension-to-suspension reactions (*see* **Note 1**). Examples of all of those, catalyzed by hydrolytic enzymes (lipases, proteases, or glycosidases) in organic media, are given in the tables in the chapter. These reactions can be further divided into three groups: reverse hydrolysis, synthetic (kinetically controlled) reactions, and kinetic resolutions.

Reverse hydrolysis reactions are thermodynamically controlled reactions that are performed in anhydrous media (**Table 1**). In early research, it was noted that the conversion of the substrate (with dissolution) can continue even when it is present in quantities significantly exceeding its solubility limit in the medium, as long as water formed in the reaction is removed *(7–9)*. Later, it was shown that the undesirable formation of polyol diesters can be suppressed by precipitating the monoester product during the reaction *(10–12)*. According to Halling et al. *(16)* for a given water activity, the equilibrium yield in a reaction involving solid substrates should be higher than that in the analogous transformation in solution, with a maximum obtained in a medium where the solubility of the starting materials is minimal. There is an additional reason for using a suspended substrate in the amidation reaction to be described. Moderately soluble ammonium salts (carbamate or hydrogen carbonate) can be used to allow the addition of an excess of ammonia in a closed system without precipitating the carboxylic acid as its ammonium salt *(14)*.

Kinetically controlled synthetic reactions have been carried out in organic solvents to suppress undesired hydrolysis of substrates, which occurs in the presence of water and leads to lower product yields. There are several examples in the literature where this problem was circumvented by suspending the reactants in an organic solvent (**Table 2**). Recent research on the enzymatic peptide synthesis has shown that similar yields can be obtained in the absence of organic solvent when a suitable for biocatalysis eutectic phase is used as a reaction medium *(18)*.

Kinetic resolutions have also been carried out as suspension reactions in organic solvents (**Table 3**). This approach allows an easy recovery of the product and/or the remaining enantiomer of the substrate. Furthermore, the use of suspended reactants leads to increased productivity and decreases the undesirable chemical hydrolysis of substrates. If the enantioselectivity of the enzyme is insufficient, the use of racemic substrate in suspension can, in principle, lead to the improved yield and/or enantiomeric purity of the remaining solid enantiomer *(21)*.

Thus, biotransformations in suspensions can be beneficial for the following reasons:

1. Shift of equilibrium for thermodynamically controlled reaction to obtain better yields
2. Suppression of enzymatic and/or chemical side reactions involving substrates and/or products

Table 1
Reverse Hydrolysis Reactions with Solid Substrates and/or Products in Organic Media

Substrate 1	Substrate 2	Product	Solvent/adjuvant	Ref.
Octanoic acid	*Glucose*	Mono- and diester	Acetronitrile	*7*
Oleic acid	*Sorbitol*	Mono-, di-, and triester	—	*8*
Fatty acid	Glycerol	*Monoester*	—	*10,11*
Fatty acid	*Glucose*	*Monoester*	Acetone	*12*
Glucuronic acid	Butanol	*Ester*	*tert*-Butanol	*13*
Oleic acid	*N-Me-Glucamine*	Amide	Hexane	*15*
Carboxylic acid	*Ammonium salt*	Amide	MIBK	*14*
Glucose	Alcohol	Glycoside	Several	*9*

Note: The solid reactants are in italics.

Table 2
Kinetically Controlled Synthesis Reactions with Solid Substrates and/or Products in Organic Media

Acyl donor	Acyl acceptor	Product	Solvent/adjuvant	Ref.
Z-Ala-Phe-OMe	*LeuNH₂*	*Peptide*	Hexane	*3*
TFAc-TyrOEt	*GlyNH₂*	*Peptide*	2-MeO-EtAc	*5*
Phe-OMe	*LeuNH₂*	*Peptide*	None	*4*
Et-butanoate	*Sucrose*	Ester mixture	*tert*-Butanol	*17*

Note: The solid reactants are in italics.

Table 3
Kinetic Resolutions Involving Suspended Substrate and/or Product in Organic Media

Racemic substrate	Product	Solvent/adjuvant	Ref.
Ac-DL-Phe-OMe	*Ac-L-Phe*	CH_2Cl_2	*19*
(+/–)-MPGM[a]	*(+)-MPG*	Toluene	*20*

Note: The solid reactants are in italics.
[a]MPGM = 3-(4-methoxyphenyl) glycidic acid methyl ester. The reaction water was introduced into the reactor via a membrane and the product was removed likewise after decarboxylation, whereas (–)-MPGM remained suspended.

3. Improved volumetric productivity resulting from the use of reactant(s) at the concentration exceeding their solubility limit in the medium
4. Facilitation of product recovery

5. Controlled addition/concentration of substrates because of their gradual dissolution
6. Improvement in the yield and/or enantiomeric purity of the remaining solid substrate

Although the above are valid for aqueous as well as nonaqueous media, reasons 1 and 2 are particularly important for suspension reactions in organic solvents because the latter are often used to reverse and/or suppress hydrolysis.

2. Materials

Generally there are no specific reagent requirements for performing enzymatic reactions in suspension as compared to other biotransformations. In the example given here, the following materials were used:

1. Oleic acid and oleamide were purchased from Sigma Chemical Co. and were stored under conditions that prevent oxidation.
2. Immobilized *Candida antarctica* lipase B (11200 U/g based on the esterification of dodecanoic acid with 1-propanol at 60°C) was a kind gift of Novo-Nordisk, Denmark. Presently, this enzyme is marketed by Roche Diagnostics GmbH, Mannheim, Germany.
3. Hexane and di-isopropylether was purchased from Merck and Aldrich Chemical Co., respectively. The latter solvent was dried over molecular sieves before use.
4. Ammonium carbamate was obtained from Fluka and it was ground to a fine powder before use.

3. Methods
3.1. Performing Reactions in Suspension: General Comments

As biotransformations in suspension are very dependent on mass transfer, it is crucial to ensure adequate stirring of the reaction mixture. Also, the separation of liquid and solid phases is required for sampling during the reaction and for the recovery of the final product. Conventional laboratory equipment and techniques are sufficient. In specific situations, the precipitation of product on the particles of substrate or enzyme may occur leading to inhibition. In the latter case, the use of enzyme immobilized on a support with relatively small pore size has been advocated *(22)*.

3.2. Particle Size Calculations

It has been long assumed that enzymatic transformations with suspended substrates are too slow for practical purposes. However, this is only the case when the dissolution of solid substrate(s) is not fast enough to maintain their concentration in solution at or close to saturation. Provided that the dissolution of substrates takes place at a sufficient rate to maintain the saturation, the enzyme should keep "working" at a rate that is defined by the intrinsic properties of the enzyme (and the solvent) irrespective of how much of the solid substrate is added or whether it is present at all. Thus, to obtain the optimal reaction rate, the rate of dissolution of substrates should be compared to the intrinsic rate of the enzymatic reaction and, if necessary, enhanced.

For pseudo-first-order reactions, the dissolution can "keep up" with the reaction if $k_L a \gg V_{max}/K_m$ *(18)*. In this formula, k_L is the mass transfer coefficient, which is typically in the range of 10^{-4} m/s in a well-stirred system, and a is the volume-specific surface area of the solid substrate particles (in m^2/m^3). It can be calculated then how small the substrate particles should be in a specific situation. For instance, suppose that the concentration of enzyme is chosen so that $V_{max}/K_m = 10^{-1}$ s^{-1}. Assuming $k_L = 10^{-4}$ m/s, this suggests that a should be $\gg 1000$ m^2/m^3. If 20 kg/m^3 of a poorly soluble substrate is to be converted (density of 2000 kg/m^3) in 1 m^3, there will be 0.01 m^3 of particles that should have a total surface area of $\gg 1000$ m^2. This means that the volume-to-area ratio of the particles should be $\ll 0.01$ $m^3/1000$ m^2. If, for simplicity, the particles are assumed to be spherical, their volume to area ratio should be $[(4/3 \, \pi r^3)]/(4 \, \pi r^2) \ll 10^{-5}$ m^3/m^2, so their radii should be $r \ll 3 \times 10^{-5}$ m. It can be easily checked under a microscope whether the particle size is in the correct range. To decrease the particle size, the substrate can be recrystallized or ground (*see* **Note 2**).

3.3. Enzymatic Amidation of Fatty Acids

This example is chosen because of the practical importance of this reaction (*see* **Note 3**). This transformation is a reverse hydrolysis in an anhydrous organic solvent. Like the conventional chemical process, it is carried out using free carboxylic acids as substrates *(14)*. The use of free carboxylic acids at low temperatures requires the controlled addition of ammonia in order to prevent precipitation of the fatty acid as its ammonium salt. A suspension of ammonium carbamate can be used as a convenient source of ammonia, because it spontaneously decomposes to give two molecules of NH_3 and one molecule of CO_2. The overall reaction is as follows:

$$\text{Fatty acid} + 1/2 \, NH_4^+(NH_2CO_2^-) = \text{Fatty amide} + 1/2 \, CO_2 + H_2O$$

The gradual dissolution of solid ammonium carbamate ensures that a suitable concentration of ammonia is maintained in the reaction mixture. Good equilibrium yields of the amide can be obtained in anhydrous solvents; the removal of water is required to improve the yield further.

3.4. Reaction Conditions

Oleic acid (4.80 g, 17.0 mmol), ammonium carbamate (0.80 g, 10.2 mmol), dry di-isopropylether (25 mL), immobilized *Candida antarctica* lipase B (600 mg) and molecular sieve 3 Å (3 g) are stirred in a closed reaction flask with a septum in the lid at 60°C. The progress of the reaction is monitored by gas chromatography (GC). After 6 h, when the conversion reaches 93%, the reaction is stopped by filtration of the solids (*see* **Note 4**). The filter cake is washed with solvent to dissolve the solidifying product and the solvent is evaporated. The final product, oleamide (3.79 g, 79% yield), is obtained as a white powder after recrystallization from hexane. The purity is 99% as determined by GC.

4. Notes

1. If a solid substrate is present without a bulk liquid phase, the expressions "solid substrate reaction" or "solid-to-solid reaction" are more suitable *(24)*.
2. Note that during the reaction the particles get smaller and the surface area decreases. Thus, even if dissolution initially is not rate limiting, it may become so at the final stage of the reaction *(25)*.
3. Fatty acid amides are produced from fatty acids in thousands of tons per annum by reaction with anhydrous ammonia at approximately 200°C and 345–690 kPa *(26)*. Oleamide (*cis*-9-octadecenamide) and erucamide (*cis*-13-docosenamide) dominate the market for refined fatty acid amides. These unsaturated amides are used as lubricants in the plastics industry. Oleamide is also known to have a physiological (sleep-inducing) activity *(27)*. As degradation of these amides occurs during the conventional manufacture because of their sensitivity to heat, additional distillation steps are necessary to meet the product specifications. To overcome this problem, De Zoete et al. *(28)* proposed a low-temperature and pressure enzymatic synthesis of the unsaturated amides from the glycerol triesters of the unsaturated acids via lipase-catalyzed ammoniolysis.
4. The reaction vessel should be opened with care because of buildup of pressure.

References

1. Inada, Y., Takahashi, K., Yoshimoto, T., Ajima, A., Matsushima, A., and Saito, Y. (1986) Application of polyethylene glycol-modified enzymes in biotechnological processes: organic solvent-soluble enzymes. *Trends Biotechnol.* **4,** 950–953.
2. Okahata, Y. and Mori, T. (1997) Lipid-coated lipase as efficient catalyst in organic media. *Trends Biotechnol.* **15,** 50–53.
3. Kuhl, P., Halling, P. J., and Jakubke, H. D. (1990) Chymotrypsin suspended in organic solvents with salt hydrates is a good catalyst for peptide synthesis from mainly undissolved reactants. *Tetrahedron Lett.* **31,** 5213–5216.
4. Gill, I. and Vulfson, E. (1993) Enzymatic synthesis of short peptides in heterogeneous mixtures of substrates. *J. Am. Chem. Soc.* **115,** 3348–3348.
5. López-Fandiño, R., Gill, I., and Vulfson, E. (1994) Enzymatic catalysis in heterogeneous mixtures of substrates: the role of the liquid phase and the effects of "adjuvants." *Biotechnol. Bioeng.* **43,** 1016–1023.
6. Erbeldinger, M., Ni, X., and Halling, P. J. (1998) Effect of water and enzyme concentration on thermolysin-catalyzed solid-to-solid peptide synthesis. *Biotechnol. Bioeng.* **59,** 68–72.

7. Ljunger, G., Adlercreutz, P., and Mattiasson, B. (1994) Lipase catalyzed acylation of glucose. *Biotechnol. Lett.* **16,** 1167–1172.
8. Ducret, A., Giroux, A., Trani, M., and Lortie, R. (1995) Enzymatic preparation of biosurfactants from sugars or sugar alcohols and fatty acids in organic media under reduced pressure. *Biotechnol. Bioeng.* **48,** 214–221.
9. Van Rantwijk, F., Woudenberg-van Oosterom, M., and Sheldon, R. A. (1999) Glycosidase-catalysed synthesis of alkyl glycosides. *J. Mol. Catal. B. Enzym.* **6,** 511–532, and references cited therein.
10. Weiss, A. (1990) Enzymatische Herstellung von festen Fettsäuremonoglyceriden. *Fat Sci. Technol.* **10,** 392–396.
11. Bornscheuer, U. T. (1995) Lipase-catalyzed syntheses of monoacylglycerols. *Enzyme Microb. Technol.* **17,** 578–586.
12. Cao, L., Fischer, A., Bornscheuer, U. T., and Schmid, R. D. (1997) Lipase-catalyzed solid phase synthesis of sugar fatty acid esters. *Biocatal. Biotransform.* **14,** 269–283.
13. Otto, R. T., Bornscheuer, U. T., Scheib, H., Pleiss, J., Syldatk, C., and Schmid, R. D. (1998) Lipase-catalyzed esterification of unusual substrates: synthesis of glucuronic acid and ascorbic acid (vitamin C) esters. *Biotechnol. Lett.* **20,** 1091–1094.
14. Litjens, M. J. J., Straathof, A. J. J., Jongejan, J. A., and Heijnen, J. J. (1999) Synthesis of primary amides by lipase-catalyzed amidation of carboxylic acids with ammonium salts in an organic solvent. *Chem. Commun.* 1255,1256.
15. Maugard, T., Remaud-Simeon, M., Petre, D., and Monsan, P. (1997) Enzymatic synthesis of glycamide surfactants by amidification reaction. *Tetrahedron* **53,** 5185–5194.
16. Halling, P. J., Eichhorn, U., Kuhl, P., and Jakubke, H. D. (1995) Thermodynamics of solid-to-solid conversion and application to enzymic peptide synthesis. *Enzyme Microb. Technol.* **17,** 601–606.
17. Woudenberg-van Oosterom, M., Van Rantwijk, F., and Sheldon, R. A. (1996) Regioselective acylation of disaccharides in tert-butyl alcohol catalyzed by Candida antarctica lipase. *Biotechnol. Bioeng.* **49,** 328.
18. Gill, I. and Vulfson, E. (1994) Enzymic catalysis in heterogeneous eutectic mixtures of substrates. *Trends Biotechnol.* **12,** 118–122.
19. Ricca, J. M. and Crout, D. H. G. (1993) Selectivity and specificity in substrate binding to proteases: novel hydrolytic reactions catalysed by α-chymotrypsin suspended in organic solvents with low water content and mediated by ammonium hydrogen carbonate. *J. Chem. Soc. Perkin Trans.* **1,** 1225–1233.
20. Furui, M., Furtani, T., Shibatani, T., Nakamoto, Y., and Mori, T (1996) A membrane reactor combined with crystallizer for production of optically active (2R,3S)-3-(4-methoxyphenyl)glycidic acid methyl ester. *J. Ferm. Bioeng.* **81,** 21–25.
21. Wolff, A., van Asperen, V., Straathof, A. J. J., and Heijnen, J. J. (1999) Potential of enzymatic kinetic resolution using solid substrates suspension: improved yield, productivity, substrate concentration, and recovery. *Biotechnol. Prog.* **15,** 216–227.
22. Kasche, V. and Galunsky, B. (1995) Enzyme catalyzed biotransformations in aqueous two-phase systems with precipitated substrate and/or product. *Biotechnol. Bioeng.* **45,** 261–267.

23. Wolff, A., Zhu, L., Wong, Y. W., Straathof, A. J. J., Jongejan, J. A., and Heijnen, J. J. (1999) Understanding the influence of temperature change and cosolvent addition on conversion rate of enzymatic suspension reactions based on regime analysis. *Biotechnol. Bioeng.* **62,** 125–134.

24. Erbeldinger, M., Ni, X., and Halling, P. J. (1998) Enzymatic synthesis with mainly undissolved substrates at very high concentrations. E*nzyme Microb. Technol.* **23,** 141–148.

25. Wolff, A., Zhu, L., Kielland, V., Straathof, A. J. J., Jongejan, J. A., and Heijnen, J. J. (1997) Simple dissolution-reaction model for enzymatic conversion of suspension of solid substrate. *Biotechnol. Bioeng.* **56,** 433–440.

26. Opsahl, R. (1992) Amides, fatty acid, in *Kirk–Othmer Encyclopedia of Chemical Technology*, 4th ed., vol. 2 (Howe-Grant, M., ed.), Wiley, New York, pp. 346–356.

27. Cravatt, B. F., Prospero-Garcia, O., Siuzdak, G., Giulia, N. B., Henriksen, S. J., Boger, D. L., et al. (1995) Chemical characterization of a family of brain lipids that induce sleep. *Science* **268,** 1506–1509.

28. De Zoete, M. C., Kock-van Dalen, A. C., Van Rantwijk, F., and Sheldon, R. A. (1996) Lipase-catalyzed ammoniolysis of lipids. A facile synthesis of fatty acid amides. *J. Mol. Catal. B. Enzym.* **1,** 109–113.

48

Characterization and Operation of a Micellar Membrane Bioreactor

Cristina M. L. Carvalho, Maria Raquel Aires-Barros, and Joaquim M. S. Cabral

1. Introduction

Recombinant cutinase from *Fusarium solani pisi* has been used in a wide variety of catalytic processes because of its high versatility as a lipolytic and estereolytic enzyme. Among the reaction systems, reverse micelles is an attractive medium to perform synthetic reactions because it allows accurate control of the water content. Furthermore, this system permits the dissolution of both hydrophobic and hydrophilic substrates and the solution remains isotropic, which facilitates the use of spectroscopic techniques for monitoring the reaction and obtaining structural data on the protein. High cutinase activity in reverse micelles was, for example, obtained in the reaction of transesterification of butyl acetate with hexanol yielding hexyl acetate *(1)*. The synthesis of this important flavor ester has been optimized with regard to the enzyme stability, water content of the medium, and substrate concentrations using the factorial design methodology *(2–4)*.

The development of reactor designs that enable continuous operation and product separation, as an integrated process, is still one of the actively investigated issues in reverse micellar enzymology. Ultrafiltration membrane bioreactors are probably the most appropriate for achieving the confinement of encapsulated enzymes at the retentate side of the membrane and simultaneous product separation in good yields. The choice of a membrane to use in a particular biotransformation is dictated by its nominal molecular-weight cutoff (NMWCO), which is a function of the average pore size and pore size distribution. The NMWCO is defined as the molecular weight of a macrosolute that yields a specified rejection, normally 90%, under defined filtration conditions

From: *Methods in Biotechnology, Vol. 15: Enzymes in Nonaqueous Solvents: Methods and Protocols*
Edited by: E. N. Vulfson, P. J. Halling, and H. L. Holland © Humana Press Inc., Totowa, NJ

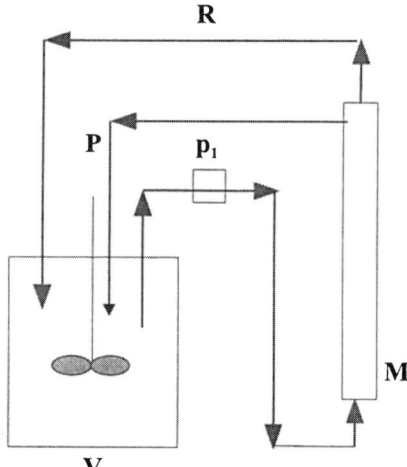

Fig. 1. Scheme of MBR operating in a total recirculation mode (*see* caption for **Fig. 2**).

(5). Other properties of the membrane such as chemical resistance to organic solvents, temperature, pH, and pressure should also be considered depending on the operational conditions and on the cleaning/sterilization process employed *(5)*.

Despite their relatively high cost, ultrafiltration ceramic membranes are highly resistant to aggressive chemicals and drastic operational conditions. Therefore, a membrane bioreactor (MBR), proposed by Prazeres et al. *(6)*, was chosen as a model system to perform the continuous alcoholysis of butyl acetate with hexanol (*see* **Subheading 3.5.** and also **refs.** *1* and *4*). Currently, the interest in membrane bioreactors is rising and numerous reports in the literature deal with their applications to the hydrolysis of triglycerides and lecithin *(7,8)*, synthesis of peptides *(9)*, pharmaceuticals *(10)*, perfusion culture of micro-organisms *(11,12)*, and transformation of hydrophobic substrates in a two-phase system *(13,14)*.

The reactor setup includes an ultrafiltration module with a tubular ceramic membrane. This anisotropic membrane is based on a carbon matrix coated with a zirconium layer and has a NMWCO of 15,000 Daltons. A scheme of the MBR operating in a total recirculation (TR) mode is depicted in **Fig. 1**, whereas the continuous operation is shown in **Fig. 2**. It should be noted that this type of reactor can be operated continuously if water-containing micelles are supplied to the reactor together with the substrates solution *(7)* in order to maintain the concentration of the constituents and the hydration level of the enzyme.

The presence of a surfactant (e.g., AOT) in the permeate is usually perceived as the major drawback of reverse-micelles-based reactors because it could hamper the purification of the final product(s). However, the possibility of

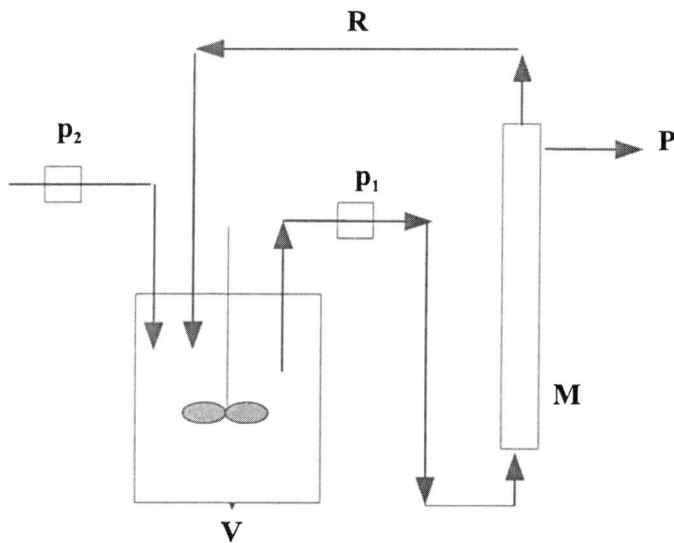

Fig. 2. Setup of MBR for continuous operation: p1—recirculating pump; p2—feeding pump; V—recirculation vessel; M—membrane module; F—feed solution; P—permeate; R—retentate.

coupling the biotransformation and separation processes in a single bioreactor *(15,16)* makes a strong case in their support.

2. Materials

2.1. MBR Setup

1. Gear pump P142-0893, Ismatec-SA (Zurich, Switzerland).
2. Peristaltic pump P-1, Pharmacia (Uppsala, Sweden).
3. Viton solvent resistant tubes (Cole Parmer, USA).
4. Carbosep® membrane M2 from Tech-Sep (Toulouse, France) with 15,000 Daltons NMWCO; 20 cm length and internal permeation area of 38 cm^2.
5. Stainless-steel monochannel ultrafiltration module and connections.
6. Thermostated recirculation vessel.

2.2. Preparation of Reverse Micelles

1. Purified *Fusarium solani pisi* cutinase cloned and expressed in *Escherichia coli* (WK-6 recombinant strain).
2. Surfactant *bis*(2-ethylhexyl) sodium sulfosuccinate, AOT, (99%) supplied by Sigma Chemical Company (St. Louis, MO, USA) used without any purification step.
3. Isooctane (99.5%), from Riedel-de Häen (Seelze, Germany).
4. NaH$_2$PO$_4$ and Na$_2$HPO$_4$ pro analysis (p.a.) from Merck (Darmstadt, Germany).
5. pH meter 691 Metrohom (Herisan, Switzerland).

2.3. Protein Determination

1. Folin–Ciocalteau Phenol Reagent 2.0 N ($C_4H_4KNaO_6$) (Sigma).
2. Bovine Albumin Fraction V (99%) (Sigma).
3. Copper (II) sulfate anhydrous, p.a. (99%) (Merck).
4. Sodium carbonate, p.a., ACS (Riedel-de Häen).
5. Potassium sodium tartrate tetrahydrate, p.a. (Merck).
6. Sodium hydroxide p.a. (Merck).
7. Spectrophotometer U-2000 Hitachi (Tokyo, Japan).
8. Centrifuge Sigma 2-15, B. Braun (Melsungen, Germany).

2.4. Determination of Water Content

1. Methanol p.a. (Merck).
2. Hydranal (Riedel-de Häen).
3. Karl–Fisher titrator Mettler DL18 (Zurich, Switzerland).

2.5. Gas Chromatograph Analysis

1. GC Hewlett Packard (Avondale, USA) 5890 HP series II.
2. HP-5 column (crosslinked 5% PM ME siloxane).
3. Flame-ionization detector (FID) using H_2.
4. Personal computer HP Vectra Pentium from Intel and software HP Chemstation from Hewlett–Packard.

2.6. Determination of Enzyme Activity and Sample Analysis

1. Butyl acetate (99.7%) (Sigma).
2. 1-Hexanol (98%) (Sigma).
3. Methanol (99.8%) (Riedel-de Häen)
4. Acetonitrile gradient grade (99.8%) (Merck).
5. Thermostated bath Thermomix MM B. Braun.
6. Magnetic stirrer IKAMAG (Staufen, Germany).
7. Vortex VF2 IKA (Staufen).
8. Sonicator Transsonic 460 Elma.
9. High-performance liquid chromatograph:
   - Waters ((Milford, USA) Model 481 LC Spectrophotometer
   - Waters 600 E System controller
   - Waters 712 Wisp Automatic Injector
   - 3390 A integrator from Hewlett-Packard
   - RP column Lichrocart® 250-4, RP-select B (5 µm) Lichrospher® 60

2.7. Reactor Cleaning

1. Distilled water.
2. Sodium hydroxide, p.a. (Merck).
3. Nitric acid (65%), p.a. (Riedel-de Häen).
4. 2-Propanol (99.7%) (Merck).

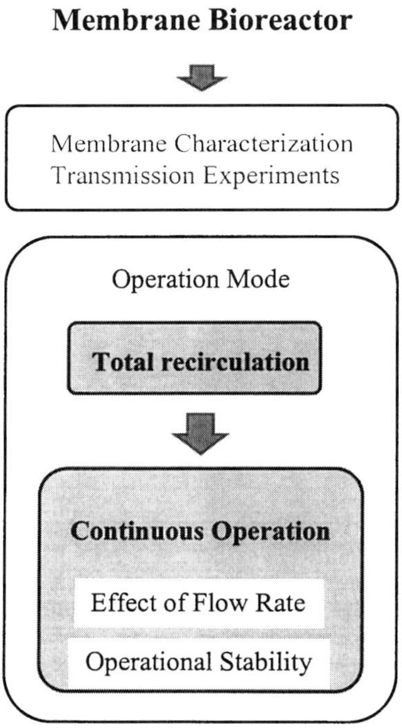

Membrane Bioreactor

Fig. 3. Plan course of the reactor characterization and subsequent operation.

3. Methods
3.1. Membrane Characterization (see *Fig. 3*)

1. Measure the permeate and the recirculation flow rates as a function of transmembrane pressure (ΔPTM) (*see* **Note 1**) using water, organic solvent, and the whole micellar system (without protein) as standard fluids.
2. Calculate the filtration flux *(J)* and the membrane permeability (L_p) for each solution tested (*see* **Note 2**).
3. Repeat the same operations with reverse micelles containing microencapsulated cutinase after 24 h in a total recirculation mode (*see* **Note 3**).
4. Calculate the membrane porosity (ε_m) (*see* **Note 4**).

3.2. Preparation of Reverse Micelles and Reactor Loading

1. Set the temperature of the thermostatted bath at 30°C.
2. Check the cleaning of MBR by recirculating 100 mL of iso-octane. (The solvent would become turbid in the presence of impurities and/or water.) If necessary substitute the solvent with a fresh portion.
3. Verify the residual iso-octane volume that remains in MBR, using the difference between the initial volume (100 mL) and the volume of solvent recovered after recirculation.

4. Dissolve AOT in iso-octane to obtain a final concentration of 150 mM (discount the volumes of aqueous phase, substrates, and iso-octane, previously retained in MBR according to **step 1**).
5. Dissolve 10 mg of lyophilised cutinase (*see* **Note 5**) in 720 µL of 200 mM phosphate buffer, pH 8.0, using vigorous vortex mixing.
6. Prepare reverse micelles by adding dropwise the aqueous phase (as prepared in **step 5**) to the AOT/iso-octane solution referred to in **step 4**. Vigorous magnetic stirring is maintained for 1 min while adding the aqueous phase (the final cutinase concentration is 0.1 mg/mL). The resulting solution should be clear, colorless, and isotropic.
7. Add hexanol (*see* **Note 6**) and stir the solution for 1–2 min.
8. Load the reactor by adding the micelles in the recirculating vessel while turning on the gear pump.
9. Allow the air to be released and start the reaction by adding the second substrate, butyl acetate (time zero).

3.3. Transmission Experiments

3.3.1. Evaluation of Protein, AOT, and Water Rejection (see **Note 7**)

1. Load the MBR as explained earlier, but without substrates (substitute their volume with iso-octane), maintaining the concentration of water and AOT as earlier.
2. Start the TR mode and take 1-mL samples, from permeate and retentate in order to quantify the different components of the system as referred to in **step 1** (*see* **Note 8**).

3.3.1.1. QUANTIFICATION OF PROTEIN

3.3.1.1.1. Construction of Calibration Curve

1. Dissolve 10 mg of bovine serum albumin (BSA) in 200 mM phosphate buffer, pH 8.0 (720 µL), to obtain solution A.

Concentration of standard (mg/L)	Solution A (µL)	Distilled water (µL)
0.1	36	—
0.075	54	18
0.050	36	36
0.025	18	54
0.0125	9	63
0	—	36

2. Use solution A to make dilutions (as indicated) and prepare 5-mL standards with different concentrations of BSA.
3. Take 36 µL of each solution and add dropwise to 4.34 mL of AOT in iso-octane. Add 1 M hexanol (624 µL). The final concentration of AOT is 150 mM.
4. Prepare samples in triplicates for each protein concentration.
5. For sample analysis, follow the procedure B below.

3.3.1.1.2. Sample Analysis (see **Note 9**)

1. Wash 10-mL centrifuge tubes with ethanol to eliminate any protein residues and dry the tubes at 50°C.
2. Add 1 M NaOH (200 µL) to a 500-µL sample in the centrifuge tube and mix twice on a vortex. The resulting solution is allowed to stand for 0.5 h at room temperature (*see* **Note 10**).
3. Prepare Lowry's reagent in the following proportions:

2% Na_2CO_3	5 mL
1% $CuSO_4$	0.1 mL
2% $C_4H_4KNaO_6$	0.1 mL

4. Add 1 mL of the Lowry reagent to each tube.
5. Stir and allow to stand for 10 min.
6. Add 200 µL of Folin's reagent diluted 1:1 with distilled water.
7. Mix with vortex and wait for 30 min; then centrifuge for 5 min at 3034g.
8. Carefully remove the top phase and discard.
9. Withdraw 700–1000 µL of the bottom phase into a glass cuvet. Use a clean pipet tip to cross the interface by releasing an air bubble to avoid phase contamination.
10. Measure the adsorbance at 750 nm against the control prepared without protein.

3.3.1.2. DETERMINATION OF WATER CONTENT

1. Set "zero" on a Karl Fischer titrator.
2. Calibrate the titrator by repeated injections (three to five times) of 10 µL of water.
3. Determine the amount of water in duplicate samples of 150 µL and take the average value.

3.3.1.3. AOT QUANTIFICATION

3.3.1.3.1. Calibration Curve

1. Prepare reverse micelles in iso-octane to obtain different concentrations of AOT as indicated:

AOT amount in 10 mL (total volume) (g)	AOT concentration (mM)
0.2221	50
0.4442	100
0.6663	150
0.8884	200
1.1105	250
1.3326	300

2. Add both substrates to the micelles (1M final concentration) and adjust the volume to 10 mL.

3.3.1.3.2. Sample Analysis

1. Follow the procedure described in **Subheading 3.3.2.1.** The retention time for AOT is 2.7 min.
2. Verify the value of water to surfactant molar ratio, W_0 (*see* **Note 11**).

3.3.2. Determination of Rejection Coefficients for Substrates and Products (see **Note 7**)

3.3.2.1. QUANTIFICATION OF ESTERS BY HPLC

3.3.2.1.1. Calibration Curve

1. Prepare the AOT/iso-octane reverse micelles, S and P, containing substrates and products respectively, as follows:

   Solution S: W_0 2.7 with 400 mM water and 150 mM AOT
                  1 M butyl acetate
                  1 M hexanol

   Solution P: W_0 2.7 with 400 mM water and 150 mM AOT
                  1 M hexyl acetate
                  1 M butanol

2. Mix both solutions to obtain 1-mL standards with the following product concentrations: 0, 20, 40, 60, 80, 100, 200, 300, 500, 700, and 1000 mM. The same curve is valid for the complementary (1000, 980, 960, 940, 920, 900, 800, 700, 500, 300, and 0 mM) substrate concentrations. The retention times for butyl acetate and hexyl acetate are 4.6 and 7.0 min, respectively.

3.3.2.1.2. Sample Analysis

1. Withdraw 200-µL samples into Eppendorf tubes containing 800 µL methanol; use vortex mixing for 40 s.
2. Store at 0°C to analyze within 24 h.
3. Add 300 µL of the above solution to an HPLC vial.
4. Analyze using the following conditions:

Eluent:	40% water: 60% acetonitrile
Injector:	injection volume: 20 µL
	run time: 10 min
System controller:	isocratic mode
	1 mL/min flow rate
	average pressure: 2700 psi
UV detector:	λ = 220 nm
	0.05 range
Integrator:	zero = 0
	attenuation = 7
	chart speed = 0.1 mm/min
	peak width = 0.64
	threshold = 5
	area rejection = 0
	stop time = 10

3.3.2.2. QUANTIFICATION OF ALCOHOLS BY GAS CHROMATOGRAPHY (GC)

1. Prepare calibration curves as described in **Subheading 3.3.2.1**.
2. Quantify the alcohols independently to relate the peak areas to concentrations.

The retention time for butanol and hexanol are 5.6 and 11.7 min, respectively. The following conditions are used for the GC analysis:

Carrier (1 mL/min) and auxiliary gas (30 mL/min): N_2 with 1 : 50 split
Injection temperature: 250°C
Detector temperature: 250°C
Oven temperature: 37°C for 7 min, then increased 30°C/min until the final temperature of 280°C is reached.

3. Inject 0.1 µL of each sample (*see* **Note 12**).

3.4. Operating the MBR in a Total Recirculation Mode

3.4.1. Determination of Cutinase Activity

1. Setup the reactor according to **Fig. 1**. The operation of the reactor in a total recirculation mode involves recycling of the permeate stream to the recirculation vessel.
2. Take samples at 1, 2, 3, 4, 5, 10, 15, and 20 min and use the procedure described in **Subheading 3.3.2.1** for analysis. The product concentration obtained after the analysis of these first eight samples is used to calculate the initial velocity and specific activity (*see* **Note 13**).

3.4.2. Determination of the Degree of Conversion

1. Withdraw samples at 0.5, 1, 2, 3, 5, 7, 10, 22, and 24 h to determine the time-course of the reaction.
2. Plot the degree of conversion (*see* **Note 14**) as a function of normalized residence time (*see* **Note 15**) to compare the results obtained in MBR to other reactor configurations.

3.5. Continuous Operation of the MBR

Change the setup and collect the permeate instead of recirculating it. At the same time, feed the system with the micelles solution at an adequate flow rate in order to maintain the total reaction volume constant (*see* **Notes 16** and **17**).

3.5.1. Procedure for Evaluating the Effect of Flow Rate on Conversion (Fig. 4)

1. Set a reference value of flow rate (*see* **Note 18**) in the appropriate range and evaluate in detail the degree of substrate conversion obtained in the permeate and recirculating solutions.
2. Change the flow rate and, at the same time, adjust the feed solution to maintain the constant volume of liquid in the reactor. Allow the system to reach steady state (i.e., when the values of product concentration stabilize).
3. Take a few more samples and calculate the standard deviation.
4. To verify that there was no loss of the enzyme activity (or to determine the degree of denaturation) return to the reference flow rate. The return to the reference flow rate is especially important when a range of different flow rate values are used in the study.

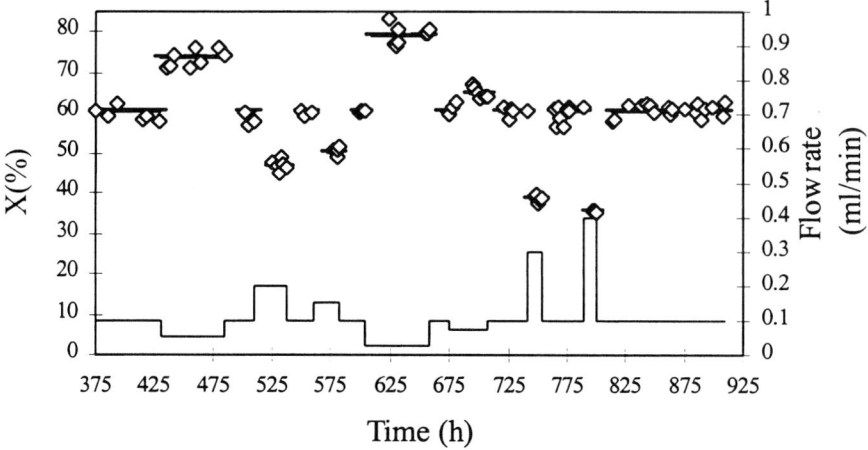

Fig. 4. Effect of the flow rate (line) on the substrate conversion degree (symbols).

5. Always start and end the assay with the reference flow rate.
6. With the results obtained in **steps 1–5**, model the reactor performance in order to estimate the substrate conversion at different operational flow rate conditions (*see* **Note 19**).

3.5.2. Determination of Operational Stability (see also **Fig. 4**)

1. Take sufficient number of samples under standard conditions (using the selected reference flow rate), depending on the kinetics of deactivation, to evaluate the degree of protein denaturation.
2. Use the degree of substrate conversion measured in the permeate and retentate to calculate the enzyme activity and plot the corresponding values against time (*see* **Note 20**).
3. Adjust the values with an appropriate model (exponential decay or other) to obtain the enzyme deactivation constant (k_d).

3.5.3. Productivity Determination

See **Note 21**.

3.6. Cleaning of the Membrane

1. Empty the bioreactor.
2. Wash with 1000 mL + 1000 mL distilled water.
3. Prepare 500 mL of 0.5 *M* NaOH and divide the solution into four portions to wash the membrane.
4. With the last portion, partially close the valve placed after the membrane to increase the pressure to force the solution through the pores of the membrane (*see* **Note 22**).

5. Wash with distilled water (1000 mL).
6. Wash with 0.03 M nitric acid (500 mL divided into four portions) (*see* **Note 23**).
7. Wash with distilled water (1000 mL).
8. Wash with propanol (130 mL).
9. Dry the membrane module at 130°C and the tubes and connectors at 50°C for at least 12 h.
10. Wash the peristaltic tubes separately with propanol to prevent AOT precipitation.

4. Notes

1. The transmembrane pressure, ΔPTM, is $\Delta PTM = [(P_i + P_0)/2]$, where P_i is the feed inlet pressure, P_0, is the retentate outlet pressure, and P_f is the permeate outlet pressure (usually $P_f = 0$, as normally the work is performed at atmospheric pressure). In order to measure P_i and P_0, it is necessary to install two manometers, before and just after the membrane module, with a sufficiently sensitive scale for the pressure values expected.
2. The filtration flux *(J)* was calculated using the relationship: $J = Q/A$, where Q is the flow rate and A is the membrane filtration area. If the flux is known, the membrane permeability, L_p, can be obtained from $J = L_p \Delta PTM$.
3. Enzymes tend to adsorb onto ceramic membranes and this leads to changes in the filtration characteristics. Twenty-four hours is assumed to be sufficient to reach a steady state, but this must be verified for each particular protein (*see* details about protein transmission in **Subheading 3.3.1.**).
4. The membrane porosity, ε_m, was calculated as $\varepsilon_m = 32 L p \mu l_p / d_p 2$, where μ is the viscosity, l_p is the pore length, and d_p is the pore diameter.
5. Cutinase was produced according to the procedure developed by Lauwereys and co-workers *(17)*.
6. It is very important to add hexanol immediately after cutinase encapsulation in order to avoid the enzyme denaturation. It was shown that the half-life time of cutinase is only 2.7 h in the absence of hexanol as compared to 39 d in the presence of 400 mM of this alcohol *(4)*.
7. The rejection coefficients (σ) are calculated using the formula: $\sigma_{obs} = (C_r - C_p)/C_r$, where C_r is the retentate concentration and C_p is the permeate concentration.
8. In parallel with protein quantification, the permeate flow rate should be measured and recorded to relate its variation to protein resistance effects (deposition or adsorption of proteins may cause a reduction in flow rate).
9. The protein determination procedure described in **Subheading 3.3.1.1.** is based on the method of Lowry et al. *(18)*, which was modified for reverse micelles by Pires et al. *(19)*. The principle modification is the extraction with ethanol (2.5% v/v) and the centrifugation step to separate the organic from aqueous phase. In the method described here, the extraction step was unnecessary, as the concentration of protein was high enough to perform the measurements accurately. To improve the accuracy of the assay, calibration curves were constructed using essentially the same technique as applied to the preparation of microencapsulated cutinase.

10. To quantify the adsorbed protein, after 48 h the micellar solution was removed from MBR and substituted by the same volume of 0.5 M NaOH, and the recirculation was continued for 1 h. Standards were prepared in aqueous phase.

11. The W_0 is given by the expression

$$W_0 = \frac{[\text{Water}]\ (\text{m}M)}{[\text{Surfactant}]\ (\text{m}M)}$$

or in the present conditions,

$$\text{Water (m}M) = \frac{(X \times 10^6\ \text{mg})/150\ \mu\text{L}}{18\text{g}} = \frac{X \times 10^4\ \text{mg}}{27}$$

as the [AOT] = 150 mM, the experimental value is obtained by using the expression $W_0 = X$ mg/0.405.

12. The HP-5 column may not be suitable for routine analysis of AOT/reverse micelles, as the surfactant peak was not identified under the chromatographic conditions used. It seems that this component is retained in the column. The supplier was unable to clarify the matter.

13. The determination of specific activity is based on the initial velocity. The specific activity is defined as millimolars of product (hexyl acetate) formed per minute and milligram of protein to facilitate the comparison of results obtained with different enzyme preparations and the data reported elsewhere.

14. The degree of conversion (X) is calculated as the ratio of the substrate concentration transformed into product (P) to the initial substrate present (S_0). In the case of two substrates present in different concentrations, the concentration of the limiting substrate is used for calculating the maximum conversion. The expression for the degree of conversion is $X(\%) = ([P]/[S_0] \times 100$.

15. Normalized residence time is defined as $\tau = Et/V$ for a batch reactor, or $\tau = E/Q$ for a continuous reactor, using mg·min/mL as units in both cases. The application of normalized residence time concept is very useful for comparison of the enzyme performance in different type of reactors and/or using different concentrations of biocatalyst.

16. Because of the protein adsorption to the membrane (as verified in transmission experiments), the continuous operation with total recirculation was started after 24 h, despite the system reaching equilibrium, in terms of conversion, after 5 h.

17. The feed solution is prepared in the same way as the reverse micelles solution described in **Subheading 3.2.** but with no enzyme. The concentrations of AOT and water are 150 mM and 400 mM, respectively. The concentrations of substrates are maintained at their initial values and there is no need to correct the iso-octane volume.

18. A typical reference feed/permeate flow rate for this reactor is 0.1 mL/min. This value corresponds to a recirculation flow rate of 1020 mL/min. Permeate flow rates could be varied in the range of 0.025–0.4 mL/min for the micellar system described.

19. Continuous reactors are usually subdivided into two major groups according to the mode of agitation and to the flow characteristics of substrate and product *(20)*. A reactor is classified as a CSTR (continuous stirred tank reactor) if mixing is accomplished by stirring and the degree of conversion is essentially the same in any part of the reactor. In a PFR (plug flow reactor), there is no agitation of the reaction mixture and, therefore, the conversion depends on the reactor length and is different for each particular section. Consequently there is an increase in the product concentration along the reactor. Despite the difficulties in obtaining the ideal PFR, this type of reactor is kinetically more efficient than a CSTR, as for the same residence time, it requires less enzyme to obtain the same conversion *(20)*.

The following equations describe the performance of a CSTR and a PFR, for a reversible Michaelis–Menten kinetics.

CSTR

$$\frac{k_{cat}}{Q} = \frac{X_e \cdot X [K_m + S_0 - S_0 \cdot X + (K_m/K_p) \cdot S_0 \cdot X]}{X_e - X}$$

PFR

$$(k_{cat} \cdot E)/Q = X_e \{ S_0 X \cdot [1 - (K_m/K_p)] + [K_m + S_0 - S_0 X_e + S_0 X_e \cdot (K_m/K_p)] \cdot \ln [X_e/(X_e - X)] \}$$

where $K_{cat} = V_{max}/[E]$ (mmol·min^{-1}g^{-1}), K_m and K_p, the Michaelis-Menten constants for substrate and product respectively, and, X_e, is the equilibrium conversion.

A continuous membrane reactor with a high recirculation flow rate behaves as a CSTR, since the recycling flow rate is much higher than the feed flow rate, and this results in the homogeneous distribution of the reaction components in the medium.

20. Assuming a CSTR behaviour, the activity retention, which is given by the ratio E/E_0, is related to the deactivation constant (k_d) by the following expression:

$$\frac{E}{E_0} = e^{-k_d \cdot (t_f - t_i)} = \frac{[K_m + S_0 - S_0 \cdot X_f + S_0 X_f (K_m/K_p)] \cdot (X_e - X_i)}{[K_m + S_0 - S_0 \cdot X_i + S_0 X_i (K_m/K_p)] \cdot (X_e - X_f)}$$

where X_i and t_i are the initial conversion and the initial operation time, respectively, after the continuous reactor had reached the steady state (16 h for $Q = 0.1$ mL/min). By plotting the activities against time, the expression $Y = Ae^{-kdt}$ describes the exponential deactivation model.

21. The productivity expresses the overall process yield using a defined set of conditions. The productivity can be related to the enzyme concentration (specific productivity, P_e) or to the reactor volume (volumetric productivity, P_v). Their definitions are

$$P_e = \frac{S_0 X Q}{E} \qquad\qquad P_v = \frac{S_0 X Q}{V}$$

where E the total amount of enzyme, S_0, is the initial substrate concentration, X is the conversion degree, and V is the reactor volume. P_v is the average production capacity per unit volume of the bioreactor and time; P_e is defined as the total amount of product (moles) produced by 1 mol of enzyme during its operational lifetime *(20)*.

22. The sodium hydroxide is used to denature and hydrolyze the protein adsorbed to the membrane. As a part of the protein may be entrapped inside the pores, it is very important to obtain a good permeation flux through the inner pores to ensure adequate cleaning efficiency.

23. Nitric acid neutralizes the remaining basic residues. At this step, a significant increase in the permeation flow rate should occur. Proceed as in the case of the NaOH cleaning, using the reduction valve to assure that all base traces are removed.

References

1. Carvalho, C. M. L., Serralheiro, M. L. M., Cabral, J. M. S., and Aires-Barros, M. R. (1997) Application of factorial design to the study of transesterification reactions using cutinase in aot reversed micelles. *Enzyme Microb. Technol.* **21,** 117–123.

2. Box, G. E. P., Hunter, W. G., and Hunter, J. S. (1985) *Statistics for Experimenters.* Wiley, New York.

3. Barker, T. B. (1985) *Quality by Experimental Design.* Marcel Dekker, New York.

4. Carvalho, C. M. L., Cabral, J. M. S., and Aires-Barros, M. R. (1999) Cutinase stability in AOT reversed micelles: system optimization using the factorial design methodology. *Enzyme Microb. Technol.* **24,** 569–576.

5. Hildebrandt, J. R. (1991) Membranes for bioprocessing: design considerations, in *Chromatographic and Membrane Processes in Biotechnology.* (Costa, C. A. and Cabral, J. M. S., eds.), Kluwer, Dordrecht, The Netherlands, pp. 363–378.

6. Prazeres, D. M. F., Garcia, F. A. P., and Cabral, J. M. S. (1993) An ultrafiltration membrane bioreactor for the lipolysis of olive oil in reversed micellar media. *Biotechnol. Bioeng.* **41,** 761–770.

7. Prazeres, D. M. F., Garcia, F. A. P., and Cabral, J. M. S. (1994) Continuous lipolysis in a reversed micellar membrane bioreactor. *Bioprocess. Eng.* **10,** 21–27.

8. Morgado, M. A Suzuki, T. (1998) Improvement of filtration performance of stirred ceramic membrane reactor and its application to rapid fermentation of lactic acid by dense cell culture of lactococcus lactis. *J. Ferment. Bioeng.* **85,** 422–427.

9. Serralheiro, M. L. M., Prazeres, D. M. F., and Cabral, J. M. S. (1994) Dipeptide synthesis and separation in a reversed micellar membrane reactor. *Enzyme Microb. Technol.* **16,** 1064–1073.

10. Satory, M., Furlinger, M., Haltrich, D., Klube, K. D., Pittner, F., and Nidetzky, B. (1997) Continuous enzymatic productio of lactobionic acid using glucose–fructose oxireductase in an ultrafiltration membrand reactor. *Biotechnol. Lett.* **19,** 1205–1208.

11. Kamoshita, Y., Ohashi, R., and Suzuki, T. (1998) Improvement of filtration performance of stirred ceramic membrane reactor and its application to rapid fermentation of lactic acid by dense cell culture of lactococcus lactis. *J. Ferment. Bioeng.* **85,** 422–427.

12. Zhang, W., Park, B. G., Chang, Y. K., Chang, H. N., Yu, X. J., and Yuan, Q. (1998) Factors affecting membrane fouling in filtration of saccharomyces cerevisiae in an internal ceramic filter bioreactor. *Bioprocess Eng.* **18,** 317–322.

13. Doig, S. D., Boam, A. T., Leak, D. I., Livingston, A. G., and Stuckey, D. C. (1998) A membrane bioreactor for biotransformations of hydrophobic molecules. *Biotechnol. Bioeng.* **58,** 587–594.
14. Bouwer, S. T., Cuperus, F. P., and Derksen, J. T. P. (1997) The performance of enzyme-membrane reactors with immobilized lipase. *Enzyme Microb. Technol.* **21,** 291–296.
15. Nakamura, K. and Hakoda, H. (1992) Electro-ultrafiltration bioreactor for enzymatic reaction in reverse micelles, in *Proceedings of Asia–Pacific Biochemical Engineering Conference 1992, Biochemical Engineering for 2001* (Furusaki, S., Endo, I., Matsuno, R., eds.), Springer-Verlag, Tokyo, pp. 433–436.
16. Hakoda, M., Enomoto, A., Hoshino, T., and Shiragami, N. (1996) Electro-ultrafiltration bioreactor for enzymatic reaction in reversed micelles, *J. Ferment. Bioeng.* **82,** 361–365.
17. Lauwereys, M., De Geus, P., De Meutter, J., Stanssens, P., and Matthyssens, G. (1990), in *Lipases: Structure, Mechanism and Genetic Engineering*, (Alberghina, R.D., Verger, R., eds.), VCH, New York, pp. 243–251.
18. Lowry, O. H., Rosenbrough, N. J., Farr, A. L., and Randall, R. J. (1951) *J. Biol. Chem.* **193,** 265–275.
19. Pires, M. J., Prazeres, D. M. F., and Cabral, J. M. S. (1993) Protein assay in reversed micelle solutions. *Biotechnol. Tech.* **7,** 293,294.
20. Cabral, J. M. S. and Tramper, J. (1994) Bioreactor design, in *Applied Biocatalysis* (Cabral, J. M. S., Best, D., Boross, L., Tramper, J., eds.), Harwood, Basel, Switzerland, pp. 333–370.

49

Immobilization of Lipase Enzymes and Their Application in the Interesterification of Oils and Fats

Alan D. Peilow and Maha M. A. Misbah

1. Introduction

Lipase enzymes occur widely in microorganisms, plants, and animals. The function of lipase is to hydrolyze triacylglycerol to diacylglycerol, mono-acylglycerol, fatty acid, and glycerol and therefore provides energy and essential fatty acids for the organism. The hydrolysis reaction is easily reversible and, consequently, lipases catalyze the reverse reaction to synthesize fatty acid esters under low water (microaqueous) conditions. This has led to their extensive use in lipid modification *(1–5)*.

Lipase enzymes can show specificity toward both fatty acid and acylglycerol parts of their substrate and this selectivity can be exploited to synthesize esters difficult to obtain via conventional chemistry. For example, a number of lipases show regio- (or sn –1,3) specificity, where lipase catalyse reactions at the outer (sn-1 and sn-3) positions of the triacylglycerol, but cannot catalyze reactions at the inner (or sn-2) position. This allows the synthesis of new triacylglycerols in which the fatty acid profile of the outer (sn-1,3) positions can be modified without changing the fatty acid profile of the inner (sn-2) position. Regio (or sn-1,3) specificity has been exploited in the manufacture of triacylglycerols used in the confectionery fats industry by modifying cheaper oils via interesterifcation reactions to prepare higher value cocoa butter equivalents *(6)*.

Industrial applications have, in general, been developed using lipase obtained from microbial organisms. The enzyme is produced extracellularly during fermentation and is easily recovered by conventional downstream processing techniques. The application of modern recombinant DNA techniques has led to a much more efficient production and the resultant lower cost

From: *Methods in Biotechnology, Vol. 15: Enzymes in Nonaqueous Solvents: Methods and Protocols*
Edited by: E. N. Vulfson, P. J. Halling, and H. L. Holland © Humana Press Inc., Totowa, NJ

has enabled the application of lipase in laundry products *(7)*. Although the availability of cheaper lipases has widened the potential for applications in synthetic reactions, the reuse of lipase is fundamental in the economics of interesterification processes. Crude suspensions of enzyme powders have been used for these reactions *(8)*, their effectiveness (efficiency) is in general much lower than for immobilized lipase biocatalysts. In order to fully exploit lipases under essentially nonaqueous conditions, the use of immobilization technology offers a number of important benefits, including enzyme reuse, easy separation of product from enzyme, and the potential to run continuous processes via packed-bed reactors. In some cases the activity and stability of the enzyme is also improved *(9)*.

In this chapter we describe both the issues of lipase immobilization (**Subheading 2.**) and the use of immobilized lipase in interesterification reactions (**Subheading 3.**).

2. Immobilization of Lipases

2.1. Background

There is a large amount of literature describing the many methods for the immobilization of lipase enzymes, much of which has been reviewed *(10,11)*. These methods can be split into the six broad groups:

1. Drying, precipitation, or adsorption on hydrophilic particles (e.g., Celite) *(12,13)*.
2. Adsorption on hydrophobic surfaces *(9,14–16)*.
3. Adsorption on ion-exchange resins *(17–19)*.
4. Covalent attachment to suitable ligands *(20–22)*.
5. Crosslinked enzyme crystals *(23)*.
6. Physical entrapment in polymer *(24)* or hydrophobic silica gels *(25)*.

It is important to choose a support material that permits the immobilized lipase to perform at maximum efficiency *(26)*. To function effectively a potential support material must satisfy a number of criteria:

1. It must allow easy enzyme immobilization without appreciable losses of activity.
2. It must have appropriate pore and particle size so it does not limit diffusion of substrate *(14,27)*.
3. It must contain suitable mechanical properties for the desired process.
4. It must contain no extractable materials likely to contaminate the product stream *(18)*.
5. Ideally it should be inexpensive.

The most widely used immobilization methods are those based on adsorption processes. These are generally simple, effective, do not require the use of potentially harmful chemicals, and do not result in large losses of lipase activity. Consequently, the method described in this chapter describes the adsorption of lipase onto a hydrophobic porous solid. Lipase enzymes hydrolyze triacylglycerol by adsorbing to the lipid substrate at the fat/water interface and are, therefore, very stable when adsorbed to hydrophobic surfaces. We have found the adsorption of lipase (from a wide variety of sources) onto Accurel EP100 to be easy, reproducible, and efficient *(15)*. Accurel EP100® is a porous polypropylene, which has a wide distribution of pore size (10 nm–20 µm) and a large internal surface area (40 m²/g) and is capable of adsorbing high loading of lipase.

To immobilize lipase onto Accurel EP100, the hydrophobic polypropylene requires the addition of a wetting agent to allow the hydrophilic lipase solution to contact the polymer surface in order for adsorption to occur. The adsorption process is monitored by following the loss of lipase activity from the aqueous phase using an hydrolysis assay, described in detail in **Subheading 2.3.1.1.** *(28)*. On completion of the adsorption process, which can take between a few minutes and many hours depending on the amount of protein present, the catalyst is collected by filtration, washed with water to remove unadsorbed protein, and dried.

2.1.1. Measurement of Catalyst Activity

Activity of lipase solutions is usually determined by measuring the rate of release of fatty acid from a dispersion of oil in water. This is due to the fact that most lipase enzymes have maximum activity in the presence of an oil/water interface. For lipase immobilized in porous materials, this is an inappropriate method because of the problem of oil droplets accessing the lipase in the interior of porous particles. For example, it has been shown that for *Rhizomucor miehei* lipase adsorbed onto Accurel EP100, the efficiency of activity (activity/loading), as measured in the tributyrin assay, was found to decline rapidly with increasing lipase loading *(15)*. This is due to dominance of the mass transfer limitations imposed on the heterogeneous substrate by the porous structure of Accurel EP100. It is expected that these mass transfer problems are particularly severe in this system as the substrate is in the form of a suspension of small droplets in water (approx 10 µm), most of which are larger than the mean pore diameter of Accurel EP100 (approx 0.2 µm). The use of heterogeneous substrate mixtures is therefore generally unsuitable for the assay of enzymes immobilized in porous materials.

To help in understanding the factors that affect catalyst activity (e.g., immobilization method, comparing different lipases, lipase content, etc.), we have found an esterification assay to be most useful. The assay uses an equimolar mixture of lauric acid and octan-1-ol containing dissolved water, giving a single-phase system and therefore minimizing mass transfer problems. The assay procedure is described in detail in **Subheading 2.3.1.2.**

The activity of lipase catalysts has been shown to vary with water activity (a_w) in the system *(29–31)*. Water plays two roles, as a product of the synthesis reaction and stabilising the tertiary structure of the protein. To obtain reliable estimates of catalyst activity it is vital to control a_w. This is achieved by controlling the water content of both substrate and catalyst at the start of the assay.

2.1.2. Relationship Between Catalyst Activity and Lipase Content

The effect of lipase loading on catalyst activity in synthetic reactions is not a simple relationship. A plot of esterification activity for *Thermomyces* (formerly known as *Humicola*) lipase adsorbed onto Accurel EP100 is shown in **Fig. 1**. At intermediate lipase loading (50–200 kLU/g) [LU = lipase unit], activity of the catalyst increases as the lipase loading is increased. At low loading (10–50 kLU/g), activity appears to be suppressed, giving an apparent lag phase in the plot of activity versus lipase loading. This "spreading" effect is more obvious when the activity data are plotted in the form of efficiency (activity/loading) against loading, shown in **Fig. 2**. The plot shows that at low loading, efficiency rapidly increases as loading is increased until it reaches a maximum at a loading of approximately 150 kLU/g.

The suppression of activity at low loading is possibly the result of conformational changes in the enzyme molecule. Lipase has a strong affinity for the surface (this is the basis of the immobilization method), and at low loading the enzyme attempts to maximize its contact with the surface. This results in a loss of conformation, and consequently a reduction in efficiency. As loading is increased, the area available for the lipase to adsorb decreases, and therefore the lipase is less able to spread itself, leading to an increase in the amount of active conformation retained, and hence the efficiency increases. An alternative explanation is that the surface contains areas of high affinity ("hot spots"), which lead to conformational change on adsorption and these areas adsorb lipase preferentially.

The depression of activity at low loading can be overcome by treating Accurel EP100 with a nonlipase protein prior to the adsorption of lipase. The exact mechanism by which the addition of other proteins improves lipase activity is not clear. It is possible that the proteins occupy sites of high affinity on the support or they may simply reduce the excess surface area available and therefore inhibit lipase deactivation. A number of proteins have been shown to be effective in stopping the deactivation at low loading for a range of lipases *(32)*. A plot of efficiency of activity versus loading for *R. miehei* lipase immobilized onto Accurel EP100, (with and without ovalbumin treatment) is shown in **Fig. 3**. At low loading (10–100 kLU/g) the deactivation process is inhibited and a large increase in efficiency over the untreated sample is observed. This suggests that *R. miehei* lipase is prone to unfolding on adsorption to hydropho-

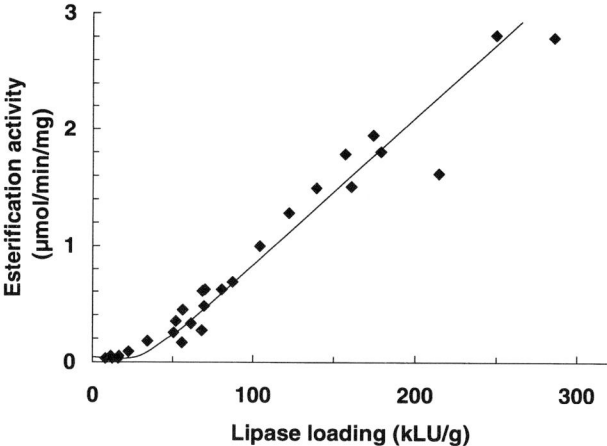

Fig. 1. Plot of activity versus lipase loading for *Thermomyces* sp. lipase absorbed onto Accurel EP 100.

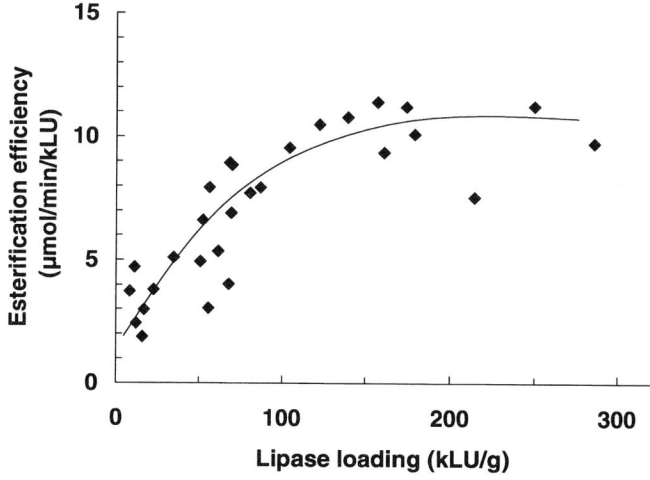

Fig. 2. Effect of lipase loading on the efficiency of esterification activity for *Thermomyces* lipase absorbed onto Accurel EP100.

bic surfaces in the absence of protein treatment. At high lipase loading (200–600 kLU/g) the efficiency declines with increased lipase loading. This is presumably due to mass transfer limitations of the substrate unable to diffuse fast enough into the particles at the high lipase loading. The mass transfer limitation was not seen in the *Thermomyces* data, because of the much lower reaction rates achieved in comparison to *R. miehei* lipase.

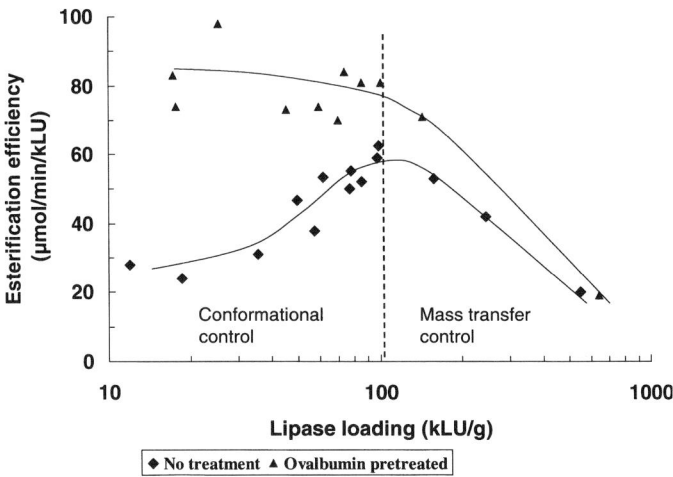

Fig. 3. Efficiency of esterification activity versus lipase loading for *Rhizomucor miehei* lipase absorbed onto Accurel EP100

Not all lipases suffer from loss of activity on adsorption at low loading. The B lipase from *Candida antarctica* does not show increased activity when adsorbed onto Accurel EP100 treated with ovalbumin (*see* **Fig. 4**). The plot of efficiency versus loading shows a plateau at low loading and thereafter decreases in efficiency with increasing loading as the mass transfer limitations are encountered. This infers that the structure of *C. antarctica* B is less prone to unfolding as a result the adsorption process, this could be the result of it having a weaker affinity for the support surface or a more "rigid" conformational structure, or both.

2.2. Materials

2.2.1 Lipase Assays

2.2.1.1. TRIBUTYRIN HYDROLYSIS ASSAY

1. Lipase enzyme (E.C. 3.1.1.3) dissolved in deionized water to contain 100–1000 LU/mL.
2. Water, deionized (Milli-Q plus), Millipore, UK.
3. Tributyrin (glycerol tributyrate), purum, Fluka Chemicals, UK.
4. Emulsification reagent (gum arabic [6 g], NaCl [17.9 g], KH_2PO_4 [0.41 g], glycerol [540 mL], and deionized water [500 mL]).
5. Homogenizer, Ultra-Turrax Model T25, Janke & Kunkel GmbH, Germany.
6. VIT 90 autotitrator in pHstat mode (Radiometer, Denmark). Set point pH 7.0. Titrant 0.100 *M* NaOH solution. 1.0 mL buret. Titration cup temperature is maintained at 30°C using a water bath.

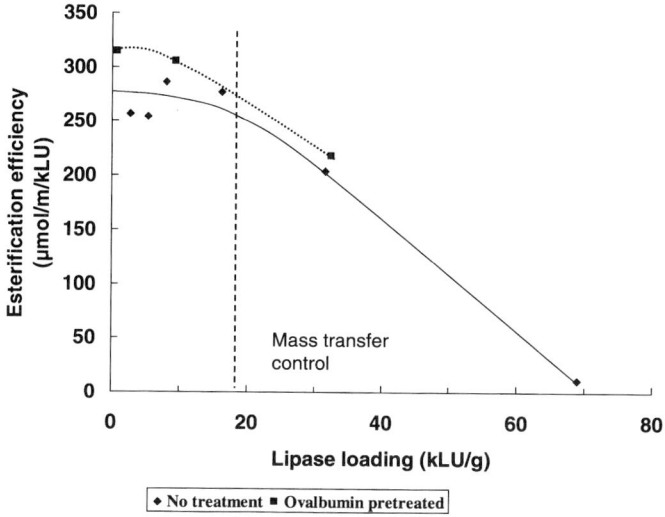

Fig. 4. Efficiency of esterification activity versus lipase loading for *C. antarctica* B lipase absorbed onto Accurel EP100.

2.2.1.2. Octyl Laurate Esterification Assay

1. Immobilized lipase (5–100 mg), depending on lipase content.
2. Mg (NO₃)₂ solution (saturated solution).
3. Lauric acid, purum (Fluka Chemicals).
4. Octan-1-ol, puriss (Fluka Chemicals).
5. Water, deionized (Milli-Q plus), Millipore.
6. Crimp-top vials and caps, 20 mL (Chromacol, UK).
7. OLS 200 shaking water bath (Grant Instruments, UK).
8. Aluminium oxide, basic, activity II (Fisons Scientific Equipment, UK).
9. Ethyl palmitate 99%, (Aldrich, UK).
10. Diethyl ether A.R., Petroleum ether A.R. (bp 100–120) (Fisher Scientific Chemicals, UK).
11. Glass-distilled hexane (Rathburn Chemicals, Scotland).
12. Gas chromatograph (GC) for fatty acid ester analysis. A GC fitted with a capillary column such as BP20 phase, 25 m × 0.53 mm, 1 μm film (SGE, Milton Keynes, UK) is suitable.

2.2.2. Lipase Immobilization: Adsorption onto Accurel EP100

1. Accurel EP100 (macroporous polypropylene) was obtained from Akzo Nobel, Germany.
2. Ethanol; 100 % v/v (Fisher Scientifi)c.
3. Deionized water (Milli-Q plus).
4. Ovalbumin, grade V (Sigma).

5. Lipase solution (50 – 200 kLU), e.g., 1 mL lipolase 100L (*Thermomyces* sp., Novo-Nordisk, Bagsvërd, Denmark.)

2.3. Methods

2.3.1. Lipase Assays

2.3.1.1. TRIBUTYRIN HYDROLYSIS ASSAY

The activity of lipase solutions is determined using an assay adapted from the method described by Novo-Nordisk *(28)*. The assay measures the rate of release of butyric acid from a suspension of tributyrin at 30°C.

1. The emulsification reagent is prepared by dissolving gum arabic (6.0 g), glycerol (540 mL), NaCl (17.9 g), and KH_2PO_4 (0.41 g) in distilled water (400 mL) and then making up the volume to 1000 mL with distilled water. The solution is stable up to 1 mo at room temperature.
2. The tributyrin suspension is prepared by homogenising a mixture of tributyrin (9 mL), distilled water (141 mL), and emulsification reagent (30 mL) at 24,000 rev/min^{-1} for 15 s. The emulsion should be prepared daily. Transfer 20 mL suspension to the titration cup and allow it to reach the assay temperature (30°C). The volume of emulsion is enough for nine tributyrin assays. The volume can be scaled up or down to the number of assays required for the day.
3. The initial pH of the tributyrin emulsion should be about 4.0. Prior to the start of the assay the pH should be raised to 6.8–7.0 by predosing an aliquot of titrant.
4. The autotitrator should be started and the lipase solution added after the set pH has been reached (pH 7.0). The addition of lipase solution may lead to an increase or decrease in pH. The action of lipase on the substrate leads to the release butyric acid reducing the pH, which, on detection by the autotitrator, will add titrant. During this initial phase the rate of addition will increase to match the rate of release of butyric acid. This should take 30–60 s. Thereafter, the rate of addition should settle to a constant figure.
5. The rate of addition of titrant is determined between 2 and 6 min.
6. One lipase unit is defined as the amount of lipase that liberates 1 μmol butyric acid per minute under these conditions.

2.3.1.2. OCTYL LAURATE ESTERIFICATION ASSAY

The following method is written for the simultaneous assay of four catalysts (each in duplicate). The assay measures the initial rate of esterification of an equimolar mixture of octan-1-ol and lauric acid.

1. The substrate solution is prepared by mixing lauric acid (48.0 g), octan-1-ol (31.0 g) and water (0.95 g). The mixture is heated to 50°C in a stoppered flask to melt the lauric acid. The liquid is stirred vigorously until the water is fully dissolved in the organic phase.
2. Samples of lipase catalyst are preweighed (5–100 mg depending on lipase content) into crimp-top vials (8 × 20 mL). An empty vial is used for a control sample.

3. Add 10.0 mL substrate solution to each vial (at 1 min intervals), cap vial, and place in shaking water bath for 60 min at 50°C. Vials are shaken at 200 strokes/min.
4. Remove sample from vial (100 μL) and immediately place on an alumina column (basic, 2 g), together with ethyl palmitate solution (100 μL, 25 mg/mL). Elute the fatty acid esters using diethyl ether (4 mL).
5. Collect the eluent and remove the diethyl ether by evaporation.
6. Dissolve fatty acid esters in hexane (4 mL) and measure the ratio of octyl laurate to ethyl palmitate by GC. (Calibrate GC using known ratios of ethyl palmitate and octyl laurate.)
7. The activity of the lipase catalyst is expressed as micromole ester formed per minute per milligram of catalyst. The efficiency of esterification activity is calculated by dividing the activity by the lipase loading, expressed as micromole ester per minute per kilo lipase unit.

2.3.2. Immobilization Method: Adsorption onto Accurel EP100

1. Wet a sample of Accurel EP100 (2.0 g) by adding absolute ethanol (10 mL) and then add deionized water (30 mL). Mix the suspension gently on an orbital shaker for 15 min.
2. Dissolve ovalbumin (0.26 g) in water (20 mL) and add to the Accurel EP100 suspension. Continue gentle mixing for 1 h.
3. Dissolve the required amount of lipase in water (40 mL) and add to the wetted Accurel EP100. Leave the sample mixing on an orbital shaker at room temperature (approx 20°C).
4. Monitor the adsorption of lipase onto the support by measuring loss of activity from the solution using the tributyrin hydrolysis assay (*see* **Subheading 2.3.1.1.**). Remove a sample (0.50 mL) at about hourly intervals and determine its lipase activity. The adsorption can take between 1 and 16 h to reach completion (>90 % of activity adsorbed), depending on the lipase preparation and the required loading.
5. Collect the catalyst by vacuum filtration using filter paper (Whatman No. 1), wash twice in water (100 mL), and rinse with deionized water several times to remove unadsorbed material. Leave the sample under suction on the filter paper for 1 h to remove the majority of the water.
6. Complete drying by placing the sample in a vacuum oven at room temperature until no further weight loss is recorded (normally overnight). Store at 4°C.
7. Equilibrate lipase catalyst over saturated $Mg(NO_3)_2$ solution (relative humidity = 52%) at room temperature for at least 48 h before use.
8. Lipase loading is calculated from the loss of activity from solution, expressed per weight support (kLU/g).

2.4. Notes

2.4.1. Tributyrin Assay

The first assay of the day can be unreliable. The cause is not obvious, but subsequent assays are reproducible within the normal scope of an enzyme assay. The first assay should be carried out in triplicate, and thereafter assays should be carried out in duplicate.

2.4.2. Esterification Assay

1. The activity of lipase catalysts is strongly influenced by the water content (a_w). The control of water activity prior to an assay is best carried out using the minimum amount of catalyst required for the assay, and maximizing the contact with the controlled humidity atmosphere by using a thin layer of catalyst. This is achieved by placing samples (200 mg) in Petri dishes and incubating the dishes in a desiccator containing saturated $Mg(NO_3)_2$ solution (a_w approx 0.5).
2. When measuring the esterification activity of a number of catalysts, it is best to start each assay at minute intervals. This allows sufficient time between sampling each vial to place the sample immediately onto the small alumina column and elute with diethyl ether to terminate the reaction.
3. It is important to carry out a control esterification reaction to measure the background esterification rate, which can be subtracted from the lipase catalyzed rate. Up to 0.5% of the substrate can be esterified by nonenzymatic reactions. Consequently, for reliable estimates of catalyst activity the reactions should exceed 2% conversion. Each catalyst should be assayed in duplicate. Repeated assays indicate the standard deviation for the assay is about 6% of the mean.
4. Due to the viscosity of the octanol/lauric acid mixture, sampling is best carried out with a positive displacement pipet. Avoid sampling of catalyst particles by placing the tip next to the wall of the vial.

2.4.3. Lipase Immobilization

1. In order to accurately measure lipase loading it is essential that the lipase solution is stable over the required period. Apparent loss of activity can arise from proteolysis; some lipase preparations contain significant protease activity. If proteolysis is thought to be a problem it can often be minimized by lowering the immobilization temperature (e.g., to approx 5°C). Monitor the activity of the lipase solution prior to immobilization to confirm stability.
2. Some lipase preparations (powdered forms) can contain significant levels of insoluble material. Centrifugation and/or filtration before immobilization should remove this, as the material can lead to blockage of pores, resulting in poor immobilization and/or low catalyst activity.
3. Care should be taken to ensure the support is fully wetted both before the lipase solution is added and during the immobilization process. This can be difficult with Accurel EP100, which tends to float and can form a dry "crust" on the enzyme solution. A small amount of mechanical agitation should be used to maintain good contact between enzyme solution and support.
4. Accurel EP100 contains small residues of vegetable oil from the manufacturing process. On addition of lipase the oil is hydrolyzed and can lead to a fine dispersion of fatty acid in the water phase, giving a cloudy appearance. This does not affect the adsorption process but can lead to blockage of filter paper during the

washing process. Changing the filter paper after the first wash significantly improves the rates of filtration and drying.
5. Using of strongly buffered solutions in immobilization can lead to problems of salt precipitating into the porous structure during drying. Using low concentrations of buffer combined with extra washes with deionized water should overcome the problem if control of pH is required during immobilization.

3. Interesterification of Oils and Fats

3.1. Background

The properties of fat depend on its fatty acid structure, i.e., the type of fatty acid and their position on the glycerol backbone. The functional and commercial value of fat is based on its structure. Upgrading of fats is mainly changing the relative amounts of the triacylglyceride molecules in fats that contain particular fatty acid structures. Fat and oil processors have changed the fatty acid structure of their materials either by blending different triacylglyceride mixtures or by chemical modification of the fatty acids by hydrogenation, or by rearrangement of the fatty acids on the glyceride backbone (interesterification). In many instances, interesterification is complementary or used in combination with hydrogenation or fractionation.

Chemical interesterification is characterized by a randomization in the distribution of fatty acid moieties in the triacylglycerol molecules by applying a catalyst such as sodium methoxide or glycerolate *(33–35)*. If chemical catalysts are used to promote the migration of fatty acid groups between triacylglycerol molecules, the products consist of triacylglycerol mixtures in which the fatty acid groups are randomly distributed among the glycerol molecules.

Great interest has been directed toward the use of lipases in interesterification reactions because of some advantages over chemical catalysis. Enzyme reactions can be highly specific (depending on type of enzyme used) and occur under mild conditions of temperature and pH. The use of regiospecific lipase gives the possibility of tailor-made products, which are unobtainable by chemical interesterification methods, and some of these products have properties of considerable value to the edible oils and fats industry. Regiospecificity is an important aspect of lipase applications and provides a means of classifying them. The properties and applications of lipases have been described in a vast number of articles *(36–47)*.

The enzymatic interesterification reaction system consists of the lipase catalyst together with a very small amount of water dissolved in a continuous organic phase comprising the reactants. The water is dissolved in the organic phase and equilibrates with the enzyme catalyst particles; there is no distinct aqueous phase. Lipases catalyze reactions at the interface, and to obtain high interesterification reaction rates a large interfacial area between the reactants and the more hydrophilic enzyme phase is required. This can be achieved by producing a fine dispersion of lipase in the organic phase or by immobilizing the enzyme on the internal surface of macroporous support particles. Immobilized lipases are usually preferred for interesterification (*see* **Subheading 2.**).

Immobilized enzyme catalysts can be used in both stirred-batch and packed-bed reactors. The properties of the immobilized enzyme are governed by the properties of the enzyme and the carrier material. The criteria for selection of the lipases as interesterification catalysts are specificity, activity, and stability in the reaction system and the clearance for use in food processing.

The following paragraphs describe the practical procedures applied on a laboratory scale for both continuous and batch processes.

3.2. Materials

3.2.1. Oils

The type of oil used in the experiment depends mainly on the objective of the investigation and the type of properties required in the product. The use of refined oils is recommended.

3.2.2. Immobilized Enzyme

An example of a commercially available immobilized enzyme cleared for food use is Lipozyme IM, e.g., Novo-Nordisk. Lipase IM is *Rhizomucor miehei* lipase immobilized onto a macroporous ion-exchange resin (Duolite ES568N).

Other lipase preparations can be used after immobilization (as described in **Subheading 2.**).

3.3. Methods

3.3.1. Packed-Bed Reactor

A number of studies have been published that cover the application of immobilized enzymes in packed-bed reactors in enzymatic interesterification reactions *(48–57)*. The work reported investigated mainly the following topics:

1. Effect of reaction conditions (water content, temperature, flow) on process.
2. Kinetics of interesterification reactions.
3. Production of structured triacylglycerides and acyl migration.
4. Pressure drop.
5. Mass transfer.

The parameters used to control the enzymatic reaction are temperature, water level in the feed stream, and the flow rate through the reactor.

The reaction system must contain a small amount of water. Water has two functions. First, it is essential for the enzyme catalyst to maintain an active hydrated state. Second, it is a reactant that generates free fatty acids and partial acylglycerides from the triacylglycerides by hydrolysis. The diacylglycerides are essential intermediates, which accelerate the interesterification reaction.

The flow rate is used to control predominantly the degree of conversion. Conversion can be based on the target triacylglycerides being produced (measured by silver-phase high-performance liquid chromatography [HPLC] or carbon number gas–liquid chromatography [GLC]).

The catalyst activity begins to decay as a result of thermal effects and poisoning of the lipase from minor components in the feed oils. The activity is monitored with time and as the conversion drops below the set levels the flow rate through the reactor is reduced. The residence time in a packed column can vary between 10 min to a few hours, depending on the concentration and activity of the catalyst. At a certain level the catalyst in the reactor needs to be replaced. High catalyst concentrations and short contact times between catalyst and reactants are characteristics of packed-bed reactors, whereas lower catalyst concentrations and longer reaction times are used in stirred-batch reactors.

The rate of catalyst decay is a crucial parameter in the process economics, and therefore accurate temperature and water control as well as adequate feed quality control is a must.

In the following paragraphs we describe typical enzymatic interesterification laboratory facilities along with details of operating conditions, daily monitoring, and analytical requirements.

3.3.1.1. EQUIPMENT

The lab scale rig is illustrated in **Fig. 5** and consists of:

1. A heated stainless steel cabinet containing the feed tank, fan for circulating hot air, the packed column, product collection point, flow controller, and a heater.
2. A 10-L double-walled glass feed tank equipped with a stirrer. The tank is filled through a funnel fitted in the lid and kept under nitrogen pressure. The tank has a drain point.
3. A double-walled glass column having a diameter of 20 mm, a length of 200 mm, and containing a section filled with immobilized enzyme (*see* **Fig. 6**).
4. A product collection vessel kept heated and under nitrogen.
5. All pipelines are double walled to prevent cold spots occurring in the system.
6. A flow controller providing a maximum flow of 50 g/h.
7. A water bath to supply hot water to the jackets of the feed tank, column, and pipe lines.
8. Control panel for setting flows and temperatures.

3.3.1.2. OPERATING PROCEDURES

1. *Startup:*
   a. Switch on the water bath to heat feed tank and lines.

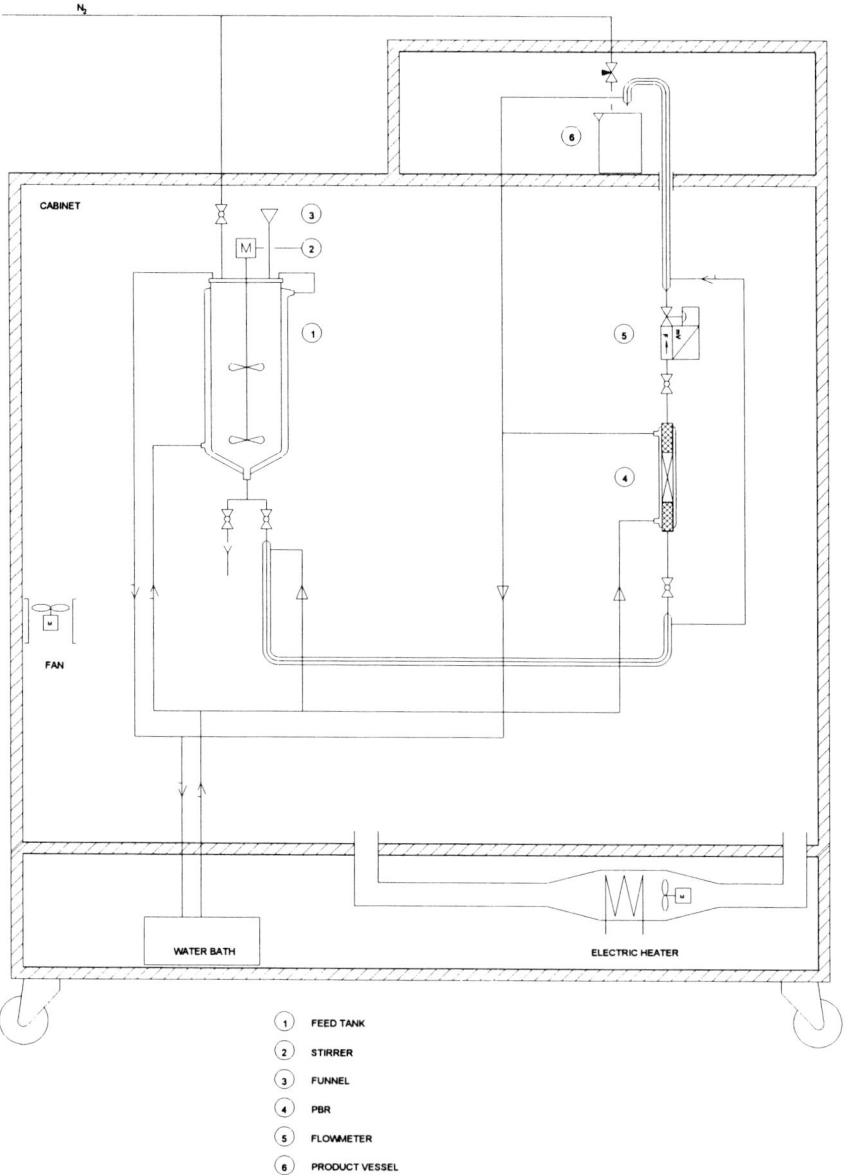

Fig. 5. Packed-bed reactor experimental setup. 1. Feed tank. 2. Stirrer. 3. Funnel. 4. Packed-bed reactor. 5. Flow meter. 6. Product vessel.

    b. Fill feed tank with the oil mixture required (described later).

    c. Switch on nitrogen for the overpressure in the feed tank.

    d. Allow 2 h for the cabinet to reach required temperature.

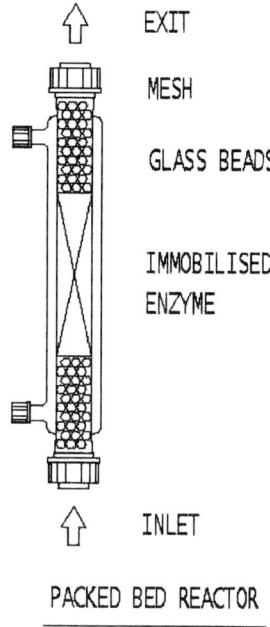

EXIT

MESH

GLASS BEADS

IMMOBILISED
ENZYME

INLET

PACKED BED REACTOR

Fig. 6. Packed-bed reactor.

e. Fill reactor with the catalyst system required (*see* **step 3**).
f. When conditions are right (i.e., constant temperature in the whole system/required saturation of the feed mixture) the feed is passed through the reactor. The flow required to achieve the desired conversion is set by adjusting the flow controller. The product is collected under nitrogen in a vessel.

2. *Loading feed tanks:*
   a. The tank is filled with the oil mixture under nitrogen.
   b. The calculated amount of deionized water (i.e., usually 0.1–0.2% wt) is added using a pipet over a 1-h time period with vigorous mixing to dissolve the water. Water content is thereafter checked by using Karl Fischer titration. Stirring speed is reduced after the water is fully dissolved.

3. *Packing the column:* A metal mesh is placed in the bottom part of the column and the reactor is filled with 10 g glass beads (3 mm diameter), the required amount of the immobilized lipase (2–8 g), glass beads, and a metal mesh.

4. *Monitoring the experiment:*
   a. Operating temperature depends on the melting point of the oil mixture chosen and the stability of the lipase catalyst. It is recommended to operate at least 2°C above the melting point in order to prevent crystallization occurring in the experimental setup. The operating temperature should also be kept as low as possible to minimize thermal deactivation of the lipase catalyst.
   b. Samples for analysis are taken on a daily basis for water content measure-

ments (feed stream) and triglyceride analysis (product stream). During the first day of starting up, the sampling is more frequent. Throughout the experiment the water content in the feed stream has to be constant. The triglyceride analysis determines whether any adjustments have to be made to the flow to achieve the conversion degree required.

    c. Flow is checked by collecting a product sample over a 1-h period.

    d. Refilling the feed tank before it is completely empty (i.e., approx 2 L left).

5. *Shutdown*: Switch off nitrogen, thermostat bath, stirrer, drain feed tank, and dismantle reactor.

6. *Cleaning*: After each experiment the feed tank, reactor, and pipelines are cleaned using acetone and dried with compressed air.

7. *Analytical requirements:*

    a. Karl Fischer (water content).

    b. Triglyceride analysis (measured by silver-phase HPLC or carbon number GLC).

3.3.2. Stirred-Batch Reactor

3.3.2.1. DESCRIPTION OF SETUP

A lab scale stirred batch vessel is illustrated in **Fig. 7**. It consists mainly of:

1. One liter double-walled stirred glass vessel.
2. Filter to allow simple separation of oil/catalyst.

3.3.2.2. OPERATING PROCEDURE

The vessel is filled with the required oil mixture and brought to operating temperature. Operating temperature is at least a couple of degrees above the melting point of the mixture, and the vessel is constantly kept under nitrogen. Deionized water is added using a pipet to bring the mixture to the required moisture level. The immobilized lipase is added and the reaction followed with time.

When the required conversion is reached, the vessel is drained and the new feed mixture is loaded.

3.3.3. Enzyme Reaction Kinetics

Several kinetic models have been developed to describe the enzymatic interesterification process *(54,58–60)*. These models can describe the effects of reaction parameters (catalyst activity, water, diacylglycerides) on reaction rate and product composition.

The following pseudo first-order kinetic model can be used to calculate the catalyst activity:

$$A = \ln (1 - X) \cdot \frac{Q}{W}$$

where

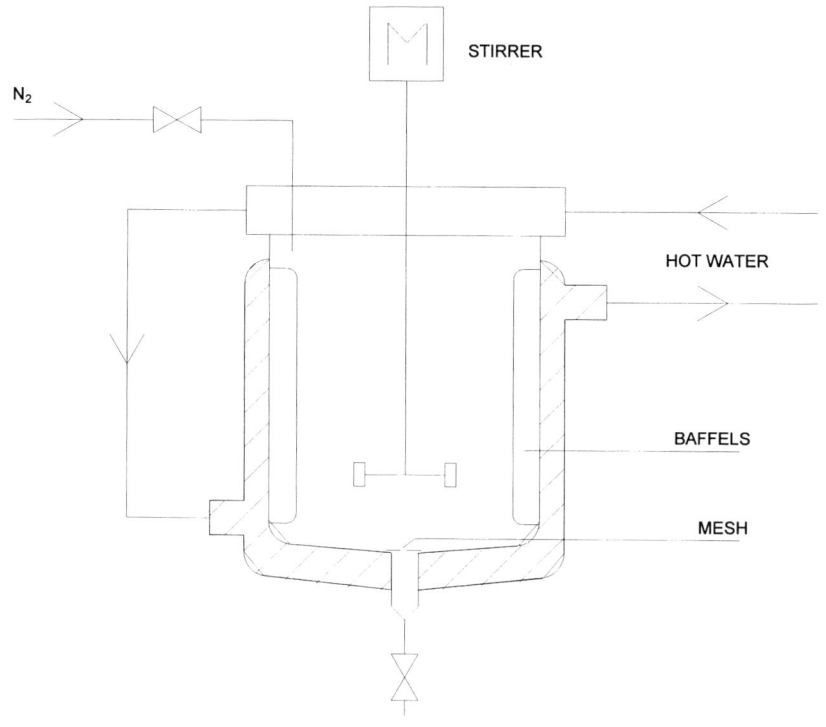

N₂ → STIRRER

HOT WATER

BAFFELS

MESH

BATCH VESSEL

Fig. 7. Stirred-batch reactor.

A = catalyst activity (kg$_{oil}$/kg$_{cat}$/h)
Q = oil flow rate in reactor (kg$_{oil}$/h)
W = weight of catalyst in reactor (kg)
X = degree of conversion (–)

$$X = \frac{TAGt - TAGi}{TAGe - TAGi}$$

where TAG = selected range of triacylglycerides (HPLC or GLC analysis) and subscript t reflects product TAG at any moment in time, i reflects TAG at start of reaction, and e reflects TAG at equilibrium conditions.

The enzymatic interesterification of palm oil (PO) (fractions) and palm kernel oil (PK) (fractions) was described in detail *(61)*. In this invention the reaction was carried out at 70°C using packed-bed reactors filled with SP-392. A number of productions were carried out varying the ratio of feed oils, the amount of catalyst, and the flow rates.

Based on the carbon number analysis using GLC, the conversion degrees were calculated. In the aforementioned invention the analysis and calculations were listed in detail. To illustrate this, **Table 1** lists the carbon number analysis of palm kernel oil and palm stearin separately, as a mixture (50:50 w/w) and after interesterification. The reaction was carried out at 70°C in a column packed with 75 g of SP-392 (*R. miehei* lipase immobilized on Duolite, e.g., Novo-Nordisk) at a flow rate of 50 g/h.

The carbon numbers obtained for the 100% converted mixture are consistent with the estimates calculated from the composition of the starting material assuming complete exchange at the sn-1and sn-3 positions. The table shows that CN44/CN46 change the most between the feed and the interesterified mixture. Therefore, they have been used for these feed oils for calculating the degree of conversion.

3.4. Notes

The biocatalyst performance is greatly affected by the temperature, water content in the system, support properties, and oil quality. Therefore attention has to be given to the following aspects.

1. *Temperature:* Heat inactivation is a factor that contributes to the loss of catalyst activity *(45,49)*. Thus the lipase used has to be resistant to heat inactivation. Reaction temperatures between 60 and 70°C ensure that the reactants and products are fully liquid.
2. *Water content:* Water is essential for the activity of the biocatalyst *(50,56)*. Some immobilized lipases have good catalytic activity at low water activity (a_w), whereas others are only active at high a_w *(30,32)*. An unexpected drop in activity could be caused by loss of water in the system. Thus, water content in the feed oils has to be checked on a regular basis. A drop in water content is usually due to leaks present in the system.
3. *Support properties:* The mechanical properties and particle size of the support have to be taken in consideration when choosing the enzymatic interesterification catalyst.

   For packed-bed reactors, avoid using small particles (<0.5 mm), as high-pressure drops will occur. Fouling in the system can also cause blockage and increase the pressure drop.

   For stirred-batch reactors, use catalysts prepared from strong support material with some elasticity, as brittle material is quickly damaged by attrition.

   Novo-Nordisk produces a highly active interesterification catalyst (*R. miehei* on ion-exchange resin) suitable for packed-bed reactors *(48)*.

   Very effective interesterification catalysts suitable for both stirred-tank and packed-bed reactors can be prepared immobilizing lipase onto Accurel EP100 (hydrophobic macroporous polypropylene particles), as described in **Subheading 2**.
4. *Oil quality:* Minor components, such as oxidation products and phospholipids, can inactivate lipase catalysts *(62)*. Thus the use of refined oils and operating

Table 1
Carbon Number Analysis

Carbon number	Palm kernel	Palmstearin	1:1 feed	Product	100% converted product
CN28	0.2	—	0.1	0.1	0.1
CN30	1.1	<0.05	0.6	0.2	0.2
CN32	6.0	0.1	3.1	0.7	0.7
CN34	8.2	0.1	4.2	1.1	1.1
CN36	21.0	0.1	10.6	3.7	3.7
CN38	15.9	0.1	8.0	4.9	4.9
CN40	9.6	0.1	4.9	10.2	10.2
CN42	9.6	0.1	4.8	13.4	13.4
CN44	7.1	0.2	3.6	14.4	14.4
CN46	5.6	0.6	3.6	18.7	18.7
CN48	6.6	23.4	14.9	13.7	13.7
CN50	2.8	38.7	20.6	10.1	10.1
CN52	2.9	27.4	15.0	7.0	7.0
CN54	3.3	7.5	5.4	1.8	1.8
CN56	0.2	0.5	0.3	0.1	0.1
CN58	<0.05	0.1	0.1	—	—
CN44+46	—	—	7.2	32.4	33.1
Conversion	—	—	0%	97%	100%

under nitrogen is essential. The peroxide values of the oil can be used as an indication of oil quality. An increase in this value indicates deterioration in the oil quality used. Care has to be taken when using oils rich in unsaturated fatty acid, as they are vulnerable to oxidation.

References

1. Vulfson, E. N. (1994) Industrial applications of lipases, in *Lipases—Their Structure, Biochemistry and Application* (Woolley, P. and Petersen, S. B., eds.), Cambridge University Press, Cambridge, UK, pp. 271–288.
2. Schmid, R. D. and Verger, R (1998) Lipases: interfacial enzymes with attractive applications. *Angew. Chem. Int. Ed.* **37,** 1608–1633.
3. Yahya, A. R. M., Anderson, W. A., and Moo-Young, M. (1998) Ester synthesis in lipase-catalyzed reactions. *Enzyme Microb. Technol.* **23,** 438–450.
4. Benjamin, S. and Pandey, A. (1998) Candida rugosa lipases: molecular biology and versatility in biotechnology. *Yeast* **14,** 1069–1087.
5. Bockisch, M. (1998) Modification of fats and oils, in *Fats and Oils Handbook,* AOCS, Champaign, IL, pp. 540–545.
6. Coleman, M. H. and Macrae, A. R. (1980) Fat process and composition. British patent specification 1,577,933.

7. Boel, E. and Huge-Jensen, B. (1989) Recombinant humicola lipase and process for the production of recombinant humicola lipases. European patent application EP 305 216.

8. Wu, X. Y., Jääskeläinen, S., and Linko, Y.-Y. (1996) An investigation of crude lipases for hydrolysis, esterification, and transesterification. *Enzyme Microb. Technol.* **19**, 226–231.

9. Pencreach, G. and Baratti, J. C. (1997) Activity of *Pseudomonas cepacia* lipase in organic media is greatly enhanced after immobilization on a polypropylene support. *Appl. Microbiol. Biotechnol.* **47**, 630–635.

10. Malcata, F. X., Reyes, H. R., Garcia, H. S., Hill, C. G. Jr., and Amundsen, C. H. (1990) Immobilized lipase reactors for modification of fats and oils — a review. *J. Am. Oil Chem. Soc.* **67**, 890–910.

11. Balcão, V. M., Paiva, A. L., and Malcata, F. X. (1996) Bioreactors with immobilized lipases: state of the art. *Enzym. Microb. Technol.* **18**, 392–416.

12. Wisdom, R. A., Dunnill, P., Lilly, M. D., and Macrae, A. R. (1984) Enzymic interesterification of fats: factors influencing the choice of support for immobilized lipase. *Enzyme Microb. Technol.* **6**, 443–446.

13. Oladepo, D. K., Halling, P. J., and Larsen, V. F. (1995) Effect of different supports on the reaction rate of *Rhizomucor miehei* lipase in organic media. *Biocatal. Biotrans.* **12**, 47–54.

14. Xu, H., Li, M., and He, B. (1995) Immobilization of *Candida cylindracea* lipase on methyl acrylate-divinyl benzene copolymer and its derivatives. *Enzyme Microb. Technol.* **17**, 194–199.

15. Bosley, J. A. and Peilow, A. D. (1997) Immobilization of lipases on porous polypropylene: reduction in esterification efficiency at low loading. *J. Am. Oil Chem. Soc.* **74**, 107–111.

16. Gitlesen, T., Bauer, M., and Adlercreutz, P. (1997) Adsorption of lipase on polypropylene powder. *Biochim. Biophys. Acta* **1345**, 188–196.

17. Jensen, B. F. and Eigtved, P. (1990) Safety aspects of microbial enzyme technology, exemplified by the safety asessment of an immobilized lipase preparation, Lipozyme™. *Food Biotechnol.* **4**, 699–725.

18. Kosugi, Y., Takahashi, K., and Lopez, C., (1995) Large-scale immobilization of lipase from *Pseudomonas fluorescens* biotype I and an application for sardine oil hydrolysis. *JAOCS* **72**, 1281–1285.

19. Gandhi, N. N., Vijayalakshmi, V., Sawant, S. B., and Joshi, J. B. (1996) Immobilization of *Mucor miehei* lipase on ion-exchange resins. *Chem. Eng. J. (Lausanne)*, **61**, 149–156.

20. Stark, M.-B. and Holmberg, K. (1989) Covalent immobilization of lipase in organic solvents. *Biotech. Bioeng.* **34**, 942–950.

21. Cho, S.-W. and Rhee, J. S. (1993) Immobilization of lipase for effective interesterification of fats and oils in organic solvent. *Biotech. Bioeng.* **41**, 204–210.

22. Sato, S., Murakata, T., Suzuki, T., Chiba, M., and Goto, Y. (1997) Esterification activity in organic medium of lipase immobilized on silicas with differently controlled pore size distribution. *J. Chem Eng Jpn.* **30**, 654–661.

23. Lalonde, J., Govardhan, C., Khalaf, N., Martinez, A. G., Visuri, K., and Margolin, A. L. (1995) Cross-linked crystals of candida rugosa lipase: highly efficient catalysts for resolution of chiral esters. *J. Am. Chem. Soc.* **117,** 6845–6852.

24. Yokozeki, K., Yamanaka, S., Takinami, K., Hirose, Y., Tanaka, A., Sonomoto, K., et al. (1982) Application of immobilized lipase to regio-specific interesterification of triglyceride in organic solvent. *Eur. J. Appl. Microbiol. Biotechnol.* **14,** 1–5.

25. Reetz, M. T. (1997) Entrapment of biocatalysts in hydrophobic sol-gel materials for use in organic chemistry. *Adv. Mater.* **9,** 943–954.

26. Adlercreutz, P., Barros, R., and Wehtje, E. (1996) Immobilization of enzymes for use in organic media. *Ann. NY Acad. Sci.* **799,** 197–200.

27. Bosley, J. A. and Clayton, J. C. (1994) Blueprint for a lipase support: use of hydrophobic controlled-pore glasses as model systems. *Biotech. Bioeng.* **43,** 934–938.

28. *Analytical Method AF 95.1/3-GB* (1983) Novo-Nordisk, Copenhagen, Denmark.

29. Valivety, R. H., Halling, P. J., Peilow, A. D., and Macrae, A. R. (1992) Lipases from different sources vary widely in dependence of catalytic activity on water activity. *Biochim. Biophys Acta.* **1122,** 143–146.

30. Oladepo, D. K., Halling, P. J., and Larsen, V. F. (1995) Effect of different supports on the reaction rate of *Rhizomucor miehei* lipase in organic media. *Biocatal. Biotrans.* **12,** 47–54.

31. Valivety, R. H., Halling, P. J., and Macrae, A. R. (1992) Water effects on suspended lipase in organic solvents: better characterised by thermodynamic activity rather than content. *Indian J. Chem.* **31B,** 914–916.

32. Bosley, J. A. and Peilow, A. D. (1991) Supported enzyme. European Patent Application EP 424,130.

33. Rozendaal, A. (1990) Interesterification of oils and fats, in *World Conference Proceedings: Edible Fats and Oil Processing: Basic Principle and Modern Practices* (Erickson, D. R., ed.). AOCS, Champaign, IL, pp. 152–157.

34. Rozendaal, A. (1992) Interesterification of oils and fats. *Inform,* **3,** 1232–1237.

35. Rozendaal, A. and Macrae, A. R. (1997) Interesterification of Oils and Fats, in *Lipid Technologies and Applications* (Gunstone, F. D. and Padley, F. B., eds.), Dekker, New York, pp. 223–263.

36. Macrae, A. R. and Hammond, R. C. (1985) Present and future applications of lipases, *Biotechnol. Genet. Eng. Rev.* **3,** 193–217.

37. Wooley, P. and Petersen, S. F. (1994) *Lipases—Their Structure, Biochemistry and Application,* Cambridge University Press, Cambridge, UK.

38. Godtfredsen, S. E (1990) Microbial lipases, in *Microbial Enzymes and Biotechnology,* 2nd ed. (Fogarty, W. M., Kelly, C. T., eds.), Elsevier Applied Science, London, pp. 255–274.

39. Egloff, M. P., Ransac, S., Marguet, F., Rogaslska, E., van Tilbeurgh, G., Buono, C., et al. (1995) Enzymes lipolytiques et lipolyse. *Oleagineux Corps Gras Lipides* **2,** 52–56.

40. Adlercreutz, P. (1994) Enzyme-catalysed lipid modification. *Biotechnol. Genet. Eng. Rev.* **12,** 231–254.

41. Gunstone, F. D. (1998) Movements towards tailor-made fats. *Prog. Lipid Res.* **37,** 277–305.

42. Matori, M., Asahara, T., and Ota Y. (1991) Positional specificity of microbial lipases. *J. Ferment. Bioeng.* **72,** 397–398.

43. Macrae, A. R. (1985) Interesterification of fats and oils, in *Biocatalysis in Organic Synthesis* (Tramper, J., van der Plas, H. C., Linko, P., eds.), Elsevier, Amsterdam, pp. 195–208.

44. Bloomer, S., Adlercreutz, P., and Mattiasson, B. (1990) Triglyceride interesterification by lipases. *J. Am. Oil Chem. Soc.* **67,** 519–534.

45. Heldt-Hansen, H. P., Ishii, M., Patkar, S. A., Hansen, T. T., and Eigtved, P. (1989) A new immobilized positional non-specific lipase for fat modification and ester synthesis. *ACS Symp. Ser.* **389,** 158–172.

46. Charton, E. and Macrae, A. R. (1992) Substrate specificities of lipases A and B from Geotrichum candidum AMICC 335426. *Biochim. Biophys. Acta* **1123,** 59–64.

47. Macrae, A. R. (1983) Lipases catalysed interesterification of oils and fats. *J. Am. Oil Chem. Soc.* **60,** 291–294.

48. Posorske, L. H., LeFebvre, G. K., Miller, C. A., Hansen, T. T., and Glenvig, B. L. (1988) Process considerations of continuous fat modification with an immobilized lipase. *J. Am. Oil Chem. Soc.* **65,** 922–926.

49. Eigtved, P. (1992) Enzymes and lipid modification. *Adv. Appl. Lipid Res.* **1,** 1–64.

50. Forssell, P., Parovuori, P., Linko, P., and Poutnaen, K. (1993) Enzymatic transesterification of Rapeseed oil and lauric acid in continuous reactor. *J. Am. Oil Chem. Soc.* **70,** 1105–1109.

51. Luck, T. and Bauer, W. (1991) Lipase catalyzed interesterification of triglycerides in solvent-free process I, Analysis and Kinetics of the Interesterification. *Fett Wiss. Technol.* **93,** 41–49.

52. Luck, T. and Bauer, W. (1991) Lipase catalyzed interesterification of triglycerides in solvent-free process II, Engineering parameters for the application of continuous process. *Fett Wiss. Technol.* **93,** 197–203.

53. Macrae, A. R. (1992) Modifying oils—enzymatic methods, in *Oils and Fats in the Nineties* (Shukla, V. K. S. and Gunstone, F. D., eds.), International Food Science Centre, Lystrup, Denmark, pp. 199–208.

54. Jung, H. J. and Bauer, W. (1992) Determination of process parameters and modelling of lipase-catalyzed transesterification in a fixed bed reactor. *Chem. Eng. Technol.* **15,** 341–348.

55. Mu, H., Xu, X., and Hoy, C. E. (1998) Production of specific structured triacylglycerols by lipase-catalyzed interesterification in a laboratory scale continuous reactor. *JAOCS* **75,** 1187–1193.

56. Macrae, A. R. (1989) Tailored triacylglycerols and esters. *Biochem. Soc. Trans.* **17,** 1146–1148.

57. Quinlan, P. and Moore, S. R. (1993) Modification of triglycerides by lipases: process technology and its application to the production of nutritionally improved fats. *Inform* **4,** 580–585.

58. Basheer, S., Mogi, K., and Nakajima, M. (1995) Interesterification kinetics of trig-lycerides and fatty acids with modified lipase in n-hexane. *J. Am. Oil Chem. Soc.* **72,** 511–518.
59. Reyes, H. R. and Hill, C. G. (1994) Kinetic modelling of interesterification reactions catalysed by immobilized lipase. *Biotech. Bioeng.* **43,** 171–182.
60. Moore, S. R. and Davies, J. (1994) The kinetics of lipase catalysed interesterification of fats and oils. *Inst. Chem. Eng. Res. Event* **1,** 526–530.
61. Huizinga, H., Sassen, C., and Vermaas, L. (1998) Process for preparing a fat blend and plastic spread comprising the fat blend obtained. European patent application EP 0792 106.
62. Wang, Y. Q. and Gordon, M. H. (1991) Effect of lipid oxidation-products on the transesterification activity of an immobilized lipase. *J. Agric. Food. Chem.* **39,** 1693–1695.

Index